EARTH, MOON, AND SUN

Earth

Mass	$M_E = 5.98 \times 10^{24}$ kg
Equatorial radius	$R_E = 6.378 \times 10^6$ m
Polar radius	$R_E' = 6.357 \times 10^6$ m
Mean density	5520 kg/m³
Surface gravity	$g = 9.81$ m/s² $= 32.2$ ft/s²
Period of rotation	1 sidereal day = 23 h 56 min 4 s $= 8.616 \times 10^4$ s
Moment of inertia:	
about polar axis	$I = 0.331 M_E R_E^2$
about equatorial axis	$I' = 0.329 M_E R_E^2$
Mean distance from Sun	1.50×10^{11} m
Period of revolution (period of orbit)	1 year = 365 days 6 h
	$= 3.16 \times 10^7$ s
Orbital speed	29.8 km/s

Moon

Mass	7.35×10^{22} kg
Radius	1.74×10^6 m
Mean density	3340 kg/m³
Surface gravity	1.62 m/s²
Period of rotation	27.3 days
Mean distance from Earth	3.84×10^8 m
Period of revolution	1 sidereal month = 27.3 days

Sun

Mass	$M_S = 1.99 \times 10^{30}$ kg
Radius	6.96×10^8 m
Mean density	1410 kg/m³
Surface gravity	274 m/s²
Period of rotation	~26 days
Luminosity	3.9×10^{26} W

GREEK ALPHABET

A	α	alpha	N	ν	nu
B	β	beta	Ξ	ξ	xi
Γ	γ	gamma	O	o	omicron
Δ	δ	delta	Π	π	pi
E	ε	epsilon	P	ρ	rho
Z	ζ	zeta	Σ	σ	sigma
H	η	eta	T	τ	tau
Θ	θ	theta	Y	υ	upsilon
I	ι	iota	Φ	ϕ	phi
K	κ	kappa	X	χ	chi
Λ	λ	lambda	Ψ	ψ	psi
M	μ	mu	Ω	ω	omega

Physics

Hans C. Ohanian

UNION COLLEGE AND

RENSSELAER POLYTECHNIC INSTITUTE

VOLUME TWO

 W · W · NORTON & COMPANY

NEW YORK · LONDON

To Susan Farnsworth Ohanian, writer,
who gently tried to teach me some of her craft.

Published simultaneously in Canada by Penguin Books Canada Ltd.,
2801 John Street, Markham, Ontario L3R 1B4.
Printed in the United States of America.
First Edition

Book design by Antonina Krass
Makeup by Roberta Flechner
Production editor Frederick E. Bidgood
Picture research by Amy Boesky and Natalie Goldstein

Cover photo by Photo Researchers 1977 © Robert Houser
Photograph credits appear on page 571

Library of Congress Cataloging in Publication Data
Ohanian, Hans C.
  Physics. Vol. II
  Includes index.
  1. Physics. I. Title.
QC21.2.037 1985 530 84-25540
ISBN 0-393-95407-2

W. W. Norton & Company, Inc., 500 Fifth Avenue, New York, N.Y. 10110
W. W. Norton & Company Ltd., 37 Great Russell Street, London WC1B 3NU

1 2 3 4 5 6 7 8 9 0

Contents

24 | Gauss' Law   565

25 | The Electrostatic Potential   580

26 | Electric Energy   601

* This section is optional.

* This section is optional.

* This section is optional.

* This section is optional.

---

* This section is optional.

40 | Quanta of Light   900

41 | Atomic Structure and Spectral Lines   920

Preface

This is a textbook for a two- or three-semester calculus-based physics course for science and engineering students. My main objectives in writing this book were to present a contemporary, modern view of classical mechanics and electromagnetism, and to offer the student a glimpse of what is going on in physics today. Thus, throughout the book, I encourage students to keep in mind the atomic structure of matter and to think of the material world as a multitude of restless electrons, protons, and neutrons. For instance, in the mechanics chapters, I emphasize that all macroscopic bodies are systems of particles; and in the electricity chapters, I introduce the concepts of positive and negative charge by referring to protons and electrons, not by referring to the antiquated procedure of rubbing glass rods with silk rags (which, according to experts on triboelectricity, can give the wrong sign if the silk has been thoroughly cleaned). I try to make sure that students are always aware of the limitations of the nineteenth-century fiction that matter and electric charge are continua. Blind reliance on this fiction has often been justified by the claim that engineering students need physics as a tool, and that the atomic structure of matter is of little concern to them. But if physics is a tool, it is also a work of art, and its style cannot be dissociated from its function. In this book I give a physicist's view of physics, because I believe that it is fitting that all students should gain some appreciation for the artistic style of the tool-maker.

The book contains two kinds of chapters: *core* chapters and *interlude* chapters. The 41 core chapters cover the essential topics of introductory physics: mechanics of particles, rigid bodies, and fluids; oscillations; wave motion; heat and thermodynamics; electricity and magnetism; optics; and quanta. They also include a chapter on special relativity, but this chapter is optional, as are a few sections of some other chapters (these are clearly marked in the table of contents and in the text). The 12 interludes present some of the fascinating discoveries and applications of physics today: crystal structure and symmetry, ionizing radiation, elementary particles, the expansion of the universe, general relativity, energy resources, fields and quanta, fission, atmospheric electricity, plasmas, superconductivity, and lasers. All of these interludes are optional — they rely on the core chapters, but the core chapters do not rely on them.

In order to accommodate students who are taking an introductory calculus course concurrently, derivatives are used slowly and hesitantly at first (Chapter 2), and routinely later on. Likewise, the use of integrals is postponed as far as possible (Chapter 7), and they come into heavy use only in the second volume (after Chapter 21). For students who need a review of calculus, Appendix 5 contains a concise primer on derivatives and integrals.

The organization of the core chapters is fairly traditional, with some innovations. I deviate from tradition by an early introduction of angular momentum, starting with the angular momentum of a particle instead of the angular momentum of a rigid body (Chapter 5). In my experience, for many students the first chapters are an expanded review of high-school physics, and students can profit from some new concepts at this stage (teachers who prefer to introduce angular momentum in conjunction with rigid bodies may postpone Section 5.6 until later). More notably, I start the study of magnetism (Chapter 30) with the law for the magnetic force between two moving point charges; this magnetic force is no more complicated than the force between two current elements, and the crucial advantage is that magnetism can be developed from the magnetic-force law in much the same way as electricity is developed from Coulomb's Law. This approach is consistent with the underlying philosophy of the book: particles are primary entities and should always be treated first, whereas macroscopic bodies and currents are composite entities which should be treated later. As another innovation, I include a simple derivation of the electric radiation field of an accelerated charge (Chapter 35); this calculation relies on Richtmeyer and Kennard's clever analysis of the kinks in the electric field lines of an accelerated charge (a set of computer-generated film loops available from the Educational Development Center shows how such kinks propagate along the field lines; these film loops tie in very well with the calculations of Chapter 35).

The chapters include generous collections of solved examples (about 275 altogether) and of problems (about 1700 altogether). Answers to the even-numbered problems are given in Appendix 11. The problems are grouped by sections, with the most difficult problems at the end of each section (exceptionally challenging problems are marked with an asterisk). I have tried to make the problems interesting to the student by drawing on realistic examples from technology, sports, and everyday life. Many of the problems are based on data extracted from engineering handbooks, car-repair manuals, *Jane's Book of Aircraft, The Guinness Book of World Records,* newspaper reports, etc. Many other problems deal with atoms and subatomic particles; these are intended to reinforce the atomistic view of the material world. In some cases, cognoscenti will perhaps consider the use of classical physics somewhat objectionable in a problem that really ought to be handled by quantum mechanics. But I believe that the advantages of familiarization with atomic quantities and magnitudes outweigh the disadvantages of a naïve use of classical mechanics.

Each chapter also includes a collection of qualitative questions intended to stimulate thought and to test the grasp of basic concepts (some of these questions are discussion questions that do not have a unique answer). Moreover, each chapter contains a brief summary of the main physical quantities and laws introduced in it. The virtue of these summaries lies in their brevity. They include essential definitions and equations, but no restatements of arguments or additional explanations, because the statements in the body of each chapter are adequate.

The inspiration for the 12 interlude chapters grew out of my unhappiness over a paradox afflicting the typical undergraduate physics curriculum: liberal-arts students in a nonmathematical physics course often get to see more of the beauty and excitement of the physics of today than science and engineering students in a calculus-based physics course. While liberal-arts students get a glimpse of quarks, black holes, or the Big Bang, science and engineering students are expected to calculate the motion of blocks on top of other blocks sliding down an inclined plane, or the motion of a baseball thrown in some direction or another by a man (or woman) riding in an elevator. To some extent this is unavoidable — science and engineering students need to learn and practice classical mechanics and electromagnetism, and they have little time left for dabbling in the arcane mysteries of contemporary physics. Nevertheless, most teachers will occasionally find an hour or two to tell their students a little of what is going on in physics today. I wrote the interludes to lend encouragement and support to such excursions to the frontiers of physics.

My choices of topics for the 12 interludes reflect the interests expressed by my students. Over the years, I have often been asked: When will we get to black holes? or Are you going to tell us about quarks? and I came to feel that such curiosity must not be allowed to whither away. Obviously, in the typical introductory course it will be impossible to cover all of the interludes (I have usually covered two per term), but the broad range of topics will permit teachers to select according to their own tastes. The interludes are mainly descriptive rather than analytic. In them, I try to avoid formulas and instead give the students a qualitative feeling for the underlying physics, keeping the discussion simple so that students can read them on their own. Thus, the interludes could be used for supplementary reading, not necessarily accompanied by lectures. For the inquisitive student, each interlude includes a collection of qualitative questions and an extensive annotated list of further readings.

The predominant system of units used in this text is SI. However, since American engineers continue to work with the British system, the text also includes examples and problems with these units. In the abbreviations for the units, I follow the dictates of the Conférence Générale des Poids et Mesures of 1971, although I deplore the majestic stupidity of the decision to replace the old, self-explanatory abbreviations amp, coul, nt, sec, °K by an alphabet soup of cryptic symbols A, C, N, s, K, etc. For the sake of clarity, I spell out the names of units in full whenever the abbreviations are likely to lead to ambiguity and confusion.

An excellent study guide for this book has been written by Professors Van E. Neie (Purdue University) and Peter J. Riley (University of Texas, Austin). This guide includes for every chapter a brief introduction laying out the objectives; a list of key terms for review; detailed commentaries on each of the main ideas; and a large collection of interesting sample problems, which alternate between worked problems (with full solutions) and guided problems (which provide step-by-step schemes that lead students to the solutions).

The book has been seven years in the making. The original manuscript went through several revisions, based both on my own experience with students at Union College and on extensive reviews and class testing at several other institutions. A preliminary edition of this book, reproduced photographically from typescript, was used in classes by Professors John R. Boccio (Swarthmore College), A. Douglas Davis (Eastern Illinois University), J. David Gavenda, (University of Texas, Austin), Frank Moscatelli (Swarthmore College), Harvey S. Picker (Trinity College), Peter J. Riley (University of Texas, Austin), Kenneth L. Schick (Union College), and Mark P. Silverman (Trinity College). The experience gained in these class tests permitted me to make many improvements. I am grateful to both teachers and students for sharing their reactions to the book with me.

I have greatly benefited from many comments by reviewers who carefully read my manuscript and suggested corrections and alterations. I am indebted to Professors John R. Boccio (Swarthmore College), Roger W. Clapp, Jr. (University of South Florida), A. Douglas Davis (Eastern Illinois University), Anthony P. French (Massachusetts Institute of Technology), J. David Gavenda (University of Texas, Austin), Roger D. Kirby (University of Nebraska), Roland M. Lichtenstein (Rensselaer Polytechnic Institute), Richard T. Mara (Gettysburg College), John T. Marshall (Louisiana State University), Harvey S. Picker (Trinity College), and Peter J. Riley (University of Texas, Austin) for very detailed, comprehensive reviews; and I am indebted to Professors John R. Albright (Florida State University), Frank A. Ferrone (Drexel University), James R. Gaines (Ohio State University), Michael A. Guillen (Harvard University), Walter Knight (University of California, Berkeley), Jean P. Krisch (University of Michigan), Hermann Nann (Indiana University), Norman Pearlman (Purdue University), P. Bruce Pipes (Dartmouth College), Jack Prince (Bronx Community College), Gerald A. Smith (Michigan State University), Julia A. Thompson and David Kraus (University of Pittsburgh), and Gary A. Williams (University of California, Los Angeles) for briefer reviews. I am also indebted to the experts who reviewed the interludes: Professors Edmond Brown (Rensselaer Polytechnic Institute; "The Architecture of Crystals"), Priscilla W. Laws (Dickinson College; "Radiation and Life"), Stephen Gasiorowicz (University of Minnesota; "Elementary Particles"), Alan H. Guth (Massachusetts Institute of Technology; "The Big Bang and the Expansion of the Universe"), Donald F. Kirwan (University of Rhode Island; "Energy, Entropy, and the Environment"), Malvin Ruderman (Columbia University; "Forces, Fields, and Quanta"), Irving Kaplan (Massachusetts Institute of Technology; "Nuclear Fission"), Bernard Vonnegut (State University of New York at Albany; "Atmospheric Electricity"), Sam Cohen (Princeton University Plasma Physics Laboratory; "Plasma"), and Margaret L. A. MacVicar (Massachusetts Institute of Technology; "Superconductivity").

I thank Dr. Richard D. Deslattes and Dr. Barry N. Taylor of the National Bureau of Standards for information on units, standards, and precision measurements. I thank some of my colleagues at Union College: Professor C. C. Jones prepared the beautiful photographs of interference and diffraction by light (Chapters 38 and 39); Professor Barbara C. Boyer gave valuable advice and assistance on photographs dealing with biological material; and the staff of the library helped me find many a tidbit of information needed for an example or a problem. Michael Rooks and Ben Hu independently checked the solutions to the problems and proposed corrections for some problems they found insoluble.

I thank the editorial staff of W. W. Norton & Co., who worked long and hard to bring all the pieces of this book together: Drake McFeely, editor, who gave the manuscript the most meticulous attention and with sharp eyes spotted many a sentence that required rephrasing and clarification; Christopher Lang, who guided the project through its early stages; Alicia Salomon, copy editor, who corrected subtle errors of grammar and enforced stylistic consistency; and Amy Boesky and Natalie Goldstein, who searched out many of the fascinating photographs that add interest to this book. And it is a pleasure to thank Terry Hynes for the typing of the first draft of the text from my handwritten manuscript; she was always eager to improve my style and skillfully supplied all those words and endings of words that my pen, somehow, left out when my thoughts raced ahead of my hand.

H. C. O.
February 1985

CHAPTER 22

Electric Force and Electric Charge

Ordinary matter — solids, liquids, and gases — consists of atoms, each with a nucleus surrounded by a swarm of electrons. For example, Figure 22.1 shows the structure of an atom of neon. At the center of this atom there is a nucleus made of ten protons and ten neutrons packed very tightly together — the diameter of the nucleus is only about 6×10^{-15} m. Moving around this nucleus there are ten electrons; these electrons are confined to a roughly spherical region about 1×10^{-10} m across. Figure 22.1 shows the electrons at one instant of time. Since the electrons move around the nucleus very quickly, for most purposes the average distribution of electrons is more relevant than the instantaneous distribution. Figure 22.2 is a picture of the average distribution of electrons in a neon atom. This picture gives a good impression of the spherical shape of this atom.

The atom somewhat resembles the Solar System, with the nucleus as Sun and the electrons as planets. In the Solar System the force that

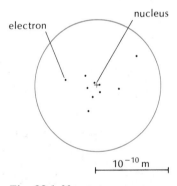

Fig. 22.1 Neon atom as seen at one instant of time with a (hypothetical) gamma-ray microscope.

Fig 22.2 Neon atom as seen with an electron-holography microscope. The atom looks like a spherical cloud. The density, or brightness, at any point of this cloud is roughly proportional to the probability for finding an electron at this point. The magnification is $2 \times 10^8 \times$. (Courtesy L. S. Bartell, University of Michigan.)

holds a planet near the Sun is the gravitational force. What is the force that holds an electron near the nucleus? The force is the **electric force** of attraction between the electron and the protons in the nucleus. The

Electric force

531

electric force not only holds electrons inside atoms, but also holds atoms together in molecules or crystals, and holds these building blocks together in large-scale macroscopic structures — a rock, a tree, a human body, a skyscraper, or a supertanker. All the mechanical "contact" forces of everyday experience — the push of a hand against a door, the pull of an elevator cable, the pressure of water against the hull of a ship — are nothing but the combined electric forces of many atoms. Thus our immediate environment is dominated by electric forces.

Electrostatic force

In the following chapters we will study electric forces and their effects. For a start (Chapters 22–29) we will assume that the particles exerting these forces are at rest or moving only very slowly (quasistatic). The electric forces are then called **electrostatic forces.** Later on (Chapters 30–34) we will consider the forces when the particles are moving with uniform velocity or nearly uniform velocity. Under these conditions the electric forces are modified — besides the electrostatic force there arises a **magnetic force.** The magnetic force may be regarded as a supplement to the electric force, a supplement depending on the velocity of the particles. The combined electrostatic and magnetic forces are called **electromagnetic forces.** Finally, we will consider the forces when the particles are moving with accelerated motion (Chapters 35 and 36). The electromagnetic forces are then further modified with a drastic consequence, that is, the emission of electromagnetic waves.

Electricity was first discovered through friction. The ancient Greeks noticed that, when rubbed, rods of amber gave off sparks and attracted small bits of straw or feathers. In the nineteenth century technical applications of electricity were gradually developed, but it was only in the twentieth century that the pervasive presence of electric forces holding together all the matter of our environment was recognized.

We will begin our study of electricity with the fundamental electric force between charged particles rather than following the historical route, because the origin of frictional electricity remains rather mysterious. Even now, physicists have no precise understanding of the detailed mechanism that generates electric charge on rubbed objects or why some rubbed objects acquire positive charge and some negative charge.

22.1 The Electrostatic Force

The electric force between, say, an electron and a proton resembles gravitation in that it decreases in proportion to the inverse square of the distance. But the electric force is much stronger than the gravitational force. The electric attraction between an electron and a proton (at any given distance) is about 2×10^{39} times as strong as the gravitational attraction. Thus the electric force is by far the strongest force felt by an electron in an atom. The other great difference between the gravitational and the electric force is that gravitation is always attractive, whereas electric forces can be attractive or repulsive. The electric forces between electron and proton, electron and electron, and proton and proton all have the same magnitude (for the same distance); the electron–proton force is attractive, but the electron–electron and proton–proton forces are repulsive. Table 23.1 gives a qualitative summary of the electric forces between the particles in an atom.

Table 22.1 ELECTRIC FORCES (QUALITATIVE)

Particles	Force
Electron and proton	Attractive
Electron and electron	Repulsive
Proton and proton	Repulsive
Neutron and anything	Zero

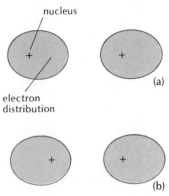

Fig. 22.3 (a) Two neighboring distorted atoms. The colored regions represent the average distributions of the electrons. The electrons of the left atom are closer to the nucleus of the right atom than to its electrons. (b) The nucleus of the left atom is closer to the nucleus of the right atom than to its electrons.

The "contact" force between two atoms close together actually arises from a counterplay of the attractive and repulsive electric forces between the electrons of the atoms and the protons in their nuclei. The force between two neighboring atoms depends on the exact location of the electrons and the nuclei. If the distribution of the electrons around the nucleus is somewhat distorted so that the electrons in one atom are closer to the nucleus of the neighboring atom than to its electrons, then the force between these atoms will be attractive. Figure 22.3a shows a distortion that leads to an attractive force; the distortion may either be intrinsic to the structure of the atom or induced by the presence of the neighboring atom. Figure 22.3b shows a distortion that leads to a repulsive force.

If two macroscopic bodies are separated by some distance, the electric attractions and repulsions between them tend to cancel, provided the bodies are neutral, i.e., provided their number of protons matches their number of electrons. For example, if the macroscopic bodies are a man and a woman separated by a distance of 10 m, then each electron in the man will be strongly attracted by the protons in the woman, but simultaneously it will be strongly repelled by the electrons in the woman; these forces cancel each other. Only when the surfaces of the two macroscopic bodies are very near one another ("touching") will the atoms in one surface exert a net force on those in the other surface.

22.2 Coulomb's Law

As stated in the preceding section, the electric forces between electron and proton, electron and electron, and proton and proton all have the same magnitude. Obviously, this implies that the force does not depend on the mass of these particles. Instead, it depends on a new quantity: the **electric charge.** For the mathematical formulation of the law of electric force between electrons, protons, and other particles we assign to each particle an electric charge which is either positive, negative, or zero. The charge on the proton is taken as positive and that on the electron negative. Experiments show that the absolute values of these charges are exactly the same. We will designate the charges of the proton and electron by $+e$ and $-e$, respectively. The charge on the neutron is zero. Table 22.2 summarizes these values of the charges.

Benjamin Franklin, *1706–1790, American scientist, statesman, and inventor. Although he is most often remembered for his hazardous experiments with a kite in a thunderstorm, which demonstrated that lightning is an electric phenomenon, and for his invention of lightning rods, Franklin also made other significant contributions to the experimental and theoretical studies of electricity, and he was admired and honored by the leading scientific associations in Europe. Among these contributions was his formulation of the law of conservation of electric charge and his introduction of the modern notation for plus and minus charges, which he regarded as an excess or deficiency of "electric fluid."*

Table 22.2 ELECTRIC CHARGES OF PROTONS, ELECTRONS, AND NEUTRONS

Particle	Charge
Proton, p	$+e$
Electron, e	$-e$
Neutron, n	0

Charles Augustin de Coulomb (koolom), *1736–1806, French physicist. A military engineer by profession, he retired at age 53 from his post as superintendent of waters and fountains so that he could fully pursue scientific research. With the torsion balance, which he invented, he established that the electric force between small charged balls obeys an inverse-square law.*

In terms of these electric charges, we can then state that the electric force between charges of like sign is repulsive and that the electric force between charges of unlike sign is attractive.

The precise value of the electric force that one charged particle exerts on another is given by **Coulomb's Law:**

> *The magnitude of the electric force that a particle exerts on another particle is directly proportional to the product of their charges and inversely proportional to the square of the distance between them. The direction of the force is along the line joining the particles.*

Thus, the magnitude of the force that a particle of charge q' exerts on a particle of charge q at a distance r is

$$F = [\text{constant}] \times \frac{|q'q|}{r^2} \tag{1}$$

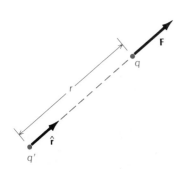

To express the direction of the force vectorially, we will use the unit vector $\hat{\mathbf{r}}$ pointing from the charge q' to the charge q (Figure 22.4). In terms of this unit vector, we can write the following equation for the force that q' exerts on q:

$$\mathbf{F} = [\text{constant}] \times \frac{q'q}{r^2} \hat{\mathbf{r}} \tag{2}$$

Fig. 22.4 Two point charges q' and q. The unit vector $\hat{\mathbf{r}}$ points from q' to q.

This vector equation gives us both the magnitude and the direction of **F.** If the charges are of like sign, then the product $q'q$ is positive and **F** is parallel to $\hat{\mathbf{r}}$ (repulsive force). If the charges are of unlike sign, then the product $q'q$ is negative and **F** is antiparallel to $\hat{\mathbf{r}}$ (attractive force).

The numerical value of the charge of the proton and the numerical value of the constant in Eqs. (1) and (2) depend on the system of units. In the SI system of units the electric charge is measured in **coulombs** (C) and the corresponding numerical values are

coulomb, C

Charge of proton, e

$$\boxed{e = 1.60 \times 10^{-19} \text{ C}} \tag{3}$$

and

$$[\text{constant}] = 8.99 \times 10^9 \text{ N} \cdot \text{m}^2/\text{C}^2 \tag{4}$$

For obscure historical reasons this constant is traditionally written in the complicated form

$$[\text{constant}] = \frac{1}{4\pi\varepsilon_0} \tag{5}$$

with

$$\boxed{\varepsilon_0 = 8.85 \times 10^{-12} \text{ C}^2/(\text{N} \cdot \text{m}^2)} \tag{6}$$ *Permittivity constant*

The quantity ε_0 is called the **permittivity constant.**[1]

Using the permittivity constant, Coulomb's Law for the force that a particle of charge q' exerts on a particle of charge q then becomes

$$\boxed{\mathbf{F} = \frac{1}{4\pi\varepsilon_0} \frac{q'q}{r^2} \hat{\mathbf{r}}} \tag{7}$$ *Coulomb's Law in vector notation*

This equation applies to particles — electrons and protons — and also to any small charged bodies, provided that the sizes of these bodies are much less than the distance between them; such bodies are called **point charges.** Equation (7) obviously resembles Newton's Law for the gravitational force (see Section 13.1); the constant $1/4\pi\varepsilon_0$ is analogous to the gravitational constant G and the charge is analogous to the mass. *Point charge*

In the SI system the coulomb is defined in terms of a standard electric current: one coulomb is the amount of electric charge that a current of one ampere delivers in one second. Unfortunately the definition of the standard current involves magnetic fields and we will therefore have to postpone the question of the precise definition of ampere and coulomb to a later chapter. For the time being, we may define the coulomb by the electric repulsive force between two equal charges at a standard distance from one another: the charge on each of two equally charged small bodies is one coulomb if they repel with a force of 8.99×10^9 N when their distance is one meter. (Incidentally, this shows that charge, like everything else, can be defined in terms of the fundamental units of mass, length, and time.)

EXAMPLE 1.   Compare the magnitudes of the gravitational force of attraction and of the electric force of attraction between the electron and the proton in a hydrogen atom. According to Newtonian mechanics, what is the acceleration of the electron? Assume that the distance between the two particles is 0.53×10^{-10} m.

SOLUTION: The gravitational force is

$$\frac{GmM}{r^2} = \frac{6.67 \times 10^{-11} \text{ N} \cdot \text{m}^2 \cdot \text{kg}^{-2} \times 9.11 \times 10^{-31} \text{ kg} \times 1.67 \times 10^{-27} \text{ kg}}{(0.53 \times 10^{-10} \text{ m})^2}$$

$$= 3.6 \times 10^{-47} \text{ N}$$

The electric force is

---

[1] The values of e and ε_0 in Eqs. (3) and (6) have been rounded off to three significant figures. The best available values of these constants are listed in Appendix 8.

$$\frac{1}{4\pi\varepsilon_0}\frac{e^2}{r} = 8.99 \times 10^9 \; \text{N} \cdot \text{m}^2/\text{C}^2 \times \frac{(1.60 \times 10^{-19} \; \text{C})^2}{(0.53 \times 10^{-10} \; \text{m})^2} = 8.2 \times 10^{-8} \; \text{N}$$

The ratio of these forces is $8.2 \times 10^{-8} \; \text{N}/3.6 \times 10^{-47} \; \text{N} = 2.3 \times 10^{39}$.

Since the gravitational force is very small compared to the electric force, it can be neglected. The acceleration of the electron is then

$$a = \frac{F}{m} = \frac{8.2 \times 10^{-8} \; \text{N}}{9.1 \times 10^{-31} \; \text{kg}} = 9.0 \times 10^{22} \; \text{m/s}^2$$

EXAMPLE 2. How much negative and how much positive charge is there on the electrons and protons of a cup of water (0.25 kg)?

SOLUTION: The molecular mass of water is 18 g; hence 250 g of water amounts to 250/18 moles. Each mole has 6.0×10^{23} molecules, giving $6.0 \times 10^{23} \times (250/18)$ molecules in a cup. Each molecule consists of two hydrogen atoms (one electron apiece) and one oxygen atom (8 electrons). Thus, there are 10 electrons in each molecule and the total negative charge on all the electrons is $-6.0 \times 10^{23} \times 250/18 \times 10 \times 1.6 \times 10^{-19} \; \text{C} = -1.3 \times 10^7 \; \text{C}$. The positive charge on the protons is the opposite of this.

EXAMPLE 3. What is the magnitude of the attractive force exerted by the electrons in a cup of water on the protons in a second cup of water at a distance of 10 m?

SOLUTION: According to the preceding example, the charge on the electrons is $-1.3 \times 10^7 \; \text{C}$, and the charge on the protons is $+1.3 \times 10^7 \; \text{C}$. If we treat both of these charges (approximately) as point charges, the force is

$$F = \frac{1}{4\pi\varepsilon_0}\frac{qq'}{r^2}$$

$$= 9.0 \times 10^9 \; \text{N} \cdot \text{m}^2 \cdot \text{C}^{-2} \times \frac{(-1.3 \times 10^7 \; \text{C}) \times (+1.3 \times 10^7 \; \text{C})}{(10 \; \text{m})^2}$$

$$= -1.5 \times 10^{22} \; \text{N}$$

This is the weight of 10^{18} tons! This enormous attractive force is precisely canceled by an equally large repulsive force exerted by the protons in one cup on the protons in the other cup. Thus, the cups exert no net force on each other.

Coulomb's Law of Force is an empirical fact about the mutual interactions of charges. This law was first discovered in the eighteenth century by simple and direct experiments with small balls and globes on which electric charges were placed. Nowadays, the strongest observational evidence for Coulomb's Law comes from atomic mechanics, just as the strongest observational evidence for Newton's law comes from celestial mechanics. Of course, the motion of an electron inside an atom must be described by quantum mechanics rather than classical mechanics. The electron has no well-defined orbit and it is not possible to check on the force law by direct measurement of orbital positions; however, orbital energies can be measured and compared with theoretical values calculated from Coulomb's Law. In the best of these calculations, which take into account not only quantum theory but also relativity theory and many subtle effects arising from the interplay of these theories, the orbital energies of the hydrogen atom have been calculated to *nine significant figures*. To within this precision the theoretical and experimental values agree. This represents one of the greatest triumphs of modern theoretical physics and of Coulomb's Law — even the legendary precision of celestial mechanics and of Newton's Law of Gravitation pale by comparison. There is no question that our understanding of the electric force surpasses our understanding of any other force in nature.

22.3 Charge Quantization and Charge Conservation

Not only electrons and protons exert electric forces on one another, but so do many other particles. The magnitudes of these electric forces are given by Eq. (7) with the appropriate values of the electric charges. Table 22.3 lists the electric charges of some particles; a more complete list will be found in Section C.3. The charges of antiparticles are always opposite to those of the corresponding particles; e.g., the anti-electron (or positron) has charge $+e$, the antiproton has charge $-e$, the antineutron has charge 0, etc.

All the particles that have been found in nature have charges that are some integer multiple of the fundamental charge e, i.e., the charges are always 0, $\pm e$, $\pm 2e$, $\pm 3e$, etc. Why no other charges exist is a mystery for which classical physics offers no explanation. (Recent investigations suggest that the unified theory of fields may supply an explanation; see Interlude G.) Much effort has been expended on experimental searches for charges of $\frac{1}{3}e$ and $\frac{2}{3}e$ (these are the charges of the quarks, the elementary constituents supposedly found inside protons and neutrons; see Interlude C), but neither such fractional charges nor any other fractional charges have ever been found.

Table 22.3 ELECTRIC CHARGES OF SOME PARTICLES

Particle	Charge
Photon, γ	0
Neutrino, ν	0
Electron, e	$-e$
Muon, μ	$-e$
Pion, π^0	0
Pion, π^+	$+e$
Pion, π^-	$-e$
Proton, p	$+e$
Neutron, n	0
Delta, Δ^{++}	$+2e$
Delta, Δ^+	$+e$
Delta, Δ^0	0
Delta, Δ^-	$-e$

Since charges only exist in discrete packets, we say that **charge is quantized** — the fundamental charge e is called the quantum of charge. However, in a description of the charge distribution on macroscopic bodies, the discrete nature of charge can often be ignored and it is usually sufficient to treat the charge as a continuous "fluid" with a charge density (C/m^3) that varies more or less smoothly as a function of position. This is analogous to describing the mass distribution of

Quantization of charge

a solid, liquid, or gas by a mass density (kg/m³) which ignores the fact that on a microscopic scale the mass is concentrated in atoms. A solid, liquid, or gas seems smooth because there are very many atoms in each cubic millimeter (about 10^{16} atoms/mm³ in a gas under standard conditions) and the distances between atoms are very small. Likewise, charge distributions placed on wires or other conductors will seem smooth because they are due to very many electrons (or protons) in each cubic millimeter.

Conservation of charge

The electric charge is a **conserved quantity:** in any reaction involving charged particles, the total charges before and after the reaction are always the same. Here are some examples of reactions in which particles are destroyed, yet the charge remains constant:

matter–antimatter annihilation,

$$[\text{electron}] + [\text{antielectron}] \rightarrow 2[\text{photons}]$$
$$\text{charges:} \quad -e \quad + \quad e \quad \rightarrow \quad 0 \qquad (8)$$

radioactive disintegration of a neutron,

$$[\text{neutron}] \rightarrow [\text{proton}] + [\text{electron}] + [\text{antineutrino}]$$
$$\text{charges:} \quad 0 \quad \rightarrow \quad e \quad + \quad (-e) \quad + \quad 0 \qquad (9)$$

pion creation in a high-energy proton–proton collision,

$$[\text{proton}] + [\text{proton}] \rightarrow [\text{neutron}] + [\text{proton}] + [\text{pion}, \pi^+]$$
$$\text{charges:} \quad e \quad + \quad e \quad \rightarrow \quad 0 \quad + \quad e \quad + \quad e \qquad (10)$$

No reaction that creates or destroys electric charge has ever been found in nature. In an effort to test the law of charge conservation, physicists have looked for hypothetical reactions such as

$$[\text{electron}] \rightarrow [\text{neutrino}] + [\text{photon}] \qquad (11)$$

that would destroy negative charge. There is strong experimental evidence that such reactions never happen.

Charge is of course also conserved in chemical reactions. For instance, in a lead–acid battery (automobile battery), plates of lead and of lead dioxide are immersed in an electrolytic solution of sulfuric acid (Figure 22.5). The reactions that take place on these plates involve sulfate ions (SO_4^{--}) and hydrogen ions (H^+); the reactions release electrons at the lead plate and absorb electrons at the lead-dioxide plate:

$$Pb + SO_4^{--} \rightarrow PbSO_4 + 2[\text{electrons}] \qquad (12)$$
$$\text{charges:} \quad 0 + (-2e) \quad \rightarrow \quad 0 \quad + \quad (-2e)$$

$$PbO_2 + 4H^+ + SO_4^{--} + 2[\text{electrons}] \rightarrow PbSO_4 + 2H_2O \qquad (13)$$
$$\text{charges:} \quad 0 \quad + 4e + (-2e) + \quad (-2e) \quad \rightarrow \quad 0 \quad + \quad 0$$

The plates of such a battery are connected by an external circuit (e.g., a wire) and the electrons released by the reaction (12) travel from one plate to the other via this external circuit, forming an electric current (Figure 22.5).

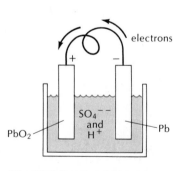

Fig. 22.5 Lead–acid battery.

EXAMPLE 4. A fully "charged" battery contains a large amount of sulfuric acid in the electrolytic solution (H_2SO_4 in the form of SO_4^{--} ions and H^+ ions). As the battery delivers charge to the external circuit connecting its poles, the amount of sulfuric acid in solution gradually decreases. Suppose that while discharging completely, the positive pole of an automobile battery delivers a charge of 1.8×10^5 C through the external circuit. How many grams of sulfuric acid will be used up in this process?

SOLUTION: Since each electron has a charge of -1.6×10^{-19} C, the number of electrons in -1.8×10^5 C is $1.8 \times 10^5 / 1.6 \times 10^{-19}$, or 1.1×10^{24}. According to the reactions (12) and (13), whenever two electrons are transferred from the lead to the lead-dioxide plate, two sulfate ions are absorbed (one at each plate). Thus 1.1×10^{24} sulfate ions will be used up, i.e., 1.1×10^{24} molecules of sulfuric acid will be used up. The required number of moles of sulfuric acid is therefore $1.1 \times 10^{24} / 6.02 \times 10^{23} = 1.9$ and, since the molecular mass of sulfuric acid is 98 g per mole, the required mass of sulfuric acid is 1.9×98 g $= 183$ g.

The conservation of charge in chemical reactions is a trivial consequence of the conservation of electrons and protons. All such reactions involve nothing but a rearrangement of electrons and protons; during this, the numbers of electrons and protons remain constant. Obviously, the net charge must then also remain constant.

The same argument applies to all macroscopic electric processes, such as the operation of electrostatic machines and generators, the flow of currents on wires, the storage of charge in capacitors, the electric discharge of thunderclouds, etc. All such processes involve nothing but a rearrangement of electrons and protons. Consequently, the net charge must remain constant.

22.4 Conductors and Insulators; Frictional Electricity

A conductor — such as copper, aluminum, or iron — is a material that permits the motion of electric charge through its volume. An insulator — such as glass, porcelain, rubber, or nylon — is a material that does not permit the motion of electric charge. Thus, when we place some electric charge on one end of a conductor, it immediately spreads out over the entire conductor until it finds an equilibrium distribution[2]; when we place some charge on one end of an insulator, it stays in place.

Conductors and insulators

All metals are good conductors. The motion of charge in metals is due to the motion of electrons. In a metal, some of the electrons of each atom are **free,** that is, they are not bound to any particular atom although they are bound to the metal as a whole. The free electrons come from the outer parts of the atoms. The outer electrons of the atom are not very strongly attached and readily come loose; the inner electrons are firmly bound to the nucleus of the atom and are likely to

Free electrons

---

[2] We will study the conditions for the equilibrium of electric charge on a conductor in Chapter 24. It turns out that when the charges finally reach equilibrium, they will all sit on the surface of the conductor (provided the conductor is homogeneous).

stay put. The free electrons wander through the entire volume of the metal, suffering occasional collisions, but they only experience a restraining force when they hit the surface of the metal. The electrons are held inside the metal in much the same way as the particles of a gas are held inside a container — the particles of gas can wander through the volume of the container, but they are restrained by the walls. In view of this analogy, electrons in a metallic container are often said to form a **free-electron gas.** If one end of a metallic conductor has an excess or deficit of electrons, the motion of the free-electron gas will quickly distribute this excess or deficit to other parts of the metallic conductor.

Free-electron gas

The charging of a body of metal is usually accomplished by the removal or addition of electrons. A body will acquire a net positive charge if electrons are removed and a net negative charge if electrons are added. Thus, positive charge on a body of metal is simply a deficit of electrons, and negative charge an excess of electrons.

Liquids containing ions (atoms or molecules with missing electrons or with excess electrons) are also good conductors. For instance, a solution of common salt in water contains ions of Na^+ and Cl^-. The motion of charge through the liquid is due to the motion of these ions. Liquid conductors with an abundance of ions are called **electrolytes.**

Incidentally, very pure, distilled water is a poor conductor, but ordinary water is a good conductor because it contains some ions contributed by dissolved impurities. The ubiquitous water in our environment makes many substances into conductors. For example, earth (ground) is a reasonably good conductor, mainly because of the presence of water. Furthermore, on a humid day many insulators acquire a microscopic surface film of water and this permits electric charge to leak away along the insulator; thus, on humid days it is difficult to store electric charge on bodies supported by insulators.

Fig. 22.6 Lightning.

Ordinary gases are insulators, but ionized gases are good conductors. For example, air is an insulator, but the ionized air found in the path of a lightning bolt is a good conductor (Figure 22.6). This ionized air is a mixture of positive ions and free electrons; the motion of charge in such a mixture is due mainly to the motion of the electrons. An ionized gas is called a **plasma.**

Plasma

We will end this chapter with a few brief comments on frictional electricity. One can accumulate electric charge on a glass rod merely by rubbing it with a piece of silk. The silk becomes negatively charged and the glass positively charged — the rubbing motion between the surfaces of the silk and the glass rips charges off one of these surfaces and makes them stick to the other, but the detailed mechanism is not well understood. It is believed that what is usually involved is a transfer of ions from one surface to the other. Contaminants residing on the rubbed surfaces play a crucial role in frictional electricity. If glass is rubbed with an absolutely clean piece of silk or other textile material, the glass becomes negatively charged rather than positively charged. Ordinarily pieces of silk apparently have such a large amount of dirt on their surfaces that the charging process is dominated by the dirt rather than by the silk. Even air can act as contaminant for some surfaces; for instance, careful experiments on the rubbing of platinum with silk show that in vacuum the platinum becomes negatively charged, but in air it becomes positively charged.

The electric charge that can be accumulated on the surface of a body of ordinary size (a centimeter or more) by rubbing may be as much as 10^{-9} to 10^{-8} coulomb per square centimeter. If the charge concentration on a body is higher than that, it will cause an electrical discharge into the surrounding air (a corona discharge; see Section I.4); a higher charge concentration can subsist only on a small body or on a small spot on a large body.

Once we have accumulated some charge on, say, a rod of glass, we can produce charges on other bodies by a process of **induction,** as follows: First we bring the glass rod near a metallic body supported on an insulating stand. The positive charge on the rod will then attract free electrons to the near side of the body and leave a deficit of free electrons on the far side (Figure 22.7a); thus, the near side will acquire negative charge and the far side positive charge. If we next momentarily connect the far side to the ground, the positive charge will be neutralized by an influx of electrons from the ground (virtually, the positive charge leaks away; Figure 22.7b). This leaves the metallic body with a net negative charge. When we finally withdraw the glass rod, this charge will remain on the metallic body, distributing itself over its entire surface (Figure 22.7c).

Some old electrostatic machines accumulate large amounts of charge by a repetitive operation of induction. Figure 22.8 illustrates the principle. Two metallic bodies on insulating stands (A and A') initially carry some small positive and negative charges. In order to increase these charges, we place two small metallic bodies B and B' nearby. Charges will then be induced on B and B' as shown in Figure 22.8a. If we now momentarily connect the far sides of B and B' with a wire, the charges on these sides will cancel, leaving B with a negative charge and B' with a positive charge (Figure 22.8b). Next, we put B' in contact with A and B in contact with A' (Figure 22.8c); this transfers the charges of the small bodies almost entirely to the larger bodies. The small bodies are

Induction

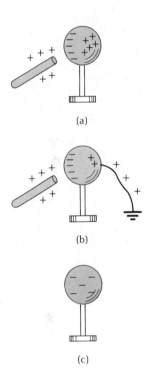

Fig. 22.7 (a) The positively charged glass rod induces a charge distribution on the metallic sphere. (b) When the far side of the sphere is connected to the ground by means of a wire, the positive charge leaks away. (c) Finally, only negative charge remains on the sphere.

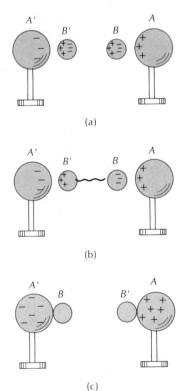

Fig. 22.8 (a) The large spheres induce charge distributions on the small spheres. (b) When the neighboring sides of the small spheres are connected by means of a wire, the charges of these sides cancel. (c) The large spheres absorb the charge of the small spheres.

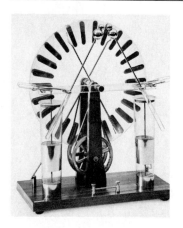

Fig. 22.9 A Wimshurst machine.

then left uncharged and we can again repeat the sequence of operations. The net result is a stepwise increase of the electric charges on *A* and *A'*. In electrostatic machines — such as Wimshurst's machine (Figure 22.9) — the entire sequence of manipulations of the bodies *B* and *B'* is carried out automatically when the experimenter turns a crank.

SUMMARY

Electric charge may be positive, negative, or zero: like charges repel, unlike charges attract.

Coulomb's Law: $F = \dfrac{1}{4\pi\varepsilon_0}\dfrac{qq'}{r^2}$

Permittivity constant: $\varepsilon_0 = 8.85 \times 10^{-12}$ C^2/N $\cdot$ m^2

$$\frac{1}{4\pi\varepsilon_0} = 8.99 \times 10^9 \text{ N} \cdot \text{m}^2/\text{C}^2$$

Charge of proton: $e = 1.60 \times 10^{-19}$ C

Charge of electron: $-e = -1.60 \times 10^{-19}$ C

Charge conservation: In any reaction the net electric charge remains constant.

Conductor: Permits motion of charge.

Insulator: Does not permit motion of charge.

QUESTIONS

1. Suppose that the Sun has a positive electric charge and each of the planets a negative electric charge, and suppose that there is no gravitational force. In what way would the motions of the planets predicted by this "electric" model of the Solar System differ from the observed motions?

2. The protons in the nucleus of an atom repel each other electrically. What holds the protons (and the neutrons) together, and prevents the nucleus from bursting apart?

3. Describe how you would set up an experiment to determine whether the electric charges on an electron and a proton are exactly the same.

4. Assume that neutrons have a small amount of electric charge, say, positive charge. Discuss some of the consequences of this assumption for the behavior of matter.

5. If we were to assign a positive charge to the electron and a negative charge to the proton, would this affect the mathematical statement [Eq. (7)] of Coulomb's Law?

6. In the cgs, or Gaussian, system of units, Coulomb's Law is written as $F = qq'/r^2$. In terms of grams, centimeters, and seconds, what are the units of electric charge in this system?

7. Could we use the electric charge of an electron as an atomic standard of electric charge to define the coulomb? What would be the advantages and disadvantages of such a standard?

8. Besides electric charge, what other physical quantities are conserved in reactions among particles? Which of these quantities are quantized?

9. Since the free electrons in a piece of metal are free to move any which way, why don't they all fall to the bottom of the piece of metal under the influence of the pull of gravity?

10. If the surface of a piece of metal acts like a container in confining the free electrons, why can't we cause these electrons to spill out by drilling hole in the surface?

11. If you rub a plastic comb, it will attract hairs or bits of paper, even though they have no net electric charge. Explain.

12. Some old-fashioned physics textbooks define positive electric charge as the kind of charge that accumulates on a glass rod when rubbed with silk. What is wrong with this definition?

13. When you rub your shoes on a carpet, you sometimes pick up enough electric charge to feel an electric shock if you subsequently touch a radiator or some other metallic body connected to the ground. Why is this more likely to happen in winter than in summer?

14. Some automobile operators hang a conducting strap on the underside of their automobile, so that this strap drags on the street. What is the purpose of this arrangement?

15. Some electric charge has been deposited on a ping-pong ball. How could you find out whether the charge is positive or negative?

16. Two aluminum spheres of equal radii hang from the ceiling on insulating threads. You have a glass rod and a piece of silk. How can you give these two spheres exactly equal amounts of electric charge?

PROBLEMS

Section 22.2

1. Within a typical thundercloud there are electric charges of -40 C and $+40$ C separated by a vertical distance of 5 km. Treating these charges as pointlike, find the magnitude of the electric force of attraction between them.

2. A crystal of NaCl (common salt) consists of a regular arrangement of ions of Na^+ and Cl^-. The distance from one ion to its neighbor is 2.82×10^{-10} m. What is the magnitude of the electric force of attraction between the two ions? Treat the ions as point charges.

3. The two protons in the nucleus of a helium atom are at a distance of 2×10^{-15} m from each other. What is the magnitude of the electric force of repulsion that they exert on each other? What would be the acceleration of each if this were the only force acting on them? Treat the protons as point charges.

4. An alpha particle (charge $+2e$) is launched at high speed toward a nucleus of uranium (charge $+92e$). What is the magnitude of the electric force on the alpha particle when it is at a distance of 5×10^{-14} m from the nucleus? What is the corresponding instantaneous acceleration of the alpha particle?

5. According to recent theoretical and experimental investigations, the subnuclear particles are made of quarks and of antiquarks (see Interlude C). For example, a positive pion is made of a u quark and a d antiquark. The electric charge on the u quark is $\frac{2}{3}e$ and that on the d antiquark is $\frac{1}{3}e$. Treating the quarks as classical particles, calculate the electric force of attraction between the quarks in the pion if the distance between them is 1.0×10^{-15} m.

6. A small charge of 2×10^{-6} C is at the point $x = 2$ m, $y = 3$ m in the x–y plane. A second small charge of -3×10^{-6} is at the point $x = 4$ m, $y = -2$ m. What is the electric force that the first charge exerts on the second? What is

the force that the second charge exerts on the first? Express your answers as vectors, with x and y components.

7. Two tiny chips of plastic of mass 5×10^{-5} g are separated by a distance of 1 mm. Suppose that they carry equal and opposite electrostatic charges. What must the magnitude of the charge be if the electric attraction between them is to equal their weight?

8. Although the best available experimental data are consistent with Coulomb's Law, they are also consistent with a modified Coulomb's Law of the form

$$F = \frac{1}{4\pi\varepsilon_0} \frac{q_1 q_2}{r^2} e^{-r/r_0}$$

where r_0 is a constant with the dimensions of a length and a numerical value which is known to be no less than 10^9 m and is probably much larger. Assuming that $r_0 = 10^9$ m, what is the fractional deviation between Coulomb's Law and the modified Coulomb's Law for $r = 1$ m? For $r = 10^4$ m? (Hint: Use the approximation $e^x \cong 1 + x$ for small x.)

9. A proton is at the origin of coordinates. An electron is at the point $x = 0.40$ Å, $y = 0.20$ Å, $z = 0.15$ Å. What are the x, y, and z components of the electric force that the proton exerts on the electron? That the electron exerts on the proton?

10. Precise experiments have established that the magnitudes of the electric charges of an electron and a proton are equal to within an experimental error of $\pm 10^{-21}e$ and that the electric charge of neutron is zero to within $\pm 10^{-21}e$. Making the worst possible assumption about the combination of errors, what is the largest conceivable electric charge of an oxygen atom consisting of 8 electrons, 8 protons, and 8 neutrons? Treating the atoms as point particles, compare the electric force between two such oxygen atoms with the gravitational force between these atoms. Is the net force attractive or repulsive?

11. Suppose that under the influence of the electric force of attraction the electron in a hydrogen atom orbits around the proton on a circle of radius 0.53×10^{-10} m. What is the orbital speed? What is the orbital period?

12. According to Bohr's theory of the atom, the electron in a hydrogen atom orbits around the nucleus in a circular orbit. The force that holds the electron in this orbit is the Coulomb force. The size of the orbit depends on the angular momentum — the smallest possible orbit has an angular momentum $\hbar = 1.05 \times 10^{-34}$ J·s; the next possible orbit has angular momentum $2\hbar$; the next, $3\hbar$, etc.
   (a) Calculate the radius of each of these three possible circular orbits.
   (b) In general, show that if a circular orbit has an angular momentum $n\hbar$ (where $n = 1, 2, 3, \ldots$), then its radius is

$$r = \frac{4\pi\varepsilon_0}{m_e e^2} n^2 \hbar^2$$

   (c) Evaluate this radius for $n = 1$.

13. A maximum electric charge of 7.5×10^{-6} C can be placed on a metallic sphere of radius 15 cm before the surrounding air suffers electric breakdown (sparks). How many excess electrons (or missing electrons) does the sphere have when breakdown is about to occur?

14. How many electrons are in a paper clip of iron of mass 0.3 g?

15. The electric charge in one mole of protons is called **Faraday's constant.** What is its numerical value?

16. What is the number of electrons and of protons in a human body of mass 73 kg? The chemical composition of the body is given in Problem 1.24.

17. Suppose that you remove all the electrons in a copper penny of mass 2.7 g and place them at a distance of 2.0 m from the remaining copper ions. What would be the electric force of attraction on the electrons?

18. How many extra electrons would we have to place on the Earth and on the Moon so that the electric repulsion between these bodies cancels their gravitational attraction? Assume that the numbers of extra electrons on the Earth and on the Moon are in the same proportion as the radial dimensions of these bodies (6.38:1.74).

19. At a place directly below a thundercloud, the induced electric charge on the surface of the Earth is $+10^{-7}$ coulomb per square meter of surface. How many singly charged positive ions per square meter does this represent? The number of atoms on the surface of a solid is typically 3×10^{20} per square meter. What fraction of these atoms must be ions to account for the above electric charge?

20. The electric charge flowing through an ordinary 110-volt, 150-watt light bulb is 1.5 C/s. How many electrons per second does this amount to?

Section 22.3

21. Consider the following hypothetical reactions involving the collision between a high-energy proton (from an accelerator) and a stationary proton (in the nucleus of a hydrogen atom serving as target):

$$p + p \rightarrow n + n + \pi^+$$

$$p + p \rightarrow n + p + \pi^0$$

$$p + p \rightarrow n + p + \pi^+$$

$$p + p \rightarrow p + p + \pi^0 + \pi^0$$

$$p + p \rightarrow n + p + \pi^0 + \pi^-$$

where the symbols p, n, π^+, π^-, and π^0 stand for proton, neutron, positively charged pion, negatively charged pion, and neutral pion. Which of these reactions are impossible?

22. Consider the reaction

$$Ni^{++} + 4H_2O \rightarrow NiO_4^{--} + 8H^+ + [electrons]$$

How many electrons does this reaction release?

23. We can silver-plate a metallic object, such as a spoon, by immersing the spoon and a bar of silver in a solution of silver nitrate ($AgNO_3$). If we then connect the spoon and the silver bar to an electric generator and make a current flow from one to the other, the following reactions will occur at the immersed surfaces (Figure 22.10):

$$Ag^+ + [electron] \rightarrow Ag_{(metal)}$$

$$Ag_{(metal)} \rightarrow Ag^+ + [electron]$$

The first reaction deposits silver on the spoon and the second removes silver from the silver bar. How many electrons must we make flow from the silver bar to the spoon in order to deposit 1 g of silver on the spoon?

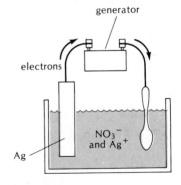

Fig. 22.10

CHAPTER 23

The Electric Field

Up to this point we have taken the view that the gravitational forces and the electric forces between particles are **action-at-a-distance**, i.e., a particle exerts a direct gravitational or electric force on another particle even though these particles are not touching. Although such an interpretation of gravitational and electric forces as a ghostly tug-of-war between distant particles is suggested by Newton's Law of gravitation and by Coulomb's Law of electric force, Newton himself expressed considerable misgivings about this interpretation. In his own words:

> It is inconceivable, that inanimate brute matter, should, without the mediation of something else, which is not material, operate upon and affect other matter without mutual contact. That Gravity should be innate, inherent and essential to Matter so that one Body may act upon another at a Distance thro' a *Vacuum* without the Mediation of anything else, by and through which their Action and Force may be conveyed from one to another, is to me so great an Absurdity that I believe no Man who has in philosophical Matters a competent Faculty of thinking can ever fall into it.

Field　　According to the modern view, there is indeed an entity that conveys the force from one particle to another by contact. This entity is the **field**. In the present chapter we will become acquainted with the electric field that conveys the electric force from one charge to another distant charge. But we must first take a look at what happens to electric forces when many electric charges interact simultaneously.

23.1 The Superposition of Electric Forces

According to Eq. (22.7), the electric force exerted by a point charge q' on a point charge q is

$$\mathbf{F} = \frac{1}{4\pi\varepsilon_0} \frac{qq'}{r^2} \hat{\mathbf{r}} \qquad (1)$$

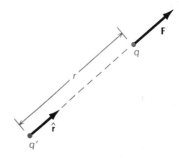

where $\hat{\mathbf{r}}$ is a unit vector pointing from the charge q' toward the charge q (Figure 23.1).

If several point charges q_1, q_2, q_3, $\ldots$, simultaneously exert electric forces on a charge q, then the collective force on q is obtained by taking the vector sum of the individual forces. Thus, the force on q is

$$\mathbf{F} = \mathbf{F}_1 + \mathbf{F}_2 + \mathbf{F}_3 + \cdots$$

$$= \frac{1}{4\pi\varepsilon_0} \frac{qq_1}{r_1^2} \hat{\mathbf{r}}_1 + \frac{1}{4\pi\varepsilon_0} \frac{qq_2}{r_2^2} \hat{\mathbf{r}}_2 + \frac{1}{4\pi\varepsilon_0} \frac{qq_3}{r_3^2} \hat{\mathbf{r}}_3 + \cdots \qquad (2)$$

Fig. 23.1 A charge q' exerts an electric force **F** on the charge q.

where r_1, r_2, r_3, $\ldots$, and $\hat{\mathbf{r}}_1$, $\hat{\mathbf{r}}_2$, $\hat{\mathbf{r}}_3$, $\ldots$, are the distances and unit vectors, respectively (Figure 23.2).

Equation (2) expresses the **principle of linear superposition** of electric forces. According to Eq. (2), the force contributed by each charge is independent of the presence of the other charges; for instance, charge q_2 does not affect the interaction of q_1 with q, it merely adds its own interaction with q. This simple combination law is an important *empirical* fact about electric forces. Since the contact forces (pushes and pulls) of everyday experience are electric forces, they will likewise obey the superposition principle and they can be combined by simple vector addition. Incidentally, the gravitational forces on the Earth and within the Solar System also obey the linear superposition principle. Thus, all the forces in our immediate environment obey this principle (see Section 5.3).

Principle of linear superposition

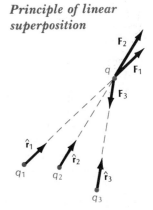

Fig. 23.2 Charges q_1, q_2, q_3 exert electric forces $\mathbf{F}_1$, $\mathbf{F}_2$, $\mathbf{F}_3$ on the charge q.

EXAMPLE 1. Point charges Q and $-Q$ are separated by a distance d. A point charge q is equidistant from these charges, at a distance x from their midpoint (Figure 23.3). What is the electric force on q?

SOLUTION: With the choice of axes shown in Figure 23.3, the distance between the charges $\pm Q$ and q is $\sqrt{x^2 + d^2/4}$. Hence the magnitudes of the Coulomb forces exerted by $+Q$ and $-Q$ are

$$F_1 = F_2 = \frac{1}{4\pi\varepsilon_0} \frac{qQ}{x^2 + d^2/4} \qquad (3)$$

The vector $\mathbf{F}_1$ points away from Q and $\mathbf{F}_2$ points toward $-Q$. In the vector sum $\mathbf{F}_1 + \mathbf{F}_2$, the x components obviously cancel and only the z component survives. The latter component has a magnitude

$$F_z = -\frac{1}{4\pi\varepsilon_0} \frac{qQ}{x^2 + d^2/4} \cos\theta - \frac{1}{4\pi\varepsilon_0} \frac{qQ}{x^2 + d^2/4} \cos\theta \qquad (4)$$

With $\cos\theta = \frac{1}{2}d/(x^2 + d^2/4)^{1/2}$, this gives

$$F_z = -\frac{1}{4\pi\varepsilon_0} \frac{qQd}{(x^2 + d^2/4)^{3/2}} \qquad (5)$$

Note that at large distance from the two charges, d^2 can be neglected com-

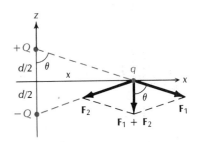

Fig. 23.3 The charges $+Q$ and $-Q$ exert forces $\mathbf{F}_1$ and $\mathbf{F}_2$ on the charge q.

pared to x^2 so that $(x^2 + d^2/4)^{3/2} \cong x^3$. The force is then proportional to $F \propto 1/x^3$, i.e., the force decreases in inverse proportion to the cube of the distance. Thus, although the force generated by each point charge is an inverse-square force, the net force has a quite different behavior because at large distance the force generated by one charge tends to cancel the force generated by the other charge.

EXAMPLE 2. Charge is distributed uniformly along a very long, thin line (say, along a string of silk). If the amount of the charge is λ coulomb per meter of line, what is the electric force on a charge q placed near the line?

SOLUTION We will pretend that the line of charge is of infinite length; this simplifies the calculation and introduces no appreciable error since, in any event, most of the force on q is contributed by those portions of the line which are nearest to q.

Figure 23.4 shows the line of charge along the z axis and the charge q on the x axis, at a distance x from the origin. To find the force, we must perform an integration, regarding the line of charge as made up of infinitesimal line elements, each of which can be treated as a point charge (Figure 23.4). Before proceeding with such an integration it is always best to determine the *direction* of the force by a preliminary, qualitative argument. The force generated by the line element dz shown in Figure 23.4 has both x and z components. Upon integration, the z component will cancel against the z component contributed by a line element lying at the same distance above the origin. Therefore the net force will only have an x component.

The line element dz carries a charge $dQ = \lambda\, dz$. Since this line element can be treated as a point charge, it generates a force

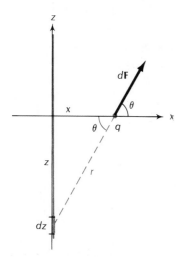

Fig. 23.4 A small segment dz of a line of charge exerts a force on the charge q. The angle θ is reckoned as positive if z is positive, and negative if z is negative.

$$dF = \frac{1}{4\pi\varepsilon_0}\frac{q\,dQ}{r^2} = \frac{1}{4\pi\varepsilon_0}\frac{q\lambda\,dz}{r^2} \tag{6}$$

This force has an x component:

$$dF_x = \frac{1}{4\pi\varepsilon_0}\frac{q\lambda\cos\theta\,dz}{r^2} \tag{7}$$

The net force is then

$$F_x = \int_{-\infty}^{+\infty}\frac{1}{4\pi\varepsilon_0}\frac{q\lambda\cos\theta}{r^2}\,dz \tag{8}$$

To evaluate this, it is convenient to express all variables in terms of θ. From Figure 23.4,

$$z = x\tan\theta \tag{9}$$

i.e.,

$$dz = x\sec^2\theta\,d\theta \tag{10}$$

Furthermore,

$$r = x\sec\theta \tag{11}$$

With these substitutions we obtain an integral over the angle θ; in terms of this angle, the limits of integration are $\theta = -90°$ and $\theta = 90°$, or, in radians, $\theta = -\pi/2$ and $\theta = \pi/2$:

$$F_x = \frac{1}{4\pi\varepsilon_0}\frac{q\lambda}{x}\int_{-\pi/2}^{\pi/2}\cos\theta\,d\theta \tag{12}$$

Since

$$\int_{-\pi/2}^{\pi/2} \cos \theta \, d\theta = \left[\sin \theta \right]_{-\pi/2}^{\pi/2} = 2$$

we find

$$F_x = \frac{1}{2\pi\varepsilon_0} \frac{q\lambda}{x} \tag{13}$$

Note that this force decreases in inverse proportion to the distance (not the square of the distance). The direction of the force is radially away from the line of charge.

23.2 The Electric Field

As we remarked in the introduction to this chapter, the simplest interpretation of Coulomb's Law is that the electric force between charges is action-at-a-distance, i.e., a charge q' exerts a direct force on a charge q even though these charges are separated by a large distance and are not touching. Such an interpretation of electric force leads to serious difficulties in the case of moving charges. Suppose that we suddenly move the charge q' somewhat nearer to the charge q; then the electric force has to increase. But the required increase cannot occur instantaneously — the increase can be regarded as a signal from q' to q and it is a fundamental rule of Special Relativity that no signal can propagate faster than the speed of light (as discussed in Section 17.3). This suggests that, when we suddenly move the charge q', some kind of disturbance travels through empty space from q' to q at the speed of light and, when this disturbance reaches q, it adjusts the electric force to the new increased value (Figure 23.5). Thus charges exert forces on one another by means of disturbances that they generate in the space surrounding them. These disturbances are called **electric fields.**

Fields are a form of matter — they are endowed with energy and with momentum and they therefore exist in a material sense. In the context of the above example it is easy to see why the disturbance, or field, generated by the sudden displacement of q' must carry momentum and energy: as we suddenly move q' toward q, the force on q' immediately increases according to Coulomb's Law, but the increase in the force on q will be delayed until a signal has had time to propagate from q' to q carrying the information regarding the changed position of q'. Thus action and reaction will be temporarily out of balance, i.e., the momentum of the two particles will not be conserved. In order to maintain an overall momentum balance, the momentum missing from the particles must be transferred to the field, i.e., the field must acquire momentum. This special case of the two charged particles is representative of the general case — it can be demonstrated rigorously that, when relativity is taken into account, the momentum and energy of a general system of interacting particles cannot be conserved by itself. An extra entity, such as the field, is needed to take up the momentum and energy missing from the particles.

Although the above arguments for the existence of fields arose from the problem of charges in motion, we will now adopt the very natural

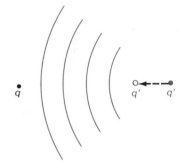

Fig. 23.5 A disturbance emanates from q' and reaches q.

Electric field

view that the forces on charges at rest involve the same mechanism. We suppose that each charge generates a permanent, static disturbance in the space surrounding it and that this disturbance exerts forces on other charges. Thus we take the view that the electric interaction between charges is **action-by-contact**: a charge q' generates a field which fills the surrounding space and exerts forces on any other charges that it touches. The field serves as the mediator of the force according to the scheme

Action-by-contact

$$\text{charge } q' \rightarrow \text{field of charge } q' \rightarrow \text{force on charge } q \qquad (14)$$

The formal definition of the electric field is as follows: To find the field at a given position, take a point charge q (a "test charge") and place it at that position. The charge will then feel an electric force **F**; the electric field is defined as the force **F** divided by the charge q,

Relationship between electric force and field

$$\boxed{\mathbf{E} = \mathbf{F}/q} \qquad (15)$$

Thus, the electric field is simply the force per unit positive charge.[1]

The unit of electric field is the newton/coulomb.[2] Table 23.1 gives the magnitudes of some typical electric fields.

Table 23.1. SOME ELECTRIC FIELDS

At surface of uranium nucleus	2×10^{21} N/C
At surface of pulsar	$\sim 10^{14}$ N/C
At orbit of electron in hydrogen atom	6×10^{11} N/C
In X-ray tube	5×10^{6} N/C
Electrical breakdown of air	3×10^{6} N/C
In Van de Graaff accelerator	2×10^{6} N/C
Within lightning bolt	10^{4} N/C
Under thundercloud	1×10^{4} N/C
Near radar transmitter (FPS-6)	7×10^{3} N/C
In sunlight (rms)	1×10^{3} N/C
In atmosphere (fair weather)	1×10^{2} N/C
In beam of small laser (rms)	1×10^{2} N/C
In fluorescent lighting tube	10 N/C
In radio wave	$\sim 10^{-1}$ N/C
Within household wiring	$\sim 3 \times 10^{-2}$ N/C
In thermal radiation in intergalactic space (rms)	3×10^{-6} N/C

The electric field surrounding a charge or a distribution of charges is a function of position. For example, according to Coulomb's law, a point charge q' exerts a force $(1/4\pi\varepsilon_0)q'q\hat{\mathbf{r}}/r^2$ on a charge q; hence the electric field generated by q' at a distance r is

---

[1] The procedure involved in this definition of the electric field implicitly assumes that all the charges that generate the field **E** remain fixed in their positions while the test charge is brought up. To avoid disturbances to these charges, it is usually convenient to take a very small charge q.

[2] Newton/coulomb is the same thing as volt/meter (see Section 25.1).

$$\mathbf{E} = \frac{1}{4\pi\varepsilon_0} \frac{q'}{r^2} \hat{\mathbf{r}} \qquad (16)$$

Note that the electric field is a vector. Its direction must be specified either by unit vectors [as in Eq. (16)] or else by components.

The net electric field generated by any collection of charges can be calculated by linear superposition of the individual fields, in much the same way as is done for electric forces.

EXAMPLE 3. A charge Q is uniformly distributed along the circumference of a thin ring of radius R. What is the electric field on the axis of the ring?

SOLUTION: We regard the ring as made up of infinitesimal line elements ds (Figure 23.6), each of which can be treated as a point charge. Before proceeding with the integration of the contributions of all such line elements, let us determine the direction of the net electric field. The field generated by the line element ds shown in Figure 23.6 has both a horizontal and a vertical component. Obviously, for any given line element there is an equal line element on the opposite side of the ring's center that contributes an electric field of opposite horizontal component, i.e., the horizontal components cancel pairwise. The net electric field is therefore vertical.

The charge per unit length along the circumference is $Q/2\pi R$; hence, the charge in the line element ds is $dQ = ds(Q/2\pi R)$. At a height z above the plane of the ring (Figure 23.6), the electric field contributed by the charge element dQ has a magnitude

$$dE = \frac{1}{4\pi\varepsilon_0} \frac{dQ}{r^2} = \frac{1}{4\pi\varepsilon_0} \frac{Q\,ds}{2\pi R} \frac{1}{z^2 + R^2} \qquad (17)$$

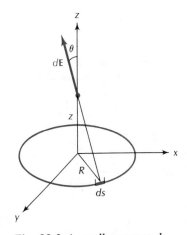

Fig. 23.6 A small segment ds of a ring of charge contributes an electric field dE.

This field has a vertical component

$$dE_z = \frac{1}{4\pi\varepsilon_0} \frac{Q\,ds}{2\pi R} \frac{\cos\theta}{z^2 + R^2} \qquad (18)$$

Consequently the net electric field is

$$E_z = \int \frac{1}{4\pi\varepsilon_0} \frac{Q}{2\pi R} \frac{\cos\theta}{z^2 + R^2} ds \qquad (19)$$

Since the integrand has the same value at all points of the circumference, the integral is of the form [constant] $\times \int ds$; this equals [constant] $\times 2\pi R$, since the length of the circumference is $2\pi R$. Hence,

$$E_z = \frac{1}{4\pi\varepsilon_0} \frac{Q\cos\theta}{z^2 + R^2} = \frac{1}{4\pi\varepsilon_0} \frac{Qz}{(z^2 + R^2)^{3/2}} \qquad (20)$$

where $\cos\theta = z/\sqrt{z^2 + R^2}$ has been inserted. In vector notation,

$$\mathbf{E} = \frac{1}{4\pi\varepsilon_0} \frac{Qz}{(z^2 + R^2)^{3/2}} \hat{\mathbf{z}} \qquad (21)$$

Note that for $z = 0$, $\mathbf{E} = 0$; and for $z \gg R$, $\mathbf{E} \cong Q\hat{\mathbf{z}}/(4\pi\varepsilon_0 z^2)$, which is the electric field of a point charge.

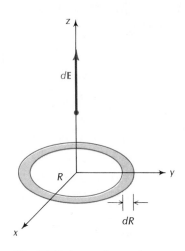

Fig. 23.7 A very large sheet of charge lies in the *x–y* plane. A thin ring of charge within this sheet produces an electric field *d***E**.

EXAMPLE 4. What is the electric field generated by a large sheet (sheet of paper) carrying a uniform charge density of σ coulomb per square meter?

SOLUTION: We will pretend that the sheet is infinitely large. The sheet can be regarded as made up of a collection of rings. Figure 23.7 shows one of these rings with radius R, width dR; this ring has an area $2\pi R\, dR$ and a charge $dQ = (2\pi R\, dR) \times \sigma$. This produces a vertical electric field [see Eq. (21)],

$$dE = \frac{1}{4\pi\varepsilon_0} \frac{2\pi R\sigma z\, dR}{(z^2 + R^2)^{3/2}} \tag{22}$$

The net electric field is therefore

$$E = \frac{2\pi\sigma z}{4\pi\varepsilon_0} \int_0^\infty \frac{R\, dR}{(z^2 + R^2)^{3/2}} \tag{23}$$

where $R = 0$ is the radius of the smallest ring and $R = \infty$ the radius of the largest.

With the substitution of $u = R^2$, the integral becomes

$$\int_0^\infty \frac{R\, dR}{(z^2 + R^2)^{3/2}} = \int_0^\infty \frac{(1/2)\, du}{(z^2 + u)^{3/2}} = \left[\frac{-1}{(z^2 + u)^{1/2}} \right]_0^\infty = \frac{1}{z}$$

and therefore

$$E = \frac{2\pi\sigma z}{4\pi\varepsilon_0} \frac{1}{z} = \frac{\sigma}{2\varepsilon_0} \tag{24}$$

In vector notation,

Electric field of flat sheet

$$\boxed{\mathbf{E} = \frac{\sigma}{2\varepsilon_0}\hat{\mathbf{z}}} \tag{25}$$

This is a constant electric field, i.e., the electric field of a uniform sheet of charge is the same at points near and at points far from the plane of the sheet. (Of course, if the sheet of charge is of finite size, then this constancy of the electric field is only true in the vicinity of the sheet; at a very large distance from a *finite* sheet of charge the electric field will resemble that of a point charge.)

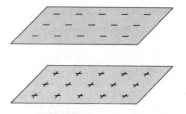

Fig. 23.8 Two very large sheets of charge.

The electric field of two parallel uniform sheets with opposite charges (Figure 23.8) can be obtained by superposing the fields generated by each sheet. In the space between the sheets, the two fields add together, giving a net field

$$E = \sigma/\varepsilon_0 \tag{26}$$

where σ is the magnitude of the charge per unit area on either sheet. Above and below the pair of sheets, the two fields cancel.

EXAMPLE 5. An electron is placed in the uniform electric field between the two charged sheets of Figure 23.8. (a) If the magnitude of the field is 3.0×10^4 N/C, what is the acceleration of the electron? (b) Suppose the electron is initially at rest on the negative sheet and then moves toward the positive sheet under the influence of the electric force. With what speed will it reach the positive sheet? The distance between the sheets is 1.0 cm.

SOLUTION: The force on the electron has a magnitude $F = eE$ and the acceleration has a magnitude

$$a = \frac{eE}{m_e} = \frac{(1.6 \times 10^{-19}\ \text{C}) \times (3 \times 10^4\ \text{N/C})}{0.91 \times 10^{-30}\ \text{kg}} = 5.3 \times 10^{15}\ \text{m/s}^2$$

For constant acceleration, $v^2 - v_0^2 = 2a(x - x_0)$ [see Eq. (2.25)]. With $v_0 = 0$, $a = 5.3 \times 10^{15}\ \text{m/s}^2$, and $x - x_0 = 0.01$ m this gives

$$v = \sqrt{2a(x - x_0)} = 1.0 \times 10^7\ \text{m/s}$$

23.3 Lines of Electric Field

The electric field can be represented graphically by drawing, at any given point of space, a vector whose magnitude and direction are those of the electric field at that point. Figure 23.9 shows the electric field vectors in the space surrounding a positive point charge.

Alternatively, the electric field can be represented graphically by **field lines.** These lines are drawn in such a way that, at any given point, the tangent to the line has the direction of the electric field. Furthermore, the density of lines is directly proportional to the magnitude of the electric field; that is, where the lines are closely spaced the electric field is strong and where the lines are far apart the field is weak. Figure 23.10 shows the electric field lines of a positive point charge and Figure 23.11 those of a negative point charge. The arrows on these lines indicate the direction of the electric field along each line.

Field lines

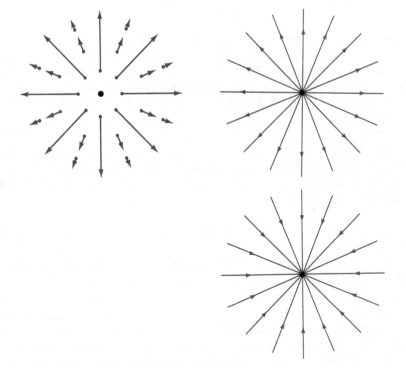

Fig. 23.9 (top left) Electric field vectors surrounding a positive point charge.

Fig. 23.10 (top right) Electric field lines of a positive point charge. Note that in three dimensions the lines spread out in all three directions of space, whereas the diagram shows the lines spreading out only in the two directions within the page. This gives a misleading impression of the density of field lines as a function of distance. This limitation of two-dimensional diagrams should always be kept in mind when looking at pictures of field lines.

Fig. 23.11 (bottom) Electric field lines of a negative point charge.

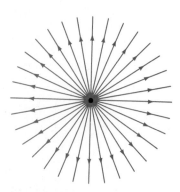

Fig. 23.12 Electric field lines of a positive point charge, twice as large as in Figure 23.10.

When we draw a pattern of field lines, we must begin each line on a positive point charge and end on a negative point charge. Since the magnitude of the electric field is directly proportional to the amount of electric charge, the number of field lines that we draw emerging from a positive charge must be proportional to the charge. Figure 23.12 shows the electric field lines of a positive charge twice as large as that of Figure 23.10. We will adopt the convention that the number of field lines emerging from a charge Q is Q/ε_0; hence the number of lines emerging from one coulomb of charge is $1/\varepsilon_0 = 1.13 \times 10^{11}$. This normalization is very convenient for making computations with field lines, but it is not always quite practical for making drawings — depending on the magnitude of the charge, it sometimes yields an enormous number of field lines so that drawing them becomes an unbearable chore, sometimes a fractional number, so that drawing them becomes altogether meaningless. For instance, according to our normalization, a proton with $e = 1.6 \times 10^{-19}$ C, has $e/\varepsilon_0 = 1.8 \times 10^{-8}$ field line. Such a number makes good sense in a computation, but no sense at all in a drawing. If we want a draftsman to prepare a drawing of the field lines of a proton, we will first have to alter our normalization. From an artistic point of view, it is desirable that the spacing between the field lines be small compared to the distance from the charge or any other relevant distance; this produces a clear picture of the spatial dependence of the electric field. In case of need, we can alter the normalization — but we must be careful to maintain a fixed normalization throughout any given computation or series of drawings.

The inverse-square law for the electric field of a point charge can be "derived" from the picture of field lines — it is easy to show that the density of lines necessarily obeys an inverse-square law. Before we do this, we must give a precise definition of density of field lines. Density of lines is the number of lines per unit area, i.e., the number of lines intercepted by a small area A erected perpendicularly to the lines (Figure 23.13) divided by the magnitude of this area,

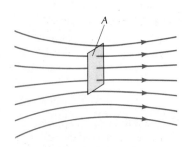

Fig. 23.13 A small area A intercepts some field lines.

$$[\text{density of lines}] = \frac{[\text{number of intercepted lines}]}{A}$$

The area A used in this equation must be small, but not too small: it must be small compared to the distance over which the electric field varies appreciably, but large compared to the spacing between the field lines, so that it intercepts a large number of field lines. If we choose A too small, the density of field lines would not be a well-defined, continuous function of position — the number of intercepted field lines and their density would jump to zero whenever the area A fits between adjacent field lines. By keeping A sufficiently large, we smooth out irrelevant, erratic fluctuations in the density of field lines. Obviously, the restrictions on the choice of the area A are analogous to the restrictions on the volume element used for the computation of the density of a compressible fluid: the volume element must be small compared to the distance over which the density varies appreciably, but large compared with the distance between the molecules.

Consider now a point charge q'. There will be q'/ε_0 lines emerging from this point charge, uniformly distributed over all radial directions. At a distance r from the point charge, these lines are uniformly distributed over the area $4\pi r^2$ of a concentric sphere, i.e., there are

$(q'/\varepsilon_0)/4\pi r^2$ lines per unit area. Hence the density of lines decreases in proportion to the inverse square of the distance. In fact, with our normalization regarding the number of lines per coulomb of charge, the density of lines is not only proportional to, but exactly *equal* to the magnitude of the electric field [see Eq. (16)]. This "derivation" of the Coulomb Law is really only a consistency check — it is because the Coulomb Law is an inverse-square law that the field can be represented by field lines; any other dependence on distance would make it impossible to draw continuous field lines that start and end only on charges.

Figure 23.14 shows the field lines generated jointly by a positive and a negative charge of equal magnitudes; in the midplane between the charges the electric force has the magnitude given by Eq. (5). Figure 23.15 shows the field lines of a pair of equal positive charges and Figure 23.16 those of a pair of unequal positive and negative charges. Figure 23.17 shows the field lines of a large, uniformly charged sheet with a positive density of charge.

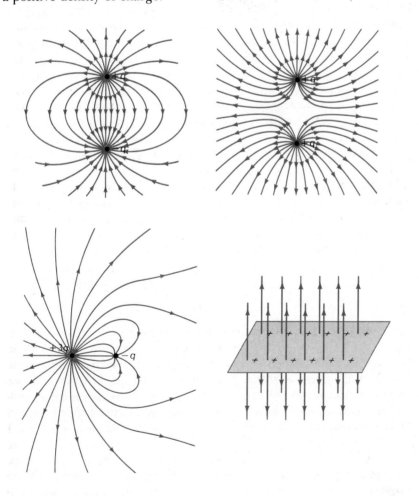

Fig. 23.14 (left) Field lines generated by positive and negative charges of equal magnitudes.

Fig. 23.15 (right) Field lines generated by two positive charges of equal magnitudes.

Fig. 23.16 (left) Field lines generated by positive and negative charge of unequal magnitudes. The positive charge has three times the magnitude of the negative charge.

Fig. 23.17 (right) Field lines of a very large sheet of charge.

Note that in all cases, the field lines start on positive charges and end on negative charges — the positive charges are **sources** of field lines and the negative charges are **sinks.** Also, note that field lines never intersect (except where they start or end at point charges). If the lines ever were to intersect, the electric field would have *two* directions at the point of intersection; this is impossible.

The above pictures of field lines help us to develop some intuitive feeling for the spatial dependence of the electric fields surrounding diverse arrangements of electric charges. But we must not fall into the trap of thinking of the field lines as physical objects. The field lines are merely mathematical crutches to aid our imagination.

23.4 Electric Dipole in an Electric Field

If a neutral body is placed in a given electric field, we might expect that the body experiences no force. However, this expectation is not always realized. A neutral body may contain within it separate positive and negative charges (of equal magnitudes) and it is possible that the electric force on one of these charges is larger than on the other; the body then experiences a net force. Such an imbalance of the forces on the negative and positive charges will happen if the electric field is stronger at the location of one kind of charge than at the location of the other. For example, the body shown in Figure 23.18 with positive charges on one end and negative charges on the other end will be pushed to the left because the electric field that acts on the body is stronger at the location of the negative charges. Note that this electric field — indicated by field lines in Figure 23.18 — is not the field generated by the body; rather it is an electric field generated by some other charges (not shown in Figure 23.18). The electric field that acts on a body is often called the **external field;** in contrast, the field generated by the body itself is called the **self-field.** The latter field only exerts internal forces within the body and does not contribute to the net force acting on the body from the outside.

If the external electric field is uniform, then the forces on the positive and negative charges in a neutral body cancel and there is no net force. However, there may still be a torque. Figure 23.19 shows a neutral body in a uniform electric field. The body carries equal positive and negative charges $\pm Q$, with the average positions of these charges separated by a distance l. Such a body is called an **electric dipole.** Obviously there is a torque on the body. The torque of each force about the center is $-\frac{1}{2}lQE \sin \theta$ and the total torque is

$$\tau = -lQE \sin \theta \tag{27}$$

where θ is the angle between the direction of the electric field and the line from the negative charge to the positive charge. The minus sign in Eq. (27) indicates that the torque is clockwise, in the sense of negative θ; the torque tends to align the body with the electric field.

One can write Eq. (27) as

$$\tau = -pE \sin \theta \tag{28}$$

where

$$\boxed{p = lQ} \tag{29}$$

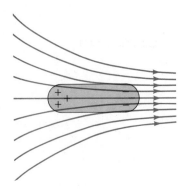

Fig. 23.18 An elongated body with positive charge at one end, negative charge on the other end. The body is placed in a nonuniform electric field.

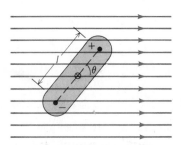

Fig. 23.19 Electric dipole in a uniform electric field. The dots indicate the average positions of the positive and the negative charges, respectively.

Electric dipole

Dipole moment

The quantity p is called the **dipole moment** of the body; it is simply the charge multiplied by the separation between the charges. The units of dipole moment are meter · coulomb.

In vector notation, we can write the torque as

$$\boxed{\tau = \mathbf{p} \times \mathbf{E}}$$ (30)   *Torque on dipole*

where vector **p** is directed from the negative charge toward the positive charge.

Corresponding to the torque (28) there exists a potential energy that equals the amount of work that *you* must do against the electric forces to twist the dipole through some angle. If you want to twist the dipole through some angle, you must supply a torque $-\tau = +pE \sin\theta$ [opposite to the torque (28)] and do work[3]

$$U = \int_{\theta_0}^{\theta} -\tau \, d\theta' = \int_{\theta_0}^{\theta} pE \sin\theta' \, d\theta'$$

$$= pE\left[-\cos\theta'\right]_{\theta_0}^{\theta} = -pE\cos\theta + pE\cos\theta_0$$

It is customary to take the starting angle as $\theta_0 = 90°$. Then

$$U = -pE\cos\theta$$ (31)

or

$$\boxed{U = -\mathbf{p} \cdot \mathbf{E}}$$ (32)   *Potential energy of dipole*

This potential energy has a minimum when **p** and **E** are parallel, a maximum when they are antiparallel. Figure 23.20 is a plot of the potential energy U vs. the angle θ.

Many asymmetric molecules have **permanent dipole moments** due to an excess of electrons on one end of the molecule and a corresponding deficit on the other. This means that the molecule has a negative charge on one end and positive charge on the other. For example, Figure 23.21 shows a water molecule. In this molecule the electrons tend to concentrate on the oxygen atom; in Figure 23.21, the left side of the molecule is negatively charged and the right side positively charged. Since the average positions of the positive and negative charges do not coincide, the water molecule has a dipole moment. For a water molecule in water vapor, $p = 6.1 \times 10^{-30} \ \text{C} \cdot \text{m}$.

Molecules of atoms that do not have a permanent dipole moment may acquire a temporary dipole moment when placed in an electric field. The opposite electric forces on the negative and positive charges can distort the molecule and produce a charge separation. Such a dipole moment, which only lasts as long as the molecule is immersed in the electric field, is called an **induced dipole moment**. The magnitude of the induced dipole moment is directly proportional to the magnitude of the electric field. (This proportionality holds for electric fields of the strengths we will be concerned with in this text; however, it fails for electric fields of extreme strength.)

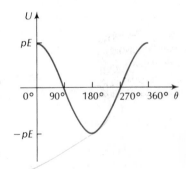

Fig. 23.20 Potential energy of an electric dipole as a function of angle.

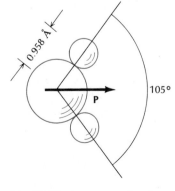

Fig. 23.21 A water molecule.

---

[3] The variable of integration in this integral has been written θ' in order to distinguish it from the limit of integration θ.

EXAMPLE 6. A molecule of water with a dipole moment of 6.1×10^{-30} C·m is placed in an electric field of 2.0×10^5 N/C. What is the difference between the potential energies for parallel and antiparallel orientations?

SOLUTION: When the dipole moment is parallel to the electric field, the potential energy is

$$U = -pE \cos 0° = -6.1 \times 10^{-30} \text{ C·m} \times 2.0 \times 10^5 \text{ N/C}$$

$$= -1.2 \times 10^{-24} \text{ J}$$

When the dipole moment is antiparallel,

$$U = -pE \cos 180° = +6.1 \times 10^{-30} \text{ C·m} \times 2.0 \times 10^5 \text{ N/C}$$

$$= 1.2 \times 10^{-24} \text{ J}$$

Hence the energy difference between the parallel and antiparallel orientations is 2.4×10^{-24} J.

The tendency for alignment of a dipole with an electric field can be exploited to make the field lines visible. For this purpose, small bits of thread are suspended in oil in a container placed in the electric field; alternatively, small grass seeds are scattered on a sheet of paper placed in the electric field. The electric field induces a dipole moment along the long axis of the bit of thread or the grass seed, and the torque on the dipole then aligns the bit of thread or the grass seed with the electric field. Figures 23.22 and 23.23 show two photographs of field lines made visible by bits of thread (Figures 28.1 and 28.2 show photographs of field lines made visible with grass seeds).

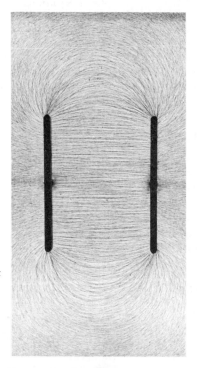

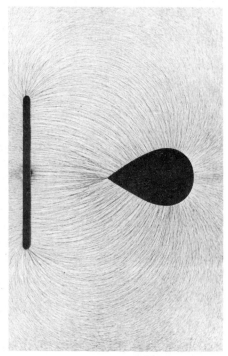

Fig. 23.22 (left) Field lines between a pair of charged parallel plates, made visible by small bits of thread aligned with the field lines. Note the fringing, or spreading, of the field lines near the edges of the plates. (Courtesy H. Waage, Princeton University.)

Fig. 23.23 (right) Field lines between a charged pointed body and a charged flat plate, made visible by small bits of thread aligned with the field lines. Note the strong concentration of field lines at the sharp point, indicating a strong electric field. (Courtesy H. Waage, Princeton University.)

SUMMARY

Superposition principle: Electric fields combine additively (as vectors).

Definition of electric field: $\mathbf{E} = \mathbf{F}/q$

Electric field of point charge: $\mathbf{E} = \dfrac{1}{4\pi\varepsilon_0}\dfrac{q'}{r^2}\,\hat{\mathbf{r}}$

Electric field of uniformly charged plane: $E = \sigma/2\varepsilon_0$

Electric dipole moment: $p = lQ$

Torque on dipole: $\boldsymbol{\tau} = \mathbf{p} \times \mathbf{E}$

Potential energy of dipole: $U = -\mathbf{p}\cdot\mathbf{E}$

QUESTIONS

1. Does it make any difference whether the value of the charge q in the equation defining the electric field [Eq. (15)] is positive or negative?

2. In the cgs, or Gaussian, system of units, Coulomb's Law is written as $F = qq'/r^2$. In terms of grams, centimeters, and seconds, what are the units of the electric field in this system?

3. How would you formally define a gravitational field vector? Is the unit of gravitational field the same as the unit of electric field? According to your definition, what are the magnitude and the direction of the gravitational field at the surface of the Earth?

4. During days of fair weather, the Earth has an atmospheric electric field that points vertically down. This electric field is due to charges on the surface of the Earth. What must be the sign of these charges?

5. A large flat sheet measures $L \times L$; the sheet carries a uniform distribution of charge. Roughly how far from the center of the sheet would you expect the electric field to be markedly different from the uniform electric field of an infinitely large sheet?

6. Figure 23.24 shows diagrams of hypothetical field lines corresponding to some static charge distributions, which are beyond the edge of the diagram. What is wrong with these field lines?

7. If a positive point charge is released from rest in an electric field, will its orbit coincide with a field line? What if the point charge has zero mass?

8. A **tube of force** is the volume enclosed between a bundle of adjacent field lines (Figure 23.25; such a tube of force is analogous to a flow tube in hydrodynamics, and the field lines are analogous to stream lines). Along such a tube of force, the magnitude of the electric field varies in inverse proportion to the cross-sectional area of the tube. Explain.

9. A negative point charge $-q$ sits in front of a very large flat sheet with a uniform distribution of positive charge. Make a rough sketch of the pattern of field lines. Is there any point where the electric field is zero?

10. A very long straight line of positive charge lies along the z axis. A very large flat sheet of positive charge lies in the x–y plane. Sketch a few of the field lines of the net electric field produced by both of these charge distributions acting together.

11. A large, flat, thick slab of insulator has positive charge uniformly distrib-

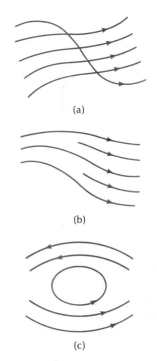

(a)

(b)

(c)

Fig. 23.24

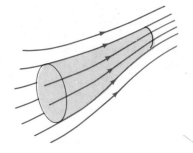

Fig. 23.25

uted over its volume. Sketch the field lines on both sides and inside the slab; pay careful attention to the starting points of the field lines.

12. If our universe is topologically closed, so that a straight line drawn in any direction returns on itself from the opposite direction, can the net charge in the universe be different from zero?

13. How could you build a "compass" that indicates the direction of the electric field?

14. When a neutral metallic body, insulated from the ground, is placed in an electric field, it develops a charge separation, acquiring positive charge on one end and negative charge on the other. This means the body acquires an induced dipole moment. How is the direction of this dipole moment related to the direction of the electric field?

15. By inspection of Figure 23.22, make a rough, qualitative plot of the electric field strength as a function of position along a line parallel to the plates, midway between the plates.

16. One electric dipole is at the origin, oriented parallel to the z axis. Another electric dipole is at some distance on the x axis. The electric field of the first dipole then exerts a torque on the second dipole. For what orientation of the second dipole is the potential energy minimum?

PROBLEMS

Sections 23.1 and 23.2

1. Electric fields as large as 3.4×10^5 N/C have been measured by airplanes flying through thunderclouds. What is the force on an electron exposed to such a field? What is its acceleration?

2. The Earth has not only a magnetic field, but also an atmospheric electric field. During days of fair weather (no thunderclouds), this atmospheric electric field has a strength of about 100 N/C and it points down. Taking into account this electric field and also gravity, what will be the acceleration (magnitude and direction) of a grain of dust of mass 1.0×10^{-18} kg carrying a single *electron* charge?

3. **Millikan's experiment** measures the elementary charge e by the observation of the motion of small oil droplets in an electric field. The oil droplets are charged with one or several elementary charges and, if the (vertical) electric field has the right magnitude, the electric force on the droplet will balance its weight, holding the drop suspended in midair. Suppose that an oil droplet of radius 1.0×10^{-4} cm carries a single elementary charge. What electric field is required to balance the weight? The density of oil is 0.80 g/cm³.

4. In an X-ray tube (see Figure B.1), electrons are exposed to an electric field of 8×10^5 N/C. What is the force on an electron? What is its acceleration?

5. According to a theoretical estimate, at the surface of a neutron star of mass 1.4×10^{30} kg and radius 1.0×10^4 m there is an electric field of magnitude 6×10^3 N/C pointing vertically up. Show that the corresponding electric force on a proton more than balances the gravitational force on the proton.

6. A long hair, taken from a girl's braid, has a mass of 1.2×10^{-3} g. The hair carries a charge of 1.3×10^{-9} C distributed along its length. If we want to suspend this hair in midair, what (uniform) electric field do we need?

7. The electric field in the electron gun of a TV tube is supposed to accelerate electrons uniformly from 0 to 3.3×10^7 m/s within a distance of 1.0 cm. What electric field is required?

8. Electric breakdown (sparks) occurs in air if the electric field reaches

3×10^6 N/C. At this field strength, free electrons present in the atmosphere are quickly accelerated to such large velocities that upon impact on atoms they knock electrons off the atom and thereby generate an avalanche of electrons. How far must a free electron move under the influence of the above electric field if it is to attain a kinetic energy of 3×10^{-19} J (that is sufficient to produce ionization)?

9. The nuclei of the atoms in a chunk of metal lying on the surface of the Earth would fall to the bottom of the metal if their weight were the only force acting on them. Actually, within the interior of any metal exposed to gravity there exists a very small electric field that points vertically up. The corresponding electric force on a nucleus just balances the weight of the nucleus. Show that for a nucleus of atomic number Z, mass m, the required field has a magnitude $mg/(Ze)$. What is the numerical value of this electric field in a chunk of iron?

10. The hydrogen atom has a radius of 0.53×10^{-10} m. What is the magnitude of the electric field that the nucleus of the atom (a proton) produces at this radius?

11. What is the strength of the electric field at the surface of a uranium nucleus? The radius of the nucleus is 7.4×10^{-15} m and the electric charge is $92e$. For the purposes of this problem the electric charge may be regarded as concentrated at the center.

12. Figure 23.26 shows the arrangement of nuclear charges (positive charges) of a KBr molecule. Find the electric field that these charges produce at the center of mass at a distance of 0.93×10^{-10} m from the Br atom.

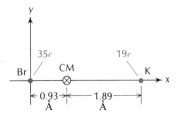

Fig. 23.26 The positive (nuclear) charges in a KBr molecule.

13. The distance between the oxygen nucleus and each of the hydrogen nuclei in an H_2O molecule is 0.958 Å; the angle between the two hydrogen atoms is 105° (Figure 23.27). Find the electric field produced by the nuclear charges (positive charges) at the point P at a distance of 1.2 Å to the right of the oxygen nucleus.

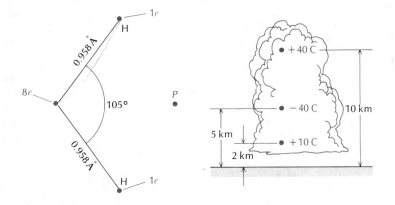

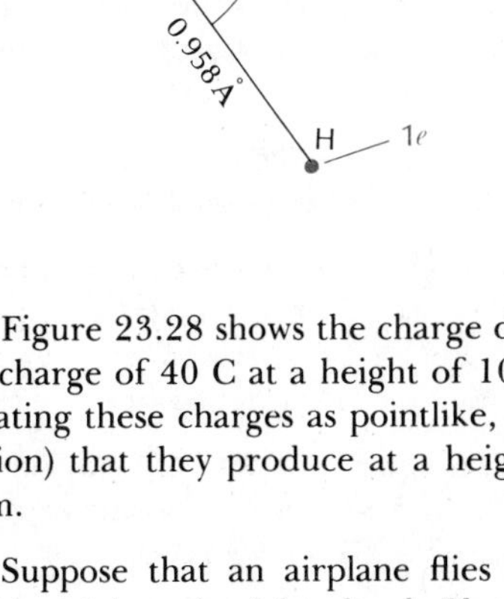

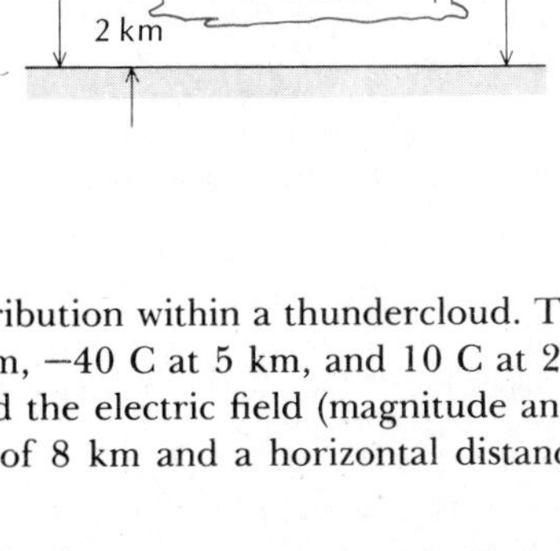

Fig. 23.27 (left) The positive (nuclear) charges in a water molecule.

Fig. 23.28 (right) Charges in a thundercloud.

14. Figure 23.28 shows the charge distribution within a thundercloud. There is a charge of 40 C at a height of 10 km, −40 C at 5 km, and 10 C at 2 km. Treating these charges as pointlike, find the electric field (magnitude and direction) that they produce at a height of 8 km and a horizontal distance of 3 km.

15. Suppose that an airplane flies through the thundercloud described in Problem 14 at the 8-km level. Plot the magnitude of the electric field as a function of position along the path of the airplane; start with the airplane 10 km away from the thundercloud.

16. Suppose that the charge distribution of a thundercloud can be approxi-

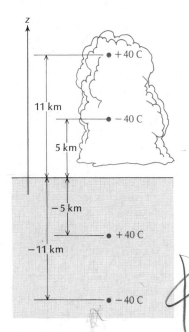

Fig. 23.29 Charges in a thundercloud and their images in the ground.

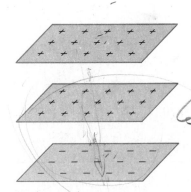

Fig. 23.30 Three parallel sheets.

mated by two point charges, a negative charge of −40 C at a height of 5 km (above ground) and a positive charge of 40 C at a height of 11 km. To find the electric field strength at the ground we must take into account that the ground is a conductor and that the charge of the thundercloud induces charges on the ground. It can be shown that the effect of the induced charges can be simulated by a point charge of 40 C at 5 km *below ground* and a point charge of −40 C at 11 km below ground (Figure 23.29); these fictitious charges are called *image charges* (described in Section I.4). By adding the electric field of the image charges to that of the two real charges in the thundercloud, calculate the magnitude of the electric field at a point on the ground directly below the thundercloud charges. Similarly, calculate the magnitude of the electric field at horizontal distances of 2, 4, 6, 8, and 10 km. Plot the field vs. the distance.

17. Consider eight of the ions of Cl^- and Na^+ in a crystal lattice of common salt. The ions are located at the vertices of a cube measuring 2.82×10^{-10} m on an edge. Calculate the magnitude of the electric force that seven of these ions exert on the eight.

18. Each of two very long, straight, parallel lines carries a positive charge of λ coulomb per meter. The distance between the lines is d. Find the electric field at a point equidistant from the lines, with a distance $2d$ from each line. Draw a diagram showing the direction of the electric field.

19. Two infinite lines of silk with uniform charge distributions of λ coulomb per meter lie along the x and the y axes, respectively. Find the electric field at a point with coordinates x, y, z; assume that $x > 0, y > 0, z > 0$.

20. A semi-infinite line carrying a uniform charge distribution of λ coulomb lies along the positive x axis from $x = 0$ to $x = \infty$. Find the components of the electric field at the point with coordinates x, y, with $z = 0$; assume $x > 0, y > 0$.

21. A semi-infinite line with a uniform charge distribution of $+\lambda$ coulomb per meter lies along the positive x axis from $x = 0$ to $x = \infty$. Another semi-infinite line with a charge distribution of $-\lambda$ coulomb per meter lies along the negative x axis from $x = 0$ to $x = -\infty$. Find the electric field at a point on the y axis.

22. Electric charge is uniformly distributed over each of three large, parallel sheets of paper (Figure 23.30). The charges per unit area on the sheets are 2×10^{-6} C/m², 2×10^{-6} C/m², and -2×10^{-6} C/m², respectively. The distance between one sheet and the next is 1.0 cm. Find the strength of the electric field **E** above the sheets, below the sheets, and in the space between the sheets. Find the direction of **E** at each place.

23. Each of two very large, flat sheets of paper carries a uniform positive charge distribution of 3.0×10^{-4} C/m². The two sheets of paper intersect at an angle of 45° (Figure 23.31). What are the magnitude and the direction of the electric field at a point between the two sheets?

24. Two large sheets of paper intersect at right angles. Each sheet carries a uniform distribution of positive charge (Figure 23.32). The charge per unit

Fig. 23.31 (left) Two sheets intersecting at 45°.

Fig. 23.32 (right) Two sheets intersecting at right angles.

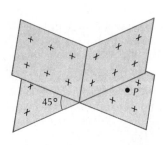

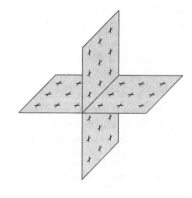

area on the sheets is 3×10^{-6} C/m². Find the magnitude of the electric field in each of the four quadrants. Draw the field lines in each quadrant.

25. The electric field within a chunk of metal exposed to the Earth's gravity (see Problem 9) is due to a distribution of surface charge. Suppose that we have a slab of iron oriented horizontally (Figure 23.33). What must be the surface charge densities on the upper and lower surfaces?

26. A total amount of charge Q is uniformly distributed along a thin, straight plastic rod of length l.
   (a) Find the electric field at the point P, at a distance x from one end of the rod (Figure 23.34).
   (b) Find the electric field at point P', at a distance y from the midpoint of the rod (see Figure 23.34).

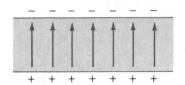

Fig. 23.33 A horizontal slab of iron.

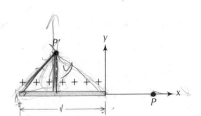

Fig. 23.34 A thin rod, with a uniform distribution of charge.

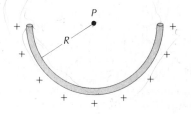

Fig. 23.35 A semicircular rod, with a uniform distribution of charge.

27. A thin, plastic rod is bent so that it has the shape of a semicircle of radius R (Figure 23.35). An amount of charge Q is uniformly distributed along the rod. What is the electric field at the center of the circle?

28. A Plexiglas square of dimension $l \times l$ has a uniform charge density of magnitude λ coulomb per meter along its edges. Two of the edges are positive and two are negative (Figure 23.36). Find the electric field at the center of the square.

29. Three thin glass rods carry charges Q, Q, and $-Q$, respectively. The length of each rod is l and the charge is uniformly distributed along each rod. The rods form an equilateral triangle. Calculate the electric field at the center of the triangle.

30. A disk of radius R carries an amount of charge Q uniformly distributed over its surface. Find the electric field at a point on the axis of the disk at a distance z from the center. Show that in the limiting case $z \gg R$, the result reduces to that for a point charge.

*31. Two thin, semi-infinite rods lie in the same plane. They make an angle of 45° with each other and they are joined by another thin rod bent along an arc of circle of radius R, with center at P (see Figure 23.37). All the rods carry a uniform charge distribution of λ coulomb per meter. Find the electric field at the point P.

*32. A thin, semi-infinite rod with a uniform charge distribution of λ coulomb per meter lies along the positive x axis from $x = 0$ to $x = \infty$; a similar rod lies along the positive y axis from $y = 0$ to $y = \infty$. Calculate the electric field at a point in the x–y plane in the first quadrant.

*33. A cylindrical Plexiglas tube of length l, radius R carries a charge Q uniformly distributed over its surface. Find the electric field on the axis of the tube at one of its ends.

*34. Two thin rods of length L carry equal charges Q uniformly distributed over their lengths. The rods are aligned, and their nearest ends are separated

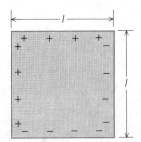

Fig. 23.36 A square, with charge along its edges.

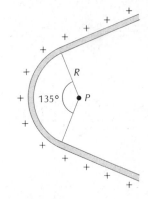

Fig. 23.37 Rods with uniform distributions of charge.

by a distance x (Figure 23.38). What is the electric force of repulsion between these rods?

Section 23.4

35. The two charges of ± 40 C in the thundercloud of Figure 23.28 form a dipole. What is the dipole moment?

36. In a hydrogen atom, the electron is at a distance of 0.53×10^{-10} m from a proton.
    (a) What is the instantaneous dipole moment of this system?
    (b) Taking into account that the electron moves around the proton on a circular orbit, what is the time-average dipole moment of this system?

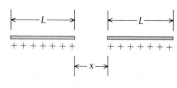

Fig. 23.38

37. (a) Pretend that the HCl molecule consists of (pointlike) ions of H^+ and Cl^- separated by a distance of 1.0×10^{-10} m. If so, what would be the dipole moment of this system?
    (b) The observed dipole moment is 3.4×10^{-30} C · m. Can you suggest a reason for this discrepancy?

38. The dipole moment of a HCl molecule is 3.4×10^{-30} C · m. Calculate the magnitude of the torque that an electric field of 2.0×10^6 N/C exerts on this molecule when the angle between the electric field and the longitudinal axis of the molecule is $45°$.

39. The dipole moment of the water molecule is 6.1×10^{-30} C · m. In an electric field, a molecule with a dipole moment will tend to settle into an equilibrium orientation such that the dipole moment is parallel to the electric field. If disturbed from this equilibrium orientation, the molecule will oscillate like a torsional pendulum. Calculate the frequency of small oscillations of this kind for a water molecule about an axis through the center of mass (and perpendicular to the plane of the three atoms) when the molecule is in an electric field of 2.0×10^6 N/C. The moment of inertia of the molecule about this axis is 1.93×10^{-47} kg · m².

FORCES, FIELDS, AND QUANTA[1]

As noted in Chapter 6, the bewildering varieties of forces that we find in nature — gravitational attractions, elastic forces, viscous forces, friction forces, electric forces, intermolecular and interatomic forces, nuclear forces, etc. — all result from just four fundamental forces: the gravitational force, the electromagnetic force, the "strong" force, and the "weak" force.

The gravitational force acts on everything — it acts on all forms of matter. One usually says it acts on the masses of particles, but it would be more accurate to say that it acts on both mass and energy, i.e., not only are particles under the sway of gravitational attractions, but so is energy. For example, the thermal energy of the gas in the Sun attracts the Earth — about 1 part in 10^6 of the Sun's pull is due to its thermal energy. Since energy has mass (see Chapter 17), it should come as no surprise that energy, and hence any form of matter, is subject to gravitational interactions.

The electric force acts between electric charges, and the magnetic force acts between magnets and between electric currents. However, a closer look reveals that the force between, say, two magnets is really nothing but a kind of extra electric force between the moving charges inside the magnets. Thus, the magnetic force is not fundamentally different from the electric force; these two forces are merely two aspects of a single force called the electromagnetic force, just as space and time are two aspects of a single entity called spacetime. This unification of electricity and magnetism is a consequence of the theory of relativity, just as is the unification of space and time. When Einstein formulated the theory of relativity and proved that space and time transform into one another under Lorentz transformations, he also proved that electric and magnetic fields transform into one another. In Section 30.4 we will examine a special instance of such a transformation; we will see that an electric field transforms into a combination of electric and magnetic fields when we go from one inertial reference frame to another. Electromagnetism is the best-known example of a **unified field theory,** that is, a theory that treats two fields as two aspects of a single underlying field.

The "strong" force acts on the protons and neu-

trons in nuclei and it acts on all the baryons and mesons. The particles that interact via the strong force are often called **hadrons;** thus, baryons and mesons are hadrons, but leptons are not. (Interlude C contains tables of baryons, mesons, and leptons.)

The "weak" force is important for neutrinos: it is the only force that neutrinos ever experience (apart from the gravitational force, which is insignificant for them under laboratory conditions); hence this force is crucial in the collisions between neutrinos and other matter. This force also acts on all other leptons, and on baryons and mesons, but in these cases the effects of the weak force are often hidden behind the much larger effects produced by the strong or the electromagnetic force. The weak force is deeply implicated in many reactions that bring about the disruption of the unstable particles. For example, the weak force is responsible for the disruption, or decay, of the neutron.

Table G.1 lists the strength of each of the fundamental forces and also the range, or the maximum distance over which this force can reach from one particle to another. This table contains the same information as Table 6.1, but in the new table we have expressed the strengths of the forces relative to that of the strong force to which we arbitrarily have assigned a strength of 1.[2]

In this chapter we will explore the mechanism that generates these forces. To attain a full understanding of this mechanism, we would have to begin with a study of quantum theory, which we cannot do here. We will therefore have to leave out many details and deal only with the broad qualitative features of the mechanism.

G.1 REACTION RATES

A basic assumption of Newtonian physics is that the motion of a particle can be described by position as a function of time. But on a microscopic level, this assumption breaks down: in consequence of their quantum-mechanical properties, particles acquire uncertainties in position and in velocity, and their tra-

---

[1] This chapter is optional.

[2] The strengths of the forces depend on the energies of the particles. The values in the table are appropriate for low energies.

Table G.1 THE FOUR FUNDAMENTAL FORCES

Force	Acts on	Relative strength	Range
Gravitational	All forms of matter (all forms of energy)	10^{-38}	Infinite
Weak	Leptons, baryons, and mesons	10^{-6}	Less than 10^{-17} m
Electromagnetic	Charged particles	10^{-2}	Infinite
Strong	Baryons and mesons (hadrons)	1	$\sim 10^{-15}$ m

jectories are not well defined. Worse yet, in high-energy reactions among particles, the number of particles is not well defined. In such reactions particles are created, and destroyed, and created again. Thus, the particles are evanescent objects with uncertain trajectories. Under these circumstances, the concept of force ceases to have an exact meaning. Hence particle physicists prefer to speak of **interactions,** that is, the action of particles on other particles; they speak of "strong" interactions, electromagnetic interactions, "weak" interactions, and gravitational interactions. Mathematically, these interactions can be described by formulas that specify the amount of energy for each interaction — energies remain meaningful even at a microscopic level, and therefore energies are more relevant than forces. Nevertheless, for the purpose of the following qualitative discussion, let us continue to speak of forces even though this intuitive concept is ambiguous and ought to be replaced by a more precise and sophisticated mathematical concept of interaction energies.

For the experimental particle physicist the phenomena most readily accessible to observation are collisions, either collisions in which new particles are created by the destruction of some old particles or by the conversion of energy into mass (inelastic reactions), or collisions in which particles merely bounce off one another (elastic reactions). The rates of these reactions depend on the energy, momentum, and angular momentum of the colliding particles. And they also depend on the strength of the force between the particles — the stronger the force, the faster the reaction will proceed. Figure G.1 shows an extremely fast reaction recorded in a photographic emulsion. A very high-energy cosmic ray — probably a proton — smashes into the nucleus of an atom and produces a "shower" of particles. Among the particles created in such collisions are pions, kaons, protons, electrons, etc. These particles are created by the strong force between the incident proton and the protons and neutrons in the nucleus. The time scale of the reaction can be estimated as follows: the incident proton travels at

nearly the speed of light and hence passes through the nucleus very quickly; since the dimension of the nucleus is of the order of 10^{-15} m, the time available for the interaction can be calculated by dividing this distance by the speed of light; this gives about 10^{-23} s. Thus, the typical time scale for such reactions is extremely short and the reactions are extremely fast.

The strong force can bring about not only particle

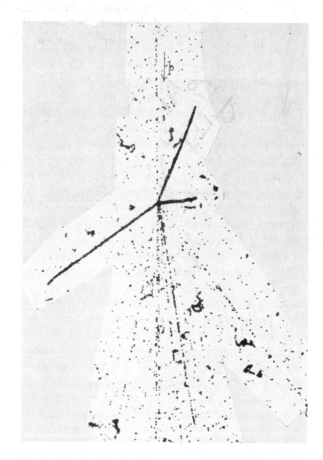

Fig. G.1 Tracks of particles in a photographic emulsion. The faint track at the top was made by a high-energy cosmic ray. This cosmic ray collided with a nucleus of an atom in the photographic emulsion. In this collision many new particles were created. Their tracks diverge from the point of collision.

creation but also particle decay. The time scale for such decays is about the same as the time scale for creation: about 10^{-23} s. Hence those particles that decay strongly lead a very ephemeral existence; they are very unstable and burst apart within an instant of their creation. For example, the ρ meson decays into two pions (π) by the strong reaction

$$\rho^- \to \pi^- + \pi^0$$

within 10^{-23} s. An object that only lasts such a short time cannot be detected directly; its existence can only be inferred from the correlations observed in the motion of the decay products (see Interlude C).

Reactions caused by electromagnetic interactions take longer. Figure G.2 shows the decay of a pion into an electron (e), antielectron ($\bar{e}$), and gamma ray (γ) by the electromagnetic reaction

$$\pi^0 \to e + \bar{e} + \gamma$$

which takes about 10^{-14} s (this meson can also decay into two gamma rays, which is a somewhat faster reaction). Although 10^{-14} s is a short time by ordinary standards, it is a long time by the standards of high-energy

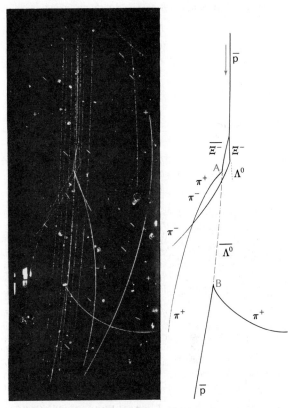

Fig. G.3 (left) Photograph of tracks of particles in a bubble chamber and (right) tracing based on the photograph. The dashed line indicates the trajectory of a $\overline{\Lambda}$ particle; it is created at A, travels a distance of about 20 cm, and then decays into a $\overline{p}$ and a π^+ at B.

physicists; a particle that takes this "long" to decay is regarded by them as stable!

The reactions involving the weak force take much longer. Figure G.3 shows the decay of an antilambda baryon ($\overline{\Lambda}$) into an antiproton ($\overline{p}$) and a pion (π^+),

$$\overline{\Lambda} \to \overline{p} + \pi^+$$

This reaction is weak and takes about 10^{-10} s. By the standards of high-energy physicists, this is a *very* long time. Moving at close to the speed of light, the $\overline{\Lambda}$ travels a distance of several centimeters between its creation and its decay.

Note that the above reactions, with time scales of 10^{-23}, 10^{-14}, and 10^{-10} s for strong, electromagnetic, and weak forces, are intended only as more or less typical examples. Rates of other reactions involving these forces can differ by several, or even by many, powers of 10 from the above. Thus, the decay of the neutron into a proton, an electron, and an antineutrino,

$$n \to p + e + \bar{\nu}$$

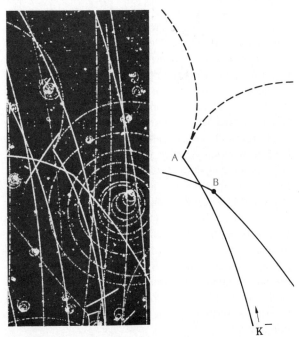

Fig. G.2 (left) Photograph of tracks of particles in a bubble chamber and (right) tracing based on the photograph. Neutral particles do not make visible tracks in the bubble chamber, but their extrapolated trajectories have been indicated by dashed lines in the tracing. A Λ baryon and a π^0 were created in a collision at A. The π^0 decayed immediately into an e, $\bar{e}$, and γ. The tracks of the e and $\bar{e}$ can be seen diverging from the point of decay, but the γ made no track in the photograph.

Table G.2 FORCES AND CONSERVED QUANTITIES

Force	Energy, momentum, and angular momentum	Charge	Baryon number	Lepton number	Strangeness	Parity	Isospin
Gravitational	✓	✓	✓	✓			
Weak	✓	✓	✓	✓			
Electromagnetic	✓	✓	✓	✓	✓	✓	
Strong	✓	✓	✓	✓	✓	✓	✓

is a weak reaction that is extremely slow — it takes about 15 min. This delay in the reaction is to be blamed on the small amount of energy available for the decay process. Although there is some overlap among the rates of the reactions generated by the different kinds of forces, as a rough rule we can state that the strong force generates the fastest reactions, the electromagnetic force slower reactions, and the weak force the slowest reactions.[3] The gravitational force between individual particles is so feeble that it is incapable of generating reactions that can be observed in our laboratories.

G.2   CONSERVATION LAWS AND SYMMETRY

All reactions conserve energy, momentum, angular momentum, and electric charge, i.e., the energy, momentum, angular momentum, and electric charge before and after any reaction among particles are exactly the same. Furthermore, all reactions ever observed conserve baryon number and lepton number (as discussed in Section C.4).

Besides these exactly conserved quantities there are some others that are only approximately conserved; these other quantities — isospin, parity, and strangeness — are conserved in some reactions but not in all (see Section C.4). Which reactions conserve what quantities depends on the type of force involved in the reaction; Table G.2 lists the four forces and some of the quantities conserved by each.

The strong force obeys all conservation laws; the electromagnetic interaction violates the conservation of isospin; the weak force violates the conservation of isospin, of parity, and of strangeness; and the gravitational force violates the same laws as the weak force. The violation of conservation laws in gravitation is an indirect effect. All forms of energy gravitate, and when the weak and electromagnetic energies gravitate, they can produce reactions that violate some of the conservation laws — the weak and electromagnetic forces contaminate gravitation with their violations.

There is a profound connection between conserva-

tion laws governing a physical system and the symmetries of that system. This connection is contained in the following general theorem: *to every symmetry there corresponds a conservation law.* A symmetry of the system is any operation that leaves the system invariant, i.e., unchanged. To understand the meaning of the general theorem, it will be best to examine a particular case, say, the conservation of momentum. This conservation law can be regarded as a consequence of the invariance of a system under translations in space. Consider a system of particles, such as a water molecule; if it makes no difference whether we place this system at one point of space or at another, then the system has invariance under translations in space. Obviously, if an external force were acting on the system, it would make a difference where we placed the system — at some points of space the potential energy associated with the external force would be high, at others it would be low. Hence invariance under translations in space demands the absence of external forces and this, in turn, implies the conservation of the momentum of the system.

By similar arguments, one can show that the conservation of angular momentum is related to invariance under rotations in space; conservation of energy is related to invariance under translations in time; and the conservation of other quantities listed in Table G.2 is related to other symmetries of a somewhat more abstruse kind, which we cannot discuss in detail here.

The exploitation of symmetries has proved very profitable in the study of the strong and the weak forces. Both of these forces are extremely complicated; the known symmetries help to place restrictions on the mathematical formulas describing these forces. High-energy physicists are engaged in a continuing search for further symmetries (and further conservation laws).

G.3   FIELDS AND QUANTA

According to classical theory, forces are mediated by fields. Distant particles do not act on one another directly; rather, each particle generates a field of force and this field acts on the other particles. As we saw in Section 23.2, the existence of fields is required by

[3] The reaction rates depend on energy. This comparison of the reaction rates is valid for the typical energies attained in laboratories.

conservation of energy and momentum. Fields play the role of storehouses of energy and momentum; the energy and momentum stored in the fields balance any excess or deficit in the energy and momentum of the interacting particles engaged in (nonuniform) motion. For example, radio waves and light waves are electromagnetic fields that store the energy and momentum lost by charges moving on radio antennas and in atoms. The most spectacular example of the conversion of particle energy into field energy occurs in the annihilation of matter with antimatter: if an electron collides with an antielectron, the two particles annihilate one another, giving off a burst of very energetic light, or gamma rays. In this annihilation the energy of the particles — including their rest-mass energy — is completely converted into field energy. The reverse reaction is also possible: if a gamma ray collides with a charged particle, it can create an electron–antielectron pair. In such a pair creation, the energy of the gamma ray is converted into the energy of the pair of particles (Figure G.4).

Each of the four fundamental forces is mediated by fields of its own. Hence there are gravitational fields, electromagnetic fields, strong fields, and weak fields. On a macroscopic scale, the gravitational and electric fields of everyday experience are smooth functions of space and time; these functions can be calculated by classical field theory. The electric-field problems that we solved in Chapter 23 are examples of such classical calculations. However, on a microscopic scale, physics is ruled by quantum theory and not by classical theory. Quantum theory introduces a new fundamental feature: the quantization of energy. It turns out that the energy stored in fields is not smoothly distributed; rather the energy is found in **quanta,** that is, small packets or lumps of energy.

Fig. G.4 The two spiraling tracks in this bubble-chamber photograph were made by an electron and an antielectron. These particles were created by a high-energy gamma ray in a collision with the electron of a hydrogen atom in the bubble chamber. The long, slightly curved downward track was made by the recoiling electron.

The quanta of electric and magnetic energy in a radio wave or a light wave are called **photons.** In an ordinary radio or light wave, the photons are not directly noticeable, just as the molecules in air are not directly noticeable. There are so many molecules in each cubic centimeter of air — about 10^{19} molecules per cubic centimeter under standard conditions — that for most practical purposes air can be described as a continuous fluid with a smoothly varying pressure and density. Likewise, there are so many photons in each cubic centimeter of a light wave — about 10^7 photons per cubic centimeter in bright sunlight — that for most purposes the wave can be described as a smoothly varying electric and magnetic field. However, the "lumps" in light can be detected by placing electrons in their path and watching for collisions; the recoil of an electron hit by a photon announces the presence of the latter. In their collisions with electrons, the photons display particle properties: they have energy, momentum, and spin-angular momentum. Their mass is zero, as is required by the theory of relativity, according to which only a zero-mass particle can move at the speed of light yet have finite energy [see Eq. (17.51)].

Table G.3 lists the quanta corresponding to the fields that mediate the four kinds of forces. Our belief in gravitons rests largely on theoretical considerations; these particles have not yet been detected and, given the extremely small strengths of gravitational forces, it is unlikely that they will be directly observed soon.

Table G.3 FIELDS AND THEIR QUANTA

Field	Quanta	Mass[a]
Gravitational	Gravitons	0 MeV
Weak	W particles	81,000
	Z particles	93,000
Electromagnetic	Photons	0
Strong	Pions and some other mesons	140

[a] The masses are expressed in energy units, as in Tables C.3–C.5.

Pions are easy to observe; they are bona fide particles, even more so than photons. Pions have energy, momentum, electric charge (0, +e, or −e, depending on the type of pion; see Table C.5). Their spin is zero, but in contrast to photons their mass is not zero. The notion that a "solid" particle such as a pion should play the role of quantum of a field seems a bit bizarre. But quantum theory teaches us that pions, like all other particles, have a dual character: they are both waves and particles (see Section 41.6). In their character of waves, pions are described by a wave field; taking into account that the energy of this field must be quantized, it turns out that the corresponding quanta are nothing but the pions themselves. This identification of particles and quanta holds in general: *all particles are quanta of fields.* Pions are quanta of the **pion field,** electrons are quanta of the **electron field,** protons are quanta of the **proton field,** etc. Unfortunately, this means that there is one kind of field for each kind of particle — there are very many kinds of fields. Since it is unlikely that the many, many particles we know of are all elementary, it is unlikely that the many, many fields are fundamental. Thus, if the proton is a composite structure made of three quarks (see Section C.6), then the proton field is also a composite structure made of a combination of three quark fields.

In view of this multitude of new fields, the question arises whether each of the new fields will act as mediator and carry some kind of force between particles. The answer is yes: all fields can play an intermediate role and mediate forces between particles. For example, the electron field can act between two photons, permitting them to exert forces on one another. Diverse meson fields (ρ-meson field, ω-meson field, etc.) can act between two protons, permitting them to exert strong forces in addition to the forces generated via the pion field; in fact, the diversity of these additional corrections is one of the reasons why the theoretical analysis of the strong force is so dreadfully difficult.

At the quantum level, we can picture the field of force generated by a particle as a swarm of quanta buzzing around the particle. For example, we can picture the electric field surrounding an electron or any other charged particle as a swarm of photons. The swarm is in a state of everlasting activity — the charged particle continually emits and reabsorbs the photons of the swarm. Emission is creation of a photon; absorption is annihilation of a photon. Hence we can say that the electric field arises from the continual interplay of three fundamental processes: creation, propagation, and annihilation of photons. The action of one charged particle on another involves a sequence of these three fundamental processes: a photon is emitted by one particle, propagates through the intervening distance, and is absorbed by the other particle. This exchange process can be represented graphically by a diagram showing the worldlines of the particles (Figure G.5); such a diagram is called a **Feynman diagram.** The photon exchanged between the two electrons is called a **virtual photon** because it lasts only a very short time and, being reabsorbed by an electron, is undetectable by any direct experiment. The steady attractive or repulsive force between two charged particles is generated by continual repetition of this exchange process. This is action-by-contact with a vengeance — at a fundamental level, all forces

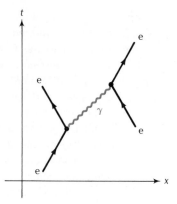

Fig. G.5 Feynman diagram representing the exchange of a virtual photon between two electrons. The solid black lines indicate the worldlines of the two electrons, with the *t* axis plotted vertically and the *x* axis plotted horizontally. The wavy colored line indicates the worldline of the photon. The electron on the left emits this photon and the electron on the right absorbs it.

reduce to local acts of creation and destruction involving particles in direct contact.

In terms of a simple analogy, we can easily understand how the exchange of particles brings about forces. Imagine two boys tossing a ball back and forth between them (Figure G.6); it is intuitively obvious that this produces a net repulsive force between the boys due to the recoil they suffer when throwing or catching the ball. Our intuition suggests that no such exchange process can ever produce attraction. However, imagine two Australian boys tossing a boomerang back and forth between them (Figure G.7); it is then obvious that this produces an attractive force between the boys.[4] Whether a photon exchanged between two charges behaves like a ball or like a boomerang depends on the signs of the charges. Quantum calculations, which take into account the wave nature of all the particles involved, show that the net force is attractive for unlike charges and repulsive for like charges, as it should be.

This theory of electric forces is called **quantum electrodynamics** (or QED); it was developed by P. A. M. Dirac, R. P. Feynman, J. Schwinger, S. Tomonaga,[5] and

---

[4] One of the flaws in this analogy is that the momentum of the boomerang changes during its flight through the air. But this is not a fatal flaw because we can assume that the initial and the final direction of motion are the same (a 180° turn), so that the boomerang merely borrows momentum from the air, without any *net* change.

[5] Paul Adrien Maurice Dirac, 1902–1984, English physicist; he received the Nobel Prize in 1933 for his contributions to quantum theory. Richard P. Feynman, 1918–, American physicist, Julian Schwinger 1918–, American physicist, and Sin-Itiro Tomonaga, 1906–1979, Japanese physicist, shared the Nobel Prize in 1965 for their work on quantum electrodynamics.

Fig. G.6 Two boys throw a ball back and forth.

Fig. G.7 Two boys throw a boomerang back and forth.

other physicists between 1930 and 1950. Quantum electrodynamics is the most accurate theory in all of physics, and in all of science. Some theoretical calculations in quantum electrodynamics have been carried out to nine significant figures and delicate experimental measurements have confirmed these calculations exactly.

The gravitational, weak, and strong forces are also generated by the exchange of virtual particles: gravitons, W and Z particles, and pions (and some other mesons). It is a general rule of quantum theory that the range of the force is inversely related to the mass of the particle that serves as the carrier of the force. Thus, the photons and the gravitons that are the carriers of the electromagnetic and the gravitational force have zero mass — the ranges of these forces are infinite. The pions that are the carriers of the strong force have a fairly large mass — the range of this force is short. The W and Z particles that are the carriers of the weak force have a very large mass — the range of this force is very short.

G.4 THE UNIFIED THEORY OF WEAK AND ELECTROMAGNETIC FORCES

It has long been one of the aspirations of theoretical physicists to achieve the unification of the forces of nature. The theory of electromagnetic forces is an example of such a unification — the electric and the magnetic forces are nothing but different aspects of a single underlying force. Theoretical physicists tend to

regard this example as rather trite — they are much more interested in a grandiose scheme for the unification of all four forces. Einstein devoted much effort to an attempt at joining the gravitational and the electromagnetic forces, but he failed to reach this goal.

Some years ago, S. Weinberg, A. Salam, and S. Glashow[6] achieved the unification of weak and electromagnetic forces. This new theory has shown that, in spite of the seemingly drastic differences between their characters, the weak and electromagnetic forces are basically the same — they are merely two aspects of a single **electroweak force.**

If we seek to unify the weak and electromagnetic forces, we must regard the carriers of these forces — the quanta whose exchange generates the forces — as closely related. Thus, the proton and the W and Z particles ought to be very similar particles that differ only in small details; all of these particles ought to belong to a family, just like, say, the proton and neutron belong to a family (see Section C.5). The trouble is that the photon and the W and Z particles do not look at all similar: the photon has zero mass while the W and Z particles have very large masses, masses much larger than that of any other particle discovered so far. The unified theory predicts the masses of the W and Z particles by the following argument: If the weak and electromagnetic forces are essentially the same, then they must also have the same strength. The fact that the experimentally observed strengths seem quite different (see Table G.1) is attributed to the masses of the W and Z particles — under certain conditions a strong force can have the appearance of a weak force if the particle that carries the force is very massive. A theoretical calculation shows that at a fundamental level the weak and the electromagnetic forces can be regarded as having the same strength provided that the W and the Z particles have masses of about 80 times the mass of a proton. These large masses explain both the strength of the weak force and its short range.

But here we run into an obstacle that would seem to defeat our efforts to construct a unified theory: on the one hand, we need very massive W and Z particles to account for the relative strengths of the weak and electromagnetic forces, and on the other hand, if these particles are very massive, then they are very different from the photon, which is massless. To overcome this obstacle we need to find a clever explanation of how particles can be related — can be members of the same family — and yet have very different masses.

[6] Stephen Weinberg, 1933–, American physicist, Abdus Salam, 1926–, Indian physicist, and Sheldon Lee Glashow, 1932–, American physicist, shared the Nobel Prize in 1979 for their unified theory of weak and electromagnetic forces.

Such an explanation is supplied by the scheme of **"broken symmetry,"** which could also be called "hidden symmetry." According to this scheme, the photon and the W and Z particles are indeed closely related, but the relationship is hidden from direct view. A discussion of "broken symmetry" is beyond the scope of this textbook, but the following is a simple example that illustrates the meaning of such a symmetry. Consider a pencil standing vertically on its tip on a table. This equilibrium configuration has rotational symmetry: the configuration is invariant under rotations about a vertical axis passing through the pencil. However, the equilibrium is unstable, and sooner or later the pencil will fall flat on the table and settle into a new, stable equilibrium configuraton. When this happens, the rotational symmetry will be destroyed because the pencil has to fall in one definite direction. Such a breakdown of symmetry is called **spontaneous** because it is not attributable to the equation of motion. In fact, the equation of motion is rotationally symmetric, i.e., it has the same form for all the directions in which the pencil can fall, and it therefore does not select any preferred directon. Generalizing from this simple example, we can say that a spontaneously broken symmetry is a symmetry that is present in the equation of motion but absent from the actual equilibrium configuration into which the system settles.

Photons and W and Z particles can be regarded as some kind of equilibrium configurations of fields. Their equations of motion (field equations) are rotationally symmetric; here, the relevant rotation is not an ordinary rotation, but rather a rotation in an abstract mathematical space, a rotation that transforms photons and W and Z particles into each other, and thereby defines their family relationships. In the abstract mathematical space, the photons and the W and Z particles can assume a symmetric equilibrium configuration, in which they all have the same mass. But this equilibrium configuration is unstable and, under normal laboratory conditions, photons and W and Z particles settle into a different, stable equilibrium configuration, with a spontaneous breakdown of the rotational symmetry. The observed mass differences between these particles are a by-product of this breakdown of symmetry. Thus, photons and W and Z particles are closely related, but under ordinary laboratory conditions, their relationship is hidden from view. The symmetry between photons and W and Z particles can be restored by giving all these particles very high energies, in excess of 100 GeV; at such high energies, the particles can attain their symmetric equilibrium configuration, and the symmetry between them would then become manifest. Such high energies are difficult to achieve in our laboratories, but they were readily available during the very early stages of the Big Bang, when the uni-verse was younger than 10^{-10} s and had a temperature in excess of 10^{15} K. It is believed that at these early times the symmetry between photons and W and Z particles was unbroken.

One notable consequence of the unified theory of weak and electromagnetic interactions is that it provides a natural explanation of the quantization of electric charge. As we saw in Section 22.3, the electric charge on all known particles is some integral multiple of the basic unit of charge, $e = 1.60 \times 10^{-19}$ C. It can be shown that this quantization rule emerges as an immediate consequence of the rotational symmetry of the unified theory.[7]

From the experimental point of view, an early success of the unified theory was its prediction of certain hitherto unknown forces between baryons and neutrinos. The weak forces between electrons and neutrinos can be accounted for by exchange of two kinds of W particles: a positively charged particle W^+ and a negatively charged particle W^-. However, the photon and these two charged W particles do not make a complete family by themselves. The scheme of broken symmetry demands that the family contain one more particle: the electrically neutral Z, also called Z^0. The force generated by the exchange of these neutral Z particles is called the neutral-current force. The existence of this new force was soon confirmed by experimenters on high-energy baryon–neutrino collisions at CERN.

The most recent and most impressive success of the unified theory was its prediction of the masses of the W and Z particles. These particles were detected in 1982 in experiments at the proton–antiproton collider at CERN. The experiments involved the observation of about a billion head-on collisions between protons and antiprotons of the same energy, 270 GeV. A few dozen W and Z particles were produced in these collisions (Figure G.8). The measured masses of the W and the Z particles are, respectively, 81 GeV and 93 GeV (expressed in energy units). These measured values are within 2 GeV of the theoretically predicted values. This excellent agreement constitutes a brilliant confirmation of the unified theory of weak and electromagnetic interactions.

G.5 GLUONS AND QUANTUM CHROMODYNAMICS

As described in Interlude C, protons, neutrons, and other hadrons are made of small fundamental particles

---

[7] The unified theory of weak and electromagnetic forces explains the quantization of electric charge only for particles within given families of particles. In order to explain the quantization of electric charge for all particles in all families, it is necessary to go one step further and unify the weak, electromagnetic, and *strong* forces (see next section).

Fig. G.8 Tracks of particles produced in a very energetic head-on collision between a proton of 270 GeV that entered from the right and an antiproton of 270 GeV that entered from the left. In this collision a W⁻ particle was created. It immediately decayed into an electron and a neutrino; the track of the former (indicated by the arrow) can be seen emerging toward the lower right.

called **quarks.** According to recent theoretical and experimental investigations, there exist six kinds, or "flavors," of quarks: **up, down, strange, charmed, top,** and **bottom;** and each of these can have any of three "colors": **red, green,** and **blue.** For example, a proton is made of two u quarks and one d quark; one of these (but not always the same one) is *red,* one *green,* and one *blue.* Furthermore, for each of the quarks there exists an antiquark. An antiproton is made of two u antiquarks and one d antiquark; one of these is anti*red,* one anti*green,* and one anti*blue.*

The quarks are confined inside the proton by very strong mutual attractive forces. These forces between quarks are color forces — the source of these forces is color just as the source of electric forces is electric charge. Each of the three varieties of color (*red, green,* and *blue*) is analogous to a kind of positive electric charge and each of the "anticolors" is analogous to a kind of negative electric charge. A body is color neutral if it contains equal amounts of all three colors or if it contains equal amounts of color and anticolor, just as a body is electrically neutral if it contains equal amounts of positive and negative charge. Thus, protons, as well as other ordinary particles, are color neutral.

The color force is a fundamental force that should be included in our table of fundamental forces instead of the strong force (Table G.1). The color force is closely related to the strong force — the latter is actually a special instance of the former. The relationship between the color force and the strong force is analogous to the relationship between the electric force and the intermolecular force. As we saw in Section 22.1, the force between two electrically neutral atoms or molecules is a residual electric force resulting from an imperfect cancellation among the attractions and repulsions of the charges in the two atoms or molecules.

Likewise, the strong force between, say, two "colorless" protons is a residual color force resulting from an imperfect cancellation among the attractions and repulsions of the quarks in the two protons. Thus, the "strong" force between protons is no more than a pale reflection of the much stronger color forces acting within each proton.

The color force has some very remarkable properties. All the forces listed in Table G.1 *decrease* with distance, but the color force remains constant as the distance between the quarks increases. This persistence of the color force brings about the confinement of quarks. If one of the quarks in, say, a proton is somewhat separated from its companions by the violent impact of a collision, the color attraction pulls it back into its original position as soon as the collision ends. The quarks behave as though linked by rubber strings. During a collision, the rubber strings stretch, but after the collision they again contract (Figure G.9). This analogy is perhaps fairly close to the truth; according to one theory, the color force is communicated from one quark to another along tightly bundled field lines concentrated in thin channels, similar to the strings shown in Figure G.9.

Although the color force keeps the quarks on a short leash and pulls them back sharply whenever they wander too far apart, the force allows the quarks to move rather freely within certain limits. This behavior is in accord with the string analogy — if the quarks stretch the strings, they experience strong restraining forces, but if they stay within the limits set by the relaxed lengths of the strings, they feel almost no force. This situation has been described by the picturesque phrases "infrared slavery" and "ultraviolet freedom" (in this context *infrared* refers to long distances and *ultraviolet* refers to short distances).

In an extremely violent collision, one of the strings

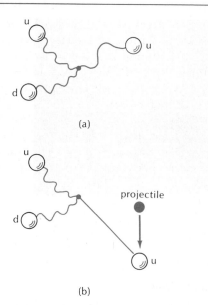

(a)

(b)

Fig. G.9 (a) The three quarks in a proton held together by "strings." (b) When a projectile collides with one of the quarks, the "string" stretches.

might break. But such a break does not result in a free quark with a dangling piece of string hanging out. Rather, the energy stored in the stretched string creates a quark–antiquark pair at the site of the break (Figure G.10). What happens in the breaking of a string is analogous to what happens in the breaking of a bar magnet. Such a magnet has a north pole and a south pole and we might hope to obtain a free north pole and south pole by simply breaking the magnet at the middle. But experience with magnets shows that this does not give us two magnets, each with a single pole. Rather, the act of breaking creates a new south pole and a new north pole, and we are left with two complete magnets (Figure G.11). The creation of new quarks during the breaking of strings makes it impossible to obtain isolated quarks; instead, the breaking of

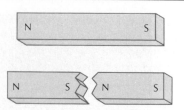

Fig. G.11 A magnet and a broken magnet.

the string results in the creation of an extra, normal particle. For example, Figure G.10 shows how the breaking of a string inside the proton leads to the creation of a pion.

At a fundamental level, the color force between quarks is due to an exchange of virtual particles between the quarks. The particle that acts as the carrier of the color force is the **gluon.** Figure G.12 shows a Feynman diagram representing the exchange of a gluon between two quarks. Such an exchange of a gluon between two colored quarks is analogous to the exchange of a photon between two charged particles (see Figure G.5). However, the gluon exchange is a rather more complicated process than photon exchange — the gluons themselves have color whereas the photons do not have any electric charge. Whenever a quark emits a gluon, this gluon takes color away from the quark and thereby changes the color of the quark. For this reason, the quark entering a vertex in Figure G.12 has a different color than the quark leaving the vertex. The color of the gluon is the difference between the colors of these quarks; for example, the gluon emerging from the left vertex in Figure G.12 has colors *green* and anti*blue*, i.e., it has two colors. Thus, the exchange of a gluon between two quarks entails an exchange of colors. The force between, say, the three quarks within a proton is due to continual, repet-

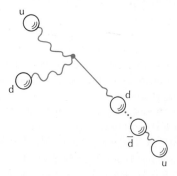

Fig. G.10 In a very violent collision, the string breaks. A quark–antiquark pair (d and d̄) is created at the site of the break. The d̄ antiquark remains tied to the u quark; this combination is a pion, π^+.

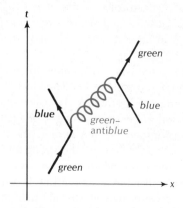

Fig. G.12 Exchange of a gluon between two quarks. The quark on the left suffers a color change from *green* to *blue;* that on the right a color change from *blue* to *green.*

itive exchanges of gluons between the quarks. During this process, the color of each quark changes again and again; for instance, the u quark within the proton is sometimes *red*, sometimes *green*, and sometimes *blue*.

The theory of the color force is called **quantum chromodynamics** (or QCD). The equations of this theory are much more complicated than those of quantum electrodynamics. Because the calculations in chromodynamics are exceedingly difficult and tedious, most of the theoretical predictions that have been obtained so far are qualitative rather than quantitative. Nevertheless, experiments performed at the DESY accelerator in 1978 did reveal some circumstantial evidence confirming the existence of the gluons. In these experiments, physicists attempted to manufacture gluons in electron–antielectron collisions of extremely high energy. Since the gluons only last for a short instant of time, there is no hope of seeing them directly. But chromodynamics predicts that in some violent electron–antielectron collisions, a quark, an antiquark, and a gluon will be produced. Each of these three particles decays into several pions; when emerging from the scene of the collision, these pions should then be arranged in three distinct jets spurting out at distinct angles. The experiments confirmed the existence of such triple jets. Figure G.13 shows a picture of such jets produced at the proton–antiproton collider at CERN.

As a next step, some theoretical physicists are speculating on a possible connection between gluons, photons, and W and Z particles. As we saw in Section G.4, photons and W and Z particles are part of a single family of particles and, consequently, the electromagnetic and weak forces are merely different aspects of a single electroweak force. If it should turn out that gluons are also part of the same family of particles, then the electromagnetic, the weak, and the strong force would all be different aspects of just one single, basic kind of force. If so, we would have achieved a deep unification of three of the fundamental forces of physics. Then there would only remain the question of how to incorporate gravitation into this grand scheme. Some clever suggestions on how to approach this question have been made, but the answer still remains hidden.

Further Reading

The World of Elementary Particles by K. W. Ford (Blaisdell, New York, 1965) gives a simple, nonmathematical introduction to fields, quanta, and the generation of force by exchange of quanta. Unfortunately, this book is by now somewhat outdated, as it was written before the discoveries of the unified theory of weak and electromagnetic forces, and of quantum chromodynamics. These recent discoveries are described in *The Moment of Creation* by J. S. Trefil (Scribner's, New York, 1983), an exceptionally lucid and readable book, which includes a discussion of the cosmological implications of the unified theory. The recent discoveries are also briefly described in several of the books already mentioned in Interlude C:

Quarks: The Stuff of Matter by H. Fritzsch (Basic Books, New York, 1983)

From Atoms to Quarks by J. S. Trefil (Scribner's, New York, 1980)

The Cosmic Code by H. R. Pagels (Simon and Schuster, New York, 1982)

What Is the World Made Of? by G. Feinberg (Doubleday, Garden City, 1977)

The following articles deal with diverse aspects of the fundamental forces:

"Field Theory," F. J. Dyson, *Scientific American*, April 1953

"Unified Theories of Elementary Particle Interactions," S. Weinberg, *Scientific American*, July 1974

"The Detection of Neutral Weak Currents," A. K. Mann and C. Rubbia, *Scientific American*, December 1974

"Light as a Fundamental Particle," S. Weinberg, *Physics Today*, June 1975

"The Confinement of Quarks," Y. Nambu, *Scientific American*, November 1976

"Supergravity and the Unification of the Laws of Physics," D. Z. Freedman and P. van Nieuwenhuizwen, *Scientific American*, February 1978

"Gauge Theories of the Forces between Elementary Particles," G. 'tHooft, *Scientific American*, June 1980

"Antiproton–Proton Colliders, Intermediate Bosons," D. Cline and C. Rubbia, *Physics Today*, August 1980

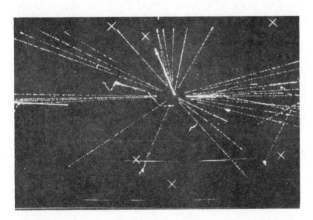

Fig. G.13 Tracks of particles produced in a very energetic head-on collision between a proton of 270 GeV that entered from the left and an antiproton of 270 GeV that entered from the right. Among the multitude of particles created in this collision, we can distinguish three jets: one jet of three particles in the lower right, one of five particles in the upper center, and one of four particles in the upper left.

"Unified Theory of Elementary-Particle Forces," H. Georgi
and S. L. Glashow, *Physics Today,* September 1980

"A Unified Theory of Elementary Particles and Forces," H.
Georgi, *Scientific American,* April 1981

"The Development of Field Theory in the Last 50 Years,"
W. F. Weisskopf, *Physics Today,* November 1981

"The Search for Intermediate Vector Bosons," D. B. Cline, C.
Rubbia, and S. van der Meer, *Scientific American,* March
1982

"Glueballs," K. Ishikawa, *Scientific American,* November 1982

"The Lattice Theory of Quark Confinement," C. Rebbi, *Scientific American,* February 1983

"The Structure of Quarks and Leptons," H. Harari, *Scientific American,* April 1983

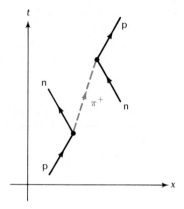

Fig. G.15

Questions

1. The strengths of the fundamental forces depend on the energies of the particles. In the case of the gravitational force, the strength increases with the energy. Why would you expect this to be true?

2. The boomerang analogy described in Figure G.7 is defective in that the boomerang requires the presence of air. What would be the motion of a boomerang in vacuum?

3. Show that if classical physics is valid, the emission of a photon by a free electron conflicts with energy conservation. (Hint: Consider the emission process in the reference frame in which the electron is initially at rest.)

4. The W particle can have either a positive charge (W^+) or a negative charge (W^-). Figure G.14 shows the Feynman diagram for the decay of the neutron (n) via exchange of a W^-; the end products are a proton (p), an electron (e), and an antineutrino ($\bar{\nu}$). Can you guess the Feynman diagram for the decay of the antineutron?

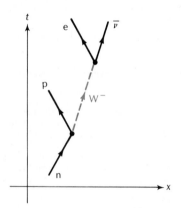

Fig. G.14

5. Figure G.15 shows the Feynman diagram for the exchange of a π^+ between a proton and a neutron; note that the proton changes into a neutron, and vice versa. Draw corresponding diagrams for the exchange of a π^- and of a π^0.

6. Give another simple example of spontaneous breaking of symmetry similar to the example of the pencil in Section G.4.

7. In Figure G.13, a large number of particles emerge in the longitudinal direction (toward the right and the left). Why is this expected, whereas the emergence of particles in the transverse direction (upward and downward) is surprising? (Hint: Consider a head-on collision between two aircraft; which way do you expect most fragments to spurt out?)

8. Figure G.10 shows the creation of a π^+ pion in a very violent collision. What is the particle (consisting of three quarks) left behind in this collision?

9. Suppose that in Figure G.10 the quark–antiquark pair created at the site of the break in the string is a u–$\bar{u}$ pair instead of a d–$\bar{d}$ pair. What kind of particle (consisting of two quarks) will then be created?

10. A pion consists of a u quark and a $\bar{u}$ antiquark tied together by a string. What kinds of particles (consisting of two quarks) can be created if the string breaks?

11. Given that each gluon has a color and an anticolor, how many different kinds of gluons are there?

12. A glueball consists of two or more gluons bound together in a composite structure (without quarks). The net color of glueballs is neutral; they always contain equal net amounts of a color and its anticolor. Suppose that a glueball consists of two gluons, one of which is *green–antiblue.* What must be the colors of the other gluon? Suppose that a glueball consists of three gluons, one *green–antiblue,* one *red–antigreen.* What must be the colors of the third gluon?

CHAPTER 24

Gauss' Law

Although the electric field of any given charge distribution can be calculated by means of Coulomb's Law, as in Examples 1–4 of the preceding chapter, this method often involves the evaluation of tedious integrals. Fortunately, there exists another method for calculating the electric field. This method relies on a theorem called Gauss' Law. That law is a consequence of Coulomb's Law and it therefore contains no new physics. It does, however, contain some new mathematics which supplies an elegant shortcut for calculating the electric field of a given charge distribution provided that this charge distribution has a certain amount of symmetry. This means that Gauss' Law does not help in every problem, but when it does help it works wonders. Before we present Gauss' Law and some examples of its use, we need to introduce the concept of electric flux.

24.1 Electric Flux and the Number of Field Lines

Consider a mathematical (i.e., imagined) surface in the shape of a rectangle of area A. Suppose that this surface is immersed in a constant electric field E (Figure 24.1). This electric field makes an angle with the surface; the electric-field vector has a component tangential to the surface and a component normal (i.e., perpendicular) to the surface. The **electric flux** Φ through the surface is defined as the product of the area A by the magnitude of the normal component of the electric field,

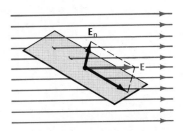

Fig. 24.1 Flat rectangular surface immersed in a uniform electric field.

$$\Phi = E_n A \qquad (1)$$

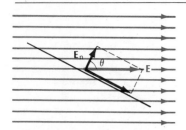

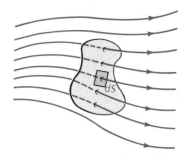

Fig. 24.2 The perpendicular to the surface makes an angle θ with the field lines. Note that the surface intercepts six field lines, i.e., the electric flux through the surface is $\Phi = 6$.

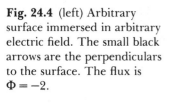

Fig. 24.3 An arbitrary surface immersed in an arbitrary electric field. The flux is $\Phi = 5$.

Electric flux

The normal component E_n can also be written as $E \cos \theta$, where θ is the angle between **E** and the perpendicular erected on the surface (Figure 24.2). Hence

$$\Phi = EA \cos \theta \tag{2}$$

The quantity $A \cos \theta$ can be interpreted as the projection of the area A onto a plane perpendicular to the electric field, that is, $A \cos \theta$ can be regarded as that part of the area A that faces the electric field. According to Section 23.3, the magnitude E of the electric field is numerically equal to the number of field lines intercepted by a unit area facing the electric field. Hence $EA \cos \theta$ must be numerically equal to the number of field lines intercepted by the area A: *the electric flux Φ through an area is equal to the number of field lines intercepted by the area.* Note that this equality hinges on the normalization adopted in Section 23.3 — flux and number of lines are equal if and only if electric field and number of lines per unit area are equal, and the latter is true if and only if we adopt the normalization that Q/ε_0 lines emerge from each charge Q.

More generally, consider a mathematical surface of arbitrary shape immersed in an electric field of arbitrary strength (Figure 24.3). Then we can define the electric flux by subdividing the surface into infinitesimal plane areas dS within each of which the electric field is nearly constant; the flux through one such infinitesimal area is $d\Phi = E \cos \theta \, dS$ and the total flux through the surface is the integral obtained by summing all these infinitesimal contributions,

$$\Phi = \int E \cos \theta \, dS \tag{3}$$

According to the above arguments, this electric flux is again numerically equal to the number of field lines intercepted by the surface. Note that lines going through the surface in one direction give a positive contribution to the flux, lines going in the opposite direction give a negative contribution (Figure 24.4). This means that the perpendiculars to the surface, with respect to which the angle θ is reckoned, must all be erected on the same side of the surface (e.g., on the right side of the surface of Figure 24.4).

Finally, consider a *closed* surface immersed in an electric field (Figure 24.5). The electric flux through this surface is given by Eq. (3) with the integration extending over all the area of the closed surface. This flux is equal to the number of field lines intercepted by the surface. The number can be positive or negative. For any closed surface we adopt the convention that the angle θ is reckoned with respect to

Fig. 24.4 (left) Arbitrary surface immersed in arbitrary electric field. The small black arrows are the perpendiculars to the surface. The flux is $\Phi = -2$.

Fig. 24.5 (right) A closed surface immersed in an electric field.

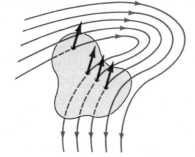

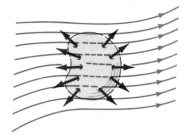

perpendiculars erected on the *outside* of the closed surface; then field lines leaving the volume enclosed by the surface make a positive contribution to the flux, and lines entering the volume make a negative contribution.

EXAMPLE 1. A point charge q is inside a closed spherical surface (Figure 24.6). What is the flux of its electric field through this surface?

SOLUTION: A positive charge q is the starting point of q/ε_0 outward field lines and all of these will pierce the surface. Hence the flux must be q/ε_0. Note that by using the concept of field lines, we are able to get the answer without explicit calculation of any integral; the explicit integration of the flux integral $\int E \cos \theta \, dS$ for the configuration shown in Figure 24.6 is rather messy (unless the charge is at the exact center of the surface).

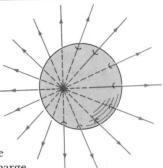

Fig. 24.6 A spherical surface surrounds a positive point charge.

24.2 Gauss' Law

We are now ready to prove an important theorem called **Gauss' Law:**

If the volume within an arbitrary closed surface holds a net charge Q, then the electric flux through the surface is Q/ε_0, that is,

Gauss' Law

$$\oint E \cos \theta \, dS = Q/\varepsilon_0 \qquad (4)$$

Here the small circle on the integral sign indicates that the integration is to be performed over a *closed* surface.

The proof of Gauss' Law is easy. The electric field appearing in Eq. (4) is a sum of the individual electric fields of some number of point charges. Some of the point charges are outside the closed surface and some are inside the closed surface. Let us consider the contribution to the flux from each individual electric field of each individual point charge. The individual electric field of a charge outside the closed surface generates no net flux through the closed surface — any field line of this field either does not touch the surface or else enters it at one point and leaves at another; neither case makes any contribution to the flux. However, the individual electric field of a positive or negative charge enclosed by the surface does contribute to the flux. For example, a positive charge q has q/ε_0 outward field lines, all of which will pierce the closed surface — such a charge therefore contributes a flux q/ε_0 (Figure 24.7). Taking into account that positive charges generate positive flux and negative charges negative flux, we see that the net flux through the surface, or the net number of lines piercing the surface, is equal to the net charge divided by ε_0. This completes the proof of Eq. (4).

With the notation E_n for the component of **E** normal to the surface, we can put Gauss' Law in the somewhat more compact form

$$\oint E_n \, dS = Q/\varepsilon_0 \qquad (5)$$

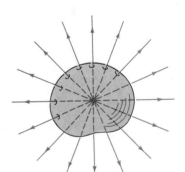

Fig. 24.7 A closed surface surrounds a positive point charge.

An alternative notation frequently used associates a vector $d\mathbf{S}$ with the area dS; this vector has the magnitude of dS and the direction of the outward perpendicular to the surface (Figure 24.8). Then $E \cos \theta \, dS = \mathbf{E} \cdot d\mathbf{S}$ and Eq. (4) takes the form

Gauss' Law in vector notation

$$\oint \mathbf{E} \cdot d\mathbf{S} = Q/\varepsilon_0 \qquad (6)$$

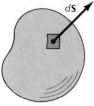

The closed mathematical surface over which the integrations in Eq. (4), (5), and (6) must be carried out is usually called a **Gaussian surface.**

Gauss' Law can be used to calculate the electric field provided the distribution of charge has a high degree of symmetry. Essentially, Gauss' Law can be regarded as a mathematical restriction imposed on the electric field. Symmetry conditions impose further restrictions on the field. By clever combination of all these restrictions one can often evaluate the electric field without the laborious process of integrating Coulomb's Law.

Fig. 24.8 A small area dS with its perpendicular vector $d\mathbf{S}$.

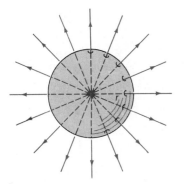

Fig. 24.9 A Gaussian spherical surface surrounds a positive point charge concentrically.

EXAMPLE 2.   Find the electric field of a point charge q using Gauss' Law.

SOLUTION: For the purpose of this example we will pretend that we do not know the electric field of a point charge. We begin with the observation that the electric field must be spherically symmetric because the point charge is spherically symmetric; hence, at all points on the surface of a mathematical sphere of radius r centered on q, the field is in the radial direction and has a constant magnitude (Figure 24.9).

If the Gaussian surface is taken to coincide with the sphere of radius r, then $\cos \theta = 1$, since $\mathbf{E}$ is perpendicular to this surface. Furthermore, $\mathbf{E}$ has a constant magnitude over this surface. Therefore

$$\oint E \cos \theta \, dS = \oint E \, dS = E \oint dS$$

But $\oint dS$ is simply the area of the Gaussian surface, i.e., $4\pi r^2$. Hence Eq. (4) becomes

$$E(4\pi r^2) = q/\varepsilon_0$$

or

$$E = \frac{1}{4\pi\varepsilon_0} \frac{q}{r^2} \qquad (7)$$

The proof of Eq. (4) shows that Gauss' Law is a consequence of the field-line concept, i.e., Gauss' Law is a consequence of Coulomb's Law. Conversely, Example 2 shows that Coulomb's Law is a consequence of Gauss' Law. These two laws are therefore equivalent, at least in regard to the electric fields of charges at rest. As we will see in a later chapter, it turns out that besides such static electric fields, there also exist time-dependent electric fields whose field lines are not attached to charges (the field lines form closed loops). Coulomb's Law does not apply to such time-dependent electric fields . . . but Gauss' Law does! The latter law is more general than the former. However, this subtle distinction need not concern us as long as all charges are static — as they are in the examples of the next section.

24.3 Some Examples

The following examples illustrate how Gauss' Law can be combined with symmetry requirements and used to evaluate the electric field of simple charge distributions. The solutions of these examples always involve two steps: first determine the *direction* of the electric field by appealing to the symmetry requirements imposed by the charge distribution, and then calculate the *magnitude* of the electric field from Gauss' Law. In all these examples the crucial trick lies in the choice of Gaussian surface — a good choice makes the calculation of the integral $\oint E \cos \theta \, dS$ easy and also permits one to "solve" Gauss' Law for the unknown value of E.

EXAMPLE 3.   Charge is distributed uniformly along a very long thin line. If the amount of charge is λ coulomb per meter of line, what is the electric field?

SOLUTION: This example has already been worked out via Coulomb's Law (see Section 23.1); the following quick and elegant method is based on Gauss' Law. First we need to determine the direction of the electric field. The only direction consistent with the symmetry of the charge distribution is the radial direction (horizontal in Figure 24.10) — if the electric field had any other direction (up or down in Figure 24.10), the field would make an unacceptable distinction between the upper and lower ends of the line of charge. Furthermore, the rotational symmetry of the line of charge tells us that the electric field has constant magnitude over the lateral curved surface of any mathematical cylinder concentric with the line of charge.

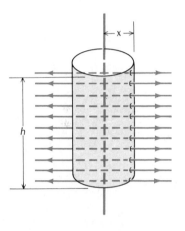

Fig. 24.10 A cylindrical Gaussian surface surrounds a line of charge.

Now take a Gaussian surface that coincides with such a cylinder, of radius x and height h (Figure 24.10). On the lateral, curved surface of the cylinder, $\cos \theta = 1$ and

$$\int_{\substack{\text{curved} \\ \text{surface}}} E \cos \theta \, dS = \int E \, dS = E \int dS = E \times (2\pi x h) \qquad (8)$$

On each of the two circular bases of the cylinder, the electric field is tangent to the surface; hence $\cos \theta = 0$ and

$$\int_{\text{base}} E \cos \theta \, dS = 0 \qquad (9)$$

The integral over the entire surface of the cylinder (curved surface plus bases) is then $2\pi x h E$. By Gauss' Law this must equal the charge contained in the cylinder divided by ε_0, i.e.,

$$2\pi x h E = Q/\varepsilon_0 = \lambda h/\varepsilon_0$$

and

$$E = \frac{1}{2\pi\varepsilon_0} \frac{\lambda}{x} \qquad (10)$$

EXAMPLE 4.   What is the electric field of a large uniform sheet of charge with a density of σ coulomb per square meter?

SOLUTION: This example also has been worked out earlier by a laborious integration procedure; this labor can be bypassed by means of Gauss' Law. Symmetry tells us that the electric field is everywhere perpendicular to the sheet of charge and it has a constant magnitude over any surface parallel to this sheet.

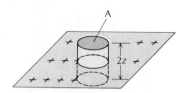

Fig. 24.11 A cylindrical Gaussian surface intersects a very large sheet of charge.

We take a Gaussian surface in the shape of a cylinder of height $2z$ and base area A (Figure 24.11). Note that since the bases are at the same distance from the sheet, the magnitude of E will be the same on both of them. On each base $\cos \theta = 1$ and, integrating over both, we find

$$\int_{bases} E \cos \theta \, dS = \int E \, dS = E \times (2A) \tag{11}$$

On the curved lateral surface of the cylinder, $\cos \theta = 0$ and therefore

$$\int_{\substack{curved \\ surface}} E \cos \theta \, dS = 0 \tag{12}$$

By Gauss' Law, the integral of $E \cos \theta$ over the complete surface must equal the charge within the cylinder divided by ε_0. Since the charge is σA, we obtain

$$2AE = Q/\varepsilon_0 = \sigma A/\varepsilon_0$$

and

$$E = \sigma/2\varepsilon_0 \tag{13}$$

EXAMPLE 5.   A spherical region of radius R has a total charge q which is uniformly distributed over the volume of this region. (a) What is the electric field at points inside the sphere? (b) What is the electric field at points outside the sphere?

SOLUTION: (a) The charge density within the sphere is

$$\rho = [\text{charge}]/[\text{volume}] = q/(4\pi R^3/3) = 3q/4\pi R^3 \tag{14}$$

Fig. 24.12 A spherical volume with a uniform distribution of charge. The Gaussian surface has a radius $r < R$.

Since the charge distribution is spherically symmetric, the electric field must be radial and of constant magnitude over any spherical mathematical surface of given radius. To find the magnitude of the electric field inside the charge distribution, take a spherical Gaussian surface of radius r, where $r \le R$ (Figure 24.12). On this surface $\cos \theta = 1$, so that

$$\oint E \cos \theta \, dS = \oint E \, dS = E \oint dS = 4\pi r^2 E \tag{15}$$

The charge inside this Gaussian surface is

$$Q = [\text{charge density}] \times [\text{volume}] = \rho \times \frac{4\pi}{3} r^3$$

$$= q \frac{r^3}{R^3} \tag{16}$$

Then Gauss' Law gives

$$4\pi r^2 E = \frac{q}{\varepsilon_0} \frac{r^3}{R^3}$$

i.e.,

$$E = \frac{1}{4\pi\varepsilon_0} \frac{qr}{R^3} \quad \text{for } r \le R \tag{17}$$

Note that this electric field increases linearly with the distance from the center and reaches a maximum value of $E = (1/4\pi\varepsilon_0)q/R^2$ when $r = R$.

(b) To find the electric field outside the sphere, take a spherical Gaussian surface of radius r (where $r \geq R$; Figure 24.13). With the usual symmetry arguments, the flux integral again has the form $4\pi r^2 E$ and Gauss' Law gives

$$4\pi r^2 E = q/\varepsilon_0$$

i.e.,

$$E = \frac{1}{4\pi\varepsilon_0} \frac{q}{r^2} \qquad \text{for } r \geq R \qquad (18)$$

This means that outside the region containing charge, the electric field is exactly the same as it would be if all the charges were located at the center. Figure 24.14 is a plot of the electric field vs. distance.

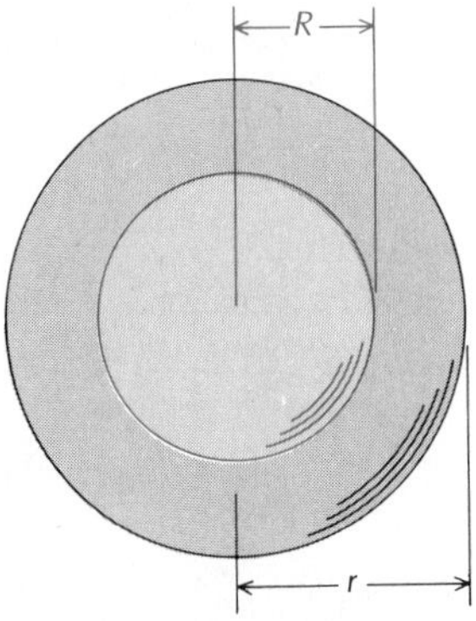

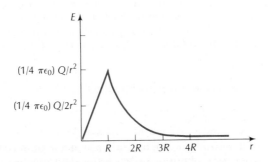

Fig. 24.13 The Gaussian surface has a radius $r > R$.

Fig. 24.14 Electric field as a function of radius for a uniformly charged sphere.

It is obvious that the argument of part (b) of the preceding example does not depend on the uniformity of the charge distribution — it only depends on its spherical symmetry. Hence outside any charge distribution that consists of a sequence of concentric shells of charge — so that the charge density is a function of the radius but not of the angular direction — the electric field will mimic that of a point charge. This result is similar to Newton's famous theorem concerning the gravitational forces exerted by a planet: the forces exerted by a spherically symmetric planet can be calculated as if all the mass were concentrated in a point at the planetary center. This similarity between electricity and gravitation reflects the similarity of the laws of force. Because of this similarity, we can use the same action–reaction argument as in Section 13.6 to show that a spherical charge distribution mimics a point charge not only in regard to the electric field that it produces, but also in regard to the force that it experiences when placed in some arbitrary external electric field. Thus, the electric force that an arbitrary electric field exerts on a spherical charge distribution can be calculated as if all of the charge were concentrated in a point at the center.

Note that Eq. (17) can be put in the form

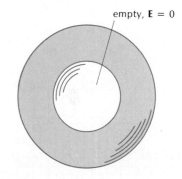

Fig. 24.15 A spherical shell (a sphere with a spherical cavity).

$$E = \frac{1}{4\pi\varepsilon_0} \frac{Q(r)}{r^2} \qquad (19)$$

Electric field of spherical charge distribution

where $Q(r)$ is the amount of charge inside the sphere of radius r. If the charge distribution is uniform, then $Q(r)$ is as given by Eq. (16); however, the expression (19) does not depend on the radial uniformity of the charge distribution — it only depends on spherical symmetry. According to Eq. (19), only the charge *inside* a radius less than r contributes to the electric field *at* the radius r. As a consequence, the electric field within the empty region inside a spherically symmetric shell of charge is exactly zero (Figure 24.15).

EXAMPLE 6.   A proton is (approximately) a spherically symmetric ball of charge of radius 1.0×10^{-15} m. What is the electric field at the surface of the proton? If a second proton is brought within touching distance of the first, what is the repulsive electric force?

SOLUTION: For $q = e = 1.6 \times 10^{-19}$ C and $r = 1.0 \times 10^{-15}$ m, the electric field is

$$E = \frac{1}{4\pi\varepsilon_0} \frac{q}{r^2}$$

$$= \frac{9.0 \times 10^9 \ \text{N} \cdot \text{m}^2 \cdot \text{C}^{-2} \times 1.6 \times 10^{-19} \ \text{C}}{(1.0 \times 10^{-15} \ \text{m})^2}$$

$$= 1.4 \times 10^{21} \ \text{N/C} \tag{20}$$

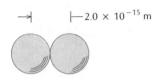

$\vdash$ 2.0×10^{-15} m

Fig. 24.16 Two protons in contact. Each proton may be regarded as a spherical ball of positive charge.

If the protons are touching (Figure 24.16), the center-to-center distance is $2 \times 1.0 \times 10^{-15}$ m. The electric field generated by one proton at this distance is one-quarter of the above value, i.e., $E = 3.6 \times 10^{20}$ N/C. The force exerted by one proton on the other is

$$F = eE = 1.6 \times 10^{-19} \ \text{C} \times 3.6 \times 10^{20} \ \text{N/C}$$

$$= 58 \ \text{N} \tag{21}$$

This is about 13 lb of force; acting on a particle of a mass of only 10^{-27} kg, this represents a gigantic force! In the nucleus of an atom, the large repulsive electric force between the tightly packed protons is more than canceled by an even larger binding force (the "strong" force) that holds the nucleus together.

24.4 Conductors in an Electric Field

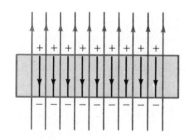

Fig. 24.17 Conductor immersed in an electric field. The charges that have accumulated on the surfaces generate an electric field (black) opposite to the original electric field (color).

Electrostatic equilibrium

As we pointed out in Section 22.4, in metallic conductors — such as copper, silver, aluminum — some of the electrons are free, that is, they can move without restraint within the volume of the metal. If such a conductor is immersed in an electric field, the free electrons move in response to the electric force. The electrons move in a direction opposite to the direction of the electric field and they continue moving until they reach the surface of the metal. As an excess of electrons accumulates on one part of the surface, a deficit of electrons will appear on another part of the surface: negative and positive charges are induced on the conductor. Within the volume of the conductor, the electric field of the induced charges tends to cancel the original electric field in which the conductor was immersed (Figure 24.17). The accumulation of negative and positive charge on the surface of the conductor continues until the electric field generated by these charges exactly cancels the original electric field. Consequently, *when the charge distribution on a conductor reaches static equilibrium, the net electric field within the material of the conductor is exactly zero.* The proof of this statement is by contradiction: if the electric field were different from zero, the free electrons would continue to move and the charge distribution would *not* (yet) be in equilibrium. For a good conductor (copper, aluminum, etc.), the equilibrium is reached in a fairly short time, a small fraction of a second.[1]

In a conductor in static equilibrium, all the (extra) electric charge resides on the surface of the conductor. One can prove this by means of Gauss' Law: Consider a small closed surface inside the conducting material (Figure 24.18). Since $\mathbf{E} = 0$ everywhere in this material, the left side of Eq. (4) vanishes and therefore the right side must also vanish — which means that the charge enclosed by *any* arbitrary small surface is zero, that is, the charge in *any* small volume of the conductor is zero. Obviously, if the charges are not in the volume of the conductor, they must be on the surface.

Finally, we can say something about the electric field just outside a conductor: *the electric field at the surface of a conductor in static equilibrium is normal to the surface.* The proof is again by contradiction: if the electric field had a component tangential to the surface of the conductor, the free electrons would move along the surface and the charge distribution would *not* be in equilibrium.

Note that this argument does not exclude an electric field perpendicular to the surface of the conductor; such an electric field merely pushes the free electrons against the surface, where they are held in equilibrium by the combination of the force exerted by the electric field and the restraining force exerted by the surface of the conductor. Figure 24.19 displays an experimental demonstration of electric fields perpendicular to the surfaces of conductors. The flat plate and the cylinder shown in this figure are conductors, and we see that the field lines meet the surfaces of these conductors at right angles.

In the preceding paragraphs we have implicitly assumed that the material of the conductor is homogeneous, i.e., the material has uniform density and uniform chemical composition. If this is not the case, then our conclusions are not quite valid. For instance, suppose that the body of the conductor consists of two dissimilar metals joined together. Figure 24.20 shows a conductor made of a block of lead in contact with a block of silver. In this conductor there will exist some electric field at and near the interface of the metals and, what is more, such an electric field will exist even if the conductor carries no net charge. This electric field is created by the contact of the two dissimilar metals. To understand how the electric field comes about, imagine that at first the blocks of lead and of silver are separated. Each metal is neutral and contains within it a gas of free electrons. If the blocks are now brought into contact, some of the free electrons from the lead block will flow into the silver block. This is so because silver exerts a slightly stronger hold on free electrons than lead (the energy required to remove a free electron from silver is larger than that from lead). As the extra free electrons accumulate in the silver, their electric repulsions gradually bring the flow to a halt. At equilibrium, the silver will have an excess of electrons (negative charge) and the lead a deficiency of electrons (positive charge). These charges will then create an electric field at and near the interface of the joined blocks.

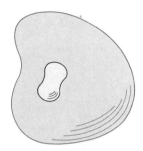

Fig. 24.18 Closed Gaussian surface inside a volume of conducting material.

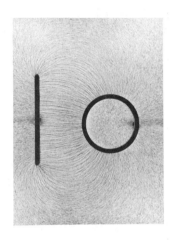

Fig. 24.19 Field lines in the space surrounding a charged flat plate and cylinder, made visible by small bits of thread suspended in oil. (Courtesy H. Waage, Princeton University.)

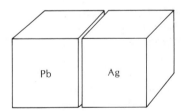

Fig. 24.20 A conductor made of two blocks of lead and silver joined together.

---

[1] The time for reaching equilibrium depends on details such as the shape and the size of the conductor and the characteristics of the conducting material. Paradoxically, a good conductor takes somewhat longer to reach equilibrium than a poor conductor. In the good conductor, the charges move very easily and they tend to overshoot their equilibrium positions — the charges slosh around on the conductor and they take a while to settle in their equilibrium positions.

The charges and the electric fields that are created by the contact of dissimilar metals are usually quite small. In the following chapters we will ignore these "contact" charges and fields. However, it is well to keep in mind that these charges and fields play a crucial role in the operation of transistors, solar cells, and other solid-state devices that consist of pieces of dissimilar conductors or semiconductors joined together.

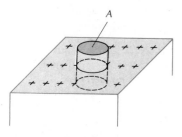

Fig. 24.21 A very large slab of conductor, with a uniform distribution of charge on its surface.

EXAMPLE 7. Find the electric field outside a very large flat conducting surface on which there is a surface charge density of σ coulomb per square meter.

SOLUTION: In view of the symmetry of the charge configuration, the electric field will be perpendicular to the conducting plane and it will have constant magnitude over any plane parallel to the conducting plane. As Gaussian surface, take the cylinder shown in Figure 24.21; the base area of the cylinder is A. The upper base contributes an amount

$$\int E \cos \theta \, dS = \int_{\substack{\text{upper} \\ \text{base}}} E \, dS = E \int dS = EA \tag{22}$$

to the flux integral. The lower base, in the conductor, does not contribute to the flux integral, since $E = 0$ in this region. Finally, the curved lateral surface does not contribute, since $\cos \theta = 0$. The charge within the Gaussian surface is σA. Hence,

$$EA = \sigma A / \varepsilon_0$$

and

$$E = \sigma / \varepsilon_0 \tag{23}$$

This is a constant electric field, independent of distance from the conducting plane. Of course, if the plane is of finite size, then the electric field will have the constant value (23) only in a region very near the plane.

Note that, for a given surface-charge density, the electric field generated by a conducting surface is twice as strong as the electric field generated by a sheet of charge [see Eq. (13)]. The reason is obvious: in the former case all field lines that originate on the charges go to the same side of the surface, in the latter case half go to each side.

Over a small region, a curved conducting surface can be approximated by a flat surface. Hence the expression (23) can be used to find the electric field in a region very near any smooth curved conducting surface [the expression (23) is *not* a good approximation near sharp edges]; it is not even necessary that σ be constant — it can be some smooth function of position.

EXAMPLE 8. At the ground directly below a thundercloud, the electric field is 2×10^4 N/C and points upward. What is the surface charge density on the ground?

SOLUTION: For the purpose of this problem, we treat the ground as a good conductor. Equation (23) gives the relation between electric field and surface-charge density for points at or near the charged surface,

$$\sigma = \varepsilon_0 E = 8.85 \times 10^{-12} \frac{\text{C}^2}{\text{N} \cdot \text{m}^2} \times 2 \times 10^4 \text{ N/C}$$

$$= +1.8 \times 10^{-7} \text{ C/m}^2$$

SUMMARY

Electric flux through surface: $\Phi = \int E_n \, dS$

$$= \int \mathbf{E} \cdot d\mathbf{S}$$

Gauss' Law (for closed surface): $\oint \mathbf{E} \cdot d\mathbf{S} = Q/\varepsilon_0$

Conductor in electrostatic equilibrium:
The electric field within the conductor is zero.
The charge resides on the surface.
The electric field at the surface is perpendicular and of magnitude $E = \sigma/\varepsilon_0$.

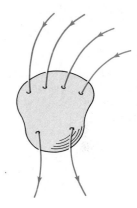

Fig. 24.22

QUESTIONS

1. A point charge Q is inside a spherical Gaussian surface, which is enclosed in a larger cubical Gaussian surface. Compare the fluxes through these two surfaces.

2. Figure 24.22 shows a Gaussian surface and lines of electric field entering and leaving this surface. Assuming that the number of lines has been normalized according to the convention of Section 23.3, what can you say about the magnitude and the sign of the electric charge within the surface?

3. Suppose that instead of $1/\varepsilon_0$ lines of force per unit charge, we had adopted some other normalization, say, k/ε_0 lines per unit charge. Would this change Gauss' Law?

4. A Gaussian surface contains an electric dipole, and no other charge. What is the electric flux through this surface?

5. Defining the gravitational field as the gravitational force per unit mass, formulate a Gauss' Law for gravity. Check that your law implies Newton's Law of universal gravitation.

6. Suppose that the electric field of a point charge were not exactly proportional to $1/r^2$, but rather proportional to $1/r^{2+a}$ where a is a small number, $a \ll 1$. Would Gauss' Law still be valid? (Hint: Consider Gauss' Law for the case of a point charge.)

7. Prove that if the electric field is uniform in some region, then the charge density must be zero in that region.

8. Problems soluble by Gauss' Law fall into three categories, according to their symmetry: spherical, cylindrical, and planar. Give some examples in each category.

9. A hemisphere of radius R has a charge Q uniformly distributed over its volume. Can we use Gauss' Law to find the electric field?

10. A spherical rubber balloon of radius R has a charge Q uniformly distributed over its surface. The balloon is placed in a uniform electric field of 120 N/C. What is the net electric field inside the rubber balloon?

11. The electric field at the surface of a conductor in static equilibrium is normal to the surface. Is the gravitational field at the surface of a mass in static equilibrium necessarily normal to the surface? What if the surface is that of a fluid, such as water?

12. Suppose we drop a charged ping-pong ball into a cookie tin and quickly close the lid. What happens to the portions of the electric field lines that are outside of the cookie tin when we close the lid?

13. When an electric current is flowing through a wire connected between a source and a sink of electric charge — such as the poles of a battery — there is an electric field inside the wire, even though the wire is a conductor. Why does our conclusion about zero electric field inside a conductor not apply to this case?

14. The free electrons belonging to a metal are uniformly distributed over the entire volume of the metal. Does this contradict the result we derived in Section 24.4, according to which the charges are supposed to reside on the surface of a conductor?

PROBLEMS

Sections 24.1 and 24.2

1. Consider the thundercloud described in Problem 23.14 and Figure 23.28. What is the total electric flux coming out of the surface of the cloud?

2. A point charge of 1.0×10^{-8} C is placed inside an uncharged metallic can (say, a closed beer can) insulated from the ground. How many flux lines will emerge from the surface of the can when the point charge is inside?

3. On a clear day, the Earth's atmospheric electric field near the ground has a magnitude of 100 N/C and points vertically down. Inside the ground, the electric field is zero, since the ground is a conductor. Consider a mathematical box of 1 m $\times$ 1 m $\times$ 1 m, half below the ground, and half above. What is the electric flux through the sides of this box? What is the charge enclosed by the box?

4. Suppose we suspend a small ball carrying a charge of 1.0×10^{-6} C in the middle of a safe and lock the door. The safe is made of solid steel; it has inside dimensions 0.3 m $\times$ 0.3 m $\times$ 0.3 m and outside dimensions 0.4 m $\times$ 0.4 m $\times$ 0.4 m. What is the electric flux through a cubical surface measuring 0.2 m $\times$ 0.2 m $\times$ 0.2 m centered on the ball? A cubical surface measuring 0.35 m $\times$ 0.35 m $\times$ 0.35 m? A cubical surface measuring 0.5 m $\times$ 0.5 m $\times$ 0.5 m?

5. Consider a mathematical surface having the shape of a cube of edge 5 cm. You do not know the electric charge or the electric field inside the cube, but you do know the electric field at the surface: at the top of the cube the electric field has a magnitude of 5×10^5 N/C and points perpendicularly out of the cube; at the bottom of the cube, the electric field has a magnitude of 2×10^5 N/C and points perpendicularly into the cube; on all other faces, the electric field is tangential to the surface of the cube.
   (a) How much electric charge is inside the cube?
   (b) Can you guess what charge distribution inside (and outside) the cube would generate this kind of electric field?

6. An electric field has the following form as a function of x, y, z:

$$E_x = 5.0x \qquad E_y = 0 \qquad E_z = 0$$

where E is in newtons per coulomb and x in meters. This represents an electric field in the x direction with a magnitude that increases in direct proportion to x. Show that such an electric field can only exist if space is filled with some electric charge density. Find the value of the required charge density as a function of x, y, z.

7. A point charge of 6.0×10^{-8} C sits at a distance of 0.30 m above the x–y plane. What is the electric flux that this charge generates through the (infinite) x–y plane?

Section 24.3

8. Charge is placed on a small metallic sphere which is surrounded by air. If the radius of the sphere is 0.5 cm, how much charge can be placed on the sphere before the air near the sphere suffers electric breakdown? The critical electric field strength that leads to breakdown in air is 3×10^6 N/C.

9. A uranium nucleus is a spherical ball of radius 7.4×10^{-15} m with a charge of $92e$ uniformly distributed over its volume. Plot the electric field produced by this charge distribution as a function of radius for $0 < r < 15 \times 10^{-15}$ m.

10. In symmetric fission, a uranium nucleus splits into two equal pieces each of which is a palladium nucleus. The palladium nucleus is spherical with a radius of 5.9×10^{-15} m and a charge of $46e$ uniformly distributed over its volume. Suppose that immediately after fission, the two palladium nuclei are barely touching (Figure 24.23). What is the value of the total electric field at the center of each? What is the repulsive force between them? What is the acceleration of each? The mass of a palladium nucleus is 1.99×10^{-25} kg.

11. Charge is distributed uniformly over the volume of a very long cylindrical plastic[2] rod of radius R. The amount of charge per meter of length of the rod is λ. Find a formula for the electric field at a distance r from the axis of the rod. Assume $r < R$.

12. What is the maximum amount of electric charge per unit length that one can place on a long, straight, human hair of diameter 8×10^{-3} cm if the surrounding air is not to suffer electrical breakdown? The air will suffer breakdown if the electric field exceeds 3×10^6 N/C.

13. A long plastic[2] pipe has inner radius a and outer radius b. Electric charge is uniformly distributed over the region $a < r < b$. The amount of charge is λ coulomb per meter of length of the pipe. Find the electric field in the regions $r < a$, $a < r < b$, and $r > b$.

14. Charge is uniformly distributed over the volume of a large, plane slab of plastic[2] of thickness d. The charge density is ρ coulomb per cubic meter. The midplane of the slab is the y–z plane (Figure 24.24). What is the electric field at a distance x from the midplane? Consider both the cases $|x| < d/2$ and $|x| > d/2$.

15. The tube of a Geiger counter consists of a thin conducting wire of radius 1.3×10^{-3} cm stretched along the axis of a conducting cylindrical shell of radius 1.3 cm (Figure 24.25). The wire and the cylinder have equal and opposite charges of 7.2×10^{-10} C distributed along their length of 9.0 cm. Find a formula for the electric field in the space between the wire and the cylinder; pretend that the electric field is that of an infinitely long wire and cylinder. What is the magnitude of the electric field at the surface of the wire?

16. A thick spherical shell of inner radius a, outer radius b has a charge Q uniformly distributed over its volume. Find the electric field in the regions $r < a$, $a < r < b$, and $r > b$.

17. According to a (crude) model, the neutron consists of an inner core of positive charge surrounded by an outer shell of negative charge. Suppose that the positive charge has a magnitude $+e$ and is uniformly distributed over a sphere of radius 0.50×10^{-15} m; suppose that the negative charge has a magnitude $-e$ and is uniformly distributed over a concentric shell of inner radius 0.50×10^{-15} m, outer radius 1.0×10^{-15} m (Figure 24.26). Find the magnitude and direction of the electric field at 1.0×10^{-15}, 0.75×10^{-15}, 0.50×10^{-15}, and 0.25×10^{-15} m from the center.

---

[2] Assume that the plastic has no effect on the electric field.

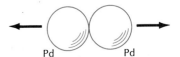

Fig. 24.23 Two palladium nuclei in contact. Each nucleus is a sphere.

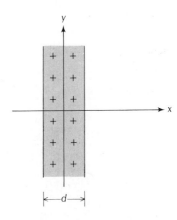

Fig. 24.24 A slab of plastic with electric charge uniformly distributed over its volume.

Fig. 24.25

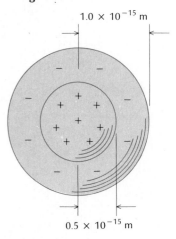

Fig. 24.26 Charge distribution inside a neutron.

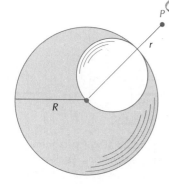

Fig. 24.27 Sphere with spherical cavity.

18. Positive charge Q is uniformly distributed over the volume of a solid sphere of radius R. Suppose that a spherical cavity of radius $R/2$ is cut out of the solid sphere, the center of the cavity being at a distance of $R/2$ from the center of the original solid sphere (Figure 24.27); the cut-out material and its charge are discarded. What new electric field does the sphere with the cavity produce at the point P at a distance r from the original center? Assume $r > R$.

19. According to an old (and erroneous) model due to J. J. Thomson, an atom consists of a cloud of positive charge within which electrons sit like plums in a pudding. The electrons are supposed to emit light when they vibrate about their equilibrium positions in this cloud. Assume that in the case of the hydrogen atom the positive cloud is a sphere of radius $R = 0.5$ Å with a charge of e uniformly distributed over the volume of this sphere. The (pointlike) electron is held at the center of this charge distribution by the electrostatic attraction.
   (a) Show that the restoring force on the electron is $e^2 r/(4\pi\varepsilon_0 R^3)$ when the electron is at a distance r from the center ($r \leqslant R$).
   (b) What is the frequency of small oscillations of the electron moving back and forth along a diameter? Give a *numerical* answer.

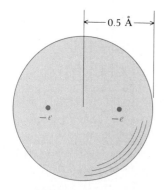

Fig. 24.28 Model of the helium atom.

20. According to the Thomson model (see also Problem 19), the atom of helium consists of a uniform spherical cloud of positive charge within which sit two electrons. Assume that the positive cloud is a sphere of radius 0.5 Å with a charge $2e$ uniformly distributed over the volume. The two electrons are symmetrically placed with respect to the center (Figure 24.28). What is the equilibrium separation of the electrons?

21. A charge distribution with spherical symmetry has a charge density ρ coulomb per cubic meter described by the formula $\rho = kr^n$, where k is a constant and $n > -3$.
   (a) What is the amount of charge $Q(r)$ inside a sphere of radius r?
   (b) What is the magnitude of the electric field as a function of r?
   (c) For what value of n is the magnitude of the electric field constant?
   (d) Why is it necessary to assume that $n > -3$?

22. The tau particle is a negatively charged particle similar to the electron, but of much larger mass — its mass is 3.18×10^{-27} kg, about 3490 times the mass of an electron. Nuclear material is transparent to the tau; thus, the tau can orbit around inside a nucleus, under the influence of the electric attraction of the nuclear charge. Suppose that a tau is in a circular orbit of radius 2.9×10^{-15} m inside a uranium nucleus. Treat the nucleus as a sphere of radius 7.4×10^{-15} m with a charge $92e$ uniformly distributed over its volume. Find the speed, the kinetic energy, the angular momentum, and the frequency of the orbital motion of the tau.

23. A very long cylinder of radius R has positive charge uniformly distributed over its volume. The amount of charge is λ coulomb per meter of length of the cylinder. A spherical cavity of radius $R' \leqslant R$, centered on the axis of the cylinder, has been cut out of this cylinder, and the charge in this cavity has been discarded.
   (a) Find the electric field as a function of distance from the center of the sphere along the axis of the cylinder.
   (b) Find the electric field as a function of distance along a line passing through the center of the sphere perpendicular to the axis of the cylinder. Consider both distances smaller and larger than R'.

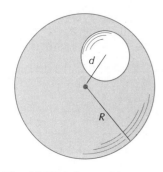

Fig. 24.29 Sphere with spherical cavity.

*24. A charge Q is uniformly distributed over the volume of a solid sphere of radius R. A spherical cavity is cut out of this solid sphere (Figure 24.29) and the material and its charge are discarded. Show that the electric field in the cavity will then be uniform, of magnitude $(1/4\pi\varepsilon_0)Qd/R^3$, where d is the distance between the centers of the spheres (Figure 24.29). Make a drawing of the lines of electric field in the cavity.

*25. When a point charge q moving at high speed passes by another stationary point charge q', the main effect of the electric forces is to give each charge a transverse impulse. Figure 24.30 shows the charge q moving at (almost) constant velocity v along the x axis in an almost straight line and shows the charge q' sitting at a distance R below the origin. The transverse impulse on q is

$$\int_{-\infty}^{\infty} F_y \, dt = \frac{q}{v} \int_{-\infty}^{\infty} E_y \, dx$$

Evaluate the integral $\int E_y \, dx$ by means of Gauss' Law and prove that

$$\int_{-\infty}^{\infty} F_y \, dt = \frac{q}{v} \frac{q'}{2\pi\varepsilon_0 R}$$

(Hint: Consider $2\pi R \int E_y \, dx$; show that this is the flux that q' produces through the infinite cylindrical surface indicated in Figure 24.30.)

26. The formula derived in Problem 25 gives the transverse momentum that a high-speed charged particle acquires as it passes by a stationary charged particle.
   (a) Calculate the transverse momentum that an electron of speed 4.0×10^7 m/s acquires as it passes by a stationary electron at a distance of 0.60×10^{-10} m.
   (b) What transverse velocity corresponds to this transverse momentum?
   (c) What will be the recoil velocity of the stationary electron (if it is free to move)?

Section 24.4

27. The surface of a long, cylindrical copper pipe has a charge of λ coulomb per meter (Figure 24.31). What is the electric field outside the pipe? Inside the pipe?

28. A solid copper sphere of radius 3 cm carries a charge of 10^{-6} C. This sphere is placed concentrically within a spherical, thin copper shell of radius 15 cm carrying a charge of 3×10^{-6} C. Find a formula for the electric field in the space between the sphere and the shell. Find a formula for the electric field outside the shell. Plot these electric fields as a function of radius.

29. A thick spherical shell made of metal has an inner radius a, an outer radius b, and is initially uncharged. A point charge of 1.0×10^{-7} C is placed at the center of the shell. Find the electric field in the regions $r < a$, $a < r < b$, and $r > b$. Find the induced surface charge densities at $r = a$ and $r = b$.

30. On days of fair weather, the atmospheric electric field of the Earth is about 100 N/C; this field points vertically downward (compare Problem 23.2). What is the surface charge density on the ground? Treat the ground as a flat conductor.

31. You wish to generate a uniform electric field of 2.0×10^5 N/C in the space between two flat, parallel plates of metal placed face to face. The plates measure 0.30 cm $\times$ 0.30 cm. How much electric charge must you put on each plate? Assume that the gap between the plates is small so that the charge distribution and the electric field are approximately uniform, as for infinite plates.

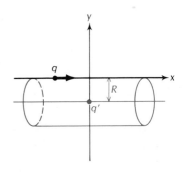

Fig. 24.30 The charge q moves along the x axis. The charge q' sits at a distance R below the x axis.

Fig. 24.31 A long pipe with positive charge on its surface.

CHAPTER 25

The Electrostatic Potential

In Chapter 8 we saw that to formulate a law of conservation of energy for a particle moving under the influence of a conservative force, we had to construct a potential energy. In this chapter we will construct the electrostatic potential energy for a charged particle moving under the influence of the electric force generated by a static charge distribution. This potential energy will help us in the calculation of the motion of the particle. Furthermore, this potential energy will give us yet another method for the calculation of the electric field of a given charge distribution. The method involves two steps: first, find the potential energy of a unit point charge placed somewhere near the charge distribution; second, find the electric field by differentiating this potential energy. Thus, we will have available three alternative methods for the calculation of the electric field: via Coulomb's Law, via Gauss' Law, and via the potential energy of a unit charge. If a problem does not yield to the simple and elegant method based on Gauss' Law, it is usually best to to use the method based on the potential energy because the mathematics are likely to be less cumbersome than with the method based on Coulomb's Law.

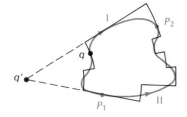

Fig. 25.1 A charge q moves from the point P_1 to the point P_2 while under the influence of the electric field of a charge q'. The path can be approximated by radial segments and circular arcs.

25.1 The Electrostatic Potential

According to the definition given in Chapter 8, a force is conservative if the work done by this force on a particle depends on the initial and final positions of the particle, but not on the shape of the path connecting these positions. The electric force exerted by a fixed point charge q' on a moving point charge q is a conservative force. The proof is the same as in the case of the gravitational force. Figure 25.1 shows a fixed point charge q' and a second point charge q that moves from P_1 to P_2; the figure also shows two alternative paths I and II from P_1 to P_2. The work done by the electric force along a path is $\int \mathbf{F} \cdot d\mathbf{l}$.[1] To evaluate this integral, we approximate each path by a sequence of infinitesimal radial segments and circular arcs (Figure 25.1). Along the circular arcs the work is zero because the force is perpendicular to the displacement. Along the radial segments the work is not zero. Since for each radial segment belonging to the first path, there is a corresponding radial segment (at the same radius) belonging to the second path, and since the magnitude of the force is the same at both these radial segments, the contributions to the work along the first and second paths are equal. This establishes that the work is independent of the shape of the path and that the force is conservative.

More generally, the net electric force exerted by any number of fixed point charges on another given point charge is conservative, since the net force is a sum of individual conservative forces.

For any conservative force we can construct a corresponding potential energy by means of the recipe given in Section 8.2: Take a reference point P_0 and at this point assign to the potential energy the value $U(P_0)$. At any other point, assign to the potential energy the value [see Eq. (8.4)]

$$U(P) = -\int_{P_0}^{P} \mathbf{F} \cdot d\mathbf{l} + U(P_0) \tag{1}$$

The **electrostatic potential** is defined as the potential energy of a charge q divided by q, i.e., it is defined as the potential energy per unit charge,

$$V(P) = \frac{U(P)}{q} \tag{2}$$

$$= -\int_{P_0}^{P} \frac{\mathbf{F}}{q} \cdot d\mathbf{l} + \frac{U(P_0)}{q} \tag{3}$$

Since the force per unit charge is the electric field ($\mathbf{E} = \mathbf{F}/q$), we can also write this as

$$\boxed{V(P) = -\int_{P_0}^{P} \mathbf{E} \cdot d\mathbf{l} + V(P_0)} \tag{4}$$

Electrostatic potential

In this equation $V(P_0)$ plays the role of an additive constant in the potential. In the calculations of potential differences between points (see below), the additive constant in the potential cancels. Hence, this constant is of no physical significance. We can make any convenient choice for this constant, but we must remain faithful to our choice throughout any subsequent calculation. We will see some examples of convenient choices of $V(P_0)$ in what follows.

---

[1] Note that we are now using the symbol $d\mathbf{l}$ for an infinitesimal displacement while in Chapters 7 and 8 we used the symbol $d\mathbf{r}$; we now want to reserve the letter $\mathbf{r}$ for the radial distance in Coulomb's Law.

Potential difference

Alessandro, Conte Volta, *1745–1827, Italian physicist, professor at Pavia. Volta established that the "animal electricity," observed by Liugi Galvani, 1737–1798, in experiments with frog muscle tissue placed in contact with dissimilar metals, was not due to any exceptional property of animal tissues, but was also generated whenever any wet body was sandwiched between dissimilar metals. This led him to develop the first "voltaic pile," or battery, consisting of a large stack of moist discs of cardboard (electrolyte) sandwiched between discs of metal (electrodes).*

The **potential difference** between two arbitrary points P_1 and P_2 is [compare Eq. (8.5)]:

$$V(P_2) - V(P_1) = -\int_{P_0}^{P_1} \mathbf{E} \cdot d\mathbf{l} + V(P_0) + \int_{P_0}^{P_2} \mathbf{E} \cdot d\mathbf{l} - V(P_0)$$

or

$$V(P_2) - V(P_1) = -\int_{P_1}^{P_2} \mathbf{E} \cdot d\mathbf{l} \tag{5}$$

Hence the potential difference is the work that *you* must do (against the electric force) in order to push one coulomb of charge from the point P_1 to the P_2.

The unit of electrostatic potential is the **volt** (V),[2]

$$1 \text{ volt} = 1 \text{ V} = 1 \text{ joule/coulomb} = 1 \text{ J/C} \tag{6}$$

The unit of electric field is 1 N/C. This can be expressed in terms of volts as follows:

$$1 \, \frac{\text{N}}{\text{C}} = 1 \, \frac{\text{N} \cdot \text{m}}{\text{C} \cdot \text{m}} = 1 \, \frac{\text{J}}{\text{C}} \frac{1}{\text{m}} = 1 \, \frac{\text{V}}{\text{m}} \tag{7}$$

Thus N/C and V/m are equal units; in practice, volt per meter is the preferred unit for the electric field.

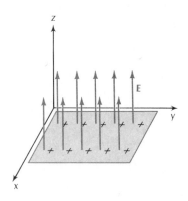

Fig. 25.2 Electric field of a very large charged surface lying in the *x–y* plane.

EXAMPLE 1.   A very large, flat conducting surface carries a uniform surface charge density which generates a constant electric field **E** (see Example 24.7). What is the electrostatic potential at some distance from this surface? Assume that the potential is zero at the surface.

SOLUTION: In Figure 25.2, the charged surface coincides with the *x–y* plane. According to Example 24.7, the electric field generated by this surface is vertical and is independent of distance:

$$E_z = E_0 = [\text{constant}] \tag{8}$$

If P_0 is a point on the surface, then $V(P_0) = 0$ and

$$V(P) = -\int \mathbf{E} \cdot d\mathbf{l} = -\int E_x \, dx - \int E_y \, dy - \int E_z \, dz$$

$$= -\int_0^z E_0 \, dz = -E_0 z \tag{9}$$

Hence the potential is directly proportional to the distance from the surface.

EXAMPLE 2.   Near the ground directly below a thundercloud the electric field is 2×10^4 V/m and points upward. What is the potential difference be-

---

[2]Note that the same letter *V* is used in physics both as a symbol for potential and as an abbreviation for *volt*. This leads to confusing equations such as $V = 3.0$ V (which means $V = 3.0$ volt). If there is a possibility of confusion, it is best not to abbreviate *volt*.

tween the ground and a point in the air, 50 m above ground? Assume that the electric field is constant.

Solution: From Eq. (9)

$$V(P) = -E_0 z = -2 \times 10^4 \text{ volt/m} \times 50 \text{ m} = -1 \times 10^6 \text{ volt}$$

Since the electric field in a conducting body in electrostatic equilibrium is zero, Eq. (5) tells us that the potential difference between any two points within a conducting body is zero. Thus, all points within a conducting body are at the same potential. For instance, since the Earth is a conductor, all points in the Earth or on the surface of the Earth are at the same electric potential. In experiments with electric circuits, it is usually convenient to adopt the convention that the potential of the Earth is zero, $V = 0$. The Earth is said to be the **electric ground,** and any conductors connected to it are said to be grounded.

Electric ground

25.2 The Electrostatic Potential of a Point Charge

To find the electrostatic potential of a fixed point charge q', we must substitute the radial electric field of such a point charge into the general formula for the potential,

$$V(P) = -\int_{P_0}^{P} \mathbf{E} \cdot d\mathbf{l} + V(P_0) \tag{10}$$

Since the integral does not depend on the path between P_0 and P, we may take any path that is convenient. Figure 25.3 shows a path consisting of one radial segment and one circular arc. Along the circular arc, the integral receives no contribution. Along the radial segment, the electric field $\mathbf{E} = (1/4\pi\varepsilon_0)q'\hat{\mathbf{r}}/r^2$ is in the same direction as the displacement, since $d\mathbf{l} = \hat{\mathbf{r}}dr$; hence $\mathbf{E} \cdot d\mathbf{l} = (1/4\pi\varepsilon_0)(q'/r^2) \, dr$ and[3]

Fig. 25.3 A path from P_0 to P consisting of a radial segment and a circular arc.

$$V(P) = -\int_{r_0}^{r} \frac{1}{4\pi\varepsilon_0} \frac{q'}{r'^2} \, dr' + V(P_0) \tag{11}$$

$$= \frac{q'}{4\pi\varepsilon_0} \left(\frac{1}{r} - \frac{1}{r_0} \right) + V(P_0) \tag{12}$$

For the reference point P_0 we choose a point at infinite distance from the fixed charge; for the value of the potential at this point we take $V(P_0) = 0$. The potential at the point P can then be written

$$\boxed{V(r) = \frac{1}{4\pi\varepsilon_0} \frac{q'}{r}} \tag{13}$$

Potential of point charge

Thus, the potential in the space surrounding a point charge is inversely proportional to the distance.

---

[3] The variable of integration in this integral has been written r' to distinguish it from the limit of integration r.

The potential energy of a charge q under the influence of the electric field of the charge q' is then

Potential energy of two point charges

$$U(r) = qV(r) = \frac{1}{4\pi\varepsilon_0}\frac{qq'}{r} \tag{14}$$

EXAMPLE 3. The electron in a hydrogen atom is at a distance of 0.53×10^{-10} m from the nucleus. What is the electrostatic potential generated by the nucleus at this position? What is the potential energy of the electron?

SOLUTION: The nucleus is (approximately) a point charge with $q' = e = 1.6 \times 10^{-19}$ C. The electrostatic potential generated by the nucleus is

$$V = \frac{1}{4\pi\varepsilon_0}\frac{q'}{r} = \frac{1}{4\pi\varepsilon_0} \times \frac{1.6 \times 10^{-19}\ \text{C}}{0.53 \times 10^{-10}\ \text{m}}$$

$$= 27\ \text{volt} \tag{15}$$

The charge of the electron is $q = -e = -1.6 \times 10^{-19}$ C. The potential energy of the electron is

$$U = qV = -e \times 27\ \text{volt} \tag{16}$$

$$= 1.6 \times 10^{-19}\ \text{C} \times 27\ \text{volt} = -4.3 \times 10^{-18}\ \text{J} \tag{17}$$

For the purposes of atomic physics, the joule is a rather large unit of energy and it is more convenient to leave the answer as in Eq. (16),

$$U = -27e \cdot \text{volt}$$

The product of the fundamental unit of atomic charge and the unit of potential, $e \cdot$ volt or eV, is a unit of energy. This unit of energy is called an **electron-volt.** It can be converted to joule by substituting the numerical value for e,

electron-volt, eV

$$1\ \text{eV} = 1 \times 1.60 \times 10^{-19}\ \text{C} \times 1\text{V} = 1.60 \times 10^{-19}\ \text{J} \tag{18}$$

In chemical reactions between atoms or molecules, the energy released or absorbed by each atom or molecule is typically 1 or 2 eV. Such reactions involve a change in the arrangement of the exterior electrons of the atoms, and the energy of 1 or 2 eV represents the typical amount of energy needed for this rearrangement.

EXAMPLE 4. An electron is initially at rest at a very large distance from a proton. Under the influence of the electric attraction, the electron falls toward the proton, which remains (approximately) at rest. What is the speed of the electron when it has fallen to within 0.53×10^{-10} m of the proton?

SOLUTION: The potential energy of the electron is $qV(r) = -eV(r)$. The total energy is the sum of the potential and kinetic energies,

$$E = \tfrac{1}{2}m_e v^2 - eV(r) \tag{19}$$

This total energy is conserved. The initial value of the energy is zero; hence the final value of the energy must also be zero:

$$\tfrac{1}{2}m_e v^2 - eV(r) = 0$$

and

$$v = \sqrt{\frac{2eV(r)}{m_e}} \qquad (20)$$

According to Example 3, $V(r) = 27$ volt for $r = 0.53 \times 10^{-10}$ m, so that

$$v = \sqrt{\frac{2 \times 1.6 \times 10^{-19} \text{ C} \times 27 \text{ V}}{9.1 \times 10^{-31} \text{ kg}}} = 3.1 \times 10^6 \text{ m/s}$$

EXAMPLE 5. A sphere of radius R carries a total charge Q uniformly distributed over its volume. Find the electrostatic potential inside and outside the sphere.

SOLUTION: Outside the sphere ($r \geq R$), the electrostatic potential is the same as for a point charge,

$$V(r) = \frac{1}{4\pi\varepsilon_0} \frac{Q}{r} \qquad r \geq R \qquad (21)$$

Inside the sphere ($r \leq R$), the electric field is [see Eq. (24.17)]

$$E(r) = \frac{1}{4\pi\varepsilon_0} \frac{Qr}{R^3} \qquad (22)$$

The potential difference between R and r can be calculated from this field [see Eq. (5)]:

$$V(r) - V(R) = -\int_R^r \frac{1}{4\pi\varepsilon_0} \frac{Qr'}{R^3} dr'$$

$$= -\frac{Q}{4\pi\varepsilon_0} \left(\frac{r^2}{2R^3} - \frac{1}{2R} \right) \qquad (23)$$

Consequently,

$$V(r) = -\frac{Q}{4\pi\varepsilon_0} \left(\frac{r^2}{2R^3} - \frac{1}{2R} \right) + V(R)$$

Since $V(R) = (1/4\pi\varepsilon_0)(Q/R)$, this gives

$$V(r) = -\frac{1}{4\pi\varepsilon_0} \frac{Qr^2}{2R^3} + \frac{1}{4\pi\varepsilon_0} \frac{3Q}{2R} \qquad r \leq R \qquad (24)$$

Figure 25.4 is a plot of the potential as a function of radius. The potential has a maximum at the center. The value of this maximum is

$$V(0) = \frac{1}{4\pi\varepsilon_0} \frac{3Q}{2R} \qquad (25)$$

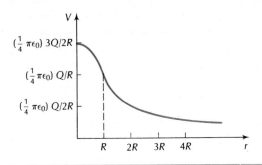

Fig. 25.4 Electrostatic potential of a uniformly charged sphere.

25.3 The Electrostatic Field as a Conservative Field

Conservative field

The electric field produced by a static charge distribution is said to be a **conservative field** because the integral $\int \mathbf{E} \cdot d\mathbf{l}$ between any two points P_1 and P_2 is independent of the path between these points. As we have seen in Section 8.1, this statement about the path independence of the integral is equivalent to the statement that the integral vanishes for any closed path, i.e.,

$$\oint \mathbf{E} \cdot d\mathbf{l} = 0 \tag{26}$$

where the circle on the integral sign indicates that the integration is around a closed path. Note that this equation implies that a field line can never form a closed loop — if it did, then the integral $\oint \mathbf{E} \cdot d\mathbf{l}$ around such a closed loop would not be zero. Consequently, any field line must start or end somewhere (on an electric charge). Since we derived the conservative property of the electrostatic field by examining the electric field produced by a point charge, it is not surprising that Eq. (26) should imply the existence of starting points and ending points for field lines. These sources and sinks of field lines are, of course, the positive and negative electric charges, and Gauss' Law tells us just how many field lines emerge from each electric charge. It can be shown that Eq. (26) together with Gauss' Law is exactly equivalent to Coulomb's Law, i.e., any electric field that satisfies both Eq. (26) and Gauss' Law is necessarily the electric field of a static distribution of point charges.

With Eq. (26) we can easily prove an interesting theorem about static electric fields: *within a closed, empty cavity inside a homogeneous conductor, the electric field is exactly zero.* Figure 25.5 shows such a cavity. If there were an electric field inside this cavity, then there would have to be field lines. Consider one of these field lines. Since the cavity is empty (contains no charge), the field line cannot end or begin within the cavity — it must therefore begin and end on the surface of the cavity (Figure 25.5). The field line cannot penetrate the conducting material, since the electric field is zero in this material. Now take a closed path consisting of one portion that follows the field line and a second portion that lies entirely in the conducting material. Along the first portion the integral $\int \mathbf{E} \cdot d\mathbf{l}$ is positive and along the second portion it is zero. Hence $\oint \mathbf{E} \cdot d\mathbf{l} \neq 0$ for this closed path. This is in contradiction to Eq. (26) and establishes the impossibility of such an electric field.

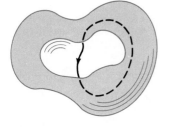

Fig. 25.5 Empty cavity in a volume of conducting material.

The absence of electrostatic fields in closed conducting cavities has important practical applications. Delicate electric instruments can be shielded from atmospheric electric fields, and other stray electric fields, by placing them in a box made of sheet metal. Such a box is called a **Faraday cage.** Often, the box is made of fine wire mesh rather than sheet metal; although such wire mesh does not have the perfect shielding properties of solid sheet metal, it provides good enough shielding for most purposes.

Faraday cage

25.4 The Gradient of the Potential

The electrostatic potential can be calculated from the electric field by integration,

$$V(P) = -\int_{P_0}^{P} \mathbf{E} \cdot d\mathbf{l} + V(P_0) \qquad (27)$$

Conversely, the electric field can be calculated from the potential by differentiation. Consider two nearby points separated by an infinitesimal displacement $d\mathbf{l}$. The change in potential between these points is then

$$dV = -\mathbf{E} \cdot d\mathbf{l} = -E \cos \theta \, dl \qquad (28)$$

where θ is the angle between the displacement $d\mathbf{l}$ and the electric field. Hence

$$\frac{dV}{dl} = -E \cos \theta \qquad (29)$$

This shows that the derivative dV/dl equals the negative of the component of E in the direction of dl.

If the displacement dl is in the x direction, $dl = dx$ and Eq. (29) becomes

$$\boxed{\frac{\partial V}{\partial x} = -E_x} \qquad (30) \qquad \textit{Derivatives of the potential}$$

Likewise

$$\boxed{\frac{\partial V}{\partial y} = -E_y} \qquad (31)$$

and

$$\boxed{\frac{\partial V}{\partial z} = -E_z} \qquad (32)$$

Thus, if the potential is a known function of position, the components of the electric field can be calculated by taking derivatives (compare Section 8.3).

In vector notation, we can express Eqs. (30)–(32) as

$$\frac{\partial V}{\partial x} \hat{\mathbf{x}} + \frac{\partial V}{\partial y} \hat{\mathbf{y}} + \frac{\partial V}{\partial z} \hat{\mathbf{z}} = -\mathbf{E} \qquad (33)$$

The quantity on the left side of this equation is called the **gradient** of the potential. The gradient is a vector with components $\partial V/\partial x$, $\partial V/\partial y$, and $\partial V/\partial z$. Thus, Eq. (33) asserts that the electric field is the negative of the gradient of the potential.

EXAMPLE 6. The electric potential generated by a uniformly charged flat conducting plate is

$$V = -E_0 z \tag{34}$$

where E_0 is a constant [see Eq. (9)]. Calculate the electric field from this potential.

SOLUTION: Equations (30)–(32) immediately give

$$E_x = -\frac{\partial V}{\partial x} = 0 \tag{35}$$

$$E_y = -\frac{\partial V}{\partial y} = 0 \tag{36}$$

$$E_z = -\frac{\partial V}{\partial z} = +E_0 \tag{37}$$

EXAMPLE 7. The electrostatic potential due to a point charge q' is

$$V = \frac{1}{4\pi\varepsilon_0} \frac{q'}{r} \tag{38}$$

Calculate the electric field from this potential.

SOLUTION: In order to use Eqs. (30)–(32) we would first have to express r in terms of x, y, z (i.e., $r = \sqrt{x^2 + y^2 + z^2}$). It is simpler to return to Eq. (29), which, with $dl = dr$, directly gives the component of **E** in the *radial direction*,

$$E_r = -\frac{dV}{dr} = -\frac{d}{dr}\left(\frac{1}{4\pi\varepsilon_0}\frac{q'}{r}\right) = \frac{1}{4\pi\varepsilon_0}\frac{q'}{r^2} \tag{39}$$

According to Eq. (29), the electric field only has components in those directions in which the potential changes. Since the potential (38) does not change in the tangential direction, the radial electric field given by Eq. (39) is the total field.

The results obtained in the two preceding examples do not tell us anything new; these calculations merely verify the consistency of our formulas.

A set of points at which the potential function has a fixed, constant value is called an **equipotential surface.** Figure 25.6 shows the equipotential surfaces belonging to the potential of a uniformly charged flat sheet — the equipotential surfaces are parallel planes. Figure 25.7

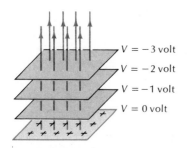

Fig. 25.6 Equipotential surfaces for a very large sheet with a uniform charge distribution.

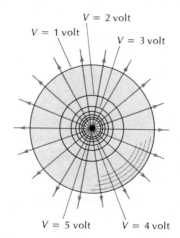

Fig. 25.7 Equipotential surfaces for a positive point charge. The equipotential surfaces are concentric spheres.

shows the equipotential surfaces belonging to the potential of a point charge — the equipotential surfaces are concentric spheres.

Note that the electric field is everywhere perpendicular to the equipotentials. This is a direct consequence of Eq. (29): along any equipotential surface the potential is constant, i.e., $dV = 0$; consequently the component of E along the surface must be zero and **E** must be perpendicular to the surface.

Conversely, if the electric field is everywhere perpendicular to a given surface, then this surface must be an equipotential. Since we already know (see Section 24.4) that along the surface of any conductor in electrostatic equilibrium **E** is perpendicular to the surface, we conclude immediately that any conducting surface is an equipotential surface. (This conclusion agrees with the general statement made at the end of Section 25.1: the potential is constant throughout any conductor.)

The calculation of the electric field of a static charge distribution can often be considerably simplified by first finding the electrostatic potential and then deriving the electric field from this. Equation (13) permits us to find directly the potential of any given charge distribution; we need only regard the charge distribution as a collection of point charges and sum, or integrate, the contributions the individual charges make to the potential. We can then calculate the components of the electric field by taking partial derivatives according to Eqs. (30)–(32). The advantage of this procedure is that it is much easier to sum the potentials of a distribution of point charges than it is to sum their electric fields; in the latter sum it is necessary to take into account the directions of the electric fields and to take into account the three electric field components — a rather messy procedure. The following is an example of a calculation of the electric field via the potential.

EXAMPLE 8. A charge Q is uniformly distributed along the circumference of a ring of radius R. Find the electric field on the axis of the ring.

SOLUTION: We have already solved this problem in Chapter 23 (see Example 23.3). Now we will solve it again by beginning with the electrostatic potential. At a height z above the plane of the ring (Figure 25.8) the potential contributed by a small charge element dQ is simply

$$dV = \frac{1}{4\pi\varepsilon_0} \frac{dQ}{r} = \frac{1}{4\pi\varepsilon_0} \frac{dQ}{(z^2 + R^2)^{1/2}} \qquad (40)$$

Since all charge elements around the ring are at the same distance from the point z, they all contribute equally and the total potential is simply

$$V = \frac{1}{4\pi\varepsilon_0} \frac{Q}{(z^2 + R^2)^{1/2}} \qquad (41)$$

Consequently, the z component of the electric field is

$$E_z = -\frac{\partial V}{\partial z} = \frac{1}{4\pi\varepsilon_0} \frac{Qz}{(z^2 + R^2)^{3/2}} \qquad (42)$$

and the x and y components are zero. This result agrees with Eq. (23.20), but we now obtained it a bit more quickly.

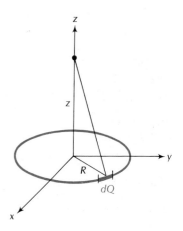

Fig. 25.8 A uniformly charged ring.

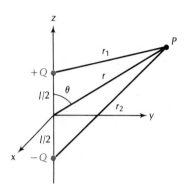

Fig. 25.9 Charges Q and $-Q$ on the z axis at $z = l/2$ and $z = -l/2$.

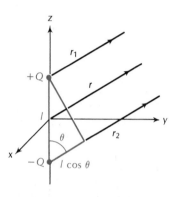

Fig. 25.10 If the point P is at a large distance (beyond the diagram), then the lines QP and $-QP$ are nearly parallel and the difference between their lengths r_1 and r_2 is $l \cos \theta$.

Potential of dipole

25.5 The Potential and Field of a Dipole

We will now calculate the electric field generated by an electric dipole. As in the above example, we will do this calculation via the potential.

Figure 25.9 shows an electric dipole consisting of charges $\pm Q$ on the z axis at $\pm l/2$. The net potential generated by this pair of charges is just the sum of the individual potentials,

$$V = \frac{1}{4\pi\varepsilon_0} \frac{Q}{r_1} - \frac{1}{4\pi\varepsilon_0} \frac{Q}{r_2} \tag{43}$$

$$= \frac{Q}{4\pi\varepsilon_0} \frac{r_2 - r_1}{r_1 r_2} \tag{44}$$

Where r_1 and r_2 are the lengths of the lines QP and $-QP$. From this potential function we can calculate the electric field by taking derivatives. Before we do this, we will make the simplifying assumption that r_1 and r_2 are much larger than l, i.e., we will make the assumption that the field point P is at a large distance from the electric dipole. Figure 25.10 shows that under these conditions r_1 and r_2 are approximately equal,

$$r_1 \cong r_2 \cong r \tag{45}$$

and their difference is a small quantity,

$$r_2 - r_1 \cong l \cos \theta \tag{46}$$

This permits us to write the following approximation for V:

$$V \cong \frac{Q}{4\pi\varepsilon_0} \frac{l \cos \theta}{r^2} \tag{47}$$

The product of Q and l is the dipole moment

$$p = lQ$$

and hence our approximation has the form

$$V = \frac{p}{4\pi\varepsilon_0} \frac{\cos \theta}{r^2} \tag{48}$$

To calculate the components of **E,** it is convenient to express everything in rectangular coordinates,

$$r = \sqrt{x^2 + y^2 + z^2} \tag{49}$$

$$\cos \theta = \frac{z}{\sqrt{x^2 + y^2 + z^2}} \tag{50}$$

so that

$$V = \frac{p}{4\pi\varepsilon_0} \frac{z}{(x^2 + y^2 + z^2)^{3/2}} \tag{51}$$

The components of the electric field are then

$$E_x = -\frac{\partial V}{\partial x} = \frac{p}{4\pi\varepsilon_0} \frac{3zx}{(x^2 + y^2 + z^2)^{5/2}} \qquad (52)$$

$$E_y = -\frac{\partial V}{\partial y} = \frac{p}{4\pi\varepsilon_0} \frac{3zy}{(x^2 + y^2 + z^2)^{5/2}} \qquad (53)$$

$$E_z = -\frac{\partial V}{\partial z}$$

$$= -\frac{p}{4\pi\varepsilon_0} \left(\frac{1}{(x^2 + y^2 + z^2)^{3/2}} - \frac{3z^2}{(x^2 + y^2 + z^2)^{5/2}} \right) \qquad (54)$$

These expressions for the electric field are only approximations, but they are very good at large distances from the dipole. For instance, if the dipole is within a molecule, then l is very small — 10^{-10} m or less — and our approximation is valid whenever the distance is large compared to 10^{-10} m. Figure 25.11 shows the field lines for this electric field.

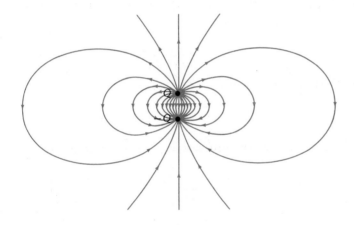

Fig. 25.11 Field lines of an electric dipole.

Equations (52)–(54) describe the electric field surrounding a water molecule, with a dipole moment $p = 6.1 \times 10^{-30}$ C·m (see Section 23.4). By means of this electric field the water molecule can act on other molecules — it can exert forces on the electric charges in other molecules. This is what makes water such a good solvent.

25.6 The Mean-Value Theorem[4]

There exists a very beautiful theorem concerning the electrostatic potential in a region free of electric charge. This theorem, called the **mean-value theorem,** states the following: *If* S *is the surface of a (mathematical) sphere whose interior is free of charge, then the potential at the center of the sphere equals the mean value of the potential over the surface.*

The proof of the theorem is simple. Suppose the sphere has a radius R. Then the mean value of the potential over the surface of the sphere

Mean-value theorem

---

[4] This section is optional.

is obtained by integrating the potential over the surface and dividing by the area of the surface,

$$\overline{V} = \frac{1}{4\pi R^2} \int V \, dS \tag{55}$$

The essential step of the proof is demonstrating that $\overline{V}$ is independent of R, i.e., $\overline{V}$ is independent of the size of the sphere. Consider the derivative of the integral (55),

$$\frac{d\overline{V}}{dR} = \frac{d}{dR} \left(\frac{1}{4\pi R^2} \int V \, dS \right) \tag{56}$$

Figure 25.12 shows an area element dS. This area is approximately the base of a cone of half-angle $\Delta\theta$; the base has an area $\pi(R \, \Delta\theta)^2$. The contribution to the integral (56) from this cone is

$$\frac{d}{dR} \left(\frac{1}{4\pi R^2} V \, dS \right) = \frac{d}{dR} \left[\frac{1}{4\pi R^2} V\pi(R \, \Delta\theta)^2 \right]$$

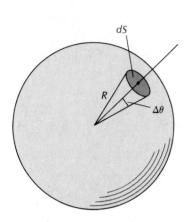

Fig. 25.12 Small circular area on surface of a sphere. The radius of this circular area is approximately $R \, \Delta\theta$.

$$= \frac{d}{dR} \left[\frac{(\Delta\theta)^2}{4} V \right] = \frac{(\Delta\theta)^2}{4} \frac{dV}{dR}$$

$$= \frac{1}{4\pi R^2} \frac{dV}{dR} \pi(R \, \Delta\theta)^2$$

$$= \frac{1}{4\pi R^2} \frac{dV}{dR} \, dS \tag{57}$$

The entire sphere can be regarded as a collection of such cones. Hence

$$\frac{d\overline{V}}{dR} = \frac{1}{4\pi R^2} \int \frac{dV}{dR} \, dS \tag{58}$$

But, according to Eq. (29), dV/dR is the negative of the component of **E** in the radial direction, i.e., it is the component of **E** in a direction perpendicular to the surface of integration. By Gauss' Law, the surface integral of this component is zero since the sphere S is free of charge, i.e.,

$$\frac{d\overline{V}}{dR} = \frac{1}{4\pi R^2} \int \frac{dV}{dR} \, dS = \frac{-1}{4\pi R^2} \int E_n \, dS = 0 \tag{59}$$

This establishes that $\overline{V}$ *is independent of R*. But if so, then the sphere of radius R and any smaller sphere must have the same value for the corresponding mean potential. In particular, since the potential at the center of the sphere can be regarded as the mean potential for a sphere of infinitesimal radius, it follows that the potential at the center must have the same value as the mean potential over the sphere of radius R. This concludes the proof.

The mean-value theorem has an important corollary: in a region free of charge, the electrostatic potential cannot have a minimum or a maximum. For suppose it had a minimum at a point P. This would mean that at *all* points in the immediate vicinity the potential is higher.

Hence, if we draw a small sphere centered on *P*, the mean value of the potential over the surface would have to be larger than the potential at the center. This would contradict the mean-value theorem and is therefore impossible. A similar argument shows that a maximum is also impossible. The absence of minima and maxima means that if the potential increases in some directions, it must decrease in others in such a way that the average of the changes in all directions is zero.

Finally, we will describe a very useful numerical method for the approximate calculation of the electrostatic potential. It will be easiest to describe this method in the context of a special example. Figure 25.13 shows a pair of large conducting plates with a kink. Figure 25.14 shows the plates edge on and gives the relevant dimensions. Suppose that the potential of the lower plate is 0 volt and that of the upper plate 3.0 volt. What is the potential in the space between the plates?

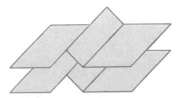

Fig. 25.13 Very large parallel plates with a kink.

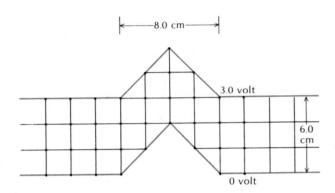

Fig. 25.14 Plates viewed edge on. A coordinate grid has been superimposed on the diagram; the squares of this grid are 2 cm × 2 cm.

In order to apply numerical methods to this problem, we must represent the space between the plates by a discrete, and finite, set of points. Figure 25.14 shows a coordinate grid superimposed on the picture of the plates. We will try to find the potential at the intersection points of this grid. This will give us only a rather rough description of the potential; for a more precise description we would have to take a finer grid. Note that the grid is two dimensional. The third dimension (out of the plane of the page in Figure 25.14) can be ignored since the potential at all points above or below the plane of the page is the same as that at the points shown in Figure 25.14.

To obtain the potential at the grid points, we proceed by the following method of successive approximations: We begin by making some reasonable first guess for the potential. Since for flat plates the potential in the space between would increase regularly from 0 volt to 3.0 volt, the values given in Figure 25.15a are a reasonable first approximation. To find the second approximation to the potential, we rely on the mean-value theorem. According to this theorem, the potential at any point should equal the average of the potentials of all the points that are at a distance of, say, 2 cm from the given point. In our grid, this average is approximated by an average over the four nearest neighbor points. We therefore obtain a second approximation by replacing the potential at each point by this average over the potentials of the four nearest neighbor points. This averaging procedure yields the values in Figure 25.15b. Next we obtain a third approximation by again replacing each of the potentials of Figure 25.15b with the average over the potentials of the four nearest neighbor points. This yields the values in Figure 25.15c, etc.

Fig. 25.15 The numbers at the grid points give the potential in volts.

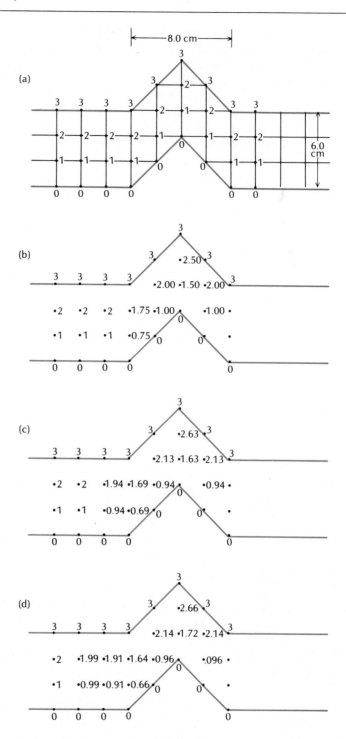

After a few successive steps, this iteration procedure yields self-consistent values for the potential, i.e., it yields values that do not change appreciably from one step to the next. These values are an approximate answer to the problem. The components of the electric field can then be obtained from the values of the potential by numerical calculation of the derivatives of the potential in the horizontal and vertical directions.

As a check on the approximation, it is a good idea to repeat the calculation with the finer grid size. This tends to get tedious, but the method lends itself very well to programming on a digital computer.

SUMMARY

Definition of electrostatic potential:

$$V(P) = -\int_{P_0}^{P} \mathbf{E} \cdot d\mathbf{l} + V(P_0)$$

Potential of point charge: $V = \dfrac{1}{4\pi\varepsilon_0}\dfrac{q'}{r}$

Potential energy of two point charges: $U = \dfrac{1}{4\pi\varepsilon_0}\dfrac{qq'}{r}$

Derivatives of the potential: $\dfrac{dV}{dl} = -E\cos\theta$

$$\frac{\partial V}{\partial x} = -E_x, \quad \frac{\partial V}{\partial y} = -E_y, \quad \frac{\partial V}{\partial z} = -E_z$$

Mean-value theorem (for charge-free region): potential at center of any sphere equals mean potential on surface.

QUESTIONS

1. The potential difference between the poles of an automobile battery is 12 volt. Explain what this means in terms of the definition of potential as work per unit charge.

2. An old-fashioned word for electrostatic potential is electrostatic *tension*. Is it reasonable to think of the potential as analogous to mechanical tension?

3. If the electric field is zero in some region, must the potential also be zero? Give an example.

4. A bird sits on a high-voltage power line which is at a potential of 345,000 volt. Does this harm the bird?

5. How would you define the gravitational potential? Are the units of gravitational potential the same as the units of electric potential? According to your definition, what is the gravitational potential difference between the ground and a point 50 m above the ground?

6. Consider an electron moving in the vicinity of a proton. Where is the electrostatic potential produced by the proton highest? Where is the potential energy of the electron highest?

7. Suppose that the electrostatic potential has a minimum at some point. Is this an equilibrium point for a positive charge? For a negative charge? Is the equilibrium stable?

8. Consider a sphere of radius R with a charge Q uniformly distributed over its volume. Where does the potential have a maximum? Where does the magnitude of the electric field have a maximum?

9. Give an example of a conductor that is not an equipotential. Is this conductor in electrostatic equilibrium?

10. If the potential in a three-dimensional region of space is known to be constant, what can you conclude about the electric field in this region? If the potential on a two-dimensional surface is known to be constant, what can you conclude about the electric field on this surface?

11. In many calculations it is convenient to assign a potential of 0 volt to the ground. If so, what is the potential at the top of the Eiffel Tower? What is the potential at the top of your head? (Hint: Your body is a conductor.)

12. Is it true that the surface of a mass in static mechanical equilibrium is a gravitational equipotential surface? What if the surface is that of a fluid, such as water?

13. If a high-voltage power cable falls on top of your automobile, you will probably be safest if you remain inside the automobile. Why?

14. Suppose that several separate solid metallic bodies have been placed near a charge distribution. Is it necessarily true that all of these bodies will have the same potential?

15. If we surround some region with a conducting surface, we shield it from external electric fields. Why can we not shield a region from gravitational fields by a similar method?

16. The interior of an automobile provides good protection against lightning. Explain.

17. A cavity is completely surrounded by conducting material. Can you create an electric field in this cavity?

18. Show that different equipotential surfaces cannot intersect.

19. Consider the patterns of field lines shown in Figures 23.14 and 23.15. Roughly, sketch some of the equipotential surfaces for each case.

20. Sketch the equipotential surfaces for the potential described numerically in Figure 25.15d.

PROBLEMS

Section 25.1

1. In order to charge a typical 12-volt automobile battery fully, the charging device must force $+2.0 \times 10^5$ coulombs from the negative terminal of the battery to the positive terminal. How much work must the charging device do during this process?

2. A proton sits at the origin of coordinates. How much work (in electron-volts) must you do to push an electron from the point $x = 1.0$ Å, $y = 0$, $z = 0$ to the point $x = 0.5$ Å, $y = 0.5$ Å, $z = 0$?

3. On days of fair weather, the atmospheric electric field of the Earth is about 100 V/m; this field points vertically downward (compare Problem 23.2). What is the electric potential difference between the ground and an airplane flying at 600 m (2000 ft)? What is the potential difference between the ground and the tip of the Eiffel Tower? Treat the ground as a flat conductor.

4. Consider the arrangement of parallel sheets of charge described in Problem 23.22. Find the potential difference between the upper sheet and the lower sheet.

5. Suppose that, as a function of x, y, z, an electric field has components

$$E_x = 6x^2y \qquad E_y = 2x^3 + 2y \qquad E_z = 0$$

where E is measured in volts per meter and the distances are measured in meters.
   (a) Find the potential difference between the origin and the point $x = 3$, $y = 0$, $z = 0$.
   (b) Find the potential difference between the origin and the point $x = 0$, $y = 2$, $z = 0$.

6. At the Stanford Linear Accelerator (SLAC), electrons are accelerated from an energy of 0 eV to 20×10^9 eV as they travel in a straight evacuated tube 1600 m in length. The acceleration is due to a strong electric field pushing the electrons along. Assume that the electric field is uniform. What must be its strength?

7. The potential difference between the two poles of an automobile battery is 12.0 V. Suppose that you place such a battery in empty space and you release an electron at a point next to the negative pole of the battery. The electron will then be pushed away by the electric force and move off in some direction.
   (a) If the electron strikes the positive pole of the battery, what will be its impact speed?
   (b) If instead the electron moves away toward infinity, what will be its ultimate speed?

8. The gap between the electrodes of a spark plug in an automobile is 0.025 in. In order to produce an electric field of 3×10^6 V/m (required to initiate an electric spark), what minimum potential difference must you apply to the spark plug?

9. Prove that the plane midway between a positive and a negative point charge of equal magnitudes is an equipotential surface. Is this also true if both charges are positive?

Section 25.2

10. The nucleus of lead has a charge of $82e$ uniformly distributed over a spherical region of radius 7.1×10^{-15} m. What is the electrostatic potential at the nuclear surface? At the center?

11. An alpha particle of kinetic energy 1.7×10^{-12} J is shot directly toward a platinum nucleus. What will be the distance of closest approach? The electric charge of the alpha particle is $2e$ and that of the platinum nucleus is $78e$. Treat the alpha particle and the nucleus as spherical charge distributions and disregard the motion of the nucleus.

12. What is the minimum kinetic energy with which an alpha particle must be launched toward a plutonium nucleus if it is to make contact with the nuclear surface? The plutonium nucleus is a sphere of radius 7.5×10^{-15} m with a charge of $94e$ uniformly distributed over the volume. For the purpose of this problem, the alpha particle may be regarded as a particle (of negligible radius) with a charge of $2e$.

13. A thorium nucleus emits an alpha particle according to the reaction

$$\text{thorium} \rightarrow \text{radium} + \text{alpha}$$

Assume that the alpha particle is pointlike and that the residual radium nucleus is spherical with a radius of 7.4×10^{-15} m. The charge on the alpha particle is $2e$ and that on the radium nucleus is $88e$.
   (a) At the instant the alpha particle emerges from the nuclear surface, what is its electrostatic potential energy?
   (b) If the alpha particle has no initial kinetic energy, what will be its final kinetic energy and speed when far away from the nucleus? Assume that the radium nucleus does not move. The mass of the alpha particle is 6.7×10^{-27} kg.

14. Consider again the arrangement of charges within the thundercloud of Figure 23.28. Find the electric potential due to these charges at a point which is at a height of 8 km and on the vertical line passing through the charges. Find the electric potential at a second point which is at the same height and has a horizontal distance of 5 km from the first point.

15. In a helium atom, at some instant one of the electrons is at a distance of 0.3×10^{-10} m from the nucleus and the other electron is at a distance of

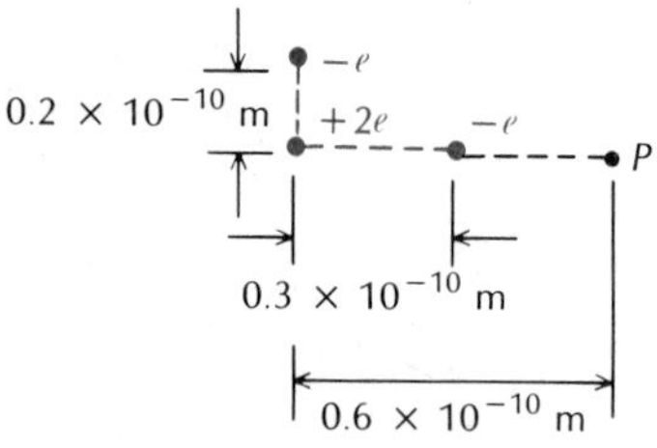

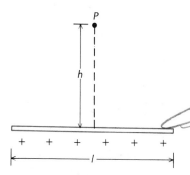

Fig. 25.16 Nucleus (charge $+2e$) and electrons (charge $-e$) of helium atom at one instant of time.

0.2×10^{-10} m, $90°$ away from the first (Figure 25.16). Find the electric potential produced jointly by the two electrons and the nucleus at a point P beyond the first electron and at a distance of 0.6×10^{-10} m from the nucleus.

*16. According to Bohr's theory of the atom (see also Problem 22.12), the electron in a hydrogen atom orbits around the nucleus in a circular orbit. The force that holds the electron in this orbit is the Coulomb force. The size of the orbit depends on the angular momentum — the smallest possible orbit has an angular momentum $\hbar = 1.05 \times 10^{-34}$ J · s; the next possible orbit has angular momentum $2\hbar$; the next $3\hbar$, etc.

(a) Show that if a circular orbit has angular momentum $n\hbar$ (where $n = 1$, 2, 3, . . .), then its radius is

$$r = \frac{4\pi\varepsilon_0}{m_e e^2} n^2 \hbar^2$$

(b) Show that the orbital energy (kinetic and potential) of the electron in such an orbit is

$$E = -\frac{m_e e^4}{2(4\pi\varepsilon_0)^2 \hbar^2} \frac{1}{n^2}$$

(c) Evaluate this energy for $n = 1$; express your answer in electron-volts.

Section 25.3

17. A total charge Q is distributed uniformly along a straight rod of length l. Find the potential at a point P at a distance h from the midpoint of the rod (Figure 25.17).

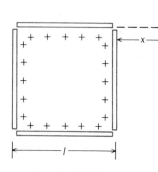

Fig. 25.17 A charged rod of length l.

18. Three thin rods of glass of length l carry charges uniformly distributed along their lengths. The charges on the three rods are $+Q$, $+Q$, and $-Q$, respectively. The rods are arranged along the sides of an equilateral triangle. What is the electrostatic potential at the midpoint of this triangle?

19. A uniformly charged sphere of radius a is surrounded by a uniformly charged concentric spherical shell of inner radius b and outer radius c. The total charge on the sphere is Q and that on the shell $-Q$. Find the potential at $r = b$, at $r = a$, and at $r = 0$.

20. Four rods of length l are arranged along the edges of a square. The rods carry charges $+Q$ uniformly distributed along their lengths (Figure 25.18). Find the potential at the point P at a distance x from one corner of the square.

21. Two semicircular rods and two short straight rods are joined in the configuration shown in Figure 25.19. The rods carry a charge of λ coulomb per meter. Calculate the potential at the center of this configuration.

22. A long straight wire of radius 0.80 mm is surrounded by an evacuated concentric conducting shell of radius 1.2 cm. The wire carries a charge of -5.5×10^{-8} coulomb per meter of length. Suppose that you release an electron at the surface of the wire. With what speed will this electron hit the conducting shell?

23. A long plastic pipe has an inner radius a and an outer radius b. Charge is uniformly distributed over the volume $a < r < b$. The amount of charge is λ coulomb per meter of length of the tube. Find the potential difference between $r = b$ and $r = 0$. Assume that the plastic has no effect on the electric field.

24. A flat disk of radius R has charge Q uniformly distributed over its surface. Find a formula for the potential along the axis of the disk.

25. The tube of a Geiger counter consists of a thin straight wire surrounded by a coaxial conducting shell. The diameter of the wire is 0.001 in. and that of

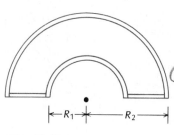

Fig. 25.18

Fig. 25.19

the shell is 1.0 in. The length of the tube is 4.0 in; however in your calculation use the formula for the electric field of an infinitely long line of charge. If the potential difference between the wire and the shell is 1.0×10^3 volt, what is the electric field at the surface of the wire? At the cylinder?

*26. An infinite charge distribution with spherical symmetry has a charge density ρ coulomb per cubic meter given by the formula $\rho = kr^{-5/2}$ where k is a constant. Find the potential as a function of the radius. Assume $V = 0$ at $r = \infty$.

*27. A point charge Q is on the positive z axis at the point $z = h$. A point charge $-Q \times R/h$ (where R is a positive length, $0 < R < h$) is on the z axis at the point $z = R^2/h$. Show that the surface of the sphere of radius R about the origin is an equipotential surface.

Sections 25.4 and 25.5

28. In some region of space the electrostatic potential is the following function of x, y, and z:

$$V = x^2 + 2xy$$

where the potential is measured in volts and the distances in meters. Find the electric field at the point $x = 2$, $y = 2$.

29. In terms of x, y, and z, the potential of a point charge is

$$V(x, y, z) = \frac{1}{4\pi\varepsilon_0} \frac{q'}{\sqrt{x^2 + y^2 + z^2}}$$

    (a) By differentiating this potential function, calculate the components E_x, E_y, and E_z of the electric field.
    (b) Show that the magnitude $\sqrt{E_x^2 + E_y^2 + E_z^2}$ agrees with the usual expression for the electric field of a point charge.

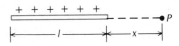

Fig. 25.20

30. A rod of length l has a charge Q uniformly distributed along its length (Figure 25.20). Find the potential at the point P at a distance x from one end of the rod. Find the electric field at this point.

31. Two rods of equal length l form a symmetric cross. The rods carry charges $\pm Q$ uniformly distributed along their lengths. Calculate the potential at the point P at a distance x from one end of the cross (Figure 25.21). Calculate the electric field at this point.

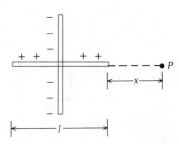

32. An annulus (a disk with a hole) made of paper has an outer radius R and an inner radius $R/2$ (Figure 25.22). An amount Q of electric charge is uniformly distributed over the paper.
    (a) Find the potential as a function of distance on the axis of the annulus.
    (b) Find the electric field on the axis of the annulus.

Fig. 25.21

33. A nucleus of carbon (charge $6e$) and one of helium (charge $2e$) are separated by a distance of 1.2×10^{-13} m and instantaneously at rest. The center of mass of this system is at a distance of 0.4×10^{-13} m from the carbon nucleus. Take this point as origin and take the x axis along the line joining the nuclei, with the carbon nucleus on the negative x axis.
    (a) Find the potential V as a function of x, y, and z.
    (b) Find E_x and E_y as a function of x, y, and z.

34. The water molecule has a dipole moment of 6.1×10^{-30} C·m.
    (a) Find the magnitude and direction of the electric field at a point on the axis of the dipole at a distance of 12.0 Å from the molecule.
    (b) Find the magnitude and direction of the electric field at a point on a line transverse to the axis of the dipole at the same distance.

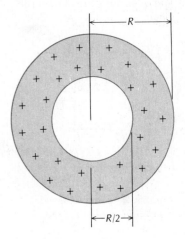

*35. A thin cylindrical cardboard tube has a charge Q uniformly distributed over its surface. The radius of the tube is R and the length l.

Fig. 25.22

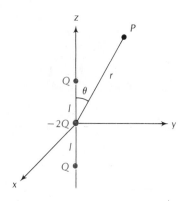

Fig. 25.23 Charges − 2Q, Q, and Q on the z axis.

(a) Find the potential at a point on the axis of the tube at a distance x from the midpoint. Assume $x > l$.

(b) Find the electric field at this point.

*36. A point charge $-2Q$ is at the origin of coordinates; two point charges $+Q$ are on the z axis, at $z = \pm l$, respectively (Figure 25.23).

(a) Show that, for $r \gg l$, the net potential of these charges is approximately

$$V = \frac{2Ql^2}{4\pi\varepsilon_0 r^3} \frac{(3\cos^2\theta - 1)}{2}$$

(b) Calculate E_x, E_y, and E_z, expressing these in terms of the coordinates x, y, and z.

Section 25.6

37. Figure 25.24 shows two large, parallel, conducting plates seen in cross section. One of the plates has a kink. The potential difference between them is 2.0 V. Use the mean-value theorem to find the potential at the points of the grid in Figure 25.24 to at least two significant figures.

38. Figure 25.25 shows two large, parallel conducting plates seen in cross section. Both plates have rectangular kinks. The potential difference between the plates is 6.0 V. Use the mean-value theorem to find the potential at the grid points to within at least two significant figures.

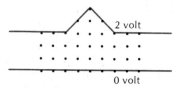
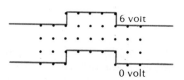

Fig. 25.24 Very large parallel plates, viewed edge on. One of the plates has a kink.

Fig. 25.25 Very large parallel plates, viewed edge on. Both plates have rectangular ridges.

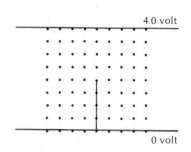

Fig. 25.26 Two very large parallel plates, seen edge on. The lower plate has a protruding thin ridge; the potential is $V = 0$ on this plate and on the ridge.

39. Two large, parallel conducting plates have a potential difference of 4.0 V between them. One of the plates carries a thin vertical ridge (Figure 25.26). Use the mean-value theorem to find the potential at the grid points to within two significant figures or better.

40. Given that each box of the grid of Figure 25.15d is 2 cm × 2 cm, evaluate numerically the derivative $\Delta V / \Delta z$ at each of the grid points and thereby find E_z. The z direction is the vertical direction (perpendicular to the flat portion of the plate).

41. Two long concentric tubes of sheet metal have a square cross section; Figure 25.27 shows their cross section. The outer tube is at a potential of 10 V; the inner tube is at a potential of 0 V. Use the mean-value theorem to find the potential at all the grid points shown to within two significant figures.

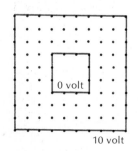

Fig. 25.27 Two concentric tubes (ducts) of sheet metal, seen in cross section.

42. When calculating the average potential in the method of successive approximations described in Section 25.6, we only took into account the nearest four points surrounding the given point in a *two-dimensional* grid. Since the mean-value theorem applies to a sphere surrounding the given point, we should actually take into account the six nearest points surrounding the given point in a *three-dimensional* grid, i.e., we should take into account an extra point in front of and an extra point behind the plane shown in Figure 25.14. Prove that for the problem discussed in Section 25.6, in which the potential in front of the given point and the potential behind the given point are the same as the potential at the given point, the average over the four nearest points in the two-dimensional grid coincides with the average over the six nearest points in the three-dimensional grid.

CHAPTER 26

Electric Energy

In the preceding chapter we calculated the electric potential energy of a charge — a test charge — placed in the electric field of a given charge distribution. Now we will calculate the potential energy of the charge distribution by itself, without the test charge. The charge distribution can be regarded as a collection of point charges; since all of these point charges exert forces on one another, it requires a certain amount of work to bring them together into their final configuration if they are initially separated by large distances. This amount of work is the potential energy of the charge distribution.

We will see that this potential energy is stored in the electric field. The energy is concentrated in those regions of space where the electric field is strong. We will see that the distribution of energy in the electric field can be described by an energy density. Since the electric field is endowed with energy, we must regard the field as a material object, a fifth state of matter.

26.1 Energy of a System of Point Charges

The electric potential energy of two point charges q_1, q_2 separated by a distance r is [see Eq. (25.14)]

$$U = \frac{1}{4\pi\varepsilon_0} \frac{q_1 q_2}{r} \tag{1}$$

This potential energy can be regarded as the work required to move q_1 from infinity to within a distance r of q_2 or, alternatively, the work required to move q_2 to within a distance r of q_1. It is a *mutual* potential

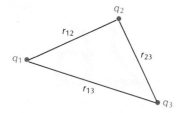

Fig. 26.1 Three point charges.

energy which belongs to both q_1 and q_2, i.e., it is an energy associated with the pair q_1, q_2.

For configurations consisting of more than two charges, the net potential energy can be calculated by writing down a term similar to that in Eq. (1) for *each pair* of charges. For instance, if we are dealing with three charges (Figure 26.1), we have three possible pairs, (q_1, q_2), (q_2, q_3), and (q_1, q_3), so that the net potential energy is

$$U = \frac{1}{4\pi\varepsilon_0} \frac{q_1 q_2}{r_{12}} + \frac{1}{4\pi\varepsilon_0} \frac{q_2 q_3}{r_{23}} + \frac{1}{4\pi\varepsilon_0} \frac{q_1 q_3}{r_{13}} \tag{2}$$

This is the work required to assemble the charges in the final configuration shown in Figure 26.1 starting from an initial condition of infinite separation.

Note that Eq. (2) is identically equal to

$$U = \frac{1}{2} \left(\frac{1}{4\pi\varepsilon_0} \frac{q_2}{r_{12}} + \frac{1}{4\pi\varepsilon_0} \frac{q_3}{r_{13}} \right) q_1 + \frac{1}{2} \left(\frac{1}{4\pi\varepsilon_0} \frac{q_1}{r_{12}} + \frac{1}{4\pi\varepsilon_0} \frac{q_3}{r_{23}} \right) q_2$$

$$+ \frac{1}{2} \left(\frac{1}{4\pi\varepsilon_0} \frac{q_1}{r_{13}} + \frac{1}{4\pi\varepsilon_0} \frac{q_2}{r_{23}} \right) q_3 \tag{3}$$

Here the potential energy of each pair of charges appears twice, each time with a factor of $\frac{1}{2}$. Equation (3) therefore leads to the following expression for the energy in terms of potentials:

$$U = \tfrac{1}{2} V_{\text{other}}(1) q_1 + \tfrac{1}{2} V_{\text{other}}(2) q_2 + \tfrac{1}{2} V_{\text{other}}(3) q_3 \tag{4}$$

where $V_{\text{other}}(1)$ is the electric potential produced at the position of charge 1 by the *other* charges (charges 2 and 3), i.e.,

$$V_{\text{other}}(1) = \frac{1}{4\pi\varepsilon_0} \frac{q_2}{r_{12}} + \frac{1}{4\pi\varepsilon_0} \frac{q_3}{r_{13}} \tag{5}$$

and similarly for $V_{\text{other}}(2)$ and $V_{\text{other}}(3)$.

By means of a generalization of this argument we can easily show that for a configuration consisting of any number of point charges, the electric potential energy is the sum

Energy of system of point charges

$$\boxed{U = \tfrac{1}{2} V_{\text{other}}(1) q_1 + \tfrac{1}{2} V_{\text{other}}(2) q_2 + \tfrac{1}{2} V_{\text{other}}(3) q_3 + \tfrac{1}{2} V_{\text{other}}(4) q_4 + \cdots} \tag{6}$$

This expression gives the work that must be done to bring the point charges to their final positions starting from initial positions at very large distances from each other. However, this expression is not the total potential energy because it does not take into account the energy that a point charge has when it is by itself, at a large distance from all other point charges. Such an isolated point charge has potential energy because it takes work to assemble the point charge out of infinitesimal pieces of charge. The energy needed to assemble the point charge is called the **self-energy** of the point charge.

The calculation of the self-energy of point charges, such as electrons, is one of the unsolved problems of physics. A straightforward calculation of the self-energy of an electron yields the absurd result that this energy is infinite [essentially, the calculated energy is infinite because the potential at the position of the point charge is infinite: $(1/4\pi\varepsilon_0)\,(q/r) \to \infty$ as $r \to 0$]. Although up to now physicists have found no satisfactory way to calculate the self-energy, they have invented several rules for bypassing this problem. These rules, called renormalization rules, give prescriptions on how to extract experimentally meaningful numbers from the theoretical calculations. In essence, the rules assert that the self-energy is a constant quantity that never has any effect on energy conservation and can therefore be ignored. The theoretical calculations based on this scheme have been extremely successful. For instance, by combining electromagnetic theory, relativity theory, and quantum theory, the electric energy of an electron in a hydrogen atom has been calculated to nine significant figures. But from the mathematical point of view this scheme has some shady aspects.

In the following we will always ignore the electric self-energy of point charges and pretend that Eq. (6) is the total electric energy.

26.2 Energy of a System of Conductors

Equation (6) permits us to evaluate the electric energy of a system of charged conductors. Figure 26.2 shows several conductors of arbitrary shapes. Suppose that the charges on these conductors are Q_1, Q_2, Q_3, . . . , and that their potentials are V_1, V_2, V_3, . . . , respectively. The charge Q_1 consists of many point charges distributed over conductor 1. Each of these point charges is at potential V_1. This potential acting on a given point charge dQ_1 can be regarded as due to the *other* point charges (on conductor 1 and on other conductors) because the given point charge makes only an insignificant contribution to V_1. Hence, according to Eq. (6), the electric energy associated with a point charge dQ_1 on conductor 1 is $\frac{1}{2}(dQ_1)V_1$ and the potential energy associated with all the point charges on conductor 1 is simply $\frac{1}{2}Q_1V_1$. Since similar arguments apply to the other conductors, we conclude that net electric energy is

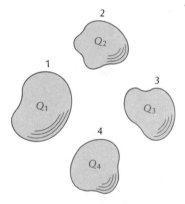

Fig. 26.2 Four conducting bodies carrying electric charges.

$$\boxed{U = \tfrac{1}{2}Q_1V_1 + \tfrac{1}{2}Q_2V_2 + \tfrac{1}{2}Q_3V_3 + \cdots} \qquad (7)$$

Energy of system of conductors

EXAMPLE 1. A metallic sphere of radius R carries a charge Q uniformly distributed over its surface. How much electric energy is stored in this charge distribution?

SOLUTION: The potential outside a spherically symmetric charge distribution is given by Eq. (25.21). At $r = R$, this potential is

$$V = \frac{1}{4\pi\varepsilon_0}\frac{Q}{R}$$

According to Eq. (7), the electric energy is then

$$U = \tfrac{1}{2}QV = \tfrac{1}{2}Q\,\frac{1}{4\pi\varepsilon_0}\frac{Q}{R} = \frac{1}{8\pi\varepsilon_0}\frac{Q^2}{R} \qquad (8)$$

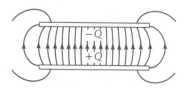

Fig. 26.3 Two very large parallel conducting plates with opposite electric charges.

EXAMPLE 2. Two large, parallel metallic plates of area A are separated by a distance d. Charges $+Q$ and $-Q$ are placed on the plates, respectively (Figure 26.3). What is the electric energy?

SOLUTION: The electric field between the plates is approximately the field of an infinite plate, i.e., the electric field is constant,

$$E = \frac{\sigma}{\varepsilon_0} = \frac{Q}{\varepsilon_0 A} \tag{9}$$

This expression fails near the edges of the plates, where there is an electric fringing field that is not constant (Figure 26.3). We will investigate the energy in such a fringing field in the next section. If the plates are large, then the edge region is only a very small fraction of the total region between the plates, and we can ignore this edge region without introducing excessive errors in our calculation.

The potential difference between the plates is

$$V_2 - V_1 = -Ed = -\frac{Qd}{\varepsilon_0 A} \tag{10}$$

and the electric potential energy is

$$U = \tfrac{1}{2}Q_1 V_1 + \tfrac{1}{2}Q_2 V_2 = \tfrac{1}{2}QV_1 - \tfrac{1}{2}QV_2$$

$$= \tfrac{1}{2}Q(V_1 - V_2) = \tfrac{1}{2}\frac{Q^2 d}{\varepsilon_0 A} \tag{11}$$

Note that Eq. (11) can be rewritten in the following interesting way:

$$U = \tfrac{1}{2}\varepsilon_0 \left(\frac{Q}{\varepsilon_0 A}\right)^2 Ad$$

$$= \tfrac{1}{2}\varepsilon_0 E^2 \times [\text{volume}] \tag{12}$$

where the "volume" is the volume between the plates, i.e., the volume of the region in which there is an electric field. The energy per unit volume of electric field is therefore $\tfrac{1}{2}\varepsilon_0 E^2$. This suggests that the electric energy is distributed over space, being concentrated in those regions where the electric field is strong. In the next section we will confirm that this is indeed the case.

26.3 The Energy Density

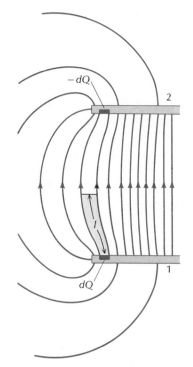

Fig. 26.4 Field lines start on the charge dQ on the lower plate and end on the charge $-dQ$ on the upper plate.

For the special case of the constant electric field in the region between parallel conducting plates, we found that the energy per unit volume of field is $\tfrac{1}{2}\varepsilon_0 E^2$. We will now prove that this result is also true for an electric field that is not constant, such as the fringing electric field at the edges of the conducting plates. Figure 26.4 shows this field at the edges. Consider a narrow bundle of field lines that start on one conductor and end on the other conductor. The bundle of field lines intercepts two small areas on the surfaces of the two conductors. The electric charges on these small areas are dQ and $-dQ$, respectively. The contribution to the net electrostatic energy from these charges $\pm dQ$ is

$$dU = \tfrac{1}{2}dQ\,V_1 - \tfrac{1}{2}dQ\,V_2 = \tfrac{1}{2}dQ(V_1 - V_2)$$

$$= \tfrac{1}{2}\,dQ \int_1^2 E(l)\,dl \qquad (13)$$

where l is the length measured along the bundle of field lines starting at conductor 1, and $E(l)$ is the strength of the electric field at the distance l. [Note that $E(l)\,dl = \mathbf{E} \cdot d\mathbf{l}$ because the path of integration coincides with a field line whose direction is always the same as that of the electric field.] Since dQ is constant, it can be placed inside the integral sign,

$$dU = \tfrac{1}{2} \int dQ\,E(l)\,dl \qquad (14)$$

Next we apply Gauss' Law to the tubular volume shown in Figure 26.5. The sides of the tube run along field lines; one cap of the tube is just inside the conductor, the other cap is perpendicular to field lines. The charge in the tube is dQ and the flux through the surface of the tube is entirely due to the field lines that emerge through the perpendicular cap $dS(l)$ at a distance l. Gauss' Law then tells us

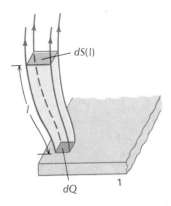

Fig. 26.5 The field lines that start on dQ form a tube of rectangular cross section.

$$\frac{dQ}{\varepsilon_0} = E(l)\,dS(l) \qquad (15)$$

Substituting this expression for dQ into Eq. (14), we obtain

$$dU = \tfrac{1}{2} \int \varepsilon_0 E(l)E(l)\,dS(l)\,dl \qquad (16)$$

But $dS(l)\,dl$ is the amount of volume in our tube between l and $l + dl$. Hence Eq. (16) can be written

$$dU = \tfrac{1}{2} \int \varepsilon_0 E^2\,dv \qquad (17)$$

where the integration extends over the volume of the tube from conductor 1 to conductor 2. This shows that the potential energy of the charges $\pm dQ$ on the conductors equals the integral of $\tfrac{1}{2}\varepsilon_0 E^2$ over the volume of the tube that lies within the bundle of field lines originating on these charges. Therefore the potential energy of all the charges on the conductors will equal the integral of $\tfrac{1}{2}\varepsilon_0 E^2$ over the volume of all the regions in which there is electric field.

Clearly, the above argument in no way depends on the shape of the conductors. Hence the result applies not only to parallel conducting plates, but to any number of conductors of arbitrary shape. Furthermore, although we have obtained this result for the case of field lines that start on one conductor and end on another, it is easy to see that we can obtain the same result for field lines that start at one conductor and go on to infinity. This establishes that for any arbitrary system of charged conductors the electric energy can be expressed as a volume integral of $\tfrac{1}{2}\varepsilon_0 E^2$, i.e.,

$$\boxed{U = \int \tfrac{1}{2}\varepsilon_0 E^2\,dv} \qquad (18)$$

Energy in electric field

where the integration extends over all the regions where there is electric field. Incidentally, this equation for the electric energy is also valid for charges placed on nonconductors; we will not deal with the proof of this assertion but we will take for granted that Eq. (18) is a general result for the energy associated with an electric field in vacuum.

Note that Eq. (7) expresses the energy as a sum over the electric charges, whereas Eq. (18) expresses the energy as an integral over the electric field. Thus, the former equation suggests that the energy is in the charges while the latter suggests it is in the field. To decide which of these alternatives is correct, we need some extra information. The clue is the existence of electric fields that are independent of electric charges. As we will see in Chapter 36, radio waves and light waves consist of electric and magnetic fields traveling through space. Although these fields are originally generated by electric charges, they persist even when the charges disappear. For example, a radio wave or a light beam continues to travel through space long after the radio transmitter has been shut down or the candle has been snuffed out. Obviously, the energy of a radio wave resides in the radio wave itself, in its electric and magnetic fields, and not in the electric charges in the antenna of the radio transmitter. But if energy is associated with the traveling electric fields of a radio wave, then energy should also be associated with the electric fields of a static charge distribution. We will therefore suppose that the energy (18) is in the electric field. The energy density, or energy per unit volume is

Energy density in electric field

$$u = \tfrac{1}{2}\varepsilon_0 E^2 \tag{19}$$

EXAMPLE 3. Since energy has mass, the electric field should have not only an energy density but also a mass density. What is the corresponding mass density in a thundercloud where $E = 2 \times 10^6$ V/m?

SOLUTION:

$$u = \tfrac{1}{2}\varepsilon_0 E^2 = \tfrac{1}{2}\varepsilon_0 \times (2 \times 10^6 \text{ V/m})^2 = 18 \text{ J/m}^3$$

The mass density is the energy density divided by c^2, the square of the speed of light,

$$\frac{u}{c^2} = \tfrac{1}{2}\varepsilon_0 E^2/c^2 = (18 \text{ J/m}^3)/(3 \times 10^8 \text{ m/s})^2$$

$$= 2.0 \times 10^{-16} \text{ kg/m}^3$$

This is obviously much too small to be detectable.

EXAMPLE 4. To a good approximation, a uranium nucleus can be regarded as a sphere with charge uniformly distributed over its volume. The radius of the nucleus is 7.4×10^{-15} m and the electric charge is $92e$. What is the electric energy of the nucleus?

SOLUTION: According to Example 24.5, the electric field outside of the nucleus and inside the nucleus is, respectively,

$$E = \frac{1}{4\pi\varepsilon_0} \frac{q}{r^2} \qquad r \geq R \tag{20}$$

and

$$E = \frac{1}{4\pi\varepsilon_0} \frac{qr}{R^3} \qquad r \leq R \tag{21}$$

The energy in the volume outside the nucleus is then

$$U_{\text{ext}} = \tfrac{1}{2}\varepsilon_0 \int E^2 \, dv = \tfrac{1}{2}\varepsilon_0 \int \left(\frac{1}{4\pi\varepsilon_0} \frac{q}{r^2}\right)^2 dv \tag{22}$$

The amount of volume in a radial interval dr is $dv = 4\pi r^2 \, dr$ and hence

$$U_{\text{ext}} = \tfrac{1}{2}\varepsilon_0 \int_R^\infty \left(\frac{1}{4\pi\varepsilon_0} \frac{q}{r^2}\right)^2 4\pi r^2 \, dr$$

$$= \frac{q^2}{8\pi\varepsilon_0} \int_R^\infty \frac{1}{r^2} \, dr = \frac{q^2}{8\pi\varepsilon_0}\left[-\frac{1}{r}\right]_R^\infty = \frac{1}{8\pi\varepsilon_0} \frac{q^2}{R} \tag{23}$$

The energy in the volume inside the nucleus is

$$U_{\text{int}} = \tfrac{1}{2}\varepsilon_0 \int E^2 \, dv = \tfrac{1}{2}\varepsilon_0 \int_0^R \left(\frac{1}{4\pi\varepsilon_0} \frac{qr}{R^3}\right)^2 4\pi r^2 \, dr$$

$$= \frac{q^2}{8\pi\varepsilon_0} \frac{1}{R^6} \int_0^R r^4 \, dr = \frac{q^2}{8\pi\varepsilon_0} \frac{1}{R^6}\left[\frac{r^5}{5}\right]_0^R$$

$$= \frac{1}{8\pi\varepsilon_0} \frac{q^2}{5R} \tag{24}$$

The total electrostatic energy is the sum of Eqs. (23) and (24),

$$U = U_{\text{ext}} + U_{\text{int}} = \frac{1}{8\pi\varepsilon_0} \frac{q^2}{R} + \frac{1}{8\pi\varepsilon_0} \frac{q^2}{5R} = \frac{1}{4\pi\varepsilon_0} \frac{3q^2}{5R} \tag{25}$$

Inserting numerical values, we obtain

$$U = \frac{1}{4\pi\varepsilon_0} \frac{3 \times (92 \times 1.6 \times 10^{-19} \text{ C})^2}{5 \times 7.4 \times^{-15} \text{ m}}$$

$$= 1.6 \times 10^{-10} \text{ J} \tag{26}$$

Expressed in electron-volts, this energy amounts to about 9.8×10^8 eV. Compared to the typical electric energy of an electron in an atom (about 27 eV for a hydrogen atom, see Example 25.3), this is a very large amount of energy. The energy released in nuclear fission arises from this large electric energy of the nucleus (as described in Interlude H).

SUMMARY

Energy of a system of point charges:

$$U = \tfrac{1}{2}q_1 V_{\text{other}}(1) + \tfrac{1}{2}q_2 V_{\text{other}}(2) + \tfrac{1}{2}q_3 V_{\text{other}}(3) + \cdots$$

Energy of a system of conductors:

$$U = \tfrac{1}{2}Q_1 V_1 + \tfrac{1}{2}Q_2 V_2 + \tfrac{1}{2}Q_3 V_3 + \cdots$$

Energy density in electric field:

$$u = \tfrac{1}{2}\varepsilon_0 E^2$$

(a)

(b)

(c)

(d)

(e)

(f)

Fig. 26.6

QUESTIONS

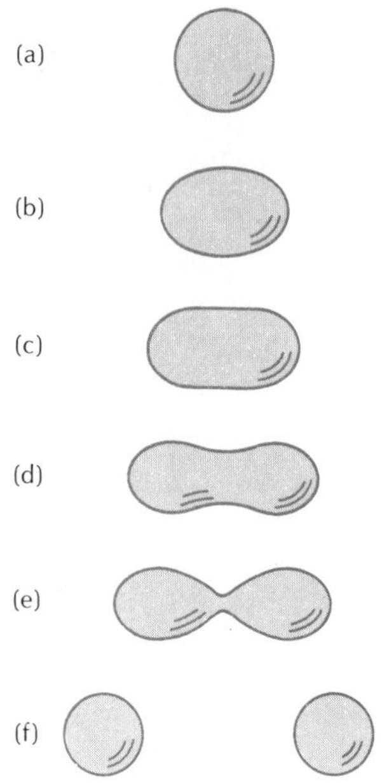

1. Suppose we have a system of electric point charges with the electrical potential energy given by Eq. (6). By what factor will this energy change if we increase the values of all the electric charges by a factor of 2?

2. Consider a metallic sphere carrying a given amount of charge. Explain why the electric energy is large if the radius of the sphere is small. Would you expect a similar inverse proportion between the electric energy and the size of a conductor of arbitrary shape?

3. Suppose that we increase the separation between the metallic plates described in Example 2. Does this change the electric energy density? The net electric energy?

4. Consider the electric fields shown in Figures 23.14 and 23.15. In what regions of the latter figure is the electric energy density larger than in the former? Can you guess which of these electric fields has a larger energy density on the average?

5. Equation (6) suggests that the electric energy is located at the charges, whereas Eq. (18) suggests it is located in the field. How could we perform an experiment to test where the energy is located? (Hint: Energy gravitates.)

6. Figure 26.6 shows a sequence of deformations of a nucleus about to undergo fission. The volume of the nucleus and the electric charge remain constant during these deformations. Which configuration has the highest electric energy? The lowest?

7. Consider a sphere with a uniform distribution of charge over its volume. Where is the energy density within this sphere highest? Lowest?

8. What fraction of the electric energy of a sphere with a uniform distribution of charge is inside the sphere? Outside the sphere?

9. Suppose that a nucleus of charge Q, radius R, and electric energy $(1/4\pi\varepsilon_0)$ $(3q^2/5R)$ fissions into two equal parts of charge $q/2$ each. The nuclear material in the original nucleus and in the final two nuclei has the same density. What is the radius of each of the two final nuclei? How does the sum of the individual electric energies of the two final nuclei compare with the initial electric energy?

10. Since the electric energy density is never negative, how can the mutual electric potential energy of a pair of opposite charges be negative?

PROBLEMS

Section 26.1

1. Consider once more the distribution of charges within the thundercloud shown in Figure 23.28. What is the electric potential energy of this charge distribution?

2. Problem 23.13 describes the arrangement of nuclear charges (positive charges) in a water molecule. Treating the nuclei as point charges, calculate the electric potential energy of this arrangement of three charges.

3. Suppose that at one instant the electrons and the nucleus of a helium atom occupy the positions shown in Figure 26.7; at this instant, the electrons are at a distance of 0.20×10^{-10} m from the nucleus. What is the electric potential energy of this arrangement? Treat the electrons and the nucleus as point charges.

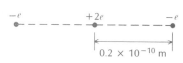

Fig. 26.7 The nucleus $(2e)$ and the electrons $(-e)$ of an atom of helium at an instant of time.

4. According to the alpha-particle model of the nucleus, some nuclei consist of a regular geometric arrangement of alpha particles. For instance, the nucleus of ^{12}C consists of three alpha particles arranged on an equilateral triangle (Figure 26.8). Assuming that the distance between pairs of alpha particles is 3.0×10^{-15} m, what is the electric energy (in eV) of this arrangement of alpha particles? Treat the alpha particles as pointlike.

5. According to the alpha-particle model (see also the preceding problem), the nucleus of ^{16}O consists of four alpha particles arranged on the vertices of a tetrahedron (Figure 26.9). If the distance between pairs of alpha particles is 3.0×10^{-15} m, what is the electric energy (in eV) of this configuration of alpha particles? Treat the alpha particles as pointlike.

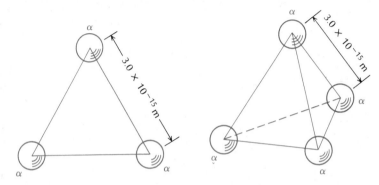

Fig. 26.8   **Fig. 26.9**

6. Problem 24.20 describes the Thomson model of the helium atom. The equilibrium separation of the electrons is 0.50 Å. Calculate the electric energy of this configuration. Take into account both the electric energy between the electrons and the positive charge, and the electric energy between the electrons; ignore the energy of the cloud and of the electrons by themselves.

7. Four equal particles of positive charges q and masses m are initially held at the four corners of a square of side L. If these particles are released simultaneously, what will be their speeds when they have separated by a very large distance?

*8. Two thin rods of length l carry equal charges Q uniformly distributed over their lengths. The rods are aligned, and their nearest ends are separated by a distance x (Figure 26.10). Calculate the mutual electric potential energy. Ignore the self-energy of each rod.

Section 26.2

9. A pair of parallel plates, each measuring 30 cm $\times$ 30 cm, are separated by a gap of 1.0 mm. How much work must you do against the electric forces to charge these plates with $+1.0 \times 10^{-6}$ C and -1.0×10^{-6} C, respectively?

10. A charge of 7.5×10^{-6} C can be placed on a metallic sphere of radius 15 cm before the surrounding air suffers electrical breakdown. What is the electric energy of the sphere with this charge?

11. A sphere of radius R has a charge Q uniformly distributed over its volume. A thin conducting shell of radius $2R$ surrounds the sphere concentrically. The shell carries a charge $-Q$ on its interior surface. What is the electric energy of this system?

12. Pretend that an electron is a conducting sphere of radius R with a charge e distributed uniformly over its surface. In terms of e and the mass m_e of the electron, what must be the radius R if the electric energy is to equal the rest-mass energy $m_e c^2$ of the electron? Numerically, what is the value of R?

Fig. 26.10

*13. Consider the Geiger-counter tube described in Problem 25.25. If the tube is initially uncharged, how much work must be done to bring the tube to its operating voltage of 1.0×10^3 V?

Section 26.3

14. Near the surface of the nucleus of a lead atom, the electric field has a strength of 3.4×10^{21} V/m. What is the energy density in this field?

15. The atmospheric electric field near the surface of the Earth has a strength of 100 V/m.
    (a) What is its energy density?
    (b) Assuming that the field has the same magnitude everywhere in the atmosphere up to a height of 10 km, what is the corresponding total energy?

16. Calculate the energy density in each of the electric fields, listed in the first four entries of Table 23.1. Calculate the mass densities that correspond to these energy densities.

17. The nuclei of ^{235}Pu, ^{235}Np, ^{235}U, and ^{235}Pa all have the same radii, about 7.4×10^{-15} m, but their electric charges are $94e$, $93e$, $92e$, and $91e$, respectively. Treating these nuclei as uniformly charged spheres, calculate their electric energies; express your answers in electron-volts.

18. One method for the determination of the radii of nuclei makes use of the known difference of electric energy between two nuclei of the same size but different electric charges. For instance, the nuclei ^{15}O and ^{15}N have the same size but their charges are $8e$ and $7e$, respectively. Given that the difference in electric energy is 3.7×10^6 eV, what is the nuclear radius?

19. A solid sphere of copper of radius 10 cm with a charge of 1.0×10^{-6} C is placed at the center of a thin, spherical copper shell of radius 20 cm with a charge of -1.0×10^{-6} C. Find a formula for the energy density in the space between the solid sphere and the shell. Find the total electric energy.

20. A proton can be crudely described as a uniformly charged sphere of charge e and radius 1.0×10^{-15} m. Find the electric self-energy of the proton. Express your answer in eV.

21. A spherical shell of inner radius a, outer radius b carries a charge Q uniformly distributed over its volume. What is the electric energy of this charge distribution?

22. In analogy to the electric field **E** (eleric force per unit mass) we can define a gravitational field **g** (gravitational force per unit mass).
    (a) The energy density in the electric field is $\varepsilon_0 E^2/2$. By analogy, show that the energy density in the gravitational field is $g^2/(8\pi G)$.
    (b) Calculate the gravitational field energy of the Moon due to its own gravity; treat this body as a sphere of uniform density. What is the ratio of the gravitational field energy to the rest-mass energy of the Moon?[1]

23. In symmetric fission, the nucleus of uranium (^{238}U) splits into two nuclei of palladium (^{119}Pd). The uranium nucleus is spherical with a radius of 7.4×10^{-15} m. Assume that the two palladium nuclei adopt a spherical shape immediately after fission; at this instant, the configuration is as shown in Figure 26.11. The size of the nuclei in Figure 26.11 can be calculated from the size of the uranium nucleus because the nuclear material maintains a constant density (the initial nuclear volume equals the final nuclear volume).

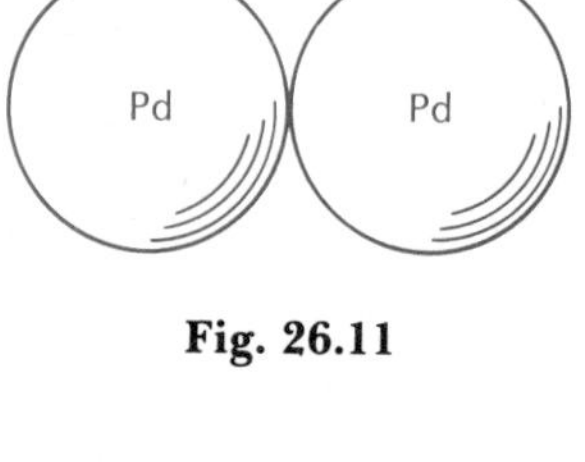

Fig. 26.11

---

[1] Warning: This problem does not take the gravitational *interaction* energy into account. The gravitational interaction energy density is [mass density] × [gravitational potential]. The total gravitational energy is the sum of the field energy and the interaction energy; this total gravitational energy is always negative.

(a) Calculate the electric energy of the uranium nucleus before fission.

(b) Calculate the total electric energy of the palladium nuclei in the configuration shown in Figure 26.11, immediately after fission. Take into account the mutual electric potential energy of the two nuclei and also the individual electric energies of the two palladium nuclei by themselves.

(c) Calculate the total electric energy a long time after fission when the two palladium nuclei have moved apart by a very large distance.

(d) Ultimately, how much electric energy is released into other forms of energy in the complete fission process (a) through (c)?

(e) If 1 kg of uranium undergoes fission, how much electric energy is released?

*24. Using the model described in Problem 24.17 for the charge distribution of a neutron, calculate the electric self-energy of a neutron. Express your answer in eV.

*25. A long rod of plastic of radius a has a charge of λ coulomb per unit length uniformly distributed over its volume. The rod is surrounded by a concentric cylinder of sheet metal of radius b with a charge of $-\lambda$ coulomb per unit length on its interior surface.

(a) What is the energy density (as a function of radius) in the space between the rod and the cylinder?

(b) What is the energy density in the volume of the rod?

(c) What is the total electric energy per unit length?

NUCLEAR FISSION[1]

The nucleus of the atom is very small and very dense. It consists of protons and neutrons closely packed together. Since the protons in the nucleus carry a positive electric charge, they exert large repulsive electric forces on one another; correspondingly, the electric energy stored within the nucleus is large. In a heavy nucleus, such as the nucleus of uranium, the disruptive electric forces that push the protons apart will sometimes overcome the cohesive "strong" forces that pull the protons and neutrons together — the nucleus splits or fissions into two or more fragments. When this happens, the large electric energy stored within the nucleus will be (partially) converted into kinetic energy of the fission fragments. Nuclear bombs and nuclear reactors derive almost all their power from this conversion of electric energy into kinetic energy. Thus, the "nuclear" energy released in fission devices is primarily electric energy.

H.1 THE NUCLEUS

Figure H.1 shows a nucleus of uranium (the isotope ^{238}U) consisting of 92 protons and 146 neutrons packed together in a spherical region. This is one of the largest, heaviest nuclei and yet its diameter is only 15×10^{-13} cm. Figures H.2–H.4 show some of the smallest, lightest nuclei: ordinary hydrogen (the isotope ^1H), heavy hydrogen or deuterium (the isotope

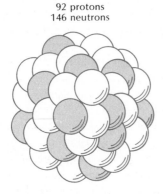

Fig. H.1 Uranium nucleus consisting of 92 protons (color) and 146 neutrons (white).

92 protons
146 neutrons

[1] This chapter is optional.

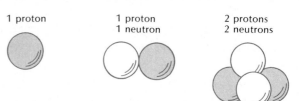

1 proton

1 proton
1 neutron

2 protons
2 neutrons

Fig. H.2 Hydrogen nucleus: one proton.

Fig. H.3 Deuterium nucleus: one proton and one neutron.

Fig. H.4 Helium nucleus: two protons and two neutrons.

^2H, or D), and helium (the isotope ^4He); all of these are less than 4×10^{-13} cm across.

The constituents of a nucleus — protons and neutrons — are usually called **nucleons.** Thus, the uranium nucleus of Figure H.1 has 238 nucleons. In general, the size of a nucleus depends on the number of nucleons that it contains — the more nucleons, the larger the size. Scattering experiments (such as the Rutherford experiment, described in Section C.1) as well as other experiments indicate that the radius of any given nucleus is proportional to the cube root of the number of its nucleons:

$$R = 1.2 \times 10^{-13}A^{1/3} \text{ cm} \qquad (1)$$

where R is the radius and A the number of nucleons in the nucleus, often called the **mass number.**

The proportionality between R and $A^{1/3}$ implies that the number of nucleons per unit volume is the same for all nuclei,

$$\frac{A}{(4\pi/3)R^3} = \frac{A}{(4\pi/3)(1.2 \times 10^{-13}A^{1/3})^3 \text{ cm}^3}$$

$$= 1.38 \times 10^{38} \text{ nucleons/cm}^3 \qquad (2)$$

Since the mass of each nucleon is about 1.7×10^{-24} g, the mass density of the nuclear material is $1.7 \times 10^{-24} \times 1.38 \ 10^{38}$ g/cm^3 = 2.3×10^{14} g/cm^3 — one cubic centimeter of pure nuclear material would weigh 230 million metric tons! Note that the volume per nucleon is $1/(1.38 \times 10^{38})$ cm^3 and hence the average distance from one nucleon to its nearest neighbor is $1/(1.38 \times 10^{38})^{1/3}$ cm $\cong 2 \times 10^{-13}$ cm. By comparing this with the radius of a proton or neutron, about 1×10^{-13} cm, we can see that inside the nucleus the nucleons are so tightly packed together that they almost touch.

Since the protons within a nucleus are at such short distances from one another, they exert very large repulsive electric forces on one another. Two neighboring protons, separated by a center-to-center distance of $\sim 2 \times 10^{-13}$ cm, experience a repulsive force of

$$F = \frac{1}{4\pi\varepsilon_0} \frac{e^2}{r^2} = 9.0 \times 10^9 \times \frac{(1.6 \times 10^{-19})^2}{(2 \times 10^{-15})^2} \text{ N}$$

$$= 58 \text{ N} = 13 \text{ lb} \tag{3}$$

Acting on a mass of (only) 10^{-24} g, this represents a colossal force.

Obviously, some extra force must be present in the nucleus to prevent it from instantaneously bursting apart under the influence of the mutual Coulomb repulsions of the protons. This extra force is the **strong force,** already mentioned in Section 6.1. This force acts equally between any two nucleons, regardless of whether they are protons or neutrons (the force is "charge independent"). Figure H.5 is a rough plot of the potential energy associated with the strong nucleon–nucleon force. For internucleon distances between $\sim 1 \times 10^{-13}$ cm and $\sim 2 \times 10^{-13}$ cm, the force is attractive and much larger than the Coulomb force — it can be more than 1000 lb. For internucleon distances smaller than $\sim 1 \times 10^{-13}$ cm, the force becomes repulsive, i.e., the nucleons have a hard core which resists interpenetration. For distances larger than $\sim 2 \times 10^{-13}$ cm, the force vanishes. Thus, in contrast to the Coulomb force, which fades only gradually and reaches out to large distances, the strong force cuts off sharply and has only a short range. In order to feel the strong force the nucleons must be touching or almost

touching, i.e., the force acts only between nearest neighbors.

As a consequence of the short-range character of the strong force, a nucleon deep inside the nucleus does not experience any net force; the nucleon only interacts with its nearest neighbors, and since these pull it with equal force in almost all directions, the net force on the nucleon is zero or nearly zero (Figure H.6). However, a nucleon at the nuclear surface only has neighbors on the side that is toward the interior and hence these will exert a net force pulling the nucleon inward (Figure H.6). Altogether this means that

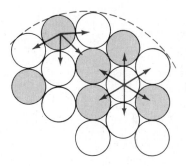

Fig. H.6 Forces on a nucleon at the nuclear surface, and forces on a nucleon in the nuclear interior.

nucleons are more or less free to wander about the interior of the nucleus, but whenever they approach the nuclear surface, the strong forces pull them back and prevent their escape.

This suggests that the nucleons in the nucleus behave somewhat like the water molecules in a drop of water; such molecules are free to wander through the volume of the drop, but when they approach the water surface, intermolecular forces hold them back. This similarity between nuclei and drops of water rests on a similarity of the laws of force. The intermolecular force has general features rather similar to those displayed in Figure H.5: it is attractive over a short range and then becomes strongly repulsive when the molecules begin to interpenetrate. The hard repulsive core of the potential makes the water nearly incompressible, while the short-range attraction provides a cohesive force that prevents water droplets from falling apart. The balance of attraction and repulsion encourages water molecules to stay at a particular distance from one another and this gives water a particular, uniform density.

Because of the similarities between a liquid and nuclear material, the nucleus can be crudely regarded as a droplet of incompressible "nuclear fluid" of uniform density. The fluid is, of course, made of nucleons, but for some purposes we can ignore the individual nucleons and calculate the properties of nuclei in terms of the gross properties of a liquid. For example, the

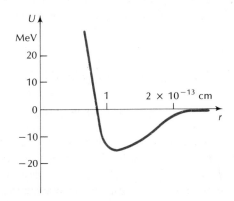

Fig. H.5 Potential energy as a function of the distance between two nucleons (for nucleons of parallel spin).

spherical shape adopted by most nuclei can be easily understood as follows: any nucleon located on the surface of a globule of nuclear fluid experiences an inward force pulling it back into the volume and consequently the fluid tends to shrink its exposed surface to the smallest value compatible with its (fixed) volume. Since a sphere has the least surface area for a given volume, the globule of fluid will take the shape of a spherical droplet.

In a stable nucleus, the repulsive Coulomb forces between the protons are held in check by the attractive strong forces. To achieve this balance of forces, the presence of neutrons is an advantage: a nucleus with more neutrons will have a larger size and therefore a larger average distance between pairs of protons — the neutrons in the nucleus dilute the repulsive effect of the Coulomb forces. Consequently, all stable nuclei, with the exception of hydrogen and one isotope of helium, contain at least as many neutrons as protons; heavy nuclei, such as uranium, contain substantially more neutrons than protons.

Figure H.7 is a plot of the number of protons vs. the number of neutrons for all known nuclei. The number of protons is represented by Z and the number of neu-

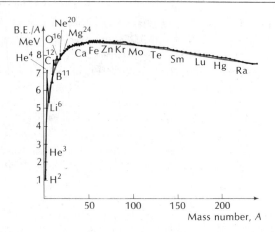

Fig. H.8 Average binding energy per nucleon vs. mass number.

trons by N (so that $A = Z + N$). On this plot, the black circles indicate the stable nuclei; the colored circles indicate the unstable nuclei, i.e., the radioactive isotopes. Note that there is no stable nucleus beyond bismuth ($Z = 83$). However, several elements beyond bismuth have some isotopes with very long half-lives; these are therefore almost stable and they occur naturally.

The energy stored in a nucleus is a sum of the potential energies contributed by electric and strong forces and the kinetic energy of the nucleons. The negative of this energy is called the **binding energy** (B.E.); this is the amount of energy released when the nucleus is assembled out of its constituent nucleons. Figure H.8 is a plot of B.E./A, the binding energy divided by the number of nucleons, or the average binding energy per nucleon. The curve plotted in Figure H.8 is called the **curve of binding energy.** Incidentally, in this figure the energy unit is the MeV,

$$1 \text{ MeV} = 10^6 \text{ eV} = 1.6 \times 10^{-13} \text{ J} \qquad (4)$$

This unit is widely used in nuclear physics.

The binding energy of a typical nucleus is a rather large amount of energy. As may be seen from Figure H.8, the average binding energy per nucleon is in the vicinity of 8 MeV for almost all nuclei; thus, a nucleus with a mass number A has a binding energy of about $A \times 8$ MeV. To put this number in perspective, we may compare it with the rest-mass energy of the nucleons. Each nucleon (neutron or proton) has a rest-mass energy $m_n c^2$; hence A nucleons have a rest-mass energy $A \times m_n c^2 = A \times 1.5 \times 10^{-10} \text{ J} = A \times 9.4 \times 10^2$ MeV. Thus the ratio of binding energy to rest-mass energy is about $8/(9.4 \times 10^2)$, i.e., the binding energy is almost 1% of the rest-mass energy! The mass associated with the binding energy is B.E./c^2; this mass is carried away by the energy released during the assem-

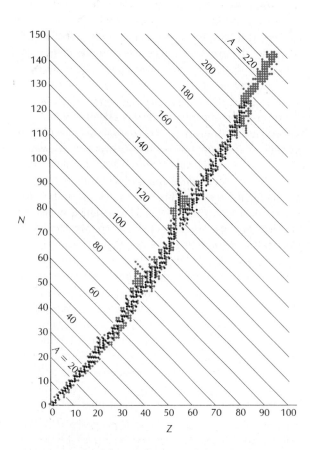

Fig. H.7 Number of protons (Z) vs. number of neutrons (N) for stable nuclei (black dots) and unstable nuclei (red dots).

neutron + ^{238}U → fission fragments

$$+ \text{ 2 or 3 neutrons} \quad (10)$$

The neutrons released in this first fission can then strike other uranium nuclei and induce their fission, and the neutrons released there can induce further fissions, etc. If no neutrons, or only a few neutrons, are lost from this avalanche, then the result is a self-sustaining **chain reaction** (Figure H.12). The rate of fission and the rate of release of energy grow drastically with time. For example, if on the average two neutrons released in each fission reaction succeed in generating further fission reactions and further neutrons, then the numbers of fission reactions in successive steps of the chain will be 2, 4, 8, 16, 64, ... If this geometric growth continues unchecked, the rate of release of energy will become explosive.

In the case of ^{238}U, conditions are not favorable for sustaining a chain reaction. Uranium-238 is a fairly stable nucleus; it will fission only if struck a hard blow by an energetic neutron — the kinetic energy of the incident neutron must be at least 1.2 MeV. The neutrons released by the fission of a uranium nucleus are energetic enough (see Table H.1), and when they collide with other nuclei they will occasionally induce a fission; but by far the most likely outcome of a collision is inelastic scattering, i.e., the neutron bounces off the nucleus with some loss of kinetic energy. Successive collisions of this kind gradually reduce the kinetic energy of the neutrons and remove their ability to induce fission — the neutrons are effectively lost from the fission chain.

In the case of ^{235}U, conditions for sustaining a chain reaction are much more favorable. This nucleus is less stable than ^{238}U; it will fission even if the kinetic energy of the incident neutron is very low. In fact, at low kinetic energy the fission reaction is more likely than at high kinetic energy; the reason for this is that the binding energy released when a neutron is captured by a ^{235}U nucleus is by itself quite sufficient to trigger the fission and, the kinetic energy of the neutron not being needed, it is advantageous for the neutron to move at low speed, since this lengthens the time it spends in the vicinity of the nucleus and therefore enhances the probability that a reaction will take place.

Besides ^{235}U, two other isotopes exist in which chain reactions are practicable. Both resemble ^{235}U in that the spontaneous decay rate is low — so that they can be held in storage without serious loss — and in that the rate of induced fission favors chain reactions: one is ^{233}U and the other is ^{239}Pu, an isotope of plutonium. The latter is very fissionable, but it is not found in ores on the Earth; it can only be obtained by artificial means, by nuclear alchemy.

In a given mass of ^{235}U or ^{239}Pu, neutrons produced by spontaneous fission or stray neutrons coming from

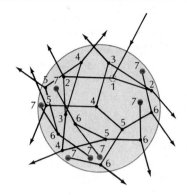

Fig. H.12 A chain reaction in a supercritical mass of fissionable material. The reaction starts at (1) with the arrival of a neutron from the outside. The diagram shows the first seven steps of the chain reaction. Each fission absorbs one neutron and releases two; occasionally a neutron escapes from the surface of the mass and is lost.

elsewhere can initiate the first step of the chain reaction. Whether the reaction keeps going depends on the **multiplication factor,** i.e., the factor by which the number of neutrons increases between one step and the next along the fission chain. If no neutrons are lost from the fission chain, then the multiplication factor is simply equal to the average number of neutrons released per fission; but if some neutrons are absorbed by impurities or escape without inducing a fission, then the multiplication factor will be smaller. The mass of fissionable material is said to be in a **critical** condition if the multiplication factor is unity; in this case the chain reaction merely proceeds at a constant rate — as in a nuclear reactor. The mass is **supercritical** if the multiplication factor exceeds unity; in that case the chain reaction proceeds at an ever-increasing runaway rate leading to an explosion — as in a nuclear bomb.

Neutrons can be lost from the fission chain by several mechanisms. For example, if the ^{235}U is not pure but contains an admixture of ^{238}U, then neutrons will be lost when they are absorbed by ^{238}U nuclei in transmutation reactions without fission. Even if the ^{235}U is pure, neutrons will still be lost when they escape from the surface of the piece of ^{235}U material. The rate of escape of the neutrons depends on the size and shape of the piece of ^{235}U material. The best shape is a sphere since in this case the surface is as distant as possible from the bulk of the material. Furthermore, neutrons are less likely to escape from a large sphere than from a small sphere — in the former a neutron released at some average point within the bulk of the material has a longer distance to travel to the surface and is therefore more likely to trigger a fission reaction while on the way. This indicates that a sphere of fissionable material must have a minimum size if it is to maintain fission. For ^{235}U, the sphere of minimum size has a diameter of 18 cm; the corresponding minimum mass — called the **critical mass** — is 53 kg.

We can significantly diminish the critical mass if we surround the sphere of fissionable material by a neutron "reflector" that prevents the loss of neutrons. The reflector must be made of some substance whose nuclei strongly scatter neutrons, but do not absorb them. In a nuclear bomb, a thick shell of beryllium metal makes a good neutron reflector; neutrons that escape from the sphere of fissionable material collide with the beryllium nuclei and bounce back into the fissionable material. In a nuclear reactor, the moderator (see Section H.6) acts as neutron reflector.

We can further diminish the critical mass by compressing the fissionable material to higher than normal densities. For example, compression of the material to twice its normal density diminishes the critical mass by a factor of 4. In a denser material, a neutron seeking to escape encounters more nuclei along its path to the surface and therefore is more likely to trigger a fission on the way.

H.4   THE BOMB

The simplest fission bomb, or **A-bomb**, consists of two pieces of ^{235}U such that separately their masses are less than the critical mass, but jointly their masses add up to more than the critical mass. To detonate such a bomb, the two pieces of ^{235}U, initially at a safe distance from one another, are suddenly brought closely together. The assembly of the two subcritical masses into a single supercritical mass must be carried out very quickly; if the two masses are brought together slowly, a partial explosion (predetonation) will push them apart prematurely, before the chain reaction can release its full energy — the explosion fizzles. The device commonly used for the assembly of the two pieces of uranium consists of a gun which propels one piece of uranium toward the other at high speed (Figure H.13); the propellant is an ordinary chemical high explosive.

A more sophisticated fission bomb consists of a (barely) subcritical mass of ^{239}Pu; if this is suddenly compressed to a higher than normal density, it will become supercritical. The sudden compression is

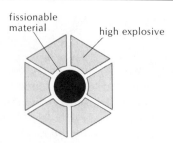

Fig. H.14 An implosion device.

achieved by the preliminary explosion of a chemical high explosive such as TNT. If this explosive has been carefully arranged in a shell around a sphere of ^{239}Pu (Figure H.14), then its detonation will crush the sphere of ^{239}Pu into itself; this implosion of the plutonium very suddenly brings its density to the supercritical value and triggers the chain reaction. The implosion technique is used with ^{239}Pu because this isotope has a strong tendency to predetonate; if one were to use the gun technique to bring together two subcritical masses of ^{239}Pu, the chain reaction would start while the masses were still moving toward one another; the consequent premature explosion would push the masses apart and prevent a full development of the chain reaction. The implosion technique assembles the supercritical mass much faster and therefore avoids the problem of a premature explosion.

During World War II, a scientific–military–industrial complex known as the Manhattan District produced three A-bombs: one plutonium bomb exploded at Alamogordo, New Mexico, on July 16, 1945, another plutonium bomb exploded at Nagasaki, Japan, on August 8, 1945, and one uranium bomb exploded at Hiroshima, Japan, on August 5, 1945. All of these bombs had yields of about 20 kilotons, i.e., an explosive energy equivalent to that of 20,000 tons of TNT. This is the energy released by the fission of 1 kg of uranium (or plutonium). Hence these devices were quite inefficient — only a small fraction of the total mass of fissionable material actually underwent fission; the rest was merely scattered in all directions, blown apart before it had a chance to react.

The uranium used for bombs has to be highly purified, "weapons-grade" ^{235}U. Uranium ores contain a mixture of 99.3% of the undesirable isotope ^{238}U and only 0.7% of the isotope ^{235}U. Since these isotopes are chemically identical, their separation is very laborious. The separation process depends on the small difference in the masses: in a gaseous compound of uranium, such as UF_6, at a given temperature, the molecules containing ^{235}U have a slightly higher average speed than the molecules containing ^{238}U, and they will diffuse slightly faster through a porous membrane; hence such a membrane acts as a (partial) filter that separates ^{235}U from ^{238}U. This is the basis of the

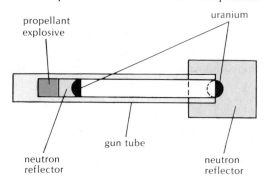

Fig. H.13 A fission bomb using a gun device.

Fig. H.15 The bomb dropped on Hiroshima. This device was 120 in. long and weighed 7000 lb.

Fig. H.16 The bomb dropped on Nagasaki. This device was 128 in. long and weighed 10,000 lb.

gaseous-diffusion process which is still the main source of highly enriched ^{235}U.

Plutonium is not found in nature, except in insignificant trace amounts. It is made artificially, by transmutation of uranium in a nuclear reactor.

As can be seen from Figures H.15 and H.16, the bombs used in World War II were rather cumbersome. Advances in the technology of shaping chemical high explosives used for the implosion of plutonium have made it possible to construct bombs in which the lump of plutonium is only about as large as a golf ball; the overall diameter of such a bomb is only about a foot, and yet it releases an amount of energy in the kiloton range.

Much higher yields can be achieved by taking advantage of nuclear **fusion.** As Figure H.8 shows, the binding energy of light nuclei is relatively low — when two such light nuclei are made to fuse together to

form a heavier nucleus, energy will be released. Fusion is the opposite of fission: in the former two nuclei merge into one, in the latter one nucleus splits into two. Furthermore, the energy released in fusion is strong energy while the energy released in fission is Coulomb energy. The strong force favors fusion since, by merging, the nuclei reduce their surface area. The Coulomb repulsion between the two nuclei opposes fusion, but in the case of light nuclei the strong force overcomes this opposition.

The heat given off by the Sun is due to a fusion reaction called hydrogen burning: hydrogen nuclei fuse together to make helium nuclei. This reaction involves several intermediate steps, which were discovered by theoretical calculations by H. Bethe.[3] This reaction cannot be duplicated on Earth because it will only proceed at extremely high temperatures and pressures, such as are found near the center of the Sun. However, some fusion reactions involving deuterium and tritium (the isotopes ^{2}H and ^{3}H) can be made to work on Earth:

$$^2\text{H} + {}^2\text{H} \rightarrow {}^3\text{He} + \text{n} \qquad (11)$$

$$^2\text{H} + {}^2\text{H} \rightarrow {}^3\text{H} + \text{p} \qquad (12)$$

$$\underline{^2\text{H} + {}^3\text{H} \rightarrow {}^4\text{He} + \text{n}} \qquad (13)$$

$$5\ {}^2\text{H} \rightarrow {}^3\text{He} + {}^4\text{He} + \text{p} + 2\text{n} + 24.3\ \text{MeV}$$

The net result of the three reactions taken together is the disappearance of five ^{2}H nuclei and the formation of ^{3}He, ^{4}He, one free proton, and two neutrons, with the release of 24.3 MeV. The amount of energy released per nucleon of reactant is 24.3 MeV per 10 nucleons = 2.43 MeV per nucleon, whereas for the fission of uranium the energy released per nucleon of reactant is 200 MeV per 235 nucleons = 0.85 MeV per nucleon. Thus the fusion of a given mass of ^{2}H will yield about three times as much energy as the fission of an equal mass of ^{235}U.

The reactions (11)–(13) are called **thermonuclear** because they will proceed only at very high temperature and pressure. The requisite temperatures and pressures are attained at the place of explosion of an A-bomb. Hence, the fusion reactions can be initiated by exploding a fission bomb next to a mass of heavy hydrogen; this results in self-sustained explosive "burning" of the hydrogen nuclei. This so-called **H-bomb** is really a fission–fusion device in which fission triggers fusion. What is more, the fusion reactions release a large number of energetic neutrons [see Eqs. (11) and (13)] which can be used to further enhance the vio-

[3] Hans A. Bethe, 1906–, German, later American, physicist. He formulated the theory of nuclear fusion in stars in 1937 and received the Nobel Prize for this in 1967; during the war he was director of the theoretical-physics division at Los Alamos.

lence of the explosion: the trick is to surround the fusion bomb by a blanket of cheap, natural uranium, consisting of mainly ^{238}U; although this isotope will not maintain a chain reaction, it will fission when exposed to the large flux of neutrons from the fusion. This kind of H-bomb is a fisson–fusion–fission device; typically, one-half of the total energy yield is due to fusion, one-half is due to fission. The fission of a large amount of uranium leaves behind a residue of highly radioactive fission products; hence fission–fusion–fission bombs are **dirty,** i.e., they generate a large volume of radioactive fallout.

The total energy yield of an H-bomb is of the order of one or several megatons[4] — roughly a thousand times the yield of an A-bomb. Explosions of up to 60 megatons have been tried with great success, and there seems to be no limit to the suicidal madness that nature will let us get away with.

H.5   THE EFFECTS OF NUCLEAR WEAPONS

The energy released by a nuclear bomb takes several forms as it emerges from the place of explosion: blast and shock in air, thermal radiation (including light), immediate nuclear radiation (neutrons and gamma rays), and delayed nuclear radiation (residual radioactivity, fallout). The exact distribution of the energy among these several forms depends on the type of bomb and on the altitude at which it explodes. For an air burst — an explosion in air at low or medium altitude — the energy distribution is typically as shown in Figure H.17.

The chronological sequence of phenomena in the air burst of a 1-megaton bomb is as follows (Figures H.18–H.23).

Within a few microseconds after the initiation of the explosion, the nuclear reactions will have released their energy. These reactions also release a large number of neutrons and gamma rays which immedi-

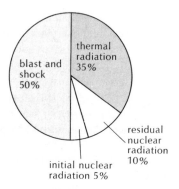

Fig. H.17 Distribution of energy released in a typical air burst at an altitude below 30,000 m. (From S. Glasstone, ed., *The Effects of Nuclear Weapons.*)

[4] One megaton = 1000 kilotons.

ately escape from the scene of the explosion (Figure H.18). Although these neutrons and gamma rays are gradually absorbed by the air, their energy is high and they penetrate the air for well over a mile with lethal intensity.

The energy released in the explosion raises the fragmented residues of the bomb to an enormous temperature — tens of millions of degrees — and vaporizes them instantly. The incandescent vapor radiates X rays which are quickly absorbed by the surrounding air, heating it. The heated air and the vaporized bomb residues form a spherical, hot, luminous mass of gas — the **fireball.** Figure H.24 is a high-speed photograph of such a fireball. For a 1-megaton explosion, at 1.8 s after time zero, the fireball is 1.2 mi across (Figure H.19); it continues to grow to a maximum diameter of 1.4 mi. The fireball attains a temperature of 6000 to 7000 K — hotter than the surface of the Sun. It emits intense thermal radiation, including intense light. Meanwhile a **blast wave,** caused by the sudden increase of pressure, travels outward from the center of explosion.

At 5 s after time zero, the blast wave has spread to a radius of almost 2 mi and begins to reflect from the ground (Figure H.20). The direct and the reflected wave merge and reinforce each other near the ground — they form a single, strong wave front called the **Mach front.** The overpressure at the wave front is 16 lb/in.² at this time. The emission of thermal radiation gradually decreases as the fireball cools, but significant amounts continue to be emitted until 10 s after time zero.

At 11 s after time zero, the blast wave has spread out to a radius of 3 mi and the overpressure at the wave front has dropped to 6 lb/in.² (Figure H.21). Behind the front, surface winds of 180 mi/h blow radially outward for a few seconds. Most of the thermal radiation from the fireball (about 80%) will already have been emitted by this time.

At 37 s after time zero, the blast wave reaches out to 10 mi, with an overpressure of about 1 lb/in.² (Figure H.22). Meanwhile, the hot gas of the fireball begins to rise like a hot-air balloon, at about 250 mi/h. The rising gas generates a vertical updraft as well as horizontal afterwinds that rush radially inward along the surface. These winds push dirt and dust into a column or stem below the fireball. As the fireball expands and cools, the vaporized bomb residues condense and form a large radioactive cloud. This cloud and the stem below take on the characteristic mushroom shape (Figure H.25).

The mushroom grows, reaching a height of 7 mi 110 s after the explosion (Figure H.23). At this time, the mushroom cloud is still shooting upward at about 150 mi/h while along the surface the afterwinds reach 200 mi/h or more. Ultimately, the mushroom cloud at-

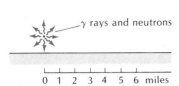

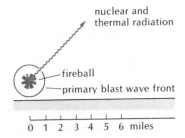

Fig. H.18 Chronological development of an air burst of 1 megaton; time 5×10^{-6} s.

Fig. H.19 Chronological development of an air burst of 1 megaton; time 1.8 s. (Figures H.19–H.23, based on S. Glasstone, ed., *The Effects of Nuclear Weapons*.)

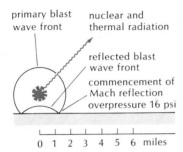

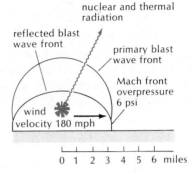

Fig. H.20 Chronological development of an air burst of 1 megaton; time 4.6 s.

Fig. H.21 Chronological development of an air burst of 1 megaton; time 11 s.

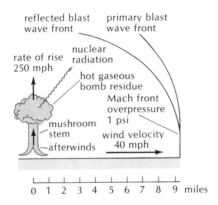

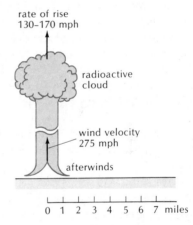

Fig. H.22 Chronological development of an air burst of 1 megaton; time 37 s.

Fig. H.23 Chronological development of an air burst of 1 megaton; time 110 s.

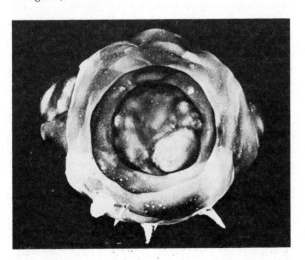

Fig. H.24 Fireball of a nuclear explosion of low yield. This high-speed photograph was taken from a large distance by means of a telescope. (Courtesy H. E. Edgerton, MIT.)

Fig. H.25 The mushroom cloud of a nuclear explosion of low yield.

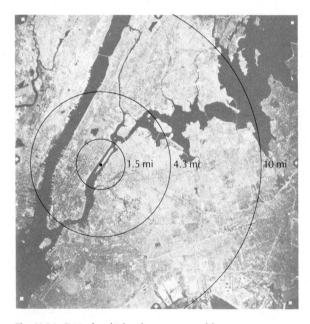

1.5 mi   4.3 mi   10 mi

Fig. H.26 Critical radii for damage caused by 1-megaton explosion over New York City.

tains a height of 14 mi. Then it gradually begins to dissipate.

For a 1-megaton bomb bursting at an altitude of, say, 6500 ft, the violent effects of thermal radiation, nuclear radiation, and blast cover an area of radius 10 mi around ground zero. Figure H.26 shows a circle of radius 10 mi superimposed on an aerial photo of New York City; circles of radii 1.5 and 4.3 mi are also shown in this photo. Within a radius of 1.5 mi, the flux of the neutrons emitted by the initial nuclear reactions is so intense that any exposed victim absorbs a lethal

dose of ionizing radiation. However, the physiological effects of overexposure to ionizing radiation are delayed; and the victims within this distance are more likely to suffer almost instantaneous death from the blast wave, which reaches them in a few seconds. Out to a radius of 4.3 mi the blast wave has an overpressure of more than 5 lb/in.², enough to level most buildings and crush the victims with flying and falling debris. Beyond 4.3 mi, the blast wave damages buildings more or less severely, but usually does not destroy them completely.

The flash of thermal radiation emitted by the fireball poses the most frightful danger in the zone from 4.3 to 10 mi from ground zero. The radiant heat is so intense that even at a distance of 10 mi, exposed combustible material will be ignited. At 10 mi, exposed human skin suffers second-degree burns (blisters); at 6 mi human skin is charred throughout its entire thickness. A victim caught in the open will be burnt fatally if at a distance of less than 8 mi; beyond this distance survival depends on the area of unprotected skin exposed to the flash of heat and light.

Since the lens of the eye focuses light on the retina, any victim looking at the fireball will suffer severe retinal damage. At distances of up to 20 mi, the result is total blindness; even at distances of 200 or 300 mi, retinal burns and partial loss of vision are likely.

Besides these primary effects of a nuclear explosion, there are secondary effects. A 1-megaton air burst over a city will set off numerous flash fires within a circle of a 10-mi radius. Such a large number of fires ignited over a wide area may cause a **firestorm:** hot air from the fires rises and creates both an updraft and strong surface winds (50 to 100 mi/h) which fan the flames. Consequently, annihilation of life and destruction of buildings may be total, or nearly total, within a 10-mi radius.

Another delayed effect is **fallout,** which consists of the radioactive residues in the mushroom cloud mixed with the dirt and dust of the mushroom stem. Wherever this contaminated dust falls on the surface of the Earth, the ground will become radioactive. Winds may carry the contaminated dust hundreds of miles as it gradually settles; the result is a fallout plume which covers a large area with dangerous radioactive contamination. The details of the distribution of fallout depend on the wind speed and direction; they also depend on the type of bomb and on the altitude of explosion — fallout is more intense for a surface burst (at ground level) than for an air burst. For a 1-megaton surface burst, whose yield is one-half from fission and one-half from fusion, fallout will be lethal over a downwind area about 150 mi long and 20 mi wide (Figure H.27). This assumes that the wind is steady at about 15 mi/h and that the victims are not sheltered from the ionizing radiation.

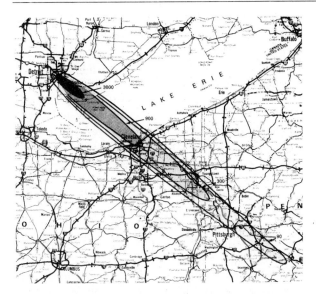

Fig. H.27 Fallout pattern for 1-megaton surface burst in Detroit, with 15-mi/h northwesterly wind. The radiation dose accumulated in a 7-day period is lethal within the second largest gray area. (From *The Effects of Nuclear War,* Office of Technology Assessment, Congress of the United States.)

Fig. H.29 Fallout pattern for a massive nuclear attack on military installations in the United States, assuming typical westerly winds. More than one-half of the inhabitants of the colored areas would die. (Based on *Sci. Am.,* November 1976.)

The basement of a house offers some shelter from the radiation emitted by the radioactive dust blanketing the ground and the roof of the house — the radiation penetrating an underground basement is $\frac{1}{25}$ to $\frac{1}{100}$ of the outside radiation. In general, a massive layer of any material provides shielding against the radiation. Crude **fallout shelters** can be improvised with a layer of any dense material; 4 in. of concrete, 7 in. of earth, or 14 in. of books all give about the same protection (Figure H.28). The radioactivity of the fallout decreases with time (as a rough rule, any sevenfold increase in time reduces the intensity of the radiation by a factor of 10); but, depending on details of the fallout pattern, it may not be safe to leave the shelter for several weeks.

In view of the awesome immediate effects of a nuclear explosion, it is easy to underrate the effects of fallout. It should be kept in mind that even in the case of a limited nuclear attack on the United States — in which bombs are only directed against military installa-

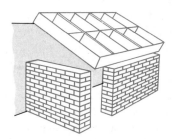

Fig. H.28 A simple fallout shelter in a basement. The bins are to be filled up with bricks or sand. (From *Radiological Defense Textbook,* Defense Civil Preparedness Agency.)

tions rather than cities — there might be more than 12 million civilian casualties from the effects of fallout alone (Figure H.29). Besides these casualties from early, local fallout, there would be some further casualties from long-term, worldwide fallout involving isotopes of strontium (^{90}Sr) and cesium (^{137}Cs). These isotopes are produced in abundance by the fission of uranium and they have very long half-lives; therefore, they remain a hazard for a long time.

H.6　NUCLEAR REACTORS

In a nuclear reactor, a fission chain reaction takes place under controlled conditions. The condition of the fission material in the reactor is *critical* rather than supercritical, i.e., the number neutrons and the number of fissions along each step of the chain reaction remains constant instead of increasing geometrically.

The most common type of reactor operates with "enriched" uranium consisting of a few percent ^{235}U mixed with 90-odd percent ^{238}U. Such a uranium mixture cannot by itself maintain a chain reaction — the ^{238}U soaks up too many of the fission neutrons. However, if the uranium is surrounded by a substance capable of slowing down the (fast) fission neutrons to a very low speed, then a chain reaction becomes viable. Slow neutrons are much more efficient at maintaining a chain reaction than fast neutrons. There are two reasons for this: slow neutrons are less likely to be absorbed by ^{238}U (the reaction rate for the capture of neutrons by ^{238}U is low at low neutron energies) and slow neutrons trigger the fission of ^{235}U more readily than fast neutrons (the reaction rate for induced fission of ^{235}U is high at low neutron energies).

The substance that slows down the neutrons is called the **moderator**. Inside the reactor the uranium is usually placed in long fuel rods and these are immersed in the bulk of the moderator (Figure H.30). Fast neutrons released by fissions travel from the fuel rods into the moderator; there they lose their kinetic energy by collisions with the moderator's nuclei; and then they wander back into one or another of the fuel

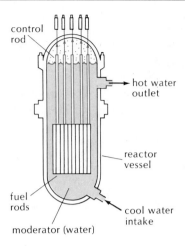

Fig. H.30 Nuclear reactor (schematic).

rods and induce further fissions. An efficient moderator must absorb the kinetic energy of the neutrons, but not absorb the neutrons themselves. The three best moderators are ordinary water (H_2O), heavy water (D_2O), and graphite (pure carbon). These materials are very efficient at slowing down neutrons because they contain nuclei of low mass, i.e., nuclei of mass equal to, or not much larger than, the mass of the neutron. As we saw in Section 10.2, the collision between two particles of equal mass transfers all of the kinetic energy to the initially stationary particle, whereas the collision between particles of unequal mass transfers only a fraction of the kinetic energy. After a few collisions with the moderator nuclei, the neutrons will have lost almost all their kinetic energy; they will only retain a kinetic energy equal to the kinetic energy of the random thermal motion of the moderator nuclei, i.e., a kinetic energy of $\frac{3}{2}kT$ [where T is the temperature of the moderator; see Eq. (19.23)]. Such neutrons are called **thermal neutrons.** Although some of these neutrons will escape from the reactor, in a critical reactor enough return to the uranium to keep the chain reaction going.

The configuration of the reactor — the size, number, and location of uranium rods, and the amount and shape of moderator — must be designed so that the reactor is critical or just barely supercritical. The number of neutrons can be finely adjusted to the point of criticality by control rods made of boron or cadmium; these substances readily soak up neutrons, and, by pushing the control rods in or pulling them out of the reactor, more or fewer neutrons can be removed from the fission chain, bringing the reactor from subcritical, to critical, or even to supercritical (brief instants of supercriticality are called **excursions**).

Both heavy water and graphite are such good moderators that in their presence even natural uranium,

with its small percentage of ^{235}U, can maintain a chain reaction. The first nuclear reactor was built at the University of Chicago in 1942 under the direction of Enrico Fermi[5]; it contained natural uranium as fuel and graphite as moderator. Several similar reactors were built at Hanford, Washington, shortly afterward, as part of the Manhattan Project. These reactors were used as **converters,** i.e., they were used to convert ^{238}U into ^{239}Pu by the following sequence of reactions: Within the rods of natural uranium, some of the neutrons from the fission of ^{235}U are soaked up by ^{238}U, transmuting it into ^{239}U. The isotope ^{239}U then spontaneously undergoes two successive beta decays, which transmute it into ^{239}Pu. The few kilograms of ^{239}Pu for the Alamogordo and Nagasaki A-bombs were obtained by these means.

Nowadays reactors are extensively used to provide intense beams of neutrons for research; to produce radioisotopes for scientific, industrial, and medical applications (see Interlude B); and to generate mechanical or electric power. The latter reactors are called **power reactors.** In them, the heat energy released by the fission reactions is removed from the reactor core by a circulating coolant; this heat is then transferred to steam, which drives a steam turbine. Thus, the nuclear reactor serves the same purpose as the furnace of a conventional steam engine — and uranium replaces coal or oil.

Most power reactors in operation or under construction in the United States have water-filled cores; the water acts simultaneously as coolant and as moderator. Figure H.31 shows an outline of a **pressurized-water reactor** (PWR), the most common type of power reactor. The reactor core with its fuel rods is enclosed

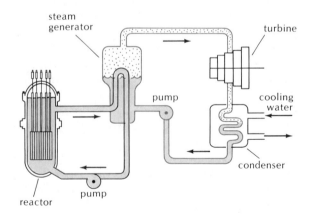

Fig. H.31 Schematic diagram of a nuclear power plant with a pressurized-water reactor.

---

[5] Enrico Fermi, 1901–1954, Italian, later American physicist. He was awarded the Nobel Prize in 1938 for his discovery of nuclear reactions initiated by neutron bombardment. The element *fermium* was named in his memory.

• Nuclear Reactors   **H-14**

Fig. H.32 Reactor vessel.

in a massive reactor vessel. Water, driven by powerful pumps, circulates through the reactor vessel; this water is kept at a high pressure (2200 lb/in.²) which prevents it from boiling even though the temperature within the reactor vessel is 600°F. The hot water circulates from the reactor vessel to a steam generator where it passes through a system of pipes immersed in water at a lower pressure. The low-pressure water boils off as steam, which drives a turbine; after the steam has done its work, it condenses and returns to the steam generator. Meanwhile the high-pressure water circulates back to the reactor vessel in a closed loop.

In a **boiling-water reactor** (BWR) the coolant is allowed to boil directly within the reactor vessel. Steam emerges from the top of the reactor and is fed into a turbine; it then condenses and circulates back to the reactor.

Figure H.32 shows the reactor vessel for a large power plant capable of generating 1200 MW of electricity; the height of the reactor vessel is 60 ft, the thickness of the steel walls is 10 in. at its thinnest, and the weight of the vessel is 1500 tons when empty. Smaller reactors are used for the propulsion of submarines, aircraft carriers, and icebreakers. For a submarine, a reactor power plant has the obvious advantage that it does not require oxygen to "burn" its fuel; besides, the reactor only needs to be refueled at very long intervals.

Both pressurized-water reactors and boiling-water reactors operate with ordinary water; they are **light-water reactors.** By contrast, **heavy-water reactors** operate with heavy water (D$_2$O) as moderator. Although heavy water is very expensive — the 750 tons of heavy water required for a single large reactor cost $80 million — it is such a good moderator that cheap, natural uranium can be used as fuel (see above). A type of heavy-water reactor developed in Canada (CANDU), contains heavy water as moderator and uses ordinary water, in a separate system of pipes, as coolant.

Another type of reactor, first developed in the United Kingdom and in France, is the **gas-cooled reactor** (GR). It contains graphite as moderator and uses gas as coolant. Gases that suit this purpose are carbon dioxide (low-temperature operation, 710°F) and helium (high-temperature operation, 1400°F). Since graphite is a good moderator, such reactors can be fueled by natural uranium; however, structural problems arising from the large size of these reactors (70-ft diameter) make it awkward to rely on natural uranium, and modern high-temperature gas-cooled reactors are fueled with highly enriched uranium, which permits a more compact design.

Power reactors now in operation obtain their energy from the fission of the isotope ^{235}U. Since the world's supply of this isotope is limited (see Interlude F) it would be desirable to build reactors that consume some other fissionable nucleus. One obvious choice is the very fissionable isotope ^{239}Pu. Although this isotope does not occur naturally, it can be readily manufactured by transmutation of the very abundant isotope ^{238}U. In fact, since the fuel of all power reactors is a mixture of ^{238}U and ^{235}U, the manufacture of ^{239}Pu is an automatic side effect of the operation of power reactors; the ^{239}Pu can subsequently be extracted by chemical reprocessing of the spent uranium fuel. A reactor fueled with ^{239}Pu not only makes good use of a material that would otherwise go to waste, but, if the reactor is supplied with a quantity of ^{238}U, it can also manufacture its own ^{239}Pu. What is more, the number of neutrons released in the fusion of ^{239}Pu is sufficiently large so that in an efficiently designed reactor slightly more than one of the neutrons released in an average fission reaction can be diverted from the fission chain to the transmutation of ^{238}U. This implies that the reactor produces *more* ^{239}Pu (from ^{238}U) than it consumes (from its original supply).

A reactor that produces more fissionable material than it consumes is called a **breeder.** Once the fuel cycle of breeder reactors has been started with an initial load of ^{239}Pu, only the abundant and cheap ^{238}U needs to be supplied to keep the fuel cycle going. Essentially, breeders extract energy indirectly from ^{238}U;

breeders would therefore be able to generate power for as long as our (abundant) supply of ^{238}U lasts (see Interlude F). Much effort has been expended on the development of breeders and a few experimental reactors have already been built. However, these reactors are afflicted with design and safety problems that have not yet been satisfactorily resolved.[6]

Further Reading

Nuclear Science and Society by B. L. Cohen (Doubleday, New York, 1974) is an excellent introduction to nuclear energy and its applications.

A concise and clear survey of nuclear physics will be found in *Secrets of the Nucleus* by J. S. Levinger (McGraw-Hill, New York, 1967). Very brief discussions of nuclear physics can be found in chapters in the books *Energy, Ecology, and the Environment* by R. Wilson and W. J. Jones (Academic Press, New York, 1974) and *The Atom and Its Nucleus* by G. Gamow (Prentice-Hall, Englewood Cliffs, 1965). *The Atomic Energy Deskbook,* edited by J. F. Hogerton (Reinhold, New York, 1963), and *Sourcebook on Atomic Energy,* edited by S. Glasstone (Van-Nostrand Reinhold, Princeton, 1967), are encyclopedias containing technical information.

The standard reference on nuclear explosions is *The Effects of Nuclear Weapons,* edited by S. Glasstone (United States Atomic Energy Commission, 1962). This contains a wealth of detail on nuclear explosions, their thermal radiation, blast, and radioactivity, and the injuries and damage wrought on people and buildings. *The Effects of Nuclear War* by the Office of Technology Assessment (Congress of the United States, Washington, D.C., 1979) presents case studies of hypothetical nuclear attacks on Soviet and U.S. industrial and civilian targets. *Radiological Defense* by the Defense Preparedness Agency (Department of Defense, 1974) is a textbook on protective measures against fallout. *Arsenal: Understanding Weapons in the Nuclear Age* by K. Tsipis (Simon and Schuster, New York, 1983) is a lucid survey of the principles and effects of nuclear bombs, the technology of delivery systems, and strategic implications.

The Cold and the Dark: The World After Nuclear War by P. R. Ehrlich, C. Sagan, D. Kennedy, and W. O. Roberts (W. W. Norton, New York, 1984) discusses the long-term worldwide consequences of nuclear war, especially the severe climatic changes ("nuclear winter") caused by the reduction of sunlight by the dust and smoke accumulated in the atmosphere, and the catastrophic effects of this on plants and animals. *Last Aid,* edited by E. Chivian, S. Chivian, R. J. Lifton, and J. E. Mack (Freeman and Co., San Francisco, 1982), is a chilling examination of the medical implications of nuclear war. *The Fate of the Earth,* by J. Schell, gives a frightening description of the global disasters that would follow a nuclear war and makes an impassioned plea for total disarmament. *Nuclear Nightmares* by N. Calder (Viking Press, New York, 1979) is a journalist's investigation of the threat of nuclear war; it deals with weapons technology, proliferation, and scenarios that might lead to war. *Survival and the Bomb,* edited by E. P. Wigner (Indiana University Press, Bloomington, 1969), presents the arguments for the implementation of a civil-defense program.

Manhattan Project by S. Groueff (Little, Brown, and Co., Boston, 1967) is a historical account of the development of the first A-bombs. *Enrico Fermi* by E. Segré (University of Chicago Press, Chicago, 1970) is a biography of the eminent physicist who designed and built the first nuclear reactor and triggered the first chain reaction. *Hans Bethe: Prophet of Energy* by J. Bernstein (Basic Books, New York, 1979) is a brilliant biography of another eminent physicist who discovered the cycles of nuclear reactions that release energy in stars. *The Curve of Binding Energy* by J. McPhee (Ballantine Books, New York, 1973) is a splendidly written account of the design and manufacturing problems involved in the production of small nuclear weapons.

The following is a list of magazine articles dealing with nuclear reactors, nuclear weapons, and related questions:

"Energy from Breeder Reactors," F. L. Culler and W. O. Harms, *Physics Today,* May 1972

"Natural-Uranium Heavy-Water Reactors," H. C. McIntyre, *Scientific American,* October 1975

"Civil Defense in Limited War — A Debate," A. A. Broyles and E. P. Wigner vs. S. D. Drell, *Physics Today,* April 1976

"A Natural Fission Reactor," G. A. Cowan, *Scientific American,* July 1976

"Limited Nuclear War," S. D. Drell and F. von Hippel, *Scientific American,* November 1976

"Superphénix: A Full-Scale Breeder Reactor," G. A. Vendryes, *Scientific American,* March 1977

"Nuclear Power and Nuclear-Weapons Proliferation," E. J. Moniz and T. L. Neff, *Physics Today,* April 1978

"Enhanced-Radiation Weapons," F. M. Kaplan, *Scientific American,* May 1978

"The Prompt and Delayed Effects of Nuclear War," K. L. Lewis, *Scientific American,* July 1979

"Catastrophic Releases of Radioactivity," S. A. Fetter and K. Tsipis, *Scientific American,* April 1981

"Gas-Cooled Nuclear Power Reactors," H. M. Agnew, *Scientific American,* June 1981

"Freeze on Nuclear Weapons Development and Deployment: Pro–Con," H. Feiveson and F. von Hippel vs. H. W. Lewis, *Physics Today,* January 1983

"Arms Limitation Strategies," H. F. York, *Physics Today,* March 1983

"Effects of Nuclear Weapons," L. Sartori, *Physics Today,* March 1983

"The Nuclear Arsenals of the US and the USSR," B. G. Levi, *Physics Today,* March 1983

"The Uncertainties of a Preemptive Nuclear Attack," M. Bunn and K. Tsipis, *Scientific American,* November 1983

"Weapons and Hope," F. J. Dyson, *The New Yorker,* February 6, 13, 20, and 27, 1984

"The Climatic Effects of Nuclear War," R. P. Turco, O. B. Toon, T. P. Ackerman, J. B. Pollack, and C. Sagan, *Scientific American,* August 1984

---

[6] For a discussion of reactor safety, see Section F.6.

Questions

1. According to Figure H.8, which isotope has the largest binding energy per nucleon? Which has the least binding energy and is therefore capable of releasing the most energy in a nuclear reaction?

2. By what factor is the amount of energy released per kilogram of reactant in a typical fission reaction larger than in a typical chemical reaction? (Hint: Compare fission of uranium with explosion of TNT.)

3. In 1933, Ernest Rutherford, the discoverer of the nucleus, declared that "the energy produced by the breaking down of the atom is a very poor kind of thing. Anyone who expects a source of power from the transformation of these atoms is talking moonshine." Taking into consideration that fission was unknown at the time, do you think Rutherford was making a fair assessment? What about heat released by natural radioactivity?

4. Free neutrons are unstable; they decay with an average lifetime of about 15 min. What effect does this have on a fission chain reaction?

5. Why is a critical mass needed for fission but not for fusion?

6. George Gamow, author of many delightful books on physics, was fond of saying that natural uranium is just as useless for carrying out a nuclear chain reaction as soaking-wet logs are for building a campfire. Explain this analogy.

7. A widely used modern method for the separation of the isotopes ^{235}U and ^{238}U relies on high-speed centrifuges. How does centrifugation affect a test tube full of a gaseous compound with molecules containing these isotopes?

8. Since the critical mass of ^{235}U is 53 kg, the Hiroshima bomb must have contained about that much uranium. If all of this uranium had undergone fission, what would have been the energy released?

9. Compare the shapes of the bombs in Figures H.15 and H.16. Why does the Hiroshima bomb have an elongated shape, and the Nagasaki bomb a rounded shape?

10. Why are high temperatures required to initiate fusion, but not to initiate fission?

11. Per kilogram of reactant, fusion releases about three times as much energy as fission. But the energy yield of a typical H-bomb is about a hundred times larger than that of a typical A-bomb. How can this be?

12. If an H-bomb is not surrounded by a blanket of natural uranium, it is a cleaner bomb, producing a smaller amount of radioactive fallout and a smaller energy yield. However, such a bomb, called a neutron bomb, releases a large number of energetic neutrons. Compare the effects of a neutron bomb with those of an ordinary H-bomb.

13. If a blast wave of overpressure 16 lb/in.² strikes you, what is the force on the front of your body?

14. If you are at a distance of 8 mi from the place of a nuclear explosion, you have available about 30 s between the flash of light and the arrival of the blast wave. Look around the room you are in. Where could you take shelter in 30 s?

15. Contamination of food with radioactive strontium poses a severe hazard because strontium is chemically similar to calcium. Explain.

16. One of the effects of a large-scale nuclear war would be the accumulation in the atmosphere of dust and smoke from explosions and fires. Such a blanket of dust and smoke, covering the entire Earth for several months, would reduce the amount of sunlight reaching the ground, perhaps by 50% or more, so that the climate would become much cooler. Recent calculations suggest that even in summer the temperature would remain below freezing, a phenomenon that has been called the **nuclear winter**. What agricultural and other problems would arise from such a drastic modification of the climate?

17. Helium should make a good moderator, since it does not absorb neutrons and has a low mass. Why do we not use helium as moderator in a nuclear reactor?

18. Why must the fuel rods in a nuclear reactor be thin?

19. The uranium in the fuel rods in a nuclear reactor is enclosed in a metal pipe (cladding). What are the requirements that the material of the pipe must meet?

20. In a nuclear power plant, the water that passes through the reactor core flows through one loop, and the water that passes through the turbine flows in a separate loop (Figure H.31). Why do we not use a single loop that directly connects the reactor core to the turbine, as in an ordinary coal-burning power plant?

21. If you wanted to build a nuclear reactor of very small size, say, to power an artificial satellite, what isotope would you use?

22. The following question was asked by a reader of the *New York Times* (February 22, 1983): "Why is it less dangerous if the nuclear reactor of a falling satellite (such as the Soviet Cosmos 1402) burns up in space than if it falls to earth in one piece? Aren't radioactive atoms just as present one way as another?" How would you answer?

23. Breeder reactors generate more fuel than they consume. Does this violate the law of conservation of energy?

24. One of the disadvantages of breeder reactors is that their fuel can be diverted to the manufacture of bombs. Why is this not a problem with ordinary reactors?

CHAPTER 27

Capacitors and Dielectrics

Capacitor Any arrangement of conductors that is used to store electric charge is called a **capacitor**, or condenser. Since work must be done during the charging process, the capacitor will also store electric potential energy. In our electric technology, capacitors find widespread application — they are part of the circuitry of radios, electronic calculators, automobile ignition systems, etc.

The first part of this chapter deals with the properties of capacitors. The second part deals with the properties of electric fields in regions *Dielectric* of space filled with an insulating material, or **dielectric.** Since many capacitors are filled with such a dielectric material, the study of the mutual effects between the electric field and the dielectric material is closely linked to the study of capacitors. But the effects of electric fields and dielectric materials upon one another are also interesting in their own right. For instance, air is a dielectric material and we ought to inquire how the electric field in air differs from that in vacuum.

27.1 Capacitance

As a first example of a capacitor, consider an isolated metallic sphere of radius R. Obviously, charge can be stored on this sphere. If the amount of charge placed on the sphere is Q, then the potential of the sphere will be

$$V = \frac{1}{4\pi\varepsilon_0}\frac{Q}{R} \tag{1}$$

Thus, the amount of charge stored on the sphere is directly proportional to the potential.

This proportionality holds in general for any conductor of arbitrary shape. The charge on the conductor produces an electric field whose strength is directly proportional to the charge (twice the charge gives twice the field) and the electric field yields a potential which is directly proportional to the field strength (twice the field strength yields twice the potential); hence charge and potential are proportional. We write this relationship as

$$Q = CV \qquad (2)$$

Capacitance of a single conductor

where C is the constant of proportionality. This constant is called the **capacitance** of the conductor. The capacitance is large if the conductor is capable of storing a large amount of charge at a low potential. For instance, the capacitance of a sphere is

$$C = \frac{Q}{V} = \frac{Q}{(1/4\pi\varepsilon_0)(Q/R)} = 4\pi\varepsilon_0 R \qquad (3)$$

Thus, the capacitance of a sphere increases with its radius.

The unit of capacitance is the **farad** (F),

$$1 \text{ farad} = 1 \text{ F} = 1 \text{ coulomb/volt} \qquad (4)$$

farad, F

This unit of capacitance is rather large; in practice, electrical engineers prefer the **microfarad** and the **picofarad**. A microfarad equals 10^{-6} farad ($1 \ \mu F = 10^{-6}$ F) and a picofarad equals 10^{-12} farad ($1 \text{ pF} = 10^{-12}$ F).

EXAMPLE 1. What is the capacitance of an isolated metallic sphere of radius 20 cm?

SOLUTION: According to Eq. (3),

$$C = 4\pi\varepsilon_0 R = 4\pi \times 8.85 \times 10^{-12} \frac{(\text{coulomb})^2}{\text{N} \cdot \text{m}^2} \times 0.20 \text{ m}$$

$$= 2.2 \times 10^{-11} \frac{\text{coulomb}}{\text{volt}} = 2.2 \times 10^{-11} \text{ F} = 22 \text{ pF}$$

Note that $1 \text{ F} = 1 \text{ C/V} = 1 \text{ C}^2/\text{N} \cdot \text{m}$ so that the constant ε_0 can be written

$$\varepsilon_0 = 8.85 \times 10^{-12} \frac{\text{C}^2}{\text{N} \cdot \text{m}^2} = 8.85 \times 10^{-12} \text{ F/m} \qquad (5)$$

The latter expression is the one usually listed in tables of physical constants.

EXAMPLE 2.   What is the capacitance of the Earth, regarded as a conducting sphere?

SOLUTION: The radius of the Earth is 6.4×10^6 m so that

$$C = 4\pi\varepsilon_0 R = 4\pi \times 8.85 \times 10^{-12} \text{ F/m} \times 6.4 \times 10^6 \text{ m}$$

$$= 7.1 \times 10^{-4} \text{ F}$$

As capacitances go, this is a rather large capacitance. However, note that it only takes a charge of $\sim 10^{-3}$ coulomb to alter the potential of the Earth by 1 volt.

In electrostatic experiments, the Earth is often used as a dump for unwanted positive or negative charge, which alters the potential of the Earth relative to infinity. This alteration conflicts with the convention adopted in Section 25.1, where we treated the ground as a body at a fixed potential, $V = 0$. But this conflict need not trouble us: for most terrestrial electric experiments, only the potential difference between the apparatus and the ground is relevant, and the alteration of the potential difference between the apparatus and infinity has no immediate effect.

The most common variety of capacitor consists of *two* metallic conductors, insulated from one another and carrying opposite amounts of electric charge $\pm Q$. The capacitance of such a pair of conductors is defined in terms of the *difference* of potential between the two conductors:

Capacitance of a pair of conductors

$$\boxed{Q = C\,\Delta V} \tag{6}$$

In this expression, both Q and ΔV are taken as positive quantities. Note that the quantity Q is not the net charge in the capacitor, but the magnitude of the charge on each plate; the net charge in the two-conductor capacitor is zero.

Figure 27.1 shows such a two-conductor capacitor consisting of two large, parallel metallic plates, each of area A, separated by a distance d. The plates carry charges $+Q$ and $-Q$, respectively. The electric field in the region between the plates is (neglecting edge effects)

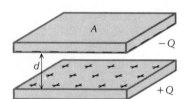

Fig. 27.1 Two parallel plates, with charges $+Q$ and $-Q$.

$$E = \frac{\sigma}{\varepsilon_0} = \frac{Q}{\varepsilon_0 A}$$

and the potential difference is

$$\Delta V = Ed = \frac{Qd}{\varepsilon_0 A} \tag{7}$$

Hence the capacitance of this configuration is

Capacitance of parallel plates

$$\boxed{C = \frac{Q}{\Delta V} = \frac{Q}{Qd/\varepsilon_0 A} = \frac{\varepsilon_0 A}{d}} \tag{8}$$

Thus, in order to store a large amount of charge at a low potential, we want a large plate area A, but a small plate separation d. Parallel-plate capacitors are usually manufactured out of two parallel sheets of aluminum foil, a few centimeters wide and several meters long. The sheets are placed very close together, but kept from contact by a thin sheet of plastic sandwiched between (Figure 27.2). For convenience, the entire sandwich is covered with another sheet of plastic and rolled up like a roll of toilet paper.

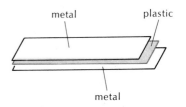

Fig. 27.2 Sheets of aluminum foil separated by a sheet of plastic.

EXAMPLE 3. A parallel-plate capacitor consists of two strips of aluminum foil, each with an area of 0.20 m², separated by a distance of 0.10 mm. The space between the foils is empty. A potential difference of 200 V is applied to this capacitor. What is the capacitance of this capacitor? What is the electric charge on each plate? What is the strength of the electric field between the plates?

SOLUTION: According to Eq. (8), the capacitance is

$$C = \frac{\varepsilon_0 A}{d} = \frac{8.85 \times 10^{-12} \text{ F/m} \times 0.20 \text{ m}^2}{1.0 \times 10^{-4} \text{ m}} = 1.8 \times 10^{-8} \text{ F}$$

$$= 0.018 \ \mu\text{F}$$

The charge on each plate is

$$Q = C \ \Delta V = 1.8 \times 10^{-8} \text{ F} \times 200 \text{ volt} = 3.5 \times 10^{-6} \text{ coulomb}$$

and the electric field between the plates is

$$E = \Delta V / d = 200 \text{ volt}/1.0 \times 10^{-4} \text{ m} = 2.0 \times 10^{6} \text{ volt/m}$$

27.2 Capacitors in Combination

Capacitors used in practical applications in electric circuitry commonly are of the two-conductor variety. Schematically, such capacitors are represented as two parallel plates with terminals emerging at their middles (Figure 27.3). In a circuit, several such capacitors are often wired together and it is then necessary to calculate the net capacitance of the combination. The simplest ways of wiring capacitors together are in **parallel** and in **series**.

Fig. 27.3 Symbol for a capacitor in a circuit diagram.

Figure 27.4 shows two capacitors connected in *parallel*. If charge is fed into this combination via the two terminals, some of the charge will be stored on the first capacitor and some on the second. The net capacitance of the combination can be found as follows. Since the corresponding plates of the capacitors are joined by a conductor, the potential differences across both capacitors are the same,

$$\Delta V = \frac{Q_1}{C_1} \quad \text{and} \quad \Delta V = \frac{Q_2}{C_2} \tag{9}$$

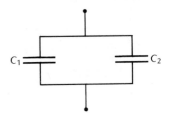

Fig. 27.4 Two capacitors connected in parallel.

Therefore the net charge can be expressed as

$$Q = Q_1 + Q_2 = C_1 \ \Delta V + C_2 \ \Delta V \tag{10}$$

i.e.,

$$Q = (C_1 + C_2)\,\Delta V \qquad (11)$$

Comparing this with the definition for capacitance given in Eq. (6), we see that the combination is equivalent to a single capacitor of capacitance

$$C = C_1 + C_2 \qquad (12)$$

Thus, the net capacitance of the parallel combination is simply the sum of the individual capacitances.

It is easy to obtain a similar result for any number of capacitors connected in parallel (Figure 27.5). The net capacitance is

$$\boxed{C = C_1 + C_2 + C_3 + \cdots} \qquad (13)$$

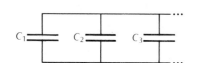

Fig. 27.5 Several capacitors connected in parallel.

Parallel combination of capacitors

Figure 27.6 shows two capacitors connected in *series*. Any charge fed into this combination via the two outside terminals will have to remain on the outside plates (the lower plate of the first capacitor (C_1) and the upper plate of the second (C_2), see Figure 27.6). Thus, the lowest plate will have a charge Q and the highest plate a charge $-Q$. But these charges on the outside plates will induce charges on the inside plates (the upper plate of the first capacitor and the lower plate of the second). The charge Q on the lowest plate will attract electrons to the facing plate and a charge $-Q$ will accumulate on this plate. Corresponding to the excess electrons on the upper plate of the first capacitor, there will be a deficit of electrons on the lower plate of the second capacitor and a charge $+Q$ will accumulate there. The capacitance of the combination can then be found as follows. The potential differences across the two capacitors are

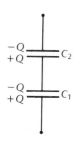

Fig. 27.6 Two capacitors connected in series.

$$\Delta V_1 = \frac{Q}{C_1} \quad \text{and} \quad \Delta V_2 = \frac{Q}{C_2} \qquad (14)$$

The net potential difference between the external terminals is the sum of these,

$$\Delta V = \Delta V_1 + \Delta V_2 = \frac{Q}{C_1} + \frac{Q}{C_2} \qquad (15)$$

i.e.,

$$\Delta V = Q\!\left(\frac{1}{C_1} + \frac{1}{C_2}\right) \qquad (16)$$

From this it is clear that the combination has a net capacitance C given by

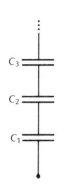

Fig. 27.7 Several capacitors connected in series.

$$\frac{1}{C} = \frac{1}{C_1} + \frac{1}{C_2} \qquad (17)$$

Thus, the net capacitance of the series combination is obtained by taking a sum of inverses. Note that the net capacitance is *less* than the individual capacitances. For example, if $C_1 = C_2$, then $C = \frac{1}{2}C_1 = \frac{1}{2}C_2$.

A similar result applies to any number of capacitors connected in series (Figure 27.7). The net capacitance is given by

$$\frac{1}{C} = \frac{1}{C_1} + \frac{1}{C_2} + \frac{1}{C_3} + \cdots$$

(18)   *Series combination of capacitors*

27.3 Dielectrics

So far, in dealing with problems of electrostatics we have assumed that the space surrounding the electric charge consisted of a vacuum, which has no effect on the electric field, or of air, which has only an insignificant effect on the electric field. However, in dealing with capacitors, we must take the effects of the medium into account. The space between the plates of a capacitor is usually filled with an insulator, or **dielectric**, which drastically changes the electric field from what it would be in a vacuum: the dielectric reduces the strength of the electric field.

To understand this, consider a parallel-plate capacitor whose plates carry some charge per unit area. Suppose that a slab of dielectric, such as glass or polyethylene, fills most of the space between the plates (Figure 27.8). This dielectric contains a large number of atomic nuclei and electrons but, of course, these positive and negative charges balance each other so that the material is electrically neutral. In an insulator, all the charges are **bound** — the electrons are confined within their atoms or molecules and they cannot wander about as in a conductor. Nevertheless, in response to the force exerted by the electric field, the charges will move very slightly without leaving their atoms. The electrons move in a direction opposite to that of the protons; consequently, atoms or molecules acquire a dipole moment in the direction of the original electric field. In most dielectrics, the magnitude of this dipole moment is directly proportional to the strength of the electric field; such dielectrics are said to be **linear**.

In some dielectrics the dipole moment results from a distortion of the molecules or atoms. By tugging on the electrons and protons in opposite directions, the electric field stretches the molecule and produces a small charge separation within it (Figure 27.9). The magnitude of the induced dipole moment is approximately proportional to the strength of the applied electric field.

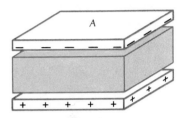

Fig. 27.8 A slab of dielectric between the plates of a capacitor.

Bound charges

Linear dielectric

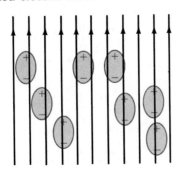

Fig. 27.9 The electric field produces a distortion of molecules.

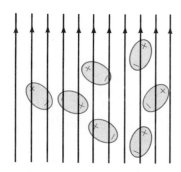

Fig. 27.10 The electric field produces a (partial) alignment of already distorted molecules.

In other dielectrics the dipole moment results (mainly) from a re-alignment of existing dipoles. In such dielectrics, usually gases or liquids, the molecules have permanent dipole moments that are randomly oriented when the dielectric is left by itself. But when the dielectric is placed in an electric field, these permanent dipoles experience a torque that tends to align them with the field (Figure 27.10). Random thermal motions oppose this alignment and the average amount of dipole moment in the direction of the applied electric field is again approximately proportional to the strength of the field.

In any case, by the action of the electric field, the average positions of the positive and negative charges in the dielectric material become displaced relative to one another — the positive and negative charge distributions cease to overlap precisely (Figure 27.11). There will be an excess of positive charge on one surface of the slab of dielectric, and an excess of negative charge on the opposite surface, i.e., the slab of dielectric acquires layers of surface charge. The slab of dielectric is

Polarization

then said to be **polarized.** These surface charges act just like a pair of parallel planes of positive and negative charge; between the planes these charges generate an electric field that is *opposite* to the original applied electric field. The total electric field, consisting of the sum of the field of the free charges on the conducting plates plus the bound charges on the dielectric surfaces is therefore smaller than the field of the free charges alone (Figure 27.12).

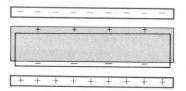

Fig. 27.11 The distributions of positive charge (color) and of negative charge (black) of the slab of dielectric do not overlap precisely.

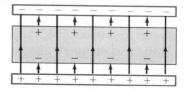

Fig. 27.12 Some electric field lines stop on the negative charges at the bottom of the slab of dielectric. The density of field lines is smaller in the dielectric than in the empty gaps adjacent to the plates.

Dielectric constant

In a linear dielectric, the amount by which the dielectric reduces the strength of the electric field can be characterized by the **dielectric constant** κ. This constant is merely the factor by which the electric field in the dielectric between the parallel plates is reduced, i.e., if E_{free} is the electric field that the free charges produce by themselves and E the electric field that the free charges and the bound charges produce together, then

Electric field in dielectric

$$E = \frac{1}{\kappa} E_{\text{free}} \tag{19}$$

where $\kappa \geq 1$.

Table 27.1 lists the values of the dielectric constant of some materials. Note that air has a value very near $\kappa = 1$, i.e., the dielectric properties of air are not very different from those of a vacuum.

Table 27.1. DIELECTRIC CONSTANTS OF SOME MATERIALS[a]

Material	κ
Vacuum	1
Helium	1.000068
Air	1.00054
Carbon dioxide	1.00098
Carbon tetrachloride	2.2
Paraffin	~2
Polyethylene	2.3
Rubber, hard	2.8
Transformer oil	~3
Plexiglas	3.4
Nylon	3.5
Epoxy resin	3.6
Paper	~4
Bakelite	~5
Pyrex glass	~5
Glass	~6
Porcelain	~7
Water, distilled	80

[a] At room temperature (20° C) and 1 atm.

Incidentally, a metal can be regarded as a dielectric with an infinite dielectric constant — if we substitute $\kappa = \infty$ into Eq. (19), we find that $E = 0$, as it should be inside the metal. This large value of the dielectric constant simply indicates that the material becomes very strongly polarized, i.e., the charges in the metal respond very strongly to an electric field.

If the slab of dielectric entirely fills the space between the plates, then the formula (19) for the reduction of the strength of the electric field applies throughout all of this space. Since the potential difference between the capacitor plates is directly proportional to the strength of the electric field, it follows that, for a given amount of free charge on the plates, the presence of the dielectric also reduces the potential difference by the factor κ,

$$\Delta V = \frac{1}{\kappa} \Delta V_0 \qquad (20)$$

where ΔV_0 is the potential difference in the absence of the dielectric. Consequently, the presence of the dielectric increases the capacitance by a factor κ,

$$C = \frac{Q}{\Delta V} = \kappa \frac{Q}{\Delta V_0} = \kappa C_0 \qquad (21)$$

where C_0 is the capacitance in the absence of dielectric. For example, the capacitance of a parallel-plate capacitor filled with dielectric is

$$C = \kappa C_0 = \kappa \varepsilon_0 A / d \qquad (22)$$

By filling the space between the capacitor plates with dielectric, we can therefore obtain a substantial gain in capacitance. Furthermore, the dielectric can prevent electric breakdown in the space between the plates. If this space contains air, sparking will occur between the plates when the electric field reaches a value of about 3×10^6 V/m and the capacitor will discharge spontaneously. Some dielectrics are better insulators than air and they will tolerate an electric field that is appreciably larger than 3×10^6 V/m. For instance, Plexiglas will tolerate an electric field of up to 40×10^6 V/m before it suffers electric breakdown (Figure 27.13).

Fig. 27.13 Electric breakdown of a Plexiglas block in a very strong electric field caused minute perforations in the block and created this beautiful arboreal pattern.

The magnitude of the surface charge density on the slab of dielectric is given by the simple formula

Surface charge on dielectric

$$\sigma_{\text{bound}} = -\frac{\kappa - 1}{\kappa} \sigma_{\text{free}} \tag{23}$$

Here the negative sign on the right side of the formula indicates that the negatively charged surface of the dielectric adjoins the positively charged plate of the capacitor (see Figure 27.12). This formula is a consequence of Eq. (19). The total electric field E in the dielectric is the sum of the fields of the free charges and the bound charges,

$$E = E_{\text{free}} + E_{\text{bound}} \tag{24}$$

Hence

$$\frac{E_{\text{free}}}{\kappa} = E_{\text{free}} + E_{\text{bound}} \tag{25}$$

which, expressed in terms of charge densities, becomes

$$\frac{\sigma_{\text{free}}}{\varepsilon_0 \kappa} = \frac{\sigma_{\text{free}}}{\varepsilon_0} + \frac{\sigma_{\text{bound}}}{\varepsilon_0} \tag{26}$$

This is equivalent to Eq. (23).

Although we derived the results of this section for the special case of a parallel-plate capacitor, Eqs. (19), (20), (21), and (23) also hold for cylindrical or spherical conductors and concentric shells of dielectric filling the space between the conductors (Figure 27.14). If the dielectric does not entirely fill the space between the conductors, then our results must be modified.

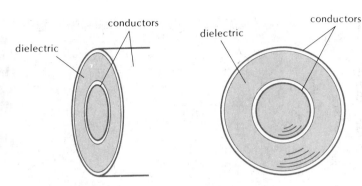

Fig. 27.14 A cylindrical capacitor and a spherical capacitor.

EXAMPLE 4.   A parallel-plate capacitor, such as found in a radio, is made of two strips of aluminum foil with a plate area of 0.75 m². The plates are separated by a layer of polyethylene 2×10^{-5} m in thickness. Suppose that a potential difference of 30 V is applied to this capacitor. What is the magnitude of the free charge on each plate? What is the magnitude of the bound charge on the surface of the dielectric? What is the electric field in the dielectric?

SOLUTION: With $\kappa = 2.3$ (see Table 27.1), the capacitance is

$$C = \kappa \varepsilon_0 \frac{A}{d} = \frac{2.3 \times 8.85 \times 10^{-12} \text{ F/m} \times 0.75 \text{ m}^2}{2 \times 10^{-5} \text{ m}}$$

$$= 7.6 \times 10^{-7} \text{ F}$$

The free charge on each plate is then

$$Q_{\text{free}} = C \, \Delta V = 7.6 \times 10^{-7} \text{ F} \times 30 \text{ V} = 2.3 \times 10^{-5} \text{ coulomb}$$

The bound charge on the surfaces of the dielectric can be found from Eq. (23):

$$Q_{\text{bound}} = A\sigma_{\text{bound}} = -\frac{\kappa - 1}{\kappa} A\sigma_{\text{free}} = -\frac{\kappa - 1}{\kappa} Q_{\text{free}}$$

$$= -\frac{2.3 - 1}{2.3} \times 2.3 \times 10^{-5} \text{ coulomb}$$

$$= -1.3 \times 10^{-5} \text{ coulomb}$$

Here, as in Eq. (23), the negative sign indicates that the bound charge on the dielectric is negative on the surface adjacent to the positive plate of the capacitor. The electric field in the dielectric is

$$E = \frac{1}{\kappa} E_{\text{free}} = \frac{1}{\kappa} \frac{\sigma_{\text{free}}}{\varepsilon_0} = \frac{1}{\kappa} \frac{Q_{\text{free}}}{\varepsilon_0 A}$$

$$= \frac{1}{2.3} \times \frac{2.3 \times 10^{-5} \text{ coulomb}}{8.85 \times 10^{-12} \text{ F/m} \times 0.75 \text{ m}^2} = 1.5 \times 10^6 \text{ volt/m}$$

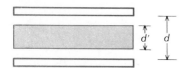

Fig. 27.15 A slab of dielectric between the plates of a capacitor.

EXAMPLE 5.   The space between two large parallel conducting plates is partially filled with a parallel slab of dielectric (Figure 27.15). The area of the plates is A, their separation is d, and the thickness of the slab is d'. The dielectric constant of the slab is κ. What is the capacitance of this arrangement?

SOLUTION: Consider a straight path from one plate to the other. A length $d - d'$ of this path is in a vacuum and a length d' is in the dielectric; the electric field has the value E_{free} in vacuum and E_{free}/κ in the dielectric. Hence the potential difference between the plates is

$$\Delta V = E_{\text{free}}(d - d') + \frac{1}{\kappa} E_{\text{free}} d'$$

$$= \left[1 - \frac{d'}{d}\left(1 - \frac{1}{\kappa}\right)\right] E_{\text{free}} d$$

$$= \left[1 - \frac{d'}{d}\left(1 - \frac{1}{\kappa}\right)\right] \Delta V_0 \qquad (27)$$

The factor in brackets is less than one; hence ΔV is smaller than ΔV_0 and, correspondingly, the capacitance C is larger than C_0,

$$C = C_0 \left/ \left[1 - \frac{d'}{d}\left(1 - \frac{1}{\kappa}\right)\right]\right.$$

$$= \left(\frac{\varepsilon_0 A}{d}\right) \left/ \left[1 - \frac{d'}{d}\left(1 - \frac{1}{\kappa}\right)\right]\right. \qquad (28)$$

27.4 Gauss' Law in Dielectrics

The electric field produced by the bound charges in a dielectric does, of course, obey Gauss' Law. Hence the *total* electric field $\mathbf{E}$ will obey Gauss' Law:

$$\varepsilon_0 \int \mathbf{E} \cdot d\mathbf{S} = Q_{\text{total}} = Q_{\text{free}} + Q_{\text{bound}} \qquad (29)$$

Here the total charge is the sum of the free charge plus the bound charge.

Unfortunately, Q_{bound} is usually not known beforehand because the amount of polarization in the dielectric depends on the (unknown) strength of the electric field. Thus, the above form of Gauss' Law is not very helpful.

Since the free charge Q_{free} is usually known, it is better to devise a modified form of Gauss' Law that depends only on this free charge. For simplicity's sake, we will assume throughout the following discussion that the conductors, the distribution of free charge on them, and the dielectrics between them have enough symmetry so that Gauss' Law suffices for the calculation of the electric field. We can then proceed as follows.

Imagine that at first the dielectrics are absent, so that only the free charge produces an electric field. For this situation, Gauss' Law is

$$\varepsilon_0 \int \mathbf{E}_{\text{free}} \cdot d\mathbf{S} = Q_{\text{free}} \qquad (30)$$

Next, imagine that the dielectrics are inserted into their proper places while the amount of free charge on the conductors is held constant. In general, we would expect that this will mess up the electric field because the bound charges on the dielectrics act back on the free charges and affect the distribution of these charges on the conductors. But if the arrangement of conductors has a high symmetry and the dielectrics have the *same* symmetry (parallel flat conductors with parallel flat slabs of dielectric; concentric cylindrical conductors with concentric cylindrical dielectrics; concentric spherical conductors with concentric spherical dielectrics), then the distribution of free charge is determined by the symmetry and the dielectric cannot disturb this distribution of free charge. Thus, the direction of the electric field at each point is unchanged. Only the strength of the electric field is altered: the new electric field will be smaller than the original one by a factor of κ,

$$\mathbf{E} = \mathbf{E}_{\text{free}}/\kappa \qquad (31)$$

Substituting this into Eq. (30), we obtain

$$\varepsilon_0 \int \kappa \mathbf{E} \cdot d\mathbf{S} = Q_{\text{free}} \qquad (32)$$

Gauss' Law in dielectrics

This is **Gauss' Law in dielectrics.** It relates the total electric field **E** to the *free* charge Q_{free}. The effect of the bound charge is implicitly contained in the factor κ appearing on the left side of the equation. Although the preceding discussion has focused on arrangements of conductors and dielectrics with high symmetry, the above modified version of Gauss' Law turns out to be valid for conductors and dielectrics of any shape whatsoever.

EXAMPLE 6. Two concentric spheres of sheet metal have radii r_1 and r_2 respectively. The space between these is filled with gas of dielectric constant κ (Figure 27.16). What is the capacitance of this contraption?

SOLUTION: Suppose that the free charge on the inner sphere is Q_{free} and on the outer $-Q_{\text{free}}$. As Gaussian surface, take a sphere of radius r ($r_2 > r > r_1$, see Figure 27.16). Equation (32) then becomes

$$\varepsilon_0 \kappa E \times 4\pi r^2 = Q_{\text{free}}$$

or

$$E = \frac{1}{4\pi\kappa\varepsilon_0} \frac{Q_{\text{free}}}{r^2} \qquad (33)$$

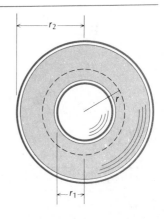

Fig. 27.16 Concentric conducting spheres.

Note that this electric field differs from that of a point charge in vacuum by the factor $1/\kappa$.

The potential difference between r_1 and r_2 is then

$$\Delta V = \int_{r_1}^{r_2} \frac{1}{4\pi\kappa\varepsilon_0} \frac{q_{\text{free}}}{r^2} \, dr = \frac{Q_{\text{free}}}{4\pi\kappa\varepsilon_0} \left(\frac{1}{r_1} - \frac{1}{r_2} \right)$$

and the capacitance

$$C = \frac{Q_{\text{free}}}{\Delta V} = \frac{4\pi\kappa\varepsilon_0}{1/r_1 - 1/r_2} \qquad (34)$$

In a more advanced study of dielectrics, it is useful to introduce a quantity **D** called the **electric displacement field,**

Electric displacement field

$$\mathbf{D} = \varepsilon_0 \kappa \mathbf{E}$$

In terms of this quantity, Gauss' Law becomes [see Eq. (32)]

$$\int \mathbf{D} \cdot d\mathbf{S} = Q_{\text{free}} \tag{35}$$

This version of Gauss' Law is of more general validity than Eq. (32); for instance, it remains valid even in nonlinear dielectrics, where Eq. (32) fails.

27.5 Energy in Capacitors

Capacitors not only store electric charge, but also electric energy. Consider a two-conductor capacitor with charges $\pm Q$ on its plates. If the capacitor contains no dielectric, then the electric potential energy can be calculated directly from Eq. (26.7):

$$U = \tfrac{1}{2}QV_2 + \tfrac{1}{2}(-Q)V_1 = \tfrac{1}{2}Q(V_2 - V_1)$$

where V_1 and V_2 are the potentials of the plates. Thus, the potential energy can be expressed in terms of charge and potential difference,

Energy in capacitor

$$\boxed{U = \tfrac{1}{2}Q \, \Delta V} \tag{36}$$

By means of the definition of capacitance, $Q = C \, \Delta V$, this can be put in the alternative forms

$$U = \tfrac{1}{2}C(\Delta V)^2 \tag{37}$$

or

$$U = \tfrac{1}{2} Q^2/C \tag{38}$$

If the capacitor contains a dielectric, then the calculation of the energy is a bit more involved. The trouble is that the dielectric, with its bound charges, contributes to the electric potential energy. However, in practice we are usually not interested in the total potential energy, but only in that part of the potential energy that changes as we charge (or discharge) the capacitor, i.e., we are interested only in the amount of work required to charge (or discharge) the capacitor. It turns out that this amount of work is correctly given by Eqs. (36)–(38), regardless of whether the capacitor contains a dielectric or not. The quantity Q in these equations is the charge on the plates, i.e., it is the *free* charge. To see this, let us derive Eq. (38) from a different starting point. Imagine that we charge the capacitor gradually, starting with an initial charge $q = 0$ and ending with a final charge $q = Q$. When the plates carry charges $\pm q$, the potential difference between them is q/C

and the work that we must perform to increase the charge on the plates by $\pm dq$ is

$$dU = \frac{q}{C}\, dq \qquad (39)$$

The total work that we must perform to charge the capacitor is then

$$U = \int_0^Q \frac{q}{C}\, dq = \frac{1}{C} \int_0^Q q\, dq = \tfrac{1}{2} Q^2/C \qquad (40)$$

This agrees with Eq. (38) and establishes its general validity. Note that as the free charges on the capacitor plate increase, the bound charges within the dielectric rearrange themselves, becoming more strongly polarized; the energy required for this rearrangement is already included in Eq. (40) (the properties of the dielectric enter into this formula via the capacitance).

In Section 26.3 we obtained a formula for the energy density in an electric field in a vacuum. That calculation was based on an examination of the electric field in the space between the plates of a parallel-plate capacitor. By repeating this calculation for a parallel-plate capacitor with dielectric, we readily find that the **energy density in the dielectric** is

$$\boxed{u = \tfrac{1}{2}\kappa\varepsilon_0 E^2} \qquad (41)$$ *Energy density in dielectric*

Formally, this differs from Eq. (26.19) by an extra factor of κ. Note, however, that the electric field **E** in Eq. (41) is the actual electric field in the dielectric, which already contains an implicit dependence on κ.

EXAMPLE 7. Consider the parallel-plate capacitor of Example 4. What is the stored potential energy? What is the energy density in the dielectric?

SOLUTION: By Eq. (36),

$$U = \tfrac{1}{2}Q\,\Delta V = \tfrac{1}{2} \times 2.3 \times 10^{-5} \text{ coulomb} \times 30 \text{ volt} = 3.4 \times 10^{-4} \text{ J}$$

The energy density can be calculated either from Eq. (41),

$$u = \tfrac{1}{2}\kappa\varepsilon_0 E^2 = \tfrac{1}{2} \times 2.3 \times 8.85 \times 10^{-12} \text{ F/m} \times (1.5 \times 10^6 \text{ volt/m})^2$$

$$= 23 \text{ J/m}^3$$

or else by taking the ratio of energy to volume,

$$u = \frac{U}{Ad} = \frac{3.4 \times 10^{-4} \text{ J}}{0.75 \text{ m}^2 \times 2.0 \times 10^{-5} \text{ m}} = 23 \text{ J/m}^3$$

SUMMARY

Capacitance of a pair of conductors: $C = Q/\Delta V$

Capacitance of parallel plates: $C = \varepsilon_0 A/d$

Parallel combination: $C = C_1 + C_2 + C_3 + \cdots$

Series combination: $1/C = 1/C_1 + 1/C_2 + 1/C_3 + \cdots$

Electric field in dielectric between parallel plates: $E = \dfrac{1}{\kappa} E_{\text{free}}$

Gauss' law in dielectric: $\int \kappa \mathbf{E} \cdot d\mathbf{S} = Q_{\text{free}}/\varepsilon_0$

Energy in capacitor: $U = \frac{1}{2} Q \, \Delta V$

Energy density in dielectric: $u = \frac{1}{2}\kappa\varepsilon_0 E^2$

QUESTIONS

1. Commercially available large capacitors have a capacitance of 1000 μF. How is it possible that the capacitance of such a device is larger than the capacitance of the Earth?

2. A single-conductor capacitor may be regarded as a two-conductor capacitor with the second plate consisting of a very large conducting shell of infinite radius. Show that for $r \to \infty$, Eq. (34), with $\kappa = 1$, reduces to Eq. (3).

3. Suppose we enclose the entire Earth in a conducting shell of a radius slightly larger than the Earth's radius. Explain why this would make the capacitance of the Earth much larger than the value calculated in Example 2.

4. Equation (8) shows that $C \to \infty$ as $d \to 0$. In practice, why can we not construct a capacitor of arbitrarily large C by making d sufficiently small? (Hint: What happens to E as $d \to 0$ while ΔV is held constant?)

5. If you put more charge on one plate of a parallel-plate capacitor than on the other, what happens to the extra charge?

6. Taking the fringing field into account, would you expect the capacitance of a parallel-plate capacitor to be larger or smaller than the value given by Eq. (8)? (Hint: How does the fringing affect the density of field lines between the plates?)

7. Explain why there must be a fringing field in the region near the edges of a pair of parallel plates. (Hint: Suppose there were no fringing field, so that the field lines look as in Figure 27.17. Is $\oint \mathbf{E} \cdot d\mathbf{l} = 0$ for the path shown in this figure?)

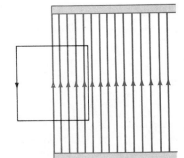

Fig. 27.17

8. Figure 27.18 shows a capacitor with a **guard rings.** These rings fit snugly around the edges of the capacitor plates, but they are not in electrical contact with the plates. In use, the potential on the rings is adjusted to the same value as the potential on the plates. Explain how the rings keep the field of the plates from fringing.

9. Figure 27.19 shows the design of an adjustable capacitor used in the tuning circuit of a radio. This capacitor can be regarded as several connected capacitors. Are these several capacitors connected in series or in parallel? If we turn the tuning knob (and the attached colored plates) counterclockwise, does the capacitance increase or decrease?

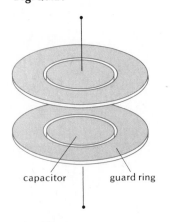

capacitor   guard ring

Fig. 27.18

Fig. 27.19 The black plates are connected together, and the colored plates are connected together.

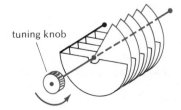

tuning knob

10. Consider a parallel-plate capacitor. Does the capacitance change if we insert a thin conducting sheet between the two plates, parallel to them?

11. Suppose we insert a thick slab of metal between the plates of a parallel-plate capacitor, parallel to the plates. Does the capacitance increase or decrease?

12. Consider a fluid dielectric that consists of molecules with permanent dipole moments. Will the dielectric constant increase or decrease as a function of temperature?

13. Figure 27.20 shows a dielectric slab partially inserted between the plates of a capacitor. Will the electric forces between the slab and the plates pull the slab into the region between the plates or push it out? (Hint: Consider the fringing field.)

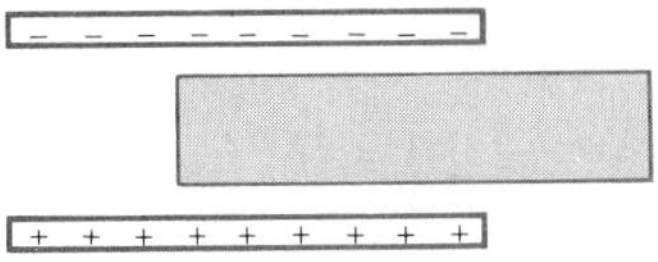

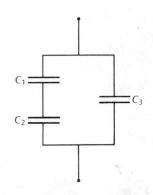

Fig. 27.20

14. If we increase the separation between the plates of a parallel-plate capacitor by a factor of 2, while holding the electric charge constant, by what factors will we change the electric field, the potential difference, the capacitance, and the electric energy?

15. Consider the parallel-plate capacitor with the slab of dielectric shown in Figure 27.15. How does the capacitance change if we move the slab up? If we move the slab to the right? If we tilt the slab?

16. Spell out the steps in the derivation of Eq. (41).

PROBLEMS

Section 27.1

1. Consider an isolated metallic sphere of radius R and another isolated metallic sphere of radius $3R$. If both spheres are at the same potential, what is the ratio of their charges? If both spheres carry the same charge, what is the ratio of their potentials?

2. The collector of an electrostatic machine is a metal sphere of radius 18 cm.
   (a) What is the capacitance of this sphere?
   (b) How many coulombs of charge must you place on this sphere to raise its potential to 2.0×10^5 V?

3. Your head is (approximately) a conducting sphere of radius 10 cm. What is the capacitance of your head? What will be the charge on your head if, by means of an electrostatic machine, you raise your head (and your body) to a potential of 100,000 V?

4. A capacitor consists of a metal sphere of radius 5 cm placed at the center of a thin metal shell of radius 12 cm. The space between is empty. What is the capacitance?

5. What is the capacitance of the Geiger-counter tube described in Problem 25.25? Pretend that the space between the conductors is empty.

Section 28.2

6. What is the combined capacitance if three capacitors of 3.0, 5.0, and 7.5 μF are connected in parallel? What is the combined capacitance if they are connected in series?

7. Three capacitors with capacitances $C_1 = 5.0$ μF, $C_2 = 3.0$ μF, and $C_3 = 8.0$ μF are connected as shown in Figure 27.21. Find the combined capacitance.

8. Two capacitors, of 2.0 and 6.0 μF, respectively, are initially charged to 24 V by connecting each, for a few instants, to a 24-V battery. The battery is then removed and the charged capacitors are connected in a closed series circuit, the positive terminal of each capacitor being connected to the negative termi-

Fig. 27.21

nal of the other (Figure 27.22). What will be the final charge on each capacitor?

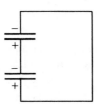

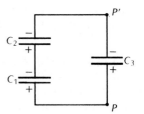

Fig. 27.22 Capacitors connected after they have been charged.

Fig. 27.23 Capacitors connected after they have been charged.

9. Three capacitors, of capacitances $C_1 = 2.0$ μF, $C_2 = 5.0$ μF, and $C_3 = 7.0$ μF, are initially charged to 36 V by connecting each, for a few instants, to a 36-V battery. The battery is then removed and the charged capacitors are connected in a closed series circuit, with the positive and negative terminals joined as shown in Figure 27.23. What will be the final charge on each capacitor? What will be the voltage across the points PP' in Figure 27.23?

Section 27.3

10. You wish to construct a capacitor out of a sheet of polyethylene of thickness 5×10^{-2} mm and $\kappa = 2.3$ sandwiched between two aluminum sheets. If the capacitance is to be 3.0 μF, what must be the area of the sheets?

11. In order to measure the dielectric constant of a dielectric material, a slab of this material 1.5 cm thick is slowly inserted between a pair of parallel conducting plates separated by a distance of 2.0 cm. Before insertion of the dielectric, the potential difference across these capacitor plates is 3.0×10^5 V. During insertion, the charge on the plates remains constant. After insertion, the potential difference is 1.8×10^5 V. What is the value of the dielectric constant?

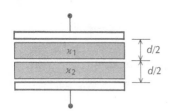

Fig. 27.24 Parallel-plate capacitor with two slabs of dielectric.

12. A parallel-plate capacitor of plate area A and spacing d is filled with two parallel slabs of dielectric of equal thickness with dielectric constants κ_1 and κ_2, respectively (Figure 27.24). What is the capacitance? (Hint: Check that the configuration of Figure 27.24 is equivalent to two capacitors in series.)

13. A capacitor with two large parallel plates of area A separated by a distance d is filled with two equal slabs of dielectric side by side (Figure 27.25). The dielectric constants are κ_1 and κ_2. What is the capacitance?

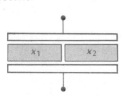

Fig. 27.25 Parallel-plate capacitor with two slabs of dielectric.

14. Show that the result of Example 5 can also be derived by regarding the capacitor partially filled with dielectric as two capacitors in series, one completely filled with dielectric, one empty.

15. A parallel-plate capacitor of plate area A and separation d contains a slab of dielectric of thickness $d/2$ (Figure 27.26) and dielectric constant κ. The potential difference between the plates is ΔV.
  (a) In terms of the given quantities, find the electric field in the empty region of space between the plates.
  (b) Find the electric field inside the dielectric.
  (c) Find the density of bound charge on the surface of the dielectric.

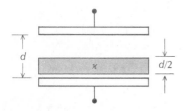

Fig. 27.26 A parallel-plate capacitor, partially filled with dielectric.

16. Within some limits, the difference between the dielectric constants of air and of vacuum is proportional to the pressure of the air, i.e., $\kappa - 1 \propto p$. Suppose that a parallel-plate capacitor is held at a constant potential difference by means of a battery. What will be the percentage change in the amount of charge on the plates as we increase the air pressure between the plates from 1.0 atm to 3.0 atm?

17. A parallel-plate capacitor is filled with carbon dioxide at 1 atm pressure. Under these conditions the capacitance is 0.5 μF. We charge the capacitor by means of a 48-V battery and then disconnect the battery so that the electric charge remains constant thereafter. What will be the change in the potential difference if we now pump the carbon dioxide out of the capacitor, leaving it empty?

*18. A parallel-plate capacitor is filled with a layer of distilled water 0.30 cm thick. The dipole moment of a water molecule is 6.1×10^{-30} C·m. Assume that the dipole moments of the water molecules are all perfectly aligned with the electric field. What is the surface charge density of bound charges on the surface of the layer of water?

Section 27.4

19. A spherical capacitor consists of a metallic sphere of radius R_1 surrounded by a concentric metallic shell of radius R_2. The space between R_1 and R_2 is filled with dielectric having a constant κ. Suppose that the free surface charge density on R_1 is $\sigma_{\text{free}(1)}$.
   (a) What is the free surface charge density on the metallic sphere at R_2?
   (b) What is the bound surface charge density on the dielectric at R_1?
   (c) What is the bound surface charge density on the dielectric at R_2?

20. A long cylindrical copper wire of radius 0.20 cm is surrounded by a cylindrical sheath of rubber of inner radius 0.20 cm and outer radius 0.30 cm. The rubber has $\kappa = 2.8$. Suppose that the surface of the copper has a free charge density of 4.0×10^{-6} C/m².
   (a) What will be the bound charge density on the inside surface of the rubber sheath? On the outside surface?
   (b) What will be the electric field in the rubber near its inner surface? Near its outer surface?
   (c) What will be the electric field just outside the rubber sheath?

21. A metallic sphere of radius R is surrounded by a concentric dielectric shell of inner radius R, outer radius $3R/2$. This is surrounded by a concentric, thin, metallic shell of radius $2R$ (Figure 27.27). The dielectric constant of the shell is κ. What is the capacitance of this contraption?

22. Two small metallic spheres are submerged in a large volume of transformer oil of dielectric constant $\kappa = 3.0$. The spheres carry electric charges of 2.0×10^{-6} C and 3.0×10^{-6} C, respectively, and the distance between them is 0.60 m. What is the force on each?

*23. A sphere of brass floats in a large lake of oil of dielectric constant $\kappa = 3.0$. The sphere is exactly halfway immersed in the oil (Figure 27.28). The sphere has a net charge of 2.0×10^{-6} C. What fraction of this electric charge will be on the upper hemisphere? On the lower? (Hint: For the path shown in Figure 27.28, $\oint \mathbf{E} \cdot d\mathbf{l} = 0$; from this prove that the electric fields in the oil and in the air above the oil will be exactly the same.)

*24. A spherical capacitor consists of two concentric spheres of metal of radii R_1 and R_2. The space between these spheres is filled with two kinds of dielectric (Figure 27.29); the dielectric in the upper hemisphere has a constant κ_1 and the dielectric in the lower hemisphere has a constant κ_2. What is the capacitance of this device? (Hint: For the path shown in Figure 27.29 $\oint \mathbf{E} \cdot d\mathbf{l} = 0$; from this prove that the electric fields in both dielectrics are exactly the same.)

*25. In a semiconductor with impurity ions, electrons will orbit around these ions. The sizes of the orbits are considerably larger than the spacings between the semiconductor atoms and hence an electron may be regarded as moving through a more or less uniform medium of a given dielectric constant.
   (a) Show that the electric force of attraction between an electron and an ion of charge e immersed in a medium of dielectric constant κ is

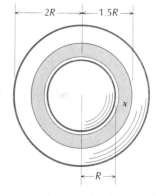

Fig. 27.27 A spherical capacitor, partially filled with dielectric.

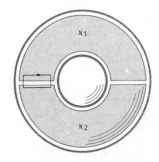

Fig. 27.28 A brass sphere afloat in a lake of oil.

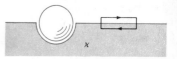

Fig. 27.29 A spherical capacitor with two kinds of dielectric.

$$F = \frac{1}{4\pi\varepsilon_0\kappa} \frac{e^2}{r^2}$$

(b) Calculate the orbital energy of an electron in a circular orbit according to Bohr's theory described in Problem 25.16.

(c) The dielectric constant of germanium is $\kappa = 15.8$. Evaluate the orbital energy of an electron moving around an ion embedded in germanium; assume that the electron is in the smallest Bohr orbit. By what factor does your result differ from what you would obtain for an electron moving around this ion in a vacuum?

Section 27.5

26. A parallel-plate capacitor has a plate area of 900 cm² and a plate separation of 0.50 cm. The space between the plates is empty.
   (a) What is the capacitance?
   (b) What is the potential difference if the charges on the plates are $\pm 6.0 \times 10^{-8}$ C?
   (c) What is the electric field between the plates?
   (d) The energy density?
   (e) The total energy?

27. Repeat Problem 26 if the space between the plates is filled with Plexiglas.

28. A TV receiver contains a capacitor of 10 μF charged to a potential difference of 2×10^4 V. What is the amount of charge stored in this capacitor? The amount of energy?

29. Two parallel conducting plates of area 0.5 m² placed in a vacuum have a potential difference of 2.0×10^5 V when charges of $\pm 4.0 \times 10^{-3}$ C are placed on them, respectively.
   (a) What is the capacitance of the pair of plates?
   (b) What is the distance between them?
   (c) What is the electric field between them?
   (d) What is the electric energy?

30. Two capacitors of 5.0 μF and 8.0 μF are connected in series to a 24-V battery. What is the energy stored in the capacitors?

31. A parallel-plate capacitor without dielectric has an area A and a charge $\pm Q$ on each plate.
   (a) What is the electric force F of attraction between the plates?
   (b) How much work must you do against this force in order to increase the plate separation by an amount Δl?
   (c) By means of Eq. (38), calculate the change of ΔU in potential energy during this change.
   (d) By comparing (a) and (c) check that $F = -\Delta U/\Delta l$.
      (Hint: One-half of the electric field between the plates is due to one plate and one-half is due to the other. Consequently, when calculating the electric force on a plate from the product of field times charge, only one-half of the field must be used; the other half gives the electric force of the plate on *itself* and is of no interest.)

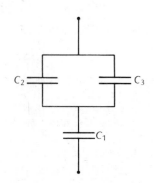

Fig. 27.30

32. Power companies are interested in the storage of surplus electric energy. Suppose we wanted to store 10^6 kW · h of electric energy (half a day's output for a large power plant) in a large parallel-plate capacitor filled with a plastic dielectric with $\kappa = 3.0$. If the dielectric can tolerate a maximum electric field of 5×10^7 V/m, what is the minimum total volume of dielectric needed to store this energy?

33. Three capacitors are connected as shown in Figure 27.30. Their capacitances are $C_1 = 2.0$ μF, $C_2 = 6.0$ μF, and $C_3 = 8.0$ μF. If a voltage of 200 V is applied to the two free terminals, what will be the charge on each capacitor? What will be the energy in each?

CHAPTER 28

Currents and Ohm's Law

Under static conditions there can exist no electric field inside a conductor. But suppose that we suddenly deposit opposite amounts of electric charge on the opposite ends of a long metallic conductor, such as a wire. The conductor will then not be in electrostatic equilibrium and the charges at the ends will generate an electric field along and inside the conductor (Figure 28.1). This electric field propels the charges toward each other. When the charges meet, they cancel. The electric field then disappears — the conductor reaches equilibrium.

For a good conductor, such as copper, the approach to equilibrium is fairly rapid; typically, the time required to achieve equilibrium is a small fraction of a second. However, we can keep a conductor in a permanent state of disequilibrium if we continually supply more electric charge to its ends. For example, we can connect the two ends of a copper wire to the terminals of a battery or of an electric generator. The terminals of such a device act as source and sink of electric charge, just like the outlet and the intake of a pump act as source and sink of water. Under these conditions electric charge will continually flow from one terminal to the other, forming an electric current.[1]

Fig. 28.1 Electric field lines in and near a straight conductor not in equilibrium. The field lines have been made visible by sprinkling grass seeds on the surface of the paper on which the conductor has been painted with conducting paint. (From O. Jefimenko, *Am. J. Phys.* **30,** 19, 1962.)

28.1 Electric Current

When a wire is connected between the two terminals of a battery or generator, the electric charges are propelled from one end of the wire to the other by the electric field that exists along and within the wire. Most of the field lines originate at the terminals of the battery or gen-

[1] We will discuss the inner workings of batteries, generators, and other "pumps" for electric charge in the next chapter.

erator, but some field lines originate at charges on the wire itself. As Figures 28.1 and 28.2 show, the field lines tend to concentrate within the conductor, and they tend to follow the conductor. If the conductor has no sharp kinks, the field lines are uniformly distributed over the cross-sectional area of the conductor. For instance, if the conductor is a more or less straight wire of constant thickness, then the electric field inside the wire will be of constant magnitude and of a direction parallel to the wire. If the length of the wire is l and if the battery or generator maintains a difference of potential ΔV across its ends, then the electric field in the wire is

Electric field in uniform wire

$$E = \Delta V / l \qquad (1)$$

This electric field causes the flow of charge, or **electric current,** from one end of the wire to the other. Before we can explore the dependence of the current on the field, we need a precise definition of the current. Suppose that an amount of charge dq flows past some given point of the wire (e.g., the end of the wire) in a time dt; then the electric current is defined as charge divided by time,

Electric current

$$I = \frac{dq}{dt} \qquad (2)$$

Note that if the sides of the wire do not leak (good insulation), then the conservation of electric charge requires that the current be the same everywhere along the wire, that is, the current is simply the rate at which charge enters the wire at one end or the rate at which charge leaves at the other end.

The SI unit of current is the **ampere** (A); this is a flow of charge of one coulomb per second,

$$1 \text{ ampere} = 1 \text{ A} = 1 \text{ C/s} \qquad (3)$$

In metallic conductors the charge carriers are electrons — a current in a metal is nothing but a flow of electrons. In electrolytes the charge carriers are positive ions, negative ions, or both — a current in such a conductor is a flow of ions. For the sake of uniformity, whenever we need to indicate the direction of the current along a conductor, we will follow the convention that the current has the direction of the positive flow of charge. This means that we pretend that the moving charges are always positive charges. Of course, in metals the moving charges are actually negative charges (electrons) and hence the above convention assigns to the current a direction opposite to that of the true motion of the charges. However, as regards the transfer of charge, the transport of negative charge in one direction is equivalent to the transport of positive charge in the opposite direction. Our convention for labeling the direction of the current takes advantage of this equivalence.

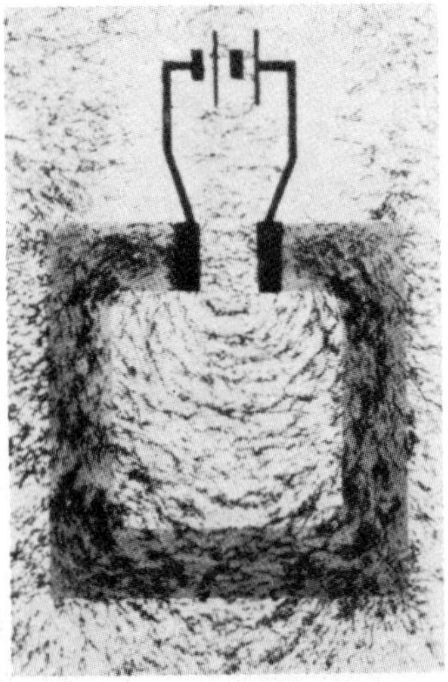

If we divide the current in a conductor by the cross-sectional area of the conductor (Figure 28.3) we obtain the **current density,**[2]

Fig. 28.2 Electric field lines in and near a rectangular conductor carrying an electric current. (From O. Jefimenko, *Am. J. Phys.* **30,** 19, 1962.)

[2] Do not confuse the symbol A for area with the abbreviation A for ampere.

$$\boxed{j = I/A} \qquad (4)$$

This is really the *average* current density over the area A. We can also define a local current density in terms of the current dI that flows across an infinitesimal portion of dA of cross-sectional area (Figure 28.3),

$$j = \frac{dI}{dA} \qquad (5)$$

Under normal conditions the current in a wire is uniformly distributed over the entire cross-sectional area of the wire. In terms of the local current density, this means that j is constant over the entire cross-sectional area.

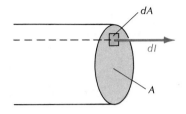

Fig. 28.3 A conductor carries a current from left to right. The conductor has been cut off on the right so as to give a clear view of its cross section.

28.2 Resistance and Ohm's Law

We will now examine in detail the behavior of a current in a metallic conductor. Such conductors contain a vast number of free electrons; for example, copper has about 8×10^{22} free electrons per cubic centimeter. These electrons form a gas which fills the entire volume of the metal. Of course, in a neutral conductor the negative charge of the free electrons is exactly balanced by the positive charge of the ions that make up the lattice of the metal. A current in a metallic conductor is simply a flow of the gas of electrons, while the ions remain at rest.

The flow of the gas of electrons along a metallic wire is analogous to the flow of water along a canal leading down a gentle slope. Under these conditions the force of gravity acting on the water has a component along the canal; this component pushes the water along. But the water does not accelerate because friction between the water and the walls of the canal opposes the motion — the water moves at a constant speed because the friction exactly matches the push of gravity.

Likewise, the electric field in the wire pushes the gas of electrons along. But the gas of electrons does not accelerate because friction between the gas and the body of the wire opposes the motion — the gas moves at a constant speed because the friction exactly matches the push of the electric field.

The analogy between the motion of water and the motion of the electron gas extends to the motion of individual water molecules and individual electrons. Although the water in a canal usually has a fairly low speed, perhaps a few meters per second, the individual molecules within the water have a rather high speed — the typical speed of the random thermal motion of water molecules is about 500 m/s at ordinary temperatures. But since this thermal motion consists of rapid zigzags which are just as likely to move the molecule backward as forward, this high speed does not contribute to the net downhill motion of the water. Figure 28.4 shows the motion of a water molecule in the canal; on a microscopic scale, this motion consists of rapid zigzags on which is superimposed a much slower "drift" along the canal.

Likewise, the electron gas moves along the wire at a rather low speed, perhaps 10^{-2} m/s, but the individual electrons have a much

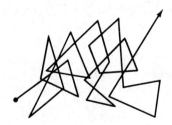

Fig. 28.4 Path of a water molecule in canal. The molecule gradually drifts from left to right.

higher speed — the typical speed of the random motion of electrons in a metal is about 10^6 m/s (this very high speed is due to quantum-mechanical effects, which we cannot discuss here). Thus, the net motion of an electron also consists of rapid zigzags on which is superimposed a much slower "drift" motion along the wire. Qualitatively, the motion resembles the path of a water molecule shown in Figure 28.4, but the amount of drift per zigzag is even less than shown in this figure.

The friction between the electron gas and the body of the wire is caused by collisions between the electrons and the ions of the crystal lattice of the wire. An electron moving through a piece of copper will suffer about 10^{14} collisions with ions per second. Each collision slows the electron down, brings it to a stop, or reverses its motion. Because of the disturbing effects of these collisions, the electron never gains much velocity from the electric field that is attempting to accelerate it. The collisions dissipate the kinetic energy that the electron receives from the electric field. The dissipated kinetic energy of the electrons remains in the crystal lattice in the form of random kinetic energy of the ions, i.e., it remains as heat.

The average velocity, or drift velocity, that an electron attains in the electric field is proportional to the strength of the electric field,

$$v_d \propto E \tag{6}$$

This proportionality merely reflects the fact that if the electric field is strong, the electron gains more velocity between one collision and the next, and therefore attains a larger average velocity. The electric current carried by the wire is proportional to the average velocity of the electrons,

$$I \propto v_d \propto E \tag{7}$$

The current is also proportional to the cross-sectional area of the wire, because a large cross-sectional area means that more electrons participate in the transport of charge. Hence

$$I \propto AE \tag{8}$$

With $E = \Delta V/l$, this proportionality becomes

$$I \propto \frac{A}{l} \Delta V \tag{9}$$

To transform this into an equality, we rewrite it as

$$I = \frac{1}{\rho} \frac{A}{l} \Delta V \tag{10}$$

Resistivity where ρ is a constant of proportionality that characterizes the material of the wire. This constant is called the **resistivity** of the material.

It is customary to define the **resistance** of the wire as

Resistance

$$\boxed{R = \rho \frac{l}{A}} \tag{11}$$

Equation (10) can then be expressed in the convenient form

$$I = \frac{\Delta V}{R}$$

(12) *Ohm's Law*

This equation is called **Ohm's Law.** It asserts that *the current is proportional to the potential difference* between the ends of the conductor. Note that in Eq. (12), the resistance plays the role of a constant of proportionality. For a wire of uniform cross section, the resistance can be calculated from the simple formula (11). But Ohm's Law is also valid for conductors of arbitrary shape — such as wires of nonuniform cross section — for which the resistance must be calculated from a more complicated formula tailored to the shape and the size of the conductor.

Ohm's Law is valid for metallic conductors and also for nonmetallic conductors (e.g., carbon) in which the current is carried by a flow of electrons.[3] It is even valid for plasmas and for electrolytes, in which the current is carried by a flow of both electrons and ions. However, we ought to keep in mind that in spite of its wide range of applicability, Ohm's Law is not a general law of nature — such as Gauss' Law — but only an assertion about the electrical properties of certain materials.

Georg Simon Ohm, *1787–1854, German physicist, professor at Munich. Ohm was led to his law by an analogy between the conduction of electricity and the conduction of heat: the electric field is analogous to the temperature gradient and the electric current is analogous to the heat flow. Equation (28.12) is then analogous to Eq. (20.5).*

28.3 The Flow of Free Electrons[4]

Before we deal with some applications of Ohm's Law, let us reexamine the derivation of this law and fill in the constants of proportionality that we left out in Eqs. (6)–(9). To do this, we need to calculate the average motion of the free electrons in a metal in some detail. The high-speed thermal motion of the electrons does not enter directly into the calculation of the average motion because it is random; however, it enters indirectly because it determines the collision rate. The large collision rate of a free electron in a metal — for instance, 10^{14} collisions per second for an electron in copper — results from the high speed of the random motion: the electron moves a large distance per second and therefore encounters many ions with which to collide. Since the extra speed that the electron gains from the electric field is very small compared with the random speed, the collision rate is nearly unaffected by the electric field. The average time per collision is therefore a constant τ that depends only on the characteristics of the metal. We can find the average motion of an electron by examining the losses and gains of momentum of this electron. If the average velocity, or drift velocity, of an electron is v_d, then the average momentum is $m_e v_d$. We expect that, on the average, a collision will absorb all of this momentum, i.e., a collision will destroy the forward drift velocity and leave the electron with only the random thermal motion. This means that, in a time interval τ, the electron loses a momentum $m_e v_d$; the

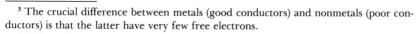

[3] The crucial difference between metals (good conductors) and nonmetals (poor conductors) is that the latter have very few free electrons.

[4] This section is optional.

average rate at which the electron loses momentum in collisions is therefore

$$\left(\frac{\Delta p}{\Delta t}\right)_{\text{loss}} = \frac{m_e v_d}{\tau} \tag{13}$$

On the other hand, the rate at which the electron gains momentum by the action of the electric force is

$$\left(\frac{\Delta p}{\Delta t}\right)_{\text{gain}} = -eE \tag{14}$$

Under steady-state conditions, the rate of loss of momentum must match the rate of gain. By setting the right sides of Eqs. (13) and (14) equal we immediately obtain

Drift velocity

$$\boxed{v_d = -eE\tau/m_e} \tag{15}$$

This is the average velocity with which the electron gas flows along the wire. As expected, this velocity is proportional to the strength of the electric field. The negative sign in Eq. (15) indicates that the direction of flow is opposite to the direction of the electric field.

EXAMPLE 1.   A potential difference of 3.0 V is applied to the ends of a copper wire 0.5 m long. What is the drift velocity of the free electrons in the wire? In copper at room temperature, the average time interval between collisions is $\tau = 2.7 \times 10^{-14}$ s.

SOLUTION: The electric field in the wire is $E = 3.0$ V$/0.5$ m $= 6.0$ V/m. Hence Eq. (15) gives

$$v_d = -\frac{1.6 \times 10^{-19} \text{ C} \times 6.0 \text{ V/m} \times 2.7 \times 10^{-14}}{9.1 \times 10^{-31} \text{ kg}}$$

$$= -2.8 \times 10^{-2} \text{ m/s}$$

Note that this speed is rather low — it takes an electron about a third of a minute to wander from one end of the wire to the other. Nevertheless, if a pulse of current (a signal) is suddenly injected into the wire at one end, a similar pulse of current will emerge from the far end *almost instantaneously*. What happens in this case is that, by means of their electric field, electrons push on neighboring electrons and a compressional wave travels through the electron gas; this ejects electrons out of the far end almost instantaneously. The phenomenon is analogous to the propagation of a wave on the water of a canal; the wave travels much faster than the flow of water.

To find the electric current that corresponds to the flow of the electron gas, we need to take into account the number of electrons. Suppose that the metal of the wire has n free electrons per unit volume. The quantity n depends on the metal; for copper $n = 8.5 \times 10^{28}/\text{m}^3$. If the wire has a cross-sectional area A and a length l, then its total number of free electrons is $n \times [\text{volume}] = nAl$ and the total charge associated with these electrons is

$$\Delta q = -enAl \tag{16}$$

It takes a time $\Delta t = l/|v_d|$ for all of these electrons to emerge at one end of the wire. Hence the current in the wire is

$$I = \frac{|\Delta q|}{\Delta t} = \frac{enAl}{l/|v_d|} = \frac{e^2 n\tau}{m_e} AE \qquad (17)$$

In this equation we have ignored the negative sign on the electric charge because the direction of the current is to be reckoned according to the convention described in section 28.1 and not according to the motion of the electrons. The current is in the same direction as the electric field.

According to Eq. (17), the current is directly proportional to the magnitude of the electric field. It is also directly proportional to the cross-sectional area of the wire. All of this agrees with our earlier, rough argument.

In terms of the potential difference $\Delta V = El$ across the ends of the wire, we can rewrite Eq. (17) as follows:

$$I = \frac{e^2 n\tau}{m_e} AE = \frac{e^2 n\tau}{m_e} \frac{A}{l} (El) \qquad (18)$$

$$= \left(\frac{e^2 n\tau}{m_e} \frac{A}{l} \right) \Delta V \qquad (19)$$

If we write the factor in parentheses as

$$R = \frac{m_e}{e^2 n\tau} \frac{l}{A} \qquad (20)$$

then Eq. (19) becomes

$$I = \frac{\Delta V}{R} \qquad (21)$$

which is Ohm's Law. Thus, Eq. (20) gives us a theoretical expression for the resistance in terms of the physical parameters associated with the free-electron gas. In practice, Eq. (20) is not very useful because there is no direct method for measuring the time per collision τ.

28.4 The Resistivity of Materials

As we saw in Section 28.2, the resistance of a wire of uniform cross section is related to the resistivity by the formula

$$R = \rho \frac{l}{A} \qquad (22)$$

We can use this formula to calculate the resistance if the resistivity of the material is known, and we can also use it to calculate the resistivity if the resistance is known. The latter calculation is important in the experimental determination of the resistivity of a material, which is done by measuring the potential difference and current in a wire of given length and cross section made of a sample of the material. This means that Ohm's Law is used both as a definition of resistance and as a law relating current, potential difference, and resistance. In the definitions of force (Section 5.2) and of temperature (Section 19.2) we have already encountered similar instances of such dual uses of laws.

As is obvious from Ohm's Law, the unit of resistance is 1 volt/ampere; this unit is called **ohm (Ω),**

ohm, Ω

$$1 \text{ ohm} = 1 \ \Omega = 1 \text{ volt/ampere} \tag{23}$$

The unit of resistivity is the ohm-meter. Table 28.1 gives the resistivities of some conducting materials.

EXAMPLE 2.   A wire commonly used for electrical installations in homes is No. 10 copper wire, which has a radius of 0.129 cm. What is the resistance of a piece of wire 30 m long? What is the potential drop along this wire if it carries a current of 10 A?

SOLUTION: The cross-sectional area of the wire is

$$A = \pi r^2 = \pi (0.129 \times 10^{-2} \text{ m})^2 = 5.2 \times 10^{-6} \text{ m}^2$$

By Eq. (22), the resistance will be

$$R = 1.7 \times 10^{-8} \ \Omega \cdot \text{m} \times 30 \text{ m} / 5.2 \times 10^{-6} \text{ m}^2 = 0.098 \ \Omega \tag{24}$$

For a current of 10 A, Ohm's Law then gives a potential drop

$$\Delta V = IR = 10 \text{ A} \times 0.098 \ \Omega = 0.98 \text{ volt} \tag{25}$$

Note that by combining Eq. (12) and (22) we readily find

$$\frac{I}{A} = \frac{1}{\rho} \frac{\Delta V}{l} \tag{26}$$

Since I/A is the current density and $\Delta V/l$ is the electric field in the conductor, Eq. (26) can be written

$$j = \frac{1}{\rho} E \tag{27}$$

i.e., the current density is directly proportional to the electric field. This is an alternative expression for Ohm's Law. Although our derivation of Ohm's Law began with a wire of uniform cross section, Eq. (27) is valid for conductors of arbitrary shape. This equation relates two *local* quantities: the current density at one point within the conductor and the electric field at that point.

Dependence of resistivity on temperature

The resistivity depends somewhat on temperature. In ordinary metals, the resistivity increases slightly with temperature. This is due to an increase in the rate of collision between electrons and atoms of the lattice — at high temperature the atoms jump violently around their positions in the lattice and they are then more likely to disturb the motion of the electrons. The numbers in the first column of Table 28.1 give the resistivity at room temperature (20°C). The numbers in the second column give the percentage increase in the resistivity per degree Celsius.

Table 28.1 RESISTIVITIES OF METALS[a]

Material	ρ	Increase in ρ per °C
Silver	$1.6 \times 10^{-8}\ \Omega \cdot m$	0.38%
Copper	1.7×10^{-8}	0.39%
Aluminum	2.8×10^{-8}	0.39%
Brass	$\sim 7 \times 10^{-8}$	0.2%
Nickel	7.8×10^{-8}	0.6%
Iron	10×10^{-8}	0.5%
Steel	$\sim 11 \times 10^{-8}$	0.4%
Constantan	49×10^{-8}	0.001%
Nichrome	100×10^{-8}	0.04%

[a] At a temperature of 20°C.

EXAMPLE 3. Suppose that because of a current overload, the temperature of the copper wire of Example 2 increases from 20°C to 90°C. How much does the resistance increase?

SOLUTION: The temperature increase is 70°C. According to Table 28.1, the resistance of copper increases by 0.39% for a temperature increase of 1°C. Hence the resistance increases by $0.39 \times 70 = 27\%$ for a temperature increase of 70°C. The change of resistance is therefore

$$\Delta R = 0.098\ \Omega \times 0.27 = 0.026\ \Omega$$

and the new resistance of the wire will be

$$0.098\ \Omega + 0.026\ \Omega = 0.124\ \Omega$$

At very low temperatures the resistivity of a metal will be substantially less than at room temperature. Some metals, such as lead, tin, zinc, and niobium, exhibit the phenomenon of **superconductivity**: their resistance vanishes completely as the temperature approaches absolute zero. For example, Figure 28.5 shows a plot of resistivity vs. temperature for tin; at a temperature of 3.72 K, the resistivity abruptly vanishes. The resistance of such a superconductor is *exactly* zero. In one experiment, a current of several hundred amperes was started in a superconducting ring; the current continued on its own with undiminished strength for over a year, without any battery or generator to maintain it. In Interlude K we will present a detailed discussion of the properties of superconductors.

According to the definition that we gave in Section 22.4, an ideal insulator is a material that does not permit the motion of electric charge. Real insulators, such as porcelain or glass, do permit some very slight motion of charge. What distinguishes them from conductors is their enormously large resistivity. Typically, the resistivity of insulators is more than 10^{20} times as large as that of conductors (see Table 28.2). This means that even when we apply a high voltage to a piece of glass, the flow of current will be insignificant (provided, of course, that the material does not suffer electrical breakdown). In fact, on a humid day it is likely that more current will flow along the microscopic film of water that tends to form on the surface of the insulator than through the insulator itself.

Superconductivity

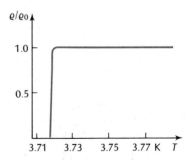

Fig. 28.5 Resistivity of tin as a function of temperature. Below 3.72 K, the resistivity is zero. The resistivity has been expressed as a fraction of the resistivity at 4.2 K, the temperature of liquefaction of helium.

Table 28.2. RESISTIVITIES OF INSULATORS

Material	ρ
Polyethylene	$2 \times 10^{11}\ \Omega \cdot m$
Glass	$\sim 10^{12}$
Porcelain, unglazed	$\sim 10^{12}$
Rubber, hard	$\sim 10^{13}$
Epoxy	$\sim 10^{15}$

28.5 Semiconductors

A semiconductor is a material with a resistivity between that of conductors and insulators. The resistivities of semiconductors vary over a wide range; the resistivities may be 10^4 to 10^{15} times a large as the resistivities of conductors (see Table 28.3).

It is a characteristic feature of semiconductors that the addition of impurities to the material has a drastic effect on the resistivity. For instance, the silicon used in electronic devices is often "doped" with small amounts of arsenic or boron; the addition of just one part per million of arsenic will decrease the resistivity of silicon by a factor of more than 10^5. The manipulation of the resistivity of materials by intentional contamination with carefully selected impurities plays a crucial role in the manufacture of semiconductor devices such as diodes and transistors. Pure semiconductor materials are hardly ever used in practical applications. It is usually the presence of impurities that gives the semiconductor materials their interesting electric properties.

n-type and p-type semiconductors

Semiconductors fall into two categories: n type and p type. In an n-type semiconductor, the carriers of current are free electrons, as in a metal. However, the resistance is higher because the semiconductor has fewer free electrons than a metal. Also, the semiconductor differs from a metal in that the resistivity *decreases* as the temperature increases. This curious behavior is due to an increase in the number of free electrons — as the temperature increases, more electrons shake loose from the atoms of the semiconductor and these extra free electrons more than compensate for the extra friction experienced by each at the higher temperature.

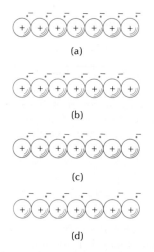

(a)

(b)

(c)

(d)

Fig. 28.6 A row of positive ions (balls marked with +) and electrons (color dots marked with −).

In a p-type semiconductor, the carriers of current are "holes" of positive charge. To understand what this means, consider Figure 28.6 showing an array of electrons and positive ions. In Figure 28.6a, these electrons and ions form neutral atoms. Suppose that the right end of this array is connected to the positive pole of a battery (not shown) and the left end to the negative pole. If the battery pulls an electron out of the right end, it will leave the array with a hole or missing electron at the position of the first atom (Figure 28.6b). The electrons will then play a game of musical chairs: the electron from the next atom will

Table 28.3. RESISTIVITIES OF SEMICONDUCTORS

Material	ρ
Silicon	$2.6 \times 10^3\ \Omega \cdot m$
Germanium	4.2×10^{-1}
Carbon (graphite)	3.5×10^{-5}

jump into this hole, leaving a hole at the position of the second atom (Figure 28.6c); and then the electron from the next atom will jump, etc. The collective motion of the electrons from left to right can be conveniently described as the motion of a hole from right to left. The hole virtually carries positive charge from the right to the left. This is essentially the mechanism for conduction in a *p*-type semiconductor. Instead of a gas of free electrons, this type of semiconductor has a gas of free holes. A flow of current is then a flow of holes and the direction of the current is the same as the direction of motion of the holes.

Semiconductors usually contain both free electrons and free holes. Whether a semiconductor is *n* type or *p* type depends on which kind of charge carrier dominates. The concentration of free electrons and of free holes is largely determined by the impurities that are present in the material. **Donor** impurities consist of atoms that release their valence electrons when placed in the semiconductor and they thereby increase the number of free electrons. **Acceptor** impurities consist of atoms that trap electrons when placed in the semiconductor and they thereby generate holes. Hence, a semiconductor with donor impurities will be *n* type and one with acceptor impurities will be *p* type. For instance, silicon doped with arsenic is an *n*-type semiconductor and silicon doped with boron is a *p*-type semiconductor. Even though the added impurity atoms may only amount to a few parts per million, they completely change the conductivity because the semiconductor has so few current carriers to start with.

Donor and acceptor impurities

28.6 Resistances in Combination

The metallic wires of any electric circuit have some resistance. But in electronic devices — radios, televisions, amplifiers — the main contribution to the resistance is usually due to gadgets that are specifically designed to have a high resistance. These gadgets are **resistors**. They are commonly made out of a short piece of pure carbon (graphite) connected between two terminals (Figure 28.7a). Carbon has a high resistivity and hence a small piece of carbon can have a higher resistance than a long piece of metallic wire. Such resistors obey Ohm's Law (current proportional to potential difference) for a wide range of currents; of course, if the resistor is overloaded with current, it will heat up, possibly even burn, and Ohm's Law will fail.

In circuit diagrams the symbol for a resistor is a zigzag line, reminiscent of the path of an electron inside a wire (Figure 28.7b).

Figure 28.8 shows two resistors connected in *series*. It is intuitively obvious that the net resistance of this combination is the sum of the individual resistances,

$$R = R_1 + R_2 \tag{28}$$

The formal derivation of this result begins with the observation that if the potential differences across the individual resistors are ΔV_1 and ΔV_2, then the net potential difference across the combination is

$$\Delta V = \Delta V_1 + \Delta V_2$$

Furthermore, the currents in both resistors must be exactly the same. Hence, by Ohm's Law

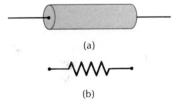

Fig. 28.7 (a) A resistor, consisting of a cylinder of carbon with two terminals attached. (b) Symbol for a resistor in a circuit diagram.

Fig. 28.8 Two resistors connected in series.

$$\Delta V = IR_1 + IR_2 = I(R_1 + R_2)$$

From this it is clear that the combination is equivalent to a single resistance given by Eq. (28).

The generalization of this result to any number of resistors in series (Figure 28.9) is obvious. The net resistance is

Series combination of resistors

$$\boxed{R = R_1 + R_2 + R_3 + \cdots} \tag{29}$$

Figure 28.10 shows two resistors in *parallel*. The potential difference across each resistor is the same as the potential difference across the combination. Hence the currents are

$$I_1 = \frac{\Delta V}{R_1} \quad \text{and} \quad I_2 = \frac{\Delta V}{R_2} \tag{30}$$

The total current through the combination is

$$I = I_1 + I_2 = \frac{\Delta V}{R_1} + \frac{\Delta V}{R_2} = \left(\frac{1}{R_1} + \frac{1}{R_2}\right)\Delta V \tag{31}$$

The combination is therefore equivalent to a single resistance given by

$$\frac{1}{R} = \frac{1}{R_1} + \frac{1}{R_2} \tag{32}$$

Note that the resistance of the combination is less than each of the individual resistances. For example, if $R_1 = R_2 = 1.0\ \Omega$, then $R = 0.5\ \Omega$.

Fig. 28.9 Several resistors connected in series.

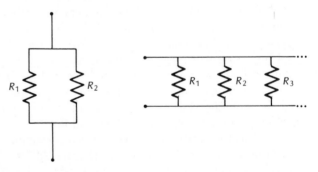

Fig. 28.10 Two resistors connected in parallel.

Fig. 28.11 Several resistors connected in parallel.

The generalization of this to any number of resistors in parallel (Figure 28.11) is again obvious. The net resistance is given by

Parallel combination of resistors

$$\boxed{\frac{1}{R} = \frac{1}{R_1} + \frac{1}{R_2} + \frac{1}{R_3} + \cdots} \tag{33}$$

EXAMPLE 4.   Two resistors, with $R_1 = 10\ \Omega$ and $R_2 = 20\ \Omega$, are connected in parallel (see Figure 28.10). A total current of 1.8 A flows through this combination. What is the potential difference across the combination? What is the current in each resistor?

SOLUTION: According to Eq. (32), the net resistance is given by

$$\frac{1}{R} = \frac{1}{R_1} + \frac{1}{R_2} = \frac{1}{10 \ \Omega} + \frac{1}{20 \ \Omega} = \frac{30}{200 \ \Omega}$$

i.e., $R = 6.67 \ \Omega$. Hence, the potential difference is

$$\Delta V = IR = 1.8 \ \text{A} \times 6.67 \ \Omega = 12.0 \ \text{volt}$$

and the individual currents are

$$I_1 = \Delta V / R_1 = 12.0 \ \text{volt} / 10 \ \Omega = 1.20 \ \text{A}$$

$$I_2 = \Delta V / R_2 = 12.0 \ \text{volt} / 20 \ \Omega = 0.60 \ \text{A}$$

SUMMARY

Electric field in uniform wire: $E = \Delta V / l$

Electric current: $I = dq/dt$

Current density: $j = I/A$

Resistance in terms of resistivity: $R = \rho \dfrac{l}{A}$

Ohm's Law: $I = \dfrac{\Delta V}{R}$

Resistivities: conductors: $\sim 10^{-8} \ \Omega \cdot \text{m}$
semiconductors: $\sim 10^{-4}$ to $10^{7} \ \Omega \cdot \text{m}$
insulators: $\sim 10^{11}$ to $10^{17} \ \Omega \cdot \text{m}$

Series combination of resistors: $R = R_1 + R_2 + R_3 + \cdots$

Parallel combination of resistors: $1/R = 1/R_1 + 1/R_2 + 1/R_3 + \cdots$

QUESTIONS

1. A wire is carrying a current of 15 A. Is this wire in electrostatic equilibrium?

2. Can a current flow in a conductor when there is no electric field? Can an electric field exist in a conductor when there is no current?

3. Figure 28.12 shows a putative mechanical analog of an electric conductor with resistance. A marble rolling down the inclined plane is stopped every so often by a collision with a pin and therefore maintains a constant average velocity v_d. Is this a good analog, i.e., is the average velocity v_d proportional to g?

4. By what factor must we increase the diameter of a wire to decrease its resistance by a factor of 2?

5. Show that for a wire of given length made of a given material, the resistance is inversely proportional to the mass of the wire.

6. What deviations from Ohm's Law do you expect if the current is very large?

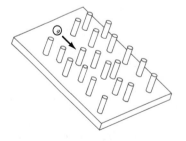

Fig. 28.12

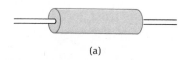

(a)

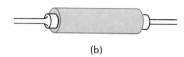

(b)

Fig. 28.13

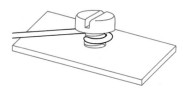

Fig. 28.14

7. Ohm's Law is an approximate statement about the electrical properties of a conducting body, just as Hooke's Law is an approximate statement about the mechanical properties of an elastic body. Is there an electric analog of the elastic limit?

8. An automobile battery has a potential difference of 12 V between its terminals even when there is no current flowing through the battery. Does this violate Ohm's Law?

9. Figure 28.13 shows two alternative experimental arrangements for determining the resistance of a carbon cylinder. A known potential difference is applied via the copper terminals, the current is measured, and the resistance is calculated from $R = \Delta V/I$. The arrangement in Figure 28.13a yields a higher resistance than that in Figure 28.13b. Why?

10. Why is it bad practice to operate a high-current appliance off an extension cord?

11. The installation instructions for connecting an outlet to the wiring of a house recommend that the wire be wrapped at least three-quarters of the way around the terminal post (Figure 28.14). Explain.

12. The **temperature coefficient of resistivity** α is defined as the fractional increase of resistivity per degree Celsius, $\alpha = (1/\rho)\, d\rho/d\mathrm{T}$. According to Table 28.1, what is the value of α for copper?

13. The resistivity of semiconductors and insulators *decreases* with temperature. Can you think of a likely explanation?

14. Figure 19.7 shows a platinum **resistance thermometer** consisting of a coil of fine platinum wire inside a glass tube that can be put in thermal contact with a body. How can this wire be used to determine the temperature of the body?

15. Aluminum wire should never be connected to terminals designed for copper wire. Why not? (Hint: Aluminum has a considerably higher coefficient of thermal expansion than copper.)

16. A wire of copper and a wire of silver are connected in parallel. Both wires have the same length and the same diameter. Which carries more current?

17. Two copper wires are connected in parallel. Both wires have the same length, but one has twice the diameter of the other. What fraction of the total current flows in each wire?

PROBLEMS

Sections 28.1–28.3

1. The effective resistance of a 150-W, 110-V light bulb is 0.73 Ω. What current passes through this light bulb when in operation? How many electrons per second does this amount to?

2. A circular loop of superconducting material has a radius of 2.0 cm. It carries a current of 4.0 A. What is the orbital angular momentum of the moving electrons in the wire? Take the center of the loop as origin.

3. A copper cable in a high-voltage transmission line has a diameter of 3.0 cm and carries a current of 750 A. What is the current density in the wire? What is the electric field in the wire?

4. When the starter motor of an automobile is in operation, the cable connecting it to the battery carries a current of 80 A. This cable is made of copper and is 0.50 cm in diameter. What is the current density in the cable? What is the electric field in the cable?

5. A table lamp is connected to an electric outlet by a copper wire of diameter 0.20 cm and length 2.0 m. Assume that the current through the lamp is 1.5 A, and that this current is steady. How long does it take an electron to travel from the outlet to the lamp?

6. A high-voltage transmission line consists of a copper cable of diameter 3.0 cm and length 250 km. Assume that the cable carries a steady current of 1500 A.
   (a) What is the electric field inside the cable?
   (b) What is the drift velocity of the free electrons?
   (c) How long does it take one electron to travel the full length of the cable?

Section 28.4

7. The electromagnet of a bell is constructed by winding copper wire around a cylindrical core, like thread on a spool. The diameter of the copper wire is 0.45 mm, the number of turns in the winding is 260, and the average radius of a turn is 5.0 mm. What is the resistance of the wire?

8. The following is a list of some types of copper wire manufactured in the United States:

Gauge No.	Diameter
8	0.3264 cm
9	0.2906
10	0.2588
11	0.2305
12	0.2053

For each type of wire, calculate the resistance for a 100-m segment.

9. To measure the resistivity of a metal, an experimenter takes a wire of this metal of diameter of 0.500 mm and length 1.10 m and applies a potential difference of 12.0 V to the ends. He finds that the resulting current is 3.75 A. What is the resistivity?

10. A high-voltage transmission line has an aluminum cable of diameter 3.0 cm, 200 km long. What is the resistance of this cable?

11. In silver, the number of free electrons is $n = 5.8 \times 10^{28}$ per cubic meter.
   (a) Using the value of the resistivity given in Table 28.1, calculate the average time interval between collisions of one of these electrons.
   (b) Assuming that the electric field in a current-carrying silver wire is 8.0 V/m, calculate the average drift velocity of an electron.

12. You want to make a resistor of 1.0 Ω out of a carbon rod of diameter 1.0 mm. How long a piece of carbon do you need?

13. The resistance of the wire in the windings of an electric starter motor for an automobile is 3.0×10^{-2} Ω. The motor is connected to a 12-V battery. What current will flow through the motor when it is stalled (does not turn)?

14. A lightning rod of iron has a diameter of 0.80 cm and a length 0.50 m. During a lightning stroke, it carries a current of 1.0×10^4 A. What is the potential drop along the rod?

15. The air conditioner in a home draws a current of 12 A.
   (a) Suppose that the pair of wires connecting the air conditioner to the fuse box are No. 10 copper wire with a diameter of 0.259 cm and a length of 25 m each. What is the potential drop along each wire? Suppose that the voltage delivered to the home is exactly 110 V at the fuse box. What is the voltage delivered to the air conditioner?
   (b) Some older homes are wired with No. 12 copper wire with a diameter of 0.205 cm. Repeat the calculation of part (a) for this wire.

16. The electromagnet of a bell is wound with 8.2 m of copper wire of diameter 0.45 mm. What is the resistance of the wire? What is the current through the wire if the electromagnet is connected to a 12-V source?

17. The copper cable connecting the positive pole of a 12-V automobile battery to the starter motor is 0.60 m long and 0.50 cm in diameter.
    (a) What is the resistance of this cable?
    (b) When the starter motor is stalled, the current in the cable may be as much as 600 A. What is the potential drop along the cable under these conditions?

18. Although aluminum has a somewhat higher resitivity than copper, it has the advantage of having a considerably lower density. Find the weight of a 100-m segment of aluminum cable 3 cm in diameter. Compare this weight with that of a copper cable of the same length and the same resistance.

19. According to the National Electrical Code, the maximum permissible current in a No. 12 copper wire (diameter 0.21 cm) with rubber insulation is 20 A.
    (a) What is the current density in a wire carrying this current?
    (b) What is the potential drop along a 1-meter segment of the wire?

20. An aluminum wire of length 15 m is to carry a current of 25 A with a potential drop of no more than 5 V along its length. What is the minimum acceptable diameter of this cable?

21. According to safety standards set by the American Boat and Yacht Council, the potential drop along a copper wire connecting a 12-V battery to an item of electrical equipment should not exceed 10%, i.e., it should not exceed 1.2 V. Suppose that a 30-ft wire (length measured around the circuit) carries a current of 25 A; what gauge of wire is required for compliance with the above standard? Use the table of wire gauges given in Problem 8. Repeat the calculation for currents of 35 A and 45 A.

22. A parallel-plate capacitor with a plate area of 8.0×10^{-2} m^2 and a plate separation of 1.0×10^{-4} m is filled with polyethylene. If the potential difference between the plates is 2.0×10^4 V, what will be the current flowing through the polyethylene from one plate to the other?

23. Consider the aluminum cable described in Problem 10. If the temperature of this cable increases from 20°C to 50°C, how much will its resistance increase?

24. What increase of temperature will increase the resistance of a nickel wire from 0.5 Ω to 0.6 Ω?

*25. A solid truncated cone is made of a material of resistivity ρ (Figure 28.15). The cone has a height h, a radius a at one end, and a radius b at the other end. Derive a formula for the resistance of this cone.

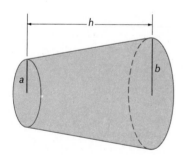

Fig. 28.15

Section 28.6

26. A brass wire and iron wire of equal diameter and of equal length are connected in parallel. Together they carry a current of 6.0 A. What is the current in each?

27. Consider the brass and iron wires described in Problem 26. What is the current in each if they are at a temperature of 90°C instead of a temperature of 20°C?

28. An electric cable of length 12.0 m consists of a copper wire of diameter 0.30 cm surrounded by a cylindrical layer of rubber insulation of thickness 0.10 cm. A potential difference of 6.0 V is applied to the ends of the cable.
    (a) What will be the current in the copper?
    (b) Taking into account the conductivity of the rubber (see Table 28.2), what will be the current in the rubber?

29. A water pipe is made of iron with an outside diameter of 2.5 cm and an inside diameter of 2.0 cm. The pipe is used to ground an electric appliance. If a current of 20 A flows from the appliance into the water pipe, what fraction of this current will flow in the iron? What fraction in the water? Assume that water has a resistivity of $0.01 \ \Omega \cdot m$.

30. A copper wire, of length 0.50 m and diameter 0.259 cm has been accidentally cut by a saw. The region of the cut is 0.4 cm long, and in this region the remaining wire has a cross-sectional area of only one-quarter of the original area. What is the percentage increase of the resistance of the wire caused by this cut?

31. The windings of high-current electromagnets are often made of copper pipe. The current flows in the walls of the pipe and cooling water flows in the interior of the pipe. Suppose the copper pipe has an outside diameter of 1.20 cm and an inside diameter of 0.80 cm. What is the resistance of 30 m of this copper pipe? What voltage must be applied to it if the current is to be 600 A?

32. An underground telephone cable, consisting of a pair of wires, has suffered a short somewhere along its length (Figure 28.16). The telephone cable is 5 km long and in order to discover where the short is, a technician first measures the resistance across the terminals *AB*; then he measures the resistance across the terminals *CD*. The first measurement yields 30 Ω; the second 70 Ω. Where is the short?

Fig. 28.16 A pair of wires with a short at the point *P* (the wires are joined at *P*).

33. The air of the atmosphere has a slight conductivity due to the presence of a few free electrons and positive ions.
   (a) Near the surface of the Earth, the atmospheric electric field has a strength of about 100 V/m and the atmospheric current density is $4 \times 10^{-12} \ A/m^2$. What is the resistivity?
   (b) The potential difference between the ionosphere (upper layer of atmosphere) and the surface of the Earth is 4×10^5 V. What is the total resistance of the atmosphere? (Hint: For the purposes of this problem you may assume that the Earth is flat.)

34. Three resistors with resistances of 3.0 Ω, 5.0 Ω, and 8.0 Ω, are connected in parallel. If this combination is connected to a 12.0-V battery, what is the current through each resistor? What is the current through the combination?

35. A flexible wire for an extension cord for electric appliances is made of 24 strands of fine copper wire, each of diameter 0.053 cm, tightly twisted together. What is the resistance of a length of 1.0 m of this kind of wire?

36. Two copper wires of diameters 0.26 cm and 0.21 cm, respectively, are connected in parallel. What is the current in each if the combined current is 18 A?

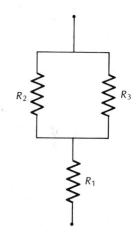

Fig. 28.17

37. Three resistors with $R_1 = 2.0$ Ω, $R_2 = 4.0$ Ω, and $R_3 = 6.0$ Ω are connected as shown in Figure 28.17.
   (a) Find the net resistance of the combination.
   (b) Find the current that passes through the combination if a potential difference of 8.0 V is applied to the terminals.
   (c) Find the potential difference and the current for each individual resistor.

38. Three resistors with $R_1 = 4.0$ Ω, $R_2 = 6.0$ Ω, and $R_3 = 8.0$ Ω are connected as shown in Figure 28.18.
   (a) Find the net resistance of the combination.
   (b) Find the current that passes through the combination if a potential difference of 12.0 V is applied to the terminals.
   (c) Find the potential difference and the current for each individual resistor.

39. Consider the combination of three resistors described in the preceding

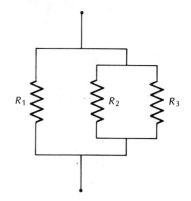

Fig. 28.18

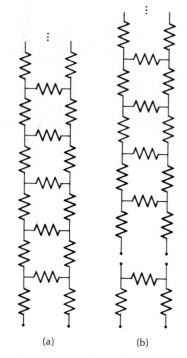

(b)

Fig. 28.19 (a) An infinite "ladder" of resistors. The terminals are marked by a pair of dots. (b) One rung has been cut off from the ladder.

problem. If we want a current of 6.0 A to flow through resistor R_2, what potential difference must we apply to the external terminals?

*40. What is the resistance of an infinite ladder of 1.0-Ω resistors connected as shown in Figure 28.19a? (Hint: The ladder can be regarded as made of two pieces connected in parallel; see Figure 28.19b.)

*41. Twelve resistors, each of resistance R, are connected along the edges of a cube (Figure 28.20). What is the resistance between diagonally opposite corners of this cube?

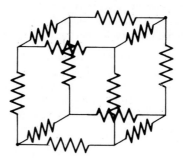

Fig. 28.20 A cube made of resistors.

CHAPTER 29

DC Circuits

If we want to keep a current flowing in a wire, we must connect its ends to a "pump of electricity," such as a battery or generator, that can continuously supply electric charges to one end of the wire and remove them from the other. Figure 29.1 shows a circuit consisting of a wire connected to the terminals of a battery. A current will flow around this circuit; in Figure 29.1 we have indicated the direction of the current according to our convention that it is in the direction of flow of (hypothetical) positive charges. As long as the "strength" of the battery and the resistance of the wire remain constant, the current will also remain constant. Such a steady, time-independent current is called a **direct current**, or DC.

Direct current, DC

29.1 Electromotive Force

The battery must do work on the charges in order to keep them flowing around the circuit shown in Figure 29.1. Suppose that a positive charge is originally at the point P, at one terminal of the battery. Pushed along by the electric field, the charge moves along the wire. On the average, the kinetic energy of the charge does not change — any kinetic energy that the charge gains from the electric field is dissipated by friction within the wire and the charge reaches the point P', at the other terminal of the battery, with its original kinetic energy.

Thus, only the potential energy changes. Since the electric field is directed along the wire, the electric potential steadily decreases with distance along the wire and the charge reaches the point P' with a potential energy lower than its original potential energy. In order to keep the current flowing, the battery must "pump" the charge from the

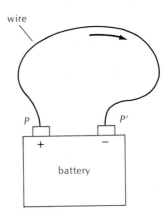

Fig. 29.1 A wire connected between the terminals of a battery.

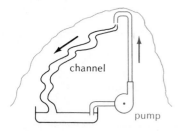

Fig. 29.2 Mechanical analog of the battery–wire circuit of Figure 29.1.

Electromotive force, emf

low-potential terminal to the high-potential terminal, i.e., the battery must supply electric potential energy to the charge. The role of the battery is analogous to that of a hydraulic pump system that lifts water from the bottom to the top of a hill (Figure 29.2). The wire is analogous to a channel by means of which the water runs down the hill returning to the pump. The water then flows in a closed hydraulic circuit, just as charge flows in a closed electric circuit. The hydraulic pump of Figure 29.2 can be regarded as a source of gravitational potential energy — it produces this energy from an external supply of chemical or mechanical energy. Likewise, the "pump of electricity" of Figure 29.1 can be regarded as a source of electric potential energy — it produces this energy from a supply of chemical energy.

To measure the "strength" of a source of electric potential energy, we introduce the concept of **electromotive force**, or **emf**. The emf of a source of electric potential energy is defined as *the amount of electric energy delivered by the source per couloumb of positive charge as this charge passes through the source from the low-potential terminal to the high-potential terminal.* Since the emf is energy per unit charge, its units are volts.

If a steady, time-independent current carries one coulomb of charge around the circuit of Figure 29.1 from P to P' along the wire and from P' to P through the source of electric potential energy, then the energy that this charge receives from the source must exactly match the energy it loses within the wire. If so, the charge returns to its starting point with exactly the same energy it had originally and it can repeat this round trip again and again in exactly the same manner. We can write this energy balance as

$$\mathcal{E} + \Delta V = 0 \tag{1}$$

where $\mathcal{E}$ represents the emf, or the increase of potential energy, due to the source and ΔV represents the decrease of potential energy along the wire ($\mathcal{E}$ is positive and ΔV is negative).

According to Eq. (1), the emf $\mathcal{E}$ has the same magnitude as the potential drop in the external circuit connected between the terminals of the source. For example, a battery with an emf of 1.5 V connected to an external circuit will do 1.5 J of work on a coulomb of positive charge that passes through the battery in a forward direction (from the − terminal to the + terminal) and the resistors and other devices in the external circuit will do −1.5 J of work on the charge as it flows around this circuit, from the + terminal to the − terminal. Because of the equality between the emf $\mathcal{E}$ and the potential drop ΔV, the emf is often *Voltage* simply called the **voltage** of the source.

EXAMPLE 1. A fresh flashlight battery with a voltage of 1.5 V will deliver a current of 1 A for about 1 h before running down. How much work does the battery do in this time interval?

SOLUTION: The battery does 1.5 J of work on each coulomb that passes through. If the current is 1 A, the charge that passes through in 1 h is 1 A × 3600 s = 3600 C, and the total work is 1.5 J/C × 3600 C = 5400 J.

Note that if one coulomb of positive charge is forced through the battery in the reverse direction (from the + terminal to the − terminal), then the battery will take electric potential energy away from the charge. The charge will then emerge from the battery at a potential that is 1.5 volt lower than the potential with which it entered. The energy taken away from the charge will either be stored within the battery (if it is a reversible battery) or else it will merely be wasted as heat within the battery (if it is an irreversible battery).

29.2 Sources of Electromotive Force

The most important kinds of sources of emf are batteries, electric generators, fuel cells, and solar cells. We will now briefly discuss each of these.

BATTERIES   These sources of emf convert chemical energy into electric energy. A very common type of battery is the **lead–acid battery** which finds widespread use in automobiles. In its simplest form, this battery consists of two plates of lead — the positive electrode and the negative electrode — immersed in a solution of sulfuric acid (Figure 29.3). The positive electrode is covered with a layer of lead dioxide, PbO_2. When the external circuit is closed, the following reactions, already mentioned in Section 22.3, take place at the immersed surfaces of the negative and positive electrodes, respectively:

$$Pb + SO_4^{--} \rightarrow PbSO_4 + 2e^- \tag{2}$$

$$PbO_2 + SO_4^{--} + 4H^+ + 2e^- \rightarrow PbSO_4 + 2H_2O \tag{3}$$

Lead–acid battery

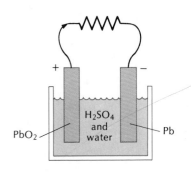

Fig. 29.3 A lead–acid battery.

These reactions deposit electrons on the negative electrode and absorb electrons from the positive electrode. Thus, the battery acts as a pump for electrons — the negative electrode is the outlet, the positive electrode is the intake, and the electrons flow from one to the other via the external circuit.

The reactions (2) and (3) deplete the sulfuric acid in the solution and they deposit lead sulfate on the electrodes. The depletion of sulfuric acid finally halts the reaction — the battery is then "discharged."

The lead–acid battery can be "charged" by simply passing a current through it in the backward direction. This reverses the reactions (2) and (3) and restores the sulfuric acid solution. Note that what is stored in the battery during the "charging" process is not electric charge, but chemical energy. The number of positive and negative electric charges (protons and electrons) in the battery remains constant; what changes is the concentration of chemical compounds. A "charged" battery contains chemical compounds (lead, lead dioxide, sulfuric acid) of relatively high internal energy; a "discharged" battery contains chemical compounds (lead sulfate, water) of lower internal energy.

The single-cell battery shown in Figure 29.3 has an emf of 2.0 V. In an automobile battery, six such cells are stacked together and connected in series to give an emf of 12.0 V. The energy stored in such a battery is typically about 0.5 kW · h.[1] Large banks of batteries, weighing several hundred tons, for the propulsion of submarines store more than 5×10^3 kW · h.

Another familiar type of battery is the **dry cell.** The positive electrode consists of maganese dioxide and the negative electrode of zinc.

Dry cell

---

[1] Recall that 1 kW · h = 1000 W × 3600 s = 3.6×10^6 J.

The electrolyte in which these electrodes are "immersed" is a moist paste of ammonium chloride and zinc chloride. The chemical reactions at the electrodes convert chemical energy into electric energy and pump electrons from one electrode to the other via the external circuit. The emf of such a dry cell is 1.5 V. Since there is no liquid to slosh around, these batteries are particularly suitable for portable devices. The energy stored in an ordinary flashlight battery is typically of the order of 2×10^{-3} kW $\cdot$ h.

ELECTRIC GENERATORS   Generators convert mechanical energy (kinetic energy) into electric energy. Their operation involves magnetic fields and the phenomenon of induction. We will leave the description of electric generators for Section 32.3.

Fuel cell   FUEL CELLS   These resemble batteries in that they convert chemical energy into electric energy. However, in contrast to a battery, neither the high-energy chemicals nor the low-energy reaction products are kept inside the fuel cell. The former are supplied to the fuel cell from external tanks and the latter are ejected. Essentially, the fuel cell acts as a combustion chamber in which controlled combustion takes place. The fuel cell "burns" a high-energy fuel, but produces electric energy rather than heat energy.

Figure 29.4a shows a fuel cell that "burns" a hydrogen–oxygen fuel. The electrodes of the fuel cell are hollow cylinders of porous carbon; oxygen at high pressure is pumped into the positive electrode and hydrogen into the negative electrode. The electrodes are immersed in a potassium-hydroxide electrolyte. The reactions at the negative and positive electrode are, respectively,

$$2H_2 + 4OH^- \rightarrow 4H_2O + 4e^- \tag{4}$$

$$O_2 + 2H_2O + 4e^- \rightarrow 4OH^- \tag{5}$$

These reactions deposit electrons on the negative electrode and remove electrons from the positive electrode. This pumps electrons from one electrode to the other via the external circuit.

Note that the net result of the sequence of reactions (4) and (5) is the conversion of oxygen and hydrogen into water. This reaction is the reverse of the electrolysis of water (decomposition of water by an electric

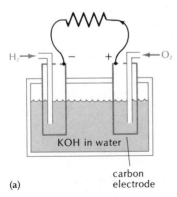

(a)

(b)

Fig. 29.4 (a) Schematic diagram of a fuel cell. (b) Fuel cell used on Skylab.

current). The excess water is removed from the cell in the form of water vapor.

All fuel cells produce a certain amount of waste heat. The best available fuel cells convert about 45% of the chemical energy of the fuel into electric energy and they waste the remainder. Fuel cells are still at an experimental stage; however, they have already been put to use as practical power sources aboard the Apollo spacecraft and on Skylab (Figure 29.4b). They are compact and clean; on Skylab the waste water eliminated from the fuel cell was used both for drinking and for showers.

SOLAR CELLS   Solar cells convert the energy of sunlight directly into electric energy. They are made of thin wafers of a semiconductor, such as silicon. Figure 29.5a shows a cross section through a solar cell. It consists of a central core of n-type silicon surrounded by an outer layer of p-type silicon. The outer layer is very thin, only about 10^{-4} cm, and when sunlight penetrates this layer and reaches the boundary between the n-type silicon and the p-type silicon, it pumps electrons from the latter into the former.

Solar cell

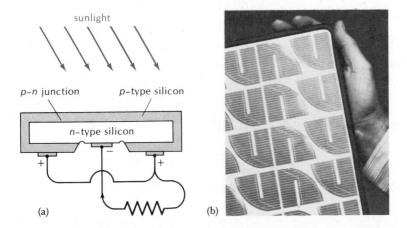

(a)   (b)

Fig. 29.5 (a) Schematic diagram of a solar cell. (b) An array of solar cells in a panel designed for charging automobile batteries.

The boundary between the two types of silicon is called a **p–n junction.** To explain why sunlight causes a current to flow across this junction, we must begin with a description of the behavior of electrons and holes. In n-type silicon the current carriers are electrons (negative charges), whereas in p-type silicon the current carriers are holes (positive charges). Figure 29.6a shows a piece of silicon of each type. When

p–n junction

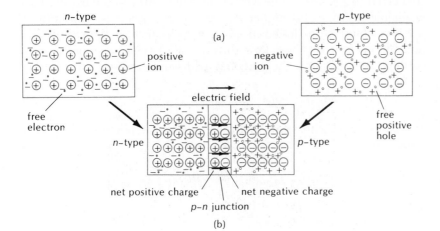

Fig. 29.6 (a) An n-type silicon and p-type silicon. (b) A p–n junction.

these two pieces are joined, some electrons will diffuse across the boundary into the *p*-type material and some holes into the *n*-type material. This diffusion comes to a halt because the accumulated charges on the two sides of the boundary repel any other charges that attempt to cross. When equilibrium is established, the *p*-type material has a layer of negative charge (electrons) and the *n*-type material a layer of positive charge (holes). Between these layers there is a strong electric field (Figure 29.6b); this is an instance of a "contact" electric field, such as mentioned in Section 24.4.

The current of the solar cell is generated by the action of sunlight on the atoms within the region of this electric field of the *p–n* junction. When the sunlight strikes one of these atoms it will ionize it, i.e., it will kick an electron out of the atom. This process amounts to the creation of a free electron and of a free hole. Under the influence of the electric field at the *p–n* junction, the electron accelerates toward the *n* side of the junction and the hole toward the *p* side. The net effect is a positive current from the *n* side toward the *p* side. In the external circuit, the current will then flow from the *p* terminal to the *n* terminal (see Figure 29.5a), i.e., the former acts as the positive pole of a battery and the latter as the negative pole.

The emf of a silicon solar cell is about 0.6 V. However, the current that can be extracted is rather small. Even in full sunlight a single solar cell of surface area 5 cm² will only deliver 0.1 ampere. Solar cells are not very efficient; only about 11% of the energy of sunlight gets converted into electric energy. Large panels, containing very many solar cells, are needed for the generation of appreciable amounts of electric power (see Figures 29.5a and F.13).

29.3 Single-Loop Circuits

In schematic diagrams of electric circuits, a source with a time-independent emf is represented by a stack of parallel thick and thin lines suggesting the plates of a lead–acid battery. The high-potential terminal carries a plus sign and the low-potential terminal a minus sign. If the terminals of such a source are connected to a network of resistances, a steady current, also called a direct current or DC, will flow through the network.

Figure 29.7 shows a very simple circuit consisting of a source of emf connected to a resistor. The wires from the resistor to the battery have negligible resistance (if higher accuracy is required, the resistance of the wires must be included in the total resistance of the circuit). To find the current that will flow through the circuit, we note that according to Ohm's Law the potential drop across the resistor must be

$$\Delta V = -IR \tag{6}$$

Fig. 29.7 A simple circuit with a source of emf and a resistor.

But, from Eq. (1), we know that the emf plus the potential drop must equal zero:

$$\mathscr{E} - IR = 0 \tag{7}$$

or

$$I = \frac{\mathscr{E}}{R} \tag{8}$$

Equation (7) is an instance of **Kirchhoff's rule** which states that *around any closed loop the sum of all the emfs and all the potential drops across resistors and other circuit elements must equal zero.*[2] In this sum, the emf of a source is reckoned as positive if the current flows through the source in the forward direction and negative if in the backward direction.

The proof of this general rule is similar to the proof of Eq. (1). If one coulomb of positive charge flows once around the closed loop, it will gain potential energy while passing through each emf and lose potential energy while passing through each resistor. Under steady conditions, the sum of gains and losses must equal zero, since the charge must return to its starting point with no change of energy. In this sum, the emf must be reckoned as negative (loss of potential energy) whenever the charge flows through the source in the backward direction (from the + terminal to the − terminal).

Kirchhoff's (second) rule

Gustav Robert Kirchhoff (keerkh-hoff), *1824–1887, German physicist, professor at Heidelberg and at Berlin. Mainly known for his development of spectroscopy, he also made many important contributions to mathematical physics, among them, his first and second rules for circuits.*

EXAMPLE 2. Figure 29.8 shows a circuit with two batteries and two resistors. The emfs of the batteries are $\mathscr{E}_1 = 12$ V and $\mathscr{E}_2 = 15$ V; the resistances are $R_1 = 4\ \Omega$ and $R_2 = 2\ \Omega$. What is the current in the circuit?

SOLUTION: To apply Kirchhoff's rule, we must decide in which direction the current flows around the loop. We will arbitrarily assume that the current flows in the clockwise direction. If this hypothesis is wrong, our calculation of the current will yield a negative value and this will indicate that the direction of the actual current is opposite to the hypothetical current.

The sum of all emfs and all potential drops across resistors is

$$\mathscr{E}_1 - IR_1 - \mathscr{E}_2 - IR_2 = 0 \tag{9}$$

Note that $\mathscr{E}_2$ enters with a negative sign into this equation since the hypothetical current passes through this source of emf in the backward direction. The solution of Eq. (9) leads to

$$I = \frac{\mathscr{E}_1 - \mathscr{E}_2}{R_1 + R_2} \tag{10}$$

or

$$I = \frac{12\ \text{V} - 15\ \text{V}}{4\ \Omega + 2\ \Omega} = -0.5\ \text{A} \tag{11}$$

The negative sign indicates that the current is *not* clockwise but counterclockwise. (We could of course have guessed that much. The stronger battery on the right will force the current backward through the weaker battery on the left.)

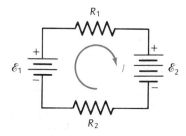

Fig. 29.8 Two sources of emf and two resistors.

Incidentally, in Example 2 we have neglected the internal resistance of the batteries. The electrolyte in a battery always has some resistance and this causes the current to suffer a voltage drop even before it leaves the external terminal of the battery. The nominal emf $\mathscr{E}$ quoted

---

[2] This is often called Kirchhoff's second rule. We will become acquainted with his first rule in Section 29.4.

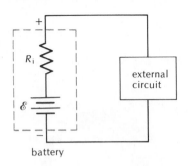

Fig. 29.9 The internal resistance is in series with the nominal emf.

on the labels of batteries refers to the potential difference between the terminals when no current, or only an infinitesimal current, is flowing. The internal resistance R_i of the battery may be regarded as connected in series with this nominal emf (Figure 29.9). When a current I is flowing, the voltage drops by $\Delta V = -IR_i$ across this internal resistance and hence the remaining voltage at the external terminals of the battery will be $\mathscr{E} - IR_i$. The internal resistance of a good battery is small and can often be neglected. But if we need to take it into account, we can do so by simply placing the appropriate internal resistance in series with each battery in the circuit diagram; then we can proceed with the usual calculation of the currents.

EXAMPLE 3.   A flashlight battery of a nominal emf 1.5 V has an internal resistance of 0.05 Ω. What will be the potential difference across its terminals if the battery is delivering a current of 2 A? A current of 10 A?

SOLUTION: For a current of 2 A, the voltage across the terminals will be

$$\mathscr{E} - IR_i = 1.5 \text{ V} - 2 \text{ A} \times 0.05 \text{ Ω} = 1.4 \text{ V}$$

For a current of 10 A, the voltage across the terminals will be

$$\mathscr{E} - IR_i = 1.5 \text{ V} - 10 \text{ A} \times 0.05 \text{ Ω} = 1.0 \text{ V}$$

29.4 Multiloop Circuits

If several sources of emf and several resistors are connected in some complicated circuit with branches and loops, then the currents will flow along several alternative paths. To find these currents, we must solve a simultaneous set of several equations with the currents as unknowns.

The procedure for obtaining the necessary equations is as follows:

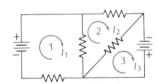

Fig. 29.10 A multiloop circuit. The currents I_1, I_2, I_3 are regarded as flowing in closed loops. Note that loops 1 and 2 share the middle resistor; and loops 2 and 3 share the diagonal resistor. The net current in the middle resistor is $I_1 - I_2$, and the net current in the diagonal resistor is $I_2 - I_3$.

Loop method

a. Regard the given circuit as a collection of closed current loops. The loops may overlap, but each loop must have at least one portion that does not overlap with other loops (Figure 29.10).
b. Label the currents in the loops $I_1, I_2, I_3, \cdots$, and arbitrarily assign a direction to each of these currents.
c. Apply Kirchhoff's rule to each loop: the sum of all the emfs and of all the potential drops across resistors must add to zero around each loop. Note that when calculating the potential drop across a resistor, we must take the product of the resistance and the *net* current through the resistor; if the resistor belongs to two adjacent loops, then the *net* current is the algebraic sum of the two loop currents (Figure 29.10).

This procedure, called the **loop method,** will result in the right number of equations for the unknown currents $I_1, I_2, I_3, \cdots$. We can then solve the equations for the unknowns by the standard mathematical methods for the solution of a system of equations with several unknowns. If a current turns out to be negative, its direction is opposite to the direction assigned in step (b).

EXAMPLE 4. Figure 29.11 shows a circuit with several batteries and resistors. Find the current in each of the resistors.

SOLUTION: Obviously, this circuit can be regarded as consisting of the two loops indicated by the arrows. We label the currents in the loops I_1 and I_2; these symbols have been written next to the arrows that indicate the directions. When calculating the potential drop in the resistor R_2 (which is included in both loops) we must take into account that both loop currents flow through R_2 simultaneously, in opposite directions. The net current through R_2 in the direction of the arrow of loop 1 is therefore $I_1 - I_2$. With this, Kirchhoff's rule for loop 1 yields

$$\mathscr{E}_1 - I_1 R_1 - (I_1 - I_2)R_2 = 0 \qquad (12)$$

Likewise, the net current through R_2 in the direction of the arrow of loop 2 is $I_2 - I_1$; and Kirchhoff's rule for loop 2 yields

$$-\mathscr{E}_2 - I_2 R_3 - (I_2 - I_1)R_2 = 0 \qquad (13)$$

This gives us two equations for the two unknowns I_1 and I_2.

Before proceeding to the solution of these equations it is convenient to substitute the numerical values for the known quantities $\mathscr{E}$ and R. For instance, if $\mathscr{E}_1 = 12.0$ V, $\mathscr{E}_2 = 8.0$ V, $R_1 = 4.0\ \Omega$, $R_2 = 4.0\ \Omega$, and $R_3 = 2.0\ \Omega$, then Eqs. (12) and (13) become

$$12 - 8I_1 + 4I_2 = 0 \qquad (14)$$

$$-8 - 6I_2 + 4I_1 = 0 \qquad (15)$$

with the solution

$$I_1 = 1.25 \text{ A} \qquad I_2 = -0.50 \text{ A} \qquad (16)$$

The negative sign on I_2 indicates that the current in the second loop is opposite to the direction shown in Figure 29.11.

The current in the resistor R_1 is then $I_1 = 1.25$ A; the current in the resistor R_2 is $I_1 - I_2 = 1.75$ A; and the current in the resistor R_3 is $I_2 = -0.50$ A.

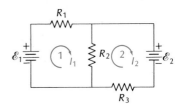

Fig. 29.11 The loops 1 and 2 share the resistor R_2. The currents I_1 and I_2 flow through R_2 in opposite directions; the net current through R_2 is therefore the difference between I_1 and I_2.

An alternative procedure for obtaining the equations for a multiloop circuit makes use of **Kirchhoff's first rule** which states that *the sum of all the currents entering any branch point of the circuit (where a wire merges with another wire) must equal the sum of currents leaving.* This rule expresses charge conservation — the amount of charge entering any branch point in some time interval equals the amount of charge leaving. Our previous procedure for circuits contained Kirchhoff's first rule implicitly, and hence we had no occasion to use this rule explicitly. The alternative procedure uses this Kirchhoff rule explicitly, as follows:

Kirchhoff's first rule

a. Regard the given circuit as a collection of closed current loops (Figure 29.12).
b. Label the currents in the distinct branches I_1, I_2, I_3, . . . , and arbitrarily assign a direction to each of these currents.
c. Apply Kirchhoff's rule, now called Kirchhoff's second rule, to each loop: the sum of all the emfs and of all the potential drops across resistors must add to zero around each loop.

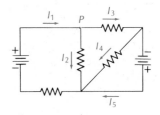

Fig. 29.12 A multiloop circuit. The currents in the distinct branches of the circuit are I_1, I_2, I_3, I_4, and I_5. The point P is a branch point, where wires merge.

d. Apply Kirchhoff's first rule to the branch points: the sum of all the currents entering a branch point must equal the sum of the currents leaving.

Branch method This procedure, called the **branch method**, will result in the right number of equations for the unknowns I_1, I_2, I_3, ...

EXAMPLE 5.   Again consider the circuit of Figure 29.11. Repeat the calculation of the current in the resistors of this circuit using the branch method.

SOLUTION: The circuit has three distinct branches; we label the currents in these branches I_1, I_2, and I_3 (Figure 29.13). Kirchhoff's second rule for loop 1 yields

$$\mathscr{E}_1 - I_1 R_1 - I_3 R_2 = 0 \tag{17}$$

Likewise, for loop 2

$$-\mathscr{E}_2 - I_2 R_3 + I_3 R_2 = 0 \tag{18}$$

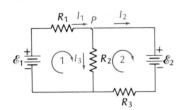

Fig. 29.13 The currents in the distinct branches of this circuit are I_1, I_2, and I_3.

The last term in Eq. (18) has been given a positive sign because the direction of the current I_3 is *opposite* to the direction of the arrow of loop 2.

Kirchhoff's first rule applied to the branch point P (Figure 29.13) yields

$$I_1 = I_2 + I_3 \tag{19}$$

Equations (17), (18), and (19) are three equations for the three unknowns I_1, I_2, and I_3. Note that in this example, the branch method gives us three equations with three unknowns, whereas the loop method gave us two equations with two unknowns. In general, the branch method always suffers from the disadvantage of giving us more unknowns and more equations. Solving Eq. (19) for I_3, we find

$$I_3 = I_1 - I_2 \tag{20}$$

When we substitute this into Eqs. (17) and (18), we obtain Eqs. (12) and (13) — the same equations for I_1 and I_2 as before. Hence the final answers obtained by the branch method are the same as those obtained by the loop method.

29.5 Energy in Circuits; Joule Heat

As we have seen in Section 29.1, to keep a current flowing in a circuit, the batteries or other sources of emf must do work. If an amount of charge dq passes through a source of an emf $\mathscr{E}$, the amount of work done will be

$$dW = \mathscr{E}\, dq \tag{21}$$

Hence the rate at which the source does work is

$$\frac{dW}{dt} = \mathscr{E}\,\frac{dq}{dt} \tag{22}$$

The rate of work is the power; the rate of flow of the charge is the current. Equation (22) therefore asserts that the electric power delivered by the source of emf to the current is

$$P = \mathscr{E}I \tag{23}$$

Power delivered by source of emf

Note that in Eq. (23) we have not yet taken into account the algebraic sign of the power. Obviously, we will have to attach a positive sign to the power if the current passes through the source in the forward direction and a negative sign if the current passes through in the backward direction. In the former case the source delivers energy to the current and in the latter case the source receives energy from the current.

EXAMPLE 6. What power do the two batteries described in Example 4 deliver?

SOLUTION: The current through the first battery is 1.25 A. This current passes through the battery in the forward direction; hence the power delivered by the battery is positive,

$$P = 1.25 \text{ A} \times 12.0 \text{ V} = 15 \text{ W}$$

Likewise, the power delivered by the other battery is

$$P = 0.50 \text{ A} \times 8.0 \text{ V} = 4.0 \text{ W}$$

The net power delivered by both batteries is 19 W.

The electric energy acquired by the charges is continually dissipated in the resistors. If within a given resistor the charge dq suffers a potential drop ΔV (regarded as a positive quantity), then the loss of potential energy is $dU = \Delta V \, dq$ and the rate at which energy is dissipated is

$$\frac{dU}{dt} = \Delta V \frac{dq}{dt} \tag{24}$$

i.e., the power dissipated in the resistor is

$$P = (\Delta V) \, I \tag{25}$$

Power dissipated in resistor

By means of Ohm's Law, $\Delta V = IR$, we can also write this power as

$$P = I^2 R \tag{26}$$

or as

$$P = (\Delta V)^2 / R \tag{27}$$

The energy lost by the charges during their passage through a resistor generates heat, i.e., it generates random microscopic motions of

Joule heat

the atoms. This conversion of electric energy into thermal energy in a resistor is called **Joule heating.**

In any circuit consisting of several sources of emf and several resistors with steady currents, the total power delivered by the sources of emf must of course equal the total power dissipated in the resistors. This equality is a direct consequence of Kirchhoff's rule. The net result of the flow of current in such a circuit is therefore a conversion of energy of the sources of emf into an equal amount of energy of heat.

EXAMPLE 7.   What is the rate at which Joule heat is produced in the resistors of Example 4?

SOLUTION: The resistances are $R_1 = 4.0\ \Omega$, $R_2 = 4.0\ \Omega$, and $R_3 = 2.0\ \Omega$; the corresponding currents are 1.25 A, 1.75 A, and 0.50 A. Equation (26) then gives the power dissipated in each resistor,

$$P_1 = (1.25\ \text{A})^2 \times 4.0\ \Omega = 6.25\ \text{W}$$

$$P_2 = (1.75\ \text{A})^2 \times 4.0\ \Omega = 12.25\ \text{W}$$

$$P_3 = (0.50\ \text{A})^2 \times 2.0\ \Omega = 0.50\ \text{W}$$

Note that the net power dissipated is 19.0 W, which agrees with the net power delivered by the batteries (see Example 6).

EXAMPLE 8.   A high-voltage transmission line that connects a city to a power plant consists of a pair of copper wires, each with a resistance of 4 Ω. (a) The transmission line delivers to the city 1.7×10^5 kW of power at 2.3×10^5 V. What is the current in the transmission line? How much power is lost as Joule heat in the transmission line? (b) If the transmission line were to deliver the same 1.7×10^5 kW of power at 110 V, how much power would be lost as Joule heat? Is it more efficient to transmit power at high voltage or at low voltage?

SOLUTION: (a) Figure 29.14 shows the circuit consisting of power plant, transmission lines, and city. In terms of the power and the voltage delivered to the city, the current through the city is

$$I = \frac{P_{\text{delivered}}}{\Delta V_{\text{delivered}}} = \frac{1.7 \times 10^8\ \text{W}}{2.3 \times 10^5\ \text{V}} = 7.4 \times 10^2\ \text{A}$$

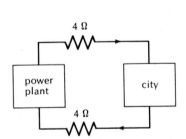

4 Ω

4 Ω

power plant

city

Fig. 29.14 Circuit diagram for a high-voltage transmission line connecting a city to a power plant.

The current in both transmission lines must of course be the same. The combined resistance of both lines is 4 Ω + 4 Ω = 8 Ω and hence the power lost in the transmission lines is

$$P_{\text{lost}} = I^2 R = (7.4 \times 10^2\,\text{A})^2 \times 8\ \Omega = 4.4 \times 10^6\ \text{W}$$

Thus, the power lost is 3% of the power delivered.

(b) For $\Delta V_{\text{delivered}} = 110$ V, the current is

$$I = \frac{1.7 \times 10^8\ \text{W}}{110\ \text{V}} = 1.6 \times 10^6\ \text{A}$$

and the power lost is

$$P_{\text{lost}} = I^2 R = (1.6 \times 10^6\ \text{A})^2 \times 8\ \Omega = 1.9 \times 10^{13}\ \text{W}$$

Thus, the power lost is much larger than the power delivered! Obviously, transmission at high voltage is much more efficient than transmission at low voltage.

29.6 The Hazards of Electric Currents

As a side effect of the widespread use of electric machinery and devices in factories and homes, each year in the United States about 1000 people die by accidental electrocution. A much larger number suffer nonfatal electric shocks. Fortunately, the human skin is a fairly good insulator, which provides a protective barrier against injurious electric currents. The resistance of a square centimeter of dry human epidermis in contact with a conductor can be as much as 10^5 Ω. However, the resistance varies in a rather sensitive way with the thickness, moisture, and temperature of the skin, and with the magnitude of the potential difference.[3]

The electric power supplied to factories and homes in the United States is usually in the form of alternating currents, or AC. These are oscillating currents, which periodically reverse direction (the standard period for the alternating current supplied by power companies is $\frac{1}{60}$ s). Since most accidental electric shocks involve alternating currents, the following discussion of the effects of currents on the human body will emphasize alternating currents. In the typical accidental electric shock, the current enters the body through the hands (in contact with one terminal of the source of emf) and exits through the feet (in contact with the ground, which constitutes the other terminal of the source of emf in most AC circuits). Thus the body plays the role of a resistor, closing an electric circuit.

The damage to the body depends on the magnitude of the current passing through it. An alternating current of about 0.001 A produces only a barely detectable tingling sensation. Higher currents produce pain and strong muscular contractions. If the victim has grasped an electric conductor with the hand, the muscular contraction may prevent the victim from releasing the hold on the conductor. The magnitude of the "let-go" current, at which the victim can barely release the hold on the conductor, is about 0.01 A. Higher currents lock the victim's hand to the conductor. Unless the circuit is broken within a few seconds, the skin in contact with the conductor will then suffer burns and blisters. Such damage to the skin drastically reduces its resistance, which can lead to a fatal increase of the current.

An alternating current of about 0.02 A flowing through the body from the hands to the feet produces a contraction of the chest muscles that halts breathing; this leads to death by asphyxiation if it lasts for a few minutes. A current of about 0.1 A lasting just a few seconds induces fibrillation of the heart. This is a rapid, uncoordinated flutter of the heart muscles, with cessation of the natural rhythm of the heartbeat and cessation of the pumping of blood. Fibrillation usually continues even when the victim is removed from the electric circuit; the

---

[3] The variation of resistance with potential difference implies that skin does not obey Ohm's Law.

consequences are fatal unless immediate medical assistance is available. The treatment for fibrillation involves the deliberate application of a severe electric shock to the heart by means of electrodes placed against the chest; this arrests the motion of the heart completely. When the shock ends, the heart usually resumes beating with its natural rhythm.

A current of a few amperes produces a block of the nervous system and paralysis of the respiratory muscles. Victims of such currents can sometimes be saved by prompt recourse to artificial respiration. At these high values of the current, the effects of AC and DC are not very different. But at lower values, a DC current poses less of a hazard than a comparable AC current, because the former does not trigger the strong muscular contractions triggered by the latter.

In the above we assumed that the path of the current through the body is from the hands to the feet. If the current enters and exits through the same arm or leg, or enters through one leg and exits through the other, no vital organs lie in its path, and the threat to life is lessened. However, an intense current through a limb tends to kill the tissue through which it passes, and may ultimately require the surgical excision of large amounts of dead tissue, and even the amputation of the limb.

Other things being equal, a higher voltage will result in a higher current. The hazard posed by contact with high-voltage sources is therefore obvious. But under exceptional circumstances, even sources of low voltage can be hazardous. Several cases of electrocution by contact with sources of a voltage as low as 12 V have been reported. It seems that in these cases death resulted from an unusually sensitive response of the nervous system; it is also conceivable that an unusually small skin resistance was a contributing factor. Thus, it is advisable to treat even sources of low voltage with respect!

First aid for electric shock   Aid to victims of electric shock should begin with switching off the current. When no switch, plug, or fuse for cutting off the current is accessible, the victim must be pushed or pulled away from the electric conductor by means of a piece of insulating material, such as a piece of *dry* wood or a rope. The rescuer must be careful to avoid electric contact. If the victim is not breathing, artificial respiration must be started at once. If there is no heartbeat, cardiac massage must be applied by trained personnel until the victim can be treated with a defibrillation apparatus.

SUMMARY

Kirchhoff's (second) rule: The sum of emfs and voltage drops around any closed loop in a circuit must equal zero.

Power delivered by source of emf: $P = I\mathscr{E}$

Power dissipated by resistor: $P = I\,\Delta V$

QUESTIONS

1. Can we use a capacitor as a pump of electricity in a circuit? In what way would such a pump differ from a battery?

2. Does a fully charged battery have the same emf as a partially charged battery?

3. The emf of a battery is sometimes called the "open-circuit voltage." Explain.

4. How would you measure the internal resistance of a battery?

5. Kirchhoff's second rule is equivalent to energy conservation in the electric circuit. Explain.

6. The **ammeter** is a device for measuring electric current. To measure the electric current at a point P in a circuit, the ammeter must be inserted as shown in Figure 29.15b. The ammeter has a very low internal resistance, so that it does not inhibit the flow of current. What would happen if we were to connect the ammeter incorrectly, as shown in Figure 29.15c?

Ammeter

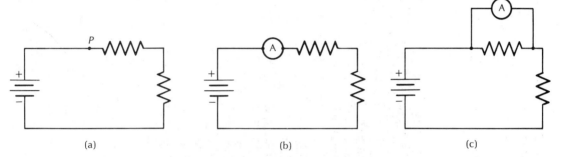

| (a) | (b) | (c) |

Fig. 29.15 (a) An electric circuit. (b) Correct connection of the ammeter. (c) Incorrect connection of the ammeter.

7. The **voltmeter** is a device for measuring differences of potential. To measure the potential difference between points P and P' in a circuit, the voltmeter must be connected as shown in Figure 29.16a. The voltmeter has a very high internal resistance, so that it does not permit the redistribution of the current in the circuit. What would happen if we were to connect the voltmeter incorrectly, as shown in Figure 29.16b?

Voltmeter

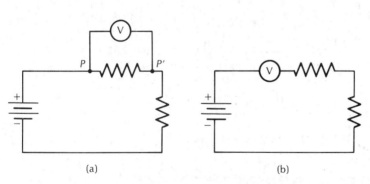

| (a) | (b) |

Fig. 29.16 (a) Correct connection of the voltmeter. (b) Incorrect connection of the voltmeter.

8. A mechanic determines the internal resistance of an automobile battery by connecting a rugged, high-current ammeter (of nearly zero resistance) directly across the poles of the battery. The internal resistance is inversely proportional to the ammeter reading, $R_i \propto 1/I$. Explain.

9. Precise comparisons of the voltages of two sources of emf can be performed with the circuit shown in Figure 29.17, called a **potentiometer.** In this circuit, $\mathcal{E}_s$ is a known, standard source of emf and $\mathcal{E}_x$ is the unknown source of emf. The resistance R is a long, uniform wire made of a metal of fairly high resistivity. The contact slider P is moved along the wire until the ammeter A

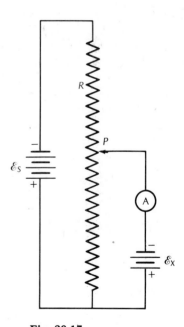

Fig. 29.17

Potentiometer

registers zero current (the potentiometer is then said to be balanced). Show that if the balance point is at the midpoint of the wire, then $\mathscr{E}_x = \frac{1}{2}\mathscr{E}_s$. What if the balance point is at three-quarters of the length of the wire?

Ohmmeter

10. An **ohmmeter** consists of a standard source of emf (a battery), connected in series with a standard resistance, and an ammeter. When the terminals of the ohmmeter are connected to an unknown resistor (Figure 29.18), the current registered by the ammeter permits the evaluation of the unknown resistance. Explain.

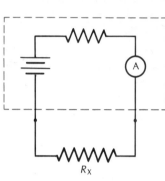

Fig. 29.18

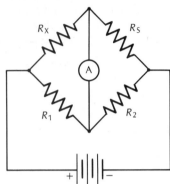

Fig. 29.19

Wheatstone bridge

11. The circuit shown in Figure 29.19 is called a **Wheatstone bridge.** It permits a very precise comparison of an unknown resistance with a known, standard resistance. In Figure 29.19 the unknown resistance is R_x, and the standard resistance is R_s. The resistances R_1 and R_2 are adjustable. In operation, the resistances R_1 and R_2 are adjusted until the ammeter A registers zero current. Show that under these conditions the ratios of the resistances are related as follows: $R_x/R_s = R_1/R_2$.

12. At $t = 0$ you connect a battery and a resistor to a capacitor, as shown in Figure 29.20. The capacitor is initially uncharged. Qualitatively, describe the current as a function of time.

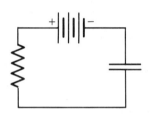

Fig. 29.20

13. A homeowner argues that he should not pay his electric bill since he is not keeping any of the electrons that the power company delivers to his home — any electron that enters the wiring of his home sooner or later leaves and returns to the power station. How would you answer?

14. The spiral heating elements commonly used in electric ranges *appear* to be made of metal. Why do they not short circuit when you place an iron pot on them?

15. What are the advantages and what are the disadvantages of high-voltage power lines?

16. In many European countries, electric power is delivered to homes at 220 V, instead of the 110 V customary in the United States. What are the advantages and what are the disadvantages of 220 V?

17. How much does it cost you to operate a 100-watt light bulb for 24 hours? The price of electric energy is 8¢ per kilowatt-hour.

PROBLEMS

Section 29.1

1. The smallest batteries weigh 0.05 ounces and store an electric energy of about 5×10^{-6} kW·h. The largest batteries (used aboard submarines) weigh

300 tons and store an electric energy of 5×10^3 kw · h. What is the amount of energy stored per pound of battery in each case?

2. A size D flashlight battery will deliver 1.2 A · h at 1.5 V (i.e., it will deliver 1.2 A for 1 h or a larger or smaller current for a correspondingly shorter or longer time). An automobile battery will deliver 55 A · h at 12 V. The flashlight battery is cylindrical with a diameter of 1.3 in., a length of 2.2 in., and a weight of 0.19 lb. The automobile battery is rectangular with dimensions 12 in. × 6.5 in. × 9 in. and a weight of 50 lb.
   (a) What is the available electric energy stored in each battery?
   (b) What is the amount of energy stored per cubic inch of battery?
   (c) What is the energy stored per pound of battery?

3. A heavy-duty 12-V battery for a truck is rated at 160 A · h, i.e., this battery will deliver 1 A for 160 h (or a larger current for a correspondingly shorter time). What is the amount of electric energy that this battery will deliver?

4. The electric starter motor in an automobile equipped with a 12-V battery draws a current of 80 A when in operation.
   (a) Suppose it takes the starter motor 3.0 s to start the engine. What amount of electric energy has been withdrawn from the battery?
   (b) The automobile is equipped with a generator that delivers 5.0 A to the battery when the engine is running. How long must the engine run so that the generator can restore the energy in the battery to its original level?

Section 29.3

5. A voltmeter of internal resistance 5.0×10^4 Ω is connected across the poles of a 12-V battery of internal resistance 0.020 Ω.
   (a) What is the current flowing through the battery?
   (b) What is the voltage drop across the internal resistance of the battery?

6. A voltmeter reads 11.9 V when connected across the poles of a battery. The internal resistance of the battery is 0.020 Ω. What must be the minimum value of the internal resistance of the voltmeter if the reading of the instrument is to coincide with the emf of the battery to within better than 1%?

Section 29.4

7. Four resistors, with $R_1 = 25$ Ω, $R_2 = 15$ Ω, $R_3 = 40$ Ω, and $R_4 = 20$ Ω, are connected to a 12-V battery as shown in Figure 29.21.
   (a) Find the combined resistance of the four resistors.
   (b) Find the current in each resistor.

8. Consider the circuit shown in Figure 29.22. Given that $\mathcal{E}_1 = 6.0$ V, $\mathcal{E}_2 = 10$ V, and $R_1 = 2.0$ Ω, what must be the value of the resistance R_2 if the current through this resistance is to be 2.0 A?

9. Find the current in the two resistors shown in Figure 29.23. Find the power delivered by the 12-V battery. The resistances are $R_1 = 10$ Ω, $R_2 = 8$ Ω; the emfs are $\mathcal{E}_1 = 10$ V and $\mathcal{E}_2 = 12$ V.

10. Two batteries of emf $\mathcal{E}$ and of internal resistances R_i and R_i', respectively, are combined in parallel. The combination is connected to an external resistance R. Find the current through each battery.

11. Consider the circuit shown in Figure 29.11 with the given resistances and emfs. Suppose we replace the emf $\mathcal{E}_1 = 12.0$ V by a larger emf. How large must we make $\mathcal{E}_1$ if the current I_2 is to charge the battery $\mathcal{E}_2$?

12. Two batteries with internal resistances are connected as shown in Figure 29.24. Given that $R_1 = 0.50$ Ω, $R_2 = 0.20$ Ω, $\mathcal{E} = 12.0$ V, $\mathcal{E}' = 6.0$ V, $R_i = 0.025$ Ω, and $R_i' = 0.020$ Ω, find the currents in the resistances R_1 and R_2.

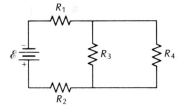

Fig. 29.21

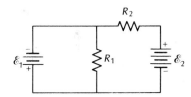

Fig. 29.22

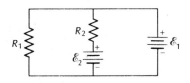

Fig. 29.23

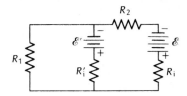

Fig. 29.24

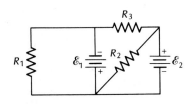

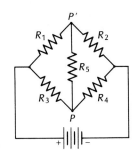

Fig. 29.25

Fig. 29.26

*13. Two batteries, with $\mathcal{E}_1 = 6.0$ V and $\mathcal{E}_2 = 3.0$ V, are connected to three resistors, with $R_1 = 6.0$ Ω, $R_2 = 4.0$ Ω, and $R_3 = 2.0$ Ω, as shown in Figure 29.25. Find the current in each resistor and the current in each battery.

*14. Five resistors, of resistances $R_1 = 2.0$ Ω, $R_2 = 4.0$ Ω, $R_3 = 6.0$ Ω, $R_4 = 2.0$ Ω, and $R_5 = 3.0$ Ω, are connected to a 12-V battery as shown in Figure 29.26.

  (a) What is the current in each resistor?
  (b) What is the potential difference between the points P and P'?

Section 29.5

15. An electric toaster uses 1200 W at 110 V. What is the current through the toaster? What is the resistance of its heating coils?

16. A cyclotron accelerator produces a beam of protons of an energy of 700 million eV. The average current of this beam is 1.0×10^{-6} A. What is the number of protons per second delivered by the accelerator? What is the corresponding power delivered by the accelerator?

17. While cranking the engine, the starter motor of an automobile draws 80 A at 12.0 V for a time interval of 2.5 s. What is the electric power used by the starter motor? How many horsepower does this amount to? What is the electric energy used up in the given time interval?

18. An air conditioner operating on 110 V uses 1500 W of electric energy. What is the electric current through the air conditioner? What is the resistance of the air conditioner?

19. A small electric motor operating on 110 V delivers 0.75 hp of mechanical power. Ignoring friction losses within the motor, what current does this motor require?

20. A 12-V battery of internal resistance 0.20 Ω is being charged by an external source of emf delivering 6.0 A.

  (a) What must be the minimum emf of the external source?
  (b) What is the rate at which heat is developed in the internal resistance of the battery?

21. An electric car is equipped with an electric motor supplied by a bank of sixteen 12-V batteries. When fully charged, each battery stores an energy of 2.2×10^6 J.

  (a) What current is required by the motor when it is delivering 12 hp? Ignore friction losses.
  (b) With the motor delivering 12 hp, the car has a speed of 65 km/h (on a level road). How far can the car travel before its batteries run down?

22. The banks of batteries in a submarine store an electric energy of 5×10^3 kW·h. If the submarine has an electric motor developing 1000 hp, how long can it run on these batteries?

23. An electric toothbrush draws 7 watts. If you use it 4 minutes per day and if electric energy costs you 8¢/kW·h, what do you have to pay to use your toothbrush for one year?

24. In a small electrostatic generator, a rubber belt transports charge from the ground to a spherical collector at 2.0×10^5 V. The rate at which the belt transports charge is 2.5×10^{-6} C/s. What is the rate at which the belt does work against the electrostatic forces?

25. A solar panel (an assemblage of solar cells) measures 58 cm × 53 cm. When facing the sun, this panel generates 2.7 A at 14 V. Sunlight delivers an energy of 1.0×10^3 W/m² to an area facing it. What is the efficiency of this panel, i.e., what fraction of the energy in sunlight is converted into electric energy?

26. A 40-m cable connecting a lightning rod on a tower to the ground is made of copper with a diameter of 7 mm. Suppose that during a stroke of lightning the cable carries a current of 1×10^4 A.
    (a) What is the potential drop along the cable?
    (b) What is the rate at which Joule heat is produced?

27. The maximum recommended current for a No. 10 copper wire of diameter 0.259 cm is 25 A. For such a wire with this current, what is the rate of production of Joule heat per meter of wire? What is the potential drop per meter of wire?

28. The cable connecting the electric starter motor of an automobile with the 12.0-V battery is made of copper with a diameter of 0.50 cm and a length of 0.60 m. If the starter motor draws 500 A (while stalled), what is the rate at which Joule heat is produced in the cable? What fraction of the power delivered by the battery does this Joule heat represent?

29. Two heating coils have resistances of 12.0 Ω and 6.0 Ω, respectively.
    (a) What is the Joule heat generated in each if they are connected in parallel to a source of emf of 110 V?
    (b) What if they are connected in series?

30. A battery of emf $\mathscr{E}$ and internal resistance R_i is connected to an external circuit of resistance R. In terms of $\mathscr{E}$, R_i, and R, what is the power delivered by the battery to the external circuit? Show that this power is maximum if $R = R_i$.

31. A 3.0-V battery with an internal resistance of 2.5 Ω is connected to a light bulb of a resistance of 6.0 Ω. What is the voltage delivered to the light bulb? How much electric power is delivered to the light bulb? How much electric power is wasted in the internal resistance?

32. Suppose that a 12-V battery has an internal resistance of 0.40 Ω.
    (a) If this battery delivers a steady current of 1.0 A into an external circuit until it is completely discharged, what fraction of the initial stored energy is wasted in the internal resistance?
    (b) What if the battery delivers a steady current of 10.0 A? Is it more efficient to use the battery at low current or at high current?

33. An electric clothes dryer operates on a voltage of 220 V and draws a current of 20 A. How long does the dryer take to dry a full load of clothes? The clothes weigh 6.0 kg when wet and 3.7 kg when dry. Assume that all the electric energy going into the dryer is used to evaporate water (the heat of evaporation is 539 kcal/kg).

34. A large electromagnet draws a current of 200 A at 400 V. The coils of the electromagnet are cooled by a flow of water passing over them. The water enters the electromagnet at a temperature of 20°C, absorbs the Joule heat, and leaves at a higher temperature. If the water is to leave with a temperature no larger than 80°C, what must be the minimum rate of flow of water (in liters per minute) through the electromagnet?

ATMOSPHERIC ELECTRICITY[1]

The atmosphere is a great electric machine. Thunderstorms are the most spectacular manifestation of electric activity in the atmosphere, but even in fair weather the atmosphere is full of electric fields and electric currents. The thunderstorms act as giant electrostatic generators, delivering negative charge to the ground, and positive charge to the upper level of the atmosphere. This upper level of the atmosphere, or ionosphere, is a good conductor and the current reaching it quickly spreads laterally, over the entire globe. In fair-weather regions, this current gradually leaks down to the ground, completing the **atmospheric electric circuit** (Figure I.1).

There are roughly 2000 thunderstorms in action all over the Earth at any given time. The time-average current generated by a thunderstorm is about 1 A (the instantaneous current can of course be much larger — up to 20,000 A in a stroke of lightning). Thus, all the thunderstorms together contribute an average of 2000 A to the current in the atmospheric circuit. Bursting air bubbles at the ocean surface also contribute to the current, by spraying small positively charged droplets into the atmosphere; but this contribution is believed to be appreciably smaller than that of thunderstorms.

Figure I.2 is a schematic diagram of the atmospheric electric circuit. Both the ionosphere and the ground are good conductors, i.e., each is an equipotential surface; the potential difference between them is about 300,000 V. The resistance of the entire fair-weather at-

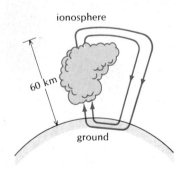

Fig. I.1 Flow of current in the Earth's atmosphere.

[1] This chapter is optional.

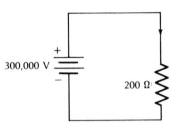

Fig. I.2 Schematic diagram of the atmospheric electric circuit.

mosphere between ionosphere and ground is about 200 Ω. Most of this resistance is concentrated in the dense, low regions of the atmosphere; correspondingly, most of the 300,000-V potential drop occurs in the low regions of the atmosphere, within a few kilometers from the ground. On the average, the total electric power delivered to the global circuit by thunderstorm activity is about 2000 A × 300,000 V = 6 × 10^8 W, i.e., nearly a million kW.

I.1 THE FAIR-WEATHER ELECTRIC FIELD

The atmospheric current that, in the fair-weather regions of the Earth, flows from the ionosphere to the ground is carried by ions — atoms and molecules that have a positive or negative charge due to a deficit or excess in the number of their electrons. Such ions are always present in the atmosphere. They are continually produced by the impacts of cosmic rays on air molecules throughout the atmosphere, by the ultraviolet irradiation of the upper atmosphere, and by the natural and man-made radioactivity near the ground. Thus the current consists of a downward motion of positive ions and an upward motion of negative ions. The driving force that maintains this motion is an **atmospheric electric field** that points vertically downward; this electric field is analogous to the electric field that exists in a current-carrying wire and drives the electrons along the wire. Near the surface of the Earth, over open ground, the strength of this atmospheric electric field is between 100 and 200 V/m in fair weather. The exact value of the strength of the field depends on local conditions, such as dust in the atmosphere, topography, time of day; the worldwide average value is 130 V/m.

The potential difference between the ground and a point 2 m above the ground (the height of a man) is therefore typically a few hundred volts. Does this mean that we can operate an appliance by plugging one terminal into the ground and the other terminal into air, 2 m above ground? Unfortunately, we cannot:

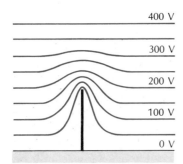

Fig. I.3 Equipotentials in air in the vicinity of a thin vertical conductor connected to the ground.

the atmosphere will deliver only an infinitesimal current to the exposed terminal and therefore that terminal (and the entire appliance) will remain at the potential of the ground. The only consequence of sticking a wire, or any conductor, into the air is to deform the atmospheric equipotentials. The 0-V equipotential will follow the shape of the conductor; successively higher equipotentials will exhibit successively less deformation (Figure I.3). Buildings, trees, or human bodies are reasonably good conductors; they will deform the equipotentials, and the electric field, in a similar manner.

The lines of the atmospheric electric field end on the surface of the Earth; thus, there must be negative charge on the surface. The amount of charge per unit area is given by Eq. (24.23):

$$\sigma = \varepsilon_0 E = -8.85 \times 10^{-12} \text{ F/m} \times 100 \text{ C/m} \cong -10^{-9} \text{ C/m}^2$$

The total negative charge on the surface of the Earth (including the oceans) is about half a million coulombs.

The electric field near the ground can be measured with a **field meter.** A simple kind of field meter is constructed with a horizontal metallic plate placed near the ground and connected to the ground by means of a wire. The lines of electric field end on the surface of the metallic plate; this surface must therefore carry an electric charge. If a second metallic plate, also connected to the ground, is suddenly placed directly above the first plate, the lines of field will end on the second plate, i.e., the second plate shields the first

plate from the electric field (Figure I.4). The charges on the first plate, released from the grip of the electric field, will then quickly flow through the wire into the ground. A sensitive ammeter connected to this wire will detect this flow charge; the total amount of charge that runs out of the plate is a measure of the strength of the electric field.

In practice, the two plates are often given the shape

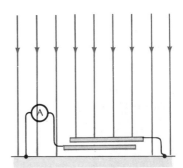

Fig. I.4 Two horizontal metallic plates, connected to the ground.

of a Maltese cross (Figure I.5). The upper plate is permanently kept above the lower plate and made to rotate in the horizontal plane. The arms of the upper plate then successively cover and uncover the arms of the lower plate; each covering and uncovering sends a pulse of current through the wire connecting the lower plate to the ground; the strength of this alternating current is a measure of the electric field. An instrument of this kind is called a **field mill.**

The strength of the atmospheric electric field decreases with altitude. This decrease is related to the decrease of the resistance of the atmosphere: at low altitude the resistance of the atmosphere is large, and at high altitude the resistance is small. Thus, at low altitude a large electric field is needed to maintain a given current, while at high altitude only a small electric field

Fig. I.5 A field mill. When in use, the cross is made to rotate at a high speed by a motor within the base.

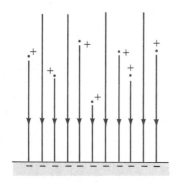

Fig. I.6 Field lines of the atmospheric electric field start on positive charges in the air and end on negative charges on the ground.

is needed. The decrease of the electric field is engendered by a positive charge density in the atmosphere (Figure I.6). Note that most of the electric field lines start on positive charges in the lower atmosphere, within a few kilometers from the ground; only very few field lines start on positive charges in the ionosphere. The total positive charge in the fair-weather at-

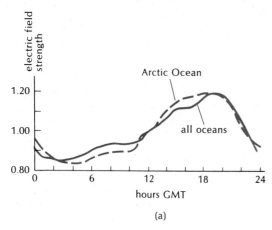

(a)

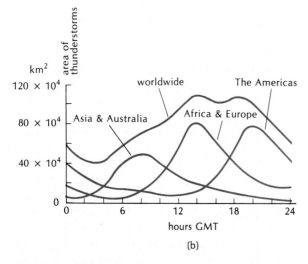

(b)

Fig. I.7 (a) Strength of electric field vs. Greenwich time. The strength of the electric field is expressed as a multiple of the average strength. (b) Thunderstorm activity vs. Greenwich time.

mosphere is of the same magnitude as the negative charge on the ground.

The strength of the fair-weather field depends to some extent on the time of day. The magnitude of the diurnal variation may be as large as 20%. Except for local effects due to comtamination of the atmosphere by smoke or dust, the variation is simultaneous at all places on the Earth; for example, the field strength is maximum at 18:00 local time (Greenwich Mean Time) in London and simultaneously at 13:00 local time (Eastern Standard Time) in New York.

Figure I.7a shows the time dependence of the strength of the fair-weather field (as a percentage of the mean strength) over the open ocean; these measurements over the open ocean give the best indication of the global time variation because they are free of spurious effects caused by local sources of pollution.

The time dependence of the field strength is due to a time dependence of thunderstorm activity. Worldwide, thunderstorm activity reaches a peak between 14:00 and 20:00 Greenwich Mean Time; this peak is mainly due to an abundance of mid-afternoon thunderstorms in the Amazon basin. Thus, the maximum in the electric field at 18:00 arises from the enhanced rate at which thunderstorms charge up the atmosphere at around this time (Figure I.7b).

I.2   THUNDERSTORMS

Thunderstorms obtain the energy for their violent mechanical and electrical activity from humid air. Thunderstorms are heat engines; their heat reservoir is the heat of evaporation stored in water vapor. One cubic kilometer of air at 17°C, at atmospheric pressure, and at 100% relative humidity contains 1.6×10^7 kg of water vapor. Since the heat of evaporation for water is 586 kcal/kg (at 17°C), condensation of all the water vapor in 1 km³ of air would release 9.2×10^9 kcal. This is equivalent to the energy released in the explosion of 9.2 kilotons of TNT. A typical thunderstorm involves many cubic kilometers of air and therefore the amount of stored energy (latent energy) is enormous. Although the efficiency for energy conversion is low, the release of just a small fraction of the energy stored in humid air suffices to account for the activity of a thunderstorm. Tornadoes and hurricanes also obtain their energy from humid air, and their destructive power reflects the large amount of available energy.

A thunderstorm is made of several **cells,** or thunderclouds, within each of which air moves upward or, in the later stages of the cell's development, downward. Each cell is some 8 km across; the base of the cell is at a height of about 1500 m and the top of the cell at a height that initially may be 7500 m but grows to 12,000 m or even 18,000 m. Within such a cell there

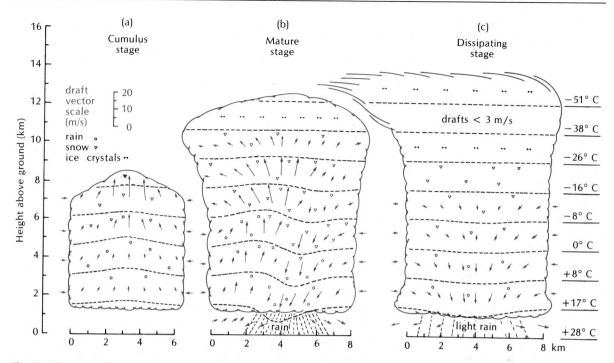

Fig. I.8 (a) A young thunderstorm cell. The vectors give the velocity of drafts within the cell. (b) A mature thunderstorm cell. (c) An old thunderstorm cell. (From A. J. Chisholm, *Meteorological Monographs,* November 1973, no. 36.)

are strong updrafts and, at later stages, strong downdrafts, commonly reaching vertical speeds of 12 m/s.

The rising motion of air in a storm cell is powered by the heat released in the condensation of water vapor. Dry air on the surface of the Earth will usually not rise; it is in stable (or neutral) equilibrium. If, because of some disturbance, a mass of dry air is pushed upward to a higher altitude, the reduced pressure of its surroundings leads to expansion of the air; the work done by the air during this expansion reduces its temperature; and this reduced temperature will match the normal reduced temperature of the environment at the higher altitude. Thus, dry air has no tendency to rise.

However, humid air is unstable with respect to vertical displacement. If a mass of humid air initially near the ground is pushed upward to a slightly higher altitude, the expansion and consequent reduction of temperature leads to condensation of a fraction of the water vapor. This supplies the air with extra heat. The air then will be warmer and less dense than the dry air that resides at the higher altitude — it continues to rise. Thus the air is unstable. Once the upward motion starts, it will continue faster and faster until all the water vapor has condensed, and even then the motion may continue, since the water can supply further energy by freezing into ice.

In a young thunderstorm cell the air rises as in a chimney, drawing in more and more humid air at the base while the top of the cell grows upward at speeds of 10 m/s or more (Figure I.8a).

A mature thunderstorm cell may extend to a height of 12,000 m or 18,000 m. Inside the cell small water droplets coalesce, forming raindrops; some of the rain freezes into ice and hail; snow may also form. Raindrops and ice particles that are too large and heavy to be supported by the updraft begin to fall. This downward motion drags some air along and starts a downdraft. Once the air begins to descend, it will continue descending — humid air is unstable with respect to downward displacement as well as upward displacement. If a mass of humid air (intermixed with water) is pushed downward to a slightly lower altitude and higher pressure, the compression tends to warm the air but the evaporation of water tends to cool it. The net result is that the air will be somewhat cooler and denser than the dry air surrounding the thunderstorm cell; therefore the mass of humid air continues to descend.

In the mature thunderstorm cell there are both strong updrafts and strong downdrafts. The top of the cell spreads out laterally, forming the characteristic anvil cloud, or **thunderhead** (Figure I.9). Observations made by looking down on thunderstorms from U-2 aircraft and satellites have revealed that above the anvil cloud, small turrets often sprout upward and reach into the stratosphere. These turrets consist of cloud masses thrown upward by violent updrafts with speeds as great as 100 m/s. During the mature stage, different regions of the thunderstorm cell acquire different electric charges, and lightning discharges ensue. Rain or hail falls out of the bottom of the cell, accom-

Fig. I.9 A thunderstorm with a well-developed anvil cloud.

panied by cool, descending air which generates a gusty wind near the ground (Figure I.8b).

In the late stage of a thunderstorm cell, the updraft ends and the downdraft takes over, covering the entire cell. Then the storm dissipates (Figure I.8c).

I.3   GENERATION OF ELECTRIC CHARGE

A mature thunderstorm cell typically has a charge distribution as shown in Figure I.10. There is a positive charge in the upper part of the cell and a negative charge in the lower part of the cell; furthermore, at the very bottom of the cell there is often an extra, small, positive charge.

Exactly how these changes are generated remains somewhat of a mystery. There are numerous theories invoking different mechanisms. Most of these theories blame the charge separation on a difference in the size of the charge carriers created in the cloud — the car-

riers of positive charge are supposed to be smaller and lighter than the carriers of negative charge. The updrafts in the cloud then blow the positive carriers upward, but the negative carriers either remain stationary or fall down. Thus the upper part of the cloud acquires a positive charge, while the lower part acquires a negative charge.

The charge carriers may be raindrops, hail, ice crystals, or ions; different theories propose different scenarios for how any or all of these particles may become charged. The conjectural charging mechanisms involve some form of electrification by friction, collision, freezing, melting, or thermoelectricity. For instance, according to one theory, the charges are created by collisions between falling hailstones and small drops of water. The vertical downward atmospheric electric field polarizes the hailstones, inducing positive charges at their bottoms and negative charges at their tops. When such a falling hailstone strikes a drop of water, pushing it aside, the drop is likely to pick up a few positive charges during the contact with the hailstone (Figure I.11); this leaves the hailstone with a net negative charge. According to an alternative theory, the charges are created by collisions between falling hailstones and ions. These ions are present in the air, with equal average concentrations of negative and

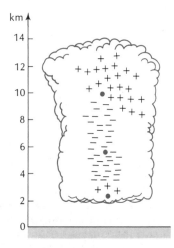

Fig. I.10 Charge distribution in a typical thundercloud.

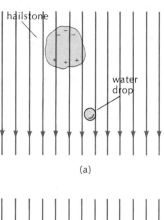

(a)

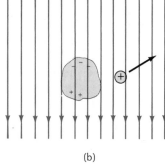

(b)

Fig. I.11 (a) A falling hailstone, polarized by the electric field, approaches a drop of water. (b) In the collision with the lower edge of the hailstone, the water drop picks up some positive charge.

positive ions. If the hailstone approaches a negative ion, it is likely to attract and capture the ion; if the hailstone approaches a positive ion, it is likely to merely repel the ion (Figure I.12). Thus the net result of such collisions is that the hailstone again tends to acquire a negative charge. The falling hailstone then carries this negative charge to the bottom of the cloud and updrafts carry the positive charge to the top of the cloud.

Another conjectural charging mechanism relies entirely on updrafts and downdrafts to accumulate charges in the cloud, after a small initial amount of charge has been generated in the cloud by other means. This mechanism can work in two ways. At the top of the cloud, the positive charge (see Figure I.10) attracts negative ions from the nearby air. If these ions become attached to droplets of water or particles of ice, downdrafts can carry them past the positive charge and deposit them in the lower portion of the cloud, adding to the negative charge already there. In a similar way, at the bottom of the cloud, the negative charge (see Figure I.10) attracts positive ions released by point discharge from sharp protuberances on the ground (see the next section). If these ions become attached to suitable water droplets or ice particles, updrafts can carry them to the upper portion of the cloud, adding to the positive charge already there.

Although all of these mechanisms, and others as well, presumably contribute to the charging of thunderclouds, it remains unclear which, if any, of the proposed mechanisms is dominant.

I.4   THE ELECTRIC FIELD OF A THUNDERCLOUD

Figure I.13a shows a crude model of the charge distributions in a thundercloud. For the purposes of this model, it has been assumed that the charge distributions are roughly spherical; the positive and negative charges can then be represented by point charges located at the centers of the charge distributions. The small positive charge at the bottom of the thundercloud has been ignored. The charges of ∓ 40 C are at heights of 5000 m and 10,000 m, respectively. At the

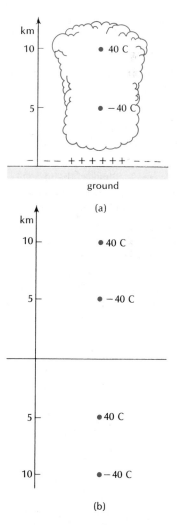

(a)

(b)

Fig. I.13 (a) In this crude model, the charge distribution has been approximated as consisting of two point charges. These point charges induce a charge distribution on the ground. (b) The effect of the induced charge distribution on the ground is equivalent to that of two virtual image charges.

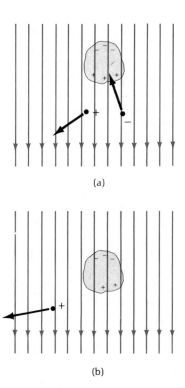

Fig. I.12 (a) A falling hailstone attracts negative ions, but repels positive ions. (b) Capture of a negative ion gives the hailstone a negative charge.

ground, directly below these charges, they produce an upward electric field of magnitude

$$E = \frac{1}{4\pi\varepsilon_0}\left(\frac{Q_1}{r_1^2} + \frac{Q_2}{r_2^2}\right)$$

$$= 9.0 \times 10^9 \text{ F/m}\left(\frac{40 \text{ C}}{(5 \times 10^3 \text{ m})^2} - \frac{40 \text{ C}}{(10 \times 10^3 \text{ m})^2}\right)$$

$$= 11 \times 10^3 \text{ V/m} \tag{1}$$

But this is not the total electric field. The charges in the cloud induce charges on the ground (a conductor), and these produce an extra electric field. The induced charge distribution on the ground will cover a large area, perhaps some 100 km²; within this area, the charge density will be concentrated directly below the charges in the cloud, and it will gradually decrease with distance.

We can calculate the effect of this induced charge distribution by a clever trick involving three steps: First, replace the ground by a thin, flat conducting plate; this does not change the electric field in the space above the ground since the thickness of the conducting plate is irrelevant. Second, place two charges of ± 40 C in the empty space below the conducting plate (Figure I.13b). These two new charges are at depths of 5000 m and 10,000 m below the conducting plate; they are called the **image** charges of the original charges in the cloud. The presence of these image charges below the plate does not change the electric field above the plate since the conductor shields these regions from each other. Third, slide the conducting plate away in a horizontal direction. This, again, does not change the electric field above the plate because, for the configuration of four charges shown in Figure I.13b, the midplane is an equipotential surface and the presence or absence of a coincident conducting surface makes no difference to the electric field.

Having taken these three steps, we recognize that the combination of cloud charges plus charges on the ground produces exactly the same electric field (above the ground) as the combination of cloud charges plus image charges; i.e., the induced charges on the ground are virtually equivalent to the image charges. Thus, we can calculate the total electric field at any given point on or above the ground by simply adding the electric fields of the four charges shown in Figure I.13b. At any given point on the ground, the image charges produce exactly the same electric field as the cloud charges — the total electric field is therefore *twice* the electric field of the cloud charges. Directly below the cloud charges, the electric field is then $2 \times$

11×10^3 V/m $= 22 \times 10^3$ V/m. Incidentally, the above trick for the evaluation of the electric field of some given point charges in the presence of a conducting surface is called the **method of images;** the method only works under rather special conditions, but when it works, it permits a very quick and elegant evaluation:

We can also use this method to calculate the potential difference between the ground and the base of the thundercloud at a height of, say, 2 km (see Figure I.10). The potential of the ground is zero and the potential of the base of the thundercloud is a sum of four terms, corresponding to the four point charges of Figure I.13b:

$$V = \frac{1}{4\pi\varepsilon_0}\left(\frac{40 \text{ C}}{8 \times 10^3 \text{ m}} - \frac{40 \text{ C}}{3 \times 10^3 \text{ m}}\right.$$

$$\left. + \frac{40 \text{ C}}{7 \times 10^3 \text{ m}} - \frac{40 \text{ C}}{12 \times 10^3 \text{ m}}\right)$$

$$= -5.4 \times 10^7 \text{ volt} \tag{2}$$

As these numbers show, the electric fields and the potential differences generated by thunderstorms are quite large.

In the above calculations we assumed a perfectly flat Earth. In reality there are protuberances on the surface of the Earth, and near these the electric field strength will be strongly enhanced.[2] It is a general rule that if a sharp conducting point protrudes from an equipotential surface, the electric field will be strong near the point. For example, Figure I.14 shows the electric field near a lightning rod; the high density of field lines near the tip of the rod indicates a strong field. It is easy to see that some enhancement of the field strength must occur for any kind of protruding point, regardless of the precise shape: the field lines must meet any conductor at right angles; hence they must bend toward the protruding conductor; hence their density increases.

The strong electric fields near any sharp spikes or edges on the ground produce **point discharge.** For example, experiments with trees have shown that a tree under a thunderstorm will carry a current of about 1 A out of the ground; the current flows into the atmosphere through the tips of the tree's leaves.

The point discharge is due to the electrical breakdown of the air — at a high electric field strength the air becomes a conductor. The sudden increase in the

[2] The field strength is also much greater in the vicinity of the charge concentrations in the cloud. On one occasion, a value of 340×10^3 V/m was measured at an airplane flying through a thunderstorm just before the airplane was struck by lightning.

conductivity of air is caused by ionization, i.e., the formation of positive ions and free electrons by the disruption of neutral atoms. There are always a few ions and free electrons present in the air, but not enough to conduct any large current. However, if the air is exposed to an intense electric field, then those few electrons are accelerated to high velocity, and when they smash into a neutral atom they can kick extra electrons out of the atom. A chain reaction sets in: each free electron generates an avalanche of free electrons. In dry air under atmospheric pressure, about 3×10^6 V/m are required to start such a discharge.

The discharges into the atmosphere can take a variety of forms: point discharge with no visible display, corona discharge (also called St. Elmo's fire) with its characteristic glow of visible light, sparks, or lightning. Although lightning is the most spectacular of these phenomena, in many thunderstorms point discharge actually plays the larger role in the transfer of charge from the thundercloud to the ground — it contributes a current to the atmospheric circuit which is several times larger than the (average) current contributed by lightning.

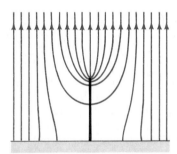

Fig. I.14 Electric field lines around a thin vertical conductor connected to the ground (lightning rod).

I.5 LIGHTNING

An electric field of 3×10^6 V/m is required to induce electric breakdown and produce small sparks in dry air at atmospheric pressure, but the presence of water drops and the lower-than-atmospheric pressure in a thundercloud favor breakdown and, therefore, sparking is likely to begin at somewhat lower electric fields. To initiate a flash of lightning, the electric field has to be very intense only in a small region near the cloud; once an avalanche of electrons starts, it can propagate into regions of less intense electric field. A lightning discharge can take place between the opposite charges inside the thundercloud, or between the thundercloud and clear air, or between the thundercloud and the ground. In the latter case the discharge usually takes place between the negative charge of the cloud

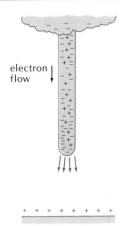

Fig. I.15 The leader. An electron avalanche from the cloud makes a channel of ionized air. Electrons fill this channel.

and the ground; discharges between the positive charge and the ground are rare.

The sequence of events in a flash of lightning occurs too fast to be resolved by the human eye; however, scientists have used techniques of high-speed photography to analyze the time development of flashes of lightning. A typical flash of lightning between a thundercloud and the ground begins with an electron avalanche in the intense electric field near the negative charge of the thundercloud. As the avalanche moves downward it leaves behind a channel of ionized air, or plasma; electrons from the cloud flow into this channel, giving it a negative charge. The concentration of negative charge near the tip of the channel generates strong electric fields which continue to push the avalanche along (Figure I.15). The channel follows a tortuous path, often with lateral branches whose twists and turns reflect random variations in the density of free electrons in the air ahead of the avalanche. The channel is called a **leader**; it has a radius of several meters (perhaps 5 m) but only its central region is (faintly) luminous. Figure I.16 shows a time sequence

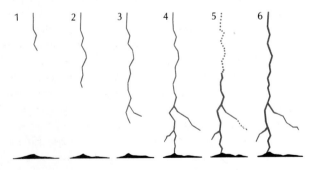

Fig. I.16 This sequence of diagrams, based on high-speed photographs, shows the downward motion of the leader (first four diagrams) and the upward motion of the return stroke (last two diagrams). The return stroke is much faster than the leader.

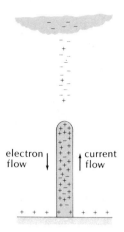

electron flow ↓   ↑ current flow

Fig. I.17 The return stroke. Electrons suddenly drain out of the bottom of the channel, leaving it with a positive charge.

of pictures of the leader. These pictures fail to illustrate one important feature of the leader: the leader does not move downward smoothly — it proceeds in steps of about 50 m with a pause of about 50 μs between steps. Because of this, it is sometimes called the **step leader.**

When the tip of the leader comes near the ground, its intense electric field initiates a discharge from the ground or from a pointed object on the ground. The upward-moving discharge meets the leader 20 m to 100 m above ground. At this instant the circuit between cloud and ground is complete — there is now a continuous conducting path and the negative charge can flow from cloud to ground with little resistance.

The electrons at the bottom of the channel are the first to move — they drain out into the ground, giving a large current. Then electrons from successively higher positions drain down. Thus, a "drainage front" moves upward toward the thundercloud. The drainage

front is the head of a tube of intense current which snakes toward the thundercloud at as much as one-half the speed of light; this tube of current is the **return stroke** (Figure I.17). The intense current heats the air and produces the bright flash of light that we see as lightning (Figure I.18). The radius of the tube of current is quite small — it ranges from less than 1 cm to a few centimeters.

The peak current of the return stroke is about 10,000 A to 20,000 A. This strong current lasts less than 100 μs. When it ends, a weaker current (a few hundred amperes) continues to flow for several milliseconds. The strong current can cause severe damage to whatever body lies in its path (Figures I.19 and I.20).

The first return stroke is often succeeded by several more return strokes. A typical flash of lightning contains three to five return strokes with intervals of 40 ms between them. Each of the latter return strokes is preceded by its own **dart leader** which travels along the channel of the first return stroke. This channel retains its conductivity for some time and enables the dart leaders to travel downward very fast, without the pauses characteristic of the first step leader.

The total charge delivered to the ground by the complete flash of lightning with its several return strokes amounts to about −25 C (see Table I.1). Since the potential difference between the base of the cloud and the ground is about 5×10^7 V [see Eq. (2)], the energy dissipated in the lightning channel is about 10^9 J. Most of this energy goes into heat (Joule heating), although a small fraction goes into emission of light and of radio waves. Immediately after the current has passed, the plasma in the lightning channel will be at an extremely high temperature (about 30,000 K) and a correspondingly high pressure. The high-pressure plasma expands explosively, generating a shock wave

Fig. I.18 Lightning.

Fig. I.19 The grass of this golf course was burnt by the currents produced by a stroke of lightning. Note that the current branches out along the ground in all directions.

Table I.1. DATA ON LIGHTNING[a]

Step leader	
Length of step	50 m
Time between steps	50 μs
Speed	1.5×10^5 m/s
Charge in channel	-5 C
Dart leader	
Speed	2×10^6 m/s
Charge in channel	-1 C
Return stroke	
Speed	5×10^7 m/s
Peak current	10^4–2×10^4 A
Charge transferred to ground	-2.5 C
Complete lightning flash	
Number of strokes	3–5
Time between strokes	40 ms
Time duration of flash	0.2 s
Charge transferred to ground (including continuing current)	-25 C

[a] Based on M. A. Uman, *Lightning*. The numbers given are average values.

in the surrounding air. The shock wave dissipates within a few meters, gradually changing into a sound wave that propagates outward as a pulse. When this sound pulse reaches our ears, we hear it as thunder. The rumbling noises in thunder are due to the successive arrivals of sound pulses from different portions of the lightning bolt, which are at different distances from our ears.

At some distance from the lightning bolt, we will hear the sound after we see the flash of lightning — because of the high speed of light, the flash arrives almost instantaneously, while the sound takes an appreciable time to cover the intervening distance. Since the speed of sound is about $\frac{1}{3}$ km/s, we can use the following rule of thumb to reckon the distance (in kilometers) to the lightning bolt: count the seconds between flash and thunder and divide by 3.

The most spectacular displays of lightning occur in tornadoes. In many instances, eyewitnesses reported that the interior of the funnel of a tornado was brightly lit by constant flashes of lightning. Measurements of the radio waves emitted by the lightning in tornadoes[3] confirm these reports; from such measurements scientists have estimated that about 20 flashes of lightning occur each second. Since the energy per flash of lightning is about 10^9 J (see above), the electric power dissipated by a tornado is about 10^9 J $\times$ 20/s = 2×10^{10} W — this is roughly $\frac{1}{20}$ of the combined output of all the electric power plants in the United States!

Fig. I.20 Tree exploded by lightning.

[3] These radio waves can be picked up on an ordinary TV set; this can give advance warning of the approach of a tornado.

Finally, some brief remarks on the weird phenomenon of ball lightning, or **Kugelblitz.** This consists of a glowing fireball that floats in midair. The sizes of these balls range from 10 cm to 100 cm, although airplane pilots have reported balls as large as 15 m to 30 m inside thunderclouds. The balls sometimes occur after a lightning stroke and sometimes they occur spontaneously, without apparent cause; they usually last only a few seconds. The balls sometimes fall down vertically from the sky, and sometimes drift horizontally near the ground. They have been known to enter buildings through doors, windows, or chimneys. Many of these balls disappear silently without a trace, but some disintegrate explosively with an ear-shattering bang.

It seems likely that the ball is created and maintained by electric effects, but we have no understanding of the mechanism involved in this phenomenon. Several theories have been proposed to explain the ball: according to one theory, it is a ball of plasma held together by magnetic fields; according to another, it is a miniature thundercloud of dust particles acting as a very efficient electrostatic generator. But owing to a lack of precise data and a lack of detailed calculations, the phenomenon remains a mystery.

Further Reading

The following books discuss thunderstorms and lightning at an introductory level:

The Lightning Book by P. E. Viemeister (Doubleday, Garden City, 1961)

The Flight of Thunderbolts by B. F. J. Schonland (Oxford, 1964)

The Nature of Violent Storms by L. J. Battan (Doubleday, Garden City, 1961)

Short but very informative articles at the same level are the following:

"Thundercloud Electricity," B. Vonnegut, *Discovery,* March 1965

"Thunder," A. A. Few, *Scientific American,* July 1975

"Thunder and Lightning," J. Latham, in *Forces of Nature,* edited by V. Fuchs (Thames and Hudson, London, 1977)

At a more technical level, the following books and articles provide a wealth of information:

Lightning by R. H. Golde (Academic Press, London, 1977)

Lightning by M. A. Uman (McGraw-Hill, New York, 1969)

Atmospheric Electricity by J. A. Chalmers (Pergamon Press, Oxford, 1967)

Physics of Lightning by D. J. Malan (English University Press, London, 1963)

"Atmospheric Electricity" by C. D. Stow, in *Reports on Progress in Physics,* Vol. 32, Part I, 1969

"Some Facts and Speculations Concerning the Origin and Role of Thunderstorm Electricity" by B. Vonnegut, in *Meteorological Monographs,* Vol. 5, No. 27, September 1963

Questions

1. It is possible to measure the potential differences between the atmosphere and the ground by shooting an arrow upward, with a fine wire trailing from its tail. A voltmeter connected between the lower end of the wire and the ground will then register the potential at the arrow. Why does this differ from sticking a stationary conductor into the air?

2. By comparing Figures I.7a and b, can you conclude that the bursting of bubbles on the ocean surface makes only a small contribution to the current in the atmospheric circuit?

3. In 1752, Benjamin Franklin flew a kite into a thundercloud and demonstrated that lightning is an electric phenomenon. A few months later, a scientist at St. Petersburg was killed by lightning while attempting to repeat this stunt. After that, scientists took the precaution of enclosing themselves in boxes of sheet metal while flying kites into thunderclouds. How does this help?

4. How could you construct a power plant that captures electric energy from a thundercloud?

5. What are the main dangers to an aircraft flying through a thundercloud?

6. A lightning rod serves not only to conduct lightning safely to the ground, but also to inhibit lightning by promoting point discharge. Explain.

7. What parts of an aircraft are most likely to be struck by lightning when flying in or near a thundercloud?

8. During a thunderstorm, would you be safer in an automobile with a sheet metal body or an automobile with a fiber glass body?

9. To protect the crew of a wood or fiber glass boat from the hazards of lightning, all metal parts on the boat should be connected to a thick conducting cable, terminating on a conducting plate on the outside of the hull, below the waterline. Explain.

10. What pattern of current flow do you expect below and on the water surface near the point of entry of lightning? Would you suffer damage if swimming nearby?

11. Is the conductivity of the ground due to a flow of electrons or a flow of ions?

12. The ancient Greeks noticed that when lightning strikes the ground, nearby cattle are killed, but nearby men often survive. The Greeks thought this indicated an affinity between men and gods. Can you think of a better explanation? (Hint: Chicken also survive.)

13. It is dangerous to take showers or baths when a thunderstorm is overhead. Why?

14. According to a familiar saying, lightning never strikes twice. According to Table I.1, how often does lightning strike, on the average?

15. Suppose you get caught in the open by a thunderstorm with severe lightning. Consider and discuss each of the following options: take shelter beneath a tree standing in isolation; lie down in the open, flat on the ground; sit in a ditch or depression, on your heels, with both feet close together.

16. When lightning strikes a tree, it often explodes the branches or the trunk. Explain. (Hint: The heating of water produces steam.)

17. Suppose that after a flash of lightning, you hear the first thunder in 3 s and the last in 9 s. What can you conclude about the distances of the nearest and the farthest portions of the lightning bolt?

CHAPTER 30

The Magnetic Force and Field

Hans Christian Oersted (örstad), *1777–1851, Danish physicist and chemist, professor at Copenhagen. He observed that a compass needle suffers a deflection when placed near a wire carrying an electric current. This discovery gave the first empirical evidence of a connection between electric and magnetic phenomena.*

The magnetic force that is most familiar in everyday experience is the force that the magnetic poles of the Earth exert on a compass needle. This magnetic force was known for many centuries, but only during the last century did experimenters discover that electric currents also exert magnetic forces on compass needles and that electric currents exert magnetic forces on one another. Finally, physicists came to understand that the magnetic force is nothing but an extra electric force acting between charges in motion. This means that between two charges in motion, there acts not only the Coulomb force, but also a force that is a function of the velocities of the charges.

In the next section we will write down an equation for the magnetic force between two moving charges. From this we will derive the equation for the magnetic force between two currents consisting of many moving charges. Our development goes counter to the historical development — the law of magnetic force between currents was discovered long before the law of magnetic force between individual charges. But our development follows the road taken in earlier chapters: we always begin with the fundamental laws that apply to *particles* and from these deduce the laws that apply to systems of particles (such as currents).

30.1 The Magnetic Force

According to Coulomb's Law, the electric force exerted by a point charge q' on a point charge q is

$$\mathbf{F} = \frac{1}{4\pi\varepsilon_0} \frac{qq'}{r^2} \hat{\mathbf{r}} \tag{1}$$

where r is the distance between the charges and $\hat{\mathbf{r}}$ a unit vector pointing from q' toward q (Figure 30.1). The force on q' exerted by q is exactly the opposite of that in Eq. (1).

Equation (1) correctly gives the force acting on charges at rest. However, when the charges are *in motion*, there is an extra force acting on these charges. This extra force is called the **magnetic force.** This force depends on the relative positions of the charges and on their velocities. If the instantaneous velocities of the charges q and q' are $\mathbf{v}$ and $\mathbf{v}'$, respectively, then the magnetic force on charge q exerted by charge q' is

$$\mathbf{F} = [\text{constant}] \frac{qq'}{r^2} \mathbf{v} \times (\mathbf{v}' \times \hat{\mathbf{r}}) \tag{2}$$

We must regard this equation for the magnetic force as a fundamental law of physics, which has the same status as Coulomb's Law or Newton's Law of Gravitation. Note that the magnetic force varies as the inverse square of the distance, like the electric force and the gravitational force. But the dependence on the velocity involves two cross products of vectors, and is rather complicated.

The direction of the force must be worked out by the right-hand rule for cross products: first we must cross-multiply $\hat{\mathbf{r}}$ by $\mathbf{v}'$ and then cross-multiply the result by $\mathbf{v}$ (see Figure 30.2). The magnetic force is always perpendicular to the velocity $\mathbf{v}$ of the charge q. Note that the magnetic force is zero unless both velocities are different from zero, i.e., the magnetic force acts only if *both* charges are in motion.

In the SI system of units, the numerical value of the constant in Eq. (2) is

$$[\text{constant}] = 1.00 \times 10^{-7} \frac{\text{N} \cdot \text{s}^2}{\text{C}^2} \tag{3}$$

This constant is conventionally written in the form

$$[\text{constant}] = \frac{\mu_0}{4\pi} \tag{4}$$

with

$$\mu_0 = 1.26 \times 10^{-6} \frac{\text{N} \cdot \text{s}^2}{\text{C}^2} \tag{5}$$

The quantity μ_0 is called the **permeability constant.**[1]

Our equation for the magnetic force on a point charge q exerted by q' then becomes

$$\boxed{\mathbf{F} = \frac{\mu_0}{4\pi} \frac{qq'}{r^2} \mathbf{v} \times (\mathbf{v}' \times \hat{\mathbf{r}})} \tag{6}$$

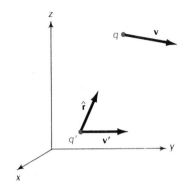

Fig. 30.1 The vectors $\mathbf{v}'$, $\hat{\mathbf{r}}$, and $\mathbf{v}$. In this example, all these vectors are in the z–y plane.

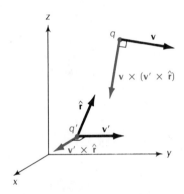

Fig. 30.2 With $\mathbf{v}'$, $\hat{\mathbf{r}}$, and $\mathbf{v}$ as in Figure 30.1, the cross product $\mathbf{v}' \times \hat{\mathbf{r}}$ is perpendicular to the z–y plane; and the cross product $\mathbf{v} \times (\mathbf{v}' \times \hat{\mathbf{r}})$ is in the z–y plane.

Permeability constant

Magnetic force on moving point charge

---

[1] The value of the constant in Eq. (3) is *exact*. Correspondingly, the value $\mu_0 = 4\pi \times 10^{-7} \text{ N} \cdot \text{s}^2/\text{C}^2$ is also exact. (These values hinge on the definitions of current and of charge in the SI system.)

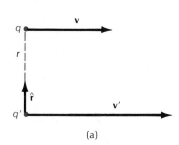

(a)

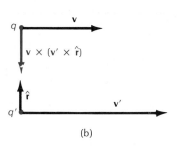

(b)

Fig. 30.3 (a) The velocity vectors **v** and **v'** are parallel. (b) The vector $\mathbf{v} \times (\mathbf{v'} \times \hat{\mathbf{r}})$ points from q toward q'.

EXAMPLE 1.   Suppose that the (instantaneous) velocities of two positive point charges are parallel (Figure 30.3a). Compare the electric and magnetic forces.

SOLUTION: For the configuration shown in Figure 30.3a, the angles in both the cross products of Eq. (6) are 90° and hence the magnitude of the magnetic force is simply

$$F_{\text{mag}} = \frac{\mu_0}{4\pi} \frac{qq'}{r^2} vv' \tag{7}$$

By working out the directions of the cross products, we find that the force is attractive, i.e., the charge q is pulled toward q' (Figure 30.3b).

The magnitude of the electric force is

$$F_{\text{el}} = \frac{1}{4\pi\varepsilon_0} \frac{qq'}{r^2} \tag{8}$$

This force is, of course, repulsive. The net force is the sum of the magnetic and electric forces.

The ratio of the magnitudes of these two forces is

$$\frac{F_{\text{mag}}}{F_{\text{el}}} = \left(\frac{\mu_0}{4\pi} \frac{qq'}{r^2} vv' \right) \bigg/ \left(\frac{1}{4\pi\varepsilon_0} \frac{qq'}{r^2} \right) = \mu_0 \varepsilon_0 vv'$$

$$= 1.26 \times 10^{-6} \frac{\text{N} \cdot \text{s}^2}{\text{C}^2} \times 8.85 \times 10^{-12} \frac{\text{C}^2}{\text{N} \cdot \text{m}^2} vv'$$

$$= (1.12 \times 10^{-17} \text{ s}^2/\text{m}^2) vv' \tag{9}$$

The numerical factor appearing in Eq. (9) is equal to $1/(3.0 \times 10^8 \text{ m/s})^2$. Hence Eq. (9) can be written

$$\frac{F_{\text{mag}}}{F_{\text{el}}} = \frac{v}{3.0 \times 10^8 \text{ m/s}} \frac{v'}{3.0 \times 10^8 \text{ m/s}} \tag{10}$$

This shows that F_{mag} will be small compared to F_{el} if v and v' are small compared to 3.0×10^8 m/s. Note that 3.0×10^8 m/s is the speed of light; why the speed of light should turn up in a calculation concerning electricity and magnetism is a puzzle which we will solve in Section 35.5.

According to Eq. (10), the magnetic force between point charges is small compared to the electric force unless the velocities of the charges approach the velocity of light. This suggests that the magnetic force only becomes significant when the velocities are so large that Newtonian physics fails. Once this happens, a variety of relativistic corrections will have to be included in the Newtonian equation of motion (see Section 17.7).[2] It would then seem that the magnetic force is no more than just another relativistic correction that has to be included in the Newtonian equation of motion of charged particles.

However, the magnetic force can become very important even when the velocities are low. This will happen whenever we are dealing with a charge distribution for which the electric forces cancel. For example, Figure 30.4 shows a moving point charge at some distance from a

Fig. 30.4 Point charge moving parallel to a current.

---

[2] If the velocities are large, some extra relativistic corrections will also be needed in Eqs. (1) and (2).

straight wire carrying a current. Under these conditions, there is no electric force on the point charge — the wire contains equal amounts of positive and negative charges and the electric attractions and repulsions on the point charge cancel. The only remaining force on the point charge is the magnetic force produced by the moving charges of the wire. Since the number of moving charges on the wire is very large, the magnetic force can be quite large.

Figure 30.5 shows two wires carrying parallel currents. The conditions here are similar to those of the preceding example. There is no electric force between these wires — the electric forces between the positive and negative charges on the two wires cancel. However, the magnetic forces do not cancel. The moving charges of one wire exert a magnetic force on the moving charges of the other wire. If the currents are parallel, as in Figure 30.5, this magnetic force is attractive (compare Example 1). If the currents are antiparallel, then the magnetic force is repulsive. The magnitude of the force between currents on wires can be calculated by integrating the forces of the individual point charges.

The magnetic force between the currents on two wires was discovered by Ampère in 1820, long before it was recognized that this force is really due to the magnetic force of Eq. (6) between individual moving charges on the wires. Experiments with individual moving charges (charged brass balls moving at high velocity) were carried out early in this century. It is very difficult to perform a precise measurement of the magnetic force between individual charges. However, it is fairly easy to perform precise measurements of the force between parallel wires carrying a current (such measurements are actually used for the *definition* of the ampere, which we will give in Section 31.4). This amounts to an indirect experimental verification of the magnetic-force law given by Eq. (6).

[As we will see in Section 30.4, the magnetic force can be regarded as generated by a transformation of reference frame applied to an electric force. Note that in a reference frame in which one of the two charges of Eq. (6) is at rest, there is no magnetic force. In this reference frame there is only an electric force. When we now transform to a different reference frame, the magnetic force suddenly makes its appearance — the transformation of reference frame generates an extra magnetic force from an electric force. In Section 30.4 we will use the principles of the theory of Special Relativity to derive the magnetic force from the electric force.]

Equation (6) tells us the force exerted by the point charge q' on the point charge q. To find the force exerted by q on q', we must exchange the velocities in Eq. (6) and we must replace $\hat{\mathbf{r}}$ by $\hat{\mathbf{r}}'$ (Figure 30.6; note that $\hat{\mathbf{r}} = -\hat{\mathbf{r}}'$). This gives the magnetic force on q,

$$\mathbf{F}' = \frac{\mu_0}{4\pi} \frac{qq'}{r^2} \mathbf{v}' \times (\mathbf{v} \times \hat{\mathbf{r}}') \tag{11}$$

As Figure 30.6 shows, in general the magnetic force $\mathbf{F}'$ exerted by q on q' is *not* opposite and equal to the magnetic force $\mathbf{F}$ exerted on q' by q. For magnetic forces, *Newton's Third Law on the equality of action and reaction fails.* Consequently, the law of conservation of momentum will also fail — when magnetic forces act, the momentum of the system of particles is not constant.

This is a disaster for Newton, but is not a disaster for physics. If we

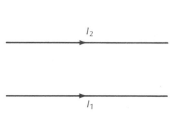

Fig. 30.5 Two parallel currents.

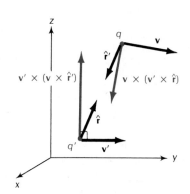

Fig. 30.6 With $\mathbf{v}'$, $\hat{\mathbf{r}}$, and $\mathbf{v}$ as in Figure 30.1, both $\mathbf{v} \times (\mathbf{v}' \times \hat{\mathbf{r}})$ and $\mathbf{v}' \times (\mathbf{v} \times \hat{\mathbf{r}}')$ are in the z–y plane, but they are not of equal magnitudes and opposite directions.

take into account the electric field, we can see why a violation of momentum conservation is quite reasonable for electric charges in motion. We know that in the space between the electric charges there is an electric field, with an energy density. As the charges move, the field continuously changes and the energy of the field must continuously redistribute itself, i.e., the energy of the field must flow through space. We know that energy has mass; hence such a changing field involves a flow of field-mass from one part of space to another. Since any moving mass has momentum, we conclude that the moving, changing electric field has momentum. Once we recognize this, we can see why the momentum of the particles is not conserved. The particles not only exchange momentum with each other, but also with the electric field. What is conserved is not the momentum of the particles, but the total momentum of *particles and field*. Although Newton's Third Law fails, the law of conservation of momentum remains valid. The explicit calculation of the momentum contained in an electric field is fairly complicated and we will not attempt it here.

Incidentally, if the system of charged particles consists of wires with steady currents flowing in closed circuits, then we can prove that the net magnetic force of one circuit on another circuit does satisfy Newton's Third Law of the equality of action and reaction. Obviously, if steady currents are flowing in closed circuits, then the fields are constant in time and their momentum must also remain constant. There is then no possibility of a momentum transfer to the fields and the conservation of momentum would fail if Newton's Third Law did not hold.

Nikola Tesla, *1856–1943, American electrical engineer and inventor. He made many brilliant contributions to high-voltage technology, ranging from new motors and generators to transformers and a system for radio transmission. Tesla designed the power-generating station at Niagara Falls.*

30.2 The Magnetic Field

In Section 23.2 we presented some arguments in favor of the view that the electric force is communicated from one charge to another by action-by-contact, through an electric field. Likewise, the magnetic force is communicated from one charge to another through a **magnetic field**.

The definition of the magnetic field is as follows: To find the magnetic field at some point in the vicinity of moving charges or currents, place a test charge q at that point and give it some velocity $\mathbf{v}$. This charge will then experience a magnetic force depending on its velocity. The magnetic field $\mathbf{B}$ is implicitly defined by the equation for the magnetic force,

Relationship between magnetic force and field

$$\boxed{\mathbf{F} = q\mathbf{v} \times \mathbf{B}} \tag{12}$$

To find the magnitude and direction of the magnetic field at some point, we place a test charge q at this point and launch it with a velocity $\mathbf{v}$ in some direction. By repeating this procedure several times, we discover how the force depends on the direction of the velocity $\mathbf{v}$. In one direction, the force will be zero; this is the direction parallel or antiparallel to $\mathbf{B}$. In the direction perpendicular to this direction, the magnitude of the force will be $F = qvB$, from which we deduce that the magnitude of the magnetic field is $B = F/(qv)$. Hence, in the special case of velocity perpendicular to the magnetic field, the magnitude of the magnetic field is the force per unit charge and unit velocity.

The SI unit of magnetic field is N/(C · m/s); this unit is called the **tesla** (T),

$$1 \text{ tesla} = 1 \text{ T} = 1 \text{ N/(C} \cdot \text{m/s)} \tag{13}$$

tesla, T

An alternative name for this unit is weber/m². For weak magnetic fields, a smaller unit is often preferred; this is the **gauss,**

$$1 \text{ gauss} \Leftrightarrow 10^{-4} \text{ T} = 10^{-4} \text{ weber/m}^2 \tag{14}$$

We have written this relationship between gauss and tesla as an equivalence (⇔) rather than as an equality (=) because the gauss is a cgs unit of magnetic field rather than an SI unit; the definitions of magnetic field in the cgs and SI systems are somewhat different and the units do not carry the same dimensions and cannot be equated. Table 30.1 gives the values of some typical magnetic fields.

Table 30.1. SOME MAGNETIC FIELDS

At surface of nucleus	$\sim 10^{12}$ T
At surface of pulsar	$\sim 10^{8}$ T
Maximum achieved in laboratory:	
Explosive compression of field lines	1×10^3 T
Steady	30 T
Large bubble-chamber magnet	2 T
In sunspot	~ 0.3 T
At surface of Sun	$\sim 10^{-2}$ T
Near small ceramic magnet	$\sim 2 \times 10^{-2}$ T
Near household wiring	$\sim 10^{-4}$ T
At surface of Earth	$\sim 5 \times 10^{-5}$ T
In sunlight (rms)	3×10^{-6} T
In Crab nebula	$\sim 10^{-8}$ T
In radio wave (rms)	$\sim 10^{-9}$ T
In interstellar galactic space	$\sim 10^{-10}$ T
Produced by human body	3×10^{-10} T
In shielded antimagnetic chamber	2×10^{-14} T

Karl Friedrich Gauss, *1777–1855, German mathematician, physicist, and astronomer. Gauss was the son of peasants, but his amazing mathematical abilities were recognized at an early age and he was granted a scholarship by the Duke of Göttingen. One of the greatest mathematicians of all time, Gauss' most celebrated work lay in number theory. Gauss was an indefatigable calculator, and he loved to perform enormously complicated computations, which today would be regarded as impossible without an electronic computer. He developed new methods for calculations in celestial mechanics, and successfully predicted the orbit of the asteroid Ceres, briefly seen and then lost by the astronomer G. Piazzi. Later, Gauss became interested in electric and magnetic phenomena, which he researched in collaboration with W. Weber. He also worked on geodetic surveys, and invented the electric telegraph.*

From Eq. (6) and (12), we see that the magnetic field generated by a point charge q' moving with velocity $\mathbf{v}'$ is

$$\boxed{\mathbf{B} = \frac{\mu_0}{4\pi} \frac{q'}{r^2} (\mathbf{v}' \times \hat{\mathbf{r}})} \tag{15}$$

Magnetic field of point charge

EXAMPLE 2.   In a hydrogen atom, an electron moves in a circular orbit of radius 5.3×10^{-11} m. The velocity of the electron is 2.2×10^6 m/s. What magnetic field does the electron produce at the center of the orbit?

SOLUTION: The velocity $\mathbf{v}'$ and the unit vector $\hat{\mathbf{r}}$ are perpendicular (Figure 30.7). Hence the magnitude of **B** is

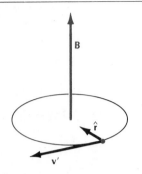

$$B = \frac{\mu_0}{4\pi} \frac{ev'}{r^2}$$

$$= \frac{1.00 \times 10^{-7} \text{ N} \cdot \text{s/C}^2 \times 1.6 \times 10^{-19} \text{ C} \times 2.2 \times 10^6 \text{ m/s}}{(5.3 \times 10^{-11} \text{ m})^2}$$

$$= 13 \text{ T} \qquad\qquad (16)$$

The direction of **B** is perpendicular to the plane of the orbit.

Fig. 30.7 Orbit of an electron around a nucleus, and magnetic field of the electron.

The magnetic field can be represented graphically by field lines. As in the case of the electric field, the tangent to the field line indicates the direction of the field and the density of field lines indicates the strength of the field. Figure 30.8 shows the pattern of magnetic field lines around a moving positive charge. The velocity of the charge is directed perpendicularly out of the plane of the page. The pattern consists of concentric circles; the density of lines becomes infinite at the position of the charge. Ahead and behind the charge, the pattern of field lines also consists of concentric circles, but without the strong concentration near the center. Note that the direction of the field lines follows a simple right-hand rule: place the thumb along the velocity of the charge; the fingers will then curl around in the direction of the field lines. For a negative charge, the direction of the field lines in Figure 30.8 must of course be reversed.

Fig. 30.8 Magnetic field lines of a positive point charge moving perpendicularly out of the plane of the page. (a) At the instant shown, the point charge is in the plane of the page. The density of field lines tends toward infinity near the point charge. (b) At the instant shown, the point charge is above the plane of the page. The density of field lines is zero at the center of the pattern, behind the point charge.

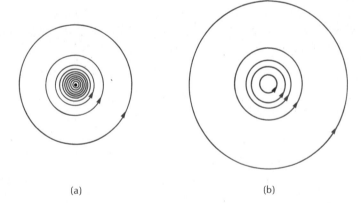

(a)   (b)

For computations with field lines we will adopt the convention that the number of lines per unit area equals the magnitude of the magnetic field. However, for the purpose of making drawings, this normalization is sometimes unwieldy and we will alter it in case of need.

The magnetic field lines of a moving charge form closed loops, i.e., the magnetic field lines do not begin or end anywhere in the way that the electric field lines begin and end on positive and negative electric charges. This indicates that there is no "magnetic charge" that acts as source or sink of magnetic field lines in the way that electric charge acts as source or sink of electric field lines.

Mathematically, we can express these features of the magnetic field lines in terms of a modified version of Gauss's Law. Consider a closed surface of arbitrary shape. The number of magnetic field lines that enter this surface is exactly equal to the number that leave, i.e., the magnetic flux through the closed surface is zero:

$$\oint \mathbf{B} \cdot d\mathbf{S} = 0 \tag{17}$$

Gauss' Law for magnetic field

This equation is not only true for the magnetic field of a single moving charge, but also for any arbitrary magnetic field produced by any arbitrary number of moving charges. The net magnetic field produced by many charges acting together is the sum of their individual magnetic fields and since each of these individual fields satisfies Eq. (17), the net field will also.

Note that the argument of the preceding paragraph hinges on the **principle of linear superposition** for the magnetic field: if several moving charges $q_1, q_2, q_3, \ldots$, simultaneously generate magnetic fields $\mathbf{B}_1, \mathbf{B}_2, \mathbf{B}_3, \ldots$, then the net magnetic field generated by all these charges acting together is

Principle of linear superposition

$$\mathbf{B} = \mathbf{B}_1 + \mathbf{B}_2 + \mathbf{B}_3 + \cdots \tag{18}$$

We will exploit this superposition principle in the next section when we calculate the magnetic field of a current, that is, the magnetic field of a large number of point charges moving along a wire.

30.3 The Biot–Savart Law

The magnetic field generated by a current is of much greater practical interest than the magnetic field generated by a single point charge. Figure 30.9 shows a thin wire of arbitrary shape carrying a steady current I. Following the usual convention, we will pretend that the current is due to a flow of positive charge. Consider a small segment dl of this wire. We can regard the moving charge dq' within dl as a point charge which produces a magnetic field

$$d\mathbf{B} = \frac{\mu_0}{4\pi} \frac{dq'}{r^2} (\mathbf{v}' \times \hat{\mathbf{r}}) \tag{19}$$

with a magnitude

$$dB = \frac{\mu_0}{4\pi} \frac{dq'}{r^2} v' \sin \theta \tag{20}$$

where θ is the angle between $\mathbf{v}'$ and $\hat{\mathbf{r}}$, i.e., it is the angle between the wire and $\hat{\mathbf{r}}$ (Figure 30.9). To relate dq' to the current, we begin with the definition of the current

$$dq' = I \, dt \tag{21}$$

Here dt is the time it takes the charge dq' to flow out of the small segment dl, i.e.,

Jean Baptiste Biot (bio), *1774–1862, French physicist, professor at the Collège de France. His most important work dealt with the refraction and polarization of light, but he was also interested in a broad range of problems in the physical sciences. With Félix Savart, 1791–1841, he confirmed Oersted's discovery of magnetic fields generated by electric currents, and formulated the equation (30.25) for the strength of the magnetic field.*

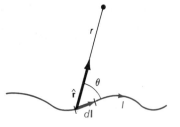

Fig. 30.9 Wire carrying a current I.

$$dt = dl/v' \tag{22}$$

Hence

$$dq' = I \, dl/v' \tag{23}$$

With this, Eq. (20) becomes

$$dB = \frac{\mu_0}{4\pi} \frac{I \, dl \sin \theta}{r^2} \tag{24}$$

To put this in vector form, we treat the segment of wire as a vector $d\mathbf{l}$ tangent to the wire and in the direction of the current (Figure 30.9). The magnetic field (19) can then be expressed as

Biot–Savart law

$$\boxed{d\mathbf{B} = \frac{\mu_0}{4\pi} \frac{I \, d\mathbf{l} \times \hat{\mathbf{r}}}{r^2}} \tag{25}$$

This equation holds true because the right side has the correct magnitude [compare with Eq. (24)] and it also has the correct direction [compare with Eq. (19) noting that $\mathbf{v}'$ and $d\mathbf{l}$ have the same direction].

Equation (25) gives us the magnetic field generated by a short segment of a wire. It is called the **law of Biot–Savart.** The magnetic field generated by a wire of any length and shape can be calculated by integrating this equation along the wire.

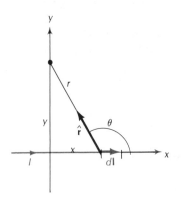

Fig. 30.10 Straight wire carrying a current I along the x axis.

EXAMPLE 3.   A very long and very thin straight wire carries a steady current I. What is the magnetic field at some distance from the wire?

SOLUTION: Figure 30.10 shows the wire lying along the x axis. The magnetic field lines are concentric circles around the wire; the lines come out of the plane of the page in the region above the wire and go into the plane of the page in the region below. We will calculate the magnetic field at a point which is at a distance y from the wire.

A small segment dx of the wire contributes a magnetic field of magnitude

$$dB = \frac{\mu_0}{4\pi} \frac{I \, dx \sin \theta}{r^2} \tag{26}$$

The directions of the contributions from all segments dx are parallel (at the point P the directions of all the contributions to the magnetic field are out of the plane of the page). Hence the magnitude of the net field is

$$B = \int_{-\infty}^{\infty} \frac{\mu_0}{4\pi} \frac{I \sin \theta}{r^2} \, dx \tag{27}$$

From Figure 30.10,

$$x = y \cot(\pi - \theta) = -y \cot \theta \tag{28}$$

so that

$$dx = y \csc^2 \theta \, d\theta \tag{29}$$

Furthermore,

$$r = y/\sin(\pi - \theta) = y/\sin \theta \qquad (30)$$

By means of Eqs. (29) and (30) we can change the variable of integration from x to θ; the new limits of integration are then $\theta = 0°$ and $\theta = 180°$, or, in radians, $\theta = 0$ and $\theta = \pi$,

$$B = \frac{\mu_0}{4\pi} \frac{I}{y} \int_0^\pi \sin \theta \, d\theta \qquad (31)$$

$$= \frac{\mu_0}{4\pi} \frac{I}{y} \left[-\cos \theta \right]_0^\pi = \frac{\mu_0}{4\pi} \frac{I}{y} \times 2$$

which yields

$$\boxed{B = \frac{\mu_0}{2\pi} \frac{I}{y}} \qquad (32)$$

Magnetic field of straight wire

Figure 30.11 shows the pattern of the magnetic field lines for this magnetic field. And Figure 30.12 shows iron filings sprinkled on a sheet of paper placed around a long straight wire carrying a strong current. The iron filings align in the direction of the magnetic field, and therefore make the pattern of field lines visible.

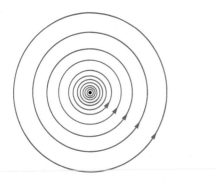

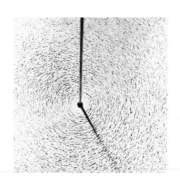

Fig. 30.11 (left) Magnetic field lines around a straight wire carrying a current. The wire is perpendicular to the plane of the page and the current emerges from this plane.

Fig. 30.12 (right) Magnetic field lines around a straight wire, made visible by iron filings sprinkled on a sheet of paper.

EXAMPLE 4.   A square loop of wire of dimension $L \times L$ carries a current I. What is the magnetic field at the center of the loop?

SOLUTION: Figure 30.13a shows the loop. Each side of the loop makes the same contribution to the magnetic field, in magnitude and in direction. Hence the net magnetic field at the center is four times the magnetic field contributed by one side.

Figure 30.13b shows one side of the loop. The calculation of the magnetic field of this side is similar to the calculation of the magnetic field of the infi-

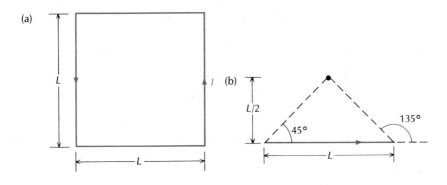

Fig. 30.13 A square loop carrying a current I.

nitely long wire, but the limits for the integration over x are now finite. In terms of the angle θ introduced in Eq. (28), the lower limit is $\theta = 45°$ (or $\theta = \pi/4$) and the upper limit is $\theta = 135°$ (or $\theta = 3\pi/4$). With these new limits, the integral in Eq. (31) becomes

$$B_1 = \frac{\mu_0}{4\pi} \frac{I}{L/2} \int_{\pi/4}^{3\pi/4} \sin\theta \, d\theta$$

$$= \frac{\mu_0}{4\pi} \frac{I}{L/2} \left[-\cos\theta \right]_{\pi/4}^{3\pi/4}$$

$$= \frac{\mu_0}{4\pi} \frac{I}{L/2} \sqrt{2} \tag{33}$$

Multiplying this by 4, we obtain the net magnetic field of the entire loop,

$$B = 4B_1 = \frac{\mu_0}{4\pi} \frac{8\sqrt{2}\,I}{L} \tag{34}$$

EXAMPLE 5. Find the magnetic field on the axis of a circular loop of wire of radius R carrying a current I.

SOLUTION: Figure 30.14 shows the loop and a point on its axis. A small segment dl of the ring produces a magnetic field of magnitude

$$dB = \frac{\mu_0}{4\pi} \frac{I \, dl}{z^2 + R^2} \tag{35}$$

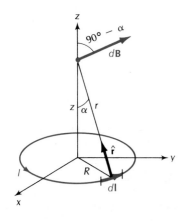

Fig. 30.14 Circular wire loop carrying a current I in the x–y plane.

The direction of this magnetic field is perpendicular to the line connecting dl and the point on the axis (Figure 30.14). Upon integration around the circle, the horizontal components of the magnetic field cancel because diametrically opposite segments dl contribute opposite horizontal components. Only the vertical component of **B** survives.

The vertical component of the magnetic field in Eq. (35) is

$$dB_z = dB\cos(90° - \alpha) = dB\sin\alpha = \frac{\mu_0}{4\pi} \frac{I \, dl}{z^2 + R^2} \sin\alpha \tag{36}$$

where, according to Figure 30.14,

$$\sin\alpha = R/\sqrt{z^2 + R^2} \tag{37}$$

Hence

$$B_z = \int \frac{\mu_0}{4\pi} \frac{I R \, dl}{(z^2 + R^2)^{3/2}} \tag{38}$$

Since all the terms in the integrand are constant, the integration amounts to multiplication of the integrand by the length of the path of integration (the circumference of the circle). Thus,

$$B_z = \frac{\mu_0}{4\pi} \frac{IR}{(z^2 + R^2)^{3/2}} \times 2\pi R \tag{39}$$

or

$$B_z = \frac{\mu_0}{2\pi} \frac{I\pi R^2}{(z^2 + R^2)^{3/2}} \tag{40}$$

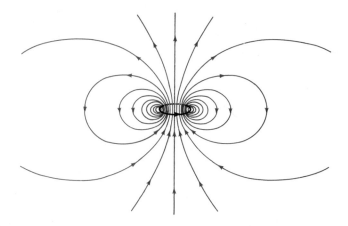

Fig. 30.15 Magnetic field lines of a circular loop of wire carrying a current.

The calculation of Example 5 only gives the magnetic field on the axis of the loop. The calculation of the magnetic field at other points is rather messy. Figure 30.15 shows the general pattern of magnetic field lines throughout space. Note that at large distance from the loop this pattern is quite similar to the pattern of electric field lines of an electric dipole (see Figure 25.11). Figure 30.16 shows the pattern of magnetic field lines, as made visible with iron filings.

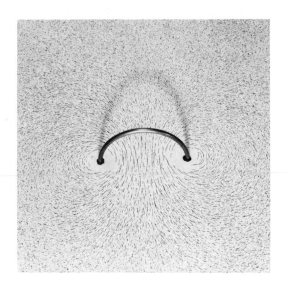

Fig. 30.16 Iron filings sprinkled on a sheet of a paper placed around a loop carrying a strong current.

A small loop of current is called a **magnetic dipole**. In this context, *small* means that the size of the loop is very small compared to the distance at which we wish to calculate the magnetic field; in Eq. (40) it means $R \ll z$. We can then approximate the expression (40) by

Magnetic dipole

$$B_z = \frac{\mu_0}{2\pi} \frac{I\pi R^2}{z^3} \qquad (41)$$

This shows that the magnetic field of a loop of current decreases as the inverse cube of the distance. It is customary to write Eq. (41) as

$$B_z = \frac{\mu_0}{2\pi} \frac{\mu}{z^3} \qquad (42)$$

where

$$\mu = I\pi R^2 \tag{43}$$

is the **magnetic dipole moment** of the ring.[3]

Note that Eq. (43) has the form

Magnetic dipole moment

$$\mu = [\text{current}] \times [\text{area of loop}] \tag{44}$$

This turns out to be a general expression for the magnetic dipole moment of a (plane) loop of arbitrary shape. For instance, the magnetic dipole moment of a square or rectangular loop of current can be calculated from Eq. (44) and the magnetic field at a large distance from the loop is then given by Eq. (42).

Electrons, protons, and many other elementary particles have magnetic dipole moments. These particles may be regarded crudely as small balls of electric charge spinning about an axis. Such a rotating charge distribution behaves like a collection of rings of current. This generates a magnetic field which, at a large distance from the particle, is essentially the dipole field of Figure 30.15. The dipole moment of an electron is 9.3×10^{-24} A $\cdot$ m². Thus, the space surrounding an electron is filled not only with electric fields, but also with magnetic fields. Figure 30.17 shows the electric and magnetic fields of an electron. Of course, if the electron also has a translational motion, then there will be an extra magnetic field of the type given by Eq. (15) around the electron.

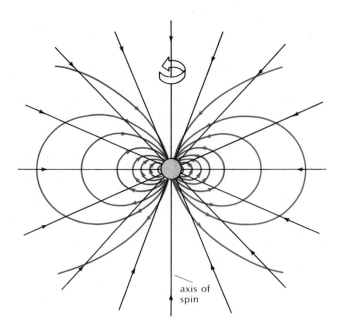

Fig. 30.17 Electric (black) and magnetic (colored) fields of an electron.

axis of spin

The Earth also has a magnetic dipole moment. This is presumably due to currents that flow around in loops deep inside the Earth, in the liquid iron of the core. Geophysicists do not yet know exactly what is

---

[3] The μ in Eq. (42) must not be confused with the μ_0 for permeability; they are not related.

the energy supply for the emf that drives these currents, but it is certain that the rotation of the Earth plays a crucial role in the generation of the currents. These currents create a magnetic field which, at large distances from the core, is nearly a dipole field. The magnetic moment of the Earth is $8.0 \times 10^{22} \ \mathrm{A \cdot m^2}$. Figure 30.18 shows the magnetic field lines. Note that the field lines emerge from the surface of the Earth near the geographic South Pole and they reenter the surface near the geographic North Pole. The force experienced by a compass needle is due to this magnetic field.

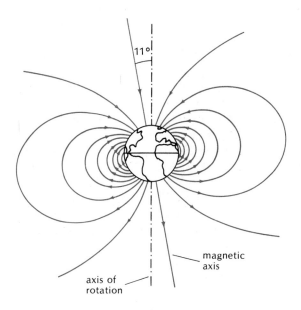

Fig. 30.18 Magnetic field of the Earth. The axis of the magnetic dipole makes an angle of 11° with the axis of rotation of the Earth.

30.4 Relativity and the Magnetic Field[4]

Since the magnetic force, $\mathbf{F} = q\mathbf{v} \times \mathbf{B}$, on a particle depends on the velocity, this force must necessarily depend on the reference frame with respect to which this velocity is reckoned. The following example illustrates this dependence in a drastic way.

Suppose that a positive charge q moves in the magnetic field generated by a current on a straight wire. We will assume that the (instantaneous) velocity $\mathbf{v}$ of the charge q is parallel to the wire. Figure 30.19 shows the situation in a reference frame in which the wire is at rest. The charge has a velocity $\mathbf{v}$ toward the right and the wire carries a current I toward the left. In this reference frame the current generates a magnetic field [see Eq. (32)]

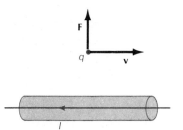

Fig. 30.19 In the rest frame of the wire, the charge q moves to the right with velocity $\mathbf{v}$.

$$B = \frac{\mu_0}{2\pi} \frac{I}{y} \tag{45}$$

and this magnetic field exerts a magnetic force

$$F = qvB = \frac{\mu_0}{2\pi} \frac{qvI}{y} \tag{46}$$

---

[4] This section is optional.

$q \bullet$

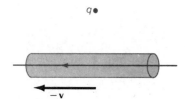

$-\mathbf{v}$

Fig. 30.20 In the rest frame of the charge q, the wire moves to the left with velocity $-\mathbf{v}$.

F ↑
•
q

Fig. 30.21 In the reference frame of the wire, the positive and the negative charge densities on the wire are equal.

on the charge q. The force points radially away from the wire (Figure 30.19).

Next, let us examine the situation in a new reference frame in which the charge q is (instantaneously) at rest. Relative to the old reference frame, this new reference frame moves toward the right with velocity $\mathbf{v}$. Figure 30.20 shows the situation in this reference frame. The charge q is at rest and the wire moves toward the left with velocity $-\mathbf{v}$. In this new reference frame there can be no magnetic force on the charge q, since the velocity of the charge is zero. We are now faced with a paradox: in the old reference frame the charge experiences a magnetic force and hence an acceleration; in the new reference frame the charge experiences no magnetic force, hence no acceleration — and yet accelerations are supposed to be independent of the reference frame!

The resolution of this paradox hinges on a subtle relativistic effect. It turns out that in the new reference frame, the wire generates an *electric field* and the corresponding electric force on the charge gives it the required acceleration.

To understand where this electric field in the new reference frame comes from, let us begin by asking why there is no electric field in the old reference frame. Obviously the answer is that in the reference frame of the wire the positive and negative charge densities on the wire are of equal magnitude; hence their electric fields cancel exactly.[5] Figure 30.21 shows these charge distributions in the reference frame of the wire. The negative charge density is due to the free electrons carrying the current; the positive charge density is due to the positive ions fixed in the lattice of the wire. The charge density of the electrons is $-\lambda$ coulomb/m and that of the ions is $+\lambda$ coulomb/m.

In the reference frame of the wire, the negative charges are in motion and the positive charges are at rest. Since the current is flowing toward the left, the free electrons carrying this current must move toward the right. For the sake of simplicity, let us assume that the velocity of the free electrons coincides with the velocity of the charge q. This is a very special case — and not very likely to happen in reality. It is, of course, possible to solve the problem in general, but it greatly helps in the solution if we have to worry about only a single velocity rather than two different velocities.

Now consider the electric charge densities in the new reference frame. The crucial point is that in this reference frame the negative and positive densities will *not be equal*. The inequality arises from the length contraction effect of special relativity. We recall from Section 17.5 that if an object has a certain length in its own reference frame, then in any other reference frame the length is shorter by a factor

$$\sqrt{1 - v^2/c^2}$$

where c is the speed of light. Thus, if a given number of positive charges sitting on the wire occupy a length of 1 m in the old reference frame (rest frame of the wire), they will occupy a shorter length of

$$1 \text{ m} \times \sqrt{1 - v^2/c^2}$$

---

[5] For the present purposes, we ignore the small electric field needed to push the current along the wire.

in the new reference frame. Correspondingly, the density of these positive charges will be larger: if the density is λ coulomb/m in the old reference frame, it will be

$$\lambda/\sqrt{1 - v^2/c^2} \text{ coulomb/m}$$

in the new reference frame. For the negative charge distribution, the length contraction has the opposite effect — the density of negative charges will be smaller: if the density of charge is $-\lambda$ coulomb/m in the old reference frame, it will be

$$-\lambda\sqrt{1 - v^2/c^2} \text{ coulomb/m}$$

in the new reference frame. This is so because in the old reference frame (rest frame of the wire) the charge distribution of electrons is in motion; it therefore is a *contracted* charge distribution. In the new reference frame, the charge distribution is at rest and it is not contracted. The transformation from the old to the new reference frame therefore is a transformation from a reference frame in which the length of the negative charge distribution is already contracted to a reference frame in which it is not contracted. Correspondingly, the density of the negative charges will be smaller, as indicated above.

In the new reference frame, the net charge per unit length of the wire is then the sum of *unequal* positive and negative contributions,

$$\lambda_{\text{new}} = \frac{\lambda}{\sqrt{1 - v^2/c^2}} - \lambda\sqrt{1 - v^2/c^2} \tag{47}$$

If the speeds are small compared to the speed of light, we can use the approximations

$$1/\sqrt{1 - v^2/c^2} \cong 1 + \tfrac{1}{2} v^2/c^2 \quad \text{and} \quad \sqrt{1 - v^2/c^2} \cong 1 - \tfrac{1}{2} v^2/c^2$$

so that Eq. (47) becomes

$$\lambda_{\text{new}} = \frac{\lambda}{\sqrt{1 - v^2/c^2}} - \lambda\sqrt{1 - v^2/c^2} \cong \lambda v^2/c^2 \tag{48}$$

Such a charge density along the wire will generate a radial electric field [see Eq. (24.10)]

$$E = \frac{1}{2\pi\varepsilon_0} \frac{\lambda_{\text{new}}}{y} = \frac{1}{2\pi\varepsilon_0 c^2} \frac{\lambda v^2}{y} \tag{49}$$

This electric field exerts an electric force

$$F_{\text{new}} = \frac{1}{2\pi\varepsilon_0 c^2} \frac{q\lambda v^2}{y} \tag{50}$$

on the charge q. This force points away from the wire (Figure 30.22).

In order to compare this electric force in the new reference frame with the magnetic force in the old reference frame, we note that the product of the velocity v of the electrons and their charge density λ is the current I on the wire. Hence Eq. (50) can be written

Fig. 30.22 In the reference frame of the charge q, the positive charge density on the wire exceeds the negative charge density.

$$F_{\text{new}} = \frac{1}{2\pi\varepsilon_0 c^2} \frac{qvI}{y} \qquad (51)$$

Now examine the ratio of the forces in the old and the new reference frames,

$$\frac{F}{F_{\text{new}}} = \left(\frac{\mu_0}{2\pi} \frac{qvI}{y}\right) \bigg/ \left(\frac{1}{2\pi\varepsilon_0 c^2} \frac{qvI}{y}\right) = \mu_0\varepsilon_0 c^2 \qquad (52)$$

Inserting numerical values for the constants on the right side of Eq. (52), we find

$$\frac{F}{F_{\text{new}}} = 1.26 \times 10^{-6} \frac{\text{N} \cdot \text{s}^2}{\text{C}^2} \times 8.85 \times 10^{-12} \frac{\text{C}^2}{\text{N} \cdot \text{m}^2} \times (3.0 \times 10^8 \text{ m/s})^2$$

$$= 1.0 \qquad (53)$$

that is, the forces are exactly equal. Note that this result hinges on the fact that the values of ε_0 and μ_0 are exactly right to cancel the factor of c^2 in Eq. (52); again, as in Section 30.1, we see that there is a deep connection between electricity, magnetism, and the speed of light.

What we conclude from the above calculation is then the following: the force that the charges on the wire exert on the positive point charge q is the same in both the old and the new reference frames. The transformation of reference frame does not change the magnitude or direction of the force (and of the acceleration) — it only changes the character of the force from purely magnetic to purely electric. Incidentally, in a reference frame moving toward the right with a speed of, say, $\frac{1}{2}v$, the force would still have the same magnitude and direction, but it would be partially magnetic and partially electric.

Although we obtained these results only for a very special and simple case, the main features have general validity. If a particle experiences a magnetic force in a given reference frame, a transformation to the rest frame of the particle will make this magnetic force disappear. But in the latter reference frame, charge distributions will appear at the locations of the currents, and the electric force of these charge distributions will replace the original magnetic force. The net force is the same in both reference frames.[6]

Electric and magnetic forces and fields transform into one another if we change the frame of reference. In the rest frame of a charged particle, only electric forces act on the particle. Hence we can regard the magnetic forces that act on the particle in any other reference frame as resulting from a transformation of the electric forces in the rest frame. In this sense, magnetic forces can be regarded as a consequence of electric forces and of relativity.

---

[6] This invariance of the force is true only if the relative speed of the reference frames is low ($v \ll c$). If the speed is high, then we must take into account the relativistic transformation law for force. It turns out that the transformation of electric and magnetic forces is consistent with that law.

SUMMARY

Magnetic force between moving point charges:

$$\mathbf{F} = \frac{\mu_0}{4\pi} \frac{qq'}{r^2} \, \mathbf{v} \times (\mathbf{v}' \times \hat{\mathbf{r}})$$

Magnetic field of point charge:

$$\mathbf{B} = \frac{\mu_0}{4\pi} \frac{q'}{r^2} \, (\mathbf{v}' \times \hat{\mathbf{r}})$$

Permeability constant: $\mu_0 = 1.26 \times 10^{-6} \, \text{N} \cdot \text{s}^2/\text{C}^2$

$$\frac{\mu_0}{4\pi} = 1.00 \times 10^{-7} \, \text{N} \cdot \text{s}^2/\text{C}^2$$

Definition of magnetic field: $\mathbf{F} = q\mathbf{v} \times \mathbf{B}$

Gauss' law for magnetism: $\oint \mathbf{B} \cdot d\mathbf{S} = 0$

Biot–Savart law: $d\mathbf{B} = \frac{\mu_0}{4\pi} I \frac{d\mathbf{l} \times \hat{\mathbf{r}}}{r^2}$

Magnetic dipole moment: $\mu = [\text{current}] \times [\text{area of loop}]$

QUESTIONS

1. List all the physical laws of force you can think of. Which of these laws are fundamental, and which can be derived from others?

2. Give an example of two moving charged particles with velocities such that the mutual magnetic forces do not obey Newton's Third Law. Give an example with velocities such that the mutual magnetic forces do obey Newton's Third Law.

3. How would the magnetic field lines shown in Figure 30.8 differ if the charge of the particle were negative instead of positive?

4. Theoretical physicists have proposed the existence of **magnetic monopoles**, which are sources and sinks of magnetic field lines, just as electric charges are sources and sinks of electric field lines. What would the pattern of magnetic field lines of a positive magnetic monople look like? Can you guess the pattern of *electric* field lines of a moving magnetic monopole?

Magnetic monopole

5. The Earth's magnetic field at the equator is horizontal, in the northward direction. What is the direction of the magnetic force on an electron moving vertically up?

6. An electron with a vertical velocity passes through a magnetic field without suffering any deflection. What can you conclude about the magnetic field?

7. At an initial time, a charged particle is at some point P in a magnetic field and it has an initial velocity. Under the influence of the magnetic field, the particle moves to a point P'. If you now reverse the velocity of the particle, will it retrace its orbit and return to the point P?

8. An electron moving northward in a region of space is deflected toward the east by a magnetic field. What is the direction of the magnetic field?

9. A Faraday cage shields electric fields. Does it also shield magnetic fields?

10. Strong electric fields are hazardous — if you place some part of your body in a strong electric field you are likely to receive an electric shock. Are strong magnetic fields hazardous? Do they produce any effect on your body?

11. Figure 30.18 shows the magnetic field of the Earth. What must be the direction of the currents flowing in loops inside the Earth to give this magnetic field?

12. The needle of an ordinary magnetic compass indicates the direction of the horizontal component of the Earth's magnetic field. Explain why the magnetic compass is unreliable when used near the poles of the Earth.

Dip needle

13. A **dip needle** is a compass needle that swings about a horizontal axis. If the axis is oriented east–west, then the equilibrium direction of the dip needle is the direction of the Earth's magnetic field. The **dip angle** of the dip needle is the angle that it makes with the horizontal. How does the dip angle vary as you transport a dip needle along the surface of the Earth from the South Pole to the North Pole?

14. Suppose we replace the single loop shown in Figure 30.14 by a coil of N loops. How does this change the formula [Eq. (40)] for the magnetic field?

15. In order to eliminate or reduce the magnetic field generated by the pair of wires that connect a piece of electric equipment to an outlet, a physicist twists these wires tightly about each other. How does this help?

16. Consider a circular loop of wire carrying a current. Describe the direction of the magnetic field at different points in the plane of the loop, both inside and outside of the loop.

17. An infinite flat conducting sheet lies in the x–y plane. The sheet carries a current in the y direction; this current is uniformly distributed over the entire sheet. What is the direction of the magnetic field above the sheet? Below the sheet?

18. Suppose that an infinitely long straight wire lies along the axis of a circular loop. Both the wire and the loop carry currents I. Draw some field lines of the net magnetic field of the wire and the loop.

19. According to Eq. (42), the magnetic field at large distance from a square loop is approximately the same as the magnetic field of a circular loop of the same area. Why is the shape unimportant? [Hint: The circular loop can be regarded as made of many small (infinitesimal) square loops.]

20. The arguments of Section 30.4 show that the Eq. (51) for the magnetic force exerted on a charged particle by the current of a straight wire is a direct consequence of the relativistic length contraction. Can we therefore regard the experimental verification of this force law as an experimental verification of the length contraction?

PROBLEMS

Section 30.1

1. A positron (charge $+e$) and an electron (charge $-e$) are moving side by side with a uniform velocity of 2.0×10^6 m/s along parallel tracks. Is the magnetic force between them attractive or repulsive? By what percentage is the total force (electric and magnetic) acting between these moving charges larger or smaller than that acting between charges at rest at the same distance?

2. Two electrons are (instantaneously) moving at right angles with speeds $v = 3.0 \times 10^6$ m/s and $v' = 1.0 \times 10^6$ m/s in the same plane (Figure 30.23). Their distance is 0.80×10^{-10} m. Calculate the magnetic force of the first electron on the second and that of the second on the first. Show the directions of the forces on a diagram.

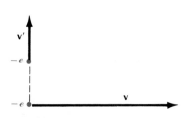

Fig. 30.23 Instantaneous positions and velocities of two electrons.

3. An electron (charge $-e$) and an antielectron (charge $+e$) are in a circular orbit about each other (Figure 30.24). The orbital radius is 0.53×10^{-10} m and the orbital speed is 1.1×10^6 m/s.
   (a) What is the magnitude of the electric force on each particle? Draw a diagram showing the direction of the electric force on each particle.
   (b) What is the magnitude of the magnetic force on each particle?

4. Two protons have equal speeds of 2.0×10^6 m/s. Their directions of motion are in the same plane and they make an angle of $60°$. The distance between the protons is 2.4×10^{-10} m, and this distance is (instantaneously) perpendicular to the velocity **v** of one of the protons.
   (a) Calculate the magnetic force on each proton due to the other proton; show the directions of the forces on a diagram.
   (b) Calculate the instantaneous rate of change of the total momentum of the two-proton system.

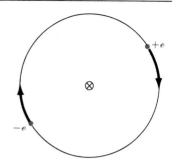

Fig. 30.24 Electron and antielectron in orbit about their common center of mass.

Section 30.2

5. Show that the magnetic field of a point charge q' moving with a velocity **v'** [Eq. (15)] can be written in terms of the electric field of this point charge as $\mathbf{B} = \mu_0 \varepsilon_0 \mathbf{v'} \times \mathbf{E}$.

6. The magnetic field surrounding the Earth typically has a strength of 5×10^{-5} T. Suppose that a cosmic-ray electron of energy 3×10^4 eV is instantaneously moving in a direction perpendicular to the lines of this magnetic field. What is the force on this electron?

7. At a location where the strength of the Earth's magnetic field is 0.60×10^{-4} T, what must be the minimum speed of an electron if the magnetic force on it is to exceed its weight?

8. At the surface of a pulsar, or neutron star, the magnetic field may be as strong as 10^8 T. Consider the electron in a hydrogen atom on the surface of such a neutron star. The electron is at a distance of 0.53×10^{-10} m from the proton and has a speed of 2.2×10^6 m/s. Compare the electric force that the proton exerts on the electron with the magnetic force that the magnetic field of the neutron star exerts on the electron. Is it reasonable to expect that the hydrogen atom will be strongly deformed by the magnetic field?

9. In New York, the magnetic field of the Earth has a vertical (down) component of 0.60×10^{-4} T and a horizontal (north) component of 0.17×10^{-4} T. What are the magnitude and direction of the magnetic force on an electron of velocity 1.0×10^6 m/s moving (instantaneously) in an east to west direction in a television tube?

Section 30.3

10. The current in a lightning bolt may be as much as 2×10^4 A. What is the magnetic field at a distance of 1.0 m from a lightning bolt? The bolt can be regarded as a straight line of current.

11. The cable of a high-voltage power line is 25 m above the ground and carries a current of 1.8×10^3 A.
    (a) What magnetic field does this current produce at the ground?
    (b) The strength of the magnetic field of the Earth is 0.60×10^{-4} T at the location of the power line. By what factor do the fields of the power line and of the Earth differ?

12. In a motorboat, the compass is mounted at a distance of 0.80 m from a cable carrying a current of 20 A from an electric generator to a battery.
    (a) What magnetic field does this current produce at the location of the compass? Treat the cable as a long, straight wire.
    (b) The horizontal (north) component of the Earth's magnetic field is 0.18×10^{-4} T. Since the compass points in the direction of the net horizontal magnetic field, the current will cause a deviation of the compass.

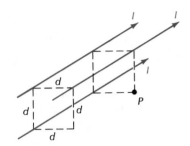

Fig. 30.25 Three long, parallel wires. The wires pass through three corners of a square.

The deviation will be maximum if the magnetic field of the current is horizontal and at right angles to the Earth's magnetic field. Under these worst circumstances, by how many degrees will the compass deviate from north?

13. A superconducting ring of diameter 3.0 cm carries a current of 12 A. What is the strength of the magnetic field at the center of the ring? Along the axis of the ring at a distance of 3.0 cm from the center?

14. A circular ring of wire of diameter 0.60 m carries a current of 35 A. What acceleration will the magnetic force generated by this ring give to an electron that is passing through the center of the ring with a velocity of 1.2×10^6 m/s in the plane of the ring?

15. Two very long, straight, parallel wires separated by distance d carry currents of magnitude I in opposite directions. Find the magnetic field at a point equidistant from the lines, with a distance $2d$ from each line. Draw a diagram showing the direction of the magnetic field.

16. Three parallel wires are spaced as shown in Figure 30.25. The wires carry equal currents in the same direction. What is the magnetic field at the point P? Draw a diagram giving the direction of the magnetic field.

17. Two very long parallel wires separated by a distance of 1.0 cm carry opposite currents of 8.0 A.
   (a) Find the magnetic field at the midpoint between the wires.
   (b) Find the magnetic field in the plane of the wires, at a distance of 2.0 cm from the midline.

18. A very long wire is bent at a right angle near its midpoint. One branch of it lies along the positive x axis and the other along the positive y axis (Figure 30.26). The wire carries a current I. What is the magnetic field at a point in the first quadrant of the x–y plane?

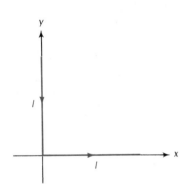

Fig. 30.26. A very long wire bent at a right angle.

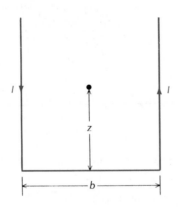

Fig. 30.27 A very long wire bent in the shape of a **U**.

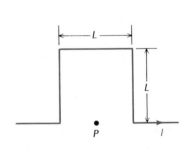

Fig. 30.28 A very long wire with square bends.

19. A very long wire carrying a current I is bent in the shape of a **U** (Figure 30.27). The two parallel segments are separated by a distance b. Find the magnetic field along the midline at a distance z from the bottom of the **U**. Consider both the case $z > 0$ and $z < 0$.

20. A very long, straight wire with a current I has a square bend near its midpoint (Figure 30.28). Each side of the square has a length L. What is the magnetic field at the point P, halfway between the two lower corners?

21. A loop of superconducting wire has the shape of a rectangle measuring $L \times 2L$. A current I flows around the wire. What is the strength of the magnetic field at the center of the rectangle?

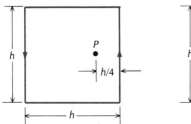

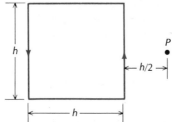

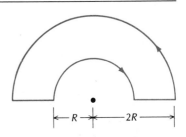

Fig. 30.29 A square loop.

Fig. 30.30 A square loop.

Fig. 30.31 A loop consisting of two semicircles and two straight segments.

22. A square loop of wire, measuring $h \times h$, carries a current I (Figure 30.29). Find the magnetic field at the point P at a distance $h/4$ from the center of the square.

23. A square loop of wire, measuring $h \times h$, carries a current I (Figure 30.30). Find the magnetic field at the point P at a distance h from the center of the square.

24. A loop of wire has the shape of two concentric semicircles connected by two radial segments (Figure 30.31). The loop carries a current I. Find the magnetic field at the center.

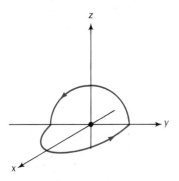

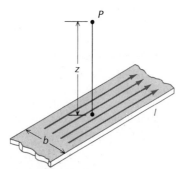

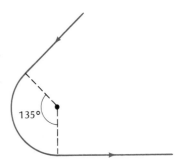

Fig. 30.32 A semicircle in the z–y plane and a semicircle in the x–y plane.

Fig. 30.33 A very long strip carrying a uniformly distributed current I.

Fig. 30.34 Two very long wires joined by an arc of circle.

25. A circular loop of wire is folded along a diameter so as to form two semicircles of radius R intersecting at right angles (Figure 30.32). A current I flows around this loop. What is the magnetic field at the center? Draw a diagram showing the direction of the magnetic field.

26. A very long strip of copper of width b carries a current I uniformly distributed over the strip. What is the magnetic field at a distance z above the midline of this strip (Figure 30.33).

27. An infinite conducting sheet occupies the x–y plane. On this sheet a current flows in the y direction; the current is uniformly distributed over the sheet with σ ampere flowing across each 1-meter segment of the x axis. Find the magnetic field at a distance z from this sheet.

28. Two semi-infinite wires are in the same plane. The wires make an angle of $45°$ with each other and they are joined by an arc of circles of radius R (Figure 30.34). The wires carry a current I. Find the magnetic field at the center of the arc of circle.

29. Helmholtz coils are often used to make reasonably uniform magnetic fields in laboratories. These coils consist of two thin circular rings of wire par-

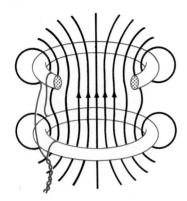

Fig. 30.35 Helmholtz coils.

allel to each other and on a common axis, the z axis (Figure 30.35). The rings have radius R and they are separated by a distance which is also R. The rings carry equal currents in the same direction.

(a) Find the magnetic field at any point of the z axis.

(b) Show that dB/dz and d^2B/dz^2 are both zero at $z = 0$.

30. The magnetic dipole moment of an electron is 9.3×10^{-24} A · m². What is the magnetic field on the axis of spin of the electron at a distance of 1.0 Å from its center?

31. The magnetic field of the Earth is approximately that of a magnetic dipole located at the center of the Earth. The strength of the magnetic field at the surface of the Earth at the magnetic north pole is 6.2×10^{-5} T. What is the strength of the magnetic field at an altitude of 1000 km above the pole? 2000 km? 3000 km? Make a plot of the strength of the magnetic field vs. altitude.

32. The magnetic field of the Earth is that of a magnetic moment of 8.0×10^{22} A · m² located deep within the Earth. Assume that the magnetic moment is due to a circular current flowing along the equator of the liquid core of the Earth; this core has a radius of 3500 km. What must be the current? What must be its direction (eastward or westward)?

33. It can be proved that the dependence on x, y, z of the magnetic field of a magnetic dipole is described by the same function as the dependence on x, y, z of the electric field of an electric dipole. Use this fact to convert the formulas of Eqs. (25.52)–(25.54) into formulas for a magnetic dipole.

34. A steady current I flows around a wire bent into a square of side L (Figure 30.36). Find the magnetic field at a distance z from the center of the square on a line perpendicular to the face of the square. Show that if $z \gg L$, your answer reduces to $B_z = (\mu_0/2\pi)\mu/z^3$, with a magnetic moment $\mu = IL^2$.

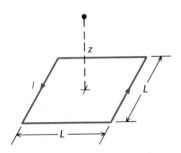

Fig. 30.36 A square loop.

35. A pipe made of a superconducting material has a length of 0.30 m and a radius of 4.0 cm. A current of 4.0×10^3 A flows around the surface of the pipe; the current is uniformly distributed over the surface. What is the magnetic moment of this current distribution? (Hint: Treat the current distribution as a large number of rings stacked one on top of another.)

36. An amount of charge Q is uniformly distributed over a disk of paper of radius R. The disk spins about its axis with angular velocity ω. Find the magnetic dipole moment of the disk. Sketch the lines of magnetic field and of electric field in the vicinity of the disk.

*37. An amount of charge Q is uniformly distributed over a disk of paper of radius R. The disk spins about its axis with angular velocity ω. Find the magnetic field produced by this disk at a point on the axis of the disk, at a distance z from the center.

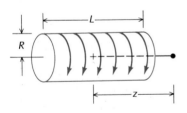

Fig. 30.37 A uniformly distributed current flows around the circumference of a copper pipe.

*38. A current I flows around the surface of a copper pipe of radius R and length L (Figure 30.37). Find a formula for the magnetic field on the axis of the pipe at a distance z from the center. Assume $z > L/2$.

*39. A semi-infinite wire lies along the positive x axis. Another semi-infinite wire lies along the y axis. The wires carry a current I from $y = \infty$ to $x = \infty$ (Figure 30.38).

(a) Find the magnetic field at a point in the first octant $(x > 0, y > 0, z > 0)$.

(b) Find the magnetic field at a point in the second octant $(x < 0, y > 0, z > 0)$.

*40. Two semi-infinite parallel wires lie in the x–y plane. The wires are joined by a semicircle of radius R whose center coincides with the origin of coordinates (Figure 30.39). The wires carry a current I in the direction shown in the figure. Find the x, y, and z components of the magnetic field at a point on the z axis.

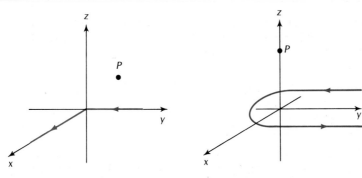

Fig. 30.38 Long straight wires along the x and the y axes.

Fig. 30.39 Two long wires in the x–y plane joined by a semicircle.

Section 30.4

41. (a) In the reference frame of the laboratory, a proton moves with velocity 1.0×10^7 m/s parallel to a long straight wire at a distance of 0.10 m. The wire carries a current of 15 A. What is the magnetic force on the proton?
    (b) In the reference frame of the proton, what is the magnetic force? What is the electric force? In this reference frame, what must be the charge density on the wire to give the correct value for the electric force?

Ampère's Law

In Chapter 24 we saw that Gauss' Law gives us a shortcut for calculating the electric field of charge distributions that have a certain amount of symmetry. In this chapter we will see that there exists a corresponding method for calculating the magnetic field of current distributions. The method is based on Ampère's law, an important relation between the magnetic field and the current.

31.1 Ampère's Law

To obtain this relation, we begin by examining the magnetic field of a very long straight wire carrying a steady current I_0. According to Eq. (30.32), the magnetic field at a distance r from this wire has a magnitude

$$B = \frac{\mu_0}{2\pi} \frac{I_0}{r} \tag{1}$$

and the magnetic field lines are concentric circles around the wire. Consider now a closed mathematical path of arbitrary shape around this wire (Figure 31.1) and evaluate the integral $\oint \mathbf{B} \cdot d\mathbf{l}$ around this path. The path can be regarded as consisting of short radial segments and circular arcs. Since $\mathbf{B}$ is in the tangential direction, the radial segments do not contribute to the integral. Since the tangential segments have lengths $r\, d\theta$, we obtain

$$\oint \mathbf{B} \cdot d\mathbf{l} = \oint Br\, d\theta = \oint \frac{\mu_0}{2\pi} \frac{I_0}{r} r\, d\theta = \frac{\mu_0}{2\pi} I_0 \oint d\theta \tag{2}$$

The integral $\oint d\theta$ over the closed path is simply 2π. Hence

$$\oint \mathbf{B} \cdot d\mathbf{l} = \mu_0 I_0 \qquad (3)$$

Although the above calculation was based on a planar path, the same result applies to nonplanar paths — any path segment parallel to the wire will be perpendicular to the magnetic field; hence, as in the above calculation, only the circular arcs of the path contribute to the integral.

It turns out that Eq. (3) is valid not only for the magnetic field produced by a current on a long wire, but also for the magnetic field produced by an arbitrary distribution of steady currents. This is **Ampère's Law:**

> *The integral of* **B** *around any closed path equals μ_0 times the current intercepted by the area spanning the path,*

$$\boxed{\oint \mathbf{B} \cdot d\mathbf{l} = \mu_0 I} \qquad (4)$$

Note that the current in Eq. (4) need not flow on thin wires; it may flow on a conductor of any shape, or it may even flow without any conductor (e.g., a beam of protons in vacuum). Furthermore, note that the current is to be reckoned as positive if it is related to the direction around the path by the right-hand rule (thumb points along current when fingers curl along path; Figure 31.2), and negative otherwise. Although Ampère's Law can be derived rigorously from our expression for the magnetic field of a small current element [see Eq. (30.25)], we will not attempt to give this general proof.

For the calculation of magnetic fields, Ampère's Law plays a role similar to that of Gauss' Law for the calculation of electric fields. Provided that a distribution of currents has sufficient symmetry, Ampère's Law completely determines the magnetic field.

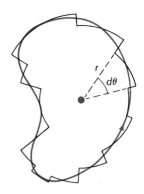

Fig. 31.1 Path of integration around a long, straight wire. The wire is perpendicular to the plane of the page. The path can be approximated by radial segments and circular arcs.

Ampère's Law

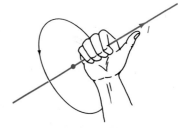

Fig. 31.2 The right-hand rule for Ampère's Law: if the fingers curl along the path of integration, the thumb points in the positive direction for the positive current.

EXAMPLE 1. Find the magnetic field of a very long, very thin, straight wire using Ampère's Law.

SOLUTION: By considerations of symmetry, the magnetic field lines of such a wire will have to be either concentric circles, or radial lines, or parallel lines in the same direction as the wire. The last two possibilities are inconsistent with the requirement that the magnetic field lines must form closed loops (see Section 30.2). Thus the field lines must be concentric circles, the magnetic field having a constant magnitude along each circle. If we integrate **B** along one of these circles we obtain

$$\oint \mathbf{B} \cdot d\mathbf{l} = \oint Br \, d\theta = 2\pi r B \qquad (5)$$

and Ampère's Law then gives us

$$2\pi r B = \mu_0 I_0 \qquad (6)$$

i.e.,

$$B = \frac{\mu_0}{2\pi} \frac{I_0}{r} \qquad (7)$$

Thus, Ampère's Law gives back the formula with which we started this section.

EXAMPLE 2. A very long, straight wire has a circular cross section of radius R. The wire carries a current I_0 uniformly distributed over the cross-sectional area of the wire. What is the magnetic field inside the wire? Outside the wire?

SOLUTION: Consider a circular path of radius r inside the wire (Figure 31.3). Symmetry tells us that the magnetic field lines are circles. Hence

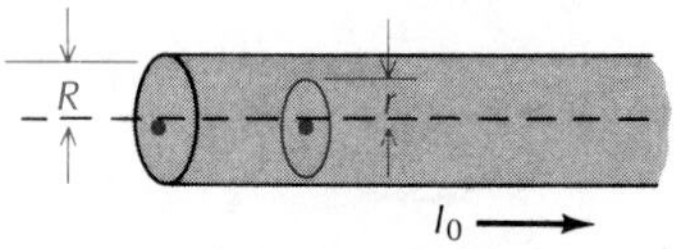

$$\oint \mathbf{B} \cdot d\mathbf{l} = 2\pi r B$$

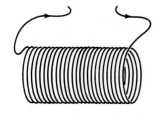

Fig. 31.3 Segment of a wire of radius R with a current I_0 uniformly distributed over the cross-sectional area.

Since the current is uniformly distributed over the volume of the wire, the amount of current I intercepted by the area within the circular path is in direct proportion to this area,

$$I = I_0 \frac{\pi r^2}{\pi R^2} = I_0 \frac{r^2}{R^2} \tag{8}$$

Ampère's Law then leads to

$$2\pi r B = \mu_0 I \tag{9}$$

or

$$2\pi r B = \mu_0 I_0 r^2 / R^2 \tag{10}$$

and

$$B = \frac{\mu_0}{2\pi} \frac{I_0 r}{R^2} \tag{11}$$

Note that the magnetic field is zero at the center of the wire ($r = 0$) and reaches a maximum value at the surface of the wire ($r = R$).

Outside the wire, we can find the magnetic field by repeating the calculation of Example 1. This calculation only depends on cylindrical symmetry and not on the thickness of the wire. The magnetic field is therefore exactly the same as for a thin wire,

$$B = \frac{\mu_0}{2\pi} \frac{I_0}{r} \tag{12}$$

31.2 Solenoids

A solenoid is a conducting wire wound in a tight helical coil (Figure 31.4). A current in this wire will produce a strong magnetic field within the coil. Because of the similarity of the current distributions, a tight coil produces essentially the same magnetic field as a collection of loops stacked next to one another.[1] The magnetic field of such a collection of loops can be calculated by summing the magnetic fields of

Fig. 31.4 A solenoid.

[1] There is, however, one small difference: in a cylindrical stack of loops the current flows in closed circles around the surface of the cylinder, whereas in a cylindrical solenoid the current has an extra component parallel to the axis of the cylinder (the current is helical). If the solenoid is tightly wound with thin wire, this axial component of the current is insignificant compared to the circular component and we can ignore it.

the individual loops. Figure 31.5 shows the pattern of field lines of the resultant magnetic field. Figure 31.6 shows the pattern of field lines of an actual solenoid with a rather loose coil, made visible with iron filings. The field is strong inside the solenoid but fairly weak outside. The detailed calculation of this magnetic field is rather messy because the individual magnetic fields of the loops of current must be added vectorially. We will not attempt this calculation here; instead, we will calculate the magnetic field of an **ideal solenoid**, that is, a very long (infinitely long) solenoid with very tightly wound coils so that the current distribution on the surface of the solenoid is nearly uniform.

Ideal solenoid

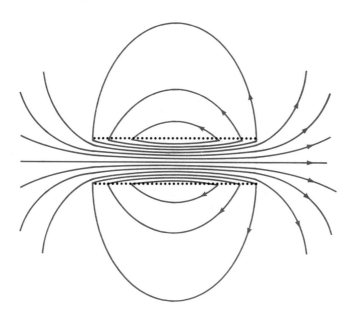

Fig. 31.5 Magnetic field lines of a solenoid.

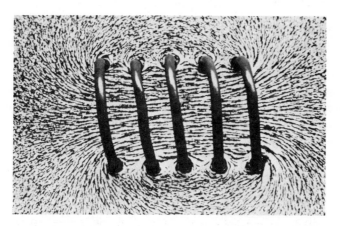

Fig. 31.6 Iron filings sprinkled on a sheet of paper inserted in a solenoid. Note that inside the solenoid, the distribution of field lines is nearly uniform.

To find this magnetic field, we begin with an appeal to symmetry, as in Example 1. The ideal solenoid has translational symmetry (along the axis of the solenoid) and rotational symmetry (around the axis). To be consistent with these symmetries, the magnetic field lines inside the solenoid will then have to be either concentric circles, or radial lines, or lines parallel to the axis. Concentric circles and radial lines are unacceptable; the former would require the presence of a current along the axis (compare Example 1) and the latter would require that the field lines start on the axis. Thus the field lines inside the solenoid must all be parallel to the axis. These field lines emerge from the end of the solenoid and return to the other end (Figure 31.5). For an ideal, very long solenoid, these external field lines will spread over a very large region of space; hence, the magnetic field outside of the solenoid is nearly zero.

We can now use Ampère's Law to find the magnitude of the magnetic field. Consider the rectangular path shown in Figure 31.7 and evaluate the integral of **B** along this path. The horizontal side external to the solenoid does not contribute to the integral (**B** is zero); the vertical sides do not contribute to the integral (**B** is perpendicular to the path). Hence only the horizontal side within the solenoid contributes. The magnetic field has some constant magnitude along this side and if the length of this side is l,

$$\oint \mathbf{B} \cdot d\mathbf{l} = Bl \tag{13}$$

The total current intercepted by the area within the rectangle is the current I_0 in the wire multiplied by the number N of wires passing through this area. Hence Ampère's Law tells us

$$Bl = \mu_0 N I_0 \tag{14}$$

or

$$B = \mu_0 I_0 (N/l) \tag{15}$$

The quantity N/l is the number of turns of wire per unit length, commonly designated by n. Thus

Field in solenoid

$$\boxed{B = \mu_0 I_0 n} \tag{16}$$

Note that this result is independent of the length of the vertical sides of the path of integration; hence **B** has the same magnitude everywhere within the solenoid. This shows that the magnetic field within an ideal solenoid is perfectly uniform.

Electromagnet

An **electromagnet** is essentially a solenoid with a gap, or, what amounts to the same thing, a pair of solenoids with their ends placed close together (Figure 31.8). Magnetic field lines come out of one sole-

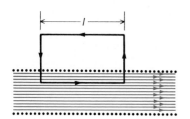

Fig. 31.7 Magnetic field lines of a very long solenoid.

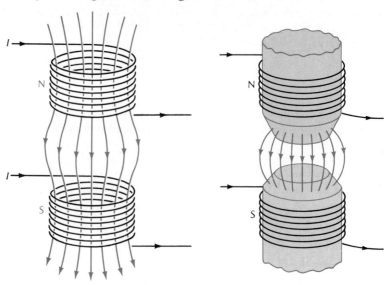

Fig. 31.8 (left) An electromagnet with two coils. The coil from which field lines emerge into the gap is called the north pole; the coil into which field lines enter is called the south pole.

Fig. 31.9 (right) An electromagnet with iron pole pieces.

noid and go into the other (of course, field lines will also have to come out of the solenoids at their other ends and close on themselves). The first solenoid is called the north pole of the electromagnet, and the second the south pole. If the gap is small, then the magnetic field in this region is almost the same as inside the solenoids. In most electromagnets the space inside the solenoids is filled with an iron core (Figure 31.9); as we will see in Section 33.3, the iron enhances the magnetic field, making it much stronger than the value given by Eq. (16).

EXAMPLE 3. A solenoid used for the investigation of the effect of magnetic fields on the propagation of light in a liquid consists of 180 turns of wire wound on a tube 19 cm long. The current in the wire is 5.0 A. What is the strength of the magnetic field within the tube?

SOLUTION: The number of turns per unit length is

$$n = 180/0.19 \text{ m} = 9.5 \times 10^2/\text{m} \qquad (17)$$

Hence, according to Eq. (16),

$$B = \mu_0 I_0 n$$

$$= 1.26 \times 10^{-6} \text{ N} \cdot \text{m}^2/\text{C}^2 \times 5.0 \text{ A} \times 9.5 \times 10^2/\text{m}$$

$$= 6.0 \times 10^{-3} \text{ T}$$

EXAMPLE 4. A **toroid** is a conducting wire wound in a tight coil in the shape of a torus (a doughnut). It may be thought of as a solenoid that has been bent so that its ends meet (Figure 31.10a). What is the magnetic field inside the toroid?

SOLUTION: By symmetry, the magnetic field lines inside the toroid are closed circles (Figure 31.10b). If we integrate **B** along one of these circular field lines, we obtain

$$\oint \mathbf{B} \cdot d\mathbf{l} = 2\pi r B \qquad (18)$$

The total current intercepted by the circular area within this path equals the total number N of wires sticking through the area multiplied by the current I_0 in each wire. Hence, from Ampère's Law,

$$2\pi r B = \mu_0 N I_0 \qquad (19)$$

or

$$B = \frac{\mu_0}{2\pi} \frac{N I_0}{r} \qquad (20)$$

Thus, the magnetic field depends on the radial distance from the axis of the torus.

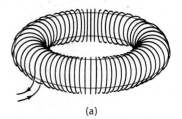

(a)

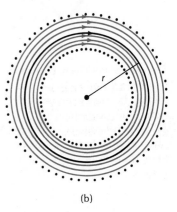

(b)

Fig. 31.10 (a) A toroid. (b) Magnetic field lines within a toroid.

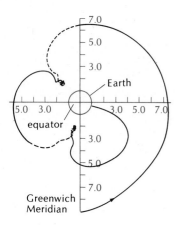

Fig. 31.11 Trajectory of a cosmic ray of momentum 5.66×10^{-19} kg · m/s in the magnetic field of the Earth. The trajectory is three dimensional; portions above the plane of the page are shown with a solid line, portions below the plane of the page are shown with a dashed line. The numbers along the axes give distances in Earth radii.

31.3 Motion of Charges in Electric and Magnetic Fields

The force exerted by a magnetic field on a charged particle is

$$\mathbf{F} = q\mathbf{v} \times \mathbf{B} \tag{21}$$

This force is always at right angles to the velocity of the particle; consequently, the force changes the *direction* of the velocity of the particle, but not the *magnitude* of the velocity. The formal proof of this is easy:

$$\frac{d}{dt} v^2 = \frac{d}{dt} \mathbf{v} \cdot \mathbf{v} = 2\mathbf{v} \cdot \frac{d\mathbf{v}}{dt} = 2\mathbf{v} \cdot \mathbf{F}/m = 0 \tag{22}$$

where the last equality expresses the fact that force and velocity are perpendicular. Thus, the magnetic force only changes the momentum of the particle, but not the kinetic energy — the magnetic force *does no work* on the particle.

In general, the motion of a particle in a magnetic field is quite complex. For example, Figure 31.11 shows the trajectory of a cosmic-ray particle approaching the Earth and being deflected by the magnetic field (the effects of gravity on such a cosmic ray are insignificant). In what follows, we will only consider the relatively simple case of motion in a uniform magnetic field.

Figure 31.12 shows a uniform magnetic field, directed into the plane of the page. Suppose that a positively charged particle has an initial velocity in the plane of the page; this initial velocity is perpendicular to the magnetic field. The magnetic force $q\mathbf{v} \times \mathbf{B}$ is perpendicular to both $\mathbf{v}$ and $\mathbf{B}$; its direction is shown in Figure 31.12. The acceleration then has a constant magnitude

$$a = F/m = qvB/m \tag{23}$$

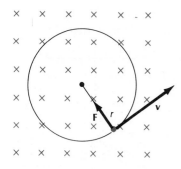

Fig. 31.12 Positively charged particle with uniform circular motion in a uniform magnetic field. The magnetic field points perpendicularly into the plane of the page; the crosses show the tails of the magnetic-field vectors.

and its direction is always perpendicular to the velocity. Such an acceleration corresponds to uniform circular motion. The particle will move in a circle of radius r and the acceleration given by Eq. (23) will play the role of centripetal acceleration, i.e.,

$$\frac{qvB}{m} = \frac{v^2}{r} \tag{24}$$

This leads to the following formula for the radius of the circular motion:

$$r = \frac{mv}{qB} \tag{25}$$

Figure 31.13 shows a beam of electrons executing such circular motion in a cathode-ray tube placed in a magnetic field.

The angular velocity of the circular motion is

$$\omega = \frac{v}{r} = \frac{qB}{m} \tag{26}$$

Fig. 31.13 Electrons moving in a circle in a cathode-ray tube in a magnetic field. The tube contains a gas at a very low pressure, and the atoms of the gas glow under the impact of the electrons; this makes the electron beam visible.

and the frequency

$$\nu = \frac{\omega}{2\pi} = \frac{qB}{2\pi m}$$
(27)   *Cyclotron frequency*

This is called the **cyclotron frequency** because the operation of cyclotrons (described in Example 5 below) involves particles moving with this frequency in a magnetic field. Note that the cyclotron frequency is independent of the speed of the circular motion — in a uniform magnetic field, slow particles and fast particles (of a given charge and mass) move around circles at the same frequency, but the slow particles move along smaller circles than the fast particles.

It is useful to write Eq. (25) in terms of the momentum $p = mv$; this leads to

$$r = \frac{p}{qB}$$
(28)   *Circular orbit in magnetic field*

The advantage of Eq. (28) is that it remains valid even when the particle moves with relativistic velocity. Thus Eq. (28) is more general than our derivation of it.

EXAMPLE 5.   A **cyclotron** is a device for acceleration of protons, deuterons, or other charged particles. It consists of an evacuated cavity placed between the poles of a large electromagnet; within the cavity there is a flat metallic can cut into two **D**-shaped pieces, or **dees** (Figures 31.14 and 31.15). An oscillating high-voltage generator is connected to the dees; this creates an oscillating electric field in the gap between the dees. The frequency of the voltage generator is adjusted so that it coincides with the cyclotron frequency of Eq. (27). An ion source at the center of the cyclotron releases protons. The electric field in the gap between the dees gives each of these protons a push and the uniform magnetic field in the cyclotron then makes the proton travel on a semicircle inside of the first dee. When the proton returns to the gap, the high-voltage generator will have reversed the electric field in the gap; the proton therefore re-

Cyclotron

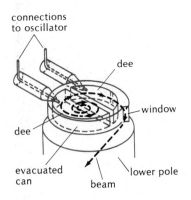

connections to oscillator

dee

window

dee

evacuated can

lower pole

beam

Fig. 31.14 Trajectory of a particle within the dees of a cyclotron. In this diagram, the upper pole of the electromagnet has been omitted for the sake of clarity.

Fig. 31.15 The dees of one of the early small cyclotrons built by E. O. Lawrence.

ceives an additional push which sends it into the second dee. There it travels on a semicircle of slightly larger radius corresponding to its slightly larger energy, etc. Each time the proton crosses the gap between the dees, it receives an extra push and extra energy. The proton travels along arcs of circles of stepwise increasing radius. When the protons reach the outer edge of the dees, they leave the cyclotron as a high-energy beam (Figure 31.14). One of the first cyclotrons, built by E. O. Lawrence at Berkeley in 1932, had dees with a diameter of 28 cm and its magnet was capable of producing a field of 1.4 T. What was the maximum energy of the protons accelerated by this cyclotron?

SOLUTION: When the proton reaches its maximum energy, its orbit has a radius of 14 cm. Since the magnetic field is 1.4 T, the momentum of such a proton is

$$p = mv = eBr$$

$$= 1.6 \times 10^{-19} \text{ C} \times 1.4 \text{ T} \times 0.14 \text{ m}$$

$$= 3.1 \times 10^{-20} \text{ kg} \cdot \text{m/s}$$

and the energy is

$$K = p^2/(2m)$$

$$= (3.1 \times 10^{-20} \text{ kg} \cdot \text{m/s})^2 / (2 \times 1.67 \times 10^{-27} \text{ kg})$$

$$= 2.9 \times 10^{-13} \text{ J} = 1.8 \times 10^6 \text{ eV}$$

Nuclear physicists like to measure energies in MeV, where

$$1 \text{ MeV} = 10^6 \text{ eV} = 1.60 \times 10^{-13} \text{ J}$$

In these units, the above kinetic energy is 1.8 MeV.

Synchrocyclotron

Cyclotrons of large size can be used to accelerate protons up to an energy of about 30 MeV. Above this energy, cyclotrons begin to fail because relativistic effects change the frequency of the orbital motion of the protons. Roughly, what happens is that at high energies the mass of the protons increases with their speed [as indicated by Eq. (17.54)] — the orbital frequency will decrease. In order to keep the pushes of the electric field in phase with the motion of the proton, we must gradually decrease the frequency of oscillation of the high-voltage generator. A modified cyclotron that automatically performs such an adjustment of frequency is called a **synchrocyclotron**. It accelerates a bunch of protons by matching its frequency of oscillation to the frequency of the orbital motion of the bunch; when it ejects the bunch, it begins to accelerate the next bunch, etc. Machines of this kind have been used to achieve energies up to several hundred MeV.

At even higher energies cyclotrons become impractical because they require excessively large magnets. It is then better to keep the protons moving around a circle of fixed radius in a large magnet of annular shape (Figure 31.16). As the protons gradually acquire more energy, the strength of the magnetic field must be gradually increased [see Eq. (28)]. Accelerator machines that automatically perform this increase of

Fig. 31.16 The Bevatron at the Lawrence Berkeley Laboratory.

magnetic field are called **synchrotrons**. The most powerful synchro- *Synchrotron*
trons built to date are the machines at Fermilab (Batavia, Illinois) and
at CERN (Centre Européenne de Recherche Nucléaire, on the Swiss–
French border near Geneva). The Fermilab machine accelerates protons up to an energy of 1,000,000 MeV (see Interlude C).

In our discussion of the motion of a charged particle in a uniform
magnetic field, we assumed that the initial velocity was perpendicular
to the magnetic field (see Figure 31.12). If this is not the case, then we
can regard the velocity as consisting of two components: one compo-
nent $v_\perp$ perpendicular to **B** and one component $v_\parallel$ parallel to **B** (Figure
31.17). Since the magnetic force is perpendicular to **B**, only the component $v_\perp$ changes, while $v_\parallel$ remains constant. The component $v_\perp$ then
gives rise to uniform circular motion with radius $r = mv_\perp/qB$ in a
direction perpendicular to **B**, while the component $v_\parallel$ gives rise to
uniform translational motion in a direction parallel to **B**. The
combination of these two motions is a helical motion with axis along
the magnetic field — the particle spirals around the magnetic field
lines (Figure 31.18). Such spiraling is a general feature of motion in a
magnetic field; it will happen even if the magnetic field is a function of

Ernest Orlando Lawrence, *1901–1958, American experimental physicist, professor at Berkeley, and director of the Radiation Laboratory (now Lawrence Berkeley Laboratory). He was awarded the Nobel Prize in 1939 for the invention and development of the cyclotron.*

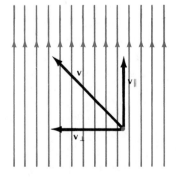

Fig. 31.17 The velocity of a particle has a component parallel to **B** and a component perpendicular to **B**.

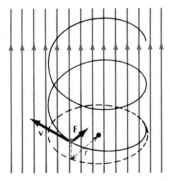

Fig. 31.18 Positively charged particle with helical motion in uniform magnetic field.

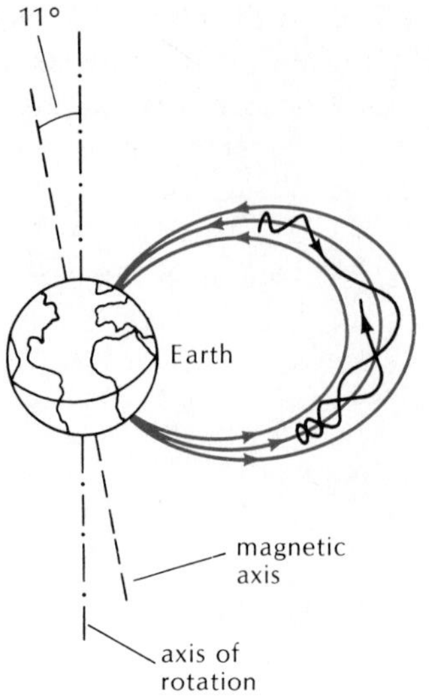

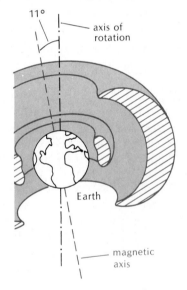

Fig. 31.19 Particle spiraling around the magnetic field lines of the Earth's field.

position. For example, Figure 31.19 shows a particle spiraling about the magnetic field lines of the field of the Earth. A detailed analysis shows that near the poles, where the field lines converge, the component $v_\parallel$ of the velocity parallel to the field lines will decrease, then vanish, and then reverse (provided $v_\parallel$ is not too large). Hence the particle spirals toward one pole, then halts its approach, and then spirals back *Van Allen belts* toward the other pole, etc. The **Van Allen belts** surrounding the Earth (Figure 31.20) consist of a large number of electrons and protons spiraling back and forth along the field lines in this manner.

Fig. 31.20 The Van Allen radiation belts. Near the poles of the Earth, where the horns of the belts are close to the upper atmosphere, charged particles sometimes spill into the atmosphere, producing the luminous glow known as the Aurora Borealis.

If electric and magnetic fields act on a particle simultaneously, then the force has both an electric and a magnetic part,

Lorentz force

$$\boxed{\mathbf{F} = q\mathbf{E} + q\mathbf{v} \times \mathbf{B}} \qquad (29)$$

This is called the **Lorentz force.**

As an example of the simultaneous action of electric and magnetic *Crossed fields* fields, let us consider **crossed fields**, that is, uniform electric and magnetic fields at right angles to one another. Figure 31.21 shows electric field lines and magnetic field lines at right angles. The motion of a particle in such fields is usually fairly complicated, but under special circumstances the motion becomes very simple. Suppose that in Figure 31.21, a positively charged particle enters the field region from the left. The magnetic force $q\mathbf{v} \times \mathbf{B}$ is then opposite to the electric force $q\mathbf{E}$ and if the velocity has just the right magnitude, these forces will cancel and the particle will continue its original motion on a straight line. The condition for this is

$$\mathbf{E} = -\mathbf{v} \times \mathbf{B} \qquad (30)$$

or, considering magnitudes,

$$E = vB$$

The "right" velocity for cancellation of the forces is then

$$\boxed{v = E/B} \qquad (31)$$

This cancellation is the basic principle behind the **velocity selectors** (or velocity filters) often used in physics laboratories in order to select particles of some desirable velocity from a beam containing particles with a large variety of velocities. It is only necessary to shoot the beam into a region containing crossed $\mathbf{E}$ and $\mathbf{B}$ fields whose magnitudes are related to the desired velocity by Eq. (31). Particles with the right velocity will then proceed undeflected, all other particles will be deflected to one side or another and they are thereby eliminated from the beam (Figure 31.22).

Crossed $\mathbf{E}$ and $\mathbf{B}$ fields also play a role whenever a wire or some other conductor carries a current in a magnetic field. Figure 31.23 shows a metallic strip carrying a current; the strip is placed in a magnetic field, perpendicular to the surface of the strip. If the current is toward the right, the motion of the free electrons in the strip must be toward the left with some average velocity $\mathbf{v}$. These free electrons experience a magnetic force $-e\mathbf{v} \times \mathbf{B}$, directed upward in Figure 31.23. This magnetic force will tend to push the electron toward the upper edge of the strip, and some electrons will accumulate there, leaving the lower edge with a deficit of electrons — the upper edge acquires a negative charge and the lower edge a positive charge. There will then exist an electric field in the strip, as shown in Figure 31.23. This electric field is perpendicular to the magnetic field, i.e., the $\mathbf{E}$ and $\mathbf{B}$ fields inside the strip are crossed fields. Under equilibrium conditions, the transverse magnetic force on an electron will be matched by the transverse electric force. The condition for this balance of forces is Eq. (30). Thus, inside the strip, there exists an electric field of magnitude $E = vB$; consequently, in a strip of width l, there is a potential difference

$$\Delta V = \int E \, dl = vBl \qquad (32)$$

between the upper and the lower edge of the strip, the lower edge being at a higher potential. This generation of potential difference between opposite edges of a conductor carrying a current in a magnetic field is called the **Hall effect**.

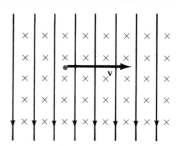

Fig. 31.21 Electric (black) and magnetic (color) fields at right angles. The crosses show the tails of the magnetic field vectors. A positively charged particle moves from left to right.

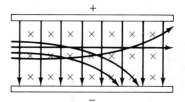

Fig. 31.22 A velocity selector with electric and magnetic fields at right angles. A beam of positively charged particles enters from the left. Particles of excessive velocity are deflected upward; particles of insufficient velocity are deflected downward.

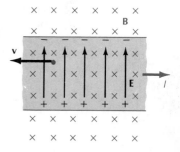

Fig. 31.23 A metallic strip carrying a current has been placed in a uniform magnetic field. The crosses show the tails of the magnetic field vectors. The current flows from left to right; the electrons making up this current move from right to left. The magnetic force on an electron is directed upward; the electric force is directed downward.

Hall effect

We can express the Hall potential difference in terms of the current by means of Eq. (28.17):

$$v = \frac{I}{enA}$$

where A is the cross-sectional area of the conductor and n the number of free electrons per unit volume. This gives

$$\Delta V = \frac{IBl}{enA} \qquad (33)$$

For example, a copper strip of cross-sectional area 5.0 cm × 0.02 cm carrying a current of 20 A in a magnetic field of 1.5 T develops a potential difference of

$$\Delta V = \frac{20 \text{ A} \times 1.5 \text{ T} \times 0.05 \text{ m}}{1.6 \times 10^{-19} \text{ C} \times 8.5 \times 10^{28}/\text{m}^3 \times 0.10 \times 10^{-4} \text{ m}^2}$$

$$= 1.1 \times 10^{-5} \text{ volt}$$

In practice, the Hall effect is often used to determine the density of free electrons n in a metal. Also, the Hall effect provides direct empirical evidence that the current carriers in metals are negative charges — if the current carriers were positive charges, then the direction of **v** in Figure 31.23 would be opposite to that shown and the positive charges would accumulate at the top of the strip rather than on the bottom.

31.4 Force on a Wire

If a wire carrying a current is placed in a magnetic field, the moving charges within the wire will experience a force; hence the wire will experience a force. According to Eq. (30.23), the amount of moving charge in a segment dl of the wire is

$$dq = I \frac{dl}{v}$$

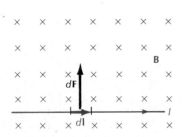

Fig. 31.24 A long straight wire in a uniform magnetic field. The force on a wire segment dl is perpendicular to dl and to **B**.

where, as always, we pretend that the moving charge is positive. If θ is the angle between the velocity and **B**, or, equivalently, the angle between the wire segment and **B** (Figure 31.24), then the force on the wire segment is

$$dF = dqvB \sin \theta = I \, dlB \sin \theta \qquad (34)$$

This force can be expressed in vector form by an argument similar to that given in connection with Eq. (30.25):

Force on wire segment

$$\boxed{d\mathbf{F} = I \, d\mathbf{l} \times \mathbf{B}} \qquad (35)$$

where the vector $d\mathbf{l}$ is along the wire in the direction of the current.

To find the net force on the entire wire we must integrate Eq. (35) along the wire.

EXAMPLE 6.   Two very long, parallel wires separated by a distance r carry currents I and I', respectively. Find the magnetic force acting between them.

SOLUTION: Each wire generates a magnetic field, which exerts a force on the other wire. Figure 31.25 shows the magnetic field $\mathbf{B}'$ that the current I' produces in the space surrounding the current I. This magnetic field has a magnitude

$$B' = \frac{\mu_0}{2\pi} \frac{I'}{r} \tag{36}$$

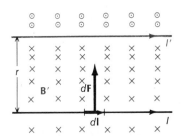

and is perpendicular to the current I. By Eq. (35), the force on a segment dl of the wire carrying the current I is

$$dF = IB' \, dl = \frac{\mu_0}{2\pi} \frac{II'}{r} \, dl \tag{37}$$

The force acting between the wires is attractive if the currents are parallel, and repulsive if they are antiparallel. Of course, the net force between two infinitely long wires is infinite. It is therefore useful to consider the force per unit length; this is finite,

$$\frac{dF}{dl} = \frac{\mu_0}{2\pi} \frac{II'}{r} \tag{38}$$

Fig. 31.25 A long straight wire carrying a current I in the magnetic field of a long straight wire carrying a current I'. The magnetic field lines of the current I' are circles perpendicular to the plane of the page. The crosses show the tails of the magnetic field vectors; the dots show the tips of the magnetic field vectors.

The official definition of the **SI unit of current** is based on the force per unit length between two long parallel wires. This force can be measured very precisely by holding one wire stationary and suspending the other from a balance; the wires are connected in series so that the currents are exactly equal ($I = I'$). The force per unit length and the distance can be measured experimentally, and the value of I calculated from Eq. (38) is then the current in amperes. The constant μ_0 appearing in Eq. (38) is given the value $\mu_0 = 4\pi \times 10^{-7}$ N · s²/C² by *definition*.[2]

SI unit of current

The official definition of the **SI unit of electric charge** is based on the unit of current. The coulomb is defined as the amount of charge that a current of one ampere will deliver in one second.

SI unit of electric charge

31.5 Torque on a Current Loop

The action of a magnetic field on a loop of wire carrying a current will in general result not only in a net force on the loop, but also in a torque. The forces and torques depend on the orientation and the shape of the loop. We will consider the special case of a rectangular loop placed in a uniform magnetic field.

Figure 31.26 shows the loop oriented perpendicular to the magnetic field. The forces on the four sides of the loop can be calculated from Eq. (35). The forces on opposite sides are opposite; hence the forces cancel in pairs and the loop will be in equilibrium.

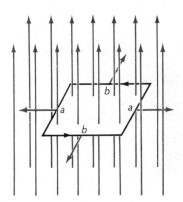

Fig. 31.26 Rectangular loop of current perpendicular to a uniform magnetic field.

---

[2] In contrast, the constant ε_0 appearing in Coulomb's Law must be determined by experiment.

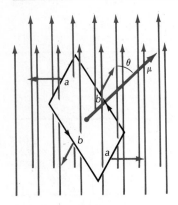

Fig. 31.27 Rectangular loop of current at an angle with a uniform magnetic field.

Figure 31.27 shows a loop oriented at an angle with the magnetic field. Again the forces on opposite sides are opposite, but the lines of action of the forces pulling to the left and the right do not coincide — these forces exert a torque which tends to rotate the loop. Since **B** is constant, Eq. (35) tells us that the magnitude of the force on the left and on the right side is

$$F = IaB \qquad (39)$$

where a is the length of the side (Figure 31.27). The torque due to this pair of forces is

$$\tau = F(b/2)\sin\theta + F(b/2)\sin\theta$$

$$= IabB \sin\theta \qquad (40)$$

where b is the length of the other side (Figure 31.27) and θ is the angle between the directions of **B** and of the normal to the loop.

According to the definition given in Section 30.3, the product of the area of the loop and the current around it is the **magnetic dipole moment**

$$\mu = Iab \qquad (41)$$

Note that if the loop consists of several turns of wire, then Eq. (41) must be evaluated with the net current flowing around the rim of the loop. This net current equals the number N of turns of wire times the current I_0 in one turn so that

$$\mu = NI_0ab$$

In any case, the torque is

$$\tau = \mu B \sin\theta \qquad (42)$$

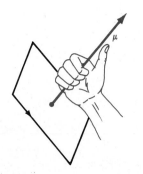

Fig. 31.28 Right-hand rule for the magnetic moment: if the fingers are curled around the loop in the direction of the current, the thumb points in the direction of μ.

Torque on current loop

This formula turns out to be valid not only for the rectangular loops, but also for loops of arbitrary shape. The direction of the torque is such that it tends to twist the loop into an orientation perpendicular to the magnetic field.

To express the torque in vector form, we must regard the magnetic dipole moment as a vector; this vector is perpendicular to the loop, in a direction given by the right-hand rule (fingers curled along current, thumb along μ; Figure 31.28). With this vector, we can write

$$\boxed{\tau = \mu \times \mathbf{B}} \qquad (43)$$

This expression for the torque on a magnetic dipole in a magnetic field is mathematically similar to the expression for the torque on an electric dipole in an electric field [see Eq. (23.30)]. The torque tends to twist the dipole so as to bring μ into alignment with **B**.

We can also write a potential energy such that changes in this potential energy represent the work that you must do against the torque (43) when changing the orientation of the loop. The expression for the potential energy is

$$U = -\mu \cdot \mathbf{B} \qquad (44)$$

Potential energy of current loop

The derivation of this expression is entirely analogous to the derivation of the expression (23.32) for the potential energy of an electric dipole in an electric field. As expected, the potential energy (44) has a minimum for μ parallel to $\mathbf{B}$, and a maximum for μ antiparallel to $\mathbf{B}$.

A loop of current suitably pivoted on an axis acts as a compass needle; the normal of the loop seeks to align itself with the magnetic field. This similarity is no accident. As we will see in Section 33.3, a compass needle actually consists of a large number of current loops. The mechanism underlying the north-seeking behavior of a compass needle is essentially as described above.

EXAMPLE 7.   A simple electric motor consists of a rectangular coil of wire that rotates on a longitudinal axle in a magnetic field of 0.50 T (Figure 31.29). The coil measures 10 cm × 20 cm; it has 40 turns of wire and the current in the wire is 8.0 A.

(a) As a function of the angle θ between the lines of $\mathbf{B}$ and the normal to the coil, what is the torque that the magnetic field exerts on the coil?

(b) In order to keep the sign of the torque constant, a switch (commutator) mounted on the axle reverses the current in the coil whenever θ passes through 0 and π radians. Plot this torque as a function of θ.

SOLUTION: (a) According to Eq. (42), the torque on the coil is

$$\tau = NI_0 abB \sin\theta$$

$$= 40 \times 8.0 \text{ A} \times 0.10 \text{ m} \times 0.20 \text{ m} \times 0.50 \text{ T} \times \sin\theta$$

$$= 3.2 \text{ N} \cdot \text{m} \sin\theta$$

(b) If the torque always has the same sign (say, positive), we can write it as

$$\tau = 3.2 \text{ N} \cdot \text{m} \, |\sin\theta|$$

Figure 31.30 shows a plot of this torque.

Practical electric motors consist of many such coils, each with its commutator, arranged at regular intervals around the axle. The plot for the net torque of this arrangement is a sum of plots such as shown in Figure 31.30, but with different starting angles. This averages out the ups and downs of Figure 31.30 and yields a torque that is nearly constant as a function of angle.

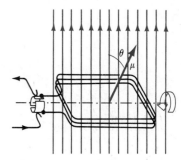

Fig. 31.29 An electric motor.

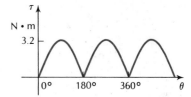

Fig. 31.30 Torque as a function of angle according to Example 7.

SUMMARY

Ampère's Law: $\oint \mathbf{B} \cdot d\mathbf{l} = \mu_0 I$

Field in solenoid: $B = \mu_0 I_0 n$

Cyclotron frequency: $v = \dfrac{qB}{2\pi m}$

Circular orbit in magnetic field: $r = \dfrac{p}{qB}$

Lorentz force: $\mathbf{F} = q\mathbf{E} + q\mathbf{v} \times \mathbf{B}$

Force on wire segment: $d\mathbf{F} = I \, d\mathbf{l} \times \mathbf{B}$

Torque on current loop: $\tau = \mu \times \mathbf{B}$

Potential energy of current loop: $U = -\mu \cdot \mathbf{B}$

QUESTIONS

1. Suppose you evaluate the integral $\oint \mathbf{B} \cdot d\mathbf{l}$ for the magnetic field of the Earth along a closed path that coincides with a meridian circle. What is the value of the integral?

2. Is the magnetic field a conservative field?

3. A long solenoid has been placed inside another long solenoid of larger radius. The solenoids are coaxial and both have the same number of turns per unit length and the same current. What is the formula for the magnetic field in the region within the smaller solenoid? Between the smaller and the larger solenoid?

4. Figure 31.31 shows a solenoid of arbitrary cross section; this solenoid is a (noncircular) cylinder. Suppose that there are n turns of wire per unit length and that the current in the wire is I. Use Ampère's Law to show that the magnetic field in the solenoid is $\mu_0 \, In$, the same as for a circular cylinder.

5. Consider a long solenoid and a long straight wire along its axis, both carrying some current. Describe the field lines within the solenoid.

6. The drawing of Figure 31.8 shows the field lines near the gap of an electromagnet. Describe the pattern of field lines beyond the edges of the drawing.

7. Cosmic rays are high-speed charged particles — mostly protons — that crisscross interstellar space and strike the Earth from all directions. Why is it easier for the cosmic rays to penetrate through the magnetic field of the Earth near the poles than anywhere else?

8. Is it possible to define a magnetic potential energy for a charged particle moving in a magnetic field?

9. If we want a proton to orbit all the way around the equator in the Earth's magnetic field, must we send it eastward or westward?

10. Consider the strip of metal placed in a magnetic field, as shown in Figure 31.23. How does the Hall potential difference between the edges of the strip change if we reverse the current? If we reverse the magnetic field? If we reverse both?

11. If we replace the metallic strip in Figure 31.23 by a semiconducting strip of p-type semiconductor, will the lower edge remain at the higher potential?

12. The Earth's magnetic field at the equator is horizontal and in the northward direction. Assume that the atmospheric electric field is downward. The electric and magnetic fields are then crossed fields. In what direction must we launch an electron if it is to move without any deflection?

13. If a strong current flows through a thick wire, it tends to cause a compression of the wire. Explain.

14. If a wire carrying a current is placed in a magnetic field, the field exerts forces on the moving electrons in the wire. How does this cause a push on the wire?

15. A horizontal wire carries a current in the eastward direction. The wire is in a uniform magnetic field. What must be the direction of this field if it is to compensate for the weight of the wire?

16. In the SI system, we first define the unit of current (by means of the mag-

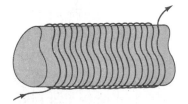

Fig. 31.31 Cylinder of arbitrary, noncircular cross section.

netic force between wires), and we then define the unit of charge in terms of the current. Could we proceed in the opposite manner: first define the unit of electric charge (by means of the electric force between charges), and then define the unit of current as a flow of charge? What would be the advantages and the disadvantages?

17. Figure 31.32 is a schematic diagram showing the mechanism of an **ammeter**. The mechanism consists of a coil suspended between the poles of a permanent magnet. The coil can rotate about an axis perpendicular to the magnetic field, and is held in its equilibrium position by a spiral spring. To operate the ammeter, the current of the external circuit is made to pass through the coil. Explain how this ammeter registers the current. Explain how a sensitive ammeter of this kind can be used as a voltmeter by connecting the coil in series with a large resistance.

Ammeter

18. The **tangent galvanometer** is an old form of ammeter, consisting of an ordinary magnetic compass mounted at the center of a coil whose axis is horizontal and oriented along the east–west line (Figure 31.33). If there is no current in the coil, the compass needle points north. Explain how the compass needle will deviate from north when there is a current in the coil.

Tangent galvanometer

19. A simple indicator of electric current, first used by H. C. Oersted in his early experiments, consists of a compass needle placed below a wire stretched in the northward direction. Explain how the angle of the compass needle indicates the electric current in the wire.

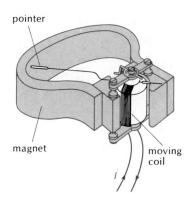

20. Positive charge is uniformly distributed over a sphere. If this sphere spins about a vertical axis in a counterclockwise direction as seen from above, what is the direction of the magnetic moment vector?

Fig. 31.32 Mechanism of the moving-coil ammeter.

PROBLEMS

Section 31.1

1. A wire of superconducting niobium, 0.20 cm in diameter, can carry a current of up to 1900 A. What is the strength of the magnetic field just outside of the wire when it carries this current?

2. A long, straight wire of copper with a radius of 1 mm carries a current of 20 A. What is the instantaneous magnetic force and the corresponding acceleration on one of the conduction electrons moving at 10^6 m/s along the surface of the wire in a direction opposite to that of the current? What is the direction of the acceleration?

3. A current I flows in a thin wire bent into a circle of radius R. The axis of the circle coincides with the z axis. What is the value of the integral $\int \mathbf{B} \cdot d\mathbf{l}$ along the z axis from $z = -\infty$ to $z = +\infty$?

4. In a proton accelerator, protons of velocity 3×10^8 m/s form a beam of current of 2×10^{-3} A. Assume that the beam has a circular cross section of radius 1 cm and that the current is uniformly distributed over the cross section. What is the magnetic field that the beam produces at its edge? What is the magnetic force on a proton at the edge of the beam?

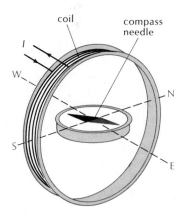

Fig. 31.33

5. A very large, thin conducting plate lies in the x–y plane. The plate carries a current in the y direction. The current is uniformly distributed over the plate with σ ampere flowing across each 1-m length along the x axis. Use Ampère's Law to find the magnetic field at some distance from the plate. (Hint: The magnetic field lines are parallel to the plate.)

6. One very large conducting plate coincides with the x–y plane. Another very large conducting plate coincides with the x–z plane. Each plate carries a uniformly distributed current with σ ampere flowing across each 1-m length per-

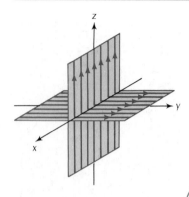

Fig. 31.34 Two large plates intersecting at right angles.

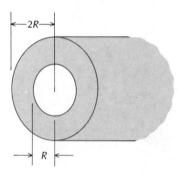

Fig. 31.35

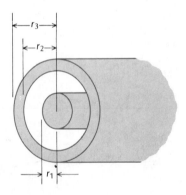

Fig. 31.36 Coaxial cylinder and cylindrical shell.

pendicular to the current (Figure 31.34). Find the magnetic field in each of the four quadrants. (Hint: See Problem 5.)

Section 31.2

7. A long solenoid has 15 turns per centimeter. What current must we put through its windings if we wish to achieve a magnetic field of 5.0×10^{-2} T in its interior?

8. The electromagnet of a small electric bell is a solenoid with 260 turns in a length of 2.0 cm. What magnetic field will this solenoid produce if the current is 8.0 A?

9. A long copper pipe with thick walls has an inner radius R and an outer radius $2R$ (Figure 31.35). A current I flows along this wall, uniformly distributed over the volume of copper. Find the magnetic fields at the radius $\frac{3}{2}R$ and at the radius $3R$.

10. A long solenoid of n turns per unit length carries a current I, and a long straight wire lying along the axis of this solenoid carries a current I'. Find the net magnetic field within the solenoid, at a distance r from the axis. Describe the shape of the magnetic field lines.

11. A coaxial cable consists of a long cylindrical wire of radius r_1 surrounded by a cylindrical shell of inner radius r_2, outer radius r_3 (Figure 31.36). The wire and the shell each carry equal and opposite currents I uniformly distributed over their volumes. Find formulas for the magnetic field in each of the regions $r < r_1$, $r_1 < r < r_2$, $r_2 < r < r_3$, $r > r_3$.

12. A toroid used in plasma research has 240 turns of wire carrying a current of 7.2×10^4 A. The inner radius of the toroid is 0.50 m and the outer radius is 1.50 m. What is the strength of the magnetic field at the inner radius? At the outer radius?

13. The coil of an electromagnet consists of a large number of layers of very thin wire wound on a cylindrical core. The inner radius of the windings is r_1 and the outer radius is r_2. The number of turns in each layer is n per unit length and the number of layers is n' per unit length in the radial direction. The current in the turns of wire is I. Find formulas for the magnetic field in the region $r < r_1$ and in the region $r_1 < r < r_2$.

Section 31.3

14. A bubble chamber, used to make the tracks of protons and other charged particles visible, is placed between the poles of a large electromagnet that produces a uniform magnetic field of 20 T. A high-energy proton passing through the bubble chamber makes a track that is an arc of a circle of radius 3.5 m. According to Eq. (28), what is the momentum of the proton?

15. A proton of energy 1.0×10^7 eV moves in a circular orbit in the magnetic field near the Earth. The strength of the field is 0.50×10^{-4} T. What is the radius of the orbit?

16. In principle, a proton of the right energy can orbit the Earth in an equatorial orbit under the influence of the Earth's magnetic field. If the orbital radius is to be 6.5×10^3 km and the magnetic field at this radius is 0.33×10^{-4} T, what must be the momentum of the proton?

17. In the Crab Nebula (the remnant of a supernova explosion), electrons of a momentum of up to about 10^{-16} kg·m/s orbit in a magnetic field of about 10^{-8} T. What is the orbital radius of such electrons? Note that it is necessary to use Eq. (28) for the calculation of the orbital radius.

18. At the Fermilab accelerator, protons of momentum 5.3×10^{-16} kg·m/s are held in a circular orbit of diameter 2.0 km by a vertical magnetic field. What is the strength of the magnetic field required for this?

19. Figure 31.37 shows the tracks of an electron (charge $-e$) and an antielectron (charge $+e$) created in a bubble chamber. When the particles made these tracks, they were under the influence of a magnetic field of a magnitude 1.0 T and of a direction perpendicular to and into the plane of the figure. What is the momentum of each particle? Assume that they are moving in the plane of the figure and that this figure is $\frac{1}{10}$ natural size. Which is the track of the electron and which is the track of the antielectron?

20. You want to confine an electron of energy 3.0×10^4 eV by making it circle inside a solenoid of radius 10 cm under the influence of the force exerted by the magnetic field. The solenoid has 120 turns of wire per centimeter. What minimum current must you put through the wire if the electron is not to hit the wall of the solenoid?

21. In the Brookhaven AGS accelerator, protons are made to move around a circle of radius 128 m by a magnetic force exerted by a vertical magnetic field. The maximum field that the magnets of this accelerator can generate is 1.3 T.
    (a) Calculate the maximum permissible momentum of the protons.
    (b) Calculate the maximum kinetic energy (use the relativistic relation between energy and momentum).
    (c) Calculate the orbital frequency of such protons.

22. Some astrophysicists believe that the radio waves of 10^9 Hz reaching us from Jupiter are emitted by electrons of fairly low (nonrelativistic) energies orbiting in Jupiter's magnetic field. What must be the strength of this field if the cyclotron frequency is to be 10^9 Hz?

23. A velocity selector with crossed **E** and **B** fields is to be used to select alpha particles of energy 2.0×10^5 eV from a beam containing particles of several energies. The electric field strength is 1.0×10^6 V/m. What must be the magnetic field strength?

24. The Earth's magnetic field at the equator has a magnitude of 5×10^{-5} T; its direction is horizontal toward the north. Suppose that the Earth's atmospheric electric field is 100 V/m; its direction is vertically down. With what velocity (magnitude and direction) must we launch an electron if the electric force is to cancel the magnetic force?

25. In a **mass spectrometer**, a beam of ions is first made to pass through a velocity selector with crossed **E** and **B** fields. The selected ions then are made to enter a uniform magnetic field **B'** where they move in arcs of circles (Figure 31.38). The radius of these circles depends on the masses of the ions. Assume that an ion has a single charge e.
    (a) Show that in terms of E, B, B', and of the impact distance l marked in Figure 31.38, the mass of the ion is

$$m = \frac{eBB'l}{2E}$$

    (b) Assume that in such a spectrometer, ions of the isotope ^{16}O impact at a distance of 29.20 cm and ions of a different isotope of oxygen impact at a distance of 32.86 cm. The mass of the ions of ^{16}O is 16.00 u. What is the mass of the other isotope?

Section 31.4

26. A straight wire is placed in a uniform magnetic field; the wire makes an angle of 30° with the magnetic field. The wire carries a current of 6.0 A and the magnetic field has a strength of 0.40 T. Calculate the force on a 10-cm segment of this wire. Show the direction of the force in a diagram.

27. Figure 31.39 shows a balance used for the measurement of a magnetic field. A loop of wire carrying a precisely known current is partially immersed in the magnetic field. The force that the magnetic field exerts on the loop can

Fig. 31.37 Tracks of an electron and an antielectron in a bubble chamber. The tracks spiral because the particles suffer a loss of energy as they pass through the liquid. For the purposes of Problem 19, concentrate on the initial portions of the tracks.

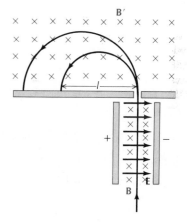

Fig. 31.38 Schematic diagram of a mass spectrometer. The magnetic fields **B** and **B'** are perpendicular to the plane of the page. The crosses show the tails of the magnetic field vectors.

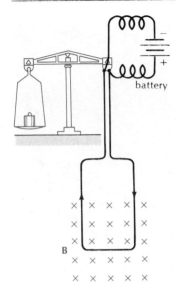

Fig. 31.39 A current balance. The magnetic field **B** is perpendicular to the plane of the page. The crosses show the tails of the magnetic field vectors.

be measured with the balance and this permits the calculation of the strength of the magnetic field. Suppose that the short side of the loop measures 10.0 cm, the current in the wire is 0.225 A, and the magnetic force is 5.35×10^{-2} N. What is the strength of the magnetic field?

28. The electric cable supplying an electric clothes dryer consists of two long straight wires separated by a distance of 1.2 cm. Opposite currents of 20 A flow on these wires. What is the magnetic force experienced by a 1.0-cm segment of wire due to the entire length of the other wire?

29. Two parallel cables of a high-voltage power line carry opposite currents of 1.8×10^3 A. The distance between the cables is 4.0 m. What is the magnetic force pushing on a 50-m segment of one of these cables? Treat both cables as very long, straight wires.

30. A rectangular loop of wire of dimensions 12 cm × 18 cm is near a long, straight wire. One of the short sides of the rectangle is parallel to the straight wire and at a distance of 6.0 cm; the long sides are perpendicular to the straight wire (Figure 31.40). A current of 40 A flows on the straight wire and a current of 60 A flows around the loop. What are the magnitude and direction of the net magnetic force that the straight wire exerts on the loop?

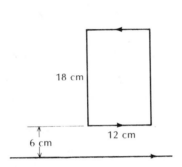

Fig. 31.40 Long, straight wire and rectangular loop.

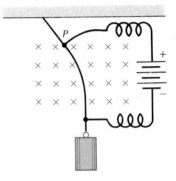

Fig. 31.41 A wire with a weight hanging in a magnetic field.

*31. A thin, flexible wire carrying a current I hangs in a uniform magnetic field **B** (Figure 31.41). A weight attached to one end of the wire provides a tension T. Within the magnetic field, the wire will adopt the shape of an arc of circle.
   (a) Show that the radius of this arc of circle is $r = T/BI$.
   (b) Show that if we remove the wire and launch a particle of charge $-q$ from the point P with a momentum $p = q\,T/I$ in the direction of the wire, it will move along the same arc of circle. (This means that the wire can be used to simulate the orbit of the particle. Experimental physicists sometimes use such wires to check the orbits of particles through systems of magnets.)

Section 31.5

32. A horizontal circular loop of wire of radius 20 cm carries a current of 25 A. At the location of the loop, the magnetic field of the Earth has a magnitude of 0.39×10^{-4} T and points down at an angle of 16° with the vertical. What is the magnitude of the torque that this magnetic field exerts on the loop?

33. The proton has a magnetic moment of 1.41×10^{-26} A · m².
   (a) If this magnetic moment makes an angle of 45° with a uniform magnetic field of 0.80 T, what is the torque on the proton?

(b) If the magnetic moment is initially oriented antiparallel to the field, how much energy will be released when the proton flips into the parallel orientation?

34. Consider the electric motor of Example 7. What is the average value of the torque over a complete rotation? If this motor rotates at the rate of 50 rev/s what is the average horsepower that it delivers to its axle?

35. A molecule with a magnetic moment $\mu = 9.3 \times 10^{-24}$ A·m² is in a uniform magnetic field $B = 0.80$ T. What is the difference in the potential energy between the parallel and antiparallel orientations for μ and **B**? Express your answer in electron-volts.

*36. A rectangular loop of wire of dimension 12 cm × 25 cm faces a long, straight wire. The two long sides of the loop are parallel to the wire and the two short sides are perpendicular; the midpoint of the wire is 8.0 cm from the wire (Figure 31.42). Currents of 95 A and 70 A flow in the straight wire and the loop, respectively.
  (a) What translational force does the straight wire exert on the loop?
  (b) What torque about an axis parallel to the straight wire and through the center of the loop does the straight wire exert on the loop?

*37. A compass needle is attached to an axle which permits it to turn freely in a horizontal plane so that only the horizontal component of the magnetic field affects its motion. The magnetic moment of the needle is 9.0×10^{-3} A·m², its moment of inertia is 2.0×10^{-8} kg·m², and the horizontal component of the magnetic field is 0.19×10^{-4} T.
  (a) What is the torque on the needle as a function of the angle θ between the needle and the north direction?
  (b) What is the frequency of small oscillations of the needle about the north direction? [Hint: Compare the equation for the rotational motion of the needle with Eq. (14.71) for the motion of a pendulum.]

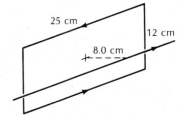

25 cm   12 cm   8.0 cm

Fig. 31.42 Long, straight wire and rectangular loop.

CHAPTER 32

Electromagnetic Induction

In this chapter we will discover that electric fields can be generated not only by charges, but also by magnetic fields. Whenever the magnetic field lines move or change in any way, they will generate an electric field, an **induced electric field.** This kind of electric field exerts electric forces on charges — in this regard the induced electric field does not differ from an ordinary electrostatic electric field. However, the induced and electrostatic electric fields differ in that the latter is conservative while the former is not. This means that the path integral of **E** around a closed path is not zero if **E** is an induced electric field. The path integral of the induced electric field is called the **induced emf.** As we will see, a famous law formulated by Michael Faraday asserts that the magnitude of this induced emf is directly related to the rate at which the magnetic field lines cut across the path.

Induced emf

We begin with a study of the induced emf produced by the motion of a path in a constant magnetic field. Such a path cuts across field lines, and thereby generates an induced emf, called a motional emf.

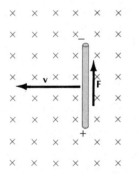

Fig. 32.1 Metallic rod moving with velocity **v** through a uniform magnetic field. A free electron within this rod experiences a force in the upward direction.

32.1 Motional Electromotive Force

Suppose that we push a rod of metal with some velocity through a uniform magnetic field, such as the magnetic field of a solenoid. If the rod and the velocity **v** of the rod are perpendicular to each other and to the magnetic field (Figure 32.1), then the free electrons in the metal will experience a magnetic force evB directed along the rod. The electrons will therefore flow along the rod, accumulating negative charge on the upper end and leaving positive charge on the lower end. This flow of charge will stop when the electric repulsion generated by the accumulated charges balances the magnetic force evB. However, if the ends of the rod are in sliding contact with a pair of long wires that provide a (stationary) return path (Figure 32.2), then the electrons will not accumulate on the ends of the rod — they will flow continuously around the circuit. Thus, the moving rod acts as a "pump of electricity." The rod is a source of emf. The upper end of the rod will act as the negative terminal of this source, and the lower end as the positive terminal.

The emf associated with the rod is the work done by the driving force on a unit positive charge that passes from the negative end of the rod to the positive. The driving force on a unit positive charge is vB and, if the length of the rod is l, the work done by this force is

$$\mathscr{E} = vBl \tag{1}$$

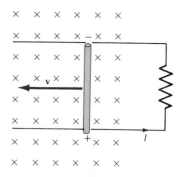

Fig. 32.2 If the ends of the rod are in sliding contact with a pair of wires, a current will flow around the circuit.

This is called a **motional emf** because it is generated by the motion of the rod through the magnetic field. Note that we have ignored the component of velocity associated with motion of the charge *along* the rod. This component leads to a magnetic force perpendicular to the length of the rod and that force is canceled by the constraining force that the sides of the rod exert on the charges, preventing their escape. The cancellation of one part of the magnetic force is crucial for the result stated in Eq. (1) — if it were not for this cancellation, the total magnetic force would do zero work [compare Eq. (31.22)]. Of course, the energy delivered to the current by the emf ultimately does not come from the magnetic field but from the mechanical device that pushes the rod. The magnetic field merely plays the role of mediator, transferring the mechanical energy into electric energy. If we did not apply an external push to the rod of Figure 32.2, it would slow down and stop as its kinetic energy is converted into electric energy and delivered to the current.

The quantity Blv appearing on the right side of Eq. (1) can be given the following interpretation: the product lv is the area that the rod sweeps across per unit time and B equals the number of lines of magnetic field per unit area;[1] hence Blv equals the number of magnetic field lines that the rod cuts across per unit time. We can then write Eq. (1) as

$$\mathscr{E} = [\text{rate of cutting of magnetic field lines}] \tag{2}$$

The advantage of Eq. (2) over Eq. (1) is that the former equation is of general validity — it holds for rods and other bodies of arbitrary shape moving through arbitrary magnetic fields (this can be proved by treating a rod of arbitrary shape as made of small straight segments chosen so that the magnetic field is approximately constant in the vicinity of each segment). Furthermore, as we will see in the next section, the relation between the induced emf and the rate of cutting of field lines holds even if the magnetic field is time dependent. Hence the concept of cutting of field lines, which is due to Faraday, not only provides us with a sharp mental picture of the mechanism of induction, but also helps us in the formulation of the general law of induction.

Motional emf

Michael Faraday, *1791–1867, English physicist and chemist. Faraday was apprentice to a bookbinder when he became interested in science. He attended lectures by Humphry Davy, the famous chemist, and prepared a set of lecture notes which so impressed Davy that he appointed Faraday his assistant at the laboratory of the Royal Institution. Ultimately, Faraday succeeded Davy as director of the laboratory. Faraday's earliest research lay in chemistry, but he soon turned to research in electricity and magnetism, making contributions of the greatest significance. His discovery of electromagnetic induction was no accident, but arose from a systematic experimental investigation of whether magnetic fields can generate electric currents. Although Faraday was essentially an experimenter, with no formal training in mathematics, he made an important theoretical contribution by introducing the concept of field lines and by recognizing that electric and magnetic fields are physical entities.*

---

[1] Note that this statement hinges on the normalization convention for magnetic field lines adopted in Section 30.2.

However, we must be careful not to interpret Faraday's vivid phrase *cutting of field lines* in a literal sense — the field lines are not physical objects, but merely mathematical constructs, and they suffer no damage when a rod passes through them.

Note that Eq. (2) only gives us the magnitude of the emf. To find the sign of the emf, we must remember that within the moving body that cuts across the field lines, a positive charge will be pushed in the direction of $\mathbf{v} \times \mathbf{B}$.

EXAMPLE 1. A straight metallic rod is rotating about its midpoint on an axis parallel to a uniform magnetic field (Figure 32.3a). The length of the rod is $2l$ and the angular velocity of rotation is ω. What is the induced emf between the midpoint of the rod and each end? What is the induced emf between the two ends?

Fig. 32.3 (a) Rod rotating in a uniform magnetic field. (b) If the ends of the rod are in sliding contact with a circular track, a current will flow as shown.

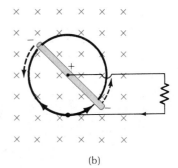

(a)   (b)

SOLUTION: Consider one-half of the rod, from the midpoint to one end. This piece takes a time $2\pi/\omega$ to sweep out the circular area πl^2. Hence the rate at which this piece sweeps out area is $\pi l^2 \omega / 2\pi = l^2 \omega / 2$ and the rate at which it cuts across magnetic field lines is $Bl^2\omega/2$. Equation (2) then leads to an induced emf

$$\mathcal{E} = Bl^2\omega/2 \tag{3}$$

between the midpoint and each end. With the direction of motion as shown in Figure 32.3a, the midpoint will act as the positive terminal of a source of emf and each moving end as the negative terminal.

The emf between the two ends is zero because both ends act as negative terminals, with the same emf (the two halves of the rod act as two batteries connected in parallel). Note that a careless calculation with Eq. (2) would have led us to the wrong conclusion that the emf for the entire rod is twice that for each half, since the entire rod cuts twice as many field lines as each half — but we must take into account the signs of these emfs!

Homopolar generator

If the moving ends of the rod are in sliding contact with a circular track connected to the midpoint via an external circuit (Figure 32.3b), then the rod will generate a current in the circuit. The device shown in Figure 32.3b acts as a generator of DC voltage; this device is called a **homopolar generator.** For practical applications, homopolar generators are constructed with a rotating disk rather than a rotating rod. This does not affect the emf of the generator, but it helps to reduce its internal resistance. Homopolar generators are used in applications requiring a large current but only a fairly small voltage, such as in electroplating.

EXAMPLE 2.   A **magnetohydrodynamic (MHD) generator** consists of a rectangular channel within which flows a hot, ionized gas, or plasma. The channel is placed in a strong magnetic field (Fig. 32.4). What emf will be induced between opposite sides of the stream of plasma? The width of the channel is 50 cm, the speed of the gas is 800 m/s, and the strength of the magnetic field is 6.0 T.

Magnetohydrodynamic generator

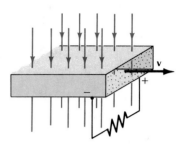

SOLUTION: Consider a straight path, fixed in the plasma, from one side of the stream of plasma to the other. This path cuts magnetic field lines, just as the straight rod discussed at the beginning of this section; hence Eq. (1) applies,

$$\mathscr{E} = vBl = 800 \text{ m/s} \times 6.0 \text{ T} \times 0.50 \text{ m} = 2.4 \times 10^3 \text{ V} \qquad (4)$$

If, as shown in Figure 32.4, the magnetic field points downward, then the left side of the stream of plasma acts as the positive terminal of a source of emf and the right side as the negative terminal. This source of emf will drive a current if we connect an external circuit, as shown in Figure 32.4. Note that the two sides of the channel must be made of a conductor, so that they are in good electrical contact with the plasma, but the top and bottom of the channel must be made of an insulator.

Fig. 32.4 A channel in which plasma flows toward the right. The channel is placed in a vertical magnetic field.

Magnetohydrodynamic generators are expected to serve as auxiliary generators in power plants burning fossil fuel (Figure 32.5). The MHD generator

Fig. 32.5 Experimental MHD generator at the Argonne National Laboratory. This generator has a circular channel of diameter ∼ 1 m, which is surrounded by a powerful superconducting magnet.

uses the exhaust gas released by the burning of this fuel. For this purpose, the fuel must be burned in a special combustion chamber, similar to that of a jet engine, so as to produce a high-speed stream of very hot exhaust gas. After the hot gas emerges from the MHD generator, it can supply heat to a boiler providing steam for a conventional power generator.

Incidentally, note that the emf induced across a stream of plasma flowing in a channel is quite analogous to the voltage produced by the Hall effect across the stream of electrons flowing in a conducting strip (see Section 31.3). The Hall voltage can be regarded as an induced, motional emf.

32.2 Faraday's Law

As we saw in the preceding section, the induced emf between the ends of a rod moving through a magnetic field equals the rate at which the rod cuts across magnetic field lines. Of course, the rod will cut across

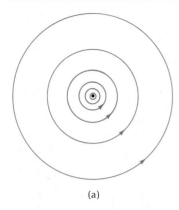

(a)

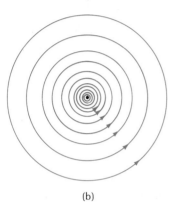

(b)

Fig. 32.6 (a) Magnetic field lines of a current on a very long wire. The wire is perpendicular to the plane of the page. (b) Magnetic field lines of a stronger current.

field lines whenever there is *relative motion* between rod and magnet. Relativity suggests that it should not make any difference whether we move the rod past a fixed magnet or move the magnet past a fixed rod — and experiments do indeed confirm that in both cases the induced emf is exactly the same.

But there is another way in which we can cut across field lines: we can hold the rod fixed and increase or decrease the strength of the magnetic field by increasing or decreasing the current. To understand why field lines will be cut under these conditions, we must first take a look at what happens to the field lines of a current when the current increases or decreases. Figure 32.6a shows the field lines produced by a current on a long, straight wire. If we increase the current, the magnetic field increases, i.e., the number of field lines increases. Figure 32.6b shows the field lines of a stronger current. Where do the extra field lines come from? Obviously, the current has to make them. When the current slowly increases, it makes new, small circles of field lines in its immediate vicinity; meanwhile the circles that already exist in Figure 32.6a expand and move outward, just as the ripples on the surface of a pond that are created when we drop a stone into the pond. Thus, the pattern shown in Figure 32.6a gradually grows into the new pattern shown in Figure 32.6b. Note that the pattern grows from the inside out.

If we decrease the current, field lines must disappear. But this is not quite the reverse of the creation of field lines — the circles of Figure 32.6b do not contract and disappear at the center. The pattern cannot change from outside in; it must first change *near* the current. When the current slowly decreases, it makes new small circles of field lines of *opposite* direction (negative field lines) in its immediate vicinity. These opposite circles expand and move outward, gradually canceling the original field lines.

In any case, a change in the strength of magnetic field involves moving field lines. If a rod is located in the vicinity of the current, the moving field lines will sweep across the rod. Hence an increase or decrease of the strength of magnetic field can cause field lines to cut across a *stationary* rod.

Does such a cutting of field lines induce an emf in the rod? This is a question that can only be resolved by experiment — and the answer is yes. The field lines cutting across the stationary rod induce an emf which is given by the same formula as before. The general statement is known as **Faraday's Law of induction:**

> *The induced emf along any moving or fixed path in a constant or changing magnetic field equals the rate at which magnetic field lines cut across the path,*

Faraday's Law

$$\mathscr{E} = [\text{rate of cutting of magnetic field lines}] \qquad (5)$$

Note that we have expressed Faraday's Law as an assertion about a mathematical *path*. The emf will exist between the ends of this path regardless of whether we place a rod or wire along this path, i.e., whenever a unit positive charge moves along this path it will gain an amount $\mathscr{E}$ of energy from the induced electric field regardless of whether the charge moves on a rod or through empty space. The rod or wire merely serves as a convenient conduit for the charge. Of course, for the practical exploitation of the emf we must provide a rod, wire, or some other conductor along which the current can flow, and we will also have to provide a return path for the current.

For a closed path, Faraday's Law can be conveniently stated in terms of the **magnetic flux.** The magnetic flux through any given area is simply the number of magnetic field lines intercepted by the area. As in the case of electric flux, we can write the flux as an integral:

$$\Phi_B = \int \mathbf{B} \cdot d\mathbf{S} \qquad (6)$$

The unit of magnetic flux is the weber (Wb),

$$1 \text{ weber} = 1 \text{ Wb} = 1 \text{ T} \cdot \text{m}^2 \qquad (7)$$

When we reckon the net rate at which field lines cut across a closed path, we must take into account that a field line can make a positive or a negative contribution to the rate of cutting, depending on whether the field line enters the area bounded by the path or leaves it. The net rate of cutting of field lines is equal to the rate of change of the number of field lines intercepted by the area bounded by the path, i.e., it is equal to the rate of change of the magnetic flux through this area. Hence, Faraday's Law can be stated as follows:

The induced emf around a closed path in a magnetic field is equal to the rate of change of the magnetic flux intercepted by the area within the path,

$$\mathscr{E} = -\frac{d\Phi_B}{dt} \qquad (8)$$

The minus sign that we have inserted in this equation indicates how the sign of the induced emf is related to the sign of the flux and of the rate of change of flux. The sign of the magnetic flux is fixed by the following **right-hand rule:** wrap the fingers around the path in the direction in which we are reckoning the emf; the magnetic flux is then positive if the magnetic field points in the direction of the thumb and negative otherwise. For example, in Figure 32.7, this right-hand rule indicates that the field lines shown make a positive contribution to the magnetic flux; consequently, if the strength of the magnetic field is increasing, $d\Phi_B/dt$ will be positive and, according to Eq. (8), the induced emf around the closed path will be negative. If the path in Figure 32.7 coincides with a conducting wire, the induced emf will therefore drive a current around this wire in a direction opposite to that indicated by the arrow.

32.3 Some Examples; Lenz' Law

In this section, we will look at some examples of the calculation of induced emfs. We will also lay down a simple rule for finding the direction of the emf.

Wilhelm Eduard Weber, *1804–1891, German physicist, professor at Göttingen. He worked on problems in magnetism, and devised a system of units for electric and magnetic quantities.*

Faraday's Law for closed loop

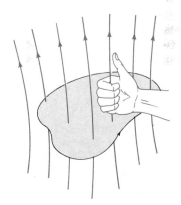

Fig. 32.7 A closed path in a changing magnetic field. The sign of the magnetic flux intercepted by the area within the path is given by a right-hand rule: if the fingers are curled around the path in the direction in which we are reckoning the emf (indicated by the black arrow), the thumb points in the direction that the magnetic field must have in order to give a positive flux.

Fig. 32.8 A rectangular coil in a uniform, decreasing magnetic field.

EXAMPLE 3. A rectangular coil of 150 loops of wire forming a closed circuit measures 0.20 m × 0.10 m. The resistance of the coil is 5.0 Ω. The coil is placed in an electromagnet, face on to the magnetic field (Figure 32.8). Suppose that when the electromagnet is suddenly switched off, the strength of the magnetic field decreases at the rate of 20 T per second. What is the induced emf in the coil? The induced current? What is the direction of the induced current?

SOLUTION: The flux intercepted by the area within the coil is $\Phi_B = NAB$, where N is the number of turns, A the area, and B the magnetic field. Hence the induced emf is

$$\mathcal{E} = -\frac{d\Phi_B}{dt} = -NA\frac{dB}{dt} = 150 \times (0.20 \text{ m} \times 0.10 \text{ m}) \times 20 \text{ T/s}$$

$$= 60 \text{ V}$$

and the induced current is

$$I = \mathcal{E}/R = 60 \text{ V}/5.0 \text{ }\Omega = 12 \text{ amperes}$$

To determine the direction of the induced current, let us reckon the emf around the loop in the direction indicated by the arrow in Figure 32.8. By the right-hand rule, the field lines shown make a positive contribution to the magnetic flux. Since the magnetic field is *decreasing*, $d\Phi_B/dt$ will then be *negative* and, according to Eq. (8), the induced emf will be positive. This induced emf will therefore drive a current in the direction of the arrow.

EXAMPLE 4. A long solenoid has a circular cross section of radius R. The solenoid is connected to a source of alternating current so that the magnetic field inside the solenoid is

$$B = B_0 \sin \omega t \tag{9}$$

where B_0 and ω are constants. What is the induced emf around a concentric circular path of radius r inside the solenoid or outside the solenoid?

SOLUTION: Inside the solenoid ($r < R$), the magnetic field is uniform. Figure 32.9 shows a circle of radius r. The area of this circle intercepts a magnetic flux $-\pi r^2 B$. [If the positive direction for **B** is into the page and the direction in which we are reckoning the emf is counterclockwise (Figure 32.9), then, by the right-hand rule discussed in connection with Eq. (8), the flux will have a sign opposite to that of **B** — at the instant shown in Figure 32.9, the flux is negative.] The rate of change of the flux is $-\pi r^2 dB/dt$. Hence Faraday's Law gives us the induced emf

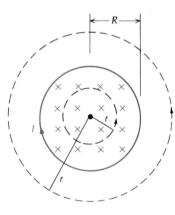

Fig. 32.9 Cross-sectional view of solenoid showing uniform magnetic field. The dashed circles are two alternative paths for reckoning the emf.

$$\mathcal{E} = -\frac{d\Phi_B}{dt} = \pi r^2 \frac{dB}{dt} = \pi r^2 \omega B_0 \cos \omega t \tag{10}$$

The emf increases in proportion to the area of the circle, reaching a maximum value if the radius of the circle coincides with the radius of the solenoid ($r = R$).

Note that as regards the time dependence, the emf is 90° out of phase with the magnetic field — when the magnetic field (9) has maximum magnitude (sin ωt = sin $\pi/2$ = 1), the emf (10) has minimum magnitude (cos ωt = cos $\pi/2$ = 0).

Outside the solenoid ($r > R$), there is no magnetic field. The magnetic flux intercepted by a circle of radius r (Figure 32.9) is therefore simply the flux $-\pi R^2 B$ of lines inside the solenoid and the rate of change in this flux is

$-\pi R^2\, dB/dt.$[2] Instead of Eq. (10) we now obtain

$$\mathscr{E} = -\frac{d\Phi_B}{dt} = \pi R^2 \frac{dB}{dt} = \pi R^2 \omega B_0 \cos \omega t \qquad (11)$$

Thus the emf remains constant as we go away from the solenoid. Figure 32.10 shows a plot of the value of the induced emf as a function of the radius of the circular path.

Note that the solenoid and its current did not directly enter into our calculation — the induced emf is entirely determined by the magnetic field. This implies that we will get the same kind of induced emf if the magnetic field is that in the gap of an electromagnet (with round pole pieces) rather than that in a solenoid. Such induced emfs in electromagnets find an important application in the design of electron accelerators. In the **betatron** (Figure 32.11), electrons travel in a circular orbit between the poles of a large electromagnet. The electrons are held in their circular orbit by the magnetic field. The acceleration of the electron is accomplished by quickly increasing the strength of the magnetic field. This change of the magnetic field induces an emf around the circular orbits of the electrons, which accelerates the electrons, increasing their energy. This acceleration continues as long as the magnetic field is increasing; at the same time, the increasing magnetic field provides just the right amount of extra centripetal force to keep the electrons in a circular orbit of constant (or nearly constant) radius. In a typical betatron, the magnetic field increases over a time interval of a few milliseconds and in this time the electrons move around their circular orbit a few hundred thousand times, acquiring a final energy of 100 MeV or so. They are then suddenly guided out of the machine (by means of deflecting electric or magnetic fields) and the magnetic field is then permitted to decrease, readying the machine for the next cycle of acceleration.

Betatron

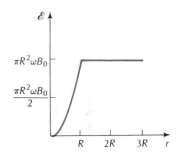

Fig. 32.10 Induced emf as a function of radius of the circular path (at time $t = 0$).

Fig. 32.11 The small machine on the cart in the center is the first betatron, built in 1940. The large machine in the background is the largest betatron. This machine, at the University of Illinois, attained an electron energy of 340 MeV.

---

[2] This leads to an intriguing question: the rate of change of flux ought to equal the rate of cutting of field lines by the circle, but if there is no magnetic field outside of the solenoid, how can the circle cut across field lines? The answer is that when we increase the current, a new field line will be created inside the solenoid and move toward the center; whenever this happens, the return portion of this field line must move from the current outward to infinity (the magnetic field lines form *closed* loops around the current). This outward-moving field line cuts across the outer circle of Figure 32.9.

The direction of the induced emf can always be figured out by an argument similar to that given in Examples 3 and 4. But there is a simple rule for finding the direction, a rule known as **Lenz' Law:**

Lenz' Law

The induced emfs are always in such a direction as to oppose the change that generated them.

Emil Khristianovich Lenz, *1804–1865, Russian physicist, rector of the University of St. Petersburg. He investigated induction and other electromagnetic phenomena. Independently of Joule, he established the law for the production of heat by a current in a resistor.*

The exact meaning of *change* depends on the problem at hand and is best made clear by some examples. For instance, the moving rod of Figure 32.12 generates a counterclockwise current. Consider now the magnetic force on this current in the rod; this force has the direction of $I\,d\mathbf{l} \times \mathbf{B}$, which points to the right in Figure 32.12. Thus, the force associated with the induced current *opposes* the motion of the rod, i.e., it opposes the change (motion) that generates the current. This is in agreement with Lenz' Law.

Likewise, if there were a circular wire loop within the solenoid of Figure 32.9, a counterclockwise current would flow in it [assuming that B of Eq. (9) is increasing, $0 < \omega t < \pi/2$]. Such a current flowing around the loop would produce a magnetic field within the loop opposite to the field B of Eq. (9). Thus, the induced current around the loop *opposes* the change of flux within the loop, i.e., it opposes the change that generates the current.

Note that Lenz' Law is consistent with energy conservation but that the contrary of Lenz' Law is not. If the contrary of Lenz' Law were true, then the force on the rod in Figure 32.12 would be toward the left so as to speed up the rod. This would develop into a runaway situation: as the rod acquires higher velocity, the force would increase and accelerate it more and more. There would then be no need for us to push the rod through the magnetic field . . . it would accelerate spontaneously! Clearly, this would be a violation of energy conservation.

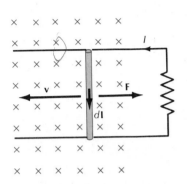

Fig. 32.12 The magnetic field exerts a force on the current in the rod.

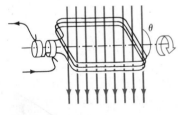

Fig. 32.13 Electromagnetic generator.

EXAMPLE 5. An electromagnetic generator consists of a coil of N loops of wire that rotates about an axis perpendicular to a constant magnetic field. Sliding contacts connect the coil to an external circuit (Figure 32.13). What emf does the coil deliver to the external circuit? The coil has an area A and rotates with an angular frequency ω.

SOLUTION: The coil makes an angle $\theta = \omega t$ with the magnetic field. The magnetic flux through the coil is $\Phi = NAB \sin \theta = NAB \sin \omega t$. By Faraday's Law [see Eq. (8)] the induced emf is

$$\mathscr{E} = -\frac{d\Phi_B}{dt} = -\frac{d}{dt}(NAB \sin \omega t)$$

$$= -NAB\omega \cos \omega t \qquad (12)$$

At the instant shown in Figure 32.13, the magnetic flux through the coil is decreasing. According to Lenz' Law, the induced current must therefore flow around the coil clockwise so as to contribute an extra magnetic flux that tends to oppose the decrease of flux.

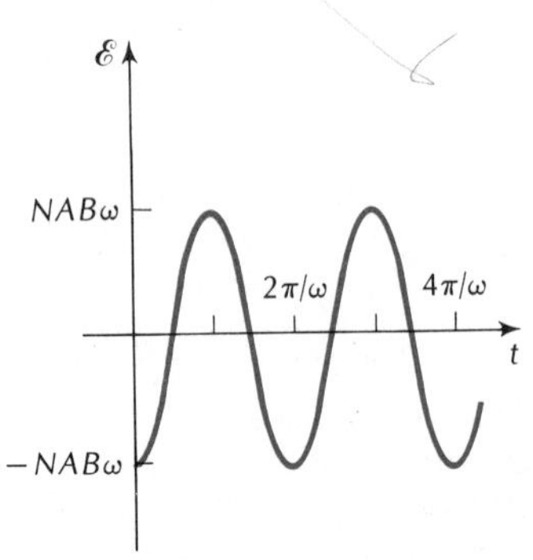

The function given by Eq. (12) is an **alternating emf,** or **AC voltage,** that oscillates between positive and negative values. A plot of this emf is shown in Figure 32.14. To obtain a constant emf, or DC voltage, from a generator, we can use a slotted sliding contact, or **commutator,** arranged so that the emf at the external terminals always has the same sign (Figure 32.15). A plot of the emf produced by such a generator is shown in Figure 32.16. This emf oscillates between zero and a maximum value but always remains positive. To elim-

Fig. 32.14 Emf of the generator; this is an alternating voltage (AC).

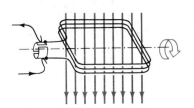

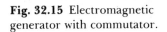

Fig. 32.15 Electromagnetic generator with commutator.

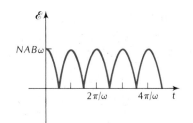

Fig. 32.16 Emf of generator with commutator.

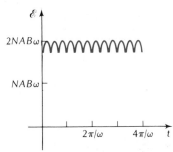

Fig. 32.17 Emf of generator with three coils 120° apart connected in series; each loop has a commutator. The ripple in the voltage is about 14% of the maximum voltage.

inate or reduce the oscillations of the emf, we can connect several generators in series, each with its coil at a slightly different initial angle to the magnetic field. In the sum of the emfs, the oscillations then tend to average out. For example, the plot of Figure 32.17 shows the combined emf of three coils whose positions are always 120° apart as they rotate. The emf delivered by this generator is fairly constant, with only a small ripple.

32.4 The Induced Electric Field

It is very instructive to reexamine the generation of the induced emf in a rod moving through a magnetic field from the point of view of a reference frame in which the rod is at rest. Consider again the rod of Figure 32.1 moving at constant velocity **v.** In a reference frame moving with the rod at the same velocity **v,** the free charges within the rod have no velocity and there is no *magnetic* force. However, the free charges must experience some other kind of force that pushes them along the rod. The question is then: What is this new kind of force in the moving reference frame that produces the same effect as the magnetic force in the stationary reference frame? Since the only kinds of force that act on electric charges are the magnetic force and the electric force, the "new" kind of force must be due to a "new" kind of electric field, an electric field that exists in the moving reference frame, but not in the stationary reference frame (Figure 32.18). This "new" electric field in the moving reference frame arises indirectly from the currents that are responsible for the magnetic field in the stationary reference frame — it turns out that although the conductors that carry these currents are electrically neutral relative to the stationary reference frame, they acquire an excess of positive or negative charge relative to the moving reference frame (for details of the mech-

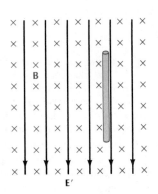

Fig. 32.18 In the moving reference frame of the rod, there is a magnetic field **B** (colored crosses) and also an electric field **E'** (black).

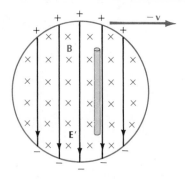

Fig. 32.19 In the moving reference frame, the solenoid has a magnetic field **B** (colored crosses) and an electric field **E'** (black). The edge of the solenoid has acquired an electric charge.

anism involved in this, see Section 30.4). For instance, if the magnetic field through which the rod moves is that of a very long solenoid (Figure 32.19), then the upper portion of the solenoid acquires a positive charge and the lower portion a negative charge, as seen in the moving reference frame.

We can determine the magnitude of the induced electric field from a consistency requirement: the "new" electric force in the moving reference frame must coincide with the magnetic force in the stationary reference frame. This tells us that the electric field in the moving reference frame must have a magnitude

$$E' = vB \tag{13}$$

This electric field does work on the free charges and therefore represents a source of emf.[3] The work done on a unit positive charge that passes through the rod is

$$\mathscr{E} = El \tag{14}$$

In view of Eq. (13), this value of the emf coincides with the value that we obtained Eq. (1). Thus, the motional emf can be calculated either in a stationary reference frame or else in a moving reference frame; in the former case it arises from a magnetic field and in the latter from an electric field.

In our simple example, the electric field **E'** is constant along the rod. More complicated problems may involve an electric field **E'** that depends on position. Equation (13) indicates that such a dependence can come about in two ways: if the magnetic field varies along the rod (a nonuniform magnetic field) or if the velocity varies along the rod (a rigid rod with rotational motion or perhaps a flexible rod undergoing a deformation). If **E'** depends on position, then we must of course replace Eq. (14) by an integral,

$$\mathscr{E} = \int \mathbf{E'} \cdot d\mathbf{l} \tag{15}$$

where the integration extends from one end of the rod to the other.

Induced electric field

The electric field **E'** that exists in the reference frame of the moving rod is called an **induced electric field.** One important feature of this kind of field is that it is *not conservative.* Figure 32.20 shows the electric field **E'** of a solenoid as seen in the reference frame of the moving rod. If we integrate **E'** around the closed path drawn in this figure, we will obviously get a result different from zero — only one side of the rectangle gives a contribution to the integral. This establishes that **E'** is not a conservative field since, for a conservative field, the integral has to be zero for *every* arbitrary closed path.[4]

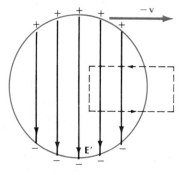

Fig. 32.20 Electric field of a long solenoid as seen in the moving reference frame. For the dashed path shown, which has one portion outside of the electric field **E'**, the integral $\oint \mathbf{E'} \cdot d\mathbf{l}$ is *not* zero.

---

[3] Note that this is an external electric field (generated by charges *not* on the rod). Besides there may be an internal electric field (generated by charges that accumulate on the ends of the rod); for now, we will not concern ourselves with this latter field.

[4] We might ask: Why can we not use a similar argument to establish that the electric field of a capacitor is nonconservative? The answer is that when we take a path that has one portion inside the capacitor and one portion outside, *both* portions contribute to the path integral because the capacitor has a fringing field that extends beyond the edges of the plates. In earlier chapters we have ignored this fringing field because it was unimportant, but for the evaluation of the path integral it is very important (see Question 27.7). In contrast, the ideal, infinitely long solenoid has no fringing field — the magnetic and the electric fields outside the solenoid are exactly zero.

In general, an induced electric field will exist in the reference frame of any path, or segment of a path, moving through a magnetic field. An induced electric field will also exist in the reference frame of a stationary path placed in a changing, time-dependent magnetic field. The induced electric field is always related to the induced emf by Eq. (15). Hence, in terms of the induced electric field, we can write Faraday's Law as

$$\int \mathbf{E}' \cdot d\mathbf{l} = [\text{rate of cutting of magnetic lines}] \qquad (16)$$

for an arbitrary path or as

$$\oint \mathbf{E}' \cdot d\mathbf{l} = -\frac{d\Phi_B}{dt} \qquad (17)$$

for a closed path. Here the induced electric field $\mathbf{E}'$ associated with some segment $d\mathbf{l}$ of the path must be reckoned in a reference frame in which the segment $d\mathbf{l}$ is (instantaneously) at rest.

EXAMPLE 6. Consider again the long solenoid with the time-dependent magnetic field described in Example 4. What is the induced electric field inside and outside the solenoid?

SOLUTION: We have already evaluated the induced emf in Example 4. Inside the solenoid $(r < R)$, its value is given by Eq. (10); and outside the solenoid $(r > R)$, its value is given by Eq. (11). The electric field is related to this emf by

$$\oint \mathbf{E} \cdot d\mathbf{l} = \mathscr{E} \qquad (18)$$

where the path of integration is a closed concentric circle of radius r. Note that the electric field $\mathbf{E}$ in this equation is an electric field in the original reference frame (there is no moving reference frame involved in this problem). We therefore are entitled to denote the electric field by $\mathbf{E}$, rather than by $\mathbf{E}'$. To evaluate the left side of Eq. (18), we observe that the field lines of the electric field must be closed concentric circles because this is the only pattern of electric field lines consistent with the rotational symmetry of the solenoid. Hence the path of integration coincides with a field line and

$$\oint \mathbf{E} \cdot d\mathbf{l} = 2\pi r E \qquad (19)$$

Combining Eqs. (10) and (19), we then obtain, inside the solenoid,

$$2\pi r E = \pi r^2 \omega B_0 \cos \omega t$$

or

$$E = \tfrac{1}{2} r \omega B_0 \cos \omega t \qquad (20)$$

The electric field is zero on the axis of the solenoid $(r = 0)$ and reaches a maximum value at the edge of the solenoid $(r = R)$.

Combining Eqs. (11) and (19), we obtain, outside the solenoid,

$$2\pi r E = \pi R^2 \omega B_0 \cos \omega t$$

or

$$E = \frac{1}{2} \frac{R^2}{r} \omega B_0 \cos \omega t \tag{21}$$

The electric field decreases in strength ($\propto 1/r$) as we go away from the solenoid.

Figure 32.21 is a plot of the strength of the electric field as a function of r. Figure 32.22 shows a picture of the field lines. According to Eqs. (20) and (21), the electric field at the initial time ($t = 0$) is positive, i.e., it has the same direction as the direction of integration around the circular path in Figure 32.9 (counterclockwise).

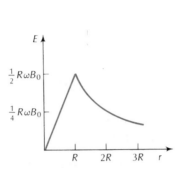

Fig. 32.21 Magnitude of the induced electric field as a function of radius (at $t = 0$).

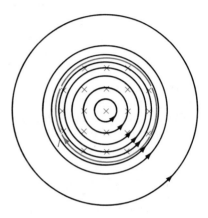

Fig. 32.22 Electric field lines (black) within solenoid and outside of solenoid.

The preceding example shows how a time-dependent magnetic field generates an electric field. This induced electric field differs from the electrostatic fields of the preceding chapters in that the field lines form closed loops — the field lines do not start on electric charges. Furthermore, the induced field is nonconservative; this becomes immediately obvious when we integrate the electric field along a closed field line [see Eq. (19)]. As we will see in Chapter 35, the electric fields of a light wave or radio wave also are instances of induced electric fields, with closed field lines.

32.5 Inductance

If a conductor carrying a time-dependent current is near some other conductor, then the changing magnetic field of the former can induce an emf in the latter. Thus, a time-dependent current in one conductor can induce a current in another nearby conductor. For instance, consider the two coils of Figure 32.23. One of the coils carries a time-dependent current, which generates a magnetic field. The changing flux Φ_{B_1} through the second coil induces an emf in this coil. The emf in the second coil is

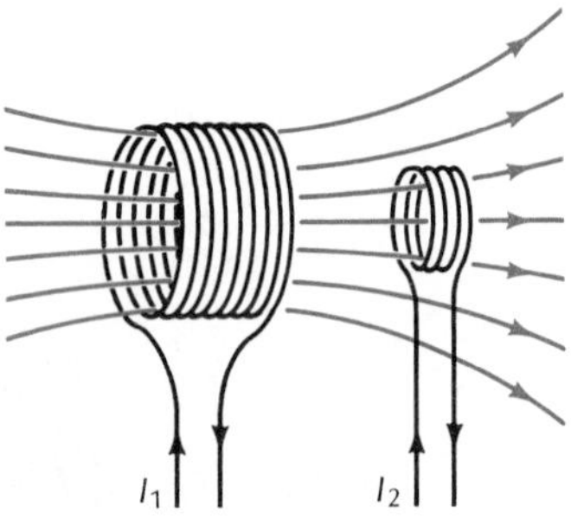

$$\mathscr{E}_2 = -\frac{d\Phi_{B_1}}{dt} \tag{22}$$

The flux Φ_{B_1} depends on the strength of the magnetic field B_1 in the

Fig. 32.23 Coil 1 creates a magnetic field. Some of the field lines pass through coil 2.

second coil produced by the current I_1 in the first coil. This field strength is directly proportional to I_1 [see Eq. (30.25)] for the dependence of magnetic field on current]; hence Φ_{B_1} is also proportional to I_1. We can write this relationship between Φ_{B_1} and I_1 as

$$\Phi_{B_1} = L_{21}I_1 \qquad (23)$$

and then Eq. (22) becomes

$$\mathscr{E}_2 = -L_{21}\frac{dI_1}{dt} \qquad (24)$$

Here L_{21} is a constant of proportionality which depends on the size of the coils, their distance, and the number of turns in each, i.e., L_{21} depends on the geometry of Figure 32.23. This constant is called the **mutual inductance** of the coils.

Equation (24) states that the emf induced in coil 2 is proportional to the rate of change of current in coil 1. But the converse is also true: if coil 2 carries a current, then the emf induced in coil 1 is proportional to the rate of change of the current in coil 2,

$$\mathscr{E}_1 = -L_{12}\frac{dI_2}{dt} \qquad (25)$$

The constants of proportionality L_{21} and L_{12} appearing in Eqs. (24) and (25) are exactly equal, i.e., $L_{21} = L_{12}$. Although we will accept this assertion without proof, we note that the result is quite reasonable: the mutual inductance reflects the geometry of the *relative* arrangement of the coils and that is of course the same in both cases.

The SI unit of inductance is called the **henry** (H),

$$1\ \text{henry} = 1\ \text{H} = 1\ \text{V}\cdot\text{s/A} \qquad (26)$$

Incidentally, the permeability constant is commonly expressed in terms of this unit of inductance,

$$\mu_0 = 1.26 \times 10^{-6}\ \text{H/m} \qquad (27)$$

Mutual inductance

Joseph Henry, *1797–1878, American experimental physicist, professor at Princeton and first director of the Smithsonian Institution. He made important improvements in electromagnets by winding coils of insulated wire around iron pole pieces, and invented an electromagnetic motor and a new, efficient telegraph. He discovered self-induction, and investigated how currents in one circuit induce currents in another.*

EXAMPLE 7.   A long solenoid has n turns per unit length. A ring of wire of radius r is placed within the solenoid, perpendicular to the axis (Figure 32.24). What is the mutual inductance of ring and solenoid?

SOLUTION: If the current in the solenoid windings is I_1, the magnetic field is $B_1 = \mu_0 n I_1$ [see Eq. (31.16)] and the flux through the ring is

$$\Phi_{B_1} = \pi r^2 \mu_0 n I_1$$

Comparing this with Eq. (23), we see that

$$L_{12} = \pi r^2 \mu_0 n \qquad (28)$$

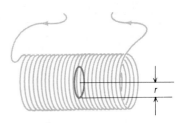

Fig. 32.24 A ring of wire inside a solenoid.

Self-inductance

A conductor by itself has a **self-inductance.** Consider a coil with a time-dependent current (Figure 32.25). The magnetic field of this coil will then produce a time-dependent magnetic flux and, by Faraday's Law, an induced emf. The net emf acting on the coil is then the sum of the external emf (applied to the terminals in Figure 32.25) and the self-induced emf. This means that whenever the current is time dependent, the coil will act back on the current and modify it (we will see how to calculate the net resulting current in Section 34.3). For this reason the self-induced emf is often called a **back emf.** From Lenz' Law we immediately recognize that the self-induced emf always acts in such a direction to *oppose* the change in the current, i.e., it attempts to maintain the current constant.

In terms of the flux through the circuit, the definition of the self-inductance is of the same form as Eq. (23),

$$\boxed{\Phi_B = LI} \tag{29}$$

and therefore the induced emf is

$$\boxed{\mathscr{E} = -L\,\frac{dI}{dt}} \tag{30}$$

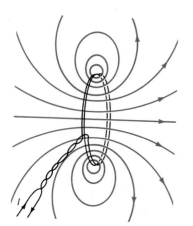

Fig. 32.25 A coil and its magnetic field.

EXAMPLE 8.   A solenoid has n turns per unit length and a radius R. What is its self-inductance per unit length?

SOLUTION: The magnetic field inside the solenoid is $B = \mu_0 nI$. The number of loops in a length l is nl; each of these loops has a flux $\pi R^2 B$. Hence the flux through all the loops in a length l is

$$\Phi_B = \pi R^2 Bnl = \pi R^2 n^2 l \mu_0 I \tag{31}$$

and

$$L = \Phi_B / I = \mu_0 n^2 \pi R^2 l \tag{32}$$

The inductance per unit length is therefore $\mu_0 n^2 \pi R^2$.

32.6 Magnetic Energy

Inductors store magnetic energy, just as capacitors store electric energy. When we connect an external source of emf to an inductor and start a current through the inductor, the back emf will oppose the increase of the current and the external emf must do work in order to overcome this opposition and establish the flow of current. This work is stored in the inductor and it can be recovered by removing the external source of emf from the circuit. As the current begins to decrease, the inductor will supply an induced emf which tends to keep the current flowing for a while (at a decreasing rate). Thus, the inductor delivers energy to the current.

To calculate the amount of stored energy, we note that when the current increases at the rate of dI/dt, the back emf is

$$\mathcal{E} = -L \, dI/dt \qquad (33)$$

The inductor does work on the current at the rate

$$I\mathcal{E} = -LI \, dI/dt \qquad (34)$$

Here the negative sign implies that the energy is delivered by the current to the inductor rather than vice versa. In a time dt, the energy stored in the inductor is therefore

$$dU = -I\mathcal{E} \, dt = LI \, dI \qquad (35)$$

If we integrate this from the initial value of the current $(I' = 0)$ to the final value $(I' = I)$, we obtain

$$U = \int_0^I LI' \, dI' = L \int_0^I I' \, dI' \qquad (36)$$

or

$$\boxed{U = \tfrac{1}{2} LI^2} \qquad (37)$$

Magnetic energy in inductor

This equation for the magnetic energy in an inductor is analogous to Eq. (27.40) for the electric energy in a capacitor.

EXAMPLE 9   A solenoid has a radius of 2.0 cm; its winding has one turn of wire per millimeter. A current of 10 A flows through the winding. What is the amount of energy stored per unit length of the solenoid?

SOLUTION: The inductance per unit length is [see Eq. (32)]

$$L/l = \mu_0 n^2 \pi R^2$$

$$= 1.26 \times 10^{-6} \text{ H/m} \times (10^3/\text{m})^2 \times \pi \times (0.020 \text{ m})^2$$

$$= 1.6 \times 10^{-3} \text{ H/m}$$

and the energy per unit length is

$$U/l = \tfrac{1}{2}(L/l)I^2$$

$$= \tfrac{1}{2} \times 1.6 \times 10^{-3} \text{ H/m} \times (10 \text{ A})^2$$

$$= 7.9 \times 10^{-2} \text{ J}$$

The energy stored in a solenoid can be expressed in terms of the magnetic field. Consider a portion of length l of the solenoid. The inductance of this portion is [see Eq. (32)]

$$L = \mu_0 n^2 \pi R^2 l \qquad (38)$$

so that

$$U = \tfrac{1}{2}\mu_0 n^2 \pi R^2 l I^2 \tag{39}$$

Since for a solenoid $B = \mu_0 n I$, we can write this as

$$U = \frac{1}{2\mu_0} B^2 \pi R^2 l \tag{40}$$

or

$$U = \frac{1}{2\mu_0} B^2 \times [\text{volume}] \tag{41}$$

where the volume $\pi R^2 l$ is the volume of magnetic field in the portion of length l of the solenoid.

According to Eq. (41), the quantity

Energy density in magnetic field

$$\boxed{u = \frac{1}{2\mu_0} B^2} \tag{42}$$

can be regarded as the magnetic energy per unit volume. Although we have only derived this equation for the special case of a long solenoid, it turns out to be generally valid (in vacuum and in nonmagnetic materials). The magnetic field, just as the electric field, stores energy. The magnetic energy density is proportional to B^2 just as the electric density is proportional to E^2 [compare Eq. (26.19)].

EXAMPLE 10.   Near the surface, the Earth's magnetic field typically has a strength of 0.3×10^{-4} T and the Earth's atmospheric electric field typically has a strength of 100 V/m. What is the energy density in each field?

SOLUTION: The magnetic energy density is

$$u_B = \frac{1}{2\mu_0} B^2 = \frac{(0.3 \times 10^{-4}\ \text{T})^2}{2 \times 1.26 \times 10^{-6}\ \text{H/m}} = 3.6 \times 10^{-4}\ \text{J/m}^3$$

and the electric energy density is

$$u_E = \frac{\varepsilon_0}{2} E^2$$

$$= \frac{(8.85 \times 10^{-12}\ \text{F/m}) \times (100\ \text{V/m})^2}{2}$$

$$= 4.4 \times 10^{-8}\ \text{J/m}^3$$

SUMMARY

Motional emf in a rod: $\mathcal{E} = vBl$

Faraday's Law: $\mathcal{E} = [\text{rate of cutting of magnetic field lines}]$

Magnetic flux: $\Phi_B = \int \mathbf{B} \cdot d\mathbf{S}$

Faraday's Law for closed loop: $\mathscr{E} = -\dfrac{d\Phi_B}{dt}$

Lenz' Law: Induced emf opposes change.

Mutual inductance: $\Phi_{B_1} = L_{21} I_1$

$$\mathscr{E}_2 = -L_{21} \frac{dI_1}{dt}$$

Self-inductance: $\Phi_B = LI$

$$\mathscr{E} = -L \frac{dI}{dt}$$

Self-inductance of solenoid: $\mu_0 n^2 \pi R^2$ per unit length

Magnetic energy in inductor: $U = \frac{1}{2} L I^2$

Energy density in magnetic field: $u = \dfrac{1}{2\mu_0} B^2$

QUESTIONS

1. At the latitude of the United States, the magnetic field of the Earth has a downward component, larger than the northward component. Suppose that an airplane is flying due west in this magnetic field. Will there be an emf between its wingtips? Which wingtip will be positive? Will there be a flow of current?

2. What is the magnetic flux that the magnetic field of the Earth produces through the surface of the Earth?

3. Is Eq. (8) valid for an open path? Is Eq. (5) valid for a closed path?

4. A long straight wire carries a steady current. A square conducting loop is in the same plane as the wire. If we push the loop toward the wire, how is the direction of the current induced in the loop related to the direction of the current in the wire?

5. A **flip coil** serves to measure the strength of a magnetic field. It consists of a small coil of many turns connected to a sensitive ammeter. The coil is placed face on in the magnetic field and then suddenly flipped over. How does this indicate the presence of the magnetic field?

6. The **magneto** used in the ignition system of old automobile engines consists of a permanent magnet mounted on the flywheel of the engine. As the flywheel turns, the magnet passes by a stationary coil, which is connected to the spark plug. Explain how this device produces a spark.

7. A long straight wire carries a current that is increasing as a function of time. A rectangular loop is near the wire, in the same plane as the wire. How is the direction of the current induced in the loop related to the direction of the current in the wire?

8. Consider two adjacent rectangular circuits, in the same plane. If the current in one circuit is suddenly switched off, describe the direction of the current induced in the other circuit.

9. Figure 32.26 shows two coils of wire wound around a plastic cylinder. If the current in the left coil is made to increase what is the direction of the current induced in the coil on the right?

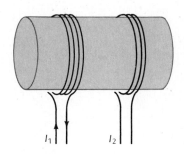

Fig. 32.26

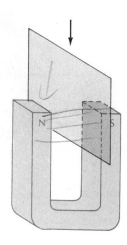

Fig. 32.27

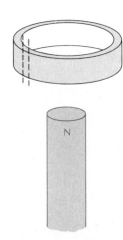

Fig. 32.28 A conducting ring falling toward a bar magnet. The dashed lines indicate where a slot will be cut through the ring.

10. A circular conducting ring is being pushed toward the north pole of a bar magnet. Describe the direction of the current induced in the ring.

11. Ganot's *Éléments de Physique*, a classic nineteenth-century textbook, states the following rules for the current induced in one loop face to face with another loop:

  I. The distance remaining the same, a continuous and constant current does not induce any current in an adjacent conductor.
  II. A current, at the moment of being closed, produces in an adjacent conductor an inverse current.
  III. A current, at the moment it ceases, produces a direct current.
  IV. A current which is removed, or whose strength diminishes, gives rise to a direct induced current.
  V. A current which is approached, or whose strength increases, gives rise to an inverse induced current.

Explain these rules on the basis of Lenz' Law.

12. A sheet of aluminum is being pushed between the poles of a horseshoe magnet (Figure 32.27). Describe the direction of flow of the induced currents, or **eddy currents**, in the sheet. Explain why there is a strong friction force that opposes the motion of the sheet.

13. Some beam balances use a magnetic damping mechanism to stop excessive swinging of the beam. This mechanism consists of a small conducting plate attached to the beam and a small magnet mounted on a fixed support near this plate. How does this damp the motion of the beam? (Hint: See the preceding question.)

14. A bar magnet, oriented vertically, is dropped toward a flat horizontal copper plate. Explain why there will be a repulsive force between the bar magnet and the copper plate. Is this an elastic force, i.e., will the bar magnet bounce if it is very strong?

15. A conducting ring is falling toward a bar magnet (Figure 32.28). Explain why there will be a repulsive force between the ring and the magnet. Explain why there will be no such force if the ring has a slot cut through it (see dashed lines in Figure 32.28).

16. You have two coils, one of slightly smaller radius than the other. To achieve maximum mutual inductance, should you place these coils face to face or one inside the other?

17. Two circular coils are separated by some distance. Qualitatively, describe how the mutual inductance varies as a function of the orientation of the coils.

18. If a strong current flows in a circuit and you suddenly break the circuit (by opening a switch), a large spark is likely to jump across the switch. Explain.

19. What arguments can you give in favor of the view that the magnetic energy of an inductor is stored in the magnetic field, rather than in the current?

PROBLEMS

Section 32.1

1. An automobile travels at 55 mi/h along a level road. The vertical downward component of the Earth's magnetic field is 0.58×10^{-4} T. What is the induced emf between the right and the left door handles separated by a distance of 7.0 ft? Which side is positive and which negative?

2. The DC-10 jet aircraft has a wingspan of 47 m. If such an aircraft is flying horizontally at 960 km/h at a place where the vertical component of the Earth's magnetic field is 0.60×10^{-4} T, what is the induced emf between its wingtips?

3. In order to detect the movement of water in the ocean, oceanographers sometimes rely on the motional emf generated by this movement of the water through the magnetic field of the Earth. Suppose that, at a place where the vertical magnetic field is 0.70×10^{-4} T, two electrodes are immersed in the water separated by a distance of 200 m measured perpendicularly to the movement of the water. If a sensitive voltmeter connected to the electrodes indicates a potential difference of 7.0×10^{-3} V, what is the speed of the water?

4. A homopolar generator consists of a metal disk rotating about a horizontal axis in a uniform horizontal magnetic field. The external circuit is connected to contact brushes touching the disk at the rim and at the axis. If the radius of the disk is 1.2 m and the strength of the magnetic field is 6.0×10^{-2} T, at what rate (rev/s) must you rotate the disk to obtain an emf of 6.0 V?

5. The rate of flow of a conducting liquid, such as detergent, tomato pulp, beer, liquid sodium, sewage, etc., can be measured with an electromagnetic **flowmeter** that detects the emf induced by the motion of the liquid in a magnetic field. Suppose that a plastic pipe of diameter 10 cm carries beer with a speed of 1.5 m/s. The pipe is in a transverse magnetic field of 1.5×10^{-2} T. What emf will be induced between the opposite sides of the column of liquid?

6. Pulsars or neutron stars rotate at a fairly high speed and they are surrounded by a strong magnetic field. The material in neutron stars is a good conductor and hence a motional emf is induced between the center of the neutron star and the rim (this is similar to the emf induced in a rotating metallic rod, see Example 1). Suppose that a neutron star of radius 10 km rotates at the rate of 30 rev/s and that the magnetic field has a strength of 10^8 T. What is the emf induced between the center of the star and a point on its equator?

7. A helicopter has blades of length 4.0 m rotating at 3 rev/s in a horizontal plane. If the vertical component of the Earth's magnetic field is 0.65×10^{-4} T, what is the induced emf between the tip of the blade and the hub?

*8. Sharks have delicate sensors on their bodies that permit them to sense small differences of potential. They can sense electrical disturbances created by other fish and they can also sense the Earth's magnetic field and use this for navigation. Suppose that a shark is swimming horizontally at 25 km/h at a place where the magnetic field has a strength of 4.7×10^{-5} T and points down at an angle of 40° with the vertical. Treat the shark as a cylinder of diameter 30 cm. What is the largest induced emf between diammetrically opposite points on the sides of the shark when heading east? North? Northeast? Make a rough plot of the induced emf vs. the direction of travel of the shark around all points of the compass. Indicate which side of the shark is positive, which negative. (Hint: The component of **B** perpendicular to the velocity **v** is proportional to $|\mathbf{v} \times \mathbf{B}|$.)

Sections 32.2 and 32.3

9. In Idaho, the magnetic field of the Earth points downward at an angle of 69° below the horizontal. The strength of the magnetic field is 0.59×10^{-4} T. What is the magnetic flux through 1 m² of ground in Idaho?

10. An electric generator consists of a rectangular loop of wire rotating about its longitudinal axis which is perpendicular to a magnetic field of 2.0×10^{-2} T. The loop measures 10.0 cm $\times$ 20.0 cm and it has 120 turns of wire. The ends of the wire are connected to an external circuit. At what speed (in rev/s) must you rotate this loop in order to induce an alternating emf of amplitude 12.0 V between the ends of the wire?

11. A very long train whose metal wheels are separated by a distance of 4 ft 9 in., is traveling at 80 mi/h on a level track. The vertical component of the magnetic field of the Earth is 0.62×10^{-4} T.
    (a) What is the induced emf between the right wheels and the left wheels?
    (b) The wheels are in contact with the rails and the rails are connected by

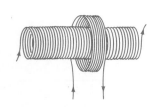

Fig. 32.29 A long solenoid surrounded by a circular coil.

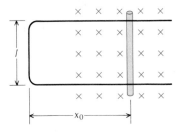

Fig. 32.30 Rod sliding on parallel tracks. The crosses show the tails of the magnetic field vectors.

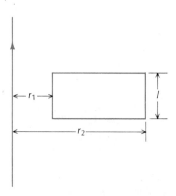

Fig. 32.31 Long straight wire and rectangular loop.

metal cross ties which close the circuit and permit a current to flow from one rail to the other. Since the number of cross ties is very large, their combined resistance is nearly zero; most of the resistance of the circuit is within the train. What is the current that flows in each axle of the train? The axles are cylindrical rods of iron of diameter 3 in. and length 4 ft 9 in.

12. (a) A long solenoid has 300 turns of wire per meter and it has a radius of 3.0 cm. If the current in the wire is increasing at the rate of 50 ampere per second, at what rate does the strength of the magnetic field in the solenoid increase?
    (b) The solenoid is surrounded by a coil of wire with 120 turns (Figure 32.29). The radius of this coil is 6.0 cm. What induced emf will be generated in this coil while the current in the solenoid is increasing?

13. A metal rod of length l and mass m is free to slide, without friction, on two parallel metal tracks. The tracks are connected at one end so that they and the rod form a closed circuit (Figure 32.30). The rod has a resistance R and the tracks have negligible resistance. A uniform magnetic field is perpendicular to the plane of this circuit. The magnetic field is increasing at a constant rate dB/dt. Initially the magnetic field has a strength B_0 and the rod is at rest at a distance x_0 from the connected end of the rails. Express the acceleration of the rod at this instant in terms of the given quantities.

14. A square loop of dimension 8.0 cm × 8.0 cm is made of copper wire of radius 1.0 mm. The loop is placed face on in a magnetic field which is increasing at the constant rate of 80 T/s. What induced current will flow around the loop? Draw a diagram showing the direction of the field and the induced current.

15. A very long solenoid with 20 turns per centimeter of radius 5.0 cm is surrounded by a rectangular loop of copper wire. The rectangular loop measures 10 cm × 30 cm and its wire has a radius of 0.05 cm. The resistivity of copper is 1.7×10^{-8} $\Omega \cdot$m. Find the induced current in the rectangular loop if the current in the solenoid is increasing at the rate of 5×10^4 A/s.

*16. A square loop of dimension $l \times l$ is moving at speed v toward a straight wire carrying a current I. The wire and the loop are in the same plane and two of the sides of the loop are parallel to the wire. What is the induced emf of the loop as a function of the distance d between the wire and the nearest side of the loop?

*17. A circular coil of insulated wire has a radius of 9.0 cm and contains 60 turns of wire. The ends of the wire are connected in series with a 15-Ω resistor closing the circuit. The normal to the loop is initially parallel to a constant magnetic field of 5.0×10^{-2} T. If the loop is flipped over, so that the direction of the normal is reversed, a pulse of current will flow through the resistor. What amount of charge will flow through the resistor? Assume that the resistance of the wire is negligible compared to that of the resistor.

*18. A flux meter used to measure magnetic fields consists of a small coil of radius 0.80 cm wound with 200 turns of fine wire. The coil is connected by means of a pair of tightly twisted trailing wires to a galvanometer that measures the electric charge flowing through the coil. The resistance of the coil–wire–galvanometer circuit is 6.0 Ω. Suppose that when the coil is quickly moved from a place outside the magnetic field to a place inside the magnetic field the galvanometer registers a flow of charge of 0.30 C. What is the magnetic flux through the coil? What is the component of **B** perpendicular to the face of the coil? Assume that **B** is uniform over the area of the coil.

*19. A long, straight wire carries a current that increases at a steady rate dI/dt.
    (a) What is the rate of increase of the magnetic field at a radial distance r?
    (b) What is the induced emf around the loop shown in Figure 32.31?

Section 32.4

20. A long solenoid of radius 3.0 cm has 2×10^3 turns of wire per meter. The wire carries a current of 6.0 A. Suppose that this solenoid is moving relative to you at a velocity of 400 m/s in a direction perpendicular to its axis.
   - (a) What is the induced electric field inside the solenoid, as seen in your reference frame?
   - (b) What is the electric field in the reference frame of the solenoid?
   - (c) What is the induced emf between two points on the edge of the solenoid at opposite ends of a diameter perpendicular to the velocity?

21. The magnetic field of a betatron has an amplitude of oscillation of 0.90 T and a frequency of 60 Hz. What is the amplitude of oscillation of the induced electric field at a radius of 0.80 m? What is the amplitude of oscillation of the induced emf around a circular path of this radius?

22. A boxcar of a train is 2.5 m wide, 9.5 m long, and 3.5 m high; it is made of sheet metal and it is empty. The boxcar travels at 60 km/h on a level track at a place where the vertical component of the Earth's magnetic field is 0.62×10^{-4} T. Assume that the boxcar is not in electrical contact with the ground.
   - (a) What is the induced emf between the sides of the boxcar?
   - (b) Taking into account the electric field contributed by the charges that accumulate on the sides, what is the net electric field inside the boxcar (in the reference frame of the boxcar)?
   - (c) What is the surface charge density on each side? Treat the sides as two very large parallel plates.

23. A disk of metal of radius R rotates about its axis with a frequency ν. The disk is in a uniform magnetic field B, parallel to the axis of the disk.
   - (a) Find the induced emf between the center of the disk and a point on the disk at a radial distance r (where $r < R$).
   - (b) Find the strength of the induced electric field at this point.

Section 32.5

24. Two coils are arranged face to face, as in Figure 32.23. Their mutual inductance is 2.0×10^{-2} H. The current in coil 1 oscillates sinusoidally with a frequency of 60 Hz and an amplitude of 12 A,

$$I_1 = 12 \sin(120\pi t)$$

where the current is measured in amperes and the time in seconds.
   - (a) What is the magnetic flux that this current generates in coil 2 at time $t = 0$?
   - (b) What is the induced emf that this current induces in coil 2 at time $t = 0$?
   - (c) What is the direction of the induced current in coil 2 at time $t = 0$, according to Lenz's Law? Assume that the positive direction for the current I_1 is as shown by the arrows in the figure.

25. A loop of wire carrying a current of 100 A generates a magnetic flux of 50 Wb through the area bounded by the loop.
   - (a) What is the self-inductance of the loop?
   - (b) If the current is decreased at the rate $dI/dt = 20$ A/s, what is the induced emf?

26. A long solenoid of radius R has n turns per unit length. A circular coil of wire of radius R' with 200 turns surrounds the solenoid (Figure 32.32). What is the mutual inductance? Does the shape of the coil of wire matter?

27. Two long concentric solenoids of n_1 and n_2 turns per unit length have radii R_1 and R_2, respectively (Figure 32.33). What is the mutual inductance per unit length of the solenoids? Assume $R_1 < R_2$.

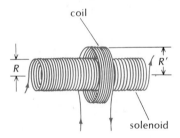

Fig. 32.32 A long solenoid and a circular coil.

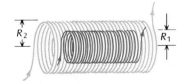

Fig. 32.33 Two long concentric solenoids.

28. A long solenoid has 2000 turns per meter and a radius of 2.0 cm.
    (a) What is the self-inductance for a 1.0-m segment of this solenoid?
    (b) What back emf will this segment generate if the current in the solenoid is changing at the rate of 3.0×10^2 A/s?

29. A solenoid of inductance 2.2×10^{-3} H in which there is initially no current is suddenly connected in series with the poles of 24-V battery. What is the instantaneous rate of increase of the current in the solenoid?

*30. A toroid with a square cross section (Figure 32.34) has an inner radius R_1, an outer radius R_2. The toroid has N turns of wire carrying a current I; assume that N is very large, so that the current can be regarded as uniformly distributed over the surface of the toroid.
    (a) Find the magnetic field as a function of radius.
    (b) Find the flux passing through one turn, and find the self-inductance of the toroid.

*31. Two inductors of self-inductances L_1 and L_2 are connected in series. The inductors are magnetically shielded from one another so that neither produces flux in the other. Show that the self-inductance of the combination is $L = L_1 + L_2$.

*32. Two inductors of self-inductance L_1 and L_2 are connected in parallel. The inductors are magnetically shielded from one another so that neither produces flux in the other. Show that the self-inductance of the combination is given by $1/L = 1/L_1 + 1/L_2$.

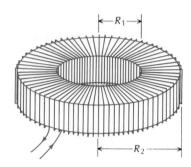

Fig. 32.34 A toroid with a square cross section.

Section 32.6

33. A ring of thick wire has a self-inductance of 4.0×10^{-8} H. How much work must you do to establish a current of 25 A in this ring?

34. The strongest magnetic field achieved in a laboratory is 10^3 T. This field can be produced only for a short instant by compressing the magnetic field lines with an explosive device. What is the energy density in this field?

35. For each of the first six entries in Table 30.1, calculate the energy density in the magnetic field.

36. For a crude estimate of the energy in the Earth's magnetic field, pretend that this field has a strength of 0.5×10^{-4} T from the ground up to an altitude of 6×10^6 m above ground. What is the total magnetic energy in this region?

37. A current of 5.0 A flows through a cylindrical solenoid of 1500 turns. The solenoid is 40 cm long and has a diameter of 3.0 cm.
    (a) Find the magnetic field in the solenoid. Treat the solenoid as very long.
    (b) Find the energy density in the magnetic field and find the magnetic energy stored in the space within the solenoid.

38. A large toroid built for plasma research in the Soviet Union has a major radius of 1.50 m and a minor radius of 0.40 m (Figure 32.35). The average magnetic field within the toroid is 4.0 T. What is the magnetic energy?

39. Consider a point charge q moving at velocity $\mathbf{v}$ through empty space. What is the energy density in the magnetic field at a distance r ahead of the point charge? At a distance r to one side of the point charge?

40. According to one proposal, the surplus energy from a power plant could be temporarily stored in the magnetic field within a very large toroid. If the strength of the magnetic field is 10 T, what volume of the magnetic field would we need to store 1.0×10^5 kW · h of energy? If the toroid has roughly the proportions of a doughnut, roughly what size would it have to be?

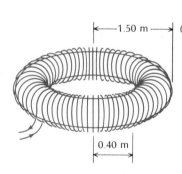

Fig. 32.35 A toroid.

*41. Two long, straight, concentric tubes made of sheet metal carry equal currents in opposite directions. The inner tube has a radius of 1.5 cm and the outer tube a radius of 3.0 cm. The current on the surface of each is 120 A. What is the magnetic energy in a 1.0-m segment of these tubes?

*42. A toroid of square cross section (Figure 32.34) has an inner radius R_1, and an outer radius R_2. The toroid has N turns of wire carrying a current I; assume that N is very large.

   (a) Find the magnetic energy density as a function of radius.

   (b) By integrating the energy density, find the total magnetic energy stored in the solenoid.

   (c) Deduce the self-inductance from the formula $U = \frac{1}{2}LI^2$.

PLASMA[1]

Plasma is a very hot, ionized gas, a gas so hot that the violent thermal collisons dissociate all or many of its atoms into positive ions and electrons. If we reckon solid, liquid, and gas as the first three states of matter, then plasma is the fourth state of matter. We might call it pure fire since an ordinary flame is part plasma and part hot gas.

Most of the matter in the universe is plasma. The Sun and all the stars are giant balls of plasma — about 99% of the total mass of the universe is in these balls of plasma. Only in planets, in pulsars, and in some clouds of interstellar gas and dust do we find solids, liquids, and gases — these bodies make up only a small fraction of all the matter in the universe. In our immediate environment, naturally occurring plasma is quite rare, so much so that it was not recognized as a separate state of matter until late in the nineteenth century. Lightning bolts, St. Elmo's fire, the Aurora Borealis, and the ionosphere are plasmas — on the Earth, these are the only forms of naturally occurring plasmas. However, modern technology makes use of many forms of artificially produced plasmas. The gas in fluorescent tubes and in neon signs is plasma; the luminous arc of an electric welder is plasma; the exhaust fire of a rocket is plasma; and the fireball of a nuclear bomb is plasma.

J.1 THE FOURTH STATE OF MATTER

A plasma contains a mixture of positive ions, electrons, and neutral atoms. The extent of ionization depends on the temperature: if the temperature is low, the plasma will contain a substantial number of neutral atoms; if the temperature is high, almost all the atoms will be ionized.

An ordinary gas also contains some ions and electrons, but not enough to make it into a plasma. If we heat a gas to higher and higher temperatures, we will gradually transform it into a plasma; however, the transition from gas to plasma is not sharply defined — there is no sudden change of phase such as in the melting of a solid or the evaporation of a liquid. For example, the flame of a candle is on the borderline

[1] This chapter is optional.

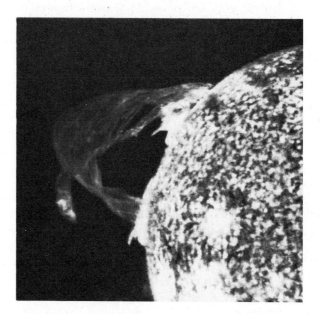

Fig. J.1 Giant prominence on the Sun photographed December 19, 1973, by an ultraviolet camera on Skylab. This prominence spans more than 588,000 km across the solar surface.

between hot gas and plasma; if it had more ions and electrons it would be a plasma, if it had less it would be an ordinary gas. The crucial distinction between an ordinary hot gas and a plasma lies in their electromagnetic properties. Plasma is an electric conductor and macroscopic electric and magnetic fields dominate its behavior, whereas an ordinary gas is an insulator and it does not respond in any drastic manner to electric and magnetic fields. Figure J.1 shows a prominence of plasma on the surface of the Sun. Obviously, this prominence has a complicated shape; it arcs up and down because its behavior is dominated by the magnetic field of the Sun rather than by gravity. In contrast, the flame of a candle has a simple shape; it streams upward because of the buoyancy of the hot gas. If we place the flame between the poles of a magnet, we find that its behavior is not changed in any noticeable way.

The criterion for a plasma is that the abundance of positive and negative charges must be so large that any local imbalance between the concentrations of

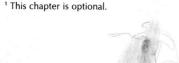

these charges is impossible. Any incipient accumulation of, say, positive charge rapidly attracts negative charges which restore the balance of charge. On a small scale, the positive and negative charges in a plasma move randomly, but on a large scale they move collectively — the positive and negative charges are strongly linked by their electric interaction and they move in unison so as to prevent any imbalance of charge. Thus, although the plasma contains a large number of free charges, it remains electrically neutral because, on the average, each unit volume contains equal amounts of positive and negative charge.

Matter in the plasma state has less order than matter in the solid, liquid, or gas state. However, the average electric neutrality of plasma represents some vestigial order. If we heat plasma to an extreme temperature, even this remaining order will be lost and the matter will turn into a chaotic mess of individual particles. This will happen when the thermal motion of the particles is so energetic that the electric forces cannot restrain them. Note that whether or not neutrality can be preserved depends on the scale of distance. On a small scale, neutrality always fails, while on a very large scale it hardly ever fails because it takes enormous amounts of energy to separate a large number of electric charges by a large distance. To gain some insight into what determines the critical scale of distance, let us examine the interaction between a plasma and a static electric field.

We know that, under static conditions, a good conductor will prevent an electric field from penetrating its interior — the conductor shields the electric field by accumulating some charge on its surface. This also happens in a plasma. If we insert a pair of electrodes into a plasma (Figure J.2), negative charges (electrons) will be drawn toward the positive electrode, and positive charges (ions) toward the negative electrode. Provided the electrodes are kept from direct contact with the plasma by a layer of dielectric insulator, the electrons and ions will accumulate, forming more or less static charge distributions around the electrodes and shielding their electric fields.

Because of thermal agitation, the charges cannot remain exactly static at the electrodes; instead, they

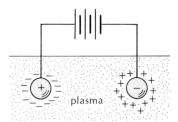

Fig. J.2 Two spherical electrodes immersed in a plasma.

form a restless cloud in the vicinity of each electrode. Only inside the cloud does the electrode disturb the plasma; outside of the cloud, the plasma hardly feels the presence of the electrode. The size of the cloud depends on the temperature — the cloud is large if the temperature is high because the random thermal back-and-forth motions of the charges will then tend to disperse them over larger distances.

The thickness of the cloud is called the shielding distance, or the **Debye distance.** It is given by the following formula:

$$D = \sqrt{\frac{\varepsilon_0 kT}{ne^2}} \qquad (1)$$

where n is the number of electrons per unit volume, T the temperature, and k Boltzmann's constant (see Chapter 19). This distance determines how far an electric field penetrates within the plasma and it also determines the magnitude of the deviations from neutrality within a plasma. On a scale large compared to the Debye distance, the plasma remains electrically neutral because any concentrations of charge are immediately hidden in a cloud of opposite charge. Consequently, an ionized gas will exhibit plasma behavior on a scale of distance that is large compared to the Debye distance. For example, the ionized gas in an ordinary neon tube has an electron density of about $10^9/cm^3$ and these electrons have a temperature of about 20,000 K. Equation (1) then gives

$$D = \sqrt{\frac{8.85 \times 10^{-12} \text{ F/m} \times 1.38 \times 10^{-23} \text{ J/K} \times 20{,}000 \text{ K}}{10^{15}/m^3 \times (1.6 \times 10^{-19} \text{ C})^2}}$$
$$= 3 \times 10^{-4} \text{ m} = 0.3 \text{ mm}$$

Hence a glob of this gas will behave as a plasma whenever it is larger than a few millimeters. Incidentally, the temperature of the electrons in a neon tube does not match the temperature of the ions — the electron temperature is about 20,000 K but the ion temperature is only about 2000 K. The reason for this discrepancy can be traced to the mechanism of energy transfer from electrons to ions. The electrons, being light and mobile, carry the electric current from one terminal of the tube to the other; the ions; being heavy and sluggish, do not contribute appreciably to this current. Thus the electrons receive all of the energy supplied to the tube by outside sources and they reach a high temperature. The ions can only acquire energy indirectly, by collisions with electrons. But in such a collision an electron will transfer only a very small fraction of its energy: when a light particle collides with a heavy particle, the light particle bounces off with next to no energy loss (see Section 10.2).

Table J.1 SOME PLASMAS

Plasma	Electron temperature	Electron density
Sun: center	2×10^7 K	10^{26}/cm³
surface	5×10^3	10^6
corona	10^6	10^5
Fusion experiments (tokamak)	1×10^8	10^{14}
Fireball of atomic bomb	$\sim 10^7$	10^{20}
Solar wind	1×10^5	5
Lightning bolt	3×10^4	10^{18}
Glow discharge (neon tube)	2×10^4	10^9
Ionosphere	$\sim 2 \times 10^3$	10^5
Ordinary flame	2×10^3	10^8

Thus, the ions only receive small amounts of kinetic energy. The ions never reach thermal equilibrium with the electrons because the rate at which they receive energy barely matches the rate at which they lose energy in inelastic processes at the wall of the neon tube.

Table J.1 gives some examples of plasmas. Note that although most of the plasmas in Table J.1 are luminous, some are not. The reason is that some plasmas are so tenuous that they do not give off an appreciable amount of light even though their temperature is very high. The ionosphere of the Earth and the corona of the Sun are examples of such very faint plasmas. Only during a total eclipse does the faint light from the solar corona become visible (Figure J.3).

J.2   PLASMAS AND THE MAGNETIC FIELD

Many plasmas — both in nature and in our laboratories — are immersed in magnetic fields. The magnetic fields may originate from external sources or else from currents flowing within the plasma itself. In contrast to static electric fields which cannot exist inside a plasma, static magnetic fields can exist. However, such magnetic fields are subject to a severe restriction: the magnetic field lines are *frozen* in the plasma. If the plasma is stationary, then the field lines within the plasma are also stationary; if the plasma flows, then the field lines flow with it, each segment of field line following the motion of the small volume element of fluid in which it is embedded. For example, Figure J.4 shows the magnetic field lines passing through a ball of plasma initially placed between the poles of an electromagnet, and what happens to the field lines as we move the ball about or compress it.

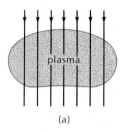

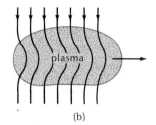

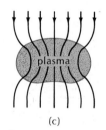

Fig. J.4 (a) Magnetic field lines in a ball of plasma. (b) If we move the ball of plasma, it drags the magnetic field lines along. (c) If we compress the ball of plasma, it compresses the field lines.

This initimate attachment of the field lines to the plasma is an immediate consequence of its high conductivity. In an ideal plasma, the conductivity is infinite and therefore the electric field in the rest frame of any given volume element of plasma must vanish. But we know, from Faraday's Law, that changes in the magnetic field induce electric fields. Since the electric field is forbidden, changes in the magnetic field are also forbidden. This means that in the rest frame of any given element of plasma, the magnetic field lines must remain fixed — they must move with the element of plasma. This freezing of the magnetic field lines within the plasma may be regarded as an extreme instance of Lenz' Law — any incipient change in the magnetic field immediately induces a current whose magnetic field combines with the original magnetic field in such a way that the net magnetic field remains constant. Note that the plasma not only keeps any interior field lines locked inside (as in Figure J.4), but it also keeps any exterior field lines outside. For exam-

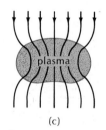

Fig. J.3 Corona of Sun visible during the eclipse of June 8, 1918.

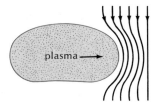

Fig. J.5 If we move a ball of plasma into a magnetic field, it pushes the field lines aside.

ple, if a ball of plasma with no field lines within it approaches a region with field lines, it will deform the latter and push them aside (Figure J.5).

This effect plays an important role in shaping the magnetic field of the Earth. If left to itself, the magnetic field of the Earth would be essentially a dipole field (Figure J.6). But this dipole field is affected by the solar wind, a stream of plasma blowing radially outward from the Sun. The solar wind consists of a neutral mixture of electrons and protons moving outward at a speed of about 400 km/s. The push of this wind against the magnetic field of the Earth causes a deformation — it compresses the field on the side facing the Sun and elongates the field on the side away from the Sun (Figure J.7). The region of space occupied by

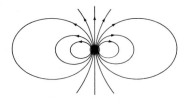

Fig. J.6 Magnetic field of an ideal dipole.

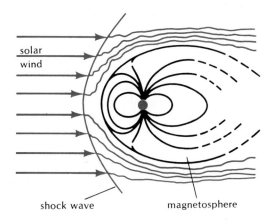

Fig. J.7 Magnetic field of Earth as distorted by the push of the solar wind. The speed of the wind is 400 m/s. At the leading edge of the magnetic field a shock wave forms.

the magnetic field of the Earth is called the **magnetosphere.** Because of the action of the solar wind, the magnetosphere acquires the shape of an elongated raindrop with a tail extending to a distance of at least several hundred thousand kilometers.

Incidentally, the speed of the solar wind past the Earth is supersonic, i.e., the speed is larger than the speed of sound waves (compressional waves) in the plasma. Consequently, the supersonic motion of the Earth relative to the solar wind generates a shock wave in the plasma, just like the motion of a supersonic aircraft generates a shock wave in air (Figure J.7).

A deformation of the magnetic field lines, such as the deformations shown in Figures J.4 and J.5, involves an increase of the density of the magnetic field lines in some regions of space; thus it involves an increase of the energy stored in the magnetic field. In order to acquire this energy from the plasma, the magnetic field must exert a force on the plasma, a force that opposes the deformation. If the magnetic field is very strong, then the force with which it opposes the motion of the plasma may be so large that it prevents this motion altogether — the magnetic field lines then confine the plasma and hold it as though it were in a cage.

The force that the magnetic field exerts on the plasma can be described (in part) as a **magnetic pressure** acting within the plasma. It turns out that if the strength of the magnetic field is B, then this extra magnetic pressure within the plasma is $B^2/2\mu_0$. The net pressure within any region of the plasma is then the sum of the ordinary kinetic pressure of the plasma particles plus the extra magnetic pressure. Note that numerically the magnetic pressure equals the energy density of the magnetic field [see Eq. (32.42)]. This is no accident — the magnetic pressure arises precisely from the changes of magnetic energy, in the following way: If we compress some volume of plasma, we will also compress the magnetic field lines frozen in this volume; this increase of density of the field lines involves an increase of the energy stored in the magnetic field. By the argument given in the preceding paragraph, the magnetic field will then exert a force that opposes the compression, that is, a pressure.

Carefully designed arrangements of magnetic fields are used for the confinement of extremely hot plasmas in experiments on thermonuclear fusion. Such arrangements of magnetic fields are called **magnetic bottles;** they can hold a plasma that is too hot to be held by an ordinary vessel of metal or glass. For instance, Figure J.8 shows a magnetic bottle that confines a plasma by means of the magnetic field produced by a solenoid. In the central region, the coils of the solenoid are uniformly spaced, but at each end an extra-large coil intensifies the magnetic field and brings the field lines together. We can understand the confine-

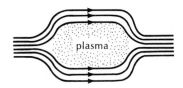

Fig. J.8 A magnetic bottle.

ment of the plasma by this bottle in terms of the magnetic pressure $B^2/2\mu_0$. This pressure is large in the strong magnetic field surrounding the plasma; hence this pressure tends to push the plasma inward, balancing the kinetic pressure that tends to push the plasma outward.

Alternatively, we can understand the confinement of the plasma in terms of the microscopic motions of the charged particles in the plasma. As we know from Section 31.3, a charged particle in a magnetic field moves in a helix around the magnetic field lines (Figure J.9). Thus, on the average, the particle gradually drifts along the direction of the field lines but does not wander off in a transverse direction. Near the ends of the bottle, where the field lines converge, the particle will be reflected. Figure J.10 shows how the converging field lines give rise to a component of force that halts the drift of the particle along the direction of the field lines and pushes it back (note that the circular motion of the particle does *not* stop or change direction; only the drift motion does). A magnetic field with convergent field lines is called a **magnetic mirror.**

A different method of confinement uses the magnetic field produced by a current flowing through the plasma, i.e., it uses the plasma's own magnetic field.

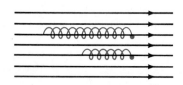

Fig. J.9 Charged particles with helical motion around the magnetic field lines.

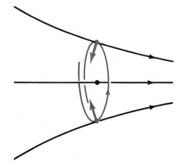

Fig. J.10 Charged particle and converging magnetic field lines. The magnetic force has a component along the axis of the helix.

Figure J.11 shows a cylindrical column of plasma, with a longitudinal current. The magnetic field produced by this current encircles the plasma and exerts a pressure on it. The magnetic field has its maximum value at the surface of the plasma; the magnetic pressure is therefore large at the surface and it pushes the plasma inward, balancing the kinetic pressure that pushes the plasma outward. The current used to hold together the very hot column of plasma in a thermonuclear fusion experiment may be as much as a million amperes.

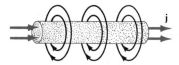

Fig. J.11 A cylindrical plasma column with an electric current passing through it.

We can also understand the confinement of such a column of plasma in terms of the attractive magnetic force between parallel current elements; obviously, this attractive force tends to compress the plasma and hold it together. The compression of a plasma by its own magnetic forces is called the **pinch effect.**

Unfortunately, the confinement provided by magnetic bottles is not perfect. The plasma tends to leak out at the ends and, what is worse, the plasma interacts with the magnetic field, giving rise to a variety of instabilities that destroy the delicate balance between the magnetic pressure and the kinetic pressure. For instance, Figures J.12 and J.13 show a kink instability and a sausage instability in a pinched plasma column. The cause of these instabilities can be qualitatively understood by examination of the field lines. In Figure J.12 the field lines are bunched together above the kink and spread apart below — hence the magnetic pressure pushes the plasma downward, further increasing the kinking of the column. In Figure J.13 the field lines are concentrated at the neck of the sausage; this squeezes the neck, further decreasing its diameter.

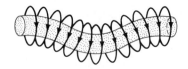

Fig. J.12 Kink instability of a plasma column.

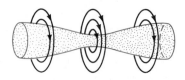

Fig. J.13 Sausage instability of a plasma column.

Both of these kinds of instability can be eliminated, or reduced, by means of an axial magnetic field in the plasma column (Figures J.14 and J.15). Such a field can be generated by placing the plasma in the core of a solenoid or by inducing an additional flow of current around the plasma column, effectively making it act as a solenoid. This magnetic field opposes the kink because deforming the axial magnetic field lines requires energy. Likewise, the magnetic field opposes the sausage constriction because compressing the magnetic field lines again requires energy.

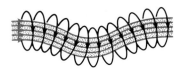

Fig. J.14 A plasma column with an extra axial magnetic field resists the kink instability.

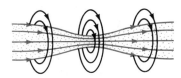

Fig. J.15 A plasma column with an extra axial magnetic field also resists the sausage instability.

Dozens of other, more complicated instabilities have been identified and studied by plasma physicists. The development of countermeasures to all these instabilities remains one of the central problems in plasma physics.

J.3   WAVES IN A PLASMA

The dynamical behavior of plasma is much more intricate than that of a gas. In the latter, the pressure completely determines the macroscopic motion, whereas in the former, electric and magnetic forces play a large role. For instance, consider the propagation of a sound wave in a plasma. This is a longitudinal wave with alternating zones of compression and rarefaction (Figure J. 16). In an ordinary gas, the restoring force in the sound wave is simply the excess pressure of the compressed zones. In a plasma, there is an additional restoring force: the concentrated positive electric charge of the ions of the plasma in the compressed zones gives rise to a repulsive electric force. Of course, the electrons of the plasma attempt to shield the ions and cancel their electric repulsion, but because the electrons have large random thermal motions (larger than those of the ions), the shielding is not quite perfect and a residual electric repulsive force remains.

Even more complicated dynamical effects can occur

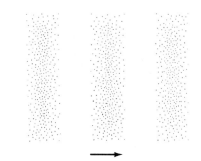

Fig. J.16 Sound wave in a gas or a plasma.

in a plasma immersed in a magnetic field. Suppose that the plasma is in an originally uniform magnetic field. A sound wave propagating in a direction perpendicular to this magnetic field will have alternating zones of compression and rarefaction and, since the magnetic field lines are frozen in the plasma, they must also have corresponding zones of compression and rarefaction (Figure J.17). The compression of the magnetic field lines generates an increase of the magnetic pressure. Thus, the restoring force that governs the propagation of this wave is due to a combination of kinetic pressure and magnetic pressure. A wave of this kind is called a **magnetosonic wave.**

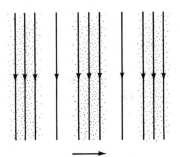

Fig. J.17 A sound wave in a plasma in the presence of a magnetic field. The compression of the plasma produces a compression of the magnetic field.

The magnetosonic wave is a longitudinal wave — it is merely a sound wave modified by magnetic effects. Surprisingly, it turns out that a plasma immersed in a magnetic field can also support a *transverse* wave. Consider a plasma in a uniform magnetic field. If one layer of this plasma is given a transverse displacement (Figure J.18), the magnetic field lines will suffer a transverse deformation. As is obvious from Figure J.18, this deformation leads to an increase of the density of field lines in the layer and hence to an increase of magnetic energy. Consequently, the embedded magnetic field exerts a restoring force on the plasma, opposing the deformation. Under the influence of the restoring force, the deformation propagates along the magnetic

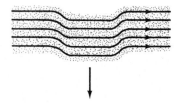

Fig. J.18 The displacement of a layer of plasma produces a deformation of the magnetic field.

field lines. This is a transverse wave, similar to a wave on a string (see Chapter 15); it propagates along the magnetic field lines just as though these lines were strings under tension. A wave of this kind is called an **Alfvén wave.**[2]

The propagation of radio waves in a plasma also has many intricate features. Since a plasma is a good conductor and shields electric fields, we expect it will shield the electric fields of a radio wave, i.e., it will reflect radio waves just as a metal does. This is true for radio waves of low frequency, but for radio waves of high frequency the shielding fails because the electrons do not have enough time to respond to the electric fields. Thus, high-frequency radio waves can penetrate into a plasma and propagate through it. The minimum frequency that can pass through a plasma is called the **cutoff frequency;** the value of this frequency depends on the electron density — it increases with the square root of the electron density. For instance, the ionosphere of the Earth will pass radio waves of frequency above about 30 MHz (radar and TV) but it will reflect waves of lower frequency (short waves, medium waves, and long waves).

The reflection of radio waves by the ionosphere is of great importance for radio communications. We can send radio waves from one end of the globe to the other by successively bouncing them back and forth between the ionosphere and the ground (Figure J.19).

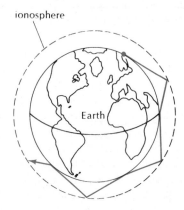

Fig. J.19 Path of a radio wave around the Earth.

[2] Hannes Alfvén, 1908–, Swedish physicist. He received the Nobel Prize in 1970 for his work in the theory of plasmas.

The ionosphere actually consists of several reflecting layers with characteristics that vary with day and night and also with the sunspot cycle (Figure J.20). Radio waves of different wavelength are reflected by different layers; for this reason, the choice of radio frequency band is critical in long-range radio communication. Incidentally, the radio blackout suffered by a space capsule during reentry of the atmosphere is also due to a plasma effect. The friction between the space capsule and the air gradually burns off the heat shield of the capsule, and the flames surround the capsule with a layer of plasma; this layer stops radio waves and prevents communication.

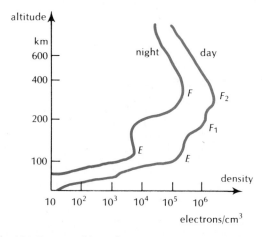

Fig. J.20 Density of free electrons in the atmosphere as a function of altitude (at a time of sunspot maximum). The peaks in the electron density at altitudes of ~120 km and ~300 km constitute the E and the F layers, respectively.

If the plasma is immersed in a magnetic field, then the behavior of radio waves depends in a complicated way on the frequency of the wave, the magnitude of the magnetic field, and the direction of propagation of the wave relative to the direction of the magnetic field. High-frequency radio waves can propagate in any direction (although their speed will depend on direction), but low-frequency radio waves can only propagate if their direction lies within a limited range of angles near the direction of the magnetic field. Hence magnetic field lines can act as guides for radio waves. This causes the weird phenomenon of whistlers that can be picked up with an audio amplifier at a few kilohertz; these are radio signals of low frequency emitted by flashes of lightning. The signals propagate in the tenuous plasma surrounding the Earth; they travel from one hemisphere to the other guided by the magnetic field lines of the Earth (Figure J.21). The high-frequency components of the signal travel faster than the low-frequency components. At the receiver, the result is a whistle consisting of a de-

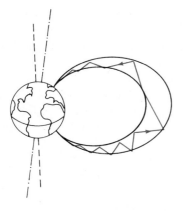

Fig. J.21 The magnetic field of the Earth acts as a waveguide for low-frequency radio waves.

scending glide tone (glissando) that begins at a high pitch and ends at low pitch.

J.4 THERMONUCLEAR FUSION

The Sun generates its heat by a nuclear fusion reaction burning hydrogen into helium. But for us on Earth, this reaction is not a viable energy source because it is a very slow reaction. A fusion reaction burning deuterium or tritium into helium is much more practical. In fact, the H-bomb derives its energy not from the fusion of hydrogen but from the fusion of deuterium and tritium (see Section H.4). Essentially, a fusion reactor brings the power of an H-bomb under control just like a fission reactor brings the power of an A-bomb under control.

Fusion is a very promising energy source because fuel for fusion is available in great abundance. Deuterium occurs naturally in molecules of heavy water (HDO) — about 0.03% of the water in the oceans is heavy water. Tritium is radioactive and does not occur naturally.[3] However, it can be easily produced by bombarding lithium with neutrons in a nuclear reactor. The available supply of deuterium is sufficient to satisfy all our energy requirements for a billion years; the supply of lithium is sufficient for a million years.

The most suitable reactions for the generation of fusion power are the **deuterium–deuterium reaction** and the **deuterium–tritium reaction.** The first of these involves the following sequence of four steps:

$$
\begin{aligned}
D + D &\rightarrow {}^3\mathrm{He} + n \\
D + D &\rightarrow T + p \\
D + T &\rightarrow {}^4\mathrm{He} + n \\
D + {}^3\mathrm{He} &\rightarrow {}^4\mathrm{He} + p \\
\hline
6D &\rightarrow 2\,{}^4\mathrm{He} + 2p + 2n + 43.1\ \mathrm{MeV}
\end{aligned}
\tag{1}
$$

[3] Deuterium (D, or ^{2}H) and tritium (T, or ^{3}H) are two isotopes of hydrogen.

The net result is the fusion of six deuterium nuclei into two helium nuclei with the release of two protons, two neutrons, and 43.1 MeV of energy. Note that the energy released for each fused deuterium nucleus is smaller than the energy released for each fissioned uranium nucleus — about 7 MeV per fused deuterium nucleus as compared to about 200 MeV per fissioned uranium nucleus. But weight for weight, the energy released in the fusion of deuterium is four times as large as in the fission of uranium. One further advantage of fusion over fission is that the reaction products from the former are harmless, whereas the reaction products from the latter are radioactive. In the reaction (1) the only potentially hazardous products are the neutrons; but these can be captured in a suitable absorber placed around the reactor vessel.

The deuterium–tritium reaction requires the presence of lithium and proceeds in two steps:

$$
\begin{aligned}
D + T &\rightarrow {}^4\mathrm{He} + n \\
n + {}^6\mathrm{Li} &\rightarrow {}^4\mathrm{He} + T \\
\hline
D + {}^6\mathrm{Li} &\rightarrow 2\,{}^4\mathrm{He} + 22.4\ \mathrm{MeV}
\end{aligned}
\tag{2}
$$

Thus, the net result is the fusion of deuterium and lithium nuclei into helium nuclei. The tritium only exists in an intermediate step in this reaction, being produced from the lithium. The technology required for the D–T reaction is much more complicated than that for the D–D reaction. The reactor chamber containing a mixture of deuterium and tritium would have to be surrounded by a blanket of lithium where neutrons can be absorbed and tritium can be generated; the tritium must afterward be extracted from the blanket and fed into the chamber. In spite of these complications, the D–T reaction is regarded as more promising than the D–D reaction because its ignition temperature is somewhat lower.

Both the D–D and D–T reactions will only proceed at extremely high temperature. The nuclei will only fuse if they are brought into contact by violent collisions — since the nuclei are positively charged, they experience a Coulomb repulsion, and to overcome this, the collision must be initiated with a large kinetic energy. The only feasible way to give the nuclei the necessary kinetic energy is by thermal motion. To attain a sufficiently large kinetic energy, a very high temperature is required — about 5×10^9 K for D–D fusion and about 1×10^8 K for D–T fusion. These temperatures are higher than at the center of the Sun. Nuclear reactions at such extreme temperatures are called **thermonuclear.** At these temperatures the deuterium or tritium atoms will be completely ionized and the nuclei and electrons will form a plasma.

To ignite the nuclear fire it is sufficient to mix the reactants and heat them to high temperature. But if the nuclear fire is to yield a profitable amount of energy, it is also necessary to keep the reactants together for some minimum length of time. Obviously, to break even, we must gain an amount of energy equal to the amount that we invested to heat the plasma. A calculation shows that for the D–T reaction this requires a minimum time (in seconds)

$$\tau = \frac{2 \times 10^{14}}{n} \qquad (3)$$

and for the D–D reaction a minimum time

$$\tau = \frac{5 \times 10^{15}}{n} \qquad (4)$$

where n is the number of ions per cubic centimeter. For instance, according to Eq. (3), a D–T plasma of density $10^{14}/cm^3$ must be confined for at least 2 s if it is to yield a profitable amount of energy. And here is the great problem of plasma physics — the plasma tends to escape and disperse. The challenge is to invent some means of confining the plasma long enough to extract significant amounts of energy.

J.5 PLASMA CONFINEMENT

Since the plasma in a thermonuclear fusion reactor must be kept very hot, we must prevent thermal contact with the walls of the reactor vessel. This can be achieved by suspending the plasma in the middle of the reactor vessel, in a magnetic field. But a magnetic field cannot confine the plasma completely. The minimum rate of leakage of the plasma is determined by diffusion caused by the random thermal motion. In practice, the leakage is much worse: the plasma tends to develop a variety of instabilities which make it squirt out of its magnetic confinement at a much faster rate than the minimum rate of leakage.

Hoping to conquer these instabilities, plasma physicists have constructed a variety of machines with fanciful names: Stellarator, Scylla, Scyllac, Tokamak, Alcator, etc. These machines are supposed to hold the plasma by means of intricate magnetic fields and they are supposed to heat the plasma to the point where fusion begins. Several of these machines have succeeded in triggering some fusion reactions, but none has generated any useful amount of power. Of these devices, the tokamak, originally developed in the Soviet Union and later copied in the United States, has so far proven the most successful.

In the **tokamak,** the plasma is confined to a toroidal region, i.e., a column bent into a ring. This avoids the loss of plasma from the ends of the column, since there are no ends. The magnetic fields confining the

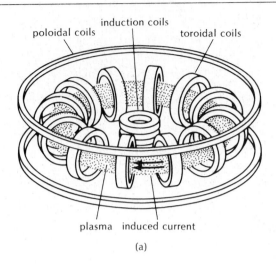

(a)

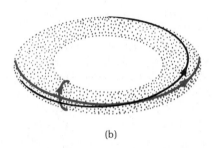

(b)

Fig. J.22 (a) Magnet coils in a tokamak. (b) The magnetic field of the toroidal coils lies along the plasma; the field lines are large horizontal circles (color). The magnetic field of the induced current flowing along the plasma column encircles the column; the field lines are small vertical circles (color). The net magnetic field consists of helical lines (black) wrapped around the plasma column. (The magnetic field of the poloidal coil has been neglected.)

plasma are in part generated by external solenoidal windings and in part by a current induced in the plasma itself. Thus, both of the simple confinement mechanisms mentioned in Section J.2 come into play. Figure J.22 shows the arrangement of magnet coils in a typical tokamak. The current along the plasma column is generated by induction, by means of a changing magnetic field provided by central, poloidal coils. The current induced by this magnetic field is usually in excess of a million amperes.

A tokamak machine at the Princeton Plasma Physics Laboratory has achieved a temperature of 7×10^7 K, four times hotter than the temperature in the center of the Sun. The plasma in the torus had a density of about $2 \times 10^{13}/cm^3$ and remained confined for about $\frac{1}{20}$ of a second. Thus, the plasma was near its thermonuclear ignition temperature, but its density and confinement time did not reach the break-even point set

Fig. J.23 The torus of the Princeton Tokamak Fusion Test Reactor (TFTR) during construction.

Fig. J.24 Neutral-beam injectors (left) attached to the Princeton TFTR (right). The torus is almost completely hidden within the support frame.

by Eq. (3). Larger versions of this machine, now in operation, are expected to do better (Figure J.23). Incidentally, the Joule heat of the current induced along the toroidal plasma provides some of the heating needed to start the thermonuclear reaction. In the Princeton machine, extra heating is provided by injectors that shoot intense beams of high-speed neutral atoms into the torus (Figure J.24).

Another promising machine is the **tandem mirror** now under construction at the Lawrence Livermore National Laboratory. This machine is a straight solenoid closed off at each end by a magnetic mirror consisting of two magnets arranged in tandem (Figure J.25). One of these magnets has an ordinary circular coil, but the other magnet has several interlocking "ying–yang" coils, shaped somewhat like the seams on a baseball. In the magnetic field of such a tandem magnet, the electrons and the positive ions of the plasma tend to separate to some extent. This charge separation generates an electric field which aids in the confinement of the plasma. Hence the action of the tandem mirror is part magnetic and part electric. Figures J.26 and J.27 show the magnet coils for Livermore under construction.

Fusion reactors based on the Tokamak design or the mirror design aim to reach the break-even point [see Eqs. (3) and (4)] by confining a low-density plasma for a fairly long time. As the fuel burns, fresh fuel would be injected into the reactor chamber, either in the form of puffs of gas or in the form of frozen pellets of a D–T mixture. In contrast, fusion reactors based on the **Scylla** design aim to reach the break-even point with a much higher-density plasma and a much shorter confinement time. Such a reactor would burn its fuel in a burst lasting perhaps 10^{-3} s and then it would receive a fresh load of fuel for the next spurt of power. The

pulsed release of heat in such a reactor is analogous to the pulsed release of heat in one of the cylinders of an internal combustion engine, which also burns its fuel in bursts and delivers spurts of power.

Instead of attempting to confine the plasma with magnetic fields, some experimenters are now exploring the possibility of a fusion reactor using the uncontrolled, explosive burning of small solid pellets of fuel. In such an **inertial confinement** reactor, one pellet at a time is dropped into a combustion chamber and suddenly vaporized into a very hot and very dense plasma by an intense pulse of light from a powerful laser or else by an intense pulse of beamed electrons or ions from an electron or ion gun. This sudden ignition will burn the deuterium and tritium before the plasma has a chance to expand and disperse. This is an extreme case of pulsed burning; it is explosive burning, lasting only about 10^{-9} s. During the explosion, the plasma is

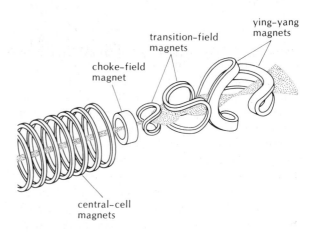

Fig. J.25 Magnet coils of the tandem mirror machine.

Fig. J.26 Yin-yang magnet coils for the tandem mirror machine under construction.

Fig. J.27 Circular magnet coils under construction.

confined to some extent by its own inertia — the outer layers of plasma do not directly participate in the fusion, but their inertia holds the core of the pellet together for a long enough time for the reactions to proceed. The success of such a fusion scheme depends on the development of very powerful and very efficient lasers (see Interlude L). Part of the output of the reactor will have to be used to energize the lasers.

Like the heat from an ordinary fission reactor, the heat from a fusion reactor can be removed from the reactor core and used to make steam to drive turbines generating mechanical and electric power. There is, however, an ingenious alternative method for the direct generation of electric power. The plasma in the combustion chamber of the fusion reactor is a mixture of electrons and positive ions. If this plasma is allowed to stream out of the chamber into a magnetic field, the negative and positive charges will separate. These opposite charges can be collected on two sets of plates; this results in a direct conversion of the plasma energy into electric energy. In principle, such a plasma generator is somewhat similar to a magnetohydrodynamic generator (see Example 32.2).

The fusion reactors of the future will probably be very large machines delivering more than 1000 MW of power (Figure J.28). But besides the confinement problem, there are several other problems that remain to be solved before we can exploit fusion power commercially. For instance, the design of the walls of the combustion chamber and the design of the magnets will require advances in technology. The walls are exposed to an abundant flux of neutrons emitted by the fusion reaction and also to the impact of charged particles escaping from the plasma. Metals exposed to such a flux of neutrons suffer from swelling and embrittlement. New, resistant alloys will have to be developed for the walls. Furthermore, the magnets surrounding

the reactor vessel will probably have to be constructed out of a superconductor; otherwise their power consumption would be prohibitive. These magnets will have to be much larger than any magnets built to date and they will have to stand up to enormous magnetic forces between the currents in their windings.

As we already pointed out in Interlude F, a fusion reactor is much cleaner than a fission reactor in that it does not produce an abundance of radioactive residues. However, the flux of neutrons released by the fusion reaction induces radioactivity in the walls of the reactor chamber, in the magnets, and wherever else it is absorbed — the impact of neutrons transmutes ordi-

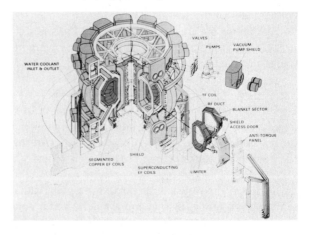

Fig. J.28 The design for the 1200-MW STARFIRE fusion reactor proposed by the Argonne National Laboratory.

nary stable nuclei into radioactive nuclei. The possible leakage of tritium from the reactor is an additional environmental hazard. Tritium is radioactive and, since it is chemically identical to hydrogen, it could easily contaminate our water supplies. If the confinement problem can be solved, these secondary problems will undoubtedly also be solved. We can then look forward to an abundant supply of relatively clean energy.

Further Reading

The Fourth State of Matter by B. Bova (New American Library, New York, 1974) is an elementary introduction to the properties and applications of plasmas. *A Physicist's ABC on Plasma* by L. A. Artsimovich (MIR Publishers, Moscow, 1978) is a very concise, slightly mathematical introduction to plasma physics, with emphasis on the magnetic confinement problem. The article "World Energy Reserves & Some Speculations on the Future of Nuclear Fusion Energy" by J. L. Tuck in *Cosmology, Fusion & Other Matters,* edited by F. Reines (Colorado Associated University Press, Boulder, 1972), gives a neat survey of available energy reserves and explains the advantages and principles of fusion reactors.

The following magazine articles deal with applications of plasma physics:

"VLF Emissions from the Magnetosphere," R. N. Sudan and J. Denavit, *Physics Today,* December 1973

"Fusion Reactors," B. B. Kadomtsev and T. K. Fowler, *Physics Today,* November 1975

"The Earth's Magnetosphere," S. Akasofu and L. J. Lanzerotti, *Physics Today,* December 1975

"Fusion Power with Particle Beams," G. Yonas, *Scientific American,* November 1978

"Recent Progress in Tokamak Experiments," M. Murakami and H. P. Eubank, *Physics Today,* May 1979

"Alternate Concepts in Magnetic Fusion," F. F. Chen, *Physics Today,* May 1979

"Progress toward a Tokamak Fusion Reactor," H. P. Furth, *Scientific American,* August 1979

"The Active Solar Corona," R. Wolfson, *Scientific American,* February 1983

"The Engineering of Magnetic Fusion Reactors," R. W. Conn, *Scientific American,* October 1983

Questions

1. Is there any plasma in the room you are in?

2. The ancient Greeks thought that there were five elements: earth, water, air, and fire. How does this list compare with the list of the four states of matter?

3. Can matter exist as a mixture of two or more of the four states? Can matter consisting of a single chemical element exist as a mixture of two or more states? Give examples.

4. According to the table of masses of the Sun and the planets printed on the endpapers of this book, roughly what fraction of the Solar System is plasma?

5. The chemical composition of the Sun is usually said to be 70% hydrogen and 30% helium. Are there actually any hydrogen or helium atoms on the Sun? What would be a more accurate description of the composition of the Sun?

6. How can the solar corona be hotter than the solar surface?

7. A neon tube contains plasma at very high temperature. Why does the tube not feel hot to the touch?

8. According to Table J.1, the temperature of the ionosphere is 2×10^3 K. Would a piece of paper burn if exposed to the ionosphere?

9. The free-electron gas in a metal has some of the features of a plasma. Does this mean we should regard a metal as plasma?

10. If you compress a gas (at fixed temperature), it becomes a liquid. If you compress a plasma, does it become a gas?

11. In the science fiction story *The Black Cloud* by the astronomer Fred Hoyle, a large cloud of plasma behaves as an intelligent being. How could the brain of such a creature store information in currents and magnetic fields? How could such a creature think?

12. Would you expect a kink or a sausage instability to develop in a plasma confined in a solenoidal magnetic field as shown in Figure J.8?

13. Explain why the magnetic mirrors shown in Figure J.8 permit some leakage along the center line.

14. Figure J.20 shows that the electron density in the ionosphere during the day is much larger than during the night (by up to a factor of 100). Can you guess why?

15. Consider a radio wave propagating around the Earth by bouncing back and forth between the ionosphere and the ground. Would you expect that a receiver at some given distance from the transmitter can be reached by two waves with different numbers of bounces?

16. What method could be used to separate the heavy water needed as fuel for a fusion reactor from ordinary water?

17. The Sun is a gigantic fusion reactor. What confines the plasma in this reactor? Why can we not use this method of confinement on Earth?

18. What would happen if the plasma in a fusion reactor were to come in contact with the wall of the reactor chamber?

19. Why is a runaway reaction possible in a fission reactor, but not in a fusion reactor?

20. Why does the reaction (1) require a higher temperature than the reaction (2)? (Hint: Which step in these reactions involves the most Coulomb repulsion?)

CHAPTER 33

Magnetic Materials

Within the atom and the nucleus, charged particles are continually moving about — electrons orbit around the nucleus and protons orbit around each other inside the nucleus. The orbital motions may be regarded as flows of electric currents within the atom, and these currents generate magnetic fields. Besides their orbital motions, the charged particles within atoms have rotational motions — electrons, protons, and neutrons all spin about their axes. The rotational motions may be regarded as flows of electric currents inside the particles, and these currents also generate magnetic fields.

The magnetic fields arising from currents flowing in loops inside atoms, nuclei, and particles can be described in terms of the corresponding magnetic dipole moments. If many of these small dipole moments within a sample of material are aligned, they will produce a strong magnetic field. Such an alignment can be achieved by placing the sample of material in the magnetic field of, say, an electromagnet. The magnetic field produced by the atomic and subatomic currents will then be strong, and it will modify the original magnetic field produced by the currents in the windings of the electromagnet. For instance, if a piece of iron is placed between the poles of an electromagnet, the magnetic field is strengthened drastically — it may become several thousand times stronger than the original magnetic field. Exactly how much the original magnetic field is increased or decreased depends on the response of the atomic and subatomic dipoles to the original magnetic field. According to the nature of their magnetic response, we can classify materials as paramagnetic, ferromagnetic, or diamagnetic. Before we discuss these types of magnetic materials, we will take a brief look at the magnitudes of the magnetic moments contributed by electrons, protons, and neutrons.

33.1 Atomic and Nuclear Magnetic Moments

An electron moving in an orbit around a nucleus produces an average current along its orbit. Strictly, the calculation of atomic orbits and currents requires quantum mechanics; but for the sake of simplicity, let us do this calculation with classical mechanics. If the electron has a circular orbit with radius r and speed v (Figure 33.1), then the time for one complete circular motion is $2\pi r/v$. The charge moved in this time is e and hence the average current along the orbit is

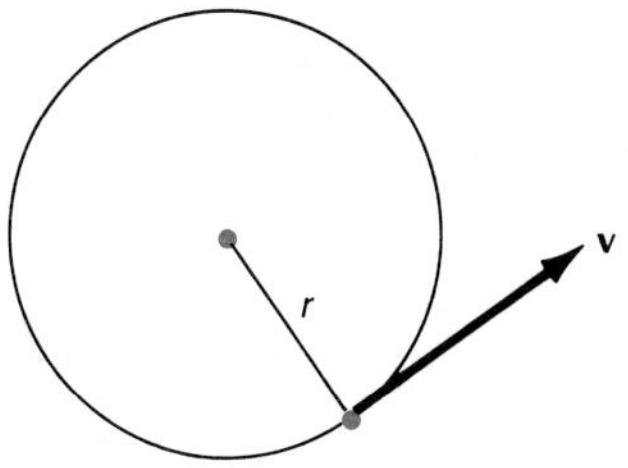

Fig. 33.1 Electron in a circular orbit around a nucleus.

$$I = \frac{e}{2\pi r/v} = \frac{ev}{2\pi r} \tag{1}$$

Such a circulating current will give rise to a magnetic moment [see Eq. (30.44)]

$$\mu = I \times [\text{area}] = \frac{ev}{2\pi r} \times \pi r^2 \tag{2}$$

$$= \frac{evr}{2} \tag{3}$$

In terms of the angular momentum $L = m_e vr$, the magnetic moment can be expressed as[1]

$$\boxed{\mu = \frac{e}{2m_e} L} \tag{4}$$

Orbital magnetic moment

This says that the magnetic moment of an orbiting electric charge is proportional to its angular momentum.

It turns out that the relation (4) is valid not only for circular orbits, but also for any other periodic orbit. Even more important: the relation (4) remains valid when we repeat the calculation using quantum mechanics. The net magnetic moment of the atom is the sum of the magnetic moments of all its electrons. Hence Eq. (4) can also be regarded as a relation between the net orbital angular momentum and the net magnetic moment of the entire atom.

It is a fundamental tenet of quantum mechanics that the magnitude of the orbital angular momentum is always some integral multiple of the constant value $\hbar = 1.06 \times 10^{-34}$ J·s.[2] Thus, the possible values of the orbital angular momentum are[3]

$$L = 0, \hbar, 2\hbar, 3\hbar, \ldots \tag{5}$$

Quantization of angular momentum

Because angular momentum only exists in discrete packets, it is said to

---

[1] In this equation, the magnetic moment μ must not be confused with the permeability μ_0.

[2] This is Planck's constant divided by 2π, i.e., $\hbar = h/2\pi = 6.63 \times 10^{-34}$ J·s/2π. See Appendix 8 for a more precise value of the constant $\hbar$.

[3] Strictly, these numbers are the possible values for the maximum component of the angular momentum in a chosen direction, say, the z direction.

be **quantized.** The constant $\hbar$ is the fundamental quantum of angular momentum, just as e is the quantum of electric charge. For instance, the oxygen atom has a net orbital angular momentum $L = 1\hbar$ and hence, according to Eq. (4), the orbital magnetic moment is

$$\mu = \frac{e}{2m_e}\hbar = 9.27 \times 10^{-24}\ \mathrm{A} \cdot \mathrm{m}^2 \tag{6}$$

Besides the magnetic moment generated by the orbital motion of the electrons, we must also take into account that generated by the rotational motion of the electrons. Crudely, an electron may be thought of as a small ball of negative charge rotating about an axis at a fixed rate. The spin, or intrinsic angular momentum, of the electron has a value of $\hbar/2 = 0.53 \times 10^{-34}\ \mathrm{J} \cdot \mathrm{s}$. This kind of rotational motion again involves circulating charge and gives the electron a magnetic moment. This intrinsic magnetic moment has a fixed magnitude

Spin magnetic moment

$$\mu_{\mathrm{spin}} = \frac{e}{2m_e}\hbar = 9.27 \times 10^{-24}\ \mathrm{A} \cdot \mathrm{m}^2 \tag{7}$$

Bohr magneton

which is called a **Bohr magneton.**[4] The direction of this magnetic moment is opposite to the direction of the spin angular momentum (Figure 33.2). Note that Eq. (4) is not valid for the spin magnetic moment. If we were to insert a spin angular momentum of $\hbar/2$ into Eq. (4), we would obtain a magnetic moment $e\hbar/(4m_e)$, one-half the value given by Eq. (7).

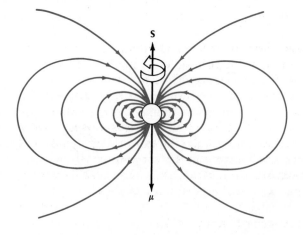

Fig. 33.2 Magnetic field lines of an electron. The magnetic moment μ is opposite to the spin angular momentum **S**.

The net magnetic moment of the atom is obtained by combining both the orbital and spin moments of all the electrons, taking into account the directions of these moments. In the oxygen atom, the net resultant magnetic moment is $13.9 \times 10^{-24}\ \mathrm{A} \cdot \mathrm{m}^2$. In some atoms — such as helium and argon — the magnetic moments cancel. But in most atoms the net magnetic moment is different from zero. Thus, most atoms behave as small magnetic dipoles. Table 33.1 gives the magnetic moments of some atoms and ions.

---

[4] See Appendix 8 for more precise values of the magnetic moments of the elementary particles.

Table 33.1 MAGNETIC MOMENTS OF SOME ATOMS AND IONS

Atom	Magnetic moment
H	9.27×10^{-24} A $\cdot$ m^2
He	0
Li	9.27×10^{-24}
O	13.9×10^{-24}
Ne	0
Na	9.27×10^{-24}
Ce^{+++}	19.8×10^{-24}
Yb^{+++}	37.1×10^{-24}

The nucleus of the atom also has a magnetic moment. This is in part due to the orbital motion of the protons inside the nucleus, and in part due to the rotational motion of individual protons and neutrons. The spin of both these particles is $\hbar/2$, the same as that of an electron. The spin magnetic moment of a proton is 1.41×10^{-26} A $\cdot$ m^2 and that of a neutron is 0.97×10^{-26} A $\cdot$ m^2 (the former magnetic moment is parallel to the axis of spin and the latter is antiparallel).[5] The magnetic moment of a proton or a neutron is small compared to that of an electron, and in reckoning the total magnetic moment of an atom, the nucleus can usually be neglected.

33.2 Paramagnetism

The behavior of the magnetic dipoles of the atoms or ions determines whether the material will be paramagnetic, ferromagnetic, or diamagnetic.

In a **paramagnetic material,** the atoms or ions have permanent magnetic dipole moments. When the material is left to itself, these dipoles are randomly oriented and their magnetic fields average to zero. However, if the material is immersed in the magnetic field of, say, an electromagnet, the torque on the dipoles tends to align them with the field (see Section 31.5). This alignment will not be perfect because of the disturbances caused by random thermal motions. But even a partial alignment of the dipoles will have an effect on the magnetic field. The material becomes **magnetized** and contributes an extra magnetic field that *enhances* the original magnetic field.

Paramagnetic material

Magnetization

To see how such an increase of magnetic field comes about, consider a piece of paramagnetic material placed between the poles of an electromagnet. Figure 33.3 shows the alignment of magnetic dipoles in this material; for the sake of simplicity, Figure 33.3b shows a case of perfect alignment. The magnetic dipoles are due to small current loops within the atoms; in Figure 33.4a we see the aligned current loops. Now look at any point inside the magnetic material where two of these current loops (almost) touch. Since the currents at this point are opposite, they cancel. Thus, everywhere inside the material the current is effectively zero. However, at the surface of the material, the current does not cancel. The net result of the alignment of current

---

[5] See Appendix 8 for more precise values of the magnetic moments of the elementary particles.

Fig. 33.3 (a) A piece of paramagnetic material. In the absence of an external magnetic field, the magnetic dipoles are oriented at random. (b) In the presence of an external magnetic field, the magnetic dipoles align with the magnetic field. The figure shows an ideal case of perfect alignment; in practice, the alignment will only be partial.

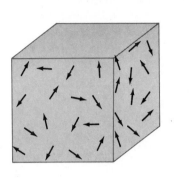
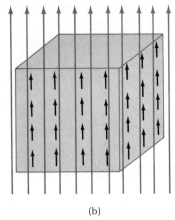

(a)   (b)

Fig. 33.4 (a) The magnetic dipoles of Figure 33.3b can be regarded as small current loops. At the point *P*, the currents of adjacent loops are opposite, and they cancel. (b) The small aligned current loops of (a) are equivalent to a current along the surface of the piece of material.

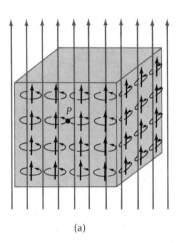
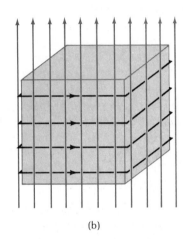

(a)   (b)

loops is therefore a current running along the surface of the magnetized material (Figure 33.4b). The material consequently behaves like a solenoid — it produces an extra magnetic field in its interior. This extra magnetic field has the *same* direction as the original magnetic field (Figure 33.3b). Hence the total magnetic field in a paramagnetic material is larger than the magnetic field produced by the free currents of the electromagnet.

The alignment of the magnetic dipoles in a magnetized paramagnetic material is analogous to the alignment of the electric dipoles in a dielectric material. However, there is a crucial difference: the alignment of magnetic dipoles *increases* the original magnetic field while the alignment of electric dipoles *decreases* the original electric field.

The increase of the strength of the magnetic field by the paramagnetic material can be described by the **relative permeability constant** κ_m. This constant is simply the factor by which the magnetic field is increased. Thus, if the free currents of the electromagnet produce a field B_{free}, then these free currents and the currents in the paramagnetic material acting together produce a field

Relative permeability constant

$$\boxed{\mathbf{B} = \kappa_m \, \mathbf{B}_{free}} \tag{8}$$

where $\kappa_m > 1$. Table 33.2 lists the values of κ_m for some paramagnetic materials.

Table 33.2 PERMEABILITIES OF SOME PARAMAGNETIC MATERIALS[a]

Material	κ_m
Air	1.000304
Oxygen	1.00133
Oxygen ($-190°C$, liquid)	1.00327
Manganese chloride	1.00134
Nickel monoxide	1.000675
Manganese	1.000124
Platinum	$1 + 13.8 \times 10^{-6}$
Aluminum	$1 + 8.17 \times 10^{-6}$

[a] At room temperature ($20°C$) and 1 atm unless otherwise noted.

Just as Gauss' Law in the presence of a dielectric material contains an extra factor κ [see Eq. (27.32)], Ampère's Law in the presence of a paramagnetic material contains extra factor κ_m. A simple argument shows that the revised form of Ampère's Law is

$$\oint \frac{1}{\kappa_m} \mathbf{B} \cdot d\mathbf{l} = \mu_0 I_{\text{free}} \tag{9}$$

Ampère's Law in magnetic materials

If the arrangement of currents and paramagnetic materials is sufficiently symmetric, then we can use Eq. (9) to calculate the magnetic field in the usual way.

EXAMPLE 1.   A solenoid filled with air has a magnetic field of 1.20 T in its core. By how much will the magnetic field decrease if the air is pumped out of the core while the current is held constant?

SOLUTION: For air, $\kappa_m = 1.00030$. Hence the magnetic field with air is larger than that without by a factor 1.00030, i.e.,

$$B_{\text{free}} = B_{\text{air}}/1.00030 \cong B_{\text{air}} (1 - 0.00030)$$

where we have used the approximation $1/(1 + x) \cong 1 - x$ for small x. The decrease of the magnetic field is then

$$\Delta B = B_{\text{free}} - B_{\text{air}} = -0.00030 \times 1.20 \text{ T}$$

$$= -3.6 \times 10^{-4} \text{ T}$$

33.3 Ferromagnetism

As is obvious from Table 33.2, the increase of magnetic field produced by a paramagnetic material is quite small. By contrast, the increase produced by a **ferromagnetic material** can be enormous. And what is more, such a material will remain magnetized even if it is not immersed in an external magnetic field. A material that retains magnetization will make a **permanent magnet.**

The intense magnetization in ferromagnetic materials is due to a strong alignment of the spin magnetic moments of electrons. In these

Ferromagnetic material

Permanent magnet

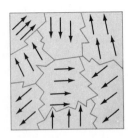

Fig. 33.5 Magnetic domains. Within each domain the magnetic dipoles have perfect alignment.

Domain

Fig. 33.6 Magnetic domains in a piece of iron. These domains are 0.1–0.3 mm across.

materials, there exists a special force that couples the spins of the electrons in adjacent atoms in the crystal, a force created by some subtle quantum-mechanical effects (which we cannot discuss here). This spin–spin force tends to lock the spins of the electrons in a parallel configuration. This force acts in the crystals of only five chemical elements: iron, cobalt, nickel, dysprosium, and gadolinium; however, it also acts in the crystals of alloys and of oxides of a large number of other elements.

Since this special spin–spin force is fairly strong, we must ask why is it that ferromagnetic materials are ever found in a nonmagnetized state? Why is it that not every piece of iron is a permanent magnet? The answer is that on a microscopic scale ferromagnetic materials are *always* magnetized. A crystal of ordinary iron consists of a large number of small **domains** within which all the magnetic dipoles are perfectly aligned. But the direction of alignment varies from one domain to the next (Figure 33.5). Hence on a macroscopic scale, there is no discernible alignment because the domains are oriented at random. The sizes and shapes of the domains depend on the crystal. Typically, the sizes of domains range from a tenth of a millimeter to a few millimeters, although in a large, uniform crystal the length of a domain may be several centimeters. Figure 33.6 shows domains in a crystal of iron.

The formation of domains results from the tendency of the material to settle into the state of least energy (equilibrium state). The state of least energy for the spins would be a state of complete alignment. But such a complete alignment would generate a large magnetic field around the material, i.e., it would make the material into a (very strong) permanent magnet. Energetically, this is an unstable configuration, because there is a very large amount of energy in the magnetic field. The domain arrangement is a compromise. The spins align within the domains, but the domains do not align — the magnetic energy is then small because there is little magnetic field and the spin–spin energy is then also reasonably small because *most* adjacent spins are aligned.

However, if the material is immersed in an external magnetic field, all dipoles tend to align along this field. The domains will then change in two ways: those domains that already are more or less aligned with the field tend to grow in size at the expense of their neighbors and, furthermore, some domains will rotate their dipoles in the direction of the field.

If *all* the magnetic dipoles in a piece of ferromagnetic material align, their contribution to the magnetic field will be very large. For example, within a piece of magnetized iron, this contribution can be as much as 2.1 T, a rather strong magnetic field. Figure 33.7 is a plot of the actual magnetic field B in a piece of iron in a solenoid as a function of the magnetic field B_{free} contributed by the free current in the solenoid [$B_{\text{free}} = \mu_0 I_0 n$; see Eq. (31.16)]. If the value of B_{free} is 2.0×10^{-4} T, the value of B is about 1.0 T — the iron increases the magnetic field by a factor of about 5000!

The solid portion of the plot in Figure 33.7 was obtained by starting with a piece of unmagnetized iron (annealed iron), and subjecting it to an increasing magnetic field B_{free}. The dashed portion of the plot was obtained by gradually reducing the magnetic field B_{free} to zero, after it had reached the value indicated by the point P. The dashed plot shows that even when B_{free} is reduced to zero, the iron retains some magnetization — the iron becomes a permanent magnet. Thus, the iron retains some memory of the magnetic field to which it was exposed earlier. Such a dependence of the state of a system on its past history is called **hysteresis.**

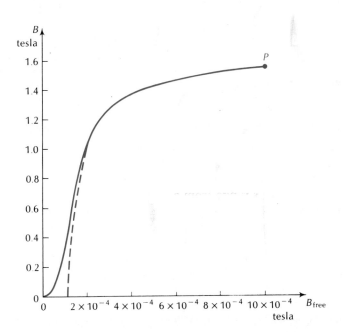

Hysteresis

Fig. 33.7 Magnetic field B in annealed iron as a function of B_{free}. The solid curve shows B if the iron is immersed in an external magnetic field B_{free} that increases from an initial value of zero to a final value of 10×10^{-4} T. The dashed curve shows B if the external magnetic field is subsequently reduced to zero.

The hysteresis of a ferromagnetic material is due to a sluggishness in the rearrangement of the domains. Once the domains have become aligned in response to a strong external magnetic field, they tend to stay that way. If we remove the ferromagnetic material from the external magnetic field, the domains will suffer some rearrangement, but they will not lose their alignment completely. The remaining alignment gives the material a permanent distribution of magnetic dipole moments over its volume. The magnetic field of a permanent magnet is produced by these remaining aligned dipole moments.

As we saw in Section 33.2, the aligned dipoles effectively amount to a current running around the surface of the magnetized material (see Figure 33.4b). In a strong permanent magnet, this surface current may amount to several hundred amperes per centimeter of length of the magnet. The magnetic field produced by such a current distribution is obviously similar to that of a solenoid of finite length — the field lines emerge from the magnet at one end (the north pole) and reenter the magnet at the other end (the south pole); see Figures 33.8 and 33.9.

Incidentally, the maximum magnetization that a ferromagnetic material will retain after it has been removed from the external magnetic field depends on the temperature. The higher the temperature, the less the remaining magnetization. Above a certain critical temperature, called the **Curie temperature**, the magnetization disappears completely. For example, iron will not retain any magnetization if the temperature is in excess of $1043\,°C$.

Curie temperature

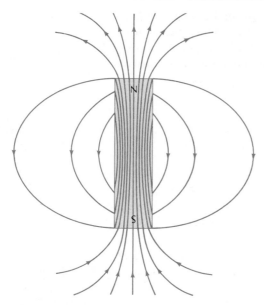

Fig. 33.8 Magnetic field lines of a permanent magnet.

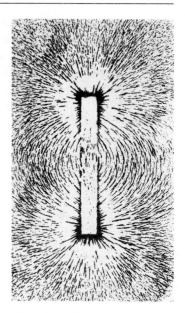

Fig. 33.9 Magnetic field lines of a permanent magnet, made visible by sprinkling iron filings on a sheet of paper.

33.4 Diamagnetism

In both paramagnetic and ferromagnetic materials the important effect is the alignment of *permanent* magnetic dipoles. This is quite analogous to the alignment of permanent electric dipoles in a dielectric. But we know that in some dielectrics the polarization is due to induced electric dipoles rather than permanent dipoles. Is it likewise possible for a material to acquire *induced* magnetic dipoles?

Diamagnetic material

In **diamagnetic materials** the magnetization arises from such induced magnetic dipoles. To see how the magnetic field can induce dipole moments in the atoms, imagine that a sample of some material is placed between the poles of an electromagnet which is initially switched off. If the electromagnet is now switched on, the external magnetic field must increase from its initial (zero) value to its final value. Thus, for a short while, the magnetic field will be time dependent and it will therefore induce an electric field within the sample. Let us consider the effect of this electric field on the motion of an electron. We will again pretend that classical mechanics applies to this problem.

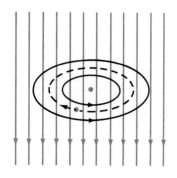

Fig. 33.10 Electron in a circular orbit around a nucleus immersed in an increasing magnetic field. The black circles indicate the induced electric field.

Figure 33.10 shows the orbit of an electron within an atom and also shows the increasing magnetic field. If the electron moves clockwise, as seen from above, the induced electric field (compare Example 32.4) will speed the electron up, giving it a larger orbital angular momentum and magnetic moment. The increment of angular momentum produces a magnetic field (within the orbit) that *opposes* the original magnetic field. This is obvious from Lenz' Law: the change in the motion of the electron amounts to an induced current, and the magnetic field associated with this current must be such as to oppose the original increasing magnetic field.

This argument establishes that induced magnetic moments reduce the strength of the magnetic field. The argument is quite general and applies to any kind of electron orbit and any kind of material. However, in paramagnetic and ferromagnetic materials the reduction of magnetic field due to induced magnetic moments is more than compensated by the increase of magnetic field due to alignment of permanent magnetic moments. If the material has no permanent magnetic moments, then the effect of the induced magnetic moments becomes noticeable and the material will be diamagnetic.

Diamagnetism is a very small effect. The following calculation gives some idea of the size of this effect. When the material is placed in a magnetic field **B**, the electrons will experience a force $-e\mathbf{v} \times \mathbf{B}$ in addition to the usual electric force acting within the atom. Figure 33.11 shows the simple case of an electron in a circular orbit of radius r around a nucleus; the orbit is perpendicular to the magnetic field **B**. If the nucleus produces an electric field **E**, then the net force on the electron is $-e\mathbf{E} - e\mathbf{v} \times \mathbf{B}$. This force must match the product of mass and centripetal acceleration,

$$eE + evB = m_e v^2/r \tag{10}$$

In terms of the angular frequency of the motion, the equation of motion becomes

$$eE + e\omega rB = m_e \omega^2 r \tag{11}$$

Let us compare this with the equation of motion in the absence of the magnetic field (undisturbed atom, $B = 0$),

$$eE = m_e \omega_0^2 r \tag{12}$$

If we subtract Eq. (12) from Eq. (11), we obtain a relation between the frequencies ω and ω_0:

$$e\omega B = m_e (\omega^2 - \omega_0^2) \tag{13}$$

It is convenient to express this in terms of the increment of frequency,

$$\Delta\omega = \omega - \omega_0 \tag{14}$$

If the magnetic field is not excessively strong, $\Delta\omega$ will be small compared to ω_0 and hence

$$\omega^2 - \omega_0^2 = (\omega_0 + \Delta\omega)^2 - \omega_0^2$$

$$= 2\omega_0 \Delta\omega + (\Delta\omega)^2 \cong 2\omega_0 \Delta\omega \tag{15}$$

so that Eq. (13) becomes

$$e\omega B \cong 2m_e\omega_0 \Delta\omega \tag{16}$$

Since ω and ω_0 are nearly equal, we can cancel them on both sides of this equation without introducing any additional errors. This leads to

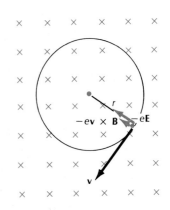

Fig. 33.11 Electron in a circular orbit around a nucleus immersed in a magnetic field. The centripetal force on the electron is the sum of the electric attraction of the nucleus and the magnetic force.

$$\Delta\omega = \frac{eB}{2m_e} \quad (17)$$

Larmor frequency This frequency is called the **Larmor frequency.** It tells us how much faster the electron will move around its orbit because of the presence of the magnetic field (of course, if the electron is initially moving in a direction opposite to that shown in Figure 33.11, then the electron will move *slower* by the same amount). Note that our calculation implicitly assumed that the orbital radius does not change as the magnetic field **B** is switched on. This assumption can be justified with an extra calculation that verifies that the work done on the electron by the induced emf changes the kinetic energy by just the amount required by the change of angular frequency at a fixed radius.

Corresponding to the change $\Delta\omega$ in the orbital frequency, there will be a change in the orbital magnetic moment. From Eq. (3),

$$\mu = \frac{evr}{2} = \frac{er^2\omega_0}{2} \quad (18)$$

Hence

$$\Delta\mu = \frac{er^2}{2}\Delta\omega \quad (19)$$

Dividing these equations into one another, we obtain an expression for the fractional change in the magnetic moment:

$$\frac{\Delta\mu}{\mu} = \frac{\Delta\omega}{\omega_0} \quad (20)$$

Let us insert some numbers. Typically the frequency of motion of an electron in an atom is $\omega_0 \cong 10^{15}$ s. If the magnetic field is fairly strong, say, $B = 1.0$ T, then

$$\Delta\omega = \frac{eB}{2m_e} = \frac{1.6 \times 10^{-19}\text{ C} \times 1.0\text{ T}}{2 \times 9.1 \times 10^{-31}\text{ kg}}$$

$$= 8.8 \times 10^{10}/\text{s} \quad (21)$$

Consequently,

$$\frac{\Delta\mu}{\mu} = \frac{\Delta\omega}{\omega_0} = \frac{8.8 \times 10^{10}/\text{s}}{10^{15}/\text{s}} \cong 10^{-4} \quad (22)$$

that is, the magnetic moment only changes by about 1 part in 10^4. This gives an indication of the small size of the diamagnetic effect.

The diamagnetic characteristics of a material can be described by a relative permeability κ_m that indicates by what factor the magnetic field is changed [compare Eq. (8)]. In the paramagnetic case $\kappa_m > 1$, but in the diamagnetic case $\kappa_m < 1$.

Table 33.3 lists some diamagnetic materials and the corresponding values of κ_m. In all cases, the value of κ_m is very near to 1.

Table 33.3 PERMEABILITIES OF SOME DIAMAGNETIC MATERIALS[a]

Material	κ_m
Bismuth	$1 - 1.9 \times 10^{-5}$
Beryllium	$1 - 1.3 \times 10^{-5}$
Methane	$1 - 3.1 \times 10^{-5}$
Ethylene	$1 - 2.0 \times 10^{-5}$
Ammonia	$1 - 1.4 \times 10^{-5}$
Carbon dioxide	$1 - 0.53 \times 10^{-5}$
Glass (heavy flint)	$1 - 1.5 \times 10^{-5}$

[a] At room temperature (20°C) and 1 atm.

SUMMARY

Magnetic moment of orbiting electron: $\mu = \dfrac{e}{2m_e} L$

Permeability constant: $B = \kappa_m B_{free}$

Magnetic materials: paramagnetic: $\kappa_m \gtrsim 1$
ferromagnetic: $\kappa_m \gg 1$
diamagnetic: $\kappa_m < 1$

Ampère's Law in paramagnetic and diamagnetic materials:

$$\oint \frac{1}{\kappa_m} \mathbf{B} \cdot d\mathbf{l} = \mu_0 I_{free}$$

Larmor frequency: $\Delta\omega = \dfrac{eB}{2m_e}$

QUESTIONS

1. Show that the SI unit of magnetic moment ($A \cdot m^2$) is the same as joule per tesla (J/T).

2. How would you measure the magnetic moment of a compass needle?

3. A bar magnet has a north pole and a south pole. If you break the bar magnet into two halves, do you obtain isolated north and south poles?

4. If we regard the Earth as a bar magnet, where is the magnetic north pole?

5. Consider a closed mathematical surface enclosing one of the poles of a bar magnet. What is the magnetic flux through this surface?

6. It is possible to magnetize an iron needle by pointing it north and giving it a few blows with a hammer. Explain.

7. If you drop a permanent magnet on a hard floor, it can become partially demagnetized. Explain.

8. Why does a magnet attract an (unmagnetized) piece of iron?

9. If you sprinkle iron filings on a sheet of paper placed in a magnetic field, the filings orient themselves along magnetic field lines (see Figure 33.9). Explain.

10. Other things being equal, a horseshoe magnet produces a stronger magnetic field than a bar magnet. Why?

11. You can make a chain of paper clips by touching one end of a clip to the pole of a magnet, then touching the free end to another paper clip, etc. Explain.

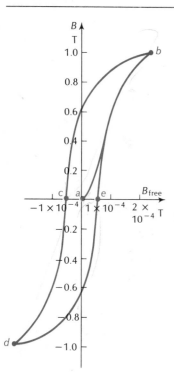

Fig. 33.12 Hysteresis loop for a piece of transformer steel. The configuration of the material initially corresponds to the point *a*. It is then successively brought to the points *b*, *c*, *d*, and *e* by suitable alterations of the external magnetic field B_{free}.

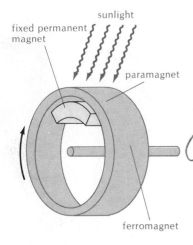

Fig. 33.13

12. When a mariner's compass is installed aboard an iron ship, it must be "adjusted" by placing several small permanent "correcting" magnets in its vicinity. What is the purpose of these magnets?

13. In the 1850s, Lord Kelvin redesigned the mariner's compass, partly by lengthy trials in his own yacht, and partly by theoretical analysis. He made the compass card (a circular disk, free to rotate, which carries the compass needles) lighter and concentrated most of its weight at the rim, and he used much smaller compass needles than had been customary before. What are the advantages of Kelvin's design?

14. The "keeper" for a horseshoe magnet is a small bar of iron that is placed across the poles of the magnet when not in use. What purpose does this serve?

15. Figure 33.12 shows a plot of B vs. B_{free} for a piece of transformer steel. This plot includes negative values of B and B_{free}. Qualitatively, explain the shape of this curve, called the **hysteresis loop.**

16. Why does the magnetization of a ferromagnet decrease with temperature?

17. Figure 33.13 shows a magnetic solar motor, called the Presnyakov wheel. The wheel is made of a ferromagnetic material. A fixed permanent magnet is installed near the top, where sunlight strikes the wheel. The heat of the sunlight converts the top portion of the ferromagnet temporarily into a paramagnet, with a much smaller value of the permeability. Explain why this wheel keeps turning after it has been given an initial push.

18. Does iron have diamagnetism?

19. The magnetic permeability of paramagnets and ferromagnets is strongly dependent on temperature, but the permeability of diamagnets is nearly independent of temperature. Why?

20. The practice of dowsing for water is still prevalent in some backward communities. The dowser holds a rod of wood, brass, or plastic in his hands in a peculiar way, and the rod supposedly dips wherever water is beneath the ground. The effect on the rod cannot be gravitational (a small body of water does not produce a perceptible disturbance of the Earth's average gravitational field). Hence, if the effect exists, it must be electric or magnetic. Discuss, keeping in mind that the ground is a conductor, and that the dowsing rod is nonmagnetic material.

PROBLEMS

Section 33.1

1. The electron in a hydrogen atom is 0.53×10^{-10} m away from the proton. What magnetic field does the magnetic moment of the proton produce at the position of the electron? Assume that the electron is instantaneously located on the axis of spin of the proton.

2. Two electrons are separated by a distance of 1.0×10^{-10} m. The first electron is on the axis of spin of the second.
   (a) What is the magnetic field that the magnetic moment of the second electron produces at the position of the first?
   (b) The potential energy of the magnetic moment of the first electron in this magnetic field depends on the orientation of the electrons. What is the potential energy (in electron-volts) if the spins of the two electrons are parallel? If antiparallel? Which orientation has the least energy?

3. Problem 22.12 gives the radii of the possible circular orbits of an electron in a hydrogen atom. For each such orbit, calculate the orbital magnetic moment.

4. The field of a fixed magnetic dipole located at the origin and oriented par-

allel to the z axis has the following components as a function of z and x in the plane $y = 0$:

$$B_x = \frac{\mu_0\mu}{4\pi} \frac{3zx}{(x^2 + z^2)^{5/2}} \qquad B_y = 0,$$

$$B_z = -\frac{\mu_0\mu}{4\pi} \left(\frac{1}{(x^2 + z^2)^{3/2}} - \frac{3z^2}{(x^2 + z^2)^{5/2}} \right)$$

Suppose that a second magnetic dipole of moment μ' is located at the point x, z (with $y = 0$); this dipole is free to rotate in the x–z plane.

(a) What is the orientation of least magnetic energy of this second dipole in the field of the first dipole, i.e., what is the angle that the second dipole will make with the z axis?

(b) What is the numerical value of the angle in the case $x = 0$, $z = r$? In the case $x = r$, $z = 0$? In the case $x = z = r/\sqrt{2}$?

5. Two magnetic dipoles are separated by a fixed distance r in the horizontal plane. The dipoles are free to rotate about a vertical axis (you may imagine that the dipoles are compass needles, but the magnetic field of the Earth is absent). If the dipoles settle into the configuration of least magnetic energy, what will be their orientation? Draw a diagram of the dipoles in this orientation. Prove your answer. (Hint: You may want to use the expression for the magnetic field given in Problem 4).

6. A compass needle has the shape of a thin rod of length 2.0 cm pivoted at its center so that it can swing freely in the horizontal plane. The compass needle has a mass of 0.12 g and a magnetic moment of 3.2×10^{-4} A · m². The compass needle is in Hawaii, where the horizontal, northerly component of the magnetic field is 2.9×10^{-5} T. What is the frequency of small rotational oscillations of the compass needle about the northerly direction?

7. Assume that the proton is a spherical ball within which the positive charge and the mass are uniformly distributed. Assume that the proton rotates rigidly with a spin angular momentum of $\hbar/2 = 0.53 \times 10^{-34}$ J · s. From this information, calculate the magnetic moment of a proton. [Hint: Use Eq. (4) with the mass of the proton. The result of this classical calculation differs by a factor of 5.6 from the actual value; this is due to a failure of classical physics when applied to subatomic particles.]

8. Assume that the charge distribution of a neutron is as indicated by the model described in Problem 24.17 and that the mass distribution is uniform. Assume that the neutron rotates rigidly with a spin angular momentum of $\hbar/2 = 0.53 \times 10^{-34}$ J · s. According to this crude model, what is the magnetic moment of a neutron? What is the direction of the magnetic moment relative to the direction of the spin angular momentum? [Hint: Apply Eq. (4) to the positive and negative charge distributions separately.]

*9. As described in Section 33.1, the proton has a magnetic moment of $\mu = 1.41 \times 10^{-26}$ A · m² parallel to the axis of its spin angular momentum. If the proton is in a magnetic field **B** and the magnetic moment makes an angle θ with this field, the torque exerted by this field on the magnetic moment will be $\tau = \mu B \sin \theta$ and the direction of this torque will be perpendicular to μ. Since the proton has a spin angular momentum **S** parallel to the magnetic moment, the torque will cause a precession of the spin about the direction of the magnetic field (see Section 12.6 for a discussion of the precession of a top under the influence of a torque).

(a) Show that the precession frequency of the proton is

$$\omega = \mu B/S$$

or, since $S = \hbar/2 = 0.53 \times 10^{-34}$ J · s,

$$\omega = 2\mu B/\hbar$$

(b) What is the precession frequency of a proton in a magnetic field of 0.20 T?

10. In a hydrogen atom, the electron orbits around the proton on a circular orbit of radius 0.53×10^{-10} m. As in Section 33.1, this orbiting electron can be regarded as a ring of current.
  (a) Calculate the magnetic field that the ring of current produces at its center.
  (b) Using the result of Problem 9, calculate the precession frequency of the proton in this magnetic field.

Section 33.2

11. A current of 25 A flows in a long solenoid of 1500 turns per meter.
  (a) If the interior of this solenoid is empty (a vacuum), what is the strength of the magnetic field?
  (b) If the interior is filled with liquid oxygen while the current stays constant, what will be the percentage change in the magnetic field?

12. Suppose that the dipole moments of all the atoms in a 20-g sample of lithium are perfectly aligned. What is the strength of the magnetic field on the axis of the dipoles at a distance of 1.0 m?

13. The space within a solenoid is to be filled with a mixture of air (paramagnetic) and methane (diamagnetic) so that the net permeability constant is exactly $\kappa_m = 1$. What percentage of air and methane should one use?

14. At a temperature of 20°C and a pressure of 1 atm the relative permeability of air is 1.000304. Calculate the relative permeability of air at the same temperature but a pressure of 3.0 atm. Assume that $\kappa_m - 1$ is proportional to the pressure.

15. Initially, the space within a long solenoid is empty. It is then filled with liquid oxygen, a paramagnetic material. What is the percentage change of the self-inductance of the solenoid? Does the self-inductance increase or decrease?

16. Show that the self-inductance per unit length of a very long solenoid filled with a paramagnetic material is $\kappa_m \mu_0 n^2 \pi R^2$, where n is the number of turns of wire per unit length and R is the radius of the solenoid.

17. Show that the energy density in the magnetic field in a very long solenoid filled with paramagnetic material is $u = B^2/(2\kappa_m \mu_0)$.

Section 33.3

18. A bar magnet of iron has a magnetic field of 0.03 T in its interior. The magnet is 15 cm long. What is the effective current running around its surface? Treat the magnet as though it were a very long cylindrical solenoid.

19. A long solenoid has 1200 turns per meter with a current of 6.0 A. The solenoid is filled with a ferromagnetic material. The value of the magnetic field B in this material is 2.0 T. What is the value of κ_m under these conditions?

20. In an iron crystal, two of the electrons of each atom participate in the alignment of spins; the magnetic field of a permanent magnet is caused by the magnetic moment of these electrons. Suppose that all of these electrons in a compass needle of mass 0.60 g are perfectly aligned. The compass needle is at a place where the horizontal, northerly component of the Earth's magnetic field is 2.4×10^{-5} T; the compass needle is free to swing in the horizontal plane. What is the torque on the compass needle when it is at an angle of 45° with the northerly component of the magnetic field? Does your answer depend on the shape of the compass needle?

21. Figure 33.7 is a plot of the magnetic field B in an iron-filled solenoid as a function of the magnetic field B_{free} that the solenoid would produce without

the iron. This plot has been prepared under the assumption that initially, when the current is zero, the iron is not magnetized. The value of the relative permeability depends on B_{free}. What is the value of κ_m when $B = 0.4$ T? When $B = 0.8$ T? When $B = 1.2$ T? Make a plot of κ_m vs. B_{free}. At what value of B_{free} is κ_m maximum?

22. Under conditions of maximum magnetization, the dipole moment per unit volume in cobalt is 1.5×10^5 A · m²/m³. Assuming that this magnetization is due to completely aligned electrons, how many such electrons are there per unit volume? How many aligned electrons per atom? The density of cobalt is 8.9×10^3 kg/m³ and the atomic mass is 58.9 g/mole.

23. In iron, two of the electrons of each atom participate in the alignment of spins that causes magnetization. Suppose that a cylindrical piece of iron, of radius 1.0 cm and length 8.0 cm, is completely magnetized along its axis, all the available electrons being in perfect alignment.
    (a) What is the number of aligned electrons?
    (b) The dipole moment of each electron is 9.27×10^{-24} A · m². What is the total dipole moment of all the aligned electrons?
    (c) What surface current running around the surface of the cylinder will give the same total dipole moment?
    (d) What magnetic field does this surface current produce in the interior of the iron?

24. The alignment of electron spins in a ferromagnetic material implies that the magnetized material has angular momentum. Suppose that a rod of iron 2.0 cm in diameter and 30 cm long is totally magnetized so that two of the electrons of each atom have their spins parallel to the axis of the rod. Suppose that the magnetization is suddenly reversed so that the spins become antiparallel to the axis. What is the change of angular momentum? The density of iron is 7.9 g/cm³ and the atomic mass is 55.8 g/mole.

*25. A long solenoid is filled with iron. The solenoid has 1800 turns per meter and the current in each turn is 50 A. Calculate the magnetic field inside the solenoid assuming that two of the electrons of each atom are completely aligned. (Hint: The magnetic field consists of two contributions: the field of the solenoid wire plus the field of the aligned electrons. The latter field can be calculated by replacing the electrons by a surface current as in Problem 23.)

Section 33.4

26. An electron in a hydrogen atom moves around a circular orbit of radius 0.53×10^{-10} m at a speed of 2.2×10^6 m/s. Suppose that the hydrogen atom is placed in a magnetic field of 0.50 T. The magnetic field is parallel to the orbital angular momentum.
    (a) What is the change of the frequency of the motion of the electron? Does the frequency increase or decrease?
    (b) What is the change of the speed of the electron? Assume the radius of the motion remains constant.
    (c) What is the change of the energy of the electron?
    (d) Check that the electric field induced by the change in magnetic flux through the orbit has the right direction to produce the change of speed and energy.

*27. A long, cylindrical bar magnet of diameter 1.0 cm has a magnetic field of 0.060 T in its interior. If you take a fine saw and cut the magnet in two pieces, the magnetic force will hold the pieces together. Estimate the magnitude of this magnetic force. [Hint: Suppose that you pull the pieces apart by a distance dx. The magnetic field in the gap between the pieces will then still be (nearly) 0.060 T. The magnetic energy in the gap equals the work that you must have done against the magnetic force while pulling the pieces apart.]

CHAPTER 34

AC Circuits

The current delivered by power companies to homes and factories is an oscillating function of time. This is called alternating current, or AC. Power companies prefer alternating currents to direct currents because of the ease with which alternating voltages can be stepped up or down by means of transformers. This makes it possible to step up the output of a power plant to several hundred thousand volts, transmit the power along a high-voltage line that minimizes the Joule losses, and finally step down the power to 220-volt AC or 110-volt AC just before delivery to the consumer.

All the appliances connected to ordinary outlets in homes therefore involve circuits with oscillating currents. Furthermore, electronic devices — such as radio transmitters and receivers — involve a variety of circuits with oscillating currents of high frequency. Many of these circuits have natural frequencies of oscillation. Such circuits exhibit the phenomenon of resonance when the natural frequency matches the frequency of a signal applied to the circuit. For instance, the tuning of a radio relies on an oscillating circuit whose frequency of oscillation is adjusted by means of a variable capacitor (attached to the tuning knob) so that it matches the frequency of the radio signal.

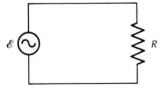

Fig. 34.1 Resistor connected to a source of alternating emf.

34.1 Simple AC Circuits with an External Electromotive Force

The simplest conceivable AC circuit consists of a pure resistor connected to an oscillating source of emf (Figure 34.1). In the circuit diagram, the source of emf is symbolized by a wavy line enclosed in a circle. This circuit might represent an electric heater or an incandescent lamp plugged into an ordinary wall outlet. In this case the emf is of the form

$$\mathscr{E} = \mathscr{E}_{max} \sin \omega t \qquad (1)$$

where $\mathscr{E}_{max}$ is the amplitude of oscillation and ω the angular frequency. In the United States, the oscillating voltage available at the outlets of private homes has an amplitude $\mathscr{E}_{max} = 156$ V and a frequency of 60 Hz, that is, an angular frequency of $\omega = 60 \times 2\pi/s$. This kind of voltage is usually called "110-volt AC" for reasons that will become clear shortly.

Applied to the circuit of Figure 34.1, Kirchhoff's Law gives, at an instant of time,

$$\mathscr{E} - IR = 0 \qquad (2)$$

or

$$\boxed{I = \frac{\mathscr{E}}{R} = \frac{\mathscr{E}_{max} \sin \omega t}{R}} \qquad (3) \qquad \textit{Current in resistor circuit}$$

Thus the current oscillates in exactly the same way as the emf (Figure 34.2).

The instantaneous electric power dissipated in the resistor is

$$P = I\mathscr{E} = \frac{\mathscr{E}_{max}^2 \sin^2 \omega t}{R} \qquad (4)$$

The power oscillates between zero and a maximum value $\mathscr{E}_{max}^2/R$ (Figure 34.3). The time-average power can be obtained by averaging

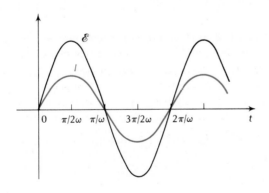

Fig. 34.2 The emf (black) and current (color) in the resistor circuit as a function of time.

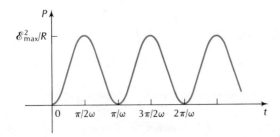

Fig. 34.3 Instantaneous power dissipated in the resistor as a function of time.

$\sin^2 \omega t$ over one cycle. We can easily verify that the average value of $\sin^2 \omega t$ is $\frac{1}{2}$,

$$\frac{1}{2\pi/\omega}\int_0^{2\pi/\omega} \sin^2 \omega t \, dt = \tfrac{1}{2} \qquad (5)$$

Hence the average power is

Average power absorbed by resistor

$$\boxed{\bar{P} = \frac{\mathscr{E}_{max}^2}{2R}} \qquad (6)$$

This is often written in the form

$$\boxed{\bar{P} = \frac{\mathscr{E}_{rms}^2}{R}} \qquad (7)$$

where the quantity $\mathscr{E}_{rms}$, called the **root-mean-square voltage,** is the square root of the time average of the square of the voltage,

Root-mean-square voltage

$$\mathscr{E}_{rms} = \sqrt{\overline{\mathscr{E}^2}} = \sqrt{\frac{\mathscr{E}_{max}^2}{2}} = \frac{\mathscr{E}_{max}}{\sqrt{2}} \qquad (8)$$

In engineering practice, an AC voltage is usually described in terms of $\mathscr{E}_{rms}$. For example, if $\mathscr{E}_{max} = 156$ V, then $\mathscr{E}_{rms} = 156/\sqrt{2}$ V $= 110$ V; an oscillating voltage with this value of $\mathscr{E}_{max}$ is described as "110-volt AC."

Comparison of Eqs. (7) and (29.23) shows that the average AC power delivered is equivalent to the DC power delivered by a steady voltage $\mathscr{E}_{rms}$. Thus, a 110-volt AC (with $\mathscr{E}_{max} = 156$ V) delivers the same average power as 110-volt DC.

EXAMPLE 1.   A 110-V AC incandescent light bulb is rated at 150 W. What is the resistance of this light bulb (when at its operating temperature)?

SOLUTION: We have $\mathscr{E}_{rms} = 110$ V and $\bar{P} = 150$ W. Hence

$$R = \mathscr{E}_{rms}^2/\bar{P} = 80.7 \ \Omega \qquad (9)$$

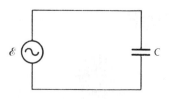

Fig. 34.4 Capacitor connected to a source of alternating emf.

Next, let us examine a circuit consisting of a capacitor connected to our oscillating source of emf (Figure 34.4). The voltage across the capacitor is Q/C and therefore Kirchhoff's Law gives

$$\mathscr{E} - Q/C = 0 \qquad (10)$$

With Eq. (1) for $\mathscr{E}$, this yields

$$Q = C\mathscr{E} = C\mathscr{E}_{max} \sin \omega t \qquad (11)$$

The current in the circuit is $I = dQ/dt$, or

$$I = \omega C\mathscr{E}_{max} \cos \omega t \qquad (12)$$

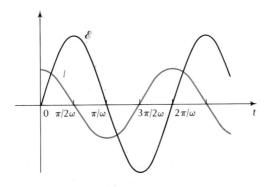

Fig. 34.5 The emf (black) and current (color) in the capacitor circuit as a function of time.

Comparison of Eqs. (12) and (1) shows that the current is a quarter cycle (90°) out of phase with the emf; for instance, at $t = 0$ the emf is minimum, and the current is maximum (Figure 34.5). Since the maxima in the current occur a quarter cycle *before* the maxima in the emf, we say that the current **leads** the emf.

It is customary to write Eq. (12) as

$$I = \frac{\mathscr{E}_{max}\cos \omega t}{X_C}$$ (13)   *Current in capacitor circuit*

where

$$X_C = \frac{1}{\omega C}$$ (14)   *Capacitive reactance*

is called the **capacitive reactance.** The quantity X_C plays roughly the same role for a capacitor in an AC circuit as does the resistance for a resistor [compare Eqs. (3) and (13)]. Note, however, that the reactance depends not only on the characteristics of the capacitor, but also on the frequency at which we are operating the circuit. The unit of reactance is the ohm, as it is for resistance.

The instantaneous power delivered to the capacitor is

$$P = I\mathscr{E} = \omega C \mathscr{E}_{max}^2 \cos \omega t \sin \omega t$$ (15)

The time dependence of this expression is contained in the factor $\cos \omega t \sin \omega t$. Since this equals $\frac{1}{2}\sin 2\omega t$, we recognize that the power oscillates at a frequency 2ω. But the important point is that the average power delivered is zero (Figure 34.6) — within one cycle, there is

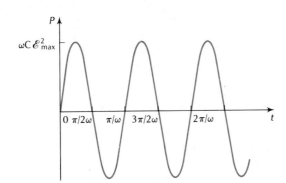

Fig. 34.6 Instantaneous power delivered to the capacitor as a function of time.

as much positive power as negative power. The source of emf does work on the capacitor during part of the cycle, but the capacitor does work on the source during other parts of the cycle so that, on the average, the *power is zero*. The ideal capacitor does not consume electric power because it has no means of dissipating electric energy.

Finally, we will examine a circuit consisting of an inductor connected to an oscillating source of emf (Figure 34.7; in this circuit diagram the inductor is represented by a coiled line). The induced emf in the inductor (back emf) is $L\, dI/dt$, and by Kirchhoff's Law this must balance the applied emf,

Fig. 34.7 Inductor connected to a source of alternating emf.

$$\mathscr{E} - L\frac{dI}{dt} = 0 \qquad (16)$$

which gives

$$\frac{dI}{dt} = \frac{\mathscr{E}}{L} = \frac{\mathscr{E}_{max}\sin\omega t}{L} \qquad (17)$$

By integrating this we obtain[1]

$$I = -\frac{\mathscr{E}_{max}\cos\omega t}{\omega L} \qquad (18)$$

Again, comparison of Eqs. (1) and (18) shows that the current is a quarter cycle (90°) out of phase with the emf (Figure 34.8). However, because of the minus sign in Eq. (18), the maxima in the current occur a quarter cycle *after* the maxima in the emf — the current **lags** the emf.

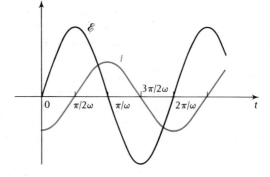

Fig. 34.8 The emf (black) and current (color) in the inductor circuit as a function of time.

We can write Eq. (18) as

Current in inductor circuit

$$I = -\frac{\mathscr{E}_{max}\cos\omega t}{X_L} \qquad (19)$$

where

Inductive reactance

$$X_L = \omega L \qquad (20)$$

---

[1] Here we assume that the constant of integration is zero. If this constant were not zero, the circuit would carry an additional time-independent current.

is the **inductive reactance.** The unit of this reactance is, again, the ohm.

The instantaneous power delivered to the inductor is

$$P = I\mathscr{E} = -\frac{1}{\omega L}\,\mathscr{E}^2_{max}\cos\omega t\sin\omega t \qquad (21)$$

As in the case of the capacitor, the average power is zero.

34.2 The Freely Oscillating LC Circuit

Figure 34.9 shows an LC circuit, which consists of an inductor and a capacitor connected in series. The circuit has no source of emf; nevertheless, a current will flow in this circuit provided that the capacitor is *initially charged.* The potential on one plate is then initially high and that on the other plate is low. A current will begin to flow around the circuit from the positive plate to the negative. If the circuit had no inductance, the current would merely neutralize the charge on the plates, i.e., the capacitor would discharge and that would be the end of the current. But the inductance makes a difference: the inductance initially opposes the buildup of the current, but once the current has become established, the inductance will keep it going for some extra time. Hence *more* charge flows from one capacitor plate to the other than required for neutrality and reversed charges accumulate on the capacitor plates. When the current finally does stop, the capacitor will again be fully charged, with reversed charges. And then a reversed current will begin to flow, etc. Thus the positive charge sloshes back and forth around the circuit.

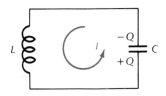

Fig. 34.9 Inductor and capacitor connected in series.

The LC system is analogous to a mass–spring system. The inductor is analogous to the mass — it tends to keep the current constant and provides "inertia." The charged capacitor is analogous to the stretched spring — it tends to accelerate the current and provides a "restoring force."

The equation of motion for the LC system follows from Kirchhoff's rule: the sum of emfs and voltage drops around the circuit must add to zero. Going around the circuit in the direction of the arrow shown in Figure 34.9, we find that the induced emf in the inductor (back emf) is

$$-L\,\frac{dI}{dt}$$

and the voltage across the capacitor is

$$-\frac{Q}{C}$$

Hence

$$-L\,\frac{dI}{dt} - \frac{Q}{C} = 0 \qquad (22)$$

Note that here Q is reckoned as positive when the charge on the lower plate is positive and I is reckoned as positive when the charge on the lower plate is increasing.

Since $I = dQ/dt$, we can also write Eq. (22) as

$$L\frac{d^2Q}{dt^2} + \frac{1}{C}Q = 0 \tag{23}$$

This equation has exactly the same mathematical form as the equation for the simple harmonic oscillator [Eq. (14.22)]:

$$m\frac{d^2x}{dt^2} + kx = 0$$

Obviously, Q plays the role of x, while L replaces m, and $1/C$ replaces k. Hence the solution of Eq. (23) can be immediately written down by recalling the solution for the simple harmonic oscillator [Eq. (14.25)]:

$$Q = Q_0 \cos\left(\frac{1}{\sqrt{LC}}t\right) \tag{24}$$

Here we have chosen the cosine solution because the initial condition of our problem specifies that at $t = 0$ the capacitor is fully charged ($Q = Q_0$) and the current is zero ($dQ/dt = 0$). Of course, other initial conditions can be accommodated by taking some suitable combination of cosine and sine solutions. From Eq. (24) we find that the current is

$$I = \frac{dQ}{dt} = -\frac{Q_0}{\sqrt{LC}}\sin\left(\frac{1}{\sqrt{LC}}t\right) \tag{25}$$

According to Eqs. (24) and (25), the charge and the current oscillate with a natural frequency

Natural frequency of LC circuit

$$\boxed{\omega_0 = \frac{1}{\sqrt{LC}}} \tag{26}$$

Figure 34.10 is a plot of the charge and the current in an LC circuit oscillating according to Eq. (25).

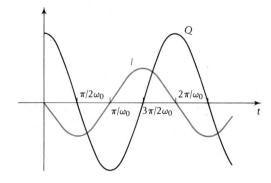

Fig. 34.10 Charge (black) on the capacitor and current (color) in the inductor as a function of time.

EXAMPLE 2.   A primitive radio transmitter, such as those built in the early days of "wireless telegraphy," consists of an LC circuit oscillating at high fre-

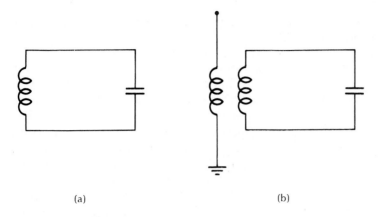

Fig. 34.11 (a) An LC circuit in a radio. (b) The LC circuit is coupled to the antenna by the mutual inductance of the two inductors.

(a)   (b)

quency (Figure 34.11a). The circuit is inductively coupled to an antenna (Figure 34.11b) so that the oscillating current in the circuit induces an oscillating current on the antenna; the latter current then radiates electromagnetic waves. Suppose that the inductance in the circuit of Figure 34.11a is 20 μH. What capacitance do we need if we want to produce oscillations of a frequency of 1500 kHz?

SOLUTION: The angular frequency is $2\pi \times 1500 \times 10^3$ radian/s. Hence, from Eq. (26)

$$C = \frac{1}{\omega_0^2 L}$$

$$= \frac{1}{(2\pi \times 1500 \times 10^3/\text{s})^2 \times 20 \times 10^{-6}\ \text{H}}$$

$$= 5.6 \times 10^{-10}\ \text{F} = 560\ \text{pF}$$

The energy of the LC system is the sum of the energies stored in the capacitor and in the inductor [see Eqs. (27.40) and (32.37)], that is,

$$U = \tfrac{1}{2}\frac{Q^2}{C} + \tfrac{1}{2}LI^2 \tag{27}$$

or

$$U = \tfrac{1}{2}\frac{Q^2}{C} + \tfrac{1}{2}L\left(\frac{dQ}{dt}\right)^2 \tag{28}$$

On physical grounds we expect that the energy remains constant during the oscillations. To prove this conservation theorem for the energy, we need only differentiate Eq. (28) with respect to time:

$$\frac{dU}{dt} = \frac{Q}{C}\frac{dQ}{dt} + L\frac{dQ}{dt}\frac{d^2Q}{dt^2} = \frac{dQ}{dt}\left(\frac{Q}{C} + L\frac{d^2Q}{dt^2}\right) \tag{29}$$

The expression on the right side is zero because of the relation between Q and d^2Q/dt^2 [see Eq. (23)].

The quantity $\tfrac{1}{2}Q^2/C$ is a potential energy (electrostatic potential energy). The quantity $\tfrac{1}{2}L(dQ/dt)^2$ may be regarded as a "kinetic" energy. Thus, the expression (28) is analogous to the expression

$$\tfrac{1}{2}kx^2 + \tfrac{1}{2}m\left(\frac{dx}{dt}\right)^2$$

for the energy of a harmonic oscillator. If the capacitor is initially charged but no current is flowing, then the energy is initially purely potential (it is stored in the electric fields in the capacitor). As the current begins to flow, the potential energy decreases and the "kinetic" energy increases. At the instant when the capacitor is completely discharged, the current reaches its maximum value. The potential energy is then zero and the energy is purely "kinetic" (it is stored in the magnetic fields in the inductor). Beyond this instant, the potential energy increases at the expense of the "kinetic" energy. When the capacitor is completely charged, with reversed charge, the current stops flowing. At this instant the energy is again purely potential. The process now repeats with the current flowing in the opposite direction. Figure 34.12 is a plot of potential energy and "kinetic" energy as a function of time for an LC circuit oscillating according to Eqs. (24) and (25).

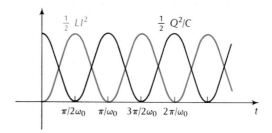

Fig. 34.12 Energy in the capacitor ($\tfrac{1}{2}Q^2/C$; black) and energy in the inductor ($\tfrac{1}{2}LI^2$; color) as a function of time.

So far we have assumed that our LC circuit contains no resistance. This is somewhat unrealistic since, at the very least, the wires connecting the circuit elements will have some resistance. The remainder of this section gives a description of the effects of resistance.

Figure 34.13 shows an LCR circuit, i.e., an LC circuit with resistance. The resistance plays a role analogous to that of the friction force in the harmonic oscillator (see Section 14.6). The resistance gradually converts electric energy into heat; hence the electric energy decreases with time. This leads to damped oscillations of gradually decreasing amplitude.

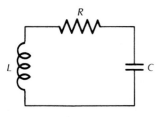

Fig. 34.13 Inductor, capacitor, and resistor connected in series.

Figure 34.14 shows the charge on the capacitor as a function of time for an LCR circuit. The charge can be described by the following equation:

$$Q = Q_0 e^{-\gamma t/2} \cos \omega_0 t \tag{30}$$

This, of course, is based on the assumption that the capacitor is fully charged at the initial time $t = 0$. The damping constant γ represents the frictional effects. It can be shown that γ is proportional to the resistance,

$$\gamma = \frac{R}{L} \tag{31}$$

As we saw in Section 14.6, the damping constant is directly related to the energy ΔU lost per period,

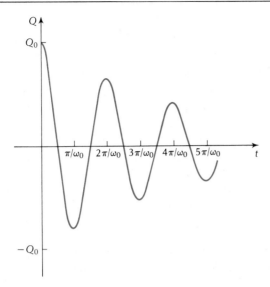

$$\frac{\Delta U}{U} = -2\pi \frac{\gamma}{\omega_0} = -\frac{2\pi R}{\omega_0 L} \tag{32}$$

Electrical engineers usually express this energy loss in terms of the **"Q"** or the **quality factor** of the circuit,[2]

$$\boxed{\text{"}Q\text{"} = 2\pi |U/\Delta U| = \omega_0 L/R} \tag{33}$$

"Q" of freely oscillating LCR circuit

Typical circuits of low resistance in radio transmitters and receivers have a *"Q"* of up to 100.

Besides damping the amplitude of the oscillations, the resistance also produces another effect: it reduces the frequency of oscillation. Intuitively, we expect such a reduction of frequency because the friction in the resistance will slow the oscillations. Mathematically, one can show that the angular frequency for the natural oscillations of the LCR circuit is

$$\omega_0 = \frac{1}{\sqrt{LC}} \sqrt{1 - \frac{CR^2}{4L}} \tag{34}$$

Here $\sqrt{1 - CR^2/4L}$ represents the factor by which the frequency is decreased as compared to an LC circuit without resistance.

34.3 The LCR Circuit with an External Electromotive Force

Figure 34.15 shows a series LCR circuit with an oscillating source of emf. This emf acts as a driving force that pushes on the charge in the circuit. As in the case of the driven harmonic oscillator (see Section

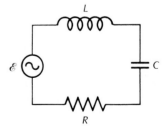

Fig. 34.15 Inductor, capacitor, and resistor connected in series to a source of alternating emf.

---

[2] This *"Q"* must not be confused with electric charge.

14.6), we expect that if the driving frequency coincides with the natural frequency ω_0 of the circuit, then the oscillations will build up to a very large value. This is the condition for **resonance.**

Suppose that the oscillating emf driving the circuit has an angular frequency ω:

$$\mathscr{E} = \mathscr{E}_{max} \sin \omega t \tag{35}$$

Under steady-state conditions the current will then oscillate with the same angular frequency. This implies that the current will be of the form

$$I = I_{max} \sin(\omega t + \phi) \tag{36}$$

We are now faced with the task of finding how the amplitude I_{max} and the phase ϕ of the current are related to the known parameters of the circuit.

We begin with Kirchhoff's rule: the external emf $\mathscr{E}$ must match the sum of the voltages across the resistor, the capacitor, and the inductor,

$$\mathscr{E} = \Delta V_R + \Delta V_C + \Delta V_L \tag{37}$$

To evaluate the three terms on the right side of this equation, we make use of the results of Section 34.1. The voltage across the resistor is in phase with the current:

$$\Delta V_R = IR = RI_{max} \sin(\omega t + \phi) \tag{38}$$

The voltage across the capacitor is one-quarter cycle behind the current,

$$\Delta V_C = -X_C I_{max} \cos(\omega t + \phi) \tag{39}$$

and the voltage across the inductor is one-quarter cycle ahead of the current,

$$\Delta V_L = X_L I_{max} \cos(\omega t + \phi) \tag{40}$$

Substitution of these into Eq. (37) yields

$$\mathscr{E}_{max} \sin \omega t = RI_{max} \sin(\omega t + \phi) - (X_C - X_L)I_{max} \cos(\omega t + \phi) \tag{41}$$

With the trigonometric identities for the sine and the cosine of the sum of two angles this becomes

$$\mathscr{E}_{max} \sin \omega t = RI_{max}(\sin \omega t \cos \phi + \cos \omega t \sin \phi)$$

$$- (X_C - X_L)I_{max} (\cos \omega t \cos \phi - \sin \omega t \sin \phi)$$

$$= [R \cos \phi + (X_C - X_L)\sin \phi]I_{max} \sin \omega t$$

$$+ [R \sin \phi - (X_C - X_L)\cos \phi]I_{max} \cos \omega t \tag{42}$$

If we examine this equation at time $t = 0$, we find that

$$0 = R \sin \phi - (X_C - X_L)\cos \phi \qquad (43)$$

or

$$\boxed{\tan \phi = \frac{X_C - X_L}{R}} \qquad (44)$$

Phase angle for series LCR circuit

From this we obtain

$$\sin \phi = \frac{1}{\sqrt{1 + 1/\tan^2 \phi}} = \frac{X_C - X_L}{\sqrt{R^2 + (X_C - X_L)^2}} \qquad (45)$$

$$\cos \phi = \frac{1}{\sqrt{1 + \tan^2 \phi}} = \frac{R}{\sqrt{R^2 + (X_C - X_L)^2}} \qquad (46)$$

With these expressions for $\sin \phi$ and $\cos \phi$, Eq. (42) reduces to

$$\mathscr{E}_{max} \sin \omega t = \frac{R^2 + (X_C - X_L)^2}{\sqrt{R^2 + (X_C - X_L)^2}} I_{max} \sin \omega t \qquad (47)$$

or

$$I_{max} = \frac{\mathscr{E}_{max}}{\sqrt{R^2 + (X_C - X_L)^2}} \qquad (48)$$

Equations (44) and (48) are the desired expressions for I_{max} and ϕ in terms of the known parameters of the circuit.

The quantity

$$Z = \sqrt{R^2 + (X_C - X_L)^2}$$

or

$$\boxed{Z = \sqrt{R^2 + (1/\omega C - \omega L)^2}} \qquad (49)$$

Impedance for series LCR circuit

is called the **impedance** of the series LCR circuit. In terms of this quantity the current is

$$\boxed{I = \frac{\mathscr{E}_{max} \sin(\omega t + \phi)}{Z}} \qquad (50)$$

Current in series LCR circuit

The relationships among the voltages and the currents in the circuit elements can be represented graphically by a **phasor diagram.** In such a diagram the amplitude of a sinusoidal function is represented by a line segment of length equal to the amplitude of oscillation, and the phase is represented by the angle between this line segment and the horizontal axis. The line segment is called a **phasor.** For instance, the function $I = I_{max} \sin(\omega t + \phi)$ is represented by the phasor shown in Figure 34.16. Obviously, the projection of this phasor on the vertical

Phasor

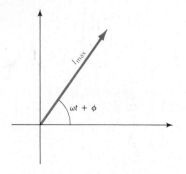

Fig. 34.16 Phasor representing the function $I = I_{max} \sin(\omega t + \phi)$. The phasor makes an angle of $\omega t + \phi$ with the horizontal axis. The projection of the phasor on the vertical axis is $I_{max} \sin(\omega t + \phi)$.

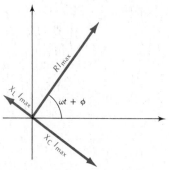

Fig. 34.17 The three phasors representing ΔV_R, ΔV_C, and ΔV_L. Their phase angles are $\omega t + \phi$, $\omega t + \phi - 90°$, and $\omega t + \phi + 90°$, respectively.

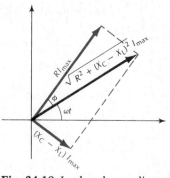

Fig. 34.18 In the phasor diagram, the sum of the voltages ΔV_R, ΔV_L, and ΔV_C is represented by the vector sum (black) of the three vectors (color) shown in Figure 34.17. From the diagram we see that this vector sum has a magnitude $\sqrt{R^2 + (X_C - X_L)^2}\, I_{max}$. The vector sum makes an angle ϕ with the phasor ΔV_R (or with the phasor representing I); from the diagram we see that $\tan \phi = (X_C - X_L)/R$.

axis is $I_{max} \sin(\omega t + \phi)$, i.e., the projection is exactly I. Likewise, we can represent each of the voltages on the right side of Eq. (37) by phasors (Figure 34.17). The vector sum of these phasors then represents the sum of these voltages. Kirchhoff's rule demands that this sum be equal to the external emf (Figure 34.18). We can then derive the expressions for the amplitude and for the phase angle of the current by direct inspection of Figure 34.18.

The amplitude of the oscillations of the current in the circuit depends critically on the frequency. With $X_C = 1/\omega C$ and $X_L = \omega L$, Eq. (48) exhibits the following dependence on frequency:

$$I_{max} = \frac{\mathscr{E}_{max}}{\sqrt{R^2 + (1/\omega C - \omega L)^2}} \tag{51}$$

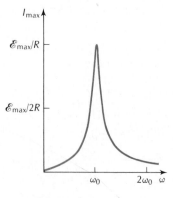

Fig. 34.19 Maximum current as a function of angular frequency.

Figure 34.19 is a plot of this maximum current as a function of frequency. The current is weak when ω is near zero, and it is also weak when ω is very large. However, when ω is near $\omega_0 = 1/\sqrt{LC}$, the current becomes very strong. This is a resonance phenomenon: the oscillations become large when the driving force pushes the circuit with the same frequency as that of the natural oscillations [see Eq. (26)]. When the frequency ω exactly matches the natural frequency ($\omega = \omega_0 = 1/\sqrt{LC}$), the amplitude of the oscillations of the current reaches the value

$$I_{max} = \frac{\mathscr{E}_{max}}{R} \tag{52}$$

Let us now examine the phase of the current given by Eq. (50). If the driving frequency is below resonance ($\omega < \omega_0$), then $X_C - X_L$ is positive, and ϕ is also positive [see Eq. (44)]. The current then leads the external emf, that is, the maxima in the current occur earlier than those in the emf (see Figure 34.20a). If the driving frequency is above resonance, ($\omega > \omega_0$), then $X_C - X_L$ is negative, and ϕ is also negative [see Eq. (44)]. The current then lags the external emf, that is, the max-

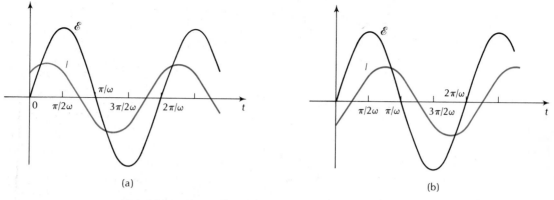

Fig. 34.20 The emf (black) and current (color) in the LCR circuit as a function of time: (a) $\omega < \omega_0$ and (b) $\omega > \omega_0$.

ima in the current occur later than those in the emf (Figure 34.20b).

Finally, what is the power delivered by the source of emf to the LCR circuit? The instantaneous power is

$$P = I\mathscr{E} = \frac{1}{Z}\mathscr{E}^2_{max} \sin(\omega t + \phi)\sin \omega t \qquad (53)$$

or

$$P = \frac{1}{Z}\mathscr{E}^2_{max}(\sin^2 \omega t \cos \phi + \cos \omega t \sin \omega t \cos \phi) \qquad (54)$$

To find the time-average power, we note that the time-average of $\sin^2 \omega t$ is $\frac{1}{2}$ [see Eq. (5)] and the time-average of $\cos \omega t \sin \omega t$ is zero. Hence

$$\bar{P} = \frac{\mathscr{E}^2_{max}}{2Z} \cos \phi \qquad (55)$$

which we can also write as

$$\boxed{\bar{P} = \frac{\mathscr{E}^2_{rms}}{Z} \cos \phi} \qquad (56) \qquad \textit{Average power absorbed by LCR circuit}$$

EXAMPLE 3. An LC circuit with $L = 3.0 \times 10^{-4}$ H and $C = 2.0 \times 10^{-6}$ F is being driven by an oscillating source delivering an AC emf of amplitude 0.40 V at a frequency of 5.0×10^4 radian/s. What is the peak instantaneous voltage across the inductor? Across the capacitor?

SOLUTION: The natural frequency of this circuit, which we will need later in our calculation, is

$$\omega_0 = \frac{1}{\sqrt{LC}}$$

$$= \frac{1}{\sqrt{3.0 \times 10^{-4} \text{ H} \times 2.0 \times 10^{-6} \text{ F}}}$$

$$= 4.1 \times 10^4 \text{ radian/s} \qquad (57)$$

The instantaneous voltage across the inductor is

$$V_L = X_L I_{max} \cos(\omega t + \phi) = \omega L I_{max} \cos(\omega t + \phi)$$

With Eq. (51) and with $R = 0$ this gives

$$V_L = \frac{\omega L \mathscr{E}_{max} \cos(\omega t + \phi)}{1/\omega C - \omega L} \tag{58}$$

This has a peak value of

$$\frac{\omega L \mathscr{E}_{max}}{1/\omega C - \omega L} = \frac{\mathscr{E}_{max}}{1/(\omega^2 LC) - 1} = \frac{\mathscr{E}_{max}}{\omega_0^2/\omega^2 - 1}$$

$$= \frac{0.40 \text{ V}}{\dfrac{(4.1 \times 10^4/\text{s})^2}{(5.0 \times 10^4/\text{s})^2} - 1} = -1.20 \text{ V}$$

The instantaneous voltage across the capacitor is

$$V_C = -X_C I_{max} \cos (\omega t + \phi)$$

$$= -\frac{1}{\omega C} \frac{\mathscr{E}_{max} \cos (\omega t + \phi)}{1/\omega C - \omega L} \tag{59}$$

This has a peak value of

$$-\frac{1}{\omega C} \frac{\mathscr{E}_{max}}{1/\omega C - \omega L} = -\frac{\mathscr{E}_{max}}{1 - \omega^2/\omega_0^2}$$

$$= -\frac{0.40 \text{ V}}{1 - \dfrac{(5.0 \times 10^4/\text{s})^2}{(4.1 \times 10^4/\text{s})^2}} = 0.80 \text{ V}$$

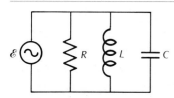

Fig. 34.21 Inductor, capacitor, and resistor connected in parallel to a source of alternating emf.

EXAMPLE 4.   Figure 34.21 shows a parallel LCR circuit with an oscillating source of emf. What is the net current delivered by the source of emf?

SOLUTION: If the emf of the source is $\mathscr{E} = \mathscr{E}_{max} \sin \omega t$, then this is also the emf acting on the inductor, the capacitor, and the resistor individually. The currents in the inductor, capacitor, and resistor are then, respectively [see Eqs. (19), (13), and (3)]:

$$I_L = -\frac{\mathscr{E}_{max} \cos \omega t}{X_L}$$

$$I_C = \frac{\mathscr{E}_{max} \cos \omega t}{X_C}$$

$$I_R = \frac{\mathscr{E}_{max} \sin \omega t}{R}$$

The net current delivered by the source of emf is the sum of these currents. To evaluate this sum, we can use a phasor diagram. Figure 34.22 shows the phasors representing the currents I_L, I_C, and I_R. The net current is represented by the vector sum of these phasors. By inspection of Figure 34.22 we see that the vector sum has a magnitude

$$I_{max} = \mathscr{E}_{max} \sqrt{\frac{1}{R^2} + \left(\frac{1}{X_C} - \frac{1}{X_L}\right)^2} \tag{60}$$

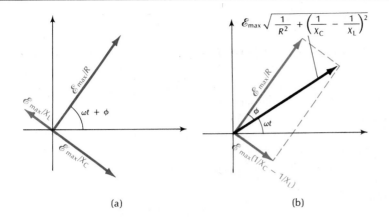

(a)   (b)

Fig. 34.22 (a) The three phasors (color) representing I_L, I_C, and I_R. (b) The vector sum (black) of these phasors represents the net current.

or

$$I_{max} = \mathscr{E}_{max}\sqrt{\frac{1}{R^2} + \left(\omega C - \frac{1}{\omega L}\right)^2} \qquad (61)$$

The angle between the phasor representing the net current and the phasor representing the current in the resistor (or the phasor representing the emf) is given by

$$\boxed{\tan\phi = \frac{1/X_C - 1/X_L}{1/R}} \qquad (62)$$

Phase angle for parallel LCR circuit

We can then write the current as

$$I = I_{max}\sin(\omega t + \phi) = \mathscr{E}_{max}\sqrt{\frac{1}{R^2} + \left(\omega C - \frac{1}{\omega L}\right)^2}\ \sin(\omega t + \phi)$$

or as

$$\boxed{I = \frac{\mathscr{E}_{max}\sin(\omega t + \phi)}{Z}} \qquad (63)$$

Current in parallel LCR circuit

where

$$\boxed{Z = 1 \Bigg/ \sqrt{\frac{1}{R^2} + \left(\omega C - \frac{1}{\omega L}\right)^2}} \qquad (64)$$

Impedance for parallel LCR circuit

is the impedance of the parallel LCR circuit.

Note that if the frequency coincides with the resonant frequency $\omega = \omega_0 = 1/\sqrt{LC}$, the current I specified by Eq. (63) is *minimum*. This is so because at this frequency, the currents in the inductor and the capacitor are of equal magnitudes; since these currents are always in opposite directions, the equality of their magnitudes implies their cancellation.

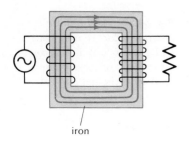

iron

Fig. 34.23 A transformer.

34.4 The Transformer

A transformer consists of two coils arranged in such a way that (almost) all the magnetic field lines generated by one of them pass through the other. This can be achieved by winding both the coils on a common iron core. As we have seen in Chapter 33, the iron increases the strength of the magnetic field in its interior by a large factor. Since the field is much stronger inside the iron than outside, the field lines will concentrate inside the iron; thus the iron tends to keep the field lines together and acts as a conduit for the field lines (Figure 34.23).

Each coil is part of a separate electric circuit (Figure 34.24). The **primary** circuit has a source of alternating emf and the **secondary** circuit has a resistance or some other load that consumes electric power. The alternating current in the primary circuit induces an alternating emf in the secondary circuit. We will show that the induced emf $\mathscr{E}_2$ in the secondary circuit is related as follows to the emf $\mathscr{E}_1$ in the primary circuit:

emf in primary and secondary circuits of transformer

$$\mathscr{E}_2 = \mathscr{E}_1 \frac{N_2}{N_1} \qquad (65)$$

where N_1 and N_2 are, respectively, the numbers of turns in the primary and secondary coils.

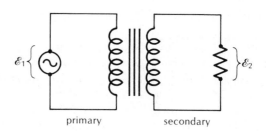

Fig. 34.24 Circuit diagram for the transformer. The parallel lines represent the mutual inductance.

primary   secondary

To prove Eq. (65), we begin with Kirchhoff's rule as it applies to the primary circuit: the emf $\mathscr{E}_1$ of the source must equal the induced emf $\mathscr{E}_{1,\text{ind}}$ across the primary coil. But by Faraday's Law, the induced emf equals the rate of change of flux,

$$\mathscr{E}_1 = \mathscr{E}_{1,\text{ind}} = -\frac{d\Phi_1}{dt} \qquad (66)$$

Likewise, the emf $\mathscr{E}_2$ delivered to the load must equal the induced emf $\mathscr{E}_{2,\text{ind}}$ in the secondary coil which, in turn, equals the rate of change of flux in that coil,

$$\mathscr{E}_2 = \mathscr{E}_{2,\text{ind}} = -\frac{d\Phi_2}{dt} \qquad (67)$$

Since the same numbers of magnetic field lines pass through both coils, the fluxes and their rates of change are necessarily in the ratio N_2/N_1,

$$\frac{d\Phi_2}{dt} = \frac{N_2}{N_1}\frac{d\Phi_1}{dt} \tag{68}$$

From Eqs. (66), (67), and (68) we obtain

$$\mathscr{E}_2 = -\frac{N_2}{N_1}\frac{d\Phi_1}{dt} = \frac{N_2}{N_1}\mathscr{E}_1 \tag{69}$$

which we wanted to prove.

If $N_2 > N_1$, we have a step-up transformer and if $N_2 < N_1$, a step-down transformer.

EXAMPLE 5. Door bells and buzzers usually are designed for 12-volt AC and they are powered by small transformers which step down 110-volt AC to 12-volt AC. Suppose that such a transformer has a primary winding with 1500 turns. How many turns are there on the secondary winding?

SOLUTION: Equation (65) applies to the instantaneous voltages. It is therefore also valid for the rms voltages. With our numerical values

$$N_2 = N_1 \frac{\mathscr{E}_2}{\mathscr{E}_1} = 1500 \times \frac{12\ \mathrm{V}}{110\ \mathrm{V}} = 164\ \text{turns}$$

As long as the secondary circuit is open and carries no current ($I_2 = 0$), an ideal transformer does not consume electric power. Under these conditions, the primary circuit consists of nothing but the source of emf and an inductance, i.e., it is a pure L circuit. In such a circuit the power delivered by the source of emf averages to zero (see Section 34.1).

If the secondary circuit is closed, a current will flow ($I_2 \neq 0$). This current contributes to the magnetic flux in the transformer and induces a current in the primary circuit. The current in the latter is then different from that in a pure L circuit and the power will *not* average to zero over a cycle. In an ideal transformer, the electric power that the primary circuit takes from the source of emf exactly matches the power that the secondary circuit delivers to the load. Good transformers approach this ideal condition fairly closely: about 99% of the power supplied to the input terminals emerges at the output terminals; the difference is lost as heat in the iron core and in the windings.

Transformers play a large role in our electric technology. As we saw in Section 29.5, transmission lines for electric power operate much more efficiently at high voltage since this reduces the Joule losses. To take advantage of this high efficiency, power lines are made to operate at several hundred kilovolt. The voltage must be stepped up to this value at the power plant and, for safety's sake, it must be stepped down just before it reaches the consumer. For these operations, large banks of transformers are needed at both ends.

SUMMARY

Average AC power absorbed by resistor: $\overline{P} = \dfrac{\mathscr{E}_{\max}^2}{2R}$

$$= \frac{\mathscr{E}_{\mathrm{rms}}^2}{R}$$

Natural frequency of LC circuit: $\omega_0 = \dfrac{1}{\sqrt{LC}}$

Energy loss in freely oscillating series LCR circuit: $Q = 2\pi |U/\Delta U|$
$$= \omega_0 L/R$$

Impedance of series LCR circuit: $Z = \sqrt{R^2 + (1/\omega C - \omega L)^2}$

Impedance of parallel LCR circuit: $Z = 1 \bigg/ \sqrt{\dfrac{1}{R^2} + \left(\omega C - \dfrac{1}{\omega L}\right)^2}$

Transformer: $\mathscr{E}_2 = \mathscr{E}_1 \dfrac{N_2}{N_1}$

QUESTIONS

1. You can perceive the 120-Hz flicker (two peaks of intensity per AC cycle) in a fluorescent light tube (by sweeping your eye quickly across the tube), but you cannot perceive any such flicker in an incandescent light bulb. Explain.

2. Do the electrons from the power station ever reach the wiring of your house?

3. Some electric motors operate only on DC, others only on AC. What is the difference between these motors?

4. If you connect a capacitor across a 110-V outlet, does any current flow through the connecting wires? Through the space between the capacitor plates? Does the outlet deliver instantaneous electric power? Average electric power?

5. It is sometimes said that a capacitor becomes a short circuit at high frequencies, and an inductor becomes an open circuit at high frequencies. Explain.

6. Can you blow a fuse by connecting a very large capacitor across an ordinary 110-V outlet?

7. How could you use an LC circuit to measure the capacitance of a capacitor?

8. In Section 34.3 we asserted that under steady-state conditions the current in an LCR circuit has the same frequency as the driving emf. Is this also true if the circuit is not in steady state? Give an example.

9. Consider a series LCR circuit. Can the voltage across the capacitor ever be larger than $\mathscr{E}_{max}$? Across the inductor? Across the resistor? (Hint: Inspect the phasor diagram.)

10. Roughly plot the phase angle ϕ given by Eq. (44) as a function of ω. What is the phase angle at resonance?

11. If you use an AC voltmeter to measure the driving emf and the voltages across the inductor, the capacitor, and the resistor in a series LCR circuit, you will find that the emf is larger than the voltage across the resistor, but smaller than the sum of the voltages across the inductor, capacitor, and resistor. Explain.

12. What is the impedance of a series LCR circuit at resonance?

13. If we substitute $R = 0$ in Eq. (50), we obtain the equation for the current in a driven LC circuit. What must we substitute to obtain the equation for a driven CR circuit? A driven LR circuit?

14. Show that the equation for the current in a series LCR circuit [Eq. (50)]

includes Eqs. (3), (13), and (19) as special cases. What are the individual impedances of a resistor, a capacitor, and an inductor?

15. Show that the equation for the current in a parallel LCR circuit [Eq. (63)] includes Eqs. (3), (13), and (19) as special cases.

16. Consider the parallel LCR circuit described in Example 4. If the driving frequency is greater than $1/\sqrt{LC}$, does the net current lead or lag the emf? If the driving frequency is smaller than $1/\sqrt{LC}$?

17. Show that the time-average power absorbed by a series LCR circuit [Eq. (56)] is maximum at resonance.

18. In a parallel LCR circuit, the time-average power absorbed is $\mathscr{E}_{max}^2/2R$. Explain.

19. Why can we not use a transformer to step up the voltage of a battery?

20. Does an electric motor absorb more electric power when pulling a mechanical load than when running freely?

PROBLEMS

Section 34.1

1. An electric heater plugged into a 110-V AC outlet uses an average electric power of 1200 W.
    (a) What is the rms current and the maximum instantaneous current through the heater?
    (b) What is the maximum instantaneous power and the minimum instantaneous power?

2. An immersible heating element used to boil water consumes an (average) electric power of 400 W when connected to a source of 110 volts AC. Suppose that you connect this heating element to a source of 110 volts DC. What power will it consume?

3. A high-voltage power line operates on an rms voltage of 230,000 volts AC and delivers an rms current of 740 A.
    (a) What are the maximum instantaneous voltage and current?
    (b) What are the maximum instantaneous power and the average power delivered?

4. An AC current of 20 A flows in a copper wire of diameter 0.30 cm connected to an electric outlet. The drift velocity of the free electrons in the copper will then oscillate at 60 Hz. What is the maximum value of the instantaneous drift velocity? What is the maximum value of the acceleration of the drift velocity?

5. An electric heater operating with a 115-V AC power supply delivers 1200 W of heat.
    (a) What is the rms current through this heater?
    (b) What is the maximum instantaneous current?
    (c) What is the resistance of this heater?

6. The GG-1 electric locomotive develops 4600 hp; it runs on an AC voltage of 1100 V.
    (a) What rms current does this locomotive draw?
    (b) Why is it advantageous to supply the electric power for locomotives at high voltage (and fairly low current)?

7. A circuit consists of a resistor connected in series to a battery; the resistance is 5 Ω and the emf of the battery is 12 V. The wires (of negligible resistance) connecting these circuit elements are laid out along a square of 20 cm × 20 cm (Figure 34.25). The entire circuit is placed face on in an oscillating mag-

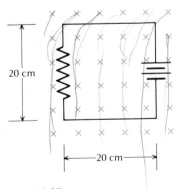

Fig. 34.25

20 cm

20 cm

netic field. The instantaneous value of the magnetic field is

$$B = B_0 \sin \omega t$$

with $B_0 = 0.15$ T and $\omega = 360$ radian/s.
(a) Find the instantaneous current in the resistor.
(b) Find the average power dissipated in the resistor.

8. A circuit consists of two capacitors of 6.0×10^{-8} F and 9.0×10^{-8} F connected in series to an oscillating source of emf (Figure 34.26). This source delivers a sinusoidal emf $\mathscr{E} = 1.8 \sin(120\pi t)$, where $\mathscr{E}$ is in volts and t in seconds.
(a) Find the charge on each capacitor as a function of time.
(b) At what time is the charge on the capacitors maximum? At what time minimum?
(c) What is the maximum energy in the capacitors? What is the time-average energy?

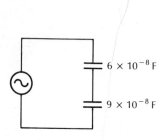

Fig. 34.26

9. An inductor of 1.6×10^{-3} H is connected to a source of alternating emf. The current in the inductor is $I = I_0 \cos \omega t$, with $I_0 = 180$ A and $\omega = 120\pi$ radian/s.
(a) What is the potential difference across the inductor at time $t = 0$? At time $t = 1/240$ s?
(b) What is the energy in the inductor at time $t = 0$? At time $t = 1/240$ s?
(c) What is the instantaneous power delivered by the source of emf to the inductor at time $t = 0$? At time $t = 1/240$ s?

10. Consider the circuit shown in Figure 34.27. The emf is of the form $\mathscr{E}_0 \sin \omega t$. In terms of this emf and the capacitance C and inductance L, find the instantaneous currents through the capacitor and the inductor. Find the instantaneous current and the instantaneous power delivered by the source of emf.

11. A capacitor with $C = 8.0 \times 10^{-7}$ F is connected to an oscillating source of emf. This source provides an emf $\mathscr{E} = \mathscr{E}_{max} \sin \omega t$, with $\mathscr{E}_{max} = 0.20$ V and $\omega = 6.0 \times 10^3$ radian/s.
(a) What is the reactance of a capacitor?
(b) What is the maximum current in the circuit?
(c) What is the current at time $t = 0$? At time $t = \pi/4\omega$?

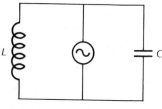

Fig. 34.27 Inductor and capacitor connected to a source of alternating emf.

12. An inductor with $L = 4.0 \times 10^{-2}$ H is connected to an oscillating source of emf. This source provides an emf $\mathscr{E} = \mathscr{E}_{max} \sin \omega t$, with $\mathscr{E}_{max} = 0.20$ V and $\omega = 6.0 \times 10^3$ radian/s.
(a) What is the reactance of the inductor?
(b) What is the maximum current in the circuit?
(c) What is the current at time $t = 0$? At time $t = \pi/4\omega$?

Section 34.2

13. What is the natural frequency for an LC circuit consisting of a 2.2×10^{-6} F capacitor and a 8.0×10^{-2} H inductor?

14. A radio receiver contains an LC circuit whose natural frequency of oscillation can be adjusted, or tuned, to match the frequency of incoming radio waves. The adjustment is made by means of a variable capacitor. Suppose that the inductance of the circuit is 15 μH. Over what range of capacitances must the capacitor be adjustable if the frequencies of oscillation of the circuit are to span the range from 530 kHz to 1600 kHz?

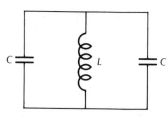

Fig. 34.28 Two equal capacitors connected to an inductor.

15. What is the natural frequency of oscillation of the circuit shown in Figure 34.28? The capacitances are 2.4×10^{-5} F each and the inductance is 1.2×10^{-3} H.

16. An LC circuit has an inductance of 5.0×10^{-2} H and a capacitance of 5.0×10^{-6} F. At $t = 0$, the capacitor is fully charged so that $Q_0 = 1.2 \times 10^{-4}$ C. What is the energy in this circuit? At what time, after $t = 0$, will the energy be purely magnetic? At what *later* time will it be purely electric?

17. Consider the LC circuit described in Problem 16. If we add a resistance of 120 Ω to this circuit (in series), what will be the frequency of the natural oscillations? By what factor is this smaller than the frequency of the circuit without resistance?

18. The circuit of Figure 34.29a is oscillating with the switch S closed. The graph of current vs. time is shown in Figure 34.29b.
   (a) At time t_1 the switch S is suddenly opened. Is the frequency of oscillation increased, decreased, or unchanged? In the space on the left in Figure 34.29b sketch the graph of current for times after t_1.
   (b) At time t_2 the switch S is closed. Sketch the graph of current after this time.

19. Consider an RC circuit consisting of a capacitor and a resistor in series (Figure 34.30). The capacitor is initially charged.
   (a) Show that Kirchhoff's rule leads to the following equation for this system:

$$R \frac{dQ}{dt} + \frac{1}{C} Q = 0$$

   (b) Verify that the solution of this equation is

$$Q = Q_0 e^{-t/RC}$$

   where Q_0 is the initial charge at time $t = 0$.
   (c) Show that the current is

$$I = -\frac{Q_0}{RC} e^{-t/RC}$$

   (d) Suppose that the resistance is 3.0 Ω. Suppose that the capacitance of the capacitor is 1.0×10^{-4} F and that the initial voltage across its terminals is 3.0 V. For this special case, plot the current as a function of time. What is the current at the initial instant? After how many seconds will the current have dropped to one-half of its initial value?

20. An LCR series circuit has $L = 0.5$ H, $C = 2.0 \times 10^{-5}$ F, and $R = 10$ Ω. At time $t = 0$, the capacitor is fully charged with a voltage of 24 V across its plates.
   (a) What is the initial energy in the circuit?
   (b) What is the percentage loss of energy per period?
   (c) At what time after $t = 0$ will the energy in the circuit have fallen to one-half of its initial value? One-tenth of its initial value?
   (d) Plot the energy as a function of time for the time interval $0 \text{ s} \leq t \leq 0.1$ s.

Section 34.3

21. A series LC circuit with $L = 3.0 \times 10^{-4}$ H and $C = 2.0 \times 10^{-5}$ F is being driven by an oscillating source of emf delivering an AC voltage of amplitude 0.40 V and frequency of 1.6×10^4 radian/s. What is the maximum instantaneous value of the current? What is the maximum instantaneous voltage across the inductor? Across the capacitor?

22. A series LC circuit is being driven by an audio generator that delivers an emf of amplitude 0.50 V. When the generator delivers the emf at a frequency of 2.0×10^3 radian/s, the maximum current is 1.0×10^{-1} A; when the generator delivers the same emf at a frequency of 1.5×10^3 radian/s, the maximum current is 2.7×10^{-2} A. From this information deduce the values of the inductance and the capacitance.

23. A capacitor of $C = 24.0$ μF and an inductor of $L = 0.180$ H are connected

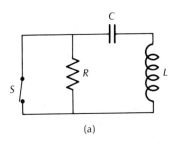

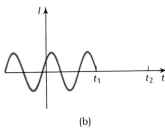

(a)

(b)

Fig. 34.29 (a) Inductor, capacitor, and resistor connected in a circuit. The switch S is initially closed. (b) Current in the circuit as a function of time. At $t = t_1$, the switch S is opened.

Fig. 34.30 Capacitor and resistor connected in series.

in series with an oscillating source that delivers an emf $\mathscr{E} = 12.0 \sin 377t$, where t is measured in seconds and $\mathscr{E}$ in volts.

(a) What is the maximum instantaneous current?

(b) What is the maximum instantaneous voltage across the capacitor?

(c) Across the inductor?

24. A series LC circuit with $C = 1.5 \times 10^{-7}$ F and $L = 2.5 \times 10^{-4}$ H is being driven by an audio generator that delivers a sinusoidal emf with an amplitude of 0.80 V and a frequency 2.2×10^4 Hz.

(a) Plot the emf as a function of time.

(b) Plot the current as a function of time.

25. A driven LC circuit consists of an inductor of 3.0×10^{-3} H, a capacitor of 2.0×10^{-8} F, and an oscillating source of emf operating at a frequency of 1.0×10^5 radian/s, all connected in series. If the amplitude of the current in the circuit is to be 6.0×10^{-3} A, what must be the amplitude of the alternating emf?

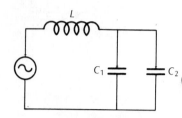

Fig. 34.31 Two capacitors and an inductor connected to a source of alternating emf.

26. Consider the circuit shown in Figure 34.31. The oscillating source delivers a sinusoidal emf of amplitude 0.80 V and frequency of 400 Hz. The inductance is 5.0×10^{-2} H and the capacitances are 8.0×10^{-7} F and 16.0×10^{-7} F. Find the maximum instantaneous current in each capacitor.

27. Consider the circuit described in Example 3. If we add a resistor with $R = 10$ Ω in series with the other elements of this circuit, what will be the peak instantaneous voltage across the inductor? Across the capacitor? Across the resistor? What will be the average power dissipated in the resistor?

28. (a) Consider the driven LCR circuit described by Eqs. (35)–(51). If $L = 6.0 \times 10^{-2}$ H, $C = 3.0 \times 10^{-6}$ F, $R = 1.2 \times 10^2$ Ω, $\mathscr{E}_{max} = 24.0$ V, and $\omega = 2.5 \times 10^3$ radian/s, calculate the maximum value of the current in this circuit under steady-state conditions.

(b) At what time after $t = 0$ does the alternating emf of the source reach its maximum value? At what time does the current reach its maximum value?

(c) Make a plot of $\mathscr{E}$ as a function of time. On top of this plot, make a plot of I as a function of time.

29. An LCR circuit consists of an inductor of 1.5×10^{-2} H, a capacitor of 2.8×10^{-6} F, and a resistor of 5.0 Ω connected in series to a source of alternating emf. The source delivers a voltage of $\mathscr{E} = \mathscr{E}_{max} \sin \omega t$, with $\mathscr{E}_{max} = 0.60$ V and $\omega = 6.0 \times 10^4$ radian/s.

(a) What is the impedance of this circuit?

(b) What is the maximum current in this circuit?

(c) What is the phase angle of the current?

30. An LCR circuit consists of an inductor of 1.2×10^{-2} H, a capacitor of 2.4×10^{-6} F, and a resistor of 2.0 Ω connected in series to a source of alternating emf with an amplitude of 0.80 V.

(a) At what frequency will this circuit be in resonance with the driving voltage?

(b) What is the maximum current in the circuit at resonance?

(c) What is the average dissipation of power in the circuit at resonance?

31. An LR circuit consists of an inductor with $L = 2.0 \times 10^{-4}$ H and a resistor with $R = 1.2$ Ω connected in series to an oscillating source of emf. This source generates a voltage $\mathscr{E} = \mathscr{E}_{max} \sin \omega t$, with $\mathscr{E}_{max} = 0.50$ V and $\omega = 3 \times 10^3$ radian/s. Find the maximum current in the circuit. Find the phase angle of the current. Find the average dissipation of power in the resistor.

32. An RC circuit consists of a resistor with $R = 0.80$ Ω and a capacitor with $C = 1.5 \times 10^{-4}$ F connected in series to an oscillating source of emf. The source generates a voltage $\mathscr{E} = \mathscr{E}_{max} \sin \omega t$, with $\mathscr{E}_{max} = 0.40$ V and $\omega = 9 \times 10^3$ radian/s. Find the maximum current in the circuit. Find the phase angle of the current. Find the average dissipation of power in the resistor.

33. Consider the circuit of Example 3. What is the maximum energy in this circuit at one instant of time? The minimum energy?

34. Derive an expression for the instantaneous power absorbed by the parallel LCR circuit of Example 4. Show that the time-average power is $\mathscr{E}_{max}^2/(2R)$.

*35. Consider an LC circuit (Figure 34.32) with a driving emf $\mathscr{E} = \mathscr{E}_{max} \sin \omega t$.
   (a) Show that if $\omega \gg 1/\sqrt{LC}$, the amplitude of the potential difference across the terminals in Figure 34.32 is $\Delta V \cong (X_C/X_L)\,\mathscr{E}_{max}$.
   (b) Show that ΔV is much smaller than $\mathscr{E}_{max}$. Thus, this circuit can be used as a **filter** that strongly attenuates high-frequency components in the driving emf.

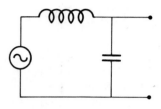

Fig. 34.32

Section 34.4

36. A transformer used to step up 110 V to 5000 V has a primary coil of 100 turns. What must be the number of turns in the secondary coil?

37. A transformer operating on a primary voltage of 110 volts AC delivers a secondary voltage of 6.0 volts AC to a small electric buzzer. If the current in the secondary circuit is 3.0 A, what is the rms current in the primary circuit? Assume that no electric power is lost in the transformer.

38. The generators of a large power plant deliver an electric power of 2000 MW at 22 kilovolts AC. For transmission, this voltage is stepped up to 400 kV by a transformer. What is the rms current delivered by the generators? What is the rms current in the transmission line? Assume that the transformer does not waste any power.

39. A power station feeds 1.0×10^8 W of electric power at 760 kV into a transmission line. Suppose that 10% of this power is lost in Joule heat in the transmission line. What percentage of the power would be lost if the power station were to feed 340 kV into the transmission line instead of 760 kV, other things being equal?

40. The largest transformer ever built handles a power of 1.50×10^9 W. This transformer is used to step down 765 kV to 345 kV. What is the rms current in the primary? What is the current in the secondary? Assume that no electric power is lost by the transformer.

SUPERCONDUCTIVITY[1]

Of all gases, helium has the lowest liquefaction temperature. All other gases had been liquified during the nineteenth century, but helium resisted the best efforts of low-temperature physicists until 1908 when it was finally liquified by H. Kammerlingh Onnes[2] at the University of Leiden. The temperature of boiling helium is 4.2 K (at atmospheric pressure).

The availability of liquid helium laid the realm of low-temperature physics open to exploration. Any material can be cooled to 4.2 K merely by immersing it in liquid helium. Furthermore, lower temperatures can be attained by pumping on the helium; this means that with a vacuum pump connected to a closed vessel containing the liquid helium, the space above the liquid is partially evacuated to a low pressure. The liquid then cools by evaporation until it reaches a lower temperature corresponding to the lower pressure (recall that the boiling point of a liquid is lowered by a reduction of pressure; see Section 20.4). By this method, temperatures slightly below 1 K can be attained.

At such extremely low temperatures, materials develop very unusual properties. Many metals and alloys become **superconductors,** i.e., their resistance to electric currents vanishes completely. Liquid helium, at a temperature of 2.2 K, becomes a **superfluid,** i.e., its internal friction disappears and it can flow without drag through very fine capillary holes. These strange properties are a manifestation of a high degree of order within the material. As we know from the Third Law of Thermodynamics, the entropy vanishes at absolute zero, i.e., the disorder caused by random thermal disturbances disappears. The material then settles into a definite microscopic state. This microscopic state is a quantum state that cannot be adequately described by classical mechanics. Thus, superconductivity and superfluidity are macroscopic manifestations of the underlying quantum behavior of matter.

---

[1] This chapter is optional.
[2] Heike Kammerlingh Onnes, 1853–1926, Dutch physicist and professor at Leiden. He was awarded the Nobel Prize in 1913 for his investigations of the properties of matter at low temperatures.

K.1 ZERO RESISTANCE

Superconductivity was discovered by Onnes in 1911 during electrical experiments with a sample of frozen mercury. Figure K.1 shows a plot of the measured values of the resistivity of mercury as a function of temperature. At a temperature of 4.15 K, the resistivity drops sharply, and below this critical temperature, the resistivity is zero.

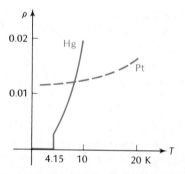

Fig. K.1 Resistivity of a sample of mercury as a function of temperature. The resistivity has been expressed as a fraction of the resistivity at 273 K. Below 4.15 K, mercury is a superconductor. For comparison, the dashed curve shows the resistivity of platinum, which is a normal conductor.

The sudden change of resistivity indicates that the material has suffered some drastic alteration of state. Experimentation reveals that in superconducting mercury, as in any metal, the carriers of electric current are free electrons. Hence, it must be that these free electrons suffer some alteration in their state. We will discuss what happens to the electrons in a later section.

Besides mercury, several other metals exhibit superconductivity at low temperatures. Table K.1 lists some superconducting metals and their transition temperatures. Furthermore, a large variety of compounds and alloys exhibit superconductivity. The highest known transition temperature is found in an alloy containing niobium and germanium (Nb_3Ge); this transition temperature is 23 K.

Table K.1 SOME SUPERCONDUCTORS

Element	T_c
Aluminum	1.20 K
Indium	3.40
Lead	7.19
Mercury	4.15
Niobium	9.26
Osmium	0.66
Tin	3.72
Tungsten	0.012
Vanadium	5.30
Zinc	0.87

A superconductor is a perfect conductor — its resistance is truly zero. Even the most precise experiments have not been able to detect any residual resistance in a superconductor. If a closed loop of superconducting wire initially has a current, then this current will keep flowing around the loop on its own accord as long as the wire is kept cold. Such a steady current that flows without any resistive loss is called a **persistent current.** In one case, a current that was started in a superconducting loop kept on flowing for two and a half years with undiminished strength, and it would probably still be flowing today if the experimenters had not run out of liquid helium for cooling their apparatus. Of course, in view of measurement errors, we cannot prove that the resistivity in superconductors is exactly zero; but if it is not zero, it is certainly extremely small — no more than 10^{-15} times the resistivity of the best normal conductors.

Persistent currents induced by changing magnetic fields bring about some spectacular levitation effects. If a small bar magnet is dropped toward a superconducting lead dish (Figure K.2), the magnetic field in-duces persistent currents along the surface of the dish; by Lenz' Law, the direction of these currents is such that their magnetic force on the magnet is repulsive. When the magnet is close to the dish, this magnetic force is large enough to support the weight of the magnet — the magnet floats forever above the lead dish.

A similar effect can be demonstrated with a superconducting lead ring and sphere (Figure K.3). The ring is stationary and it carries an initial current that has been induced by means of an external magnetic field. If now a lead sphere is dropped toward the ring, it will remain floating at some height above the ring. This is analogous to the levitation effect described in the preceding paragraph; the ring with its current and associated magnetic field plays the role of the magnet and the lead sphere plays the role of the lead dish.

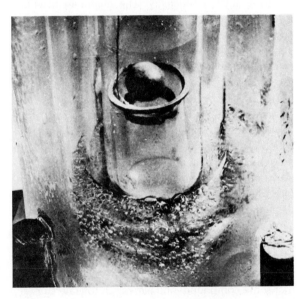

Fig. K.3 A small superconducting sphere levitated above a superconducting ring carrying a current.

Fig. K.2 A small bar magnet levitated above a superconducting dish. (Courtesy A. Leitner, Rensselaer Polytechnic Institute.)

K.2 THE CRITICAL MAGNETIC FIELD

Superconducting loops with persistent currents make magnetic fields — they are magnets. Such superconducting magnets are similar to permanent magnets in that they do not require any electric power supply to maintain their current and their magnetic field. Only an initial energy input is needed to get the persistent current started. This suggests that superconductors should permit us to produce extremely intense magnetic fields with little expenditure of energy.

Unfortunately, intense magnetic fields have an adverse effect on superconductors: intense magnetic fields destroy superconductivity. For instance, at a

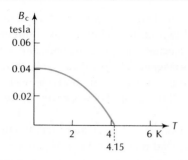

Fig. K.4 Critical magnetic field as a function of temperature for mercury.

temperature near absolute zero, a magnetic field of 0.041 T will destroy the superconductivity of mercury. At a temperature near the critical temperature (4.15 K), an even smaller magnetic field suffices to destroy the superconductivity. The minimum magnetic field that will quench the superconductivity of a material is called the **critical magnetic field,** B_c. Its strength depends on the temperature; Figure K.4 shows a plot of critical field strength for mercury as a function of temperature.

This breakdown of superconductivity imposes serious restrictions on the maximum current that can be carried by a superconductor. The current in, say, a wire will itself generate a magnetic field and, if this magnetic field is intense enough, it will cause a breakdown of the superconductivity of the wire. For example, a superconducting wire of mercury, 0.2 cm in diameter at a temperature near absolute zero, can carry a current of at most 200 A; a larger current will lead to a breakdown of the superconductivity. Such restrictions must be kept in mind in the design of superconducting magnets. We will see in Section K.4 that some compounds and alloys can tolerate substantially larger magnetic fields.

K.3 THE MEISSNER EFFECT

As we know from our study of ordinary conductors (see Chapter 28), a steady current in such a conductor requires an electric field to overcome the resistance. The electric field within a conductor carrying a given current is directly proportional to the resistance — a conductor of large resistance has a large electric field and a conductor of small resistance has a small electric field (see Ohm's Law in Section 28.2). A superconductor, with zero resistance, always has zero electric field in its interior. Furthermore, it follows from this that the rate of change of magnetic field in a superconductor must always be zero; if it were not, then the changing magnetic flux would induce an electric field, in contradiction with the requirement that the electric field remain zero. For example, if we transport a supercon-

ducting cylinder into a magnetic field (Figure K.5), it will push the magnetic field lines aside so that none of these penetrate the cylinder. What happens here is that as the cylinder touches the magnetic field, currents are induced on the surface of the cylinder and the magnetic field of these currents produces just the right deformation of the magnetic field lines to prevent their penetration into the cylinder. This behavior is characteristic of a perfect conductor; for instance, a ball of plasma exhibits the same behavior and for the same reason — a ball of plasma is an (almost) perfect conductor.

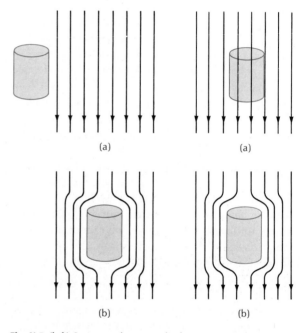

Fig. K.5 (left) Superconducting cylinder transported into a magnetic field.

Fig. K.6 (right) Conducting cylinder in a magnetic field: (a) normal metal, before T_c is reached, and (b) superconductor, after T_c is reached.

But it turns out that a superconductor is more than just a perfect conductor. A superconductor not only prevents the penetration of magnetic field lines that are initially outside of the superconducting material, but it also *expels* any magnetic field lines that are initially inside the material (before it becomes superconducting). Figure K.6 shows a cylinder of lead, above its critical temperature, placed in a magnetic field. This is an ordinary conductor and the magnetic field lines penetrate it without hindrance. However, if we cool the lead to its critical temperature, it will expel these field lines and, when the lead reaches its superconducting state, the magnetic field within it will be zero. This behavior is to be contrasted with that of a mere perfect conductor. Figure K.7 shows a ball of gas, placed in a magnetic field. If we convert this gas into a plasma (by ionizing it), the magnetic field lines are trapped (or frozen) in the plasma — when we switch the magnet off, the magnetic field lines inside the plasma remain unchanged, the induced currents in the plasma generating just enough magnetic field to keep the flux constant.

Fig. K.8 The lines with arrows show the flow of a hypothetical current within the volume of a superconductor. The closed loop is the path of integration for Ampère's Law; since some current passes through this loop, $\int \mathbf{B} \cdot d\mathbf{l}$ cannot be zero.

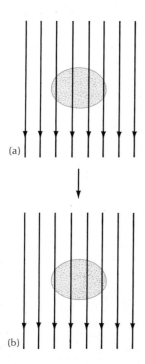

Fig. K.7 Ball of gas in a magnetic field: (a) normal gas, before ionization, and (b) plasma, after ionization.

The expulsion of magnetic flux from a metal during the transition from the normal to the superconducting state is called the **Meissner effect.** It means that a superconductor is not only a perfect conductor, but also a perfect diamagnet. (Recall that a diamagnetic material immersed in a magnetic field tends to reduce the strength of the magnetic field in its interior; see Section 33.4).

The elimination of the magnetic field from the interior of the superconductor is brought about by currents that flow along the surface of the superconductor; the magnetic field generated by these currents cancels the magnetic field generated by the external sources.

To see that the currents carried by a superconductor must always be surface currents, we note that if any currents were to flow within the volume of the superconductor, then Ampère's Law would demand a nonzero value of **B** within the volume (Figure K.8); such a nonzero value of **B** would be in contradiction to the Meissner effect. From this we can draw the general conclusion that *any* current carried by a superconductor must flow on the surface; this applies to

induced currents as well as to currents originating from other sources of emf. Although from a macroscopic point of view these superconductor currents are surface currents, from a microscopic point of view they do not flow exactly on the surface but rather in a thin layer or skin. The thickness of this layer carrying the current is typically about 10^{-5} cm. Within the surface layer the magnetic field is not quite zero; the Meissner effect is incomplete and some magnetic flux penetrates.

In Figure K.6 we illustrated the Meissner effect for a superconducting cylinder. Strictly, the ideal Meissner effect, with complete expulsion of the magnetic flux from the entire volume of the metal, occurs only if the metal has the shape of a very long cylinder (a wire) aligned with the magnetic field. For other shapes, the extent of the expulsion of the magnetic flux depends on the geometry. In general, the volume of the metal splits into domains of superconducting material and normal material. If we increase the strength of the magnetic field, the size of the normal domains increases at the expense of the superconducting domains and, when the field reaches the critical strength, the entire volume of metal becomes normal.

K.4 SUPERCONDUCTORS OF THE SECOND KIND

In most pure superconducting metals, the expulsion of magnetic flux from each of the superconducting domains in the metal is an all or nothing affair: if the metal is held at a fixed temperature and immersed in a magnetic field, it will prevent the penetration of the magnetic flux as long as the magnetic field is weaker than the critical value; but the metal will suddenly cease to be a superconductor when the magnetic field becomes stronger than the critical value and it will then freely permit the penetration of the magnetic flux.

But in niobium, in vanadium, and in alloys of other metals, the expulsion of magnetic flux from a superconducting domain is a much more complicated affair. The alloy will permit a partial penetration of flux if the magnetic field is of intermediate strength; and it will finally cease to be a superconductor and permit complete penetration of flux when the magnetic field becomes stronger. Thus, an alloy has two critical values of magnetic field strength: at a value B_{c_1} the flux begins to penetrate and at a value B_{c_2} the flux penetrates completely and superconductivity breaks down. For example, at a temperature of 4.2 K, the niobium–tin alloy Nb_3Sn has critical values $B_{c_1} = 0.019$ T and $B_{c_2} = 22$ T. The high value of B_{c_2} is of great practical importance — the alloy retains its superconductivity even in a very strong magnetic field where any pure metal would lose its superconductivity.

Materials such as Nb_3Sn that permit a partial penetration of the magnetic flux when immersed in a magnetic field of intermediate strength are called **superconductors of the second kind,** or of type II. When immersed in a magnetic field of intermediate strength, such a superconductor is in a mixed state: the bulk of the material is superconducting, but it is threaded by very thin filaments of normal material; these filaments are oriented parallel to the external magnetic field and they serve as conduits for the penetrating lines of this external magnetic field (Figure K.9). A current circulates around the perimeter of each filament; this current shields the bulk of the superconductor from the magnetic field in the filament. The flow of this current has the character of a vortex; because of this, the filaments are usually called **vortex lines.**

It turns out that the amount of flux associated with each vortex line has a fixed value related to Planck's constant and the electric charge of the electron,[3]

$$\Phi_0 = \frac{h}{2e} = 2.07 \times 10^{-15} \text{ T} \cdot \text{m}^2 \qquad (1)$$

This means that the flux is *quantized* — Φ_0 represents a **quantum of flux**[3] just as the electric charge e of an electron represents a quantum of electric charge. In a superconductor of the second kind, an increase of the strength of the external magnetic field will not cause an increase of the flux associated with each vortex line; instead it will cause an increase in the number of vortex lines threading the superconductor. The stronger the external magnetic field, the more densely will the vortex lines be packed. Figure K.10 shows the vortex lines (viewed end on) in a sample of lead–indium alloy; the vortex lines are packed in a regular triangular pattern.

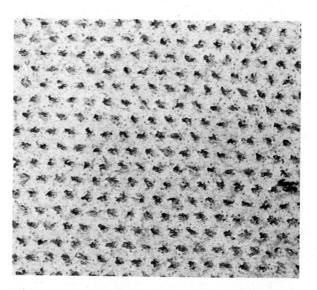

Fig. K.10 Ends of vortex lines at the surface of a sample of superconducting lead–indium. The vortex lines have been made visible by dusting with powdered iron. The separation between the vortex lines is about 0.005 cm.

K.5 THE BCS THEORY

We know from Chapter 28 that the resistivity of a metal is caused by collisions between the free electrons and the ions of the crystal lattice of the metal. The resistivity depends on temperature because the random thermal motion of the ions of the lattice increases the likelihood that an electron will suffer collisions. In fact, in a perfect lattice with no thermal vibrations an electron can travel without ever suffering any collisions. If we adopt the naive classical picture of an electron as a pointlike particle, we can see that travel without collisions is possible if the electron moves along a straight line between the rows of atoms. This classical picture of the motion of electrons has been shown to be invalid by modern quantum me-

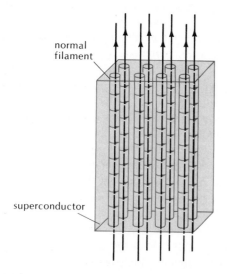

Fig. K.9 Filaments of normal material in a superconductor. The filaments serve as conduits for the magnetic field lines.

normal filament

superconductor

[3] See Appendix 8 for a more precise value of the magnetic flux quantum.

chanics, according to which electrons have wave properties (see Section 41.6); however it turns out that the quantum-mechanical picture of the motion of electron waves leads to a similar conclusion. The electron wave can travel through a perfect crystal lattice without suffering any scattering (deviation) because whatever effect is caused by one atom is canceled by the effects of other atoms. The quantum-mechanical picture indicates that in a perfect crystal lattice the electron wave can travel unhindered in *any* direction. Imperfections in the regularity of the position of the atoms of the lattice will hinder the propagation of the electron wave and give the metal a finite resistivity. Thus, the random thermal vibrations of the ions contribute to the resistivity. At low temperatures, the reduction of these thermal vibrations brings about a reduction of resistivity; at zero temperature, when the thermal vibrations disappear, the resistivity should also disappear (except for a residual contribution due to impurities and dislocations of the lattice).

It therefore came as no surprise to physicists that the experimental values of resistivity were small at low temperatures — what came as a surprise was that the resistivity of some metals vanished completely at a few degrees above absolute zero.

The details of the mechanism underlying superconductivity were finally spelled out in the **Bardeen–Cooper–Schrieffer** (BCS)[4] **theory** of superconductivity, some 50 years after the discovery of this phenomenon. The key to this mechanism is the formation of electron pairs (Cooper pairs). After many false starts, theoretical physicists recognized that the free electrons in a metal are not quite free, but they interact with one another via the lattice. The negative charge of each free electron exerts an attractive force on the positive charges of the ions of the lattice; consequently, the nearby ions contract slightly toward the electron. This slight concentration of positive charge, in turn, attracts other electrons. The net effect is that a free electron exerts a small attractive force on another free electron. Although this attractive force is too small to be of any consequence at room temperature, it is strong enough to permanently bind two electrons into a pair when the temperature is within a few degrees of absolute zero, where the thermal disturbances nearly disappear. In a superconducting metal in electrostatic equilibrium (no current), each **Cooper pair** consists of two electrons of exactly opposite momenta. Obviously such a configuration makes no sense from a classical point of view — if two particles have opposite and constant momenta, they will travel away from each

other in opposite directions; they will then cease to interact and they cannot remain bound. However, the configuration makes sense from a quantum-mechanical point of view, where each particle is described by a wave — if two waves have opposite directions of motion they can continue to overlap for a long time and they can continue to interact. Thus, the BCS theory is intrinsically quantum mechanical and the language of classical physics can only convey a crude outline of the theory.

In a superconducting metal that is carrying a current, the electron pairs have a net momentum in the direction opposite to that of the current, and they transport electric charge. The electron pairs move through the lattice without resistance because whenever the lattice scatters one of the electrons and changes its momentum, it will also scatter the other electron of the pair and change its momentum by an opposite amount. Consequently, the lattice cannot change the net momentum of a pair — it can neither slow down nor speed up the average motion of the paired electrons.

The quantum character of superconductivity shows up quite explicitly in the quantization of the current and of the magnetic flux in closed superconducting loops. Delicate experiments with small superconducting loops have shown that the current is restricted to a discrete set of values defined by the condition that the flux intercepted by the area within the loop is always a multiple of the basic quantum of flux, $\Phi_0 = 2.07 \times 10^{-15} \text{ T} \cdot \text{m}^2$. Even more spectacular quantum effects occur in a **Josephson junction**,[5] consisting of a thin layer of insulator placed between two adjacent pieces of superconductor (Figure K.11). Classically, the layer of insulator constitutes an inpenetrable barrier for electrons; but quantum-mechanically the electron waves can tunnel through this barrier. Consequently, the thin layer of insulator permits the passage of a sizable fraction of the superconducting current. The interplay of the electron waves on the two sides of the Josephson junction gives rise to some remarkable in-

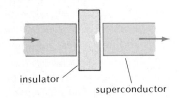

insulator

superconductor

Fig. K.11 A Josephson junction. The thickness of the layer of insulator, an oxide, is usually less than 100 Å.

[4] John Bardeen, 1908–, Leon N. Cooper, 1930–, and J. Robert Schrieffer, 1931–, American physicists. They shared the Nobel Prize in 1972 for their theory of superconductivity.

[5] Brian D. Josephson, 1940–, English physicist, and Ivar Giaever, 1929–, American physicist, shared the Nobel Prize in 1973 for the prediction of the properties of the Josephson junction and for the discovery of the tunneling of superconducting currents through insulators, respectively.

terference phenomena. For instance, if we apply a DC potential difference ΔV across the junction, the result is an AC current of frequency $2e\,\Delta V/h$. And if we apply an AC potential difference of frequency $2e\,\Delta V/h$ across the junction, the result is a combination of AC and DC currents. The measurement of the frequency of the oscillating current produced by a steady potential difference has been used for a new determination of the ratio of the fundamental constants e and h, with unprecedented precision. Such measurements also provide the most precise method for the determination of potential differences.

Other remarkable interference phenomena arise when two Josephson junctions are connected in parallel (Figure K.12), an arrangement called a **SQUID** (*superconductive quantum interference device*). The net current passing through this device depends on the magnetic flux intercepted by the area spanned by the loop: the current is zero whenever the flux is a half-integer multiple of Φ_0, and the current is maximum whenever the flux is an integer multiple of Φ_0. This sensitive dependence of the current on the magnetic flux can be exploited for extremely precise measurements of magnetic fields.

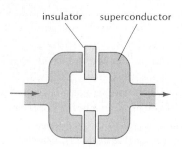

Fig. K.12 Two Josephson junctions connected in parallel form a SQUID.

K.6 TECHNOLOGICAL APPLICATIONS

Practical applications of superconductivity have focused on the development of electromagnets with superconductive coils capable of generating intense magnetic fields. Conventional electromagnets, with copper windings and iron cores, are among the least energy-efficient devices produced by our technology. They consume enormous amounts of electric power merely to keep a constant current running around in circles. Except for the magnetic energy initially stored in the electromagnet at startup, all of the electric power is wasted as heat. By contrast, a magnet with superconducting coils requires no electric power at all to keep running. Of course, the superconducting magnet must be cooled to a temperature near absolute zero and maintained at this temperature. The initial

Fig. K.13 A high-field superconducting magnet. This magnet produces a field of 15 T in a bore of 7 cm.

cooling and the continual removal of any heat that leaks into the magnet from the environment require refrigeration and a supply of liquid helium.

Figure K.13 shows a superconducting magnet capable of generating a magnetic field of 15 T; a magnetic field of this strength is very difficult to attain with a conventional magnet. In Figure K.14 we see a dramatic

Fig. K.14 Nails scattered on a board align with the field lines of the intense magnetic field generated by two superconducting magnets. The coils of these magnets are immersed in stainless-steel dewars full of liquid helium.

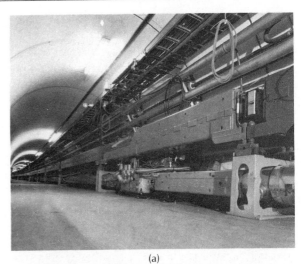

(a)

Fig. K.15 The coils of this large superconducting magnet for a bubble chamber at Argonne National Laboratory are made of niobium–titanium alloy embedded in copper.

(b)

Fig. K.16 (a) The tunnel of the main accelerator at Fermilab. The upper ring consists of ordinary magnets; the lower ring consists of superconducting magnets. (b) One of the superconducting magnets at Fermilab.

display of the strength of the magnetic fields of superconducting magnets: thousands of iron nails align along the field lines connecting two such magnets.

Figure K.15 shows the coils of a large superconducting magnet for a bubble chamber at the Argonne National Laboratory. The coils are 4.8 m across and they produce a magnetic field of 1.8 T. After the initial current has been established in this magnet, the only power required is 190 kW for a liquid-helium refrigerator; in contrast, a conventional magnet of similar size would require about 10,000 kW of electric power. The construction costs for this superconducting magnet are about the same as for a conventional magnet of similar size, but the yearly running costs are lower by a factor of 10.

Superconducting magnets have recently been installed at the Fermilab accelerator. This machine was originally built with conventional magnets, but in order to make the machine capable of handling particles of higher energy, new superconducting magnets producing more intense magnetic fields were required. Fermilab developed 7-m-long supermagnets producing a magnetic field of up to 4.5 T. Figure K.16b shows one of these magnets. The complete accelerator ring, 6.2 km in circumference, has 774 of these magnets plus 240 extra magnets used for focusing the beam of high-energy particles. The ring of supermagnets is placed just below the ring of ordinary magnets (Figure K.16a). Particles are first accelerated in the ordinary ring and then fed into the super ring, where they reach a final energy of 10^6 MeV.

The construction of superconducting coils must overcome some tricky manufacturing problems. The superconducting material commonly employed in these coils is either an alloy of niobium with titanium (NbTi) or else the compound Nb_3Sn. The latter mate-

rial can withstand the highest current densities and magnetic fields (up to 2×10^5 A/cm^2 and up to 15 T), but it is extremely brittle and cannot be drawn into wires to be wound around a magnet core. Instead, it is necessary to first wind a coil with separate strands of niobium wrapped in tin and then heat the entire coil to fuse these materials into the Nb_3Sn compound.

Figure K.17 shows a cross section through one strand of a superconducting cable of NbTi alloy. The alloy is in the form of very thin filaments embedded in a copper matrix. A bundle of thin filaments has more surface area than a single thick wire of the same cross section as the bundle; since the supercurrents always run on the surface of the superconductor, the large area of the filaments permits the flow of a large current. The complete cable consists of several strands of NbTi alloy and several wires of copper wound around

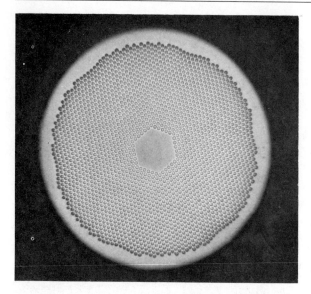

Fig. K.17 This strand of a superconducting cable consists of 2100 thin filaments of NbTi alloy embedded in a copper matrix. The diameter of this strand is about 0.7 mm.

each other. As long as the NbTi alloy is in its superconducting state, the copper plays no role; in fact, relative to the superconductor, the copper behaves as an insulator and all the current flows in the superconductor. But if the superconductor should ever fail because of inadequate cooling or excessive magnetic fields, the current can continue to flow in the parallel wires of copper. This protects the coils of the magnet from the explosive conversion of magnetic energy into heat that would occur if a large current (10^5 A or more) were suddenly halted by a large resistance.

Whereas electromagnets are among the least efficient of our technological devices, electric generators and electric motors are among the most efficient. Typically, an ordinary electric generator will convert close to 99% of the supplied mechanical power into electric power. Electric generators with superconducting coils could probably be even more efficient, but their main advantage is their smaller size and smaller weight — for a given power output, the weight of a superconducting generator can be 10 times less than that of a conventional generator. Several prototypes of superconducting generators and motors have been built (Figure K.18). Superconducting generators may find application in nuclear power plants because conventional generators capable of handling the very large power output of these plants must be of huge size and they therefore face serious structural problems because of the weight of the moving parts. Superconducting motors may also find applications in marine propulsion where, again, weight is a problem.

The transmission and storage of electric power are another important application of superconductors. Resistive losses in conventional transmission lines are of the order of 10% of the total power transmitted. A superconducting cable, with zero resistance, would of course eliminate the loss. However, from a financial point of view, the elimination of the resistive loss makes only a minor contribution to the thriftiness of superconducting cables; what makes a major contribution is their compact size — large amounts of power

Fig. K.18 Rotor (center) of an experimental superconducting electric generator (right), capable of delivering 20,000 kW.

can be carried by a single cable of fairly small dimensions that can be installed in an underground tunnel not requiring the expensive strips of real estate on which conventional transmission lines are erected. Figure K.19 shows an experimental version of a superconducting transmission line; the outside diameter of the pipe enclosing the cable is 40 cm and it can carry a power of 1000 MW, roughly the total power of a nuclear generating station.

Incidentally, an AC current in a superconductor of the second kind is not entirely free of resistive losses. The trouble is that the periodic oscillation of the current produces a similar oscillation of the strength of the magnetic field and consequently lines of magnetic flux (vortex lines) move periodically in and out of the superconductor. This motion of the vortex lines involves some "friction" and it generates some heat. To avoid this loss, the AC current must be carried by a superconductor of the first kind, without magnetic flux in its interior.

On conventional transmission lines, the electric power is carried at very high voltage to minimize the resistive loss. In a superconducting cable, high voltage is not necessary — one can dispense with the awkward step-up and step-down transformers at the generating plant and at the consumer station. Furthermore, if transformation is not required, it becomes advantageous to transmit DC power rather than AC power. A superconducting cable for DC power is cheaper than a cable for AC power; it can be made of a superconductor of the second kind which will tolerate higher current densities and, furthermore, the design can be kept simpler.

Another possible application of superconductivity is the storage of electric energy. Because of daily fluctuations in the demand for electric power by a city, it is desirable to hold large amounts of electric energy temporarily in some kind of storage system. The power can be fed into the storage system when the demand is low and released when the demand is high. Table K.2 compares the energy density in several systems. It is obvious from this table that magnetic fields constitute one of the most concentrated forms of stored energy.

Fig. K.19 Superconducting power transmission line at Brookhaven National Laboratory.

In order to store 100 MW · h, the volume of the magnetic field (at 10 T) would have to be about 10^4 m³. A large toroid about 5 m thick and 100 m across would provide this volume. The windings of the toroid would of course have to be made of superconductor. The toroid would have to withstand enormous forces due to the magnetic interaction of the currents. And it would be exposed to a severe danger: if, by some failure, one part of the winding became resistive, the sudden generation of heat would probably cause the entire toroid to become resistive and the sudden conversion of all of the magnetic energy into heat would then melt the toroid and blow it apart with great violence.

As a final example of an application of superconductivity, we briefly consider magnetic suspension of trains. In Section K.1 we saw that a permanent magnet will remain suspended over a superconducting sheet — the former induces currents in the latter and the magnetic force on the currents is repulsive. If we replace the permanent magnet by a superconducting coil with a current (a superconducting magnet), we will get the same repulsion effect. Taking advantage of this, we could attach superconducting coils to the underside of a train and levitate it on superconducting rails. The suspension is frictionless — the train sits on a cushion of magnetic fields and slides without friction. In practice, this ideal arrangement is not feasible be-

Table K.2 ENERGY STORAGE

System	Energy density
Magnetic fields, 10 T	11 kW · h/m³
Electric field, 10^5 kV/m	0.01
Water in reservoir, height 100 m	0.27
Compressed air, 50 atm	5
Hot water, 100°C	18[a]

[a] This number takes into account that the thermodynamic conversion efficiency for this heat into mechanical energy is at most 20%.

cause it would be much too expensive to lay down miles and miles of superconducting rails. However, it is also possible to obtain levitation with rails made of an ordinary conductor. In this case, there is no repulsive force between the coils and the rails when they are at rest, but there is a force when they are in motion relative to one another because the moving magnetic field induces currents in the rails. These currents suffer some resistive loss, but the effective friction would be reasonably small. Engineers estimate that at speeds in excess of 200 km/h, a train riding on superconductive coils would be much safer than a train on iron wheels.

Further Reading

The Quest for Absolute Zero by K. Mendelssohn (Halsted Press, New York, 1977) is a splendid account of the exciting discoveries in low-temperature physics; it includes a chapter on superconductivity. *Near Zero* by D. K. C. MacDonald (Doubleday, Garden City, 1961) is a much briefer and very elementary introduction. *Cryophysics* by K. Mendelssohn (Interscience Publishers, New York, 1960) is an advanced textbook that gives a concise survey of low-temperature physics; although the book is aimed a senior-level students, many sections are accessible to lower-level students.

Superconductivity by A. W. B. Taylor (Wykeham, London, 1970) is a concise introductory survey with a minimum of mathematics. *Superconductivity* by D. Schoenberg (Cambridge University Press, Cambridge, 1965) is a somewhat more advanced survey; written long before the days of the BCS theory, it is somewhat outdated, but its classic description of the phenomenological properties of superconductors remains useful.

The following magazine articles deal with applications of superconductivity:

"Large-scale Applications of Superconductivity," B. B. Schwartz and S. Foner, *Physics Today*, July 1977
"Superconductors in Electric-Power Technology," T. H. Geballe and J. K. Hulm, *Scientific American*, November 1980
"Superconducting Electronics," D. G. McDonald, *Physics Today*, February, 1981
"The Road to Superconducting Materials," J. K. Hulm, J. E. Kunzler, and B. T. Matthias, *Physics Today*, January 1981
"The Superconducting Computer," J. Matisoo, *Scientific American*, May, 1980

Questions

1. Why would you expect helium to be more difficult to liquify than any other gas?

2. The graph in Figure 28.5 gives the impression that the resistance of tin is constant for $T > 3.72$ K. Is this true?

3. Is Ohm's Law valid for a superconductor?

4. Is it reasonable to regard superconductors as another state of matter? If so, how many states of matter are there?

5. Is a ring of superconductor with a persistent current a perpetual motion machine, i.e., does it violate the law of conservation of energy?

6. Suppose that a closed ring of superconducting wire is initially outside of a magnetic field. If we push the ring into the magnetic field, the magnetic flux through the ring remains zero. Explain.

7. Figure K.6b shows a superconducting cylinder that has expelled the magnetic field by the Meissner effect. Describe the currents that must be flowing along the surface of this cylinder.

8. Suppose that a piece of metal immersed in a magnetic field is initially a normal conductor, and is then cooled so that it becomes a superconductor. To expel the magnetic field by the Meissner effect, the metal must acquire electric currents. Where does the energy for these currents come from?

9. If the resistance of a superconductor were small but not zero, could the Meissner effect be sustained for a long time?

10. The levitated magnet shown in Figure K.2 was placed near the superconducting dish *after* the dish became superconducting. Consequently, does Figure K.2 demonstrate infinite conductivity, or the Meissner effect, or both?

11. (a) If you push a ring of superconductor into a magnetic field, a current will be induced along the ring; if you then remove the ring from the magnetic field, the current disappears. Explain. (b) If you place a ring of normal conductor in a magnetic field, then cool it so it becomes a superconductor, and then remove it from the magnetic field, the ring will be left with a persistent current. Explain.

12. Suppose that a very large flat metallic plate is placed between the poles of an electromagnet. If this plate is cooled so that it becomes superconducting, it will not expel the magnetic field lines. Why not?

13. The equation describing the graph in Figure K.4 is $B_c(T) = (1 - T^2/T_c^2)B_c(0)$. At what temperature is $B_c(T) = \frac{1}{2}B_c(0)$?

14. The existence of a critical magnetic field implies the existence of a critical current. For a cylindrical wire of radius r, the critical current is given by the equation $I_c = 2\pi r B_c/\mu_0$, known as the Silsbee rule. Explain why the critical current is proportional to the radius of the wire.

15. Would you expect that a noncrystalline material, such as glass, could exhibit superconductivity?

16. Consider a pair of electrons moving around a circular superconducting ring carrying a current. According to the discussion in Section 41.4, the orbital angular momentum of the electrons is quantized. Explain how this leads to the quantization of the electric current and the magnetic flux associated with this current.

17. A superconducting cable carrying a large current and a copper cable are connected in parallel. Is it really true that the copper cable carries no current at all? (Hint: What is the potential difference from one end of the superconducting cable to the other?)

CHAPTER 35

The Displacement Current and Maxwell's Equations

James Clerk Maxwell, *1831–1879, Scottish physicist. He was professor at King's College, London, and later professor at Cambridge, where he supervised the construction of the Cavendish Laboratory. Maxwell at first attempted to explain the behavior of electric and magnetic fields in terms of a complicated mechanical model, according to which all of space was filled with an elastic medium, or ether, consisting of a multitude of small rotating vortex cells. But ultimately Maxwell discarded the mechanical model and treated the electric and magnetic fields as physical entities that exist in their own right, without any need for an underlying medium. He published his electromagnetic theory in 1873 in his celebrated* Treatise on Electricity and Magnetism. *In this book he laid down a complete and consistent set of laws for all electromagnetic phenomena, and thereby accomplished for electricity and magnetism what Newton had accomplished for mechanics.*

We already know that a changing magnetic field will induce an electric field. In this chapter we will discover that the converse is also true: a changing electric field will induce a magnetic field. The law describing this induction effect of electric fields was formulated by James Clerk Maxwell, who thereby achieved a wide-ranging unification of the laws of electricity and magnetism. The next four chapters of this book are nothing but applications of these basic laws.

The mutual induction of electric and magnetic fields gives rise to the phenomenon of self-supporting electromagnetic oscillation in empty space. If, initially, there exists an oscillating electric field, it will induce a magnetic field, and this will induce a new electric field, and this will induce a new magnetic field, etc. Thus, these fields can perpetuate each other. Of course, an oscillating charge or current is needed to get the fields started, but after this initiation the fields continue on their own. These self-supporting oscillations are **electromagnetic waves,** either traveling waves or standing waves. We will examine such electromagnetic waves in the space within a conducting cavity and in the space surrounding an accelerated electric charge.

35.1 The Displacement Current

Ampère's Law asserts that the integral of **B** around any closed path is related to the current intercepted by an arbitrary surface that spans this path:

$$\oint \mathbf{B} \cdot d\mathbf{l} = \mu_0 I \qquad (1)$$

If all the currents flow along closed, continuous circuits, then the shape of the surface used to intercept the current is irrelevant. For example, Figure 35.1 shows a circular path of integration spanned by a baglike curved surface with a circular mouth. This baglike surface intercepts the same current as a flat circular surface spanning the mouth — both surfaces give the same result in Eq. (1).

However, if the wire carrying the current is interrupted by a capacitor, then Ampère's Law develops a flaw. Suppose that a small capacitor is connected to long, straight wires as in Figure 35.2. Suppose further that the wires carry a constant current I which steadily charges up the capacitor. (This of course requires that the voltage of the source of emf driving the current steadily increases so as to allow for the steadily increasing voltage on the capacitor.) The distribution of current on the interrupted wire of Figure 35.2 differs from that on the continuous wire of Figure 35.1 only in that a short piece of current is missing (between the capacitor plates). Except in the immediate vicinity of the capacitor, the magnetic field in Figure 35.2 will therefore be about the same as in Figure 35.1. Let us now apply Ampère's Law to the baglike surface shown in Figure 35.2. This surface passes between the capacitor plates and it therefore intercepts *no current*. Hence the right side of Eq. (1) is zero, but the left side is obviously not zero — Ampère's Law fails.

The question is then, can we modify Ampère's Law so that it gives the right answer even when the surface passes between capacitor plates? To discover the required modification, we note that although the surface does not intercept current in the region between the capacitor plates, it does intercept *electric flux* (Figure 35.3). The positive plate of the capacitor is a source of electric field lines and the negative plate is a sink; field lines going from one plate to the other cross the baglike surface. This suggests that the required correction to Ampère's Law involves the electric flux Φ.

It is not hard to see that to achieve this correction for our baglike surface, the right side of Eq. (1) should be replaced by $\mu_0 \varepsilon_0 \, d\Phi/dt$,

$$\oint \mathbf{B} \cdot d\mathbf{l} = \mu_0 \varepsilon_0 \frac{d\Phi}{dt} \qquad (2)$$

To check that this works, let us calculate the amount of flux. If the electric charge on the positive plate is Q, then Q/ε_0 lines of field start on this plate. All these lines cross the baglike surface, i.e., the flux is

$$\Phi = \frac{Q}{\varepsilon_0} \qquad (3)$$

The rate of change of the flux is then

$$\frac{d\Phi}{dt} = \frac{1}{\varepsilon_0} \frac{dQ}{dt} = \frac{1}{\varepsilon_0} I \qquad (4)$$

Hence the term $\mu_0 \varepsilon_0 \, d\Phi/dt$ on the left side of Eq. (2) equals

$$\mu_0 \varepsilon_0 \frac{d\Phi}{dt} = \mu_0 \varepsilon_0 \frac{1}{\varepsilon_0} I = \mu_0 I \qquad (5)$$

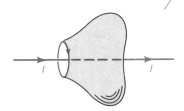

Fig. 35.1 A circular path (color) is spanned by a baglike mathematical surface. The wire carrying the current enters the "mouth" of the bag and is intercepted by the bottom of the bag. Alternatively, the circular path can be spanned by a flat circular surface. The same current will then be intercepted by this flat circular surface.

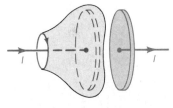

Fig. 35.2 A similar baglike surface that passes between the plates of a capacitor. The baglike surface intercepts no current.

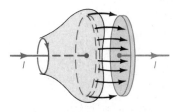

Fig. 35.3 The baglike surface intercepts electric flux.

This shows that the calculation with the baglike surface passing between the plates and intercepting the electric flux [Eq. (2)] gives the same result as a calculation with some other surface cutting across the wire and intercepting the current [Eq. (1)].

It can of course also happen that electric flux and current are present at the same place; for example, if the space between the capacitor plates is filled with some slightly conducting material (a leaky capacitor), then a surface passing between the plates will intercept both some electric flux and some current. In such a case, this electric flux and this current must be combined,

Maxwell's modification of Ampère's Law

$$\oint \mathbf{B} \cdot d\mathbf{l} = \mu_0 I + \mu_0 \varepsilon_0 \frac{d\Phi}{dt} \qquad (6)$$

where it is understood that I is the current intercepted by whatever surface is used in the calculation [thus, the current I in Eq. (6) is not necessarily the same as the current on the wires]. Equation (6) is Maxwell's modification of Ampère's Law. The great importance of this equation lies in its general validity — Maxwell boldly proposed that this equation is valid not only for a capacitor, but also for any arbitrary system of electric fields, currents, and magnetic fields. Incidentally, the sign of the electric flux in Eq. (6) is fixed by the following right-hand rule: wrap the fingers around the path of integration, in the direction of integration; the flux is to be reckoned as positive if the electric field is in the direction of the thumb.

The quantity of $\varepsilon_0 \, d\Phi/dt$ is called the **displacement current,**

Displacement current

$$I_{\rm d} = \varepsilon_0 \frac{d\Phi}{dt} \qquad (7)$$

This quantity is of course not a true current, but in Eq. (6) it has the same effect as though it were:

$$\oint \mathbf{B} \cdot d\mathbf{l} = \mu_0(I + I_{\rm d}) \qquad (8)$$

Note that the displacement current between the capacitor plates has the same direction as the true current arriving at the plates; the displacement current is the continuation of the real current "by other means."

Apart from the term $\mu_0 I$, the Maxwell–Ampère Law of Eq. (6) is analogous to the Faraday Law of Eq. (32.12). The latter relates the path integral of the electric field to the rate of change of magnetic flux. The former relates the path integral of the magnetic field to the rate of change of the electric flux. Hence these laws indicate a certain symmetry in the effects of electric and magnetic fields on one another.

Just as a time-dependent magnetic field can induce an electric field, a time-dependent electric field can induce a magnetic field. The following example is quite analogous to Example 32.3.

EXAMPLE 1.   A parallel-plate capacitor has circular plates of radius R. The capacitor is connected to a source of alternating emf so that the electric field between the plates oscillates according to

$$E = E_0 \sin \omega t \tag{9}$$

where E_0 and ω are constants. What is the induced magnetic field inside and outside the capacitor? Assume that the electric field is uniform inside the capacitor; neglect the fringing of the field at the edges.

SOLUTION: Inside the capacitor ($r < R$), the electric field is uniform. Figure 35.4 shows this electric field. To find the induced magnetic field, take a circle of radius r. The flux through this circle is $\pi r^2 E$ and the rate of change of this flux is $\pi r^2 \, dE/dt$. According to Eq. (6), we then have

$$\oint \mathbf{B} \cdot d\mathbf{l} = \mu_0 \varepsilon_0 \pi r^2 \, dE/dt \tag{10}$$

The field lines of **B** must be closed concentric circles, because this is the only pattern of field lines consistent with the rotational symmetry of the electric field shown in Figure 35.4. Since the path of integration coincides with a field line, we can immediately evaluate the left side of Eq. (10),

$$\oint \mathbf{B} \cdot d\mathbf{l} = 2\pi r B \tag{11}$$

Combining Eq. (10) and (11), we then obtain

$$2\pi r B = \mu_0 \varepsilon_0 \pi r^2 \frac{dE}{dt} \tag{12}$$

or

$$B = \frac{\mu_0 \varepsilon_0}{2} r \frac{dE}{dt} \tag{13}$$

$$= \frac{\mu_0 \varepsilon_0}{2} r \omega E_0 \cos \omega t \tag{14}$$

Outside the capacitor ($r > R$), the electric field is zero. Hence the electric flux through a circle of radius r is $\pi R^2 E$. We can then obtain the magnetic field by going through the same steps as above, with the result

$$B = \frac{\mu_0 \varepsilon_0}{2} \frac{R^2}{r} \frac{dE}{dt} \tag{15}$$

$$= \frac{\mu_0 \varepsilon_0}{2} \frac{R^2}{r} \omega E_0 \cos \omega t \tag{16}$$

The formulas (14) and (16) are obviously analogous to (32.20) and (32.21).

Figure 35.5 shows the magnetic field lines. The direction of these lines is related to that of the current flowing toward the capacitor plates by the usual right-hand rule. Note that the magnetic field given by Eq. (16) outside the capacitor is exactly that of a long, straight wire ($B \propto 1/r$). As far as this magnetic field is concerned, the removal of a short piece of wire and its replacement by a capacitor makes no difference at all.[1]

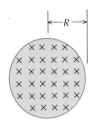

Fig. 35.4 The electric field between the capacitor plates. The crosses show the tails of the electric field vectors.

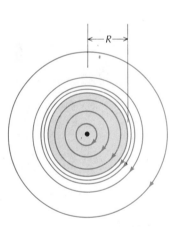

Fig. 35.5 The magnetic field between the capacitor plates.

---

[1] However, a more precise calculation, taking into account the fringing of the electric field at the edges of the capacitor plates, shows that near the plates the magnetic field differs slightly from that of a straight wire.

35.2 Maxwell's Equations

Equation (6) is the last of the fundamental laws that we need to completely describe the behavior of electric and magnetic fields. There are four fundamental laws, as follows[2]:

Gauss' Law for electricity [Eq. (24.6)]

$$\oint \mathbf{E} \cdot d\mathbf{S} = Q/\varepsilon_0 \tag{17}$$

Gauss' Law for magnetism [Eq. (30.17)]

$$\oint \mathbf{B} \cdot d\mathbf{S} = 0 \tag{18}$$

Faraday's Law [Eq. (32.12)]

$$\oint \mathbf{E} \cdot d\mathbf{l} = -\frac{d\Phi_B}{dt} \tag{19}$$

Maxwell–Ampère's Law [Eq. (6)]

$$\oint \mathbf{B} \cdot d\mathbf{l} = \mu_0 I + \mu_0\varepsilon_0\frac{d\Phi}{dt} \tag{20}$$

Maxwell's equations   Taken as a whole, these laws are known as **Maxwell's equations,** because Maxwell added the missing link between the magnetic and the electric fields [Eq. (20)], and thereby brought electromagnetic theory to completion. Maxwell recognized that these equations imply a dynamic interplay between electric and magnetic fields, an interplay that couples and unifies electric and magnetic phenomena.

The physical basis for each of these four equations may be briefly summarized as follows. Gauss' Law for electricity is based on Coulomb's Law describing the forces of attraction and repulsion between stationary charges. Gauss' Law for magnetism asserts that there exist no magnetic monopoles. Faraday's Law describes the induction of an electric field by motion or by a changing magnetic field. And finally, Maxwell–Ampère's Law is based on the law of magnetic force between moving charges and it also contains the induction of a magnetic field by a changing electric field.

The above equations must be supplemented by the equation for the **Lorentz force** on a point charge:

---

[2] These equations are valid for vacuum. The equations in the presence of dielectric or magnetic materials must be modified either by implicitly including the bound charges of the dielectric in Q and the currents of the magnetic material in I, or else by inserting the dielectric constant and the relative permeability into the appropriate terms of the equations.

$$\boxed{\mathbf{F} = q\mathbf{E} + q\mathbf{v} \times \mathbf{B}}$$ (21)   *Lorentz force*

This expression is based on the definition of **E** and **B** [see Eqs. (23.15) and (30.12)].

Maxwell's equations give a complete description of the interactions among charges, currents, electric fields, and magnetic fields. All the properties of the fields can be deduced by mathematical manipulation of these equations. If the distribution of charges and currents is known, then these equations uniquely determine the corresponding fields. Even more important, Maxwell's equations uniquely determine the time evolution of the fields, starting from a given initial condition of these fields. Thus, these equations accomplish for the dynamics of electromagnetic fields what Newton's equations of motion accomplish for the dynamics of particles.

Calculations with Maxwell's equations are often best done by converting the integral equations (17)–(20) into *differential equations;* however, the latter involve partial derivatives and for this reason we will not attempt to use them here.

Although the empirical foundation on which we based the development of Maxwell's equations was restricted to charges at rest or charges in uniform motion, these equations also govern the fields of accelerated charges and the fields of light and radio waves. In the remaining sections of this chapter and in the next chapters we will calculate these fields from our equations and we will see that the results are in agreement with the observed properties of light and radio waves.

35.3 Cavity Oscillations

As we saw in Example 1, the time-dependent electric field in a parallel-plate capacitor will induce a magnetic field. This implies that the capacitor has some self-inductance, since the changing, time-dependent magnetic field will, in turn, induce an electric field, i.e., it will induce an emf. We therefore expect that a parallel-plate capacitor placed in a circuit with a source of alternating emf will behave very much like the driven LC circuit of Section 34.3. The following is a calculation of the oscillations of a parallel-plate capacitor.[3] However, instead of treating this system as an LC circuit with some effective inductance, we will directly examine the electric and magnetic fields between the plates and their dependence on position and time.

Figure 35.6 shows a parallel-plate capacitor with circular plates connected to a source of alternating emf of frequency ω. The electric field between the plates will then oscillate at the same frequency. As a first approximation, we ignore the induction effects within the capacitor. The electric field between the plates is then uniform[4] and parallel to the axis. We use the symbol $E_{(1)}$ for this field and we write

$$E_{(1)} = E_0 \sin \omega t$$ (22)

where E_0 is a constant.

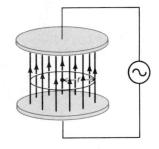

Fig. 35.6 Electric field lines in a parallel-plate capacitor connected to a source of alternating emf. The fringing of the electric field lines has been ignored.

---

[3] Based on *The Feynman Lectures on Physics* by R. P. Feynman, R. B. Leighton, and M. Sands.

[4] As always, we disregard the fringing effects at the edges of the capacitor.

According to Example 1, such a time-dependent electric field induces a magnetic field [see Eq. (14)]

$$B_{(1)} = \frac{\mu_0 \varepsilon_0}{2} \, r\omega E_0 \cos \omega t \qquad (23)$$

The magnetic field lines are closed circles (Figure 35.7).

This time-dependent magnetic field in turn induces an extra electric field, which will have to be added to the field $E_{(1)}$ of Eq. (22). The direction of the extra field is again parallel to the axis. To find an expression for this field, let us apply Faraday's Law to the path shown in Figure 35.7. If we use the symbol $\mathbf{E}_{(2)}$ for the induced electric field, then

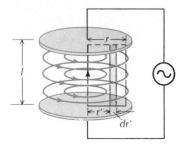

Fig. 35.7 Magnetic field lines. The path of integration is a rectangle of height *l* and width *r*. The direction of integration is clockwise.

$$\int \mathbf{E}_{(2)} \cdot d\mathbf{l} = -\frac{d\Phi_{B_{(1)}}}{dt} \qquad (24)$$

The integral on the left side does not receive any contribution from the horizontal segments (Figure 35.7), since there the field is perpendicular to the path. Furthermore, we will suppose that $E_{(2)} = 0$ along the axis of the capacitor,[5] so that the integral receives no contribution from the vertical segment along the axis (Figure 35.7). Only the other vertical segment remains. If the latter is at a distance r from the axis, the left side of Eq. (24) becomes

$$-E_{(2)} \, (r) \, l \qquad (25)$$

where l is the distance between the plates (the minus sign is needed because we reckon the electric field as positive if it is directed upward, whereas the segment of path is directed downward).

To evaluate the right side of Eq. (24), we begin by calculating the flux $\Phi_{B_{(1)}}$. Consider a strip of height l and width dr'; the flux through this strip is $B_{(1)}l \, dr'$ and the flux through all such strips is the integral

$$\Phi_{B_{(1)}} = \int_0^r B_{(1)} \, l \, dr' \qquad (26)$$

$$= \int_0^r \left(\frac{\mu_0 \varepsilon_0}{2} \, r' \, \omega E_0 \cos \omega t \right) l \, dr'$$

$$= \frac{\mu_0 \varepsilon_0}{2} \, l\omega E_0 \cos \omega t \int_0^r r' \, dr'$$

$$= \frac{\mu_0 \varepsilon_0}{4} \, lr^2 \, \omega E_0 \cos \omega t \qquad (27)$$

Hence

$$\frac{d\Phi_{B_{(1)}}}{dt} = -\frac{\mu_0 \varepsilon_0}{4} \, lr^2 \, \omega^2 E_0 \sin \omega t \qquad (28)$$

---

[5] The choice $E_{(2)} = 0$ along the axis is a matter of definition. It means that we regard $E_{(1)}$ of Eq. (22) as the *exact* electric field along the axis of the capacitor, while we regard the deviations from $E_{(1)}$ occurring at all other points as induced corrections which remain to be calculated.

Upon inserting Eqs. (25) and (28) into Eq. (24), we immediately find the induced electric field

$$E_{(2)} = -\frac{\mu_0 \varepsilon_0}{4} r^2 \omega^2 E_0 \sin \omega t \qquad (29)$$

Thus, the electric field $E_{(1)}$ must be corrected by adding to it the electric field $E_{(2)}$:

$$E = E_0 \sin \omega t - \frac{\mu_0 \varepsilon_0}{4} r^2 \omega^2 E_0 \sin \omega t + \cdots$$

$$= \left(1 - \frac{\mu_0 \varepsilon_0}{4} \omega^2 r^2\right) E_0 \sin \omega t + \cdots \qquad (30)$$

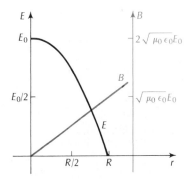

Fig. 35.8 Electric field (black) and magnetic field (color) in a parallel-plate capacitor as a function of radius.

But the correction to the electric field implies a corresponding correction to the induced magnetic field, and this implies a further correction to the electric field, etc. This calculation therefore leads to an infinite sequence of successive approximations for the fields. The result takes the form of a power series in ascending powers of the radial variable r. The two terms in parentheses in Eq. (30) are merely the first two terms of the power series; the other terms have been indicated by the three dots in Eq. (30). For a start let us ignore all the extra corrections and use Eq. (30) as it stands. One interesting feature of this solution is that the electric field vanishes at a radius R such that

$$1 - \frac{\mu_0 \varepsilon_0}{4} \omega^2 R^2 = 0 \qquad (31)$$

i.e.,

$$R = \frac{2}{\sqrt{\mu_0 \varepsilon_0}} \frac{1}{\omega} \qquad (32)$$

A more exact calculation taking into account the extra terms in Eq. (30) shows that the precise radius at which the electric field vanishes is somewhat larger,

$$R = \frac{2.405}{\sqrt{\mu_0 \varepsilon_0}} \frac{1}{\omega} \qquad (33)$$

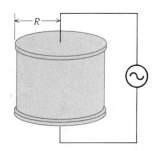

Fig. 35.9 A conducting cylindrical wall of radius R closes off the sides of the parallel-plate capacitor.

Figure 35.8 is a plot of the radial variation of the electric and magnetic fields.

The above solution for the fields between capacitor plates also gives us the solution for another problem, one of much greater practical importance: the problem of the electromagnetic oscillations of a closed conducting cavity. We can convert our pair of capacitor plates into a closed cavity by inserting a conducting cylindrical wall of radius R between the plates (Figure 35.9). The cylindrical wall does not interfere with the electric field of Eq. (30) because this electric field is zero anyhow at the radius R. Thus our solution for the fields between capacitor plates is also the solution for the fields in such a cavity. Note that here the choice of the correct radius is crucial: if the cylindrical wall had a

Cavity oscillations

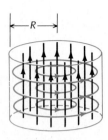

Fig. 35.10 A closed cylindrical can with the same radius R, and the electric (black) and magnetic (color) field lines in its interior. This diagram does not show the time dependence of the fields (both fields reverse direction every half cycle; when the electric field is maximum, the magnetic field is zero and conversely).

radius different from R, then the electric field would not "fit" this cylinder.

Figure 35.10 shows a closed cylindrical can and the lines of electric and magnetic field within it. The external circuit with its driving emf has been omitted in this figure. The currents can flow from one horizontal end face of the cylinder to the other via the vertical wall; the external wire of Figure 35.6 is therefore not needed. And the external source of emf is not needed either — once the oscillations of the cavity have been started, they are self-sustaining. If charge is initially placed on the end faces,[6] then the cavity begins to oscillate spontaneously just as the LC circuit of Section 34.2 begins to oscillate when charge is initially placed on the capacitor. And, just as in the LC circuit, the energy sloshes back and forth between potential energy (in electric fields) and "kinetic" energy (in magnetic fields).

The electromagnetic oscillations in a conducting cavity are analogous to the acoustic vibrations in a closed organ pipe. Equation (33) determines the resonant frequency of a cavity of given radius. This is analogous to Eq. (16.6), which determines the resonant frequency of an organ pipe. The electromagnetic oscillations in a cavity can be regarded as a standing wave (we will discuss electromagnetic waves in the next chapter) and Eq. (33) can then be regarded as the condition for a proper fit of the wave in the cavity. We know from the study of sound that an organ pipe has several possible harmonic standing waves of different resonant frequencies. It turns out that this is also true for an oscillating elecromagnetic cavity, but in order to find the higher harmonics and their frequencies, we would have to calculate the higher-order terms in Eq. (30).

Electromagnetic cavities have many practical applications in the technology of microwaves, that is, high-frequency radio and radar waves. For instance, one of the devices commonly used to generate high-frequency waves is the **klystron**, consisting of a cavity in which electromagnetic oscillations are excited by an incident beam of electrons (Figures 35.11 and 35.12). The excitation of electromagnetic oscillations in a cavity by a beam of electrons is analogous to the excitation of sound oscillations in an organ pipe by a stream of air. The

Fig. 35.11 One of the 245 large klystrons in the gallery at the Stanford Linear Accelerator Center. These klystrons generate the electromagnetic waves used in the accelerator, which is buried 25 ft below the floor.

---

[6] The charge must be placed on the *interior* of the end faces. If the charge were placed on the exterior, it would not produce any field inside the cavity.

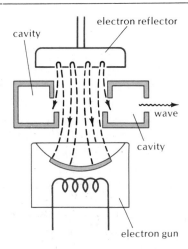

Fig. 35.12 Schematic diagram of a reflex klystron. An electron beam (dashed lines) emerges from an electron gun and moves upward, passing by the open gaps in the walls of the cavity; it is then reflected by an electric field and moves downward, again passing by the gaps. The electric fields of this beam excite oscillations in the cavity.

electrons give up their kinetic energy to the electric and magnetic fields and they thereby increase the strength of the oscillations. As we mentioned above, the oscillations can be regarded as standing waves. A small opening on the side of the klystron (Figure 35.12) permits some of these waves to spill out into space in the form of a traveling radio wave.

Incidentally, in practice the radio waves are not released into space directly from the generator, but are first guided into a dishlike antenna (a radar antenna) mounted at some suitable height. The waves are carried from the generator to the antenna by a hollow conducting tube or **waveguide** (Figure 35.13). Such a tube is nothing but a cavity with two open ends. The mathematical analysis of the electromagnetic oscillations in a waveguide is quite similar to that in a closed cavity — the electric and magnetic fields induce each other just as in the above calculation. However, these oscillations take the form of traveling waves rather than that of standing waves. This can be readily understood in terms of the analogy with sound waves. We know that a standing sound wave in, say, a closed organ pipe consists of two traveling waves of opposite directions (see Section 15.6). If we suddenly open the two ends of such an organ pipe, the standing wave will separate into its traveling parts, each spilling out of one end of the organ pipe.

Waveguides for microwaves are usually copper pipes of rectangular cross section. Figure 35.14 shows the pattern of field lines of a traveling wave in such a waveguide.

Fig. 35.13 A piece of waveguide of rectangular cross section. The flanges at the ends are used for the accurate joining of adjacent pieces.

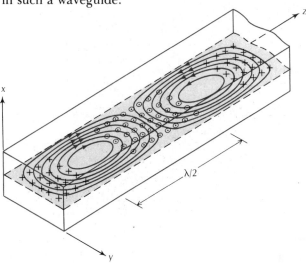

Fig. 35.14 Electric (black) and magnetic (color) field lines in a waveguide. The wave is traveling to the right. The dots and the crosses are the tips and the tails of the electric field lines, respectively. The closed loops are the magnetic field lines.

35.4 The Electric Field of an Accelerated Charge

We started our study of electromagnetic theory with the electric field of a charge at rest — this is the Coulomb field given by Eq. (23.16). Later, we set down the fields of a charge in motion with uniform velocity — in addition to the Coulomb field, such a charge has a magnetic field given by Eq. (30.15). Now we will investigate the fields of a charge with *accelerated motion*. We will find that in this case there are extra electric and magnetic fields that spread outward from the position of the charge, like ripples on a pond in which a stone has been dropped, and carry away energy and momentum. These fields are called **radiation fields.**

We can derive a formula for the electric field surrounding an accelerated charge by the following argument. Suppose that the charge is initially at rest, then is quickly accelerated for some short time interval τ, and then continues to move with a constant final velocity. We can summarize the motion thus:

$$t < 0 \qquad \text{charge is at rest}$$

$$0 \le t < \tau \qquad \text{charge has acceleration } a$$

$$t \ge \tau \qquad \text{charge moves at constant velocity } v = a\tau \qquad (34)$$

The initial electric field lines originate at the initial position of the charge. But the field lines at some later time $(t > \tau)$ must originate on the new position of the charge. The field lines cannot change from their initial configuration to their final configuration instantaneously; rather, a disturbance must travel outward from the position of the charge and gradually change the field lines. The disturbance takes the form of a kink connecting the old and the new field lines.

The disturbance travels at some speed c (we will prove below that the speed of the disturbance is actually the speed of light; our notation anticipates this result). Figure 35.15 shows the situation at some later time. The disturbance begins at the time 0 when the acceleration begins. The leading edge of the disturbance (outer edge of kink)

Fig. 35.15 Electric field lines of a charge that has suffered an acceleration. Here Q is the present position of the charge; P is the initial position of the charge. Between P and P' the charge suffered a constant acceleration. Between P' and Q the charge moved at constant velocity. The outer dashed sphere (outer edge of kink) has radius ct and is centered on P; the inner dashed sphere (inner edge of kink) has radius $c(t - \tau)$ and is centered on P'.

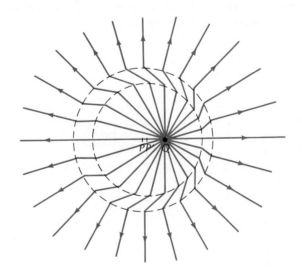

travels outward from the initial position of the charge and in a time t it reaches out to a distance ct. Beyond the sphere of radius ct, the electric field is still the old field with field lines centered on the initial position of the charge.

The disturbance ceases as soon as the acceleration ceases. The field in the vicinity of the uniformly moving charge then settles into the new radial configuration centered on the new position of the charge. The trailing edge of the disturbance (inner edge of kink) marking the cessation of the acceleration travels outward from the position that the charge has at the time τ, and by the time t it reaches out to a distance $c(t - \tau)$. Within the sphere of radius $c(t - \tau)$ the electric field is the new radial field of the uniformly moving charge.

The disturbance produced by the accelerated charge is confined to the space between the larger and smaller spheres in Figure 35.15. The field lines in this zone must connect the lines of the new field of the uniformly moving charge with the lines of the old field of the stationary charge. We have drawn the connecting line segments in Figure 35.15 as straight segments. Strictly, this is only an approximation based on the rationale that a curve can be approximated by a straight line. (A careful argument shows that the segments are actually slightly curved, but the curvature is unimportant if the speed of the charge is much smaller than the speed of light, $v \ll c$).

In the zone of the kink, the electric field has both a radial component and a tangential, or **transverse,** component. This transverse component is characteristic of the field of an accelerated charge.

Figure 35.16 shows one electric field line in detail. This figure relies on the assumption that a long time has elapsed since the acceleration interval, that is, $t \gg \tau$. Consequently, on the scale of this diagram, the distance that the charge has covered while under acceleration is very small compared to the distance that the charge has covered at uniform velocity. The inner and outer circles of Figure 35.16 are then nearly concentric and the distance from the initial position of the charge to the position at time t is nearly vt.[7] Figure 35.16 also contains the implicit assumption that the final speed of the charge is much less than the speed of light, that is, $v \ll c$. (As we remarked above, if this assumption does not hold, then the field line in the zone of the kink is appreciably curved and the straight-line approximation of Figure 35.16 is not satisfactory.)

By inspection of Figure 35.16, we see that the ratio of the transverse and radial components of the electric field in the kink is

$$\frac{E_\theta}{E_r} = \frac{vt \sin \theta}{c\tau} = \frac{at \sin \theta}{c} \tag{35}$$

where, of course, $a = v/\tau$. The radial electric field equals the number of lines passing through a unit area on a sphere; this number is not affected by the angle that the lines make with the radial direction. Thus, E_r is simply the usual Coulomb field:

$$E_r = \frac{1}{4\pi\varepsilon_0} \frac{q}{r^2} \tag{36}$$

Inserting this in Eq. (35), we obtain

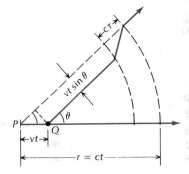

Fig. 35.16 One of the electric field lines (color). This line makes an angle θ with the direction of motion of the charge. Here we have neglected the small distance PP' (see Figure 35.15) that the charge covers while under acceleration; with this approximation, the distance PQ is $\sim vt$.

---

[7] The exact distance is $\frac{1}{2}v\tau + v(t - \tau)$ which, for $t \gg \tau$, indeed reduces to $\sim vt$.

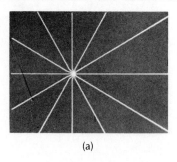

(a)

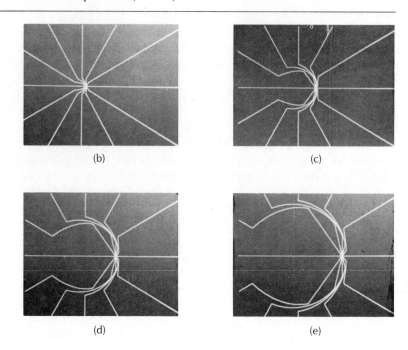

(b)

(c)

(d)

(e)

Fig. 35.17 Time-dependent electric field lines for a charge that suddenly accelerates and then moves with uniform velocity. (a) The charge is at rest. (b) The charge has begun to accelerate. (c)–(e) The charge moves with uniform velocity. Note that the field lines in the region of the kink are curved; this is due to the very high speed of the motion.

$$E_\theta = \frac{1}{4\pi\varepsilon_0} \frac{q}{r^2} \frac{at \sin\theta}{c} \tag{37}$$

Since $t \gg \tau$, the radii of the inner and outer circles in Figure 35.16 are nearly the same, that is, the difference $\Delta r = c\tau$ between them is very small compared to the radius $r = ct$. We can therefore substitute this value of the radius in Eq. (37) so that

Radiation field of accelerated charge

$$E_\theta = \frac{1}{4\pi\varepsilon_0} \frac{qa \sin\theta}{c^2 r} \tag{38}$$

This transverse electric field is the **radiation field** of the accelerated charge. Note that it is directly proportional to the acceleration a. Also note that it varies as the inverse distance, not the inverse square of the distance. The radiation field therefore decreases less sharply with distance than the Coulomb field — it remains significant at large distances where the Coulomb field practically disappears.

The radiation field produced by an acceleration a acting at time 0 reaches the distance r at a time r/c after time 0, i.e., the field needs time to propagate from the position of the charge to a remote point. Mathematically, we can express this retardation by writing the acceleration as a function of time:

$$E_\theta(t, r) = \frac{1}{4\pi\varepsilon_0} \frac{q \sin\theta}{c^2 r} a\left(t - \frac{r}{c}\right) \tag{39}$$

This formula indicates that the value of the function E_θ at time t is related to the value of the function a at the earlier time $t - r/c$. For example, if $r = 3.8 \times 10^8$ m (the distance from the Earth to the Moon), then $r/c = 3.8 \times 10^8$ m$/(3.0 \times 10^8$ m/s$) = 1.3$ s. Thus the radiation field that an accelerated charge on the Earth produces at the Moon at time t depends on the acceleration at time $t - 1.3$ s, i.e., it depends on

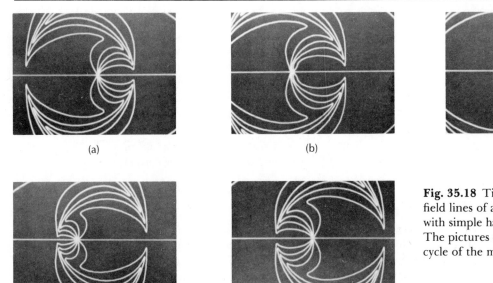

(a) (b) (c)

(d) (e)

Fig. 35.18 Time-dependent field lines of a charge moving with simple harmonic motion. The pictures display one-half cycle of the motion.

the acceleration at a time 1.3 s earlier, the radiation field needing this much time to propagate from the charge to the Moon.

If the acceleration lasts only a short time, as in the special case shown in Figure 35.15, then the electric radiation field at any point also lasts only a short time. The radiation field described by Eq. (39) then behaves like a single wave pulse that propagates outward from the place at which the acceleration occurred. Figure 35.17 shows a sequence of snapshots from a filmed computer simulation of the electric field lines of an accelerated charge, whose acceleration only lasted a short time. These snapshots give a clear impression of the outward propagation of the kinks in the field lines.

The formula (39) for the radiation field is perfectly general and applies even if the acceleration is not constant and lasts for any length of time. For instance, if the charge oscillates back and forth with simple harmonic motion of frequency ω so that

$$a = a_0 \sin \omega t \tag{40}$$

then

$$E_\theta(t,r) = \frac{1}{4\pi\varepsilon_0} \frac{q \sin\theta}{c^2 r} a_0 \sin\omega\left(t - \frac{r}{c}\right) \tag{41}$$

Radiation field of oscillating charge

This represents a (spherical) harmonic wave of frequency ω traveling in the radial direction with a speed c. The amplitude of the wave gradually decreases ($\propto 1/r$) as the wave spreads out. As we will see in the next chapter, the function (41) describes a radio wave emitted from an antenna or a light wave emitted from an atom.

Figure 35.18 shows a sequence of snapshots from a computer simulation of the electric field lines of a charge oscillating back and forth with simple harmonic motion, as indicated by Eq. (40). The folds in the field lines can be seen to propagate outward, forming a regular succession of wave peaks and wave troughs.

EXAMPLE 2.   In a head-on collision with an atom, a fast-moving electron suffers a deceleration of 4.0×10^{23} m/s². What is the electric radiation field generated by the electron at a distance of 10 cm at right angles to the direction of motion? Pretend that classical mechanics can be applied to this problem.

SOLUTION: With $\theta = 90°$, Eq. (38) gives

$$E_\theta = \frac{1}{4\pi\varepsilon_0} \frac{1.6 \times 10^{-19}\ \text{C} \times 4.0 \times 10^{23}\ \text{m/s}^2}{(3.0 \times 10^8\ \text{m/s})^2 \times 0.10\ \text{m}}$$

$$= 6.4 \times 10^{-2}\ \text{V/m}$$

Bremsstrahlung

Radiation emitted by the collision of a beam of fast-moving electrons with the atoms in a target (block of metal) is called **Bremsstrahlung**.[8] This is essentially the mechanism by means of which X rays are produced in an X-ray tube (see Figure C.1).

35.5 The Magnetic Field of an Accelerated Charge[9]

When a charge accelerates, the disturbance traveling outward consists not only of electric fields, but also of magnetic fields. The initial magnetic field of a charge at rest is zero while the final magnetic field of a moving charge is that given by Eq. (30.15). Hence a disturbance must move outward and gradually change the magnetic field from its initial to its final value.

The magnetic field can be regarded as an induced field obeying the Maxwell–Ampère Law

$$\oint \mathbf{B} \cdot d\mathbf{l} = \mu_0\varepsilon_0 \frac{d\Phi}{dt} \tag{42}$$

By symmetry, the magnetic field lines are circles about the direction of motion of the charge. Figure 35.19 shows the magnetic field lines in the zone of the kink. Note that the magnetic field is everywhere perpendicular to the electric field. Let us evaluate Eq. (42) for a circular path that follows one of the magnetic field lines (Figure 35.20). The radius of the circle is $r \sin\theta$ and hence the left side of Eq. (42) is

$$\oint \mathbf{B} \cdot d\mathbf{l} = 2\pi r(\sin\theta)B \tag{43}$$

In the evaluation of the right side of Eq. (42) we need to take into account two contributions to the flux: from the radial electric field E_r and from the transverse electric field E_θ. It turns out that the flux produced by the radial electric field induces the familiar magnetic field dependent on velocity, i.e., the magnetic field given by Eq. (30.15). We will not present the detailed calculation establishing this result; instead we leave this calculation as a problem (see Problem 29). The flux produced by the transverse electric field induces a magnetic field dependent on the acceleration. Since in the context of this section we are

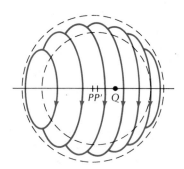

Fig. 35.19 Magnetic field lines in the zone of the kink. These lines are circles around the line of motion. At any point, the magnetic field is perpendicular to the electric field shown in Figure 35.15.

---

[8] German for "braking radiation."

[9] This section is optional.

only interested in the magnetic field characteristic of accelerated motion, we will only take into account the contribution of the flux from this transverse electric field E_θ.

Taking a flat circular area for the evaluation of the flux, we see that this area intercepts the transverse electric field E_θ only in a narrow angular region of radius $r \sin \theta$ and width $c\tau/\sin \theta$ (see Figure 35.20). The area of this region is

$$[\text{area}] \cong [\text{circumference}] \times [\text{width}] = 2\pi r \sin \theta \times \frac{c\tau}{\sin \theta} = 2\pi r c\tau \quad (44)$$

The transverse field E_θ makes an angle of θ with this area and hence the flux is

$$\Delta\Phi = E_\theta \sin \theta \times [\text{area}] = 2\pi r c\tau E_\theta \sin \theta \quad (45)$$

In a time τ, all of this flux disappears since the kink moves beyond the annular area. Hence

$$\frac{d\Phi}{dt} = \frac{\Delta\Phi}{\tau} = 2\pi r c E_\theta \sin \theta \quad (46)$$

Using Eqs. (43) and (46) in Eq. (42), we obtain

$$B = c\mu_0\varepsilon_0 E_\theta \quad (47)$$

This equation expresses the magnetic field of the pulse of radiation in terms of the transverse electric field, the speed of propagation c, and the familiar constants ε_0 and μ_0. As we have pointed out above, the speed of propagation of the pulse coincides with the speed of light,

$$c = 3.00 \times 10^8 \text{ m/s} \quad (48)$$

However, we can do better than this. In what follows we will derive a theoretical expression for the speed of propagation and will confirm that the theoretical prediction agrees with the experimental value given by Eq. (48).

The preceding calculation regarded the magnetic field **B** as induced by $\mathbf{E}_\theta$. But the converse is also true: the electric field $\mathbf{E}_\theta$ is induced by the magnetic field **B** according to Faraday's Law

$$\oint \mathbf{E} \cdot d\mathbf{l} = -\frac{d\Phi_B}{dt} \quad (49)$$

To extract some useful information from this law, it is convenient to evaluate the integral for the path shown in Figure 35.21; this path consists of two semicircles joined by short radial segments. The transverse electric field is only different from zero along the larger semicircle. Hence the left side of Eq. (49) is

$$\oint \mathbf{E} \cdot d\mathbf{l} = \int_{P_1}^{P_2} E_\theta \, dl \quad (50)$$

where the points P_1 and P_2 are shown in Figure 35.21. The magnetic

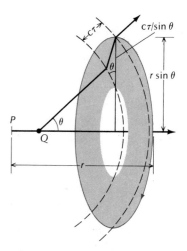

Fig. 35.20 The magnetic field line (color) is a circle of radius $r \sin \theta$ perpendicular to the plane of the page. The area bounded by this field line is a circle of the same radius perpendicular to the plane of the page. This circle intercepts the transverse electric field in the annular region marked with color.

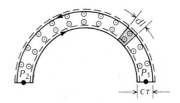

Fig. 35.21 The path of integration for Eq. (49) is marked with a solid line. As in the other figures, the boundaries of the zone of the kink are marked with dashed lines. The larger semicircle of the path is in the zone of the kink, but the smaller semicircle is not. The magnetic field is perpendicular to the plane of the page; the colored dots show the tips of the magnetic field vectors.

flux on the right side can be calculated by taking small area elements of the shape shown in Figure 35.21 with dimensions $c\tau \times dl$:

$$\Delta\Phi_B = \int \mathbf{B} \cdot d\mathbf{S} = \int Bc\tau\, dl = c\tau \int_{P_1}^{P_2} B\, dl \tag{51}$$

In a time τ, the two dashed circles of Figure 35.21, with the magnetic field between them, sweep beyond the fixed path shown in this figure. Hence

$$\frac{d\Phi_B}{dt} = \frac{-\Delta\Phi_B}{\tau} = -c \int_{P_1}^{P_2} B\, dl \tag{52}$$

Faraday's Law now becomes

$$\int_{P_1}^{P_2} E_\theta\, dl = c \int_{P_1}^{P_2} B\, dl$$

In view of Eq. (47), this can be written

$$\int_{P_1}^{P_2} E_\theta\, dl = c^2\mu_0\varepsilon_0 \int_{P_1}^{P_2} E_\theta\, dl \tag{53}$$

Obviously, the consistency of this equation requires that

$$1 = c^2\mu_0\varepsilon_0 \tag{54}$$

or

Speed of electromagnetic wave

$$\boxed{c = \frac{1}{\sqrt{\mu_0\varepsilon_0}}} \tag{55}$$

We have therefore derived a theoretical expression for the speed of propagation of an electromagnetic disturbance in vacuum. The numerical value of the right side of Eq. (55) is

$$c = 1/(1.26 \times 10^{-6} \text{ H/m} \times 8.85 \times 10^{-12} \text{ F/m})^{1/2}$$

$$= 3.00 \times 10^8 \text{ m/s} \tag{56}$$

We see that this agrees with the experimental value (48) for the speed of light (in section 36.1 we will make a more precise comparison of the theoretical and experimental values of the speed of light). The derivation of Eq. (55) was one of the great and early triumphs of the electromagnetic theory formulated by Maxwell in his Eqs. (17)–(20).

By means of Eq. (55), we can somewhat simplify the relation between B and E_θ. We have

$$B = c\mu_0\varepsilon_0 E_\theta = \frac{c}{c^2} E_\theta \tag{57}$$

or

$$B = \frac{E_\theta}{c} \qquad (58)$$

EXAMPLE 3. In Example 2 we found that the sudden deceleration of an electron during a collision produces a transverse electric field of 6.4×10^{-2} V/m at a certain distance. What is the magnetic field produced by the deceleration?

SOLUTION: According to Eq. (58),

$$B = E_\theta/c$$

$$= (6.4 \times 10^{-2} \text{ V/m})/(3.0 \times 10^8 \text{ m/s})$$

$$= 2.1 \times 10^{-10} \text{ T}$$

SUMMARY

Displacement current: $I_d = \varepsilon_0 \dfrac{d\Phi}{dt}$

Maxwell's equations: $\oint \mathbf{E} \cdot d\mathbf{S} = Q/\varepsilon_0$

$$\oint \mathbf{B} \cdot d\mathbf{S} = 0$$

$$\oint \mathbf{E} \cdot d\mathbf{l} = -\frac{d\Phi_B}{dt}$$

$$\oint \mathbf{B} \cdot d\mathbf{l} = \mu_0 I + \mu_0 \varepsilon_0 \frac{d\Phi}{dt}$$

Transverse electric field of accelerated charge:

$$E_\theta = \frac{1}{4\pi\varepsilon_0} \frac{qa \sin \theta}{c^2 r}$$

Speed of electromagnetic wave: $c = 1/\sqrt{\mu_0 \varepsilon_0}$

Transverse magnetic field: $B = \dfrac{E_\theta}{c}$

QUESTIONS

1. The displacement current between the plates of a capacitor has the same magnitude as the conduction current in the wires connected to the capacitor, and yet the magnetic field produced by the former current near and within the capacitor is much weaker than that produced by the latter current near and within the wire. Explain.

2. Consider the electric field of a single positive electric charge moving at constant velocity. What is the direction of the displacement current intercepted by a circular area perpendicular to the velocity in front of the charge? Behind the charge?

3. A capacitor is connected to an alternating source of emf. Does the displacement current between the capacitor plates lead or lag the emf?

4. Which of Maxwell's equations permits us to deduce the electric Coulomb field of a static charge? Which of Maxwell's equations permits us to deduce the magnetic field of a charge moving with uniform velocity?

5. Suppose that there exist magnetic monopoles, i.e., postive and negative magnetic charges that act as sources and sinks of magnetic field lines, analogous to positive and negative electric charges. Which of Maxwell's equations would have to be modified to take into account such monopoles? Qualitatively, what are the required modifications?[10]

6. Given Eq. (30) for the electric field between the capacitor plates described in Section 35.3, is there a well-defined potential difference between the plates?

7. Some microwave ovens have rotating turntables that continually turn the meat while cooking. What is the purpose of this arrangement? (Hint: Microwave ovens are cavities with standing electromagnetic waves.)

8. Cavities are capable of oscillating at much higher frequencies than ordinary LC circuits. What limits the maximum frequency attainable with LC circuits?

9. In practice, the electromagnetic oscillations in a cavity are damped by "frictional" losses. How does this "friction" arise?

10. Efficient waveguides are manufactured out of a very good conductor, such as copper, sometimes with a silver lining. Why is high conductivity essential for high efficiency?

11. As stated in Section 35.5, the magnetic field of a charge moving at uniform velocity can be regarded as an induced magnetic field produced by the changing electric flux in the space surrounding the charge. Check that the direction of the induced magnetic field agrees with the direction given by Eq. (30.15).

12. Since the magnetic field of a charge moving at uniform velocity can be regarded as an induced magnetic field produced by the changing electric flux, we might expect that the electric field (Coulomb field) can be regarded as an induced electric field produced by a changing magnetic flux. Is this the case?

13. Is the transverse electric radiation field E_θ a conservative field?

14. In Figure 35.16 the electric field lines have a sharp discontinuity in slope. Is this discontinuity unphysical? On what is this discontinuity to be blamed?

15. Roughly sketch the electric field lines for a positive charge that is initially moving at uniform velocity and suddenly stops.

16. In a collision with an atom, an electron suddenly stops. Describe the directions of the electric and magnetic radiation fields at some distance from the electron at right angles to the acceleration.

17. Given that the magnitudes of the magnetic and electric radiation fields of an accelerated charge are related by $B = E_\theta/c$, compare the energy densities in these fields.

18. Consider a small angular patch (Figure 35.22) in the radiation field of the accelerated charge described in Section 35.4. The volume of this small patch is [area] × [thickness] = $r^2 \, d\theta \, d\alpha \times c\tau$ and the energy in this volume is $\frac{1}{2}\varepsilon_0 E_\theta^2 \times (r^2 \, d\theta \, d\alpha \times c\tau)$. Show that this energy remains constant as the pulse of radiation propagates outward.

19. A charged particle moving around a circular orbit at uniform speed has a centripetal acceleration and therefore produces a radiation field. However, a uniform current flowing around a circular loop does *not* produce a radiation field. Is this a contradiction? Explain.

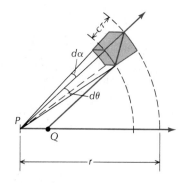

Fig. 35.22

---

[10] For a quantitative discussion of these modifications, see Problem 10.

PROBLEMS

Section 35.1

1. A parallel-plate capacitor is being charged by a current of 4.0 A.
   (a) What is the displacement current between its plates?
   (b) What is the rate of change of the electric flux intercepted by each plate?

2. A parallel-plate capacitor consists of circular plates of radius 0.30 m separated by a distance of 0.20 cm. The voltage applied to the capacitor is made to increase at a steady rate of 2.0×10^3 V/s. Assume that the electric charge distributes itself uniformly over the plates and ignore the fringing effects.
   (a) What is the rate of increase of the electric field between the plates?
   (b) What is the magnetic field between the plates at a radius of 0.15 m? At 0.30 m?

3. A parallel-plate capacitor has circular plates of radius 20 cm and a uniform electric field between the plates. The capacitor is being charged at a rate of 0.10 A.
   (a) What is the net displacement current between the plates?
   (b) What is the displacement current between the plates within the radial interval $0 \le r \le 5$ cm?

4. The space between the plates of a leaky capacitor is filled with a material of resistance 5.0×10^5 Ω. The capacitor has a capacitance of 2.0×10^{-6} F, its plates are circular, with a radius of 30 cm, and its electric field is uniform. At time $t = 0$, the initial voltage across the capacitor is zero.
   (a) What is the displacement current if we increase the voltage at the steady rate of 1.0×10^3 V/s?
   (b) At what time will the real current leaking through the capacitor equal the displacement current?
   (c) What is the magnitude of the magnetic field between the plates at radius $r = 20$ cm at $t = 0$? At $t = 1$ s? At $t = 2.0$ s?

5. A parallel-plate capacitor with circular plates of radius 25 cm separated by a distance of 0.15 cm is connected to a source of alternating emf. The voltage across the plates oscillates with an amplitude of 5000 V and a frequency of 60 Hz. What is the amplitude of the magnetic field between the plates at a distance of 20 cm from the axis of the capacitor?

6. A parallel-plate capacitor has circular plates of area A separated by a distance d. A thin straight wire of length d lies along the axis of the capacitor and connects the two plates (Figure 35.23); this wire has a resistance R. The exterior terminals of the plates are connected to a source of alternating emf with a voltage $V = V_0 \sin \omega t$.
   (a) What is the current in the thin wire?
   (b) What is the displacement current through the capacitor?
   (c) What is the current arriving at the outside terminals of the capacitor?
   (d) What is the magnetic field between the capacitor plates at a distance r from the axis? Assume that r is less than the radius of the plates.

7. Suppose that the parallel-plate capacitor discussed in Section 35.1 (see Figure 35.2) is filled with a slab of dielectric with a dielectric constant κ.
   (a) By retracing the arguments of Section 35.1, show that the displacement current must be

$$I_d = \kappa \varepsilon_0 \frac{d\Phi}{dt}$$

and Maxwell's modification of Ampère's Law must be

$$\oint \mathbf{B} \cdot d\mathbf{l} = \mu_0 I + \kappa \mu_0 \varepsilon_0 \frac{d\Phi}{dt}$$

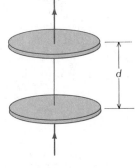

Fig. 35.23 Parallel-plate capacitor with a thin wire connecting the inside faces of the plates.

(b) If the parallel-plate capacitor described in Problem 1 is filled with a dielectric with $\kappa = 2.0$, how do the answers to that problem change?

Section 35.2

8. Prove that Maxwell's equations [Eqs. (17)–(20)] mathematically imply the conservation of electric charge, i.e., prove that if no electric current flows into or out of a given volume, then the electric charge within this volume remains constant. [Hint: Begin with Eq. (17), $Q = \varepsilon_0 \Phi$, and evaluate dQ/dt by means of Eq. (20); note that if the baglike surface used for the evaluation of Eq. (20) is actually a closed surface (a bag whose mouth has been shrunk to zero), then the path integral of **B** is zero.]

9. Write Maxwell's equations for a medium of dielectric constant κ and relative permeability κ_m.

*10. Suppose that there exist magnetic monopoles, i.e., positive and negative magnetic charges that act as sources and sinks of magnetic field lines analogous to positive and negative electric charges. The magnetic field generated by a magnetic charge q_m is an inverse-square field, $B = q_m/4\pi r^2$. Write a new set of Maxwell's equations that take into account the magnetic charge. Be careful with the constants ε_0 and μ_0, and the signs.

Section 35.3

11. An oscillating emf of amplitude 2.0 V is applied between the centers of the plates of a parallel-plate capacitor with large circular plates. The frequency of the emf is 1.5×10^{10} radian/s and the distance between the plates is 1.0 cm. Using the approximate equations derived in Section 35.3, find the amplitude of oscillation of the electric and magnetic fields at the center of the capacitor and also at a radial distance of 2.0 cm from the center.

12. (a) Find the magnetic field $B_{(2)}$ that the electric field $E_{(2)}$ of Eq. (29) will induce.
    (b) By adding $B_{(1)}$ and $B_{(2)}$, find the corrected magnetic field, $B = B_{(1)} + B_{(2)}$.
    (c) What is the value of the corrected magnetic field at the radius R given by Eq. (32)?

*13. Consider the electric and magnetic fields given by Eq. (30) and (23).
    (a) Find the energy density as a function of radius in each of these fields.
    (b) By integrating the energy density, find the electric and magnetic energy between the capacitor plates within a radius $R = 2/(\sqrt{\mu_0\varepsilon_0}\omega)$.
    (c) Plot the electric energy and the magnetic energy as a function of time. Is the sum of these energies constant?

*14. If oscillating electric and magnetic fields of the kind described by Eqs. (30) and (23) exist in a closed cylindrical cavity, currents must periodically flow back and forth between the top and the bottom of the cylinder.
    (a) Calculate the instantaneous amount of electric charge on the top and bottom faces of a cylinder of radius $R = 2/(\sqrt{\mu_0\varepsilon_0}\omega)$ if the electric field in the cylinder is given by Eq. (30).
    (b) Calculate the current that must flow between the top and bottom faces.

Section 35.4

15. Consider the electron whose transverse electric and magnetic fields we calculated in Examples 2 and 3.
    (a) For comparison, find the Coulomb field of this electron at the given point.
    (b) Assuming that the electron has a velocity of 2.0×10^7 m/s, find the magnetic field, according to Eq. (30.15), of this electron at the given point.

16. In a collision with an atom, an electron suffers a deceleration of 2.0×10^{24} m/s². What is the magnitude of the electric radiation field that this electron generates at a distance of 20 cm at an angle of 45° to the direction of

the deceleration? At what time after the instant of collision does this radiation field arrive at that distance?

17. In an X-ray tube, a beam of high-speed electrons is made to impact on a block of metal. The sudden deceleration of the electrons brings about the emission of intense electromagnetic radiation (X rays). Suppose that an electron of initial energy 2×10^4 eV decelerates uniformly and comes to rest within a distance of 5×10^{-9} m. What will be the magnitude of the electric radiation field at a distance of 0.3 m from the point of impact in a direction perpendicular to the direction of the acceleration?

18. An electron of energy 2.0×10^4 eV is in a circular orbit under the influence of a constant magnetic field of 5.0×10^{-2} T.
   (a) Consider a point of the orbit. What are the magnitude and direction of the instantaneous acceleration of the electron at this point?
   (b) Consider a point at a distance of 1.5 m from the electron in a direction perpendicular to that of the instantaneous acceleration. What is the transverse electric field due to the acceleration that reaches this point (after some suitable time delay)?

19. Repeat Problem 18, substituting a proton for the electron. By what factor does the transverse electric field generated by the proton differ from that generated by the electron?

20. Two protons of energy 1.5×10^5 eV collide head on.
   (a) What is the distance of closest approach? What is the instantaneous acceleration at this point?
   (b) What is the transverse electric field that each proton produces at a distance of 3.0×10^{-10} m from the point of collision in a direction perpendicular to the line of motion of the protons?
   (c) What is the net transverse electric field of both protons taken together?

21. On a radio antenna (a straight piece of wire), electrons move back and forth in unison. Suppose that the velocity of the electrons is $v = v_0 \cos \omega t$, where $v_0 = 8.0 \times 10^{-3}$ m/s and $\omega = 6.0 \times 10^6$ radian/s.
   (a) What is the maximum acceleration of one of these electrons?
   (b) Corresponding to this maximum acceleration, what is the strength of the transverse electric field produced by one electron at a distance of 1.0 km from the antenna in a direction perpendicular to the antenna? What is the time delay (or retardation) between the instant of maximum acceleration and the instant at which the corresponding electric field reaches a distance of 1.0 km?
   (c) There are 2.0×10^{24} electrons on the antenna. What is the collective electric field produced by all the electrons acting together? Assume that the antenna is sufficiently small so that all the electrons contribute just about the same electric field at a distance of 1.0 km.

22. The accelerated electrons in the antenna of a radio station produce an electric field of 1.0 V/m at a distance of 1.0 km from the radio station. The antenna is a straight, vertical wire made of copper, 5.0 m long with a diameter of 2.0×10^{-3} m. Assume that all the free electrons in the copper contribute equally to the field. What must be the average acceleration of each electron?

23. Consider an electron in a hydrogen atom orbiting the nucleus in a circular orbit of radius 0.53×10^{-10} m under the influence of the electrostatic force of attraction. Calculate the acceleration of this electron and then calculate the magnitude of the electric radiation field that, according to Eq. (39), the electron produces at a distance of 10 cm from the atom measured along a line tangent to the circular orbit at the position of the electron. (The result of this classical calculation does not agree with the actual behavior of the radiation emitted by the electron; quantum effects play a crucial role in this problem.)

Section 35.5

24. In a Van de Graaff accelerator, a proton is given an acceleration of

1.1×10^{14} m/s².

(a) Find the strengths of the transverse electric and magnetic fields at a distance of 0.50 m from the proton at an angle of 45° with the acceleration.

(b) Draw a diagram showing the direction of the acceleration, and the direction of the electric and magnetic fields calculated in part (a).

25. Show that the energy density in the transverse electric field of an accelerated charge equals the energy density in the corresponding magnetic field.

26. A thorium nucleus emits an alpha particle and thereby transforms itself into a radium nucleus. Assume that the alpha particle is pointlike and that the residual radium nucleus is spherical with a radius of 7.4×10^{-15} m. The charge on the alpha particle is $2e$ and that on the radium nucleus is $88e$.

(a) Calculate the acceleration of the alpha particle at the instant it leaves the surface of the nucleus.

(b) Calculate the magnitudes of the electric and magnetic radiation fields that, according to Eqs. (39) and (58), the alpha particle produces at a distance of 0.50×10^{-10} m from the nucleus along a direction perpendicular to the direction of the acceleration. (This classical calculation does not agree with the actual behavior of the radiation emitted by the alpha particle; quantum effects are important in this problem.)

27. An AC current $I = I_0 \cos 2\pi\nu t$, with $I_0 = 15.0$ A and $\nu = 60$ Hz, flows in a copper wire of diameter 0.26 cm.

(a) Find formulas for the average velocity (drift velocity) and the corresponding acceleration of the free electrons as a function of time.

(b) The acceleration reaches a maximum magnitude every $\frac{1}{120}$ s. What is this maximum magnitude?

(c) For this maximum acceleration what are the transverse and magnetic electric fields that one of the free electrons produces at a distance of 0.20 m perpendicularly away from the wire?

(d) The velocity reaches a maximum magnitude every $\frac{1}{120}$ s. What is the maximum magnitude of the velocity? What is the magnetic field that, according to Eq. (30.15), an electron moving with this velocity produces at a distance of 0.20 m perpendicularly away from the wire?

28. The electric field in a copper wire of diameter 0.26 cm carrying a current of 12.0 A is 3.9×10^{-2} V/m.

(a) What is the acceleration of one of the free electrons of copper in this electric field?

(b) What are the transverse electric and magnetic fields that the accelerated electron produces at a distance of 4.0 m perpendicularly away from the wire?

(c) Suppose that all the free electrons in a segment of this wire 5.0×10^{-2} m long simultaneously produce such transverse electric and magnetic fields. What are the net transverse electric and magnetic fields of all these electrons acting together?

(d) Compare the magnetic field calculated in part (c) with the static magnetic field associated with the current of 12.0 A in the segment of wire.

*29. Suppose that a charge moves with *uniform* velocity. By a calculation similar to that in Eqs. (44)–(47), show that the *radial* electric field (Coulomb field) of this charge induces a magnetic field. Show that the magnitude of this magnetic field coincides with the magnitude of the familiar magnetic field [see Eq. (30.15)] of a uniformly moving charge. Hence the latter can be regarded as an induced magnetic field.

*30. (a) Integrate the energy density of the electric radiation field over the volume of the zone of the kink and find the total electric energy in the radiation field of the accelerated charge described in Section 35.4.

(b) Do the same for the energy of the magnetic radiation field.

CHAPTER 36

Light and Radio Waves

As we saw in the preceding chapter, an accelerated charge creates a propagating electromagnetic wave pulse which spreads outward from the charge. In essence, this wave pulse is a disturbance of the familiar electric and magnetic fields with which we began our study of electricity and magnetism [see Eqs. (23.16) and (30.15)]. As long as the charge moves with uniform velocity, these fields accompany the charge — they move as though they were rigidly attached to the charge. But if the charge is forced to accelerate, then parts of the fields break away — they become independent of the charge and they travel outward as an electric and magnetic disturbance.

Radio waves are propagating electromagnetic disturbances of this kind. They are created by the accelerated back-and-forth motion of charges on an antenna. Light waves are also propagating electromagnetic disturbances. They are created by the oscillations of electrons within atoms. We cannot observe the creation of light waves as directly as we can that of radio waves, but we can compare the predictions of electromagnetic theory with the results of many experiments testing the interaction of light and matter, and the emission of light by matter. In the next three chapters we will deal with a few instances of the interaction between light and matter — reflection, refraction, diffraction. We will leave the study of the emission of light by matter to the last chapter because the atomic processes that lead to the emission of light must be described in terms of quantum theory.

The earliest evidence for the electromagnetic character of light waves emerged from Maxwell's theoretical calculation of the speed of electromagnetic waves. From his equations, Maxwell predicted that electromagnetic waves, consisting of dynamic electric and magnetic fields that mutually induce each other forming self-supporting electromagnetic oscillations, propagate with a speed $c = 1/\sqrt{\mu_0 \varepsilon_0}$. Numerically, this predicted speed coincides with the measured speed of light waves, a coincidence that led Maxwell to propose that light waves are electromagnetic waves. Maxwell's theory of self-supporting electromagnetic oscillations in empty space received direct experimental confirmation at the hands of Heinrich Hertz, who generated the first artificial radio waves by means of sparks triggered in a gap in a high-frequency LC circuit.

The most precise modern method for the determination of the speed of light relies on separate measurements of the wavelength and of the frequency of the light emitted by a stabilized laser. The speed can then be evaluated as the product of these, that is, $c = \lambda v$. This method was first developed by K. M. Evenson et al. at the National Bureau of Standards. The best available results of such determinations of the speed of light (in vacuum) give

Standard value of the speed of light

$$c = 299{,}792{,}458 \text{ m/s}$$

This value has an uncertainty of only ± 1 m/s. Thus, the modern method gives us the speed of light to within nine significant figures!

As stated in Chapter 1, this value of the speed of light was adopted as a standard of speed in 1983, and it is now used as the basis for the definition of the meter.

Older methods for the determination of the speed of light relied on the direct timing of a light signal traveling back and forth over a precisely measured distance. For instance, A. A. Michelson performed many careful measurements of the time required for a light signal to travel between Mt. Wilson and Mt. San Antonio in California. But uncertainties in the atmospheric conditions seriously limited the accuracy of his determination of the speed of light.

In this chapter we will be mainly concerned with the characteristics of electromagnetic waves after they have emerged from their source and are propagating on their own through empty space. By contrast, in the preceding chapter we were mainly concerned with the process of production of electromagnetic waves.

Heinrich Rudolf Hertz, *1857–1894, German physicist, and professor at Bonn. He supplied the first experimental evidence for the electromagnetic waves predicted by Maxwell's theory. Hertz generated these waves by means of an electric spark, measured their speed and wavelength, and established their similarity to light waves in the phenomena of reflection, refraction, and polarization.*

36.1 The Plane Wave Pulse

In the preceding chapter we derived formulas for the electric and magnetic radiation fields produced by an accelerated charge; we obtained these radiation fields from a careful analysis of the behavior of propagating disturbances, or "kinks," in the field lines. The most important features of the electric radiation field of an accelerated charge can be summarized as follows: The magnitude of the electric radiation field is directly proportional to the acceleration and it is inversely proportional to the radial distance from the charge. If we look at the electric radiation field in a fixed direction relative to the direction of the acceleration, then the magnitude of this electric field has the following dependence on distance and time:

$$E(r,t) = [\text{constant}] \times \frac{a(t - r/c)}{r} \tag{1}$$

where $a(t)$ represents the acceleration as a function of time, and c is the speed of propagation of the disturbance (speed of light). This says that the electric field at time t depends on the acceleration at the earlier time $t - r/c$; the time delay r/c is exactly the time required for the disturbance to travel from the charge to the distance r. By comparing Eq. (1) with Eq. (15.2), the standard equation for a traveling wave, we recognize that the former does indeed represent a traveling wave that propagates with a speed c in the radial direction. The factor $1/r$ appearing in Eq. (1) indicates how the electromagnetic wave gradually weakens as it spreads farther and farther out.

Often we will be interested in the behavior of the wave over only a limited range of distance, a range very small compared to r. For example, we might be interested in the electric and magnetic fields that the moving charges on the antenna of a radio transmitter produce in the room in which we are sitting. If the length of the room is 5 m and the distance to the transmitter is 50 km, then r only changes by 1 part in 10^4 from one end of the room to the other. Under these conditions the factor $1/r$ in Eq. (1) can be treated as approximately constant. Equation (1) then reduces to

$$E(r,t) = [\text{constant}] \times a(t - r/c) \tag{2}$$

where the constant now includes the factor $1/r$.

The electric radiation field is perpendicular to the direction of propagation. For the sake of clarity, let us assume that the direction of propagation coincides with the x axis and that the electric field is along the y axis. Then we can write Eq. (2) as

$$E_y(x,t) = [\text{constant}] \times a_y(t - x/c) \tag{3}$$

This is a plane wave. The wave fronts of this wave are flat surfaces parallel to the y–z plane. For instance, Figure 36.1 shows the electric field of such a wave if the acceleration lasts for some short time interval, and is constant during this time interval. The radiated electric field then consists of a short wave pulse within which the electric field is constant. (Note that the flat wave front shown in Figure 36.1 is simply a small patch of the spherical wave front shown in Figure 35.15; such a small patch of a spherical wave front will appear flat if its size is much smaller than the radius of the sphere.)

The time-dependent electric field induces a magnetic field. Figure 36.2 shows this induced magnetic field. The direction of the magnetic field is perpendicular to the direction of propagation and also perpendicular to the electric field. We can deduce the direction of the magnetic field from the direction of the displacement current that generates it. Figure 36.2 shows that the displacement current of our wave pulse consists of two sheets of current, one with upward current at the front edge of the pulse, and one with downward current at the rear edge of the pulse. A sheet of current can be regarded as an infinite array of parallel straight wires. Such an infinite array of wires produces a uniform magnetic field whose direction is perpendicular to the direction of the current, according to the usual right-hand rule. Since the currents in the two sheets shown in Figure 36.2 have opposite directions, their magnetic fields combine in the region between the sheets, and cancel in the region outside of the sheets.

To determine the magnitude of the uniform magnetic field in the

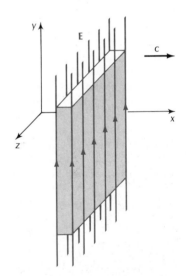

Fig. 36.1 Electric field lines of a plane wave pulse consisting of a region of uniform electric field. The wave front, or front surface of the pulse, is a flat plane. The entire pattern of field lines travels to the right with the wave speed c.

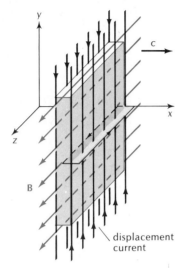

Fig. 36.2 A small stationary loop (black) perpendicular to the electric field lines registers a rate of change of electric flux at the instant the front edge of the wave pulse sweeps across the loop, and again at the instant the rear edge of the wave pulse sweeps across the loop. Hence there is a sheet of displacement current at the front edge of the wave pulse, and another sheet of displacement current at the rear edge. The black lines indicate the direction of the displacement current. The colored lines indicate the magnetic field produced by these currents.

region of the wave pulse, we apply the Maxwell–Ampère Law,

$$\oint \mathbf{B} \cdot d\mathbf{l} = \mu_0 \varepsilon_0 \frac{d\Phi}{dt} \tag{4}$$

to the stationary closed loop in the *x–z* plane shown in Figure 36.3. This loop measures $\Delta x \times \Delta z$; one edge of this loop is inside the wave pulse. The integral of **B** around this loop only receives a contribution from the edge inside the wave pulse:

$$\oint \mathbf{B} \cdot d\mathbf{l} = B_z \, \Delta z \tag{5}$$

Since in a time $\Delta t = \Delta x/c$, the wave front sweeps over the loop and fills the loop with electric flux, the rate of change of electric flux is

$$\frac{\Delta\Phi}{\Delta t} = \frac{E_y \, \Delta x \, \Delta z}{\Delta x/c} = cE_y \, \Delta z$$

Hence

$$B_z \, \Delta z = c\mu_0\varepsilon_0 E_y \, \Delta z \tag{6}$$

so that the magnitude of the magnetic field is

$$B_z = c\mu_0\varepsilon_0 E_y \tag{7}$$

This equation expresses the magnitude of the magnetic field in terms of the magnitude of the electric field, the speed of propagation of the wave pulse, and the constants ε_0 and μ_0. [Equation (7) coincides with Eq. (35.47), which we derived for the spherical wave pulse.]

Since the magnetic field of the wave pulse is a time-dependent mag-

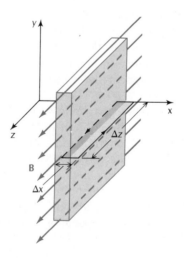

Fig. 36.3 Stationary rectangular loop in the *x–z* plane. The loop measures $\Delta x \times \Delta z$. At the instant shown, one edge of the loop is inside the front edge of the wave pulse.

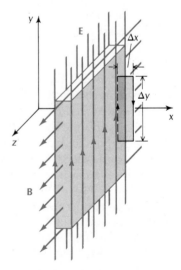

Fig. 36.4 Stationary rectangular loop in the *x–y* plane. The loop measures $\Delta x \times \Delta y$. At the instant shown, one edge of the loop is inside the front edge of the wave pulse.

netic field, it induces an electric field. For the sake of consistency, this induced electric field must coincide with the electric field E_y of the wave pulse. Thus, the electric and the magnetic fields of the wave pulse mutually induce each other. The requirement of consistency between these two mutually induced fields leads to the theoretical expression for the speed of light. To derive this expression, we apply Faraday's Law,

$$\oint \mathbf{E} \cdot d\mathbf{l} = -\frac{d\Phi_B}{dt} \tag{8}$$

to the stationary closed loop in the x–y plane shown in Figure 36.4. This loop measures $\Delta x \times \Delta y$; one edge of this loop is, again, inside the wave pulse. The integral of the electric field around this loop is $E_y \, \Delta y$; and the rate of change of magnetic flux through the loop is $\Delta\Phi_B/\Delta t = -B_z \, \Delta x \, \Delta y/(\Delta x/c)$. Hence

$$E_y \, \Delta y = cB_z \, \Delta y$$

and therefore

$$B_z = (1/c)E_y \tag{9}$$

Comparing this with Eq. (7), we see that consistency demands

$$c\mu_0\varepsilon_0 = 1/c \tag{10}$$

or

$$\boxed{c = \frac{1}{\sqrt{\mu_0\varepsilon_0}}} \tag{11}$$

Speed of electromagnetic wave

This is the theoretical expression for the speed of electromagnetic waves in vacuum first obtained by Maxwell. [Not surprisingly, Eqs. (9) and (11) coincide with Eqs. (35.58) and (35.55), derived for the spherical wave pulse.]

When comparing this theoretical value for the speed with the experimental value, we must resist the temptation of blithely inserting the best available modern values into Eq. (11). Such a procedure would be circular because the best available value for ε_0 listed in modern tables of constants — such as the table in Appendix 8 — is actually calculated from Eq. (11). For a meaningful test of Eq. (11), we must insert a value for ε_0 determined directly by electrical measurements. The best determination of this kind available, carried out by scientists at the U.S. Bureau of Standards early in this century, gave $\varepsilon_0 = 8.85433 \times 10^{-12}$ F/m.[1] With this value for ε_0 and with the (exact) value $4\pi \times 10^{-7}$ H/m for μ_0, the theoretical prediction for the speed of electromagnetic waves is

$$c = 1/(4\pi \times 10^{-7} \text{ H/m} \times 8.85433 \times 10^{-12} \text{ F/m})^{1/2}$$

$$= 2.99790 \times 10^8 \text{ m/s} \tag{12}$$

---

[1] This is the value obtained by E. B. Rosa and N. E. Dorsey, as corrected by H. L. Curtis.

This value is in excellent agreement with the best available measured value of the speed of light in vacuum,

$$c = 2.99792458 \times 10^8 \text{ m/s} \qquad (13)$$

This remarkable agreement constitutes clear evidence for the view that light is an electromagnetic wave. Other evidence for this view is found in the polarization properties of light, which agree with the polarization properties of electromagnetic waves (see Section 36.2).

36.2 Plane Harmonic Waves; Polarization

Although the above results for the magnitude of the magnetic field and the speed of electromagnetic waves were derived for the special case of a wave pulse consisting of a region of constant electric field, these results are of general validity because an arbitrary wave can be regarded as a succession of short wave pulses with piecewise constant electric fields. Accordingly, we can write Eq. (9) as a general relation between the magnitudes of the (perpendicular) electric and magnetic fields:

Relation between electric and magnetic fields of wave

$$\boxed{B(r,t) = \frac{1}{c}E(r,t)} \qquad (14)$$

The direction of the magnetic field is perpendicular to both the electric field and the direction of propagation. The directions of **E** and **B** are related by a right-hand rule: if the fingers are curled from **E** toward **B**, then the thumb lies along the direction of propagation.

A case of great practical importance involves the radiation fields generated by charges moving with simple harmonic motion. For instance, the currents and charges on the antenna of a radio transmitter move up and down the antenna with simple harmonic motion. For this kind of motion, the acceleration is a harmonic function of time,

$$a_y = [\text{constant}] \times \sin \omega t \qquad (15)$$

so that Eq. (3) becomes

$$E_y(x,t) = [\text{constant}] \times \sin(\omega t - \omega x/c) \qquad (16)$$

Obviously, the function on the right side represents a harmonic wave of angular frequency ω traveling in the x direction with a speed c [compare Eq. (15.8)]. We will write Eq. (16) as

Electric field of plane wave polarized in y direction

$$\boxed{E_y(x,t) = E_0 \sin\left(\omega t - \frac{\omega x}{c}\right)} \qquad (17)$$

where the quantity E_0 is the amplitude of the harmonic wave (compare Section 15.2). The magnetic field is perpendicular both to **E** and to the

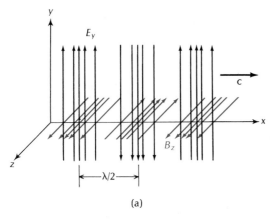

(a)

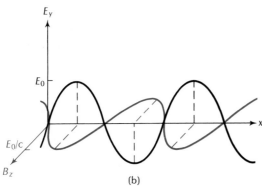

(b)

Fig. 36.5 (a) Electric (black) and magnetic (color) field lines of a plane harmonic wave traveling toward the right, shown at one instant of time. The electric field is vertical and the magnetic field is horizontal. Only the electric field lines in the x–y plane and the magnetic field lines in the x–z plane are shown. There are many more electric and magnetic field lines parallel to those shown. These lines fill slabs perpendicular to the x axis. (b) Plot of the strengths of the electric and magnetic fields as a function of x.

direction of propagation; according to Eq. (14), the magnitude of the magnetic field is E/c. Thus,

$$B_z(x,t) = \frac{E_0}{c} \sin\left(\omega t - \frac{\omega x}{c}\right)$$ (18)

Magnetic field of plane wave polarized in y direction

Equations (17) and (18) describe a **plane harmonic wave.** The wave fronts of this wave are flat surfaces parallel to the y–z plane. Note that the directions of **E** and **B** within this wave are related by the right-hand rule given at the beginning of this section. Note also that the electric and magnetic fields are in phase, i.e., **B** is at maximum wherever **E** is at maximum. Figure 36.5 shows the field lines of the plane wave.

Besides the plane wave described above, there is another possible plane wave with the same direction of propagation, but with its fields rotated by 90°. This other wave has an electric field along the z axis and a magnetic field along the y axis:

$$E_z(x,t) = E_0 \sin\left(\omega t - \frac{\omega x}{c}\right)$$ (19)

$$B_y(x,t) = -\frac{E_0}{c} \sin\left(\omega t - \frac{\omega x}{c}\right)$$ (20)

Electric and magnetic fields of plane wave polarized in z direction

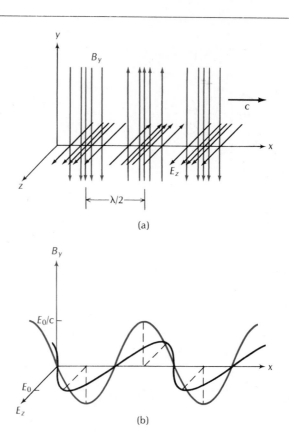

Fig. 36.6 (a) Electric (black) and magnetic (color) field lines of another plane wave traveling toward the right, shown at one instant of time. The electric field is horizontal and the magnetic field is vertical. (b) Plot of the strengths of the electric and magnetic fields as a function of x. •

Figure 36.6 shows the field lines of this wave.

Polarization

The direction of **E** is called the direction of **polarization** of the wave. The wave described by Eq. (17) is polarized in the y direction and the wave described by Eq. (19) is polarized in the z direction. It is of course also possible to construct waves polarized in some intermediate direction, say, at 45° to the y and z axes; but such waves are superpositions of those described by Eqs. (17) and (19) and therefore nothing essentially new. Thus, electromagnetic waves have only *two* independent directions of polarization.

Although an individual light wave, like any other kind of electromagnetic wave, is always polarized in some direction or another, the light beams produced by ordinary light sources — the Sun, a light bulb, a candle — do not exhibit any noticeable polarization. Such an

Unpolarized light

"unpolarized" light beam consists of a superposition of a very large number of plane waves with random directions of polarization. Hence, on the average, there is no polarization. However, the unpolarized

Polarizing filter

light can be given a polarization by passing it through a **polarizing filter**, such as a sheet of Polaroid, that only permits the passage of the electric field component lying parallel to a preferential direction and absorbs the electric field component perpendicular to the preferential direction. The Polaroid sheet contains long chains of molecules arranged parallel to each other; the preferential direction that permits passage of the electric field of a wave is *perpendicular* to the direction of alignment of these molecules. An analogous polarizing filter for microwaves, or radio waves of short wavelength, can be constructed out of a number of thin conducting rods or wires arranged parallel to each

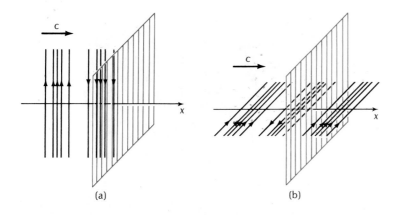

Fig. 36.7 An array of vertical wires (a) blocks the passage of a microwave of vertical polarization, but (b) permits the passage of a microwave of horizontal polarization.

other (Figure 36.7). The preferential direction of polarization that permits the passage of the electric field of a wave is then *perpendicular* to the direction of the wires because the wires have very little effect on a perpendicular electric field; on the other hand, an electric field parallel to the wires causes strong currents to flow in the wires, which both reflect the wave and dissipate its energy.

EXAMPLE 1. The wave reaching a point at some distance from a radio transmitter has an electric field with an amplitude of 2.0×10^{-3} V/m. What is the amplitude of the magnetic field?

SOLUTION: By Eq. (18) the amplitude of the magnetic field is E_0/c,

$$B_0 = \frac{E_0}{c} = \frac{2.0 \times 10^{-3} \text{ V/m}}{3.0 \times 10^8 \text{ m/s}} = 6.7 \times 10^{-12} \text{ T}$$

EXAMPLE 2. Suppose that a light wave polarized in the y direction is incident on a sheet of Polaroid whose preferential direction (direction of passage) makes an angle θ with the y axis (Figure 36.8). By what factor is the intensity of the light reduced? Pretend that the Polaroid sheet acts as an ideal polarizing filter, with 100% transmission for the component of the wave parallel to the preferential direction, and total absorption for the component perpendicular to this direction.

SOLUTION: If the magnitude of the electric field of the incident wave is E_0, then the component of the electric field parallel to the preferential direction is (Figure 36.8)

$$E' = E_0 \cos \theta \tag{21}$$

This is the only component that will pass through the sheet. Thus, the transmitted wave will be polarized at an angle θ with respect to the incident wave, and it will have the amplitude given by Eq. (21), i.e., it will have an amplitude smaller than that of the incident wave by a factor of $\cos \theta$. Since the intensity of a wave is proportional to the square of the amplitude, the intensity of the transmitted wave is smaller than that of the incident wave by a factor $\cos^2 \theta$,

$$\boxed{[\text{transmitted intensity}] = \cos^2 \theta \times [\text{incident intensity}]} \tag{22}$$

This relation between the incident and the transmitted intensities of a wave passing through a polarizing filter is called the **law of Malus**. Note that for

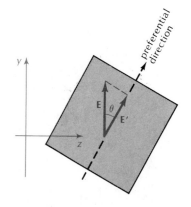

Fig. 36.8 A light wave polarized in the y direction is incident on a sheet of Polaroid whose preferential direction makes an angle θ with the y direction. The transmitted wave has a reduced amplitude and is polarized in the preferential direction of the sheet. The vectors **E** and **E'** indicate the directions of polarization of the incident and the transmitted waves.

Law of Malus

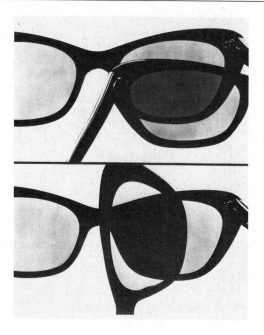

Fig. 36.9 The light that passes through the distant pair of sunglasses becomes polarized in the vertical direction, because this is the preferential direction for the Polaroid in the sunglasses. If the near pair of sunglasses is oriented parallel to the distant pair ($\theta = 0°$), it permits the passage of the polarized light. But if the near pair of sunglasses is oriented perpendicular to the distant pair ($\theta = 90°$), it stops the polarized light completely.

$\theta = 90°$, the transmitted intensity is zero, i.e., the Polaroid sheet stops the light wave completely (Figure 36.9).

36.3 The Generation of Electromagnetic Waves

Since the speed of the electromagnetic wave described by Eqs. (17)–(20) is c, the speed of light, the wavelength is

$$\lambda = \frac{2\pi}{\omega} c \tag{23}$$

or, if we replace the angular frequency ω by the ordinary frequency $\nu = \omega/2\pi$,

$$\lambda = \frac{c}{\nu} \tag{24}$$

This shows how the wavelength depends on the frequency of the source. For instance, the charges on the antenna of an FM radio station typically oscillate back and forth with a frequency of 10^8 Hz (or 100 MHz); correspondingly, the wavelength of the radiation emitted by these accelerated charges has a wavelength of

$$\lambda = \frac{c}{\nu} = \frac{3.0 \times 10^8 \text{ m/s}}{10^8/\text{s}} = 3.0 \text{ m} \tag{25}$$

The oscillations of the charges on the antenna are produced by means of a resonating LC circuit coupled to the antenna by a mutual inductance (see Figure 34.11b). This is essentially the method used to generate **long waves, medium waves** (AM), and **short waves** (including FM), as well as **TV waves;** such radio waves span a wavelength range from 10^5 m to a few centimeters.

Kinds of electromagnetic radiation

Waves of shorter wavelength, or **microwaves,** are best generated by a resonating cavity such as a klystron (see Section 35.3); the antenna in this case is merely a horn, connected to the cavity by a waveguide, that permits the waves to spill out into space. This method can be used to generate waves of wavelength as short as about a millimeter. Shorter wavelengths cannot be generated with currents oscillating on macroscopic laboratory equipment; however, short wavelengths are easily generated by electrons vibrating within molecules and atoms subjected to stimulation by heat or by an electric current. Depending on the details of the motion, the electrons in molecules and in atoms will emit **infrared** radiation, **visible** light, **ultraviolet** radiation, or **X rays;** the corresponding wavelengths range from 10^{-3} m to 10^{-11} m. X Rays can also be generated by the acceleration that high-speed electrons suffer during impact on a target; this is **Bremsstrahlung** (see Example 35.2).

Bremsstrahlung

Radiations of even shorter wavelengths are emitted by protons and neutrons moving within a nucleus; these are **gamma rays** with wavelengths as short as 10^{-13} m. Of course, the motion of subatomic particles and their emission of radiation cannot be calculated by classical mechanics or classical electromagnetic theory; such calculations require quantum mechanics and quantum electrodynamics.

In an ordinary light source, such as a neon tube, the individual atoms or molecules radiate independently. The emerging light consists of a superposition of many individual light waves with random phase differences, random directions of polarization, and diverging directions of propagation (light waves with random, unpredictable phase differences are said to be **incoherent**). In a **laser,** the atoms or molecules radiate in unison, by a quantum-mechanical phenomenon called **stimulated emission.** The emerging light is a superposition of light waves with exactly the same phases, the same directions of polarization, and the same directions of propagation (light waves with no phase differences, or with predictable phase differences, are said to be **coherent**). Since the individual light waves in this kind of light combine constructively, the light beam emerging from the laser is very intense, and it is also very sharply collimated.

Stimulated emission

Another important mechanism for the generation of electromagnetic waves is **cyclotron emission.** This involves high-speed electrons undergoing centripetal acceleration while spiraling in a magnetic field (see Section 31.3). Depending on the speed of the electron and the strength of the magnetic field, the radiation may consist of radio waves, X rays, or anything in between. Most of the radio waves reaching us from stars, pulsars, and radio galaxies are generated by this process.

Figure 36.10 displays the wavelength and the frequency bands of electromagnetic radiation. The bands overlap to some extent because the names assigned to the different ranges of wavelength depend not only on the value of the wavelength, but also on the method used to generate and/or detect the radiation. For example, radiation of a wavelength of a tenth of a millimeter will be called a radio wave (microwave) if detected by a radio receiver, but it will be called infrared radiation if detected by a heat sensor.

Visible light is electromagnetic radiation of a wavelength between

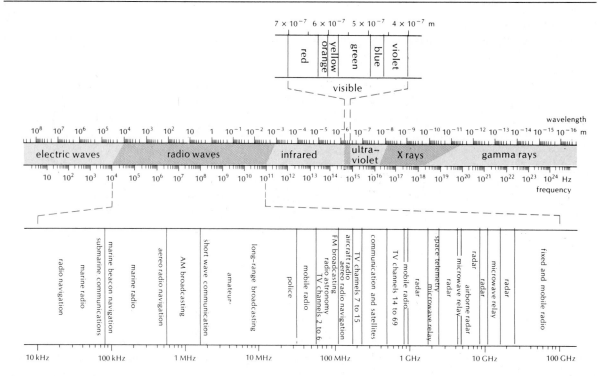

Fig. 36.10 Wavelength and frequency bands of electromagnetic radiation.

about 7×10^{-7} m and 4×10^{-7} m. This is the range of wavelengths to which our eyes are sensitive. We perceive different wavelengths within the visible region as having different colors. Figure 36.11 shows how colors are correlated with wavelength.

Incidentally, our eyes are almost completely insensitive to the polarization of light waves. We can only detect the polarization with special equipment such as Polaroid sunglasses. Radio and TV antennas are of course very sensitive to the direction of polarization of radio waves and they must have the proper orientation to pick up a strong signal.

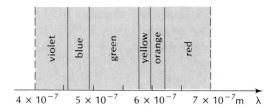

Fig. 36.11 Colors of visible light.

36.4 Energy of a Wave

The electric and magnetic fields of a wave contain energy. As the wave moves along, so does this energy — the wave transports energy.

Let us calculate the flow of energy in a plane wave. Suppose the wave is of the kind described by Eqs. (17) and (18), i.e., it is a wave propagating in the positive x direction. The densities of electric and magnetic energy are, respectively [see Eqs. (26.19) and (32.42)],

$$u_E = \frac{\varepsilon_0}{2} E^2 \quad \text{and} \quad u_B = \frac{1}{2\mu_0} B^2 \tag{26}$$

Within a thin rectangular slab of this wave, of thickness dx and frontal area A (Figure 36.12), the fields are nearly constant. The total amount of energy in this slab is

$$U = (u_E + u_B) \times [\text{volume}] \qquad (27)$$

$$= \left(\frac{\varepsilon_0}{2} E^2 + \frac{1}{2\mu_0} B^2 \right) \times A \, dx \qquad (28)$$

Since $E = cB$ and $\varepsilon_0 = 1/\mu_0 c^2$ we can write this as

$$dU = \left[\frac{1}{2\mu_0 c^2} E(cB) + \frac{1}{2\mu_0} B\left(\frac{E}{c}\right) \right] A \, dx \qquad (29)$$

$$= \frac{1}{\mu_0 c} EBA \, dx \qquad (30)$$

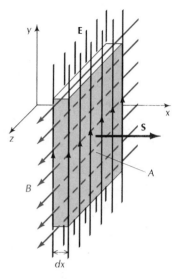

Fig. 36.12 A slab of electric (black) and magnetic (color) fields in a plane wave propagating toward the right. The slab of thickness dx and frontal area A moves with the wave.

Note that the two terms on the right side of Eq. (29) are equal, i.e., the electric and magnetic energy densities in an electromagnetic wave are equal.

The amount of energy dU moves out of the volume $A \, dx$ in a time $dt = dx/c$. Hence the rate of flow of energy is

$$\frac{dU}{dt} = \frac{1}{\mu_0} EBA \qquad (31)$$

and the rate of flow of energy per unit frontal area, or the **energy flux,** is

$$\frac{1}{A}\frac{dU}{dt} = \frac{1}{\mu_0} EB \qquad (32) \qquad \textit{Energy flux}$$

In Figure 36.12 the energy flows toward the right, along the direction of propagation of the wave. We can describe the energy flux in terms of a vector that has a magnitude equal to the right side of Eq. (32) and a direction along the flow. The vector satisfying these conditions is

$$\boxed{ \mathbf{S} = \frac{1}{\mu_0} \mathbf{E} \times \mathbf{B} } \qquad (33) \qquad \textit{Poynting vector}$$

Since $\mathbf{E}$ and $\mathbf{B}$ are perpendicular, the magnitude of the vector $\mathbf{S}$ given by Eq. (33) coincides with the magnitude given by Eq. (32). Furthermore the direction of $\mathbf{E} \times \mathbf{B}$ in Figure 36.12 is toward the right, as desired. The vector $\mathbf{S}$ is called the **Poynting vector.** It turns out that this vector gives the energy flux not only in a plane wave, but in any arbitrary electromagnetic field. The units of energy flux are $J/(m^2 \cdot s)$, or W/m^2.

Since both the electric and magnetic fields of a wave oscillate in time, so does the energy flux. If the value of x in Eqs. (17) and (18) is fixed, say, $x = 0$, we have

John Henry Poynting, *1852–1914, British physicist, professor at the University of Birmingham. He introduced the Poynting vector while presenting a proof of a general theorem on energy transfers in the electromagnetic fields.*

$$E = E_0 \sin \omega t \quad \text{and} \quad B = \frac{E_0}{c} \sin \omega t \qquad (34)$$

so that

$$S = \frac{1}{\mu_0 c} E_0^2 \sin^2 \omega t \qquad (35)$$

Thus, the energy flux oscillates between zero and a maximum value $E_0^2 / \mu_0 c$. The time-average energy flux is

Time-average magnitude of Poynting vector

$$\boxed{\overline{S} = \frac{1}{2\mu_0 c} E_0^2} \qquad (36)$$

EXAMPLE 3. At a distance of 6.0 km from a radio transmitter, the amplitude of the oscillating electric field of the radio wave is $E_0 = 0.13$ V/m. What is the time-average energy flux? What is the total power radiated by the radio transmitter? Pretend that the transmitter radiates uniformly in all directions.

SOLUTION: From Eq. (36) we find

$$\overline{S} = \frac{1}{2\mu_0 c} E_0^2$$

$$= \frac{(0.13 \text{ V/m})^2}{2 \times 1.26 \times 10^{-6} \text{ H/m} \times 3 \times 10^8 \text{ m/s}}$$

$$= 2.2 \times 10^{-5} \text{ W/m}^2$$

To obtain the total power, we must multiply the power per unit area by the area of a sphere of radius 6.0 km:

$$\overline{P} = 4\pi r^2 \overline{S} \qquad (37)$$

$$= 4\pi \times (6.0 \times 10^3 \text{ m})^2 \times 2.2 \times 10^{-5} \text{ W/m}^2$$

$$= 1.0 \times 10^4 \text{ W} = 10 \text{ kW}$$

Note that, according to Eq. (37), the energy flux $\overline{S}$ for a spherical wave spreading out from a source is inversely proportional to the square of the distance,

$$\overline{S} \propto \frac{1}{r^2} \qquad (38)$$

This decrease of the flux is consistent with Eq. (1), according to which the amplitude of the electric field is inversely proportional to the distance, $E_0 \propto 1/r$. We must take this dependence of flux and amplitude on distance into account whenever we want to investigate the spreading of the wave over a large range of distances (but we can ignore this dependence over a small range of distances).

36.5 Momentum of a Wave

An electromagnetic wave carries not only energy but also momentum. This can be readily understood in terms of Einstein's mass–energy relation, which tells us that the energy in the wave has mass. As the energy moves so does the mass, and such moving mass has momentum.

The following simple calculation permits us to derive a quantitative formula for the amount of momentum in an electromagnetic wave. Imagine that a plane electromagnetic wave polarized in the y direction and traveling in the x direction strikes a charged particle (Figure 36.13). The electric and magnetic fields of the wave will then exert forces on the particle, i.e., they will change the momentum of the particle. Let us look at the x component of the rate of change of momentum. The electric field is entirely in the y direction — it exerts no force in the x direction. The magnetic field is entirely in the z direction — it exerts the following force in the x direction:

$$\frac{dp_x}{dt} = q(\mathbf{v} \times \mathbf{B})_x = q(v_y B_z - v_z B_y) = q v_y B_z$$

Since $B_z = E_y / c$ [see Eqs. (17) and (18)], we can also write this rate of change of momentum as

$$\frac{dp_x}{dt} = \frac{q}{c} v_y E_y \qquad (39)$$

Let us compare this with the rate at which the particle acquires energy from the wave. Only the electric field does work on the particle. The rate of work, or the rate of increase of the energy of the particle, is

$$\frac{dW}{dt} = q\mathbf{E} \cdot \mathbf{v} = q E_y v_y \qquad (40)$$

(here we have used the symbol W for the energy of the particle because the symbol E is reserved for the electric field). If we compare the right sides of Eqs. (39) and (40), we recognize that

$$\frac{dp_x}{dt} = \frac{1}{c} \frac{dW}{dt}$$

i.e., the rate at which the particle acquires x momentum from the wave matches the rate at which it acquires energy. Whenever the particle absorbs energy from the wave, it will also absorb x momentum,

$$dp_x = \frac{1}{c} dW \qquad (41)$$

Any gain of energy and momentum by the particle implies a corresponding loss of energy and momentum by the wave. If we imagine that the wave loses all of its energy and momentum to the particle (and disappears),[2] we recognize that the total amount of energy and the

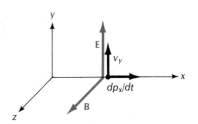

Fig. 36.13 A plane wave traveling toward the right strikes a particle. The electric and magnetic fields of the wave exert forces on the particle.

---

[2] Actually, a single charged particle cannot absorb all of a wave (the particle will scatter the wave); but a cloud of many charged particles can completely absorb a wave.

total amount of x momentum stored in the wave must be related by a formula similar to Eq. (41):

Momentum of wave

$$P_x = \frac{U}{c} \qquad (42)$$

This is the general relation between the momentum and the energy of a plane wave. This relation is also valid for the amounts of momentum and energy in some portion of the wave.

Since we already know that the energy flows with the wave, we can conclude that the momentum also flows with the wave. By means of Eq. (31), we can then express the momentum flow within a frontal area A of the wave as

$$\frac{dP_x}{dt} = \frac{1}{c}\frac{dU}{dt} = \frac{1}{c\mu_0}EBA \qquad (43)$$

This shows that the momentum flow oscillates in the same way as the energy flow. The time-average momentum flow is

$$\overline{\frac{dP_x}{dt}} = \frac{1}{2\mu_0 c^2}E_0^2 A \qquad (44)$$

Whenever a wave strikes a body and is absorbed by it, the wave will exert a force on the body and transfer momentum to it. If the body has a frontal area A facing the wave, the rate of change of momentum, or the force, will simply be given by Eq. (43),

$$F_x = \frac{dP_x}{dt} = \frac{1}{c\mu_0}EBA$$

or

$$F_x = \frac{1}{c}SA$$

where S is the magnitude of the Poynting vector. The force per unit frontal area is then

Pressure of radiation

$$\frac{F_x}{A} = \frac{S}{c} \qquad (45)$$

This is called the **pressure of radiation** or the pressure of light. Note that the formula (45) hinges on the assumption that the wave is completely absorbed. If the wave is partially or totally reflected, then the force is *larger* than given by Eq. (45). For instance, if the wave is totally reflected, then the body must not only absorb all of the initial momentum, but also supply the momentum for the reversed motion of the wave. The force is then twice as large as that given by Eq. (45).

The magnitude of the radiation-pressure force is usually insignificant compared with other ordinary forces that act on a body.

EXAMPLE 4. The average energy flux in the sunlight incident on the Earth is $\bar{S} = 1.4 \times 10^3$ W/m². What force does the pressure of light exert on the Earth? How does this compare with the gravitational force that the Sun exerts on the Earth? Pretend that all the light striking the Earth is absorbed.

SOLUTION: The cross-sectional area that the Earth offers to the stream of sunlight is πR_E^2. The total power absorbed by the Earth is then

$$\frac{dU}{dt} = \bar{S}A$$

$$= 1.4 \times 10^3 \text{ W/m}^2 \times \pi \times (6.4 \times 10^6 \text{ m})^2$$

$$= 1.8 \times 10^{17} \text{ W} \tag{46}$$

Hence

$$F = \frac{1}{c}\frac{dU}{dt} = \frac{1.8 \times 10^{17} \text{ W}}{3.0 \times 10^8 \text{ m/s}} = 6.0 \times 10^8 \text{ N}$$

The gravitational force that the Sun exerts on the Earth is much greater:

$$F = \frac{GM_S M_E}{r^2}$$

$$= \frac{6.7 \times 10^{-11} \text{ N} \cdot \text{m}^2/\text{kg}^2 \times 2.0 \times 10^{30} \text{ kg} \times 6.0 \times 10^{24} \text{ kg}}{(1.5 \times 10^{11} \text{ m})^2}$$

$$= 3.6 \times 10^{22} \text{ N}$$

The force of radiation pressure on the Earth is insignificant compared to the gravitational pull of the Sun. However, on an object of very small mass, such as a grain of dust in interplanetary space, the force of radiation pressure can be as large as, or even larger than, the gravitational pull of the Sun. A simple calculation shows that if a grain of dust is smaller than about 10^{-6} m across, then radiation pressure overcomes gravitation and blows the grain away from the Sun.[3] The tails of comets are a spectacular demonstration of this effect. These tails consist of a plume of dust that is blown away from the comet's head by the pressure of sunlight (Figure 36.14). Regardless of the direction of motion of the comet, the tail always points away from the Sun, much as a pennant fluttering from the mast of a ship always points away from the wind.

36.6 The Doppler Shift of Light

The frequency of light waves, like that of waves of any kind, suffers a Doppler shift whenever the emitter of waves is in motion relative to the receiver. However, in contrast to sound waves in air or surface

Fig. 36.14 Comet Mrkos and its tails, photographed in August 1957. This comet has both a dust tail (haze, upper right) and an ion tail (streaks, upper center). The dust tail is produced by the pressure of sunlight; the ion tail is produced by the pressure of the solar wind.

[3] For *very* small grains of dust (less than 10^{-7} m across) and for molecules this will not happen because such very small objects scatter light instead of absorbing it. This cuts down the force exerted by light.

waves in water, light waves in a vacuum are not propagating in any material medium that can serve as a preferred reference frame with respect to which we can reckon their velocity. Light waves have the same speed c with respect to *all* inertial reference frames (this well-established observational fact lies at the root of the theory of Special Relativity, as discussed in Chapter 17). Hence, for the calculation of the Doppler shift we must first decide which reference frame is the most suitable. Since we want to find the frequency as seen by the *receiver*, it is clear that the relevant reference frame is the rest frame of the receiver.

In this reference frame, the calculation of the Doppler shift can be done exactly as in Section 16.3. Since the receiver is at rest in the "medium" (reference frame) in which light has the speed c, the formula for the Doppler shift of light must be [see Equation (16.13)]

$$\nu = \frac{1}{1 \pm v/c}\, \nu_0 \tag{47}$$

where, as before, ν_0 is the frequency radiated by the emitter, ν is the frequency detected by the receiver, and v is the speed of the emitter (the positive sign is to be used if the emitter is receding, the negative sign if approaching). Since the calculations of Section 16.3 relied on Newtonian physics, which is valid only if the speed is low compared to the speed of light($v \ll c$), Eq. (47) can be replaced by the approximation

$$\nu \cong (1 \mp v/c)\nu_0 \tag{48}$$

without any significant loss of accuracy. From this we see that the frequency shift $\Delta\nu = \nu - \nu_0$ is given by

Approximate Doppler shift

$$\boxed{\frac{\Delta\nu}{\nu_0} \cong \mp\frac{v}{c}} \qquad \begin{array}{l} + \text{ for approaching emitter} \\ - \text{ for receding emitter} \end{array} \tag{49}$$

Measurements of the Doppler shift of light play a crucial role in astronomy. The velocities of remote stars relative to us can be determined via the Doppler shift of the starlight reaching us. The rotational velocities of stars can likewise be determined by a careful measurement of the difference between the Doppler shifts of the light from one edge of the star and the opposite edge.

Although Eq. (47) is adequate whenever the speed of the source is small compared to the speed of light, it must be modified for relativistic effects if the speed of the source is large. According to Section 17.4, the emitter is subject to a relativistic time-dilation effect. This, in itself, would *reduce* the frequency of the emitter by a factor $\sqrt{1 - v^2/c^2}$. We must therefore insert this extra factor in Eq. (47),

$$\nu = \sqrt{\frac{1 - v^2/c^2}{1 \pm v/c}}\, \nu_0 \tag{50}$$

This is the exact formula for the Doppler shift of light. Since $1 - v^2/c^2 = (1 - v/c)(1 + v/c)$, this formula can be simplified by appropriate cancellations:

$$v = \sqrt{\frac{1 - v/c}{1 + v/c}}\, v_0 \qquad \text{for receding emitter} \tag{51}$$

$$v = \sqrt{\frac{1 + v/c}{1 - v/c}}\, v_0 \qquad \text{for approaching emitter} \tag{52}$$

Exact Doppler shift

For low speeds there is of course no appreciable difference between these exact formulas and the approximate formula (47).

The velocities of recession of galaxies and quasars partaking of the expansion of the universe (see Interlude D) give the light and radio waves from these objects very large Doppler shifts. For instance, the light from the quasar 3C 147 (Figure 36.15) has a frequency shift amounting to a factor of 1.55. If we insert this factor in Eq. (51),

$$\frac{v}{v_0} = \frac{1}{1.55} = \sqrt{\frac{1 - v/c}{1 + v/c}} \tag{53}$$

we find a velocity of recession $v = 0.41c$, that is, 41% of the speed of light! The velocities of recession of some other quasars are even higher.

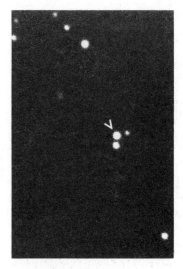

Fig. 36.15 The quasar 3C 147.

SUMMARY

Electric and magnetic fields of plane wave:

$$E_y = E_0 \sin\left(\omega t - \frac{\omega x}{c}\right)$$

$$B_z = \frac{E_0}{c} \sin\left(\omega t - \frac{\omega x}{c}\right)$$

Poynting vector: $\mathbf{S} = \dfrac{1}{\mu_0}\, \mathbf{E} \times \mathbf{B}$

Time-average energy flux of plane wave: $\overline{S} = \dfrac{1}{2\mu_0 c} E_0^2$

Momentum of wave: $P_x = \dfrac{U}{c}$

Radiation pressure: $\dfrac{F_x}{A} = \dfrac{S}{c}$

Doppler shift: $\dfrac{\Delta v}{v_0} \cong \mp \dfrac{v}{c}$ at low speed

$$v = \sqrt{\frac{1 - v/c}{1 + v/c}}\, v_0 \qquad \text{(receding)}$$

$$v = \sqrt{\frac{1 + v/c}{1 - v/c}}\, v_0 \qquad \text{(approaching)}$$

QUESTIONS

1. Since the speed of light has now been adopted as the standard of speed and has been assigned the value 2.99792458×10^8 m/s *by definition,* why is it still meaningful to compare this number with Maxwell's theoretical prediction for the speed of light?

2. The round-trip travel time for a radio signal to the Moon is 2.6 s. Does this have a noticeable effect on radio communications with astronauts on the Moon?

3. A severe limitation on the speed of computation of large electronic computers is imposed by the speed of light because the electric signals on the connecting wires within the computer are electromagnetic waves ("guided waves"), which travel at a speed roughly equal to the speed of light. If the computer measures about 1 m across, what is the minimum travel time required for a typical signal sent from one part of the computer to another? What is the maximum number of signals that can be sent back and forth (sequentially) per second? Is there any way to avoid the limitation imposed by the travel time of signals?

4. Describe the direction of polarization of the radiation field of an accelerated charge at a few typical points in the space surrounding the charge. Show that the direction of polarization is never perpendicular to the direction of acceleration.

5. Consider transverse waves on a string. How many independent directions of polarization do such waves have? How could you construct a mechanical polarization filter that only permits the passage of a wave polarized in a preferential direction?

6. If the preferential directions of two adjacent sheets of Polaroid are at right angles, no light will pass through. However, if you now slip a third sheet of Polaroid between the other two and orient its preferential direction so that it lies between the directions of the other two, then some light will pass through the three sheets. Explain.

7. It has been proposed that we could eliminate the glare of the headlights of approaching automobiles by covering the windshields and the headlights with sheets of Polaroid. What orientation should we pick for the sheets of Polaroid installed on windshields and on headlights so that the light of every approaching automobile is blocked out, but our own light is not?

8. Malus discovered that when unpolarized light is incident on any surface, the reflected light is partially polarized in a direction parallel to the surface. This phenomenon, called **polarization by reflection,** arises from a dependence of the strength of reflection on the direction of polarization. Taking this phenomenon into account, explain how Polaroid sunglasses can eliminate a large fraction of the reflected glare (Figure 36.16), and explain why it is advantageous to arrange the preferential direction in Polaroid sunglasses vertically.

9. Some small-boat sailors like to wear Polaroid sunglasses because these make disturbances of the water surface stand out with exceptional contrast, and thereby make it easier to spot approaching gusts of wind. Why are Polaroid sunglasses better for this purpose than ordinary sunglasses? (Hint: Consider polarization by reflection; see preceding question.)

10. Suppose you are given a sheet of Polaroid that has no markings identifying its preferential direction. You have available a beam of unpolarized light. How can you determine the preferential direction of the sheet? (Hint: Cut the sheet in two and place one sheet behind the other rotated by 90° around the axis of the beam, so that the light is completely blocked. What will happen if

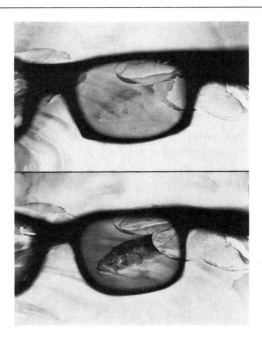

Fig. 36.16 When viewed through ordinary sunglasses (top), the fish is almost completely hidden by the glare reflected by the water. When viewed through Polaroid sunglasses (bottom), the fish becomes visible because the glare is reduced.

you now rotate one sheet about a *transverse axis,* i.e., an axis perpendicular to the beam?

11. The scattered light reaching you from the blue sky is (partially) polarized: if you look straight up, the direction of polarization is perpendicular to the direction of the Sun. How does this polarization arise? [Hint: Consider a beam of (unpolarized) sunlight passing overhead. The electric field of the light waves in this beam accelerates electrons in the molecules of air, and the radiation emitted by these electrons constitutes the scattered light of the sky. Since the direction of acceleration is perpendicular to the incident beam, what can you say about the polarization of the radiation emitted downward, toward you?]

12. Figure 36.17 shows the sensitivity of the human eye as a function of the wavelength of light. The sensitivity is maximum at about 5.5×10^{-7} m and drops to about 1% at 6.9×10^{-7} m and at 4.3×10^{-7} m. Suppose that the sensitivity of your eye were constant over the entire interval of wavelengths shown in Figure 36.17. How would this alter your visual perception of some of the things you see in your everyday life?

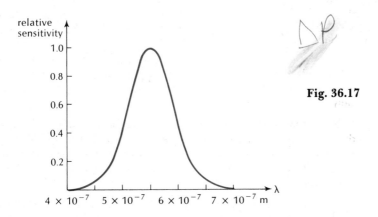

Fig. 36.17

13. An atom radiates visible light of a wavelength several thousand times longer than the size of the atom. How is this possible?

14. Why does the radio reception fade in the receiver of your automobile when you enter a tunnel?

15. Short-wave radio waves are reflected by the ionosphere of the Earth; this makes them very useful for long-range communication. Explain.

16. Why is the effect of the pressure of sunlight on a grain of dust floating in interplanetary space more significant if the grain is very small?

17. Could we propel a spaceship by shining a very intense light out of its back? What would be the advantages of such a propulsion scheme?

18. Astronauts of the future could travel all over the Solar System in a spaceship suspended from a large "sail" coated with a reflecting material. The pressure of sunlight on the sail could propel the spaceship in any direction. How would the astronauts have to orient their (flat) sail to move radially away from the Sun? To move at right angles to the radial direction? To move radially toward the Sun?

19. Suppose that a light wave is totally reflected by a moving mirror. Would you expect the energy of the light wave to change during this reflection? (Hint: Consider the work done by the radiation pressure.)

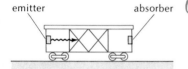

emitter   absorber

Fig. 36.18

20. Einstein's first derivation of his famous equation $E = mc^2$ relied on the equation for the momentum of light. Einstein considered a closed boxcar, free to roll on frictionless rails, with a light source at one end and an absorber at the other end (Figure 36.18). If the source at one end emits a light pulse of energy E, the recoil sets the boxcar in motion; this motion stops when the light pulse reaches the other end and is absorbed. Given that the center of mass of the closed system should remain stationary during this process, how could Einstein conclude that the transfer of the energy E from one end to the other involves a transfer of mass?

21. In the seventeenth century, the Danish astronomer Ole Roemer noticed that the orbital periods of the moons of Jupiter, as observed from the Earth, exhibit some systematic irregularities: the periods are slightly longer when the Earth is moving away from Jupiter, and slightly shorter when the Earth is moving toward Jupiter. Roemer attributed this apparent irregularity to the finite speed of propagation of light, and used it to make the first determination of the speed of light. Explain how one can use the lengthening or the shortening of the observed period to deduce the speed of light. (This shift of period may be regarded as the earliest discovery of a Doppler shift.)

22. Equations (51) and (52) give the Doppler shift for an emitter that is approaching or receding along the line of sight. Taking into account relativistic effects, would you expect a Doppler shift for an emitter moving at right angles to the line of sight? (This is called the transverse Doppler shift.)

PROBLEMS

Section 36.1

1. At a distance of 6.0 km from a radio transmitter, the amplitude of the electric radiation field of the emitted radio wave is $E_0 = 0.13$ V/m. Taking into account the decrease of the amplitude of the wave with distance, what will be the amplitude of the radio wave when it reaches a distance of 12.0 km? A distance of 18.0 km?

2. Consider the plane wave pulse shown in Figures 36.1 and 36.2. If the magnitude of the electric field in this pulse is 4×10^{-3} V/m, what is the magnitude of the displacement current flowing along the front surface of the pulse per meter of length measured perpendicularly to the current?

3. Two plane-wave pulses of the kind described in Section 36.1 are traveling

in opposite directions. Their polarizations are parallel and the magnitudes of their electric fields are 2×10^{-3} V/m.

(a) What are the electric energy density and the magnetic energy density in each pulse?

(b) Suppose that at one instant the two pulses overlap. What are the magnitudes of the electric field and the magnetic field in this superposition?

(c) What are the electric energy density and the magnetic energy density?

4. (a) One type of antenna for a radio receiver consists of a short piece of straight wire; when the electric field of a radio wave strikes this wire it makes currents flow along it, which are detected and amplified by the receiver. Suppose that the electric field of a radio wave is vertical. What must be the orientation of the wire for maximum sensitivity?

(b) Another type of antenna consists of a circular loop; when the magnetic field of a radio wave strikes this loop it induces currents around it. Suppose that the magnetic field of a radio wave is horizontal. What must be the orientation of the loop for maximum sensitivity?

Section 36.2

5. At a distance of several kilometers from a radio transmitter, the electric field of the emitted radio wave has a magnitude of 0.12 V/m at one instant of time. What is the energy density in this electric field? What is the energy density in the magnetic field of the radio wave?

6. A plane electromagnetic wave travels in the eastward direction. At one instant the electric field at a given point has a magnitude of 0.60 V/m and points down. What are the magnitude and direction of the magnetic field at this instant? Draw a diagram showing the electric field, the magnetic field, and the direction of propagation.

7. An electromagnetic wave traveling along the x axis consists of the following superposition of two waves polarized along the y and z directions, respectively:

$$\mathbf{E} = \hat{\mathbf{y}}E_0 \sin(\omega t - \omega x/c) + \hat{\mathbf{z}}E_0 \cos(\omega t - \omega x/c)$$

This electromagnetic wave is said to be **circularly polarized.**

(a) Show that the magnitude of the electric field is E_0 at all points of space at all times.

(b) Consider the point $x = y = z = 0$. What is the angle between $\mathbf{E}$ and the z axis at time $t = 0$? $t = \pi/2\omega$? $t = \pi/\omega$? $t = 3\pi/2\omega$? Draw a diagram showing the y and z axes and the direction of $\mathbf{E}$ at these times. In a few words, describe the behavior of $\mathbf{E}$ as a function of time.

8. An electromagnetic wave has the form

$$\mathbf{E} = \hat{\mathbf{x}}E_0 \sin(\omega t + \omega z/c) + 2\hat{\mathbf{y}}E_0 \sin(\omega t + \omega z/c)$$

(a) What is the direction of propagation of this wave?

(b) What is the direction of polarization, i.e., what angle does the direction of polarization make with the x, y, and z axes?

(c) Write down a formula for the magnetic field of this wave as a function of space and time.

9. The preferential directions of two adjacent sheets of Polaroid make an angle of 45°. A beam of polarized light, whose direction of polarization coincides with the preferential direction of the *second* sheet, is incident on the *first* sheet. By what factor is the intensity of the transmitted beam emerging from the second sheet reduced compared to the intensity of the incident beam? Assume that the sheets act as ideal polarizing filters.

10. Suppose that the preferential directions of two adjacent sheets of Polaroid make an angle θ with each other. If unpolarized light is incident on these

sheets, what is the transmitted intensity as a function of the angle θ? Assume that the sheets act as ideal polarizing filters.

11. If the preferential directions of two adjacent sheets of Polaroid are at right angles, they will completely block a light beam. However, if you insert a third sheet of Polaroid between the other two, then some light will pass through. Derive a formula for the dependence of the intensity of the transmitted light as a function of the angle that the preferential direction of the inserted sheet makes with that of the first sheet. Assume that the incident light is unpolarized, and that the sheets act as ideal polarizing filters. For what orientation of the inserted sheet is the transmitted intensity maximum?

*12. Example 2 shows that when a light wave passes through a sheet of Polaroid, its direction of polarization rotates, albeit with some loss of intensity. Consider an infinite number of adjacent sheets of Polaroid, each tilted by an infinitesimal angle with respect to the preceding sheet (Figure 36.19). Show that such a stack rotates the plane of polarization of the light wave through a finite angle without loss of intensity. Assume that each sheet of Polaroid acts as an ideal polarizing filter.

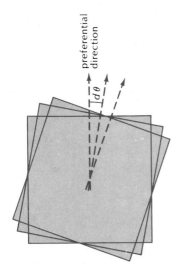

Fig. 36.19

Section 36.3

13. An ordinary radio receiver, such as found in homes across the country, has an AM dial and an FM dial. The AM dial covers a range from 530 to 1600 kHz and the FM dial a range from 88 to 108 MHz. What is the range of wavelengths for AM? For FM?

14. At many coastal locations, radio stations of the National Weather Service transmit continuous weather reports at a frequency of 162.5 MHz. What is the wavelength of these transmissions?

Section 36.4

15. A plane electromagnetic wave travels in the northward direction. At one instant, the electric field at a given point has a magnitude of 0.50 V/m and is in the eastward direction. What are the magnitude and direction of the magnetic field at the given point? What are the magnitude and direction of the Poynting vector?

16. The average energy flux of sunlight incident on the top of the Earth's atmosphere is 1.4×10^3 W/m². What are the corresponding amplitudes of oscillation of the electric and magnetic fields?

17. In the United States, the accepted standard for the safe maximum level of continuous whole-body exposure to microwave radiation is 10 milliwatts/cm².[4]
   (a) For this energy flux, what are the corresponding amplitudes of oscillation of the electric and magnetic fields?
   (b) Suppose that a man of frontal area 1.0 m² completely absorbs microwaves with an intensity of 10 milliwatts/cm² incident on this area and that the microwave energy is converted to heat within his body. What is the rate (in calories per second) at which his body develops heat?

18. A silicon solar cell of frontal area 13 cm² delivers 0.20 A at 0.45 V when exposed to full sunlight of energy flux 1.0×10^3 W/m². What is the efficiency for conversion of light energy into electric energy?

19. A magnifying glass of diameter 10 cm focuses sunlight into a spot of diameter 0.50 cm. The energy flux in the sunlight incident on the lens is 0.10 W/cm².
   (a) What is the energy flux in the focal spot? Assume that all points in the spot receive the same flux.

---

[4] It is of interest that in the Soviet Union, where many experiments on the effects of microwaves on the human body have been performed, the accepted standard is much lower, 10 microwatts/cm².

(b) Will newspaper ignite when placed at the focal spot? Assume that the flux required for ignition is 2 W/cm².

20. Binoculars are usually marked with their magnification and lens size. For instance, 7 × 50 binoculars magnify angles by a factor of 7 and their collecting lenses have an aperture of diameter 50 mm. Your pupil, when dark adapted, has an aperture of diameter 7.0 mm. When observing a distant pointlike light source at night, by what factor do these binoculars increase the energy flux penetrating your eye?

21. At night, the naked, dark-adapted eye can see a star provided the energy flux reaching the eye is 8.8×10^{-11} W/m².
  (a) Under these conditions, how many watts of power enter the eye? The diameter of the dark-adapted pupil is 7.0 mm.
  (b) Assume that in our neighborhood there are, on the average, 3.5×10^{-3} stars per cubic light-year and that each of these emits the same amount of light as the Sun (3.9×10^{26} W). If so, how many stars could we see in the sky with the naked eye? How far would the faintest visible star be?

22. The beam of a powerful laser has a diameter of 0.2 cm and carries a power of 6 kW. What is the time-average Poynting vector in this beam? What are the amplitudes of the electric and the magnetic fields?

23. Calculate the instantaneous Poynting vector for the electromagnetic wave described in Problem 7.

24. A radio transmitter emits a time-average power of 5 kW in the form of a radio wave with uniform intensity in all directions. What are the amplitudes of the electric and magnetic fields of this radio wave at a distance of 10 km from the transmitter?

25. A radio receiver has a sensitivity of 2×10^{-4} V/m. At what maximum distance from a radio transmitter emitting a time-average power of 10 kW will this radio receiver still be able to detect a signal? Assume that the transmitter radiates uniformly in all directions.

26. A TV transmitter emits a spherical wave, i.e., a wave of the form given by Eq. (1). At a distance of 5 km from the transmitter, the amplitude of the wave is 0.22 V/m. What is the time-average power emitted by the transmitter?

27. A steady current of 12 A flows in a copper wire of radius 0.13 cm.
  (a) What is the longitudinal electric field[5] in the wire?
  (b) What is the magnetic field at the surface of the wire?
  (c) What is the magnitude of the radial Poynting vector at the surface of the wire?
  (d) Consider a 1.0-m segment of this wire. According to the Poynting vector, what amount of power flows into this piece of wire from the surrounding space?
  (e) Show that the power calculated in part (d) coincides with the power of the Joule heat developed in the 1.0-m segment of wire.

28. A stationary electric point charge is located in a uniform magnetic field. Draw a diagram showing the lines of electric and magnetic field. Sketch the direction of the Poynting vector at a few places. Roughly sketch the flow lines for the energy according to the Poynting vector.

*29. A coaxial cable transmits DC power from a source of emf to a load (Figure 36.20). The cable consists of a long conducting straight wire of radius a surrounded by a conducting shell of radius b; assume that these conductors have zero resistance. The source has an emf $\mathscr{E}$ and delivers a power P.
  (a) Show that the electric field within the coaxial cable is

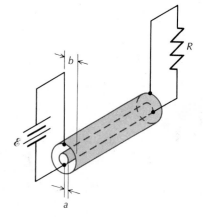

Fig. 36.20 A coaxial cable consisting of a central wire surrounded by a cylindrical shell. The radius of the wire is a; the radius of the shell is b.

---

[5] At the surface and outside the wire, the electric field also has a radial component. This contributes a *longitudinal* component to the Poynting vector, corresponding to the transport of energy *along* the wire.

$$E = \frac{\mathcal{E}}{\ln(b/a)} \frac{1}{r} \quad \text{for } a < r < b$$

(b) What is the current in each conductor of the cable? Show that the magnetic field within the coaxial cable is

$$B = \frac{\mu_0}{2\pi} \frac{P}{\mathcal{E}} \frac{1}{r} \quad \text{for } a < r < b$$

(c) What is the Poynting vector within the coaxial cable?
(d) Integrate the Poynting vector over the annular surface $a < r < b$ and show that the result equals P.

Section 36.5

30. According to a proposed scheme, solar energy is to be collected by a large power station on a satellite orbiting the Earth. The energy is then to be transmitted down to the surface of the Earth as a beam of microwaves. At the surface of the Earth, the beam is to have a width of about 10 km × 10 km and it is to carry a power of 5×10^9 W.
    (a) What would be the time-average Poynting vector in this beam?
    (b) What would be the amplitudes of the electric and magnetic fields in the beam?
    (c) What pressure would the beam exert on the surface of the Earth?

31. According to another proposal for the exploitation of solar energy, a large mirror is to be placed in orbit around the Earth. The mirror reflects sunlight and focuses it on a collector on the surface of the Earth.
    (a) One version of this proposal calls for a mirror 370 km in diameter. What force does the pressure of sunlight exert on such a mirror when face on to the Sun?
    (b) To prevent the mirror from drifting away under the influence of this force, it will have to be controlled with rocket thrusters. The mass of the mirror and its control system has been estimated as 5.7×10^9 kg. If the rocket thrusters are *not* switched on, how far would the mirror drift in one hour, starting from rest? Ignore the orbital motion of the mirror.

32. A grain of dust is spherical, with a diameter of 1.0×10^{-6} m. It consists of material with a density 2.0×10^3 kg/m³. The grain is floating in interplanetary space, its distance from the Sun equal to the Earth–Sun distance.
    (a) What is the attractive force of the Sun's gravity on the grain?
    (b) What is the repulsive force of the Sun's radiation pressure? Which force is larger? Assume that the grain completely absorbs the incident sunlight.
    (c) Repeat the above calculations if the diameter of the grain is 0.50×10^{-6} m.
    (d) Show that the ratio of the forces of gravity and radiation pressure is independent of the distance of the grain from the Sun.

33. Astronauts of the future could travel all over the Solar System in a spaceship equipped with a large "sail" coated with a reflecting material. Such a "sail" would act as a mirror; the pressure of sunlight on this mirror could support and propel the spaceship. How large a "sail" do we need to support a spaceship of 70 metric tons (equal to the mass of Skylab) against the gravitational pull of the Sun? Ignore the mass of the "sail."

34. A beam of light is incident on a mirror at an angle of 30° with the normal and is entirely reflected. The energy incident in some time interval is 10^5 J. How much momentum is transferred to the mirror in this time interval?

35. A sinusoidal electromagnetic wave travels to the right in a coaxial waveguide consisting of an inner wire surrounded by an outer shell. The radius of the inner conductor is a and the radius of the outer conductor is b. Figure

36.21 shows the electric field lines and the magnetic field lines at one instant of time.

  (a) Sketch the direction of the Poynting vector.

  (b) The electric and magnetic fields in the region between the conductors have the mathematical forms

$$E = \frac{A}{r} \sin\left(\frac{\omega}{c} x - \omega t\right)$$

$$B = \frac{A}{cr} \sin\left(\frac{\omega}{c} x - \omega t\right)$$

where A is a constant, r is the radial distance from the axis of the wire ($b > r > a$), x is the distance along the wire (direction of propagation), and the other symbols have their usual meaning. Find the instantaneous power that comes out of the coaxial waveguide at its far end at $x = L$. Find the average power.

  (c) Suppose that at the far end of the waveguide, the wave is absorbed by a perfect absorber. What is the average force exerted by the wave on the absorber?

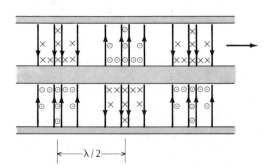

Fig. 36.21 Electric and magnetic field lines of an electromagnetic wave traveling toward the right in a coaxial waveguide. The drawing shows a longitudinal cross section of the waveguide. The electric field lines are radial lines. The magnetic field lines are circles around the central conductor; the dots and the crosses mark the heads and the tails of the magnetic field vectors.

36. Suppose that a magnetic monopole (source of magnetic field lines) and an electric point charge (source of electric field lines) are at some distance from one another.

  (a) Draw a diagram showing the overlapping electric and magnetic field lines. Sketch the direction of the Poynting vector at a few places. Describe in words or pictures along what paths the energy flows in the space around the monopole and the charge.

  (b) The Poynting vector gives not only the flow of energy, but also that of momentum. Accordingly, do you expect that the fields sketched in part (a) possess angular momentum?

*37. Question 20 describes how Einstein derived his equation $E = mc^2$, which expresses the equivalence of energy and mass. Formulating the arguments presented in this question mathematically, calculate how much mass the light pulse of energy E transports from one end of the boxcar to the other, and thereby derive Einstein's equation.

Section 36.6

38. In the spectrum of the light reaching us from the quasar PKS 0106+01 astronomers find some distinctive light emitted by hydrogen atoms. The wavelength of this light as received on Earth is 3776 Å. The wavelength as emitted by the atoms on the quasar is 1216 Å. Find the velocity of recession that will produce this shift of wavelength.

39. The light reaching us from the nearby galaxy NGC 221 has a Doppler shift, mainly due to the orbital motion of the Sun around the center of our Galaxy. A wavelength $\lambda_0 = 3968.5$ Å generated by calcium atoms in NGC 221 has shifted to $\lambda = 3965.8$ Å when detected by us on Earth. What is the radial speed of NGC 221 relative to us? Are we moving toward or away from this galaxy?

40. A spent Scout rocket, falling toward the surface of the Earth at 7.0 km/s, emits a radio signal at a frequency of 2203.08 MHz. What is the frequency of the radio signal received on the surface of the Earth?

41. The light reaching us from all the distant galaxies exhibits a red shift caused by the motion of recession of these galaxies. This is called the **cosmological red shift.** Astronomers usually measure this red shift by means of a wavelength generated by calcium atoms; in the rest frame of the atoms the wavelength is 3968.5 Å.[6] The following table lists several distant galaxies and the corresponding Doppler-shifted wavelength detected on the Earth. The table also lists the distances of these galaxies.

Galaxy	Wavelength	Distance
In Virgo Cluster	3984 Å	78×10^6 light-years
In Ursa Major Cluster	4167	980
In Corona Borealis Cluster	4254	1400
In Bootes Cluster	4485	2500
In Hydra Cluster	4776	4000

    (a) Calculate the velocities of recession of these galaxies. Make a plot of the distances vs. the velocities.

    (b) Assuming that the galaxies had the same velocities at earlier times, calculate at what time all of the galaxies were at zero distance from our Galaxy.

42. Doppler radar units, employed by police to measure the speed of automobiles, consist of a transmitter of microwaves and a receiver. The transmitter sends a wave of fixed frequency toward the target and the receiver detects the reflected wave sent back by the target. During this process the wave suffers two Doppler shifts: first when reaching the target and second when returning to the receiver.

    (a) Suppose that an automobile approaches a Doppler radar unit at a speed of 60 mi/h. Calculate the fractional change of frequency between the transmitted and the received waves.

    (b) In the radar unit, the transmitted and received frequencies are allowed to beat against one another. Suppose that the frequency of the transmitted microwaves is 8.0×10^9 Hz. What is the beat frequency?

43. Astronomers measure the velocity of rotation of stars by observing the Doppler shifts of light emitted by the approaching and the receding edge of the star. Suppose that the surface of a star emits a spectral line of a wavelength of 4101.74 Å. On the Earth, the wavelength received from the opposite edges of the surface are 4101.77 Å and 4101.71 Å.[7] What are the velocities of ap-

---

[6] This is a dark spectral line, i.e., an absorption line.

[7] The light received from other points of the surface will have a Doppler shift somewhere between the Doppler shifts of the light from the edges. Thus, the light received on Earth contains a certain continuous range of wavelengths — the sharp spectral line is *broadened* by the Doppler shift. Thermal motions of the atoms on the surface of the star contribute a further broadening.

proach and recession? What is the angular velocity of rotation of the star if its radius is 7.0×10^8 m and its axis of rotation is perpendicular to the line of sight?

44. Suppose that an astronomer attempting to measure the velocity of rotation of a star by the method described in Problem 43 finds that the wavelengths he receives from opposite edges of the star are 4103.82 Å and 4103.76 Å, respectively. As in Problem 43, the wavelength of this light is 4101.74 Å when emitted by the surface of the star. What can the astronomers conclude from this regarding the rotational and translational velocities of the star? Assume that the axis of rotation is perpendicular to the line of sight.

45. What is the percentage difference between the Doppler-shifted frequencies predicted by the formulas (48) and (51) for a receding emitter with $v/c = 0.10$? For a receding emitter with $v/c = 0.90$?

CHAPTER 37

Reflection and Refraction

So far we have examined the propagation of electromagnetic waves only in a vacuum. There, a plane wave will simply propagate in a fixed direction at the constant speed c. But if the wave encounters a region filled with matter — a sheet of metal, a pane of glass, or a layer of water — then the wave will interact with the matter and suffer changes in speed, direction, intensity, and polarization. These changes can of course be calculated from Maxwell's equations, taking into account the electric charges and currents induced in matter. Sometimes the calculations get extremely complicated and, furthermore, Maxwell's equations often tell us more than we want to know. For instance, if a wave is incident on a water surface we may wish to compute the angle at which it penetrates, but we often do not need to know the changes in intensity and polarization. We will see that a simple rule, called Huygens' Construction, permits us to discover, without complicated calculations, how the change of direction of propagation is related to the change of speed of the wave. Huygens' Construction is a geometric construction, bypassing Maxwell's equations. This construction serves as the basis for **geometric optics,** which studies the propagation of light under the assumption that light propagates in a fixed direction (rectilinearly) while in a uniform medium, and suffers changes of direction only when it encounters an interface between two different media.

Geometric optics

37.1 Huygens' Construction

The propagation of a wave can be conveniently described by means of the *wave fronts,* or wave crests, that is, the points at which the wave has maximum amplitude[1] at some instant of time. For example, Figure 37.1 shows the instantaneous wave fronts of the radio wave emitted by an antenna. With the passing of time, each of these wave fronts travels in the outward direction.

The rule governing the propagation of wave fronts is **Huygens' Construction:**

> *To find the change of position of a wave front in a small time interval* Δ*t, draw many small spheres of radius [wave speed]* × Δ*t with centers on the old wave front. The new wave front is the surface of tangency to these spheres.*

The small spheres employed in this construction are called **wavelets.** Figure 37.2 shows how Huygens' Construction applies to the propagation of one of the spherical wave fronts of Figure 37.1. The wave speed in this example is simply *c* and hence the radius of the wavelets is *c* Δ*t*. Note that the wavelets of Figure 37.2 have two tangent surfaces, one in the outward direction and one in the inward direction. Obviously, only the former surface is appropriate to our problem (the latter surface would be appropriate if we were dealing with a convergent wave sent toward, say, a radio transmitter and precisely focused on it; this wave would be the time reverse of the wave sent out by the transmitter).

Huygens' Construction applies not only to propagation of electromagnetic waves in a vacuum, but also to propagation in any transparent medium. As we will see in the following sections, this construction allows us to derive the laws of reflection and refraction. Although our emphasis will be on the propagation of light, Huygens' Construction is a general feature of wave propagation — it applies just as well to sound waves and water waves. The laws of reflection and refraction for all such waves are essentially the same and some of our examples will take advantage of this.

37.2 Reflection

When a light wave encounters a material surface — such as the surface of a pane of glass, or the surface of a pond — part of the wave penetrates the surface and part is reflected. If the surface consists of very smoothly polished metal — such as the silvered surface of a mirror — almost all of the wave is reflected. In this section we will deal with the reflected part of the wave; in the next section we will deal with the part of the wave that penetrates from one medium into the other.

The **law of reflection** for a wave incident on a flat surface is well known: *the angle of incidence equals the angle of reflection.* (The same law of reflection is valid for a wave incident on a small portion of a curved surface, which can be approximated by a flat, tangent surface.) To derive this law from Huygen's Construction, we begin with Figure 37.3a, which shows wave fronts approaching a reflecting surface; one edge of the leading wave front is barely touching the surface at the point *P*. Figure 37.3b shows some Huygens' wavelets a short time later; the portions of these wavelets below the reflecting surface have been omitted as irrelevant. The new wave front touches the reflecting surface at

Huygens' Construction

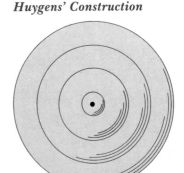

Fig. 37.1 Spherical wave fronts at one instant of time. At a later time, each of these wave fronts will have moved outward by some distance.

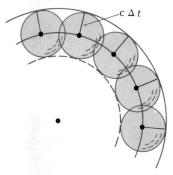

Fig. 37.2 Huygens' Construction for the propagation of a wave front. The inner solid arc shows the wave front at a time *t*; the outer solid arc shows the propagated wave front at time *t* + Δ*t*. The dashed arc shows the propagated wave front for a time-reversed (convergent) wave.

Law of reflection

---

[1] By amplitude of an electromagnetic wave we will hereafter always mean the amplitude of the electric field.

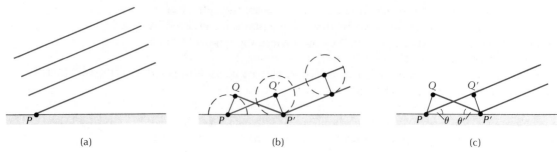

(a) (b) (c)

Fig. 37.3 (a) Wave fronts approaching a reflecting surface. The leading wave front barely touches the reflecting surface. (b) Huygens wavelets erected on the leading wave front of part (a). (c) The incident wave front PQ' makes an angle θ with the reflecting surface; the reflected wave front QP' makes an angle θ' with this surface.

the point P'. Obviously, to the right of the point P', the new wave front is simply parallel to the old wave front, i.e., this part of the wave has not yet been reflected. To find the new wave front to the left of the point P', we draw a straight line that starts at P' and is tangent to the wavelet centered on P. This straight line represents the part of the wave that has already been reflected. To see that the incident wave front and the reflected wave front make the same angle with the reflecting surface, we appeal to Figure 37.3c. The right triangles $PQ'P'$ and $P'QP$ are congruent since they have a common side (PP') and their short sides (PQ and $P'Q'$) are equal. Hence the angles θ and θ' are equal.

Rays

The direction of propagation of a wave is often described by the **rays** of the wave. These are lines perpendicular to the wave fronts. For example, Figure 37.4 shows the rays associated with the incident and reflected wave fronts.

Angle of incidence and of reflection

The angle θ (or θ') between the wave front and a reflecting surface is obviously equal to the angle between the ray and the normal to the surface. The angles θ and θ' are called the **angles of incidence and of reflection** (Figure 37.5). Thus, from the Huygens' Construction we have deduced that the angle of incidence equals the angle of reflection, which is the law of reflection. Note that in all of the above, we have implicitly taken it for granted that the incident and the reflected rays are coplanar. This is evident from considerations of symmetry, and it can also be deduced from Huygens' Construction by taking into account that, in three dimensions, the wave fronts in Figures 37.3a and b are planes and the wavelets are spheres.

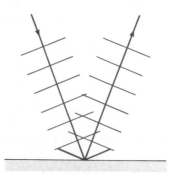

Fig. 37.4 The rays of the wave are perpendicular to the wave fronts.

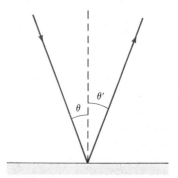

Fig. 37.5 The angle of incidence θ and the angle of reflection θ'. These angles are the same as in Figure 37.3c.

When light from some source strikes a flat mirror, the reflection of the light leads to formation of an image of the source. Figure 37.6 shows a point source of light and the rays emerging from it; it also shows the reflected rays. If we extrapolate the reflected rays to the far side of the mirror, we find that they all appear to come from a point source of light. This apparent point source is the **image.** To an eye looking into the mirror, the image looks like the original object — the eye perceives the mirror image as existing in the space beyond the mirror. But the mirror image is an illusion; the light does not come from beyond the mirror. This kind of illusory image that gives the impression that light rays emerge from where they do not is called a **virtual image.**

If, instead of a single luminous point, our light source consists of an extended object made of many luminous points, then the mirror image will also be an extended object. Note that the mirror image of an object is a mirror-reversed object. For instance, Figure 37.7 shows some written letters and their mirror images. This reversal is commonly referred to as a reversal of left and right. However, it is more accurately described as a reversal of front to back — mirror writing is ordinary writing seen from behind. And the mirror image of, say, a hand facing north is a hand facing south (the reversal is not an ordinary "about face," but involves passing the front of the hand through its back, thereby converting a right hand into a left hand, and vice versa; Figure 37.8).

Virtual image

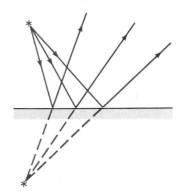

Fig. 37.6 Rays emerging from a point source are reflected by a mirror. The extrapolated rays (dashed) appear to come from a point source beyond the mirror.

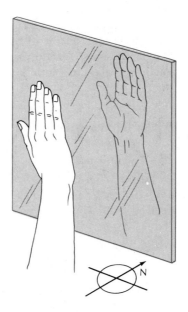

Fig. 37.7 (left) Some letters and their images in a mirror.

Fig. 37.8 (right) A hand facing north and its image in a mirror.

Two mirrors at right angles form a **corner reflector.** A ray of light incident on this reflector is sent back in exactly the same direction it came from. Figure 37.9 shows how this works in two dimensions. But the same principle also applies to three dimensions; here it involves three mirrors at right angles to each other (a corner) and the light ray is sent back upon three successive reflections. Reflectors on automobiles and bicycles make use of such reflectors arranged in an array of a large number of small corners. Radar reflectors on boats apply the same principle; Figure 37.10 shows such a reflector to be hung on the

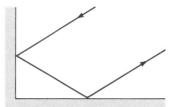

Fig. 37.9 A corner reflector.

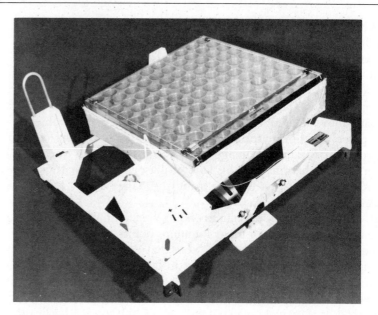

Fig. 37.10 Radar reflector for a sailboat.

Fig. 37.11 Corner reflectors for installation on the surface of the Moon.

mast of a sailboat. Finally, Figure 37.11 shows an array of corner reflectors deposited on the Moon by the Apollo 11 astronauts. This was used as a target for a laser beam sent from the Earth to the Moon, in an experiment that determined the Earth–Moon distance with very high precision (± 15 cm) by measuring the travel time of a pulse of laser light sent from the Earth to the Moon and reflected back to the Earth (see also Interlude L.3).

37.3 Refraction

The speed of light in a material medium — such as air, water, or glass — differs from the speed of light in a vacuum. We can recognize this immediately by recalling the theoretical formula for the speed of light derived from Maxwell's equations,

$$c = \frac{1}{\sqrt{\varepsilon_0 \mu_0}} \tag{1}$$

We know from Chapters 27 and 33 that in a material medium with given dielectric and magnetic characteristics, the quantities ε_0 and μ_0 in Maxwell's equations get replaced by $\kappa \varepsilon_0$ and $\kappa_m \mu_0$ [compare Eqs. (27.32) and (33.9)]. Hence, the formula (1) for the speed of light likewise gets replaced by

$$v = \frac{1}{\sqrt{\kappa \kappa_m \varepsilon_0 \mu_0}} \tag{2}$$

This is usually written as

$$\boxed{v = \frac{c}{n}} \qquad (3)$$

Speed of light in a material medium

where

$$n = \sqrt{\kappa \kappa_m} \qquad (4)$$

The quantity n is called the **index of refraction** of the material. In most materials the value of κ_m is very near 1 except, of course, in ferromagnetic materials, where light does not propagate anyhow (see Tables 33.1 and 33.3). Therefore the factor κ_m in Eq. (4) can often be omitted. This permits us to write

$$\boxed{n = \sqrt{\kappa}} \qquad (5)$$

Index of refraction

One important warning: the value of the dielectric κ depends on the frequency of the electric field. Hence the values of the dielectric constants from Table 27.1 cannot be inserted in Eq. (5), because the former values only apply to static fields whereas we are now concerned with the high-frequency fields of a light wave.

Table 37.1 gives the values of the index of refraction for a few materials. The values in this table apply to light waves of medium frequency (yellow-green light). The index of refraction is slightly larger for blue light and slightly smaller for red light; we will deal with this complication later in this section.

With the wave speed $v = c/n$, the relation between frequency and wavelength becomes

$$\lambda \nu = v = \frac{c}{n}$$

or

$$\boxed{\lambda = \frac{c}{n}\frac{1}{\nu}} \qquad (6)$$

Table 37.1 INDICES OF REFRACTION OF SOME MATERIALS[a]

Material	*n*
Air, 1 atm, 0°C	1.00029
1 atm, 15°C	1.00028
1 atm, 30°C	1.00026
Water	1.33
Ethyl alcohol	1.36
Castor oil	1.48
Quartz, fused	1.46
Glass, crown	1.52
light flint	1.58
heavy flint	1.66

[a] For light of wavelength ~5500 Å.

For example, if a wave penetrates from air into water, where $n = 1.33$, its speed is reduced by a factor of 1.33, but its frequency remains constant. Consequently, Eq. (6) shows that its wavelength will be reduced by a factor of 1.33. The fact that the frequency of the wave remains constant can be understood in terms of the atomic mechanism underlying the change of wave velocity. When the wave strikes the water surface, it shakes the electrons of the water molecules; this acceleration of electric charges produces extra waves, which combine with the original wave. The net electromagnetic wave within the water is then a superposition of the original wave plus the extra waves, and it is this superposition that makes up the refracted wave with its changed direction. The superposition has the same frequency as the original wave because the shaking of electrons proceeds at the original frequency and therefore the extra waves generated also will have exactly this frequency.

Incidentally, Eq. (6) does not imply that a light source changes color when immersed in water. The color we perceive depends on the *frequency* of the light reaching our eyes; and this frequency is independent of whether the light source, our eyes, or both are immersed in water or in air.

When a wave strikes the surface of a dielectric, part of it is reflected and part of it penetrates the dielectric. In the preceding section we have investigated the direction of propagation of the reflected wave; now let us investigate the penetrating wave. Again, we will use Huygens' Construction to find out what the wave does when it strikes the dielectric surface. In vacuum the speed of light is c; in the dielectric it is c/n. Figure 37.12a shows a wave front at one instant of time; the left edge of this wave front barely touches the surface. Figure 37.12a also shows the Huygens wavelets that determine the position of the wave front at some later time; only the forward portions of these wavelets are relevant. Above the dielectric surface, the wavelet has a radius $c\,\Delta t$; below the dielectric surface, the wavelet has a smaller radius $(c/n)\,\Delta t$. The reduction of the speed of propagation of one side of the wave front causes the wave front to swing around, changing its direction of advance. The right triangles $PP'Q$ and $PP'Q'$ have the side PP' in common (Figure 37.12b). In terms of the length l of this common side, the sines of the angles between the wave fronts and the dielectric surface are

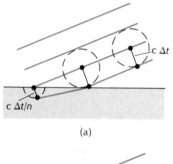

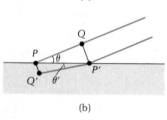

(a)

(b)

Fig. 37.12 (a) Huygens wavelets erected on a wave front whose edge barely touches the dielectric surface. (b) The incident wave front makes an angle θ with the dielectric surface; the refracted wave front makes an angle θ'.

$$\sin \theta = \frac{c\,\Delta t}{l} \tag{7}$$

and

$$\sin \theta' = \frac{(c/n)\,\Delta t}{l} \tag{8}$$

The ratio of this pair of equations is

(9)
$$\frac{\sin \theta}{\sin \theta'} = \frac{c}{c/n}$$

or

$$\boxed{\sin \theta = n \sin \theta'} \qquad (10) \qquad \textit{Law of refraction}$$

This equation describes the change of direction of a wave upon penetration into a dielectric. This change of direction is called **refraction.** Equation (10) is called the **law of refraction,** or **Snell's Law.** The angle θ is the angle of incidence, and θ' is the angle of refraction. It is usually convenient to measure these angles between the rays and the normal to the dielectric surface; Figure 37.13 shows the incident and refracted rays, and the angles of incidence and refraction. Note that the ray in the dielectric is bent toward the normal ($\theta' < \theta$). Also note that, as in the case of reflection, the incident and the refracted rays are coplanar.

Our formula (10) describing refraction at the interface between vacuum and dielectric is a special case of a general formula describing refraction at the interface of two dielectrics. If the indices of refraction are n_1, n_2 and the angles between the rays and the normals are θ_1, θ_2, then

$$n_1 \sin \theta_1 = n_2 \sin \theta_2 \qquad (11)$$

This equation can be derived by some simple changes in the argument that led to Eq. (10).

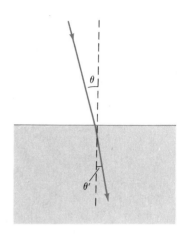

Fig. 37.13 The angle of incidence θ and the angle of refraction θ'.

EXAMPLE 1.   A small, shiny fish is in the water 1.0 m below the surface. Where does a fisherman looking down into the water see the fish; that is, where is the image of the fish?

SOLUTION: Figure 37.14 shows two light rays from the fish to the eye of the fisherman. The first light ray is perpendicular to the surface and is not bent. The second light ray is bent away from the normal. Since the angles are small, Eq. (10) gives

$$\theta = n\theta' = 1.33\theta' \qquad (12)$$

Extrapolation of the refracted ray into the water shows that it intersects the vertical ray at the point P', above the point P. Hence the image is above the object. The image distance OP' and the object distance OP are related as follows (Figure 37.14):

$$OP \tan \theta' = OP' \tan \theta \qquad (13)$$

For small angles this yields the approximation

$$\frac{OP'}{OP} = \frac{\theta'}{\theta} = \frac{1}{1.33} \qquad (14)$$

Hence the image distance is smaller than the object distance by a factor of 1.33. If $OP = 1$ m, then $OP' = 0.75$ m. The fish seems to be nearer to the surface than it is.

Note that the above calculation hinges on the assumption that the angles are small (so that $\sin \theta \cong \theta$, $\tan \theta \cong \theta$).

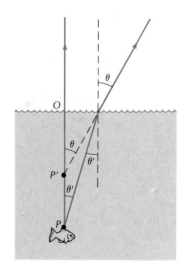

Fig. 37.14 A small shiny fish acts as a source of light. Note that the direction of propagation of the ray is opposite to that shown in Figure 37.13; but this does not affect the validity of Eq. (10). The extrapolated ray (dashed) appears to come from the point P'.

For a ray of light attempting to leave water, there is a critical angle beyond which refraction is impossible. As the light ray emerges into air it is bent away from the normal; in the extreme case it is bent so much that it lies almost along the water surface (Figure 37.15). This extreme case corresponds to $\theta = 90°$ in Eq. (10),

$$n \sin \theta' = \sin 90° = 1 \tag{15}$$

The critical angle for this extreme form of refraction is therefore

Critical angle for total internal reflection

$$\theta'_c = \sin^{-1}\left(\frac{1}{n}\right) \tag{16}$$

which for $n = 1.33$ gives

$$\theta'_c = \sin^{-1}\left(\frac{1}{1.33}\right) = 48.75° \tag{17}$$

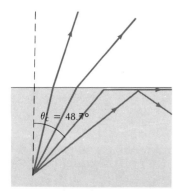

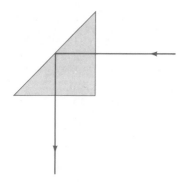

Fig. 37.15 A ray approaching a water surface from below with an angle of incidence $\theta'_c = 48.7°$ is refracted along the water surface.

If a light ray strikes a water surface from below at an angle larger than this, refraction is impossible. The only alternative is reflection — the water surface behaves as a perfect mirror. This phenomenon is called **total internal reflection.** It will occur whenever the index of refraction of the medium containing the light ray is larger than the index of refraction of the adjacent medium.

Total internal reflection has many important applications in optics. For instance, in a periscope the light is reflected down the tube by internal reflection in a prism (Figure 37.16); this gives a much better image than reflection in a mirror. In an optical fiber, light moves along a thin rod made of a transparent material; the light zigzags back and forth between the walls of the rod, undergoing a sequence of total internal reflections (Figure 37.17). Such optical fibers are being used to replace telephone cables. The electrical impulses normally carried on a cable are converted into pulses of light which can be transmitted via an optical fiber. The efficiency of optical fibers is very high because they can carry many conversations simultaneously; in modern telephone systems, single optical fibers are being used to carry 670 telephone conversations simultaneously.

In most materials the index of refraction depends somewhat on the wavelength of light. Usually, the index of refraction increases as the wavelength decreases. For instance, Figures 37.18 and 37.19 are plots of the index of refraction of light in water and in flint glass as a function of wavelength (the wavelengths plotted along the horizontal axis in these figures are measured in air, before the light penetrates the

Fig. 37.16 Internal reflection in a prism.

Fig. 37.17 (above) Internal reflection in an optical fiber. (right) Light enters the optical fiber at the top right and emerges at the center.

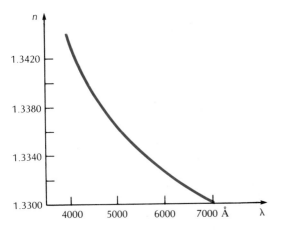

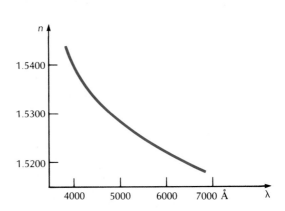

Fig. 37.18 Index of refraction of light in water as a function of wavelength. The index of refraction varies by about 1% over the range of visible wavelengths.

Fig. 37.19 Index of refraction of light in glass (Schott telescope flint glass) as a function of wavelength.

material). When a ray of light containing several wavelengths, or colors, is refracted by a medium with an index of refraction that depends on wavelength, the refracted rays of different colors will emerge at somewhat different angles. The separation of a ray by refraction into distinct rays of different colors is called **dispersion.**

Dispersion

EXAMPLE 2. The index of refraction for red light in water is 1.330 and for violet light it is 1.342. Suppose that a ray of light approaches a water surface with an angle of incidence of 80°. What are the angles of refraction for red light and for violet light?

SOLUTION: For $n = 1.330$, Eq. (10) yields

$$\sin \theta' = \frac{\sin 80°}{1.330} = 0.740 \tag{18}$$

and

$$\theta' = 47.8°$$

For $n = 1.342$, Eq. (10) yields

$$\sin \theta' = \frac{\sin 80°}{1.342} = 0.734 \tag{19}$$

and

$$\theta' = 47.2°$$

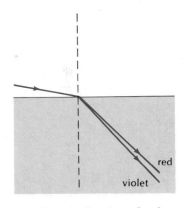

Fig. 37.20 Refraction of red and of violet light in water. The difference between the angles of the refracted rays has been exaggerated for the sake of clarity.

Thus the violet light is bent more toward the normal than the red light (Figure 37.20). For $\theta = 80°$, the difference in the angles of refraction is about 0.6°. Thus, refraction in water slightly separates a light ray according to colors. A beautiful demonstration of this effect is found in rainbows, which are produced by the refraction of sunlight in water droplets.

A **prism** is the traditional device employed to separate light rays into their constituent colors. The basic mechanism is the same as that discussed in Example 2: the glass in the prism has different indices of re-

Prism

fraction for light of different wavelengths and hence it bends rays of different colors by different amounts (Figure 37.21). In passing through a prism, the light is refracted twice: first at the air–glass interface and then at the glass–air interface. Under normal operating conditions, a good prism will introduce a difference of several degrees between the angular directions of the emerging red and violet light rays. The pattern of colors produced by the analysis of a light ray by

Spectrum

means of a prism is called the **spectrum** of the light. The white light emitted by the Sun has a continuous spectrum consisting of a mixture of all the colors. The colored light emitted by the atoms of a chemical element in an electric discharge tube, such as a neon tube, has a discrete spectrum consisting of just a few discrete colors. Each of these discrete colors is absolutely pure, i.e., it is light of a single wavelength. For example, hydrogen atoms emit the following discrete colors: red (6563 Å), blue-green (4861 Å), blue-violet (4340 Å), and violet

Spectral lines

(4102 Å). These discrete colors are called **spectral lines.**[2] Figure 37.22 shows the spectral lines of hydrogen as displayed by means of a prism illuminated with light from a fine slit. Each of the lines in this figure is a separate image of the slit made by a separate color after refraction by the prism (Figure 37.23).

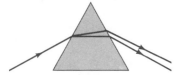

Fig. 37.21 Refraction of red and of violet light by a prism.

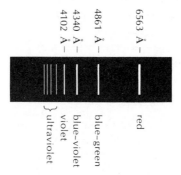

Fig. 37.22 The spectral lines of hydrogen. (A color print of these spectral lines appears between pages 850 and 851.)

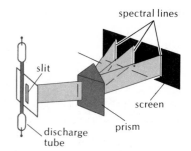

Fig. 37.23 Analysis of light by means of a prism.

37.4 Spherical Mirrors

A mirror with a surface curved like the surface of a sphere can focus a beam of light to a point. Figure 37.24 shows a *concave* spherical mirror and wave fronts of light incident and reflected on this mirror; the re-

Focal point

flected waves converge to a point, the **focal point** of the mirror. We can describe the direction of propagation of the waves by rays. Figure 37.25 shows the incident and the reflected rays. The reflected rays converge at the focal point.

The focal point of the spherical mirror is halfway between the mirror and the center of the spherical surface. To prove this, we use Figure 37.26, which shows the path of a single ray of light. The focal point is the intersection of this ray with the axial line *CA*. To find the

---

[2] Hydrogen also emits spectral lines in the ultraviolet and in the infrared; see Chapter 41.

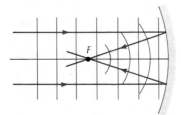

Fig. 37.24 A concave spherical mirror focuses an incident plane wave on a point.

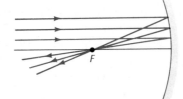

Fig. 37.25 Reflection of parallel rays by a concave spherical mirror.

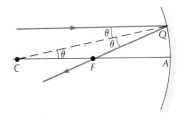

Fig. 37.26 Reflection of a single ray.

distance *FA*, called the **focal length,** we begin with the observation that in the isosceles triangle *CFQ* the length *CF* equals *FQ*. Under the assumption that the angle θ is small (equivalently, that the incident ray is near the axial line), the length *FQ* is approximately equal to *FA*. Hence

$$CF = FA \qquad (20)$$

that is, the point *F* is halfway between the mirror (*A*) and the center (*C*) of the spherical surface. The focal length *FA* is therefore one-half of the radius of the spherical surface. Designating the former by *f* and the latter by *R*, we can write

$$\boxed{f = \tfrac{1}{2}R} \qquad (21)$$

Focal length of spherical mirror

Note that this focusing of a beam of light rests on the approximation that the beam is narrow so that all the rays are near the axial line. If a ray strikes the mirror at some appreciable distance from the axial line, then the reflected ray will miss the focal point (Figure 37.27). This defect in the focusing properties of a mirror is called **spherical aberration.** For better focusing, it is necessary to replace the spherical mirror by a paraboloidal mirror (Figure 37.28).

Spherical aberration

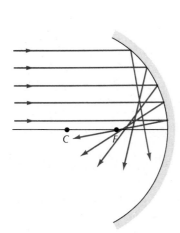

Fig. 37.27 Reflection of rays far from the axis of the mirror. Such rays miss the focal point.

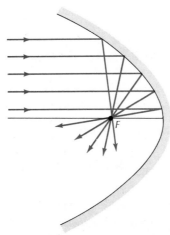

Fig. 37.28 Reflection of parallel rays by a paraboloidal mirror. All rays converge at the focal point.

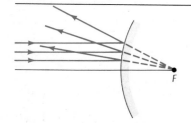

Fig. 37.29 Reflection of parallel rays by a convex spherical mirror.

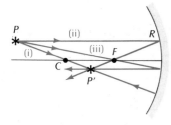

Fig. 37.30 A point source of light P in front of a concave mirror. The rays (i), (ii), and (iii) intersect at the image P'.

Figure 37.29 shows a *convex* spherical mirror. Parallel rays incident on this mirror diverge upon reflection. If we extrapolate the divergent rays to the far side of the mirror, they all seem to come from a single point, the focal point of the convex mirror. An argument similar to that given above shows that the focal length is again one-half of the radius of the spherical surface,

$$f = -\tfrac{1}{2}R \tag{22}$$

A negative sign has been inserted in Eq. (22) to indicate that the focal point is on the far side of the mirror.

Both concave and convex mirrors will form images of objects placed in front of them. Figure 37.30 shows a point source of light in front of a concave mirror. To find the position of the image, we must trace some of the rays of light. The three rays that are easiest to trace are (i) the ray PC through the center, (ii) a ray PR parallel to the axis, and (iii) a ray PF through the focal point (Figure 37.30). The first of these rays strikes the mirror perpendicularly and is therefore reflected on itself; the second ray passes through the focus after being reflected; and the third ray emerges parallel to the axis after being reflected. All these rays, and any other rays originating at P, come together at P'. This point is the image of the point source. Note that to locate the image, two out of the three rays mentioned above are already sufficient — the third is redundant but serves as a useful check.

If the source of light is an extended object, then we must find the image of each of its points. For instance, a luminous object in the shape of an arrow has an image as shown in Figure 37.31. We can easily verify this by drawing the rays that emerge from, say, the midpoint of the arrow, the tail of the arrow, etc.

The ray-tracing technique summarized in Figure 37.30 is a graphical method for finding the image of a source. This method also applies to convex mirrors; an example of this is shown in Figure 37.32.

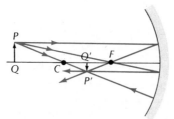

Fig. 37.31 An object PQ in the shape of an arrow and its image $P'Q'$ formed by a concave mirror.

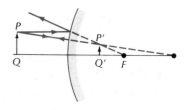

Fig. 37.32 An object PQ and its image $P'Q'$ formed by a convex mirror.

The position of the image can be calculated algebraically by means of the **mirror equation**

Mirror equation

$$\frac{1}{s} + \frac{1}{s'} = \frac{1}{f} \tag{23}$$

Here s is the distance from the object to the mirror and s' is the distance from the image to the mirror. The distance s or s' is positive if the object or image is in front of the mirror; the distance s or s' is negative if the object or image is behind the mirror.[3] As already indicated above, f is positive for a concave mirror, negative for a convex mirror.

To derive Eq. (23), we make use of Figure 37.33, which shows an object PQ, its image $P'Q'$, and two rays. The ray PCP' passes through the center of the spherical surface and is reflected on itself; the ray PAP' strikes the center of the mirror and is reflected symmetrically with respect to the axial line so that the angles θ and θ' are equal. The triangles PQA and $P'Q'A$ are similar; hence

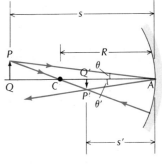

$$\frac{PQ}{P'Q'} = \frac{s}{s'} \qquad (24)$$

Fig. 37.33 The angles θ and θ' are equal, hence the right triangles PQA and $P'Q'A$ are similar.

The triangles PQC and $P'Q'C$ are also similar; hence

$$\frac{PQ}{P'Q'} = \frac{QC}{Q'C} \qquad (25)$$

or, since $QC = s - R$ and $Q'C = R - s'$,

$$\frac{PQ}{P'Q'} = \frac{s - R}{R - s'} \qquad (26)$$

Combining Eqs. (24) and (26), we find

$$\frac{s}{s'} = \frac{s - R}{R - s'} \qquad (27)$$

By a bit of algebraic manipulation, we can rearrange this equation to read

$$\frac{1}{s} + \frac{1}{s'} = \frac{2}{R} \qquad (28)$$

Obviously, Eq. (28) is equivalent to Eq. (23).

EXAMPLE 3.   A candle is placed 41 cm in front of a convex spherical mirror of radius 60 cm. Where is the image?

SOLUTION: With $s = 41$ cm and $f = 30$ cm, Eq. (23) gives

$$\frac{1}{41 \text{ cm}} + \frac{1}{s'} = \frac{1}{30 \text{ cm}} \qquad (29)$$

or $s' = 112$ cm. The positive sign indicates that the image is on the same side of the mirror as the object (Figure 37.34).

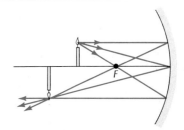

Fig. 37.34 The real image lies in the space in front of the mirror.

The image in the preceding example is a **real image.** The light rays not only seem to come from this image, but they actually do. As Figure

Real image

---

[3] The object can be behind the mirror if what serves as object for the mirror is actually an image produced by another mirror or by a lens.

37.34 shows, the light rays pass through the image and diverge from it, just as they diverge from the object. The real image is in front of the mirror. Visually, it gives the impression of a ghostly replica of the object floating in midair.

37.5   Thin Lenses

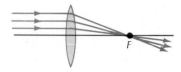

Fig. 37.35 Refraction of rays by a convex lens. The focal length is positive.

Lens-maker's formula

A lens made of a refracting material with two spherical surfaces will focus a beam of light to a point (Figure 37.35). For a thin lens, the focal length is given by the **lens-maker's formula**

$$\frac{1}{f} = (n - 1)\left(\frac{1}{R_1} + \frac{1}{R_2}\right) \tag{30}$$

where n is the index of refraction of the material of the lens and R_1, R_2 are the radii of the two spherical surfaces making up the lens. This equation is based on the assumption that the lens is thin (its thickness is small compared to R_1, R_2) and that the incident rays are near the axial line. A ray that enters the lens at some appreciable distance from the axis will miss the focus; this is a form of aberration analogous to spherical aberration of a mirror.

Equation (30) can be derived by tracing rays through the lens, taking into account their refraction at the two curved surfaces. We will not perform this tedius calculation and only note that the focusing depends on the fact that rays far from the axial line strike the surface of the lens with a larger angle of incidence than rays near the axial line. Thus, the far rays are refracted through a larger angle, i.e., they are bent more sharply toward the axis. This is of course exactly what is required to make these far rays cross the axis at the same point (the focus) as the near rays.

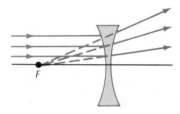

Fig. 37.36 Refraction of rays by a concave lens. The focal length is negative.

Equation (30) may also be applied to a concave lens (Figure 37.36). In this case, the radii R_1, R_2 must be reckoned as *negative* and the focal distance f is then also negative. The meaning of a negative value of f is the same as in the case of mirrors: parallel rays incident on the lens diverge when they emerge from the lens, and the focal point is the point at which the extrapolated rays appear to intersect (Figure 37.36). Furthermore, Eq. (30) can be applied to a concave–convex lens (Figure 37.37). Whether such a lens produces net convergence or divergence depends on whether the positive radius (convex) or the negative radius (concave) is smaller; for instance, the lens of Figure 37.37 will produce convergence.

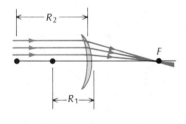

Fig. 37.37 Refraction of rays by a concave–convex lens. The convex surface (right surface) has a smaller radius than the concave surface (left surface). In Eq. (30), the radius of the former surface is reckoned positive ($R_1 > 0$) and the radius of the latter surface is negative ($R_2 < 0$). The sum $1/R_1 + 1/R_2$ is positive.

Note that a lens has *two* focal points at equal distances right and left of the lens. The point on the right of a converging lens is the focus for a parallel beam coming from the left and, conversely, the point on the left is the focus for a parallel beam coming from the right.

To find the image of an object placed near the lens, we can use a ray-tracing technique similar to that used for mirrors. Figure 37.38 shows three rays that are easy to trace: (i) the ray PQP' that starts parallel to the axis and ultimately passes through the focus F, (ii) the ray PCP' that passes undeflected through the center of the lens, and (iii) the ray $PQ'P'$ that passes through the focus F' and emerges parallel to the axis. All these rays intersect at the point P', the image point.

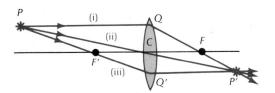 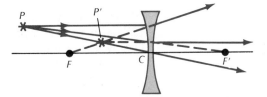

Fig. 37.38 A point source of light *P* in front of a convex lens. The rays (i), (ii), and (iii) intersect at the image *P'*.

Fig. 37.39 A point source of light *P* in front of a concave lens, and the image *P'*.

As in the case of mirrors, two of the above three rays are already sufficient to locate the image. And, of course, the same ray-tracing technique can be applied to concave lenses (Figure 37.39).

The equation to be used for the algebraic calculation of the image distances is the same as Eq. (23),

$$\boxed{\frac{1}{s} + \frac{1}{s'} = \frac{1}{f}}$$

(31)   *Lens equation*

but the sign conventions are slightly different. The object distance *s* is positive if the object is on the near side of the lens and negative if it is on the far side; the image distance *s'* is positive if the image is on the far side of the lens and negative if it is on the near side. In this context, the "near" side is the side from which the light rays are incident on the lens and the "far" side is the other side. The sign of *f* is positive for a convex lens, negative for a concave lens.

Although the derivation of the lens equation (31) can be based on a geometric argument similar to that used for the mirror formula (23), we can bypass this labor by a trick. We begin by noting that a concave mirror is equivalent to one-half of a convex lens placed directly in front of a flat mirror. If the concave mirror and the (entire) convex lens have the same value of *f*, then the two arrangements shown in Figures 37.40a and b have exactly the same optical properties — in both cases the image distances are the same. If we now remove the flat mirror in Figure 37.40a and replace the one-half lens by the entire lens, the image distance will remain the same, but the image will form on the opposite side of the lens, i.e., the sign of the image distance will be reversed. Consequently, the same equation (23) must apply to the concave mirror and the convex lens; the only difference is that the sign of the image distance is reversed. This reversal of sign has already been taken into account in our description of the sign conventions associated with Eqs. (23) and (31) — for the mirror, *s'* is taken as *positive* if on the near side of the mirror, whereas for the lens, *s'* is taken as *negative* if on the near side of the lens. There is of course a similar correspondence between a convex mirror and a concave lens.

Fig. 37.40 (a) A point source of light and its image formed by a concave mirror. (b) A similar image is formed by one-half of a convex lens placed in front of a flat mirror. Note that each ray has to pass through the one-half lens twice: once before reflection by the mirror, once after. The deflection suffered by a ray in two passages through one-half of a lens is the same as that in a single passage through the entire lens. (c) A similar image is also formed by the entire lens, but the image is now on the other side of the lens.

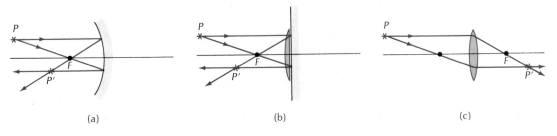

(a)   (b)   (c)

EXAMPLE 4.   A convex lens of focal length 25 cm is placed at a distance of 10 cm from a printed page. What is the image distance? How much larger is the image of the page than the page?

SOLUTION: With $s = 10$ cm and $f = 25$ cm, Eq. (31) gives

$$\frac{1}{10\text{ cm}} + \frac{1}{s'} = \frac{1}{25\text{ cm}}$$

which yields $s' = -16.7$ cm. The negative sign indicates that the image is on the near side of the lens (Figure 37.41). The image is virtual.

Since the triangles $P'Q'C$ and PQC are similar, the sizes of image and object are in the ratio

$$\frac{P'Q'}{PQ} = \frac{Q'C}{QC} = \frac{|s'|}{s} \tag{32}$$

$$= \frac{16.7\text{ cm}}{10\text{ cm}} = 1.67$$

i.e., the image is larger than the object by a factor of 1.67. This is the principle involved in the magnifying glass.

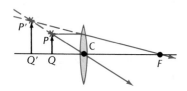

Fig. 37.41 An object PQ and its image $P'Q'$.

37.6 Optical Instruments

The simple lenses described in the preceding section suffer from diverse kinds of aberrations. Parallel rays that strike the lens at an appreciable distance from the axis will miss the focal point; this is **spherical aberration,** similar to the spherical aberration of a mirror. Rays of different colors will be refracted differently by the glass of the lens, and they will therefore be focused differently; this is **chromatic aberration.** Furthermore, the images of objects of large size will display complicated three-dimensional distortions. All these troublesome aberrations can be partially eliminated by combining several simple lenses of different shapes made of different kinds of glass into a compound lens. For instance, Figure 37.42 shows the Zeiss "Tessar" lens, one of the most best-known photographic lenses. In such a compound lens, the individual lenses are carefully designed so that their aberrations tend to cancel mutually.

Most optical instruments — cameras, magnifiers, microscopes, and telescopes — employ compound lenses. However, in the following discussion of optical instruments, we will schematically represent the compound lenses by single lenses of appropriate focal lengths.

THE PHOTOGRAPHIC CAMERA   The lens of the camera forms a real image of the object on the photographic film and thereby imprints this image on the film (Figure 37.43). The distance between the lens and the film is adjustable so that the image can always be made to fall on the film, regardless of the object distance. A shutter controls the exposure time during which light is admitted to the camera. For a given exposure time, the amount of light entering the camera is proportional to the area of the lens. Thus, a large lens permits photography with dim light. The size of a camera lens is commonly labeled by the **f**

Spherical and chromatic aberrations

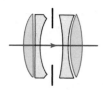

Fig. 37.42 Zeiss "Tessar" lens. The outer lenses (color) are made of crown glass, and the inner lenses (gray) of flint glass.

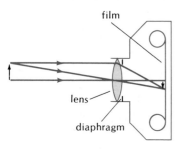

Fig. 37.43 A photographic camera.

number, which is defined as the ratio of the focal length of the lens to its diameter. For instance, a lens of focal length 55 mm and diameter 32 mm has an f number of 55:32, or 1.7. A lens of small f number is said to be "fast" because it collects sufficient light for a photograph with a short exposure time. Good cameras have an adjustable iris diaphragm that can be used to block part of the area of the lens and thereby alter the effective f number. If the diaphragm is closed down so that only a small central portion of the lens remains unblocked, the camera will require a long exposure time, but it will have a large depth of field — it simultaneously forms sharp images for objects spanning a large range of object distances. This is so because the rays emitted from any point of an object enter the camera within a narrow cone, and they therefore intersect at the image within a narrow cone; thus these rays will be close together on the photographic film even if the position of the image does not fall exactly on the film.

A **pinhole camera** can be regarded as an ordinary camera with the iris diaphragm closed down to a point. The lens can then be discarded since the rays pass through its center, where they suffer no deflection (Figure 37.44). To obtain a sharp image, we must use an extremely small pinhole; hence such a camera requires very bright light or a very long exposure time. Figure 37.45 is a picture taken with a pinhole camera.

f *number*

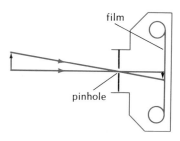

Fig. 37.44 A pinhole camera.

Pinhole camera

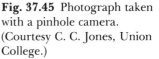

Fig. 37.45 Photograph taken with a pinhole camera. (Courtesy C. C. Jones, Union College.)

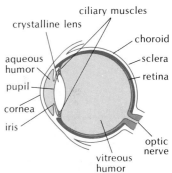

Fig. 37.46 The human eye (natural size). The space between the cornea and the crystalline lens is filled with a transparent jelly (the aqueous humor). The main body of the eye is also filled with a transparent jelly (the vitreous humor). The indices of refraction of the humors are 1.34, nearly the same as for water. The index of refraction of the crystalline lens is 1.44. The sclera is the thick white outer casing of the eye. The choroid is a pigmented black membrane that absorbs stray light, like the black paint in cameras.

THE EYE In principle, the eye is similar to a camera. The lens of the eye forms a real image on the retina, a delicate membrane packed with light-sensitive cells which send nerve impulses to the brain. Figure 37.46 shows a section through the human eye; the diameter of the eyeball is typically 2.3 cm. The cornea and the aqueous humor act as a lens; they provide most of the refraction for rays entering the eye. The crystalline lens merely provides the fine adjustment of focal length required to make the image of an object fall on the retina, regardless of the object distance. The crystalline lens is flexible; its focal length is adjusted by the ciliary muscles. If the eye is viewing a distant object, the muscles are relaxed and the lens is fairly flat, with a long focal length. If the eye is viewing a nearby object, the muscles are contracted and the lens is more rounded, with a shorter focal length. The shortest attainable focal length determines the shortest distance at which an object can be placed from the eye and still be seen sharply. This shortest distance is called the **near point.** For a normal young adult, the near point is typically 25 cm. With advancing age the lens loses its flexibility and the near point recedes; for instance, at an age of 60 years, the near point is typically around 200 cm.

Nearsightedness and farsightedness

The two most common optical defects of the eye are nearsightedness and farsightedness. In a **nearsighted** (myopic) eye, the focal length is excessively short, even when the ciliary muscles are completely relaxed. Thus, parallel rays from a distant object come to a focus in front of the retina and fail to form a sharp image on the retina — vision of distant objects is blurred. This condition can be corrected by eyeglasses with divergent lenses. In a **farsighted** (hyperopic) eye, the focal length is excessively long, even when the ciliary muscles are fully contracted (in other words, the near point of the eye is farther away than normal). Hence rays from a nearby object converge toward an image beyond the retina and fail to form a sharp image on the retina. This condition can be corrected by eyeglasses with convergent lenses. In old age, both of these conditions often occur simultaneously, through the loss of flexibility of the crystalline lens and the weakening of the ciliary muscles. The correction then requires bifocal lenses, with a lower convergent portion for near vision, and an upper divergent portion for far vision.

THE MAGNIFIER   In order to see fine detail with the naked eye, we must bring the object very close to the eye, so that the angular size of the object is large and, correspondingly, the image on the retina is large (Figure 37.47). This means we want to bring the object to the near point, at a typical distance of 25 cm for the eye of a young adult. To see finer detail, we need a magnifier. This consists of a strongly convergent lens placed adjacent to the eye (Figure 37.48).[4] Such a lens permits us to bring the object closer to the eye, and thereby increase the size of the image on the retina.

The angular magnification of the magnifier is defined as the ratio of the angular size of the image at infinity produced by the magnifier (as in Figure 37.48) to the angular size of the object seen by the naked eye at the standard distance of 25 cm (as in Figure 37.47). Since the angles in Figures 37.47 and 37.48 are small, the angles are approximately equal to the tangents:

$$\theta \cong \tan \theta \cong h/25 \text{ cm} \tag{33}$$

and

$$\theta' \cong \tan \theta' = h/f \tag{34}$$

where h is the size of the object. Taking the ratio of these angles, we find

---

[4] Note that such a magnifier is *not* the same thing as a magnifying glass. In common use, the magnifying glass is placed at an appreciable distance from the eye, near the object to be magnified, because this maximizes the magnification. A magnifying glass can be regarded as a magnifier (in the technical sense of this word) only if it is placed next to the eye.

$$[\text{angular magnification}] = \theta'/\theta = \frac{25 \text{ cm}}{f} \qquad (35)$$

Angular magnification of magnifier

This tells us the magnification relative to the (typical) naked eye, i.e., it tells us how much better the magnifier is than the naked eye. For example, a magnifier with $f = 5$ cm has an angular magnification of 25 cm/5 cm = 5. Note that this result is valid only under the assump-

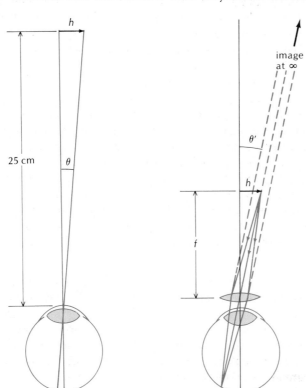

Fig. 37.47 (left) The angular size of the object determines the size of the image on the retina. Here the object has been placed at a distance of 25 cm from the eye.

Fig. 37.48 (right) The magnifier is adjacent to the eye. The object has been placed at a distance slightly shorter than the focal distance, so that the eye sees the image at infinity.

tions that the magnifier is placed adjacent to the eye, and that the object is placed near the focus of the magnifier so that the image is at infinity. The second of these assumptions is not crucial — if the object is placed closer than the focus, the magnification will be changed only slightly. But the first assumption is crucial — if the magnifier is placed at some appreciable distance from the eye, then the magnification will be very different!

THE MICROSCOPE   The microscope consists of two lenses: the objective and the ocular, or eyepiece. Both of these lenses have very short focal lengths. The objective is placed near the object, and it forms a real, magnified image of the object. This image serves as object for the ocular, which acts as a magnifier and forms a virtual image at infinity (Figure 37.49). Thus, both the objective and the ocular contribute to the magnification of the microscope. The net angular magnification of the microscope is the angular magnification of the ocular multiplied by the magnification of the objective. The angular magnification of the ocular is given by Eq. (35); and the magnification of the objective is given by Eq. (32), where s and s' are, respectively, the object and image distances for the objective (these distances are shown in Figure 37.49). Hence the net angular magnification of the microscope is

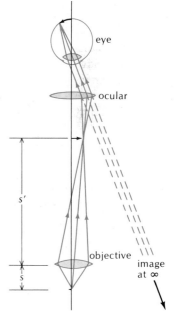

Fig. 37.49 Arrangement of lenses in a microscope. The object to be magnified is below the objective and the eye is just above the ocular. The ocular forms a virtual image at infinity, and the lens of the eye focuses on the retina the parallel rays emerging from the ocular.

Angular magnification of microscope

$$[\text{angular magnification}] = \frac{25 \text{ cm}}{f_{oc}} \times \frac{s'}{s} \qquad (36)$$

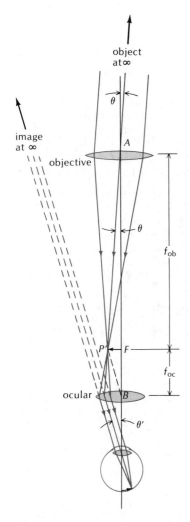

Fig. 37.50 An astronomical telescope. The object is at a large distance above. The observer's eye is below the ocular. The ocular forms a virtual image at infinity, and the lens of the eye focuses on the retina the parallel rays emerging from the ocular.

Note that, as in the case of the magnifier, this tells us the magnification relative to the (typical) naked eye. Good microscopes operate at magnifications of up to 1400. Although higher magnifications can be achieved, this serves little purpose because the diffraction of the light waves at the objective limits the detail that can be resolved. To overcome this limitation we need to use waves of shorter wavelength, such as the electron waves used in electron microscopes.

THE TELESCOPE   A simple astronomical telescope consists of an objective of very long focal length and an ocular of short focal length. These two lenses are separated by a distance (nearly) equal to the sum of their individual focal lengths, so that their focal points coincide. The objective forms a real image of a distant object. This image serves as object for the ocular, which forms a magnified virtual image at infinity (Figure 37.50). To find the angular magnification produced by this telescope, we begin by noting that the lens equation with $s = \infty$ applied to the objective gives

$$\frac{1}{\infty} + \frac{1}{s'} = \frac{1}{f_{ob}}$$

i.e.,

$$s' = f_{ob} \qquad (37)$$

This means that the image is at the focal distance, and verifies that the location of the image (FP') is correctly shown in Figure 37.50. We can therefore use the geometric relationships contained in this figure. The angular magnification is the ratio of the angles θ' and θ that represent, respectively, the angular sizes of the object and the final image viewed by the eye. Since both angles are small,

$$[\text{angular magnification}] = \frac{\theta'}{\theta}$$

$$\cong \frac{\tan \theta'}{\tan \theta} = \frac{FP'/FB}{FP'/FA} = \frac{FA}{FB} \qquad (38)$$

i.e., the angular magnification is the ratio of the focal length of the objective to that of the ocular,

Angular magnification of telescope

$$[\text{angular magnification}] = \frac{f_{ob}}{f_{oc}} \qquad (39)$$

For example, an astronomical telescope with $f_{ob} = 120$ cm and $f_{oc} = 2.5$ cm has an angular magnification of 120 cm/2.5 cm = 48.

Many astronomical telescopes are reflecting telescopes in which a concave mirror plays the role of the objective. The mirror forms a real image that serves as object for the ocular. Of course, the ocular must

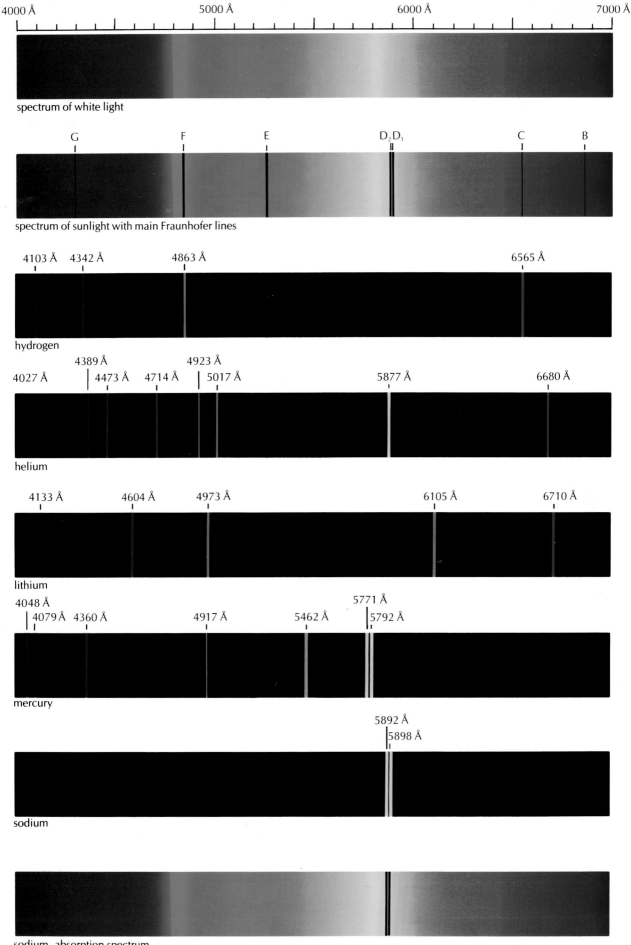

4000 Å   5000 Å   6000 Å   7000 Å

spectrum of white light

G   F   E   D₂D₁   C   B

spectrum of sunlight with main Fraunhofer lines

4103 Å   4342 Å   4863 Å   6565 Å

hydrogen

4389 Å   4923 Å
4027 Å 4473 Å 4714 Å 5017 Å   5877 Å   6680 Å

helium

4133 Å   4604 Å   4973 Å   6105 Å   6710 Å

lithium

4048 Å   5771 Å
4079 Å 4360 Å   4917 Å   5462 Å 5792 Å

mercury

5892 Å
5898 Å

sodium

sodium, absorption spectrum

(All wavelengths are measured in vacuum.)

Courtesy Eastman Kodak Company

be placed in front of the mirror (and blocks out some of the light). Figure 37.51 shows this arrangement. Note that Figure 37.51 is essentially the same as Figure 37.50 folded over at the position of the objective lens; hence the geometry of the light rays in reflecting and refracting astronomical telescopes is essentially the same. Because large mirrors of good quality, free of aberrations, are easier to manufacture than large lenses of good quality, the largest astronomical telescopes all use mirrors. For instance, the telescope on Mt. Palomar uses a mirror of diameter 510 cm (200 in.) and of a focal length 1680 cm; the mirror is aspherical for the sake of better imaging (Figure 37.52).

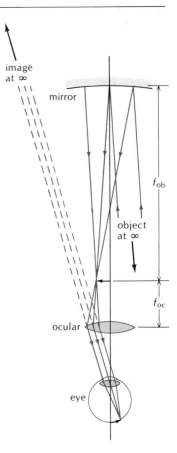

Fig. 37.51 A reflecting telescope. The object has been placed below, to facilitate comparison with Figure 37.50.

Fig. 37.52 The 200-in. telescope on Mt. Palomar. The mirror can be seen under the observation cage.

SUMMARY

Huygens' Construction: The new wave front is the surface of tangency of the wavelets erected on the old wave front.

Law of reflection: The angle of incidence equals the angle of reflection.

Index of refraction: $v = c/n$

$$n = \sqrt{\kappa}$$

Law of refraction: $\sin \theta = n \sin \theta'$

Critical angle for total internal reflection:

$$\theta'_c = \sin^{-1}\left(\frac{1}{n}\right)$$

Focal length of spherical mirror: $f = \pm\frac{1}{2}R$ (f is positive for concave mirror, negative for convex)

Mirror equation: $\dfrac{1}{s} + \dfrac{1}{s'} = \dfrac{1}{f}$ (s or s' is positive if object or image is in front of mirror, negative if behind)

Lens-maker's formula: $\dfrac{1}{f} = (n - 1)\left(\dfrac{1}{R_1} + \dfrac{1}{R_2}\right)$ (f is positive for convex lens, negative for concave)

Lens equation: $\dfrac{1}{s} + \dfrac{1}{s'} = \dfrac{1}{f}$ (s is positive if on near side of lens, negative if on far side; s' is positive if on far side, negative if on near side)

Angular magnification of magnifier: $25 \text{ cm}/f$

Angular magnification of microscope: $(25 \text{ cm}/f_{oc}) \times (s'/s)$

Angular magnification of telescope: f_{ob}/f_{oc}

QUESTIONS

1. When light is incident on a smooth surface — a glass surface, a painted surface, a water surface — the reflection is strongest if the angle of incidence is near 90° (grazing incidence). Can Huygen's Construction explain this?

2. In celestial navigation, the navigator measures the angle between the Sun, or some other celestial body, and the horizon with a sextant. If the navigator is on dry land, where the horizon is not visible, he can measure instead the angle between the Sun and its reflection in a pan full of water, and divide this angle by two. Explain.

3. Suppose we release a short flash of light in the space between two parallel mirrors placed face to face. Why does this light flash not travel back and forth between the two mirrors forever?

4. What is the minimum size of a mirror hanging on a wall such that you can see your entire body when standing in front of it?

5. Artists are notorious for making mistakes when drawing or painting mirror images. What is wrong with the position and orientation of the mirror images shown in the cartoon in Figure 37.53?

Fig. 37.53

6. Two parallel mirrors are face to face. Describe what you see if you stand between these mirrors.

7. Figure 37.54 shows spots of sunlight on a wall in the shade of a tree. The spots were made by sunlight that has passed through very small gaps between the leaves of the tree. Explain why all the spots are round and of nearly the same size, even though the gaps are of irregular shape and size. (Hint: The Sun is round.)

8. If you immerse one-half of a stick in the water, it will appear bent. Explain.

9. At sunset, the image of the Sun remains visible for some time after the actual position of the Sun has sunk below the horizon. Explain.

10. After a navigator measures the angle between the Sun and the horizon with a sextant, he must make a correction for the refraction of sunlight by the atmosphere of the Earth. Does this refraction increase or decrease the apparent angle between the Sun and the horizon?

11. Figure 37.55 shows water waves refracted in the shallows near a headland. This refraction deflects the waves toward the headland. Explain this, keeping in mind that the speed of water waves in shallow water is proportional to the square root of the depth ($v = \sqrt{gh}$).

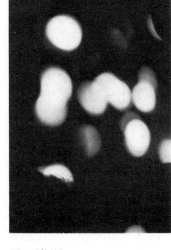

Fig. 37.54

Fig. 37.55

12. At amusement parks you find mirrors that make you look very short and fat or very tall and thin. What kinds of mirrors achieve these effects?

13. Storeowners often install convex mirrors at strategic locations in their stores to supervise the customers. What is the advantage of a convex mirror over a flat mirror?

14. You look toward a lens or a mirror and you see the image of an object. How can you tell whether this image is real or virtual?

15. Hand mirrors are sometimes concave, but never convex. Why?

16. If you place a book in front of a concave mirror, will it show you mirror writing? Does the answer depend on the distance of the book from the mirror?

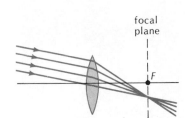

Fig. 37.56 The dashed line indicates the focal plane, seen edge on.

Fig. 37.57 Fresnel lens of the lighthouse at Point Reyes, California.

17. You place an object in front of a concave mirror, at a distance smaller than the focal length. Is the image real or virtual? Erect or inverted? Magnified or reduced? What if the distance is greater than the focal length? What if the mirror is convex?

18. How could you make a lens that focuses sound waves?

19. A convex lens is made of glass of index of refraction 1.2. If you immerse this lens in water, will it produce convergence or divergence of incident parallel rays?

20. If you place a small light bulb at the focus of a convex lens and look toward the lens from the other side, what will you see?

21. Consider Figure 37.38. How do we know that the ray $PQ'P'$ emerges parallel to the axis of the lens?

22. Are the distances to the two focal points of a thick lens necessarily the same?

23. Suppose you place a magnifying glass against a flat mirror and look into the glass. What do you see if your face is very near the glass? If it is not very near?

24. Consider a beam of parallel rays incident on a convex lens at a small angle with the axis of the lens (Figure 37.56). Show that these parallel rays will be focused at a point in a plane that is perpendicular to the axis and passes through the focal point. This plane is called the **focal plane.**

25. You place an object in front of a convex lens at a distance smaller than the focal length. Is the image real or virtual? Erect or inverted? Magnified or reduced? What if the distance is greater than the focal length? What if the lens is concave?

26. Lenses, like mirrors, suffer from aberration. Consider rays incident on a convex lens, parallel to the axis. If the point of incidence is far from the axis, would you expect the ray to pass in front of the focus or behind?

27. Figure 37.57 shows a large **Fresnel lens** used in the lantern of a lighthouse. The lens consists of annular segments, each with a curved surface similar to the curved surface of an ordinary lens. Why is this arrangement better than a single curved surface?

28. The telescopes built by Galileo Galilei consisted of a convex objective lens and a concave ocular lens. The lenses are arranged so that the focal point of the objective coincides with the focal point of the ocular (Figure 37.58). Explain how this telescope produces an angular magnification. Show that the eye sees an erect image. (Hint: The concave lens placed *before* the point F in Figure 37.58 has the same effect as the convex lens placed *beyond* the point F in Figure 37.50 — the lens produces an image at infinity. Is this image erect or inverted?)

29. Binoculars use prisms to reflect the light back and forth (Figure 37.59). What is the purpose of these prisms, and what is their advantage over mirrors?

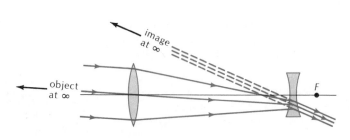

Fig. 37.58 Galilean telescope.

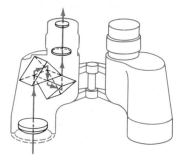

Fig. 37.59 Arrangement of prisms in a binocular.

PROBLEMS

Section 37.2

1. According to a (questionable) story, Archimedes set fire to the Roman ships besieging Syracuse by focusing the light of the Sun on them with mirrors. Suppose that Archimedes used flat mirrors. How many flat mirrors must simultaneously reflect sunlight at a piece of canvas if it is to catch fire? The energy flux of the sunlight at the surface of the Earth is 0.1 W/cm² and the energy flux required for ignition of canvas is 4 W/cm². Assume that the mirrors reflect the sunlight without loss.

2. A vertical mirror, oriented toward the Sun, throws a rectangular patch of sunlight on the floor in front of the mirror. The size of the mirror is 0.5 m × 0.5 m and its bottom rests on the floor. If the Sun is 50° above the horizon, what is the size of the patch of sunlight on the floor?

3. Two mirrors meeting at a right angle make a corner reflector (see Figure 37.9). Prove that a ray of light reflected successively by both mirrors will emerge on a path antiparallel to its original path.

Section 37.3

4. The speed of sound in air is 340 m/s and in water it is 1500 m/s. If a sound wave in air approaches a water surface with an angle of incidence of 10°, what will be the angle of refraction?

5. Make a plot of the angle of incidence vs. the angle of refraction for light rays incident on a water surface. What is the maximum angle of refraction?

6. A ship's navigator observes the position of the Sun with his sextant and measures that the Sun is exactly 39° away from the vertical. Taking into account the refraction of the Sun's light by air, what is the true angular position of the Sun? For the purpose of this problem assume that the Earth is flat and that the atmosphere can be regarded as a flat, transparent plate of uniform density and an index of refraction 1.0003.

7. A ray of light strikes a plate of window glass of index of refraction 1.5 and thickness 2.0 mm with an angle of incidence of 50°.
   (a) Show that the transmitted ray leaving the glass on the other side is parallel to the incident ray.
   (b) Find the lateral displacement between the transmitted ray and the extrapolation of the incident ray.[5]

8. A point source of light is placed above a thick plate of glass of index of refraction n (Figure 37.60). The distance from the source to the upper surface of the plate is l and the thickness of the plate is d. A ray of light from the source may suffer either a single reflection at the upper surface, or a single reflection at the lower surface, or multiple alternating reflections at the lower and upper surfaces. Thus, each ray splits into several rays, giving rise to multiple images. In terms of l and d, find the distance of the first, second, and third images below the upper surface of the plate. Assume that the angle of incidence of the ray is small.[5]

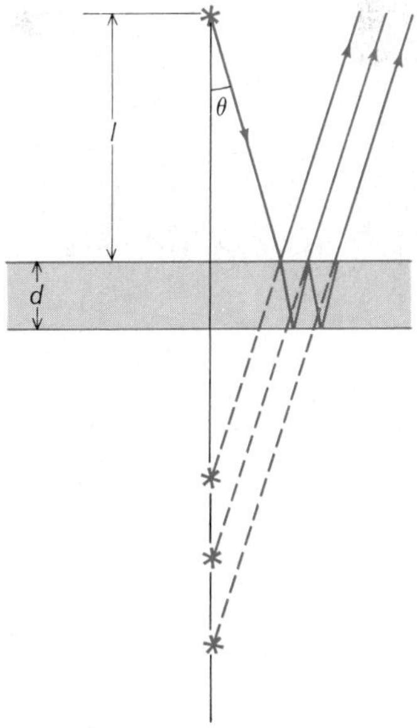

9. The bottom half of a beaker of depth 20 cm is filled with water ($n = 1.33$) and the top half is filled with oil ($n = 1.48$). If you look into this beaker from above, how far below the upper surface of the oil does the bottom of the beaker seem to be?

10. Because of refraction in the plate of glass, an object viewed through an ordinary window will seem somewhat nearer than its actual distance.
   (a) Consider the rays that strike the glass at a small angle of incidence

Fig. 37.60 Ray of light reflected and refracted by a plate of glass.

---

[5] In this problem assume that the index of refraction of air is 1.

(nearly normal). Show that if an object is at a distance l from a window, the image is at a distance that is shorter by an amount $\Delta l = (1 - 1/n)d$ where n is the index of refraction and d the thickness of the glass.

(b) What change in distance does this formula give if the windowpane is ordinary glass with $d = 2.0$ mm and $n = 1.5$?

(c) What change in distance does this formula give if the window pane is heavy plate glass with $d = 8.0$ mm and $n = 1.5$?

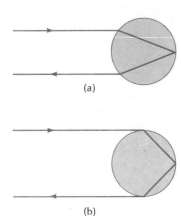

(a)

(b)

Fig. 37.61 (a) Path of a light ray that reverses direction in a transparent sphere. (b) Path of a light ray that penetrates at maximum distance from the axis of the sphere.

11. The highly reflective paint used on highway signs contains small glass beads which reverse the direction of propagation of a light ray, throwing it back toward the source of light (see Figure 37.61a). The reversal of direction will occur only for those light rays incident on the sphere of glass at a selected distance from the axis, a distance that depends on the index of refraction of the glass. Show that if the reversal of direction is to occur at all, the index of refraction of the glass must be in the range $1 < n \leq \sqrt{2}$. (Hint: Consider the light ray shown in Figure 37.61b; what index of refraction does the reversal of this light ray require?)

12. A signal rocket explodes at a height of 200 m above a ship on the surface of a smooth lake. The explosion sends out sound waves in all directions. Since the speed of sound in water (1500 m/s) is larger than the speed of sound in air (340 m/s), a sound wave can suffer total reflection at a water surface if it strikes at a sufficiently large angle of incidence. At what minimum distance from the ship will a sound wave from the explosion suffer total reflection?

13. An optical fiber is made of a thin strand of glass of index of refraction 1.5. If a ray of light is to remain trapped within this fiber, what is the largest angle it may make with the surface of the fiber?

14. (a) A transparent medium of index of refraction n adjoins a transparent medium of index of refraction n'. Assuming $n > n'$, show that the critical angle for total internal reflection of a ray attempting to leave the first medium is

$$\theta_c = \sin^{-1}\left(\frac{n'}{n}\right)$$

(b) A layer of kerosene ($n' = 1.2$) floats on a water surface. In this case, what is the critical angle for total internal reflection within the water?

15. A layer of oil, of index of refraction n', floats on a surface of water. A ray of light coming from below attempts to pass from the water to the oil and from there to the air above. What is the maximum angle of incidence of the ray on the water–oil surface that will permit the ultimate escape of the ray into the air? Does you answer depend on n'? (Hint: Use the formula derived in Problem 14.)

16. A seagull sits on the (smooth) surface of the sea. A shark swims toward the seagull at a constant depth of 5 m. How close (measured horizontally) can the shark approach before the seagull can see it?

17. To discover the percentage of sucrose (cane sugar) in an aqueous solution, a chemist determines the index of refraction of the solution very precisely and then finds the percentage in a table giving the dependence of index of refraction on sucrose concentration. He determines the index of refraction by immersing a glass prism in the sucrose solution and measuring the critical angle for total internal reflection of a light ray inside the glass prism.

(a) Suppose that with a prism of index of refraction 1.6640 the critical angle is 57.295°. Use the result of Problem 14 to find the index of refraction of the sucrose solution.

(b) Use the following table, interpolating if necessary, to find the concentration of sucrose to four significant figures.

Concentration	n
40.00%	1.3997
40.10	1.3999
40.20	1.4001
40.30	1.4003

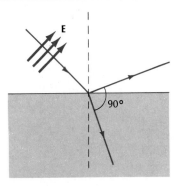

18. Consider a light wave incident on a dielectric medium of index of refraction n. It can be shown that if the light is polarized in the plane of incidence (plane of the page in Figure 37.62) and the refracted ray is perpendicular to the reflected ray, then the intensity of the reflected wave is zero. Show that the angle of incidence that makes the refracted ray perpendicular to the reflected ray is given by $\tan \theta = n$. This relation is called **Brewster's Law.**

Fig. 37.62 The light wave is polarized in the plane of the page. The refracted ray is perpendicular to the reflected ray.

*19. Rainbows are produced by the refraction of sunlight by drops of water. Figure 37.63 shows a ray of light entering a spherical drop of water. The ray is refracted at A, reflected at B, and refracted at C. The angle of incidence at A (between the ray and the normal to the surface) is θ and the angle of refraction is θ'.

   (a) By geometry, show that the angles of incidence and reflection at B coincide with θ' and that the angles of incidence and refraction at C coincide with θ' and θ, respectively.

   (b) Show that the angular deflection of the ray from its path is $\theta - \theta'$ at A, $\pi - 2\theta'$ at B, and $\theta - \theta'$ at C. These angles are all in radians and they are measured clockwise from the incident path at each point.

   (c) The total angular deflection of the ray by the raindrop is $\Delta = 2(\theta - \theta') + \pi - 2\theta'$. A rainbow will form when all the rays within an infinitesimal range $d\theta$ of angles of incidence (on different drops) suffer the same angular deflection, i.e., when the derivative $d\Delta/d\theta = 2 - 4d\theta'/d\theta$ is zero. If this condition is satisfied, the rays sent back by the raindrops are concentrated, producing a bright zone in the sky. Show that the critical angle θ_c, at which $d\Delta/d\theta = 0$, is given by

$$\cos^2 \theta_c = \tfrac{1}{3}(n^2 - 1)$$

   where n is the index of the refraction of water.

   (d) The index of refraction for red light in water is 1.330. Find θ_c and find Δ in degrees and minutes of arc. Draw a diagram showing a red ray coming from the Sun, hitting the drop, and reaching the eye of a rainbow watcher.

   (e) The index of refraction for violet light in water is 1.342. Find θ_c and find Δ. On top of the preceding diagram, draw a violet ray coming from the Sun, hitting the drop (at a different point), and reaching the eye of the rainbow watcher. Will the watcher see the red color above or below the violet?

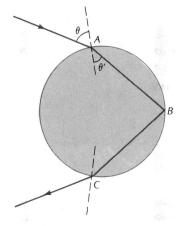

Fig. 37.63 Path of light ray in a drop of water.

Section 37.4

20. At what distance from a concave mirror of radius R must you place an object if the image is to be at the same position as the object?

21. A concave mirror has a radius of curvature R. If you want to form a real image, within what range of distances from the mirror must you place the object? If you want to form a virtual image, within what range of distances must you place the object?

22. A woman's hand mirror is to show a (virtual) image of her face magnified 1.5 times when held at a distance of 20 cm from the face. What must be the radius of curvature of a spherical mirror that will serve the purpose? Must it be concave or convex?

23. The surface of a highly polished doorknob of brass has a radius of curva-

ture of 4.5 cm. If you hold this doorknob 15 cm away from your face, where is the image that you see? By what factor does the size of the image differ from the size of your face?

24. A concave mirror of radius 30 cm faces a second concave mirror of radius 24 cm. The distance between the mirrors is 80 cm and their axes coincide. A light bulb is suspended between the mirrors, at a distance of 20 cm from the first mirror.
    (a) Where does the first mirror form an image of the light bulb?
    (b) Where does the second mirror form an image of this image?

25. A concave mirror of radius 60 cm faces a convex mirror of the same radius. The distance between the mirrors is 50 cm, and their axes coincide. A candle is held between the mirrors, at a distance of 10 cm from the convex mirror. Consider rays of light that first reflect off the concave mirror and then off the convex mirror. Where do these rays form an image?

Section 37.5

26. The crystalline lens of a human eye has two convex surfaces with radii of curvature of 10 mm and 6.0 mm. The index of refraction of its material is 1.45. Treating it as a thin lens, what is its focal length when removed from the eye and placed in air?

27. A thin lens of flint glass with $n = 1.58$ has one concave surface of radius 15 cm and one flat surface.
    (a) What is the focal length of this lens?
    (b) If you place this lens at a distance of 40 cm from a candle, where will you find the image of the candle?

28. A thin, symmetric, convex lens of crown glass with index of refraction $n = 1.52$ is to have a focal length of 20 cm. What are the correct radii of the spherical surfaces of the lens?

29. A slide projector has a lens of focal length 13 cm. The slide is at a distance of 2.0 m from the screen. What must be the distance from the slide to the lens if a sharp image of the slide is to be seen on the screen?

30. The convex lens of a magnifying glass has a focal length of 20 cm. At what distance from a postage stamp must you hold this lens if the image of the stamp is to be twice as large as the stamp?

31. If you place a convex lens of focal length 18 cm at a distance of 30 cm from a small light bulb, where will you find the image of the light bulb? Is this a real or virtual image? Is it upright or inverted?

32. Show that if a thin lens of index of refraction n is placed in a medium (e.g., water) of index of refraction n', then the lens-maker's formula (30) must be modified as follows:

$$\frac{1}{f} = \left(\frac{n}{n'} - 1\right)\left(\frac{1}{R_1} + \frac{1}{R_2}\right)$$

[Hint: In the law of refraction — Eq. (11) — only the ratio of the indices of refraction is relevant.]

33. Show that if two thin lenses of focal lengths f_1 and f_2 are placed next to one another (in contact), the net focal length f is given by

$$\frac{1}{f} = \frac{1}{f_1} + \frac{1}{f_2}$$

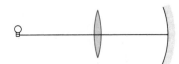

Fig. 37.64 Light bulb, convex lens, and concave mirror.

34. A convex lens of focal length 25 cm is at a distance of 60 cm from a concave mirror of focal length 20 cm. A light bulb is 80 cm from the lens (Figure 37.64).

(a) Where does the lens form an image of the light bulb?

(b) Where does the mirror form an image of this image?

35. Two lenses, one concave and one convex, have equal focal lengths of 30 cm. The lenses are separated by a distance of 10 cm. A candle is 20 cm from the convex lens (Figure 37.65).

(a) Where does the convex lens form an image?

(b) Where does the concave lens form an image of this image?

Fig. 37.65 Candle, convex lens, and concave lens.

36. A light bulb is 15 cm in front of a convex mirror of radius 10 cm. A convex lens of focal length 25 cm is 5 cm beyond the light bulb (Figure 37.66). Where do you see the light bulb if you look through the convex lens at the mirror?

Fig. 37.66 Light bulb, convex mirror, and convex lens.

*37. Figure 37.67a shows a Fresnel lens consisting of a large number of annular segments each of which is an annular portion of an ordinary lens. Fresnel lenses of diameters more than 1 m are commonly used in the lamps of lighthouses where ordinary lenses without segments would be much too thick and too heavy. Consider one of these annular segments (Figure 37.67b); for the sake of simplicity, assume that the width of the segment is infinitesimal. Derive a formula for the angle of inclination θ of the surface of this segment in terms of the index of refraction n, the distance r of the segment from the axis of the lens, and the focal length of the lens.

Section 37.6

38. The lens of a 35-mm camera has a focal length of 55 mm. The distance of the lens from the film is adjustable over a range from 55 mm to 62 mm. Over what range of object distances (measured from the lens) is this camera capable of producing sharp pictures?

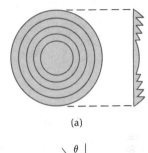

(a)

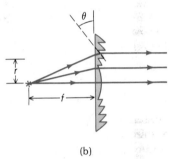

(b)

Fig. 37.67 (a) Fresnel lens, frontal view and cross section. (b) A ray passing through one of the annular segments.

39. A miniature Minox camera has a lens of focal length 15 mm. This camera can be focused on an object as close as 20 cm, or as far away as infinity. What must be the distance from the lens to the film if the camera is set for 20 cm? What if the camera is set for infinity?

40. The light meter of a 35-mm camera with a lens of f number 1.7 indicates that the correct exposure time for a photograph is $\frac{1}{250}$ s. If the iris diaphragm is closed down so that the f number becomes 4, what will be the correct exposure time?

41. Pretend that the cornea and the crystalline lens of the human eye act together as a single thin lens placed at a distance of 2.2 cm from the retina (Figure 37.68). This lens is deformable; it can change its focal length by changing its shape.

(a) What must be the focal length if the eye is viewing an object at a very large distance?

(b) What must be the focal length if the eye is viewing an object at a distance of 25 cm?

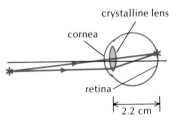

Fig. 37.68 Lens of human eye forms image on retina.

42. In a nearsighted eye the (relaxed) lens has an abnormally short focal length and consequently the eye fails to form an image of a distant object on the retina. This defect can be corrected with a contact lens. Pretend that both the lens of the eye and the contact lens are thin lenses so that the formula given in Problem 33 applies.

(a) Suppose that the focal length of the eye is 2.0 cm. What must be the focal length of the contact lens if it is to increase the net focal length to 2.2 cm? Should it be convergent or divergent?

(b) The radius of curvature of the side of the contact lens next to the eye should be −0.80 cm so as to fit tightly on the cornea. What must be the radius of curvature of the other side? The contact lens is made of plastic with an index of refraction 1.33.

43. Equation (35) gives the angular magnification for a magnifier if the object is so placed that the image is at infinity. Show that if the object is so placed

that the image is at a distance of 25 cm, then the angular magnification is $1 + 25$ cm$/f$.

44. A microscope has an objective of focal length 4.0 mm. This lens forms an image at a distance of 224 mm from the lens. If we want to attain a net angular magnification of 550, what choice must we make for the angular magnification of the ocular?

45. A microscope has an objective of focal length 1.9 mm and an ocular of focal length 25 mm. The distance between these lenses is 180 mm.
   (a) At what distance must the object be placed from the objective so that the ocular forms an image at infinity, as shown in Figure 37.49?
   (b) What is the net angular magnification of this microscope?

46. A telescope has an objective of focal length 160 cm and an ocular of focal length 2.5 cm. If you look into the *objective* (i.e., into the wrong end) of this telescope, you will see distant objects *reduced* in size. By what factor will the angular size of objects be reduced?

47. An amateur astronomer uses a telescope with an objective of focal length 90.0 cm and an ocular of focal length 1.25 cm. What is the angular magnification of this telescope?

48. The large reflecting telescope on Mt. Palomar has a mirror of focal length 1680 cm. If this telescope is operated with an ocular of focal length 1.25 cm, what is the angular magnification?

CHAPTER 38

Interference

Geometric optics relies on the assumption that the sizes of the mirrors or lenses and the separations between them are much larger than the wavelength of light. Under these conditions, we can adequately describe the propagation of light by rays that are rectilinear, except when refracted by the surfaces of dielectric media. Thus, in geometric optics, the wave properties of light do not show up explicitly (although these wave properties enter into the derivation of the laws of reflection and refraction).

Wave optics, or physical optics, deals with the propagation of light in the general case, without any restrictive assumptions as to the sizes of the bodies through or around which the light is propagating. If we let light waves interact with a body or an obstacle of a size comparable to a wavelength, the light waves will display their wave properties explicitly through the phenomena of interference and diffraction. As we saw in Chapter 16, interference is the constructive or destructive combination of two or more waves meeting at one place; diffraction is the bending and spreading of waves around obstacles.

In this chapter we will examine the interference of light waves and of other electromagnetic waves. Like all electric and magnetic fields, the fields of electromagnetic waves obey the principle of linear superposition: if two waves meet at some point, the resultant electric or magnetic field is simply the vector sum of the individual fields. If two waves of equal amplitude meet crest to crest, they combine and produce a wave of doubled amplitude; if they meet crest to trough, they cancel and give a wave of zero amplitude. The former case is called **constructive interference** and the latter **destructive interference.** We will encounter both cases of interference in the following sections. The first section involves the interference between two waves propagating in opposite directions; the other sections involve waves propagating in the same or nearly the same direction.

Thomas Young, *1773–1829, English physicist, physician, and Egyptologist. Young worked on a wide variety of scientific problems, ranging from the structure of the eye and the mechanism of vision, to the decipherment of the Rosetta stone. He revived the wave theory of light, and recognized that interference phenomena provide proof of the wave properties of light.*

Constructive and destructive interference

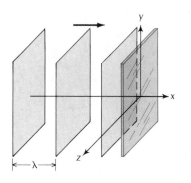

Fig. 38.1 Wave fronts of a plane electromagnetic wave striking a mirror.

38.1 The Standing Electromagnetic Wave

Suppose that an electromagnetic wave strikes a mirror perpendicularly and is totally reflected. The space in front of the mirror will then contain two overlapping waves: the incident wave and the reflected wave. Figure 38.1 shows the mirror in the *y–z* plane. We will assume that the incident wave is a plane harmonic wave, polarized in the *y* direction. The electric fields of the incident and the reflected waves are then

$$E_{\text{in}} = E_0 \cos(\omega t - \omega x/c) \tag{1}$$

$$E_{\text{ref}} = -E_0 \cos(\omega t + \omega x/c) \tag{2}$$

Both these electric fields are in the *y* direction. The first equation represents a wave of frequency ω traveling toward the right, and the second a wave traveling toward the left. A negative sign has been inserted in front of the right side of Eq. (2) so as to satisfy the boundary condition at the mirror; since the mirror at $x = 0$ is a conducting surface, the net electric field must vanish at this point at all times:

$$E_{\text{tot}} = E_{\text{in}} + E_{\text{ref}} = E_0 \cos \omega t - E_0 \cos \omega t = 0 \tag{3}$$

The negative sign in Eq. (2) means that the reflecting surface not only reverses the direction of propagation of the incident wave, but also reverses the electric field. It can be shown that this reversal of the electric field is a general feature of reflection whenever the index of refraction of the medium in which the wave is propagating is smaller than the index of refraction of the medium off which the wave reflects.[1]

In the region $x < 0$, the superposition of the incident and reflected waves gives

$$E_{\text{tot}} = E_{\text{in}} + E_{\text{ref}}$$

$$= E_0[\cos(\omega t - \omega x/c) - \cos(\omega t + \omega x/c)] \tag{4}$$

With the trigonometric identity $\cos(\alpha - \beta) - \cos(\alpha + \beta) = 2 \sin \alpha \sin \beta$, this simplifies to

$$E_{\text{tot}} = 2E_0 \sin \omega t \sin \omega x/c \tag{5}$$

Standing wave

This is a **standing wave**, i.e., a wave whose peaks do not travel right or left but remain fixed in space while the entire wave increases and decreases in unison — the wave pulsates (compare Section 15.6). The maxima of the wave are at the points where

$$\frac{\omega x}{c} = \frac{\pi}{2}, \frac{3\pi}{2}, \frac{5\pi}{2}, \cdots \tag{6}$$

Since the wavelength of the wave is $\lambda = 2\pi c/\omega$, we can also write this as

---

[1] As was pointed out in Section 27.3, a conducting medium (such as the metal of a mirror) can be regarded as having an infinite value of κ and hence, according to Eq. (37.5), an infinite value of *n*.

$$x = \tfrac{1}{4}\lambda, \ \tfrac{3}{4}\lambda, \ \tfrac{5}{4}\lambda, \ \ldots \tag{7}$$

At these points the incident and reflected waves interfere constructively during one part of the cycle and destructively during another part of the cycle so that the standing wave oscillates with an amplitude $2E_0$. The minima of the wave are at the points where

$$x = 0, \ \tfrac{1}{2}\lambda, \ \lambda, \ \tfrac{3}{2}\lambda, \ \ldots \tag{8}$$

Here the waves interfere destructively at all times so that the standing wave has zero amplitude.

The magnetic fields associated with the waves described by Eqs. (1) and (2) are in the z direction:

$$B_{\text{in}} = \frac{E_0}{c}\cos\!\left(\omega t - \frac{\omega x}{c}\right) \tag{9}$$

$$B_{\text{ref}} = \frac{E_0}{c}\cos\!\left(\omega t + \frac{\omega x}{c}\right) \tag{10}$$

Note that there is no negative sign in Eq. (10); this is so because the right-hand rule tells us that if the electric field is in the negative y direction [see Eq. (2)] and the wave is traveling in the negative x direction, the magnetic field must be in the positive z direction.

The superposition of the magnetic fields gives

$$B_{\text{tot}} = B_{\text{in}} + B_{\text{ref}} = 2\,\frac{E_0}{c}\cos \omega t \cos \omega x/c \tag{11}$$

This of course is also a standing wave. Its maxima are at the points where

$$\frac{\omega x}{c} = 0, \ \pi, \ 2\pi, \ 3\pi, \ \ldots \tag{12}$$

$$x = 0, \ \tfrac{1}{2}\lambda, \ \lambda, \ \tfrac{3}{2}\lambda, \ \ldots \tag{13}$$

and its minima at

$$x = \tfrac{1}{4}\lambda, \ \tfrac{3}{4}\lambda, \ \tfrac{5}{4}\lambda, \ \ldots \tag{14}$$

Thus the maxima of the magnetic field are displaced a distance of $\tfrac{1}{4}$ wavelength from the maxima of the electric field. Furthermore, note that the time dependence of the magnetic field ($\cos \omega t$) is a quarter of a cycle out of phase with the time dependence of the electric field ($\sin \omega t$). Figure 38.2 shows a plot of the electric and magnetic fields of the standing wave.

The interference between the incident and the reflected waves can be readily observed in a standing microwave. Such a wave can be set up by aiming a microwave generator at a metallic plate, which acts as a mirror. The positions of the maxima and minima of the electric and

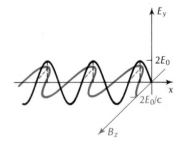

Fig. 38.2 Electric (black) and magnetic (color) fields of the standing wave plotted as a function of position (at a fixed time). The electric field is in the y direction and the magnetic field in the z direction.

magnetic fields can be investigated by probing the fields with small antennas connected to a microwave receiver.

38.2 Thin Films

The interference between a light wave incident on a mirror and the light wave reflected by the mirror is very difficult to observe. As we found in the preceding section, the two waves traveling in opposite directions make up a standing wave with destructive interference at points one-half wavelength apart [see Eq. (8)]. Since the wavelength of visible light is quite small, our eyes cannot perceive the individual minima (or maxima) of the standing wave — we only see the average intensity without noticeable interference effects.

Very spectacular interference effects may become visible when a light wave is reflected by a thin film, such as a thin film of oil floating on water. When the wave strikes the upper surface of the film (Figure 38.3), it will set up a multitude of reflected waves due to reflection on the upper surface, reflection on the lower surface, and multiple zigzags between the surfaces. These reflected waves all travel in the same direction and they can interfere destructively or constructively over a large region of space. Note that we are now interested only in the interference between the several reflected waves — the incident wave plays no direct role in this.

To find the conditions for constructive and destructive interference between the waves reflected by a thin film, let us make the simplifying assumption that the direction of propagation of the wave is nearly perpendicular to the surface of the film. The two most intense waves are those that suffer only one reflection: the wave that reflects only on the upper surface and the wave that reflects only on the lower surface. Figure 38.4 shows the rays corresponding to these two waves. Under what conditions will these waves interfere constructively in the space above the film? Obviously, the wave that is reflected at the lower surface has to travel an extra distance to emerge from the film. If the thickness of the film is d, and if the direction of propagation is nearly perpendicular to the film, then the extra distance the wave has to travel is $2d$. Provided that this extra distance is equal to one, two, three, etc., wavelengths, the wave reflected at the lower surface will meet crest to crest with the wave reflected from the upper surface. The condition for constructive interference is therefore

$$2d = \lambda,\ 2\lambda,\ 3\lambda,\ \ldots$$ (15)

Likewise, the condition for destructive interference is

$$2d = \tfrac{1}{2}\lambda,\ \tfrac{3}{2}\lambda,\ \tfrac{5}{2}\lambda,\ \ldots$$ (16)

Thus, depending on the thickness of the film and on the wavelength, the reflection can be either very strong or very weak. Note that the wavelength λ in Eqs. (15) and (16) is the wavelength of the light within the film; the wavelength outside the film will differ from this by a factor depending on the index of refraction.

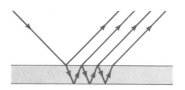

Fig. 38.3 Incident ray and multiple reflected and refracted rays produced by a thin film.

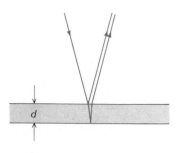

Fig. 38.4 Incident ray and reflected rays for nearly perpendicular incidence and reflection.

Constructive and destructive interference for wave reflected by thin film

EXAMPLE 1. A film of kerosene 4500 Å thick floats on water. White light, a mixture of all visible colors, is vertically incident on this film. Which of the wavelengths contained in the white light will give maximum intensity upon reflection? Which will give minimum intensity? The index of refraction of kerosene is 1.2.

SOLUTION: For maximum intensity we need

$$\lambda = 2d, \tfrac{2}{2}d, \tfrac{2}{3}d, \ldots = 9000 \text{ Å}, 4500 \text{ Å}, 3000 \text{ Å}, \ldots \tag{17}$$

These are the wavelengths in kerosene; to obtain the wavelengths in air, we must multiply by the index of refraction of kerosene, $n = 1.2$. Of the resulting wavelengths, the only one in the visible region is $1.2 \times 4500 \text{ Å} = 5400 \text{ Å}$.

For minimum intensity we need

$$\lambda = 4d, \tfrac{4}{3}d, \tfrac{4}{5}d, \ldots$$

$$= 18{,}000 \text{ Å}, 6000 \text{ Å}, 3600 \text{ Å}, \ldots \tag{18}$$

Upon multiplication by 1.2 we find that the only wavelength in air in the visible region is $1.2 \times 3600 \text{ Å} = 4320 \text{ Å}$.

A wavelength of 5400 Å corresponds to a yellow-green color. The film will therefore be seen to have this color in reflected light.

The color displays seen on oil slicks and on soap bubbles arise from such interference effects. Different portions of an oil or soap film usually have different thicknesses, and they therefore give constructive interference for different wavelengths. This results in a pattern of bright colored bands, or colored **fringes.**

Similar interference effects can also arise in a narrow gap between two adjacent glass surfaces; such a gap can be regarded as a thin film of air. For instance, Figure 38.5 shows a photograph of the interference fringes produced by the thin film of air between a flat glass plate and a lens of large radius of curvature. The convex surface of the lens is in contact with the plate at the center, but leaves a gradually widening gap for increasing distances from the center. The photograph was taken with monochromatic light. At the bright rings, the width of the gap is such as to give constructive interference of the reflected light.

Fringes

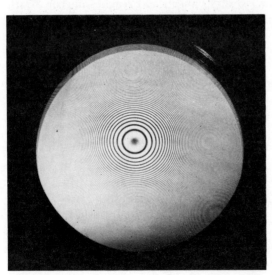

Fig. 38.5 Fringes of constructive and destructive interference seen in the light reflected in the gap between a flat glass plate and a spherical lens in contact.

At the dark rings, the width is such as to give destructive interference. The rings in Figure 38.5 are called **Newton's rings.**

Note that in the derivation of Eqs. (15) and (16) we have not taken into account the reversal, or change of phase, of the electric field upon reflection. As stated in the preceding section, this change of phase will occur if (and only if) the index of refraction of the medium in which the wave is propagating is smaller than the index of refraction of the medium off which the wave reflects. In the case of a kerosene film resting on water, both the wave reflected at the upper surface and the one reflected at the lower surface suffer this change of phase; consequently, the *relative* phase between the waves is unaffected. However, in the case of an oil film or soap film suspended in air, the wave reflected at the lower surface does not suffer a change of phase; this introduces an extra phase difference of 180° between the two waves. Such a phase difference has the same effect as an extra path difference of one-half wavelength. Consequently, Eqs. (15) and (16) must be interchanged: the former equation now applies to destructive interference and the latter to constructive interference.

Thin films are of great practical importance in the manufacture of optical instruments. The lenses of high-quality instruments are often coated with a thin film of a transparent dielectric so that undesirable reflection of light is eliminated. Of course, with one thin film we can achieve destructive interference only at one wavelength; but with several layers of thin films, we can achieve destructive interference at several wavelengths.

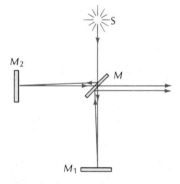

Fig. 38.6 Paths of rays in a Michelson interferometer. For the sake of clarity, rays are shown reaching the mirrors M_1 and M_2 with a small angle between them; they are actually parallel and they overlap.

38.3 The Michelson Interferometer

The Michelson interferometer takes advantage of the interference between two light waves to achieve an extremely precise comparison of two lengths. Figure 38.6 is a schematic diagram of the essential parts of such an interferometer. The apparatus consists of two arms at the ends of which are mounted mirrors M_1 and M_2. Light waves from a monochromatic source S fall on a semitransparent mirror M (a half-silvered mirror). This mirror splits the light wave into two parts: one part continues straight ahead and reaches mirror M_1, the other part is reflected and reaches mirror M_2. These mirrors reflect the waves back toward the central mirror M and, upon reflection or transmission by this mirror, the waves emerge from the interferometer. When they emerge, they interfere constructively or destructively. Suppose that the lengths MM_1 and MM_2 differ by d. Then one of the waves must travel an extra distance $2d$ and the condition for the constructive interference is

$$2d = 0, \lambda, 2\lambda, \ldots$$

(19)

and for destructive interference it is

$$2d = \tfrac{1}{2}\lambda, \tfrac{3}{2}\lambda, \tfrac{5}{2}\lambda, \ldots$$

(20)

To achieve this interference, the mirrors M_1 and M_2 must be very precisely aligned so that the image of M_1 seen in M is exactly parallel to M_2. The alignment can be achieved by means of adjusting screws on the backs of the mirrors.

The mirror M_1 is usually mounted on a carriage that can be moved along a track by means of a carefully machined screw. If the mirror is slowly moved inward or outward, the interference of the emerging waves will change back and forth between constructive and destructive whenever the mirror is displaced by $\frac{1}{4}$ wavelength — and the intensity of the emerging light will change back and forth between maxima and minima. Thus, the displacement of the mirror can be measured very precisely by counting fringes and fractions of fringes. This measurement expresses the displacement in terms of the wavelength of the light. Modern interferometers, such as that illustrated in Figure 1.10, are designed to count the fringes automatically with a photoelectric device. Such an interferometer is capable of counting 19,000 fringes per second and its mirror is capable of a displacement of up to 1 m.

As was pointed out in Section 17.1, interferometers have played an important role in the test of the dependence of the speed of light on the motion of the Earth. This test is the famous **Michelson–Morley experiment,** first performed in 1881. The principle behind this experiment is as follows: If light were to propagate in a manner analogous to sound, then we would expect that the motion of the Earth toward or away from a light wave would affect the speed of light relative to the Earth, just as the motion of a train toward or away from a source of sound affects the speed of sound relative to the train. Such an alteration of the speed of light could be detected with an interferometer by orienting one of the arms parallel to the direction of motion of the Earth and the other arm perpendicular. A difference in the speed of light c along the arms would entail a difference in the corresponding wavelengths ($\lambda = 2\pi c/\omega$) and alter the conditions (19) and (20) for bright and dark fringes. The easiest way to detect the speed difference is by rotating the interferometer so that the arm that had been parallel to the motion becomes perpendicular and vice versa. If the speed were different in the two directions, this rotation could shift the fringes from bright to dark or vice versa.

Michelson and Morley found that there was no observable fringe shift to within the accuracy of their experiment. Taking into account possible experimental errors, they established that the effect of the motion of the Earth on the speed of light was at most ± 5 km/s. This value is substantially less than the speed of the Earth around the Sun (~ 30 km/s) and proved beyond all reasonable doubt that the propagation of light through space is *not* analogous to the propagation of sound through air. Figure 38.7 shows Michelson and Morley's interferometer.

38.4 Interference from Two Slits

A very clear experimental demonstration of interference effects in light can be performed with two small light sources. The light waves spread out from the sources, run into one another, and interfere constructively or destructively, giving rise to a pattern of bright and dark zones. These interference effects were discovered by Thomas Young

Albert Abraham Michelson, *1852–1931, American experimental physicist, professor at the Case Institute of Technology and at the University of Chicago. He made precise measurements of the speed of light, and used interferometric methods to determine the length of spectral lines in terms of the standard meter. He first performed the "Michelson–Morley" experiment on his own in 1881, and then repeated it several times in collaboration with E. W. Morley, with increasing accuracy. Michelson received the Nobel Prize in 1907.*

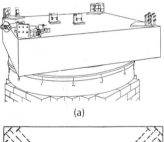

(a)

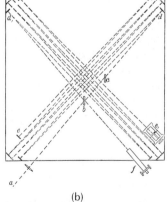

(b)

Fig. 38.7 (a) The interferometer used in the experiment of Michelson and Morley. (b) The many mirrors reflect the light beams back and forth several times, increasing the path length of the light.

around 1800 and, in conjunction with diffraction effects (see next chapter), they convinced physicists of the wave nature of light.

If the interference pattern is to remain stationary in space, the two light sources must be **coherent,** that is, they must emit waves of the same frequency and the same phase (a *constant* phase difference is also acceptable; the crucial thing is that the relative phase must not fluctuate in time). Such coherent sources are easily manufactured by aiming a monochromatic wave with plane or spherical wave fronts at an opaque plate with two small slits or holes; the waves diverging from the two slits are then coherent because they arise from a single original wave (Figure 38.8). An extended object, such as an ordinary light bulb, can be used to illuminate the slits provided it is placed at a *very large* distance. The light bulb will then effectively act as a point source, illuminating the slits with light waves that consist of a succession of nearly plane wave fronts. If the light bulb is placed too close to the slits, then light waves arrive at the slit from several directions at once and this tends to wash out the interference pattern. For convenience the light bulb is sometimes placed fairly near the slits; but it must then be covered with a shield perforated with a single pinhole so that the emerging light comes from just one point on the surface of the light bulb; this again makes the light source into a point source. In modern practice, a laser is often used to illuminate the slits because it provides a very intense plane wave, so that even very faint interference and diffraction effects become visible (lasers and the light emitted by them will be discussed in Interlude L).

Coherence

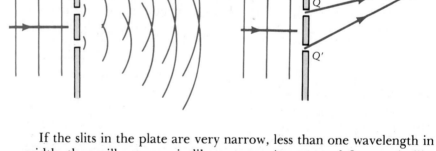

Fig. 38.8 (left) A plane light wave strikes a plate with two very narrow slits. The slits act as two coherent light sources. Light waves diverge from the two slits.

Fig. 38.9 (right) The waves reaching the point *P* have different path lengths *QP* and *Q'P*.

If the slits in the plate are very narrow, less than one wavelength in width, they will act as pointlike sources. As expected for a pointlike source, the wave diverges radially and spreads out over a wide angle beyond each slit (see Figure 38.8). Incidentally, this spreading of the wave is a diffraction effect; we will study this in detail in the next chapter.

To find the interference maxima and minima in the space beyond the slits, we need to reckon the path difference between the rays from each of the slits. Figure 38.9 shows a light wave incident on the plate from the left and the rays *QP* and *Q'P* leading from the slits to a point *P* on the right. We will assume that the light source is either a laser or else some other light source placed very far from the plate so that the incident wave is a plane wave (this assumption is implicit in all the calculations of this chapter). The waves emerging from the slits and reaching *P* interfere constructively if the difference between the lengths *QP* and *Q'P* is zero, or one wavelength, or two wavelengths, etc.; and they interfere destructively if this difference is one-half wavelength, or three-halves wavelength, etc.

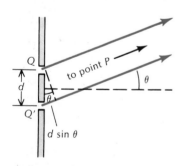

Fig. 38.10 If *P* is far away, *QP* and *Q'P* are nearly parallel. The lengths *QP* and *Q'P* then differ by *d* sin *θ*.

We can obtain a simple formula for the angular position of the maxima and minima if we make the additional assumption that the point P is at a very large distance from the plate (to be precise, QP is very large compared to QQ'). If so, then the rays QP and $Q'P$ are nearly parallel and, as Figure 38.10 shows, the difference between the lengths of these rays is approximately $d \sin \theta$, where d is the distance between the slits and θ the angle between P and the perpendicular midline.[2] Our condition for maximum intensity is then

$$d \sin \theta = 0, \lambda, 2\lambda, \ldots \qquad (21)$$

Maxima and minima for two-slit interference pattern

Likewise, the condition for minimum intensity is

$$d \sin \theta = \tfrac{1}{2}\lambda, \tfrac{3}{2}\lambda, \tfrac{5}{2}\lambda, \ldots \qquad (22)$$

These equations give the angular positions of the interference maxima and minima.

The regions of high intensity have the shape of beams fanning out in the space beyond the slits. The beams of high intensity are separated by lines of zero intensity; these are called nodal lines, analogous to the nodal points of the wave on a string (see Chapter 15). Figure 38.11 is a photograph of such a pattern of beams produced by the interference of water waves spreading out from two pointlike sources [the formulas (21) and (22) apply to water waves and to any other kinds of waves]. With light waves, we cannot photograph the entire pattern of beams at once; instead, we must be content with Figure 38.12a, which shows the pattern of bright and dark fringes recorded on a photographic film that intercepts the beams at some fixed distance beyond the slits.

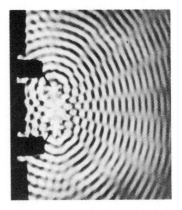

Fig. 38.11 Interference between water waves spreading out from two coherent pointlike sources in a ripple tank.

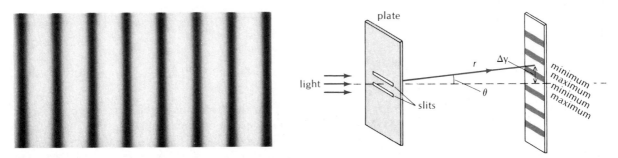

Fig. 38.12 (a) A photographic film placed beyond two illuminated narrow slits records a regular pattern of bright and dark fringes. (Courtesy C. C. Jones, Union College.) (b) Placement of the photographic film beyond the slits.

[2] This is called the **Fraunhofer approximation** for the difference between the lengths of the rays. Sometimes a lens is placed beyond the slits so as to focus their light on the point P. The approximation $d \sin \theta$ for the difference between the lengths is then justified even if the point P is not at a very large distance, because the lens brings together rays that are parallel when leaving the slits. For the sake of simpicity, we will assume throughout the following discussion that P is at a large distance and that no lens is required.

EXAMPLE 2. Two narrow slits separated by a distance of 0.12 mm are illuminated with light of wavelength 5890 Å from a sodium lamp. What is the angular position of the first lateral maximum? If the light is intercepted by a photographic film placed 2.00 m beyond the slits, what is the distance along the film between this maximum and the central maximum?

SOLUTION: With $d = 1.2 \times 10^{-4}$ m and $\lambda = 5.89 \times 10^{-7}$ m, Eq. (21) gives

$$\sin \theta = \frac{\lambda}{d} = \frac{5.89 \times 10^{-7} \text{ m}}{1.2 \times 10^{-4} \text{ m}} = 4.9 \times 10^{-3} \tag{23}$$

The corresponding angle is $\theta = 4.9 \times 10^{-3}$ radian.

The distance between the points with $\theta = 0$ and $\theta = 4.9 \times 10^{-3}$ radian on the photographic film is (see Figure 38.12b)

$$\Delta y \cong r\theta = 2.00 \text{ m} \times 4.9 \times 10^{-3} = 9.8 \times 10^{-3} \text{ m}$$

Thus the maxima are separated by nearly 1 cm.

Let us now calculate the intensity distribution as a function of angle. The electric fields of the spherical waves spreading out from points in the two slits[3] and reaching the point P are [compare Eq. (36.1)]

$$E_1 = \frac{A}{r_1} \cos\left(\omega t - \frac{\omega r_1}{c} \right) \tag{24}$$

and

$$E_2 = \frac{A}{r_2} \cos\left(\omega t - \frac{\omega r_2}{c} \right) \tag{25}$$

In these equations, A is a constant that depends on the strength of the wave incident on the slits; r_1 is the distance QP and r_2 is the distance $Q'P$. With the same large-distance approximation as above, we have

$$r_1 = r_0 - \frac{d}{2} \sin \theta \tag{26}$$

$$r_2 = r_0 + \frac{d}{2} \sin \theta \tag{27}$$

where r_0 is the distance from the midpoint of the two slits to P (Figure 38.13). We can substitute these approximations into the cosine functions in the numerators of Eqs. (24) and (25). We should also substitute these approximations into the denominators of these equations, but these denominators are not all that sensitive to small changes in the distances and the crude approximation $r_1 \cong r_2 \cong r_0$ suffices here:

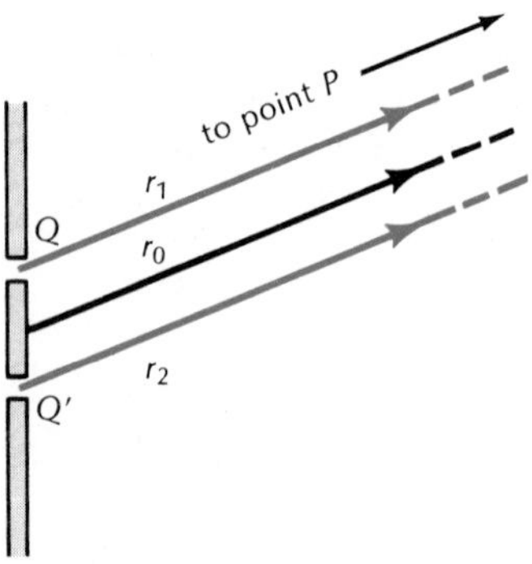

$$E_1 = \frac{A}{r_0} \cos\left(\omega t - \frac{\omega r_0}{c} + \frac{\omega d}{2c} \sin \theta \right) \tag{28}$$

$$E_2 = \frac{A}{r_0} \cos\left(\omega t - \frac{\omega r_0}{c} - \frac{\omega d}{2c} \sin \theta \right) \tag{29}$$

Fig. 38.13 Rays from the slits to the point P.

[3] In the following calculation we will only consider the waves emanating from one point in each slit (the points in the plane of Figure 38.13); other points in the slits contribute similar waves.

The superposition of the two electric fields E_1 and E_2 then gives

$$E = E_1 + E_2$$

$$= \frac{A}{r_0}\left[\cos\left(\omega t - \frac{\omega r_0}{c} + \frac{\omega d}{2c}\sin\theta\right) + \cos\left(\omega t - \frac{\omega r_0}{c} - \frac{\omega d}{2c}\sin\theta\right)\right] \quad (30)$$

With the trigonometric identity $\cos(\alpha + \beta) + \cos(\alpha - \beta) = 2\cos\alpha\cos\beta$ this becomes

$$E = \frac{2A}{r_0}\cos\left(\omega t - \frac{\omega r_0}{c}\right)\cos\left(\frac{\omega d}{2c}\sin\theta\right) \quad (31)$$

The factor $\cos(\omega t - \omega r_0/c)$ is the usual oscillating function of space and time, characteristic of a wave propagating outward. The factor $\cos[(\omega d/2c)\sin\theta]$ indicates how the amplitude of this wave depends on the position angle θ.

The intensity of the wave is proportional to E^2. Since we are only interested in the time-average intensity, we can replace the factor $\cos^2(\omega t - \omega r_0/c)$ appearing in E^2 by its time-average value $\frac{1}{2}$. Hence

$$[\text{intensity}] \propto \frac{1}{r_0^2}\cos^2\left(\frac{\omega d}{2c}\sin\theta\right) \quad (32)$$

With $\lambda = 2\pi c/\omega$, we can also write this as

$$\boxed{[\text{intensity}] \propto \frac{1}{r_0^2}\cos^2\left(\frac{\pi d}{\lambda}\sin\theta\right)} \quad (33)$$

Intensity for two-slit interference pattern

This formula describes the intensity pattern produced by the double slits. The factor $\cos^2[(\pi d/\lambda)\sin\theta]$ gives the distribution of intensity as a function of angle. Figure 38.14 is a plot of this factor. This plot represents the intensity of light that reaches different angles at some constant distance r_0 from the midpoint of the slits. If we want to measure this amount of light by intercepting it with a screen or a photographic film, we have to bend the screen along a circular arc of radius r_0 (Figure 38.15); however, the observations are usually restricted to small angles θ and then a flat screen will serve well enough.

Fig. 38.14 (below, left) Intensity as a function of the angle θ for the light arriving at a screen placed beyond two narrow slits. The distance between the slits is $d = 8\lambda$.

Fig. 38.15 (below, right) Zones of maximum intensity and minimum intensity on a curved screen. All points of this screen are at the same distance r_0.

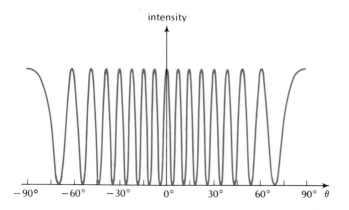

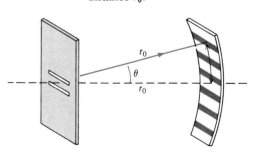

38.5 Interference from Multiple Slits

It is easy to see that Eq. (21) also applies to the case of multiple slits. Figure 38.16 shows an opaque plate with three evenly spaced slits. If waves from *adjacent* slits interfere constructively, then *all* the waves from all three slits interfere constructively. Hence the condition for maximum intensity is the same as Eq. (21).

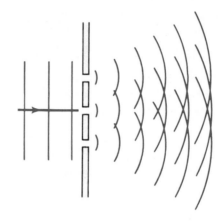

Fig. 38.16 A plane wave strikes a plate with three very narrow slits.

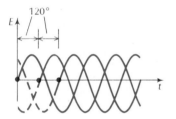

Fig. 38.17 Electric field as a function of time at the point *P*. The three waves from the three slits differ in phase by 120°; their sum is zero.

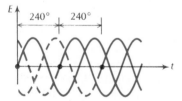

Fig. 38.18 The three waves differ in phase by 240°; their sum is zero.

However, the condition for minimum intensity is *not* Eq. (22). For instance, in the case of three slits, destructive interference of waves from adjacent slits will lead to cancellation of the waves originating from one pair of slits, but the wave from the remaining slit will not suffer cancellation. The condition for destructive interference between three waves (of equal amplitude) is that the phases differ by 120° from one wave to the next; the sum of two of the waves then always cancels against the third wave (Figure 38.17). Hence a triple slit will yield a minimum if $d \sin \theta = \lambda/3$, where, of course, d is the separation between adjacent slits. The triple slit will also yield a minimum if $d \sin \theta = 2\lambda/3$; this increased value of the path difference merely amounts to an extra shift of each wave crest by $\frac{1}{3}$ of a period and does not alter the relative distribution of these crests (compare Figures 38.17 and 38.18). Pursuing this argument, we find further minima at $d \sin \theta = 4\lambda/3$, $5\lambda/3$, etc. Figure 38.19 is a plot of the intensity as a function of θ for the case of three slits (for small values of θ). Between any two principal maxima, there are two minima. Between the two minima there is a weak maximum, a *secondary* maximum. Figure 38.20 is a photograph of this interference pattern.

Fig. 38.19 Intensity as a function of the angle θ for light arriving at a screen placed beyond three narrow slits. The distance between one slit and the next is $d = 12\lambda$.

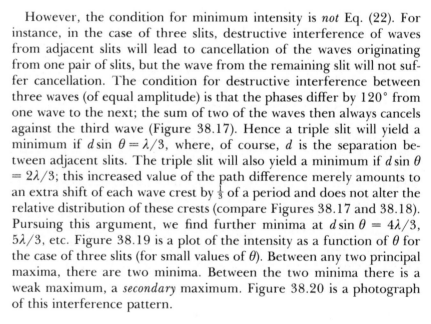

Fig. 38.20 A photographic film placed beyond the slits shows the strong principal maxima and the weaker secondary maxima. (Courtesy C. C. Jones, Union College.)

For the case of N slits, we can summarize the important features of the interference pattern as follows. The principal maxima are given by

$$d \sin \theta = n\lambda \qquad n = 0, 1, 2, \ldots \qquad (34)$$

Maxima and minima for multiple-slit interference pattern

The minima are given by

$$d \sin \theta = m\lambda/N$$

$m = 1, 2, 3, \ldots$
(with $m = N, 2N, \ldots$, excluded) $\qquad (35)$

The secondary maxima are (approximately) halfway between the minima.

Arrangements of multiple slits are commonly used to analyze light into its colors. If a light beam containing several wavelengths passes through such multiple slits, the maxima for these different wavelengths will form beams at different angles — the beams of long-wavelength light will be found at larger angles than the beams of short-wavelength light [see Eq. (34)]. Thus, the system of slits separates the light according to color and produces a spectrum in much the same way that a prism does. There is, however, one important difference between the spectra formed by a prism and by a system of slits: in the prism the long-wavelength light (red) suffers the least deflection; in the system of slits the long-wavelength light suffers the most deflection.

The system of slits will produce one complete spectrum for each value of n in Eq. (34), i.e., each principal maximum, except the central maximum, gets spread out by color. These spectra are called the first-order spectrum ($n = 1$), second-order spectrum ($n = 2$), etc. Sometimes these spectra overlap, e.g., the red end of the second-order spectrum may show up at the same angle as the blue end of the third-order spectrum (Figure 38.21).

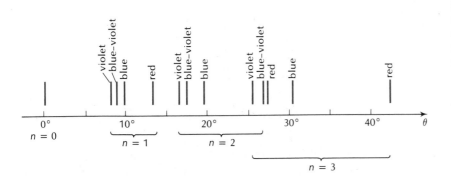

Fig. 38.21 First-, second-, and third-order spectra of hydrogen light produced by a system of N slits. The lines correspond to the principal maxima; the secondary maxima are weak and can be ignored. Each spectrum consists of a violet, blue-violet, blue, and red spectral line (compare Figure 37.22). The pattern of spectral lines for negative values of θ (negative n) is similar.

In order to achieve a very sharp separation of colors within each spectrum, we must use a very large number of slits. This makes the principal maxima very narrow and therefore reduces the overlap between maxima formed by two colors differing but little in wavelength. To understand how this helps, note that the distance from the center of a principal maximum to the next minimum is, according to Eq. (35),

Width of principal maximum

$$\Delta\theta = \frac{1}{N}\frac{\lambda}{d} \tag{36}$$

where we assume that θ is small so that $\sin\theta \cong \theta$. For a change $\Delta\lambda$ in wavelength, the change in the angular position of the principal maximum is, according to Eq. (34),

$$\Delta\theta = n\frac{\Delta\lambda}{d} \tag{37}$$

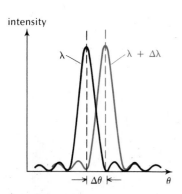

If the angular shift (37) is equal to the angular width (36), then the changed wavelength will produce a maximum at the minimum of the original wavelength. Under these conditions the two maxima are obviously well separated and can be told apart very clearly (Figure 38.22). Setting Eqs. (37) and (36) equal, we therefore find that the wavelength difference that can be clearly resolved by our arrangements of slits is given by

Fig. 38.22 Two distinct principal maxima produced by two different spectral lines of wavelengths λ and $\lambda + \Delta\lambda$. The maximum produced by one of these wavelengths coincides with the minimum produced by the other.

$$\frac{1}{N}\frac{\lambda}{d} = n\frac{\Delta\lambda}{d} \tag{38}$$

Resolving power

or

$$\frac{\lambda}{\Delta\lambda} = Nn \tag{39}$$

The ratio $\lambda/\Delta\lambda$ is called the **resolving power** of the system of slits. The formula (39) shows that if we want to detect a small wavelength difference, we need a large value of N.

EXAMPLE 3. Among the wavelengths emitted by atoms of iron are $\lambda = 5005.72$ Å and $\lambda = 5006.13$ Å. If we want to separate these two wavelengths clearly in the first-order spectrum of a system of slits, what is the minimum number of slits required?

SOLUTION: The wavelength difference is $\Delta\lambda = 0.41$ Å and $n = 1$. Hence Eq. (39) gives

$$N = \frac{\lambda}{\Delta\lambda} = \frac{5006\ \text{Å}}{0.41\ \text{Å}} = 1.2 \times 10^4$$

This means that we need at least 12,000 slits.

A system of a large number of slits used to analyze light into its colors is called a **grating.** Since it is difficult to cut a large number of slits in an opaque plate, gratings are usually manufactured by cutting fine parallel grooves in a glass or metal surface with a diamond stylus guided by a special ruling machine. When illuminated by a light wave, the edges of the grooves act as light sources in much the same way as slits. High-quality gratings used in spectroscopy have 10^5 or more lines with distances of about 10^{-6} m between them.

Radiotelescopes, consisting of a regular array of evenly spaced antennas, act as "gratings" for radio waves; the same formulas apply to these as to ordinary gratings used with light. Of course, there is a difference between the operation of a radiotelescope and that of an ordinary grating; in the former the radio waves from a distant point *enter* the antennas and interfere constructively or destructively within the radio receiver, while in the latter the light waves *emerge* from the slits and travel to a distant point where they interfere. Nevertheless, the condition for, say, maximum intensity can still be expressed in terms of the direction of the wave by Eq. (34) because this condition only hinges on the phase relationships among the waves, and these relationships are independent of the direction of propagation (for instance, in Figure 38.16 it makes no difference whether the waves are traveling toward or away from the slits).

Radiotelescope

EXAMPLE 4.   One branch of the VLA (Very Large Array) radiotelescope at Socorro, New Mexico, has nine antennas arranged on a straight line with a distance of 2.0 km between one antenna and the next (Figure 38.23). These antennas are all connected to a single radio receiver by waveguides of equal length; the receiver then registers a maximum intensity if the radio waves incident on all antennas are in phase. Radio waves of wavelength 21 cm from a pointlike source in the sky strike this radiotelescope. When a source is at the zenith, the intensity in the radio receiver is maximum. What must be the angular displacement of a source from the zenith to make the intensity minimum?

SOLUTION: According to Eq. (36), the angular distance from the center of a principal maximum to the adjacent minimum is

$$\Delta\theta = \frac{1}{N}\frac{\lambda}{d} = \frac{1}{9}\frac{0.21 \text{ m}}{2.0 \times 10^3 \text{ m}} = 1.2 \times 10^{-5} \text{ radian}$$

This angle is about 3 seconds of arc.

Fig. 38.23 The VLA radiotelescope. The branch in the center consists of nine antennas. The distance between the antennas can be adjusted by rolling them along the track.

SUMMARY

Constructive interference: Waves meet crest to crest.

Destructive interference: Waves meet crest to trough.

Change of phase by reflection: 180° if medium off which wave reflects has higher index of refraction

Two-slit interference pattern:

maxima: $d \sin \theta = 0, \lambda, 2\lambda, \ldots$

minima: $d \sin \theta = \frac{1}{2}\lambda, \frac{3}{2}\lambda, \frac{5}{2}\lambda, \ldots$

Multiple-slit interference pattern:

principal maxima: $d \sin \theta = 0, \lambda, 2\lambda, \ldots$

minima: $d \sin \theta = \lambda/N, 2\lambda/N, \ldots$

Width of principal maximum of grating: $\Delta\theta = \dfrac{1}{N}\dfrac{\lambda}{d}$

Resolving power of grating: $\dfrac{\lambda}{\Delta\lambda} = Nn$

QUESTIONS

1. According to Eq. (11), the magnetic field of the standing wave is maximum at the surface of the mirror. The magnetic field is zero inside the mirror. What is the direction of the current that must flow along the surface of the mirror to produce this sudden drop of the magnetic field?

2. The light wave inside the tube of a laser is a standing wave, reflected at both ends of the tube. If the distance between the ends is L, what condition must the wavelength of the standing wave satisfy?

3. Would you expect sound waves reflected by a wall to produce a standing wave? How could you test this experimentally?

4. The radar wave emitted by a stationary police radar unit is reflected by an approaching automobile. Does this set up a standing wave?

5. In a traveling electromagnetic wave, the maxima of the electric and magnetic fields occur at the same place. A standing wave is a superposition of two traveling waves. How, then, can the maxima of the electric and magnetic fields in this wave occur at different places?

6. The maxima and the minima of the electric and magnetic fields in a standing wave produced by a microwave generator can be detected with an electric dipole antenna or a magnetic dipole antenna connected to a radio receiver (Figure 38.24). How must you orient each of these antennas relative to the directions of the electric and magnetic fields for maximum sensitivity?

7. When two waves interfere destructively at one place, what happens to their energy?

8. When two coherent waves of equal intensity interfere constructively at one place, the energy density at that place becomes four times as large as the energy density of each individual wave. Is this a violation of the law of conservation of energy?

9. Suppose that N waves of equal intensity meet at one place. If the waves are

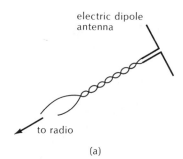

electric dipole
antenna

to radio

(a)

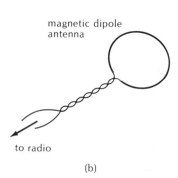

magnetic dipole
antenna

to radio

(b)

Fig. 38.24 (a) Electric dipole antenna. (b) Magnetic dipole antenna.

coherent, the net intensity is N^2 times that of each individual wave. If the waves are incoherent, the (average) intensity is N times that of each individual wave. Explain, and give an example of each case.

10. Why do we not see colored fringes in thick oil slicks?

11. When light strikes a window pane, some rays will be reflected back and forth between the two glass surfaces. Why do these reflected rays not produce visible colored interference fringes?

12. Suppose that a lens is covered with an antireflective coating that eliminates the reflection of perpendicularly incident light of some given color. Will this coating also eliminate the reflection of light incident at an angle?

13. Is it possible to cover the surface of an aircraft with an antireflective coating so that it does not reflect radar waves of wavelength, say, 5 cm?

14. Consider the light *transmitted* by a thin film. For a light wave with a direction of propagation perpendicular to the surface of the film, what is the condition for constructive interference between the direct wave and the wave that is reflected once by the lower surface and once by the upper surface? Would you expect that this condition for constructive interference in transmission coincides with the condition for destructive interference in reflection?

15. Why is the central spot in Newtons' rings (Figure 38.5) dark?

16. Explain how Newtons's rings may be used as a sensitive test of the rotational symmetry of a lens.

17. Two flat plates of glass are in contact at one edge and separated by a thin spacer at the other edge (Figure 38.25a). Explain why we see parallel interference fringes in the reflected light if we illuminate these plates from above (Figure 38.25b).

18. In the experiment that sought to test the dependence of the speed of light on the motion of the Earth, Michelson and Morley used an interferometer with very long arms (about 11 m, obtained by multiple reflections back and forth between sets of mirrors). Why does this make the instrument more sensitive?

19. If you stand next to your TV receiver, your body will sometimes affect the reception. Why?

20. Consider (1) sunlight, (2) sunlight passed through a monochromatic filter selecting one wavelength, (3) light from a neon tube, (4) light from a laser, (5) starlight passed through a monochromatic filter, and (6) radio waves emitted by a radio station. Which of these kinds of light or electromagnetic radiation is sufficiently coherent so that when it is used to illuminate two slits such as shown in Figure 38.8, it will give rise to an interference pattern?

21. When installing a pair of stereo loudspeakers, the terminals of the loudspeakers should be connected to the amplifier in the same way, so that the loudspeakers are in phase. What would happen to sound waves of long wavelength if the loudspeakers were out of phase?

22. What happens to the interference pattern plotted in Figure 38.14 if $d < \lambda$?

23. Suppose we cover the left side of one of the slits in Figure 38.10 with a glass plate. Since the wavelength in glass is shorter than that in air, the path in the glass includes more wavelengths than the path in air, and the phases of the two waves emerging from the slits will be different. If the phase difference between these waves is exactly 180°, how does this change the location of maxima and minima?

24. Consider the interference pattern shown in Figure 38.14. How would this pattern change if the entire interference apparatus were immersed in water?

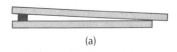

(a)

(b)

Fig. 38.25 (a) Two flat plates of glass, separated by a thin wedge of air. (b) Interference fringes.

Fig. 38.26 The plane wave is incident on the plate at a slant.

25. Suppose that a plane wave is incident at an angle on a plate with two narrow slits (Figure 38.26). In what direction will we then find the central maximum?

26. Several radio antennas are arranged at regular intervals along a straight line; the antennas are connected to the same radio transmitter, so that they radiate coherently. How can this array be used to concentrate the radio emission in a selected direction?

27. When a crystal is illuminated with X rays, each atom acts as a pointlike source of scattered X rays. The typical separation between adjacent atoms in a crystal is 1 Å. Roughly what must be the wavelength of the X rays if they are to exhibit distinct interference effects?

PROBLEMS

Section 38.1

1. A light wave of wavelength 6943 Å from a laser strikes a mirror at normal incidence. This will set up a standing wave in front of the mirror.
   - (a) At what distance from the mirror will the electric field have nodes? Antinodes?
   - (b) At what distance from the mirror will the magnetic field have nodes? Antinodes?

2. In 1887, Hertz gave the first direct experimental demonstration of the existence of the electromagnetic waves that had been predicted by Maxwell's theory. For this experiment, Hertz placed a high-frequency spark generator at some distance in front of a large vertical zinc plate. The wave emitted by the spark reflected off the zinc plate, setting up a standing wave. Hertz found that when the generator was operating at a frequency of about 4×10^7 cycles per second, the distance between a node and an antinode in the standing wave pattern was about 2 m. What value of the speed of propagation of the electromagnetic disturbances could he deduce from this?

3. With a microwave generator, you can set up a standing electromagnetic wave between two large, flat, parallel plates of metal. The standing wave must have zero electric field at the surfaces of the plates at all times. If your microwave generator produces waves of frequency 2×10^9 Hz, what minimum separation between the plates do you need for a standing wave? Where does this standing wave have a maximum amplitude of the electric field? Of the magnetic field?

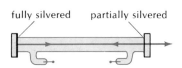

fully silvered   partially silvered

Fig. 38.27 Tube of a laser with mirrors at its ends.

4. The tube of a laser has a mirror at each end (Figure 38.27). A standing electromagnetic wave fills the space between these mirrors. This wave must satisfy the boundary condition that its electric field is zero at all times at the position of the mirrors.
   - (a) Show that the distance between the mirrors and the wavelength of the wave must be related by

$$d = \left(\frac{n+1}{2}\right)\lambda \qquad n = 0, 1, 2, \ldots$$

   - (b) A He–Ne laser with $d = 30.00$ cm operates at $\lambda = 6328$ Å. What is the value of n? How many minima and maxima does the standing wave have between the two mirrors?
   - (c) At what distance from each of the mirrors is the nearest maximum of the electric field? The nearest maximum of the magnetic field?

Section 38.2

5. The wall of a soap bubble floating in air has a thickness of 4000 Å. If sunlight strikes the wall perpendicularly, what colors in the reflected light will be

strongly enhanced as seen in air? The index of refraction of the soap film is 1.35.

6. A lens made of flint glass with an index of refraction of 1.61 is to be coated with a thin layer of magnesium fluoride with an index of refraction of 1.38.
   (a) How thick should the layer be so as to give destructive interference for the perpendicular reflection of light of wavelength 5500 Å seen in air?
   (b) Does your choice of thickness permit constructive interference for the reflection of light of some other wavelength in the visible spectrum? (If it does, you ought to make a better choice.)

7. A thin oil slick, of index of refraction 1.3, floats on water. When a beam of white light strikes this film vertically, the only colors enhanced in the reflected beam seen in air are orange-red (about 6500 Å) and violet (about 4300 Å). From this information, deduce the thickness of the oil slick.

8. Two flat, parallel plates of glass are separated by thin spacers so as to leave a gap of width d (Figure 38.28). If light of wavelength λ is normally incident on these plates, what is the condition for constructive interference between the rays reflected by the lower surface of the top plate and the upper surface of the bottom plate?

Fig. 38.28

9. A layer of oil of thickness 2000 Å floats on top of a layer of water of thickness 4000 Å resting on a flat, metallic mirror. The index of refraction of the oil is 1.24 and that of the water 1.33. A beam of light is normally incident on these layers. What must be the wavelength of the beam if the light reflected by the top surface of the oil is to interfere destructively with the light reflected by the mirror?

*10. A lens with one flat surface and one convex surface rests on a flat plate of glass (Figure 38.29). A light ray, normally incident on the lens, will be partially reflected by the curved surface of the lens and partially reflected by the flat plate of glass. The interference between these reflected rays will be constructive or destructive, depending on the height of the air gap between the lens and the plate. The interference gives rise to the pattern of Newton's rings, shown in Figure 38.5.
   (a) Why is the center of the pattern dark?
   (b) Show that the radius of the nth bright ring is

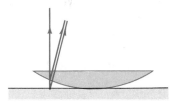

Fig. 38.29 Lens resting on flat glass plate.

$$r = \sqrt{n\lambda R - n^2\lambda^2/4}$$

   where λ is the wavelength of the light and R is the radius of the convex surface of the lens.
   (c) What is the radius of the first dark ring if $\lambda = 5000$ Å, $R = 3.0$ m?

Section 38.3

11. The **Fabry–Perot interferometer** consists of two parallel half-silvered mirrors. A ray of light entering the space between the mirrors may pass straight through, or be reflected once or several times by each mirror (Figure 38.30). Show that the condition for constructive interference of the emerging light (at large distance from the mirrors) is

$$2d \cos \theta = 0, \lambda, 2\lambda, \ldots$$

where d is the distance between the mirrors, λ the wavelength of light, and θ the angle of incidence of the light.

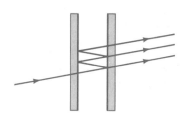

Fig. 38.30 Fabry–Perot interferometer.

12. The interferometer of the Bureau International de Poids et Mesures can count 19,000 bright fringes (maxima) per second. To achieve this count rate, what must be the speed of motion of the moving mirror (mirror M_1 in Figure 38.6)? Assume that the interferometer operates with krypton light of wavelength $\lambda = 6058$ Å.

13. An **etalon** consists of two mirrors held a fixed distance apart by means of a

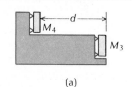

(a)

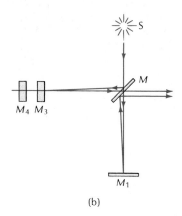

(b)

Fig. 38.31 (a) An etalon. (b) Etalon (M_3, M_4) installed in Michelson interferometer.

rigid support (Figure 38.31a). The distance between the two mirrors can serve as a standard of length. To measure this distance in terms of wavelengths of light, the etalon is installed in a Michelson interferometer, replacing the fixed mirror M_2 (Figure 38.31b). For a start, the distances MM_1 and MM_3 are made exactly equal; then the mirror M_1 is slowly moved outward, producing a sequence of interference maxima and minima until the distances MM_1 and MM_4 are exactly equal. [The exact equality of distances can be checked by replacing the usual monochromatic light source of the interferometer with a white light source. White light will give an interference maximum if and only if the distance d in Eq. (19) is exactly zero because only then will Eq. (19) be satisfied for *all* wavelengths.] Suppose that you operate a Michelson interferometer with krypton light of wavelength 6057.802 Å and you count 36,484.8 interference maxima while moving the mirror M_1 from its initial to its final position. To within six significant figures, what is the length of the etalon?

Section 38.4

14. Suppose that the two slits, the film, and the light source described in Example 2 are immersed in water. What will now be the distance between the central maximum and the first lateral maximum?

15. A piece of aluminum foil with two narrow slits is being illuminated with red light of wavelength 6943 Å from a laser. This yields a row of evenly spaced bright bands on a screen placed 3.00 m beyond the slits. The interval between the bright bands is 1.4 cm. What is the distance between the two slits?

16. Consider the water waves shown in Figure 38.11. With a ruler, measure the wavelength of the waves and the distance between the sources; with a protractor, measure the angular positions of the nodal lines (minima). Do these measured quantities satisfy Eq. (22)?

17. Two radio beacons are located on an east–west line and separated by a distance of 6.0 km. The radio beacons emit synchronous (in phase) sinusoidal waves of a frequency of 1.0×10^5 Hz. The navigator of a ship wants to determine his position relative to the radio beacons. His radio receiver indicates zero signal strength at the position of the ship. What is the angular bearing of the ship relative to the radio beacons? Assume that the distance between the ship and the radio beacons is much larger than 6 km. Note: This problem has *several* answers.

18. Light of wavelength λ is obliquely incident on a pair of narrow slits separated by a distance d. The angle of incidence of the light on the slits is ϕ (Figure 38.32).
  (a) Show that the diffracted light emerging at an angle θ interferes constructively if

$$d \sin \theta - d \sin \phi = 0, \lambda, 2\lambda, \ldots$$

and destructively if

$$d \sin \theta - d \sin \phi = \tfrac{1}{2}\lambda, \tfrac{3}{2}\lambda, \tfrac{5}{2}\lambda, \ldots$$

  (b) Show that if θ is small, the angular separation between the interference maxima and minima is independent of the angle ϕ.

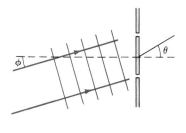

Fig. 38.32

19. When a beam of monochromatic light is incident on two narrow slits separated by a distance 0.15 mm, the angle between the central beam and the third lateral maximum in the interference pattern is 0.52°. What is the wavelength of the light?

20. An interferometric radiotelescope consists of two antennas separated by a distance of 1.0 km. The two antennas feed their signals into a common receiver tuned to a frequency of 2300 MHz. The receiver will detect a maximum (constructive interference) if the wave sent out by a radio source in the

sky arrives at the two antennas with the same phase. What possible angular positions of the radio source will result in such a maximum? Measure the angular position from the vertical line erected at the midpoint of the antennas. Treat the antennas as pointlike.

21. The radio wave from a transmitter to a receiver may follow either a direct path or else an indirect path involving a reflection on the ground (Figure 38.33). This can lead to destructive interference of the two waves and a consequent fading of the radio signal at certain locations. Suppose that a transmitter and a receiver operating at 98 MHz are both at a height of 60 m on tall buildings with bare ground between. What is the maximum distance between the buildings that will lead to destructive interference? Be careful to take into account the phase change upon reflection.

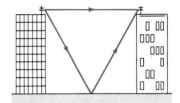

Fig. 38.33 Antennas on two buildings.

22. Two radio beacons emit waves of frequency 2.0×10^5 Hz. The beacons are on a north–south line, separated by a distance of 3.0 km. The southern beacon emits waves $\frac{1}{4}$ of a cycle later than the northern beacon. Find the angular directions for constructive interference. Measure angles relative to the east–west line and assume that the distance between the beacons and the point of observation is large.

23. A device called **Lloyd's mirror** produces interference between a ray reaching a vertical screen directly and a ray reaching the screen after reflection by a horizontal mirror. Show that in terms of the distances z, z_0, and l defined in Figure 38.34, the condition for constructive interference is

$$\sqrt{l^2 + (z + z_0)^2} - \sqrt{l^2 + (z - z_0)^2} = \tfrac{1}{2}\lambda, \tfrac{3}{2}\lambda, \tfrac{5}{2}\lambda, \ldots$$

and that for destructive interference is

$$\sqrt{l^2 + (z + z_0)^2} - \sqrt{l^2 + (z - z_0)^2} = 0, \lambda, 2\lambda, \ldots$$

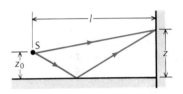

Fig. 38.34 Lloyd's mirror.

Note that in this calculation you must take into account the reversal, or change of phase, of the wave during reflection.

*24. In the first application of interferometric methods in radio astronomy, Australian astronomers observed the interference between a radio wave arriving at their antenna on a direct path from the Sun and on a path involving one reflection on the surface of the sea (Figure 38.35). Assuming that the radio waves have a frequency of 6.0×10^7 Hz and that the radio receiver is at a height of 25 m above the level of the sea, what is the least angle of the source above the horizon that will give destructive interference of the waves at the receiver?

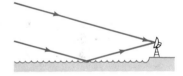

Fig. 38.35

25. Two radio beacons transmit waves of the same phase and frequency. The transmitters are on the x axis, at $\pm x_0$. Show that the interference is constructive at those points of the x–y plane satisfying the condition

$$\sqrt{(x + x_0)^2 + y^2} - \sqrt{(x - x_0)^2 + y^2} = n\lambda$$

where $n = 0, 1, 2, \ldots$. Show that for a given nonzero value of n, this is the equation of a hyperbola. (To show this, either use graphical methods to plot the curve, or else use your knowledge of analytic geometry.)

26. Light of wavelength 6943 Å from a ruby laser is incident on two narrow parallel slits cut in a thin sheet of metal. The slits are separated by a distance of 0.11 mm. A screen is placed 1.5 m beyond the slits. Find the intensity, relative to the central maximum, at a point on the screen 1.2 cm to one side of the central maximum.

Section 38.5

27. A grating has 5000 lines per centimeter. What are the angular positions of the principal maxima produced by this grating when illuminated with light of wavelength 6500 Å?

28. Sodium light with wavelengths 5889.9 Å and 5895.9 Å is incident on a grating with 5500 lines per centimeter. A screen is placed 3.0 m beyond the grating. What is the distance between the two spectral lines in the first-order spectrum on the screen? In the second-order spectrum?

29. The red line in the spectrum of hydrogen has a wavelength of 6563 Å; the blue line in this spectrum has a wavelength of 4340 Å. If hydrogen light falls on a grating of 6000 slits per centimeter, what will be the angular separation (in degrees) of these two spectral lines as seen in the first-order spectrum?

30. One of the gratings made by H. Rowland — a celebrated maker of gratings, some of which remain unsurpassed to this day for their uniformity — had 14,438 lines per inch and was 6.0 in. wide. What was the resolving power of this grating in the first order?

31. A grating has 40,000 lines uniformly spaced over a width of 8.0 cm. What is the width of the principal maxima formed by this grating with light of wavelength 5500 Å?

32. A thin curtain of fine batiste consists of vertical and horizontal threads of cotton forming a net that has a regular array of square holes. While looking through this curtain at the red (6700 Å) taillight of an automobile, a physicist notices that the taillight appears as a multiple array of images (an array of principal maxima). The angular separation between adjacent images is 2×10^{-3} radian. From this information deduce the spacing between the threads in the batiste curtain.

33. A good grating cut in speculum metal has 5900 lines per centimeter. If this grating is illuminated with white light ranging over wavelengths from 4000 Å to 7000 Å, it will produce a spectrum ranging over some interval of angles. From what angle to what angle does the first-order spectrum extend? The second-order spectrum? The third-order spectrum? Do these angular intervals overlap? Is the third-order spectrum complete?

34. Consider the Rowland grating described in Problem 30. In the interference pattern generated by this grating, how many secondary maxima are there between two adjacent principal maxima?

35. A carbon arc emits, among others, two spectral lines of wavelengths 4267.02 Å and 4267.27 Å. You wish to resolve these spectral lines in the second-order spectrum of a grating of total width 5.0 cm. How many slits per centimeter do you need in your grating?

CHAPTER 39

Diffraction

We saw in Chapter 16 that waves can bend and spread around obstacles. For instance, a water wave will spread out beyond a gap in a breakwater (see Figure 16.29) and it will spread around an island (see Figure 16.32). Such deviations from rectilinear propagation constitute **diffraction.** These deviations become quite large if the size of the gap, island, or other obstruction is of the same order of magnitude as the wavelength.

Since light is a wave, it displays diffraction effects. But the wavelength of light is very short, and hence these effects are difficult to detect (Figure 39.1). The diffraction effects become pronounced only when light passes through extremely narrow slits or when it strikes extremely small opaque obstacles. Diffraction effects with light had been noticed in the seventeenth century. Nevertheless, Newton held to the belief that light is of a corpuscular nature, and consists of a stream of particles. The importance of diffraction effects in establishing the wave nature of light was not fully appreciated until 1818, when Augustin Fresnel mathematically formulated the theory of diffraction. The brilliant success of this theory finally convinced physicists that light is a wave. In this chapter we will calculate the intensity pattern produced by a light wave diffracted by a narrow slit.

Diffraction

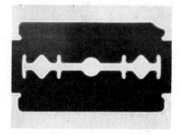

Fig. 39.1 Diffraction of light around a razor blade. The diffracted light generates a complex pattern of fine fringes at the edges of the shadow. This photograph was prepared by illuminating the razor blade with a distant light source, and throwing its shadow on a distant screen.

39.1 Diffraction by a Single Slit

Figure 39.2 shows a plane light wave approaching a slit in an opaque plate. To find the distribution of light in the space beyond the slit, we will use the following prescription, called the **Huygens–Fresnel Principle:**

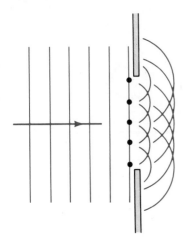

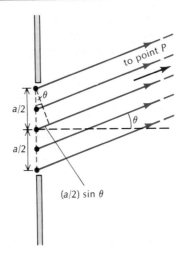

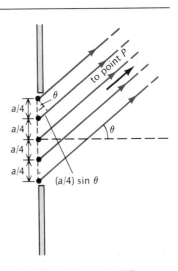

Fig. 39.2 A plane wave approaches a slit in an opaque plate. Each point on the wave front at the slit gives rise to a spherical wave that spreads out beyond the slit.

Fig. 39.3 Rays from points in the slit to the point P. The path difference between the uppermost ray and the middle ray is $(a/2)\sin\theta$.

Fig. 39.4 The path difference between the uppermost ray and the ray starting at a distance $a/4$ below the uppermost ray is $(a/4)\sin\theta$.

Huygens–Fresnel Principle

Pretend that each point of the wave front reaching the slit can be regarded as a point source of light emitting a spherical wave; the net wave in the region beyond the slit is simply the superposition of all these waves.

This prescription has some obvious similarities to Huygens' Construction. However, the latter is merely a geometric construction for finding the successive positions of the wave fronts and does not yield any information about the distribution of intensity, whereas the aim of our new prescription is precisely the calculation of this distribution of intensity. Although the new prescription has a strong intuitive appeal, it turns out that it is not all that easy to justify it rigorously. The trouble is that light waves are not really sources of light waves — only accelerated charges are sources of light waves. The intensity pattern on the far side of the slit arises from the superposition of the original wave on all the waves radiated by the electric charges sitting in the opaque plate when accelerated by the original wave. Thus, our prescription lacks a simple physical basis. Nevertheless, it gives the right answer, or almost the right answer. This can be shown by a somewhat sophisticated mathematical argument, to be presented in Section 39.3.

We begin our calculation by finding the positions of the maxima and minima in the diffraction pattern of the slit. Figure 39.3 shows some rays associated with the waves that spread out from the points of a wave front at the slit. All the rays lead to a point P in the space beyond the slit. As in our calculation of the two-slit pattern, we will assume that P is very far away so that all the rays are nearly parallel. These rays can be divided into two equal groups: those that come from the upper part of the slit and those that come from the lower. Rays from the first group have a shorter distance to travel to the point P than rays from the second group. Consider the uppermost ray from the first group and the uppermost ray from the second. The path difference between these is $(a/2)\sin\theta$, where a is the width of the slit. If this path difference is $\frac{1}{2}\lambda$, the two rays will interfere destructively. Furthermore, pairs of rays that originate at an equal distance below each of these two uppermost rays in the two groups will also interfere destructively. This shows that all the waves cancel in pairs when

$$\frac{a}{2}\sin\theta = \tfrac{1}{2}\lambda \tag{1}$$

or

$$a\sin\theta = \lambda \tag{2}$$

This is the condition for the first minimum.

To find the next minimum, we divide the rays into four equal groups and consider the uppermost ray from the first and the second groups (Figure 39.4). These rays, and other pairs of rays, will interfere destructively if their path difference is $\frac{1}{2}\lambda$:

$$\frac{a}{4}\sin\theta = \tfrac{1}{2}\lambda \tag{3}$$

or

$$a\sin\theta = 2\lambda \tag{4}$$

By continuing this argument, we find a general condition for minima:

$$\boxed{a\sin\theta = \lambda,\ 2\lambda,\ 3\lambda,\ \ldots} \tag{5}$$ *Minima for a single slit*

As regards the maxima, there is of course a strong central maximum ($\theta = 0$). The secondary maxima cannot be found by any simple argument; roughly, their position is halfway between the successive minima. On a photographic film placed at some distance, the maxima and minima show up as a pattern of light and dark fringes (Figure 39.5).

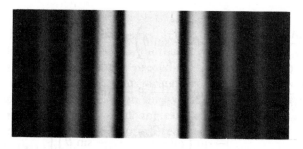

Fig. 39.5 A photographic film placed beyond an illuminated slit records a strong central maximum and successively weaker secondary maxima. (Courtesy C. C. Jones, Union College.)

EXAMPLE 1. Equation (5) applies not only to light waves, but also to radio waves and other waves. Suppose that radio waves from a TV transmitter with a wavelength of 0.80 m strike the wall of a large building. In this wall there is a very wide window with a height of 1.4 m (a horizontal slit). The wall is opaque to radio waves and the window is transparent. What is the angular

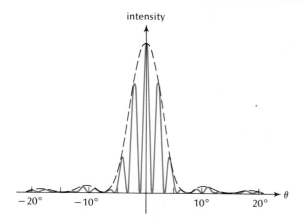

Fig. 39.8 Intensity as a function of θ for light arriving at a screen placed behind two slits of width $a = 8\lambda$ separated by a center-to-center distance of $d = 26\lambda$. The dashed curve (black) is the diffraction curve for a single slit.

Fig. 39.9 A photographic film placed beyond the two slits records a complex pattern of bands. (Courtesy C. C. Jones, Union College.)

39.2 Diffraction by a Circular Aperture; Rayleigh's Criterion

The diffraction of light by a circular aperture is in principle no different from the diffraction by a slit (a very long rectangular aperture). To calculate the light distribution we must sum the waves originating at all points of a wave front at the circular aperture. The evaluation of the resulting integral is a bit messy and we will not attempt it. Figure 39.10 shows a plot of the intensity distribution as a function of the

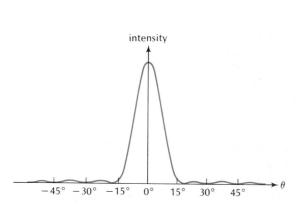

Fig. 39.10 Intensity as a function of θ for light arriving at a screen placed behind a circular aperture of diameter $a = 4\lambda$. Note that this is somewhat similar to Figure 39.7.

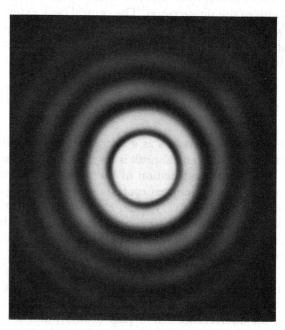

Fig. 39.11 A photographic film placed beyond a circular aperture records a strong central maximum and annular secondary maxima. (Courtesy C. C. Jones, Union College.)

usual angle θ; the distribution is of course axially symmetric. The overall shape of the curve plotted in Figure 39.10 is quite similar to that of Figure 39.7; there is a strong central maximum and a sequence of minima and secondary maxima. The angular position of the first minimum is given by

$$\sin \theta = 1.22\lambda/a \qquad (15)$$

First minimum for circular aperture

where a is the diameter of the circular aperture.

Figure 39.11 is a photograph of the diffraction pattern produced by a circular aperture.

Many optical instruments — telescopes, microscopes, cameras, etc. — have circular apertures and these will diffract light. For instance, the objective lens of an astronomical telescope will act like a circular opening in a plate; the parallel wave fronts arriving from some distant star will suffer diffraction effects and produce an intensity distribution such as that plotted in Figure 39.10. The image of the star as seen through this telescope will then not be a bright point, but a disk surrounded by concentric rings as in Figure 39.11. For example, seen through a telescope with an objective lens 6 cm in diameter, stars look like small disks, about 2×10^{-5} radian (or 4 seconds of arc) in diameter.

This spreading out of the image puts a limit on the detail that can be perceived through this telescope. If two stars are very close together, their images tend to merge and it may be impossible to tell them apart. Figures 39.12a–d show the images produced by a pair of pointlike light sources upon diffraction by a circular aperture. In the first of these figures, the angular separation of the sources is so small that the two images look like one image. In the second figure, the angular separation is large enough to give a clear indication of the existence of two separate images.

The angular separation of the sources in Figure 39.12b is such that the central maximum of the diffraction pattern of one source coincides

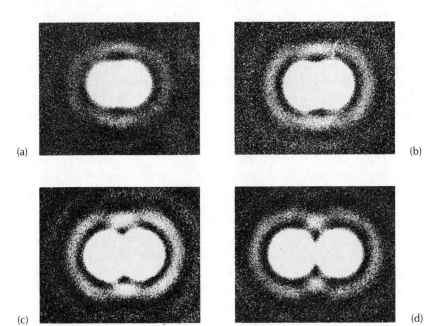

(a)

(b)

(c)

(d)

Fig. 39.12 When the light waves from two pointlike sources arrive at a circular aperture simultaneously, each set of light waves will produce its own diffraction pattern. If the angular separation between the two sources is small, the diffraction patterns overlap. In photograph (a), the angular separation is very small. In photographs (b), (c), and (d), the angular separation is progressively larger. In the first photograph, the two light sources are not resolved (not distinct). In the second photograph, they are just barely resolved, according to Rayleigh's criterion. (Courtesy C. C. Jones, Union College.)

with the minimum of the diffraction pattern of the other source. Since we are now dealing with small angles, Eq. (15) tells us that this angular separation must be

Rayleigh's criterion

$$\Delta\theta = 1.22 \, \frac{\lambda}{a} \qquad (16)$$

We will regard this as the critical angle that decides whether the two sources are clearly distinguishable: the telescope (or other optical device) can resolve the sources if their angular separation is larger than that in Eq. (16), and it cannot resolve them if the separation is smaller. This is **Rayleigh's criterion.**

EXAMPLE 2.   The star ζ Orionis is a binary star, that is, it consists of two stars very close together. The angular separation of the stars is 2.8 seconds of arc. Can the stars be resolved with a telescope having an objective lens 6 cm in diameter? Assume that the wavelength of the starlight is 5500 Å.

SOLUTION: According to Rayleigh's criterion, a telescope of this aperture can resolve stars as close as

$$\Delta\theta = 1.22 \, \frac{\lambda}{a} = \frac{1.22 \times 5.5 \times 10^{-7} \text{ m}}{0.06 \text{ m}}$$

$$= 1.1 \times 10^{-5} \text{ radian} \qquad (17)$$

This is equal to 2.3 seconds of arc. Hence the telescope can resolve the double stars.

John William Strutt, 3rd Baron Rayleigh, *1842–1919, English physicist, professor at Cambridge and at the Royal Institution. He is best known for his extensive mathematical investigations of sound and of light. He also investigated the behavior of gases at high densities, and discovered argon; for this he was awarded the Nobel Prize in 1904.*

The ability of a telescope to resolve stars or other objects of small angular separation improves with the size of the telescope — a telescope of 30-cm diameter can resolve angular separations as small as 0.5 seconds of arc. However, beyond 30 cm, the resolution of an Earthbound telescope does not increase any further with size. The trouble is that fluctuations in the density of the atmosphere produce slight perturbations in the path of light rays coming down from the sky. This causes a jitter in the apparent position of the star. Even under favorable atmospheric conditions — designated as **good seeing** by astronomers — the jitter is usually at least 0.5 seconds of arc. Thus, the very large aperture of the 5.1-m (200-in.) telescope on Mt. Palomar does not improve the resolution, but it does collect a very large amount of light and therefore permits the observation of very faint objects in the sky.

The Space Telescope, to be launched in 1986, will be placed in orbit above the atmosphere of the Earth and it will therefore not be affected by atmospheric seeing conditions. This telescope (Figure 39.13) will have an aperture of 2.4 m and it will achieve an angular resolution of about 0.1 second of arc, close to the limit set by Rayleigh's criterion. Thus, the Space Telescope will achieve higher resolution than any Earth-bound telescope. Furthermore, it will be able to observe at ultraviolet and infrared wavelengths, which are blocked by the atmosphere.

For Earth-bound radiotelescopes, the atmosphere poses no problem — radio waves do not suffer from the effects of atmospheric den-

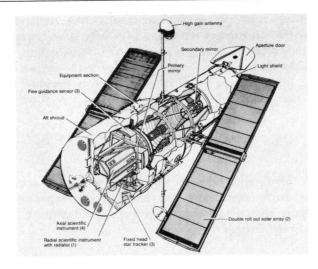

Fig. 39.13 The Space Telescope.

sity fluctuations.[2] Hence, with increasing size, the resolution of a radiotelescope improves indefinitely. Figure 39.14 shows the large radiotelescope at Arecibo, Puerto Rico. The concave "mirror" of this telescope has an aperture of 1000 ft and a radius of curvature which is also 1000 ft. The shortest wavelength at which this telescope has been operated is 4 cm. For this wavelength, Rayleigh's criterion gives a limiting angular resolution of

$$\theta = 1.22 \frac{\lambda}{a} = 1.22 \times \frac{0.04 \text{ m}}{1000 \text{ ft} \times 0.30 \text{ m/ft}}$$

$$= 1.6 \times 10^{-4} \text{ radian}$$

which is about 30 seconds of arc.

Fig. 39.14 Arecibo radiotelescope.

---

[2] The density fluctuations occur over such small distances that radio waves do not notice them.

A very substantial improvement in resolution is attainable by employing two radiotelescopes separated by a very large distance. If the signals detected by these two telescopes are brought into superposition,[3] then the system will act essentially like the two-slit system of Section 38.4. The limiting angular resolution is roughly the angle between a maximum and the adjacent minimum,

$$\Delta\theta \cong \frac{\lambda}{d} \tag{18}$$

where d is the distance between the telescopes. In one application of this method astronomers employed radiotelescopes separated by a distance of more than 10,000 km: they were in Greenbank, West Virginia, and in the Crimea.

39.3 Babinet's Principle[4]

In this last section we will present the theoretical justification for the Huygens–Fresnel Principle. According to this principle, the amplitude of the light in the region beyond an opaque plate with an aperture of a given shape (single slit, double slit, circular hole, etc.) can be calculated by pretending that the points of the aperture act as point sources of light. To gain some insight into the actual origins of the amplitude of the light, consider first an opaque plate whose aperture has been closed with a plug that exactly fits the opening (Figure 39.15). Obviously, when we place such a plugged plate in the path of a light wave, the amplitude of the light beyond the plate will be zero. It is very instructive to inquire: How does the plate stop the light wave? The absence of light beyond the plate is a cancellation effect. When the wave strikes the plate, it shakes the electrons within the atoms and causes them to radiate light of the same wavelength as that of the incident light. The original wave passes through the plate and reaches the region beyond the plate. The radiated waves from all the electrons also reach this region. The net amplitude of the light in this region is then the sum of the contribution of the original wave plus the contribution from all the electrons — in order that the plugged plate be truly opaque, these two contributions must cancel exactly.

Mathematically we can express this as follows: At some point beyond the plate, the electric field contributed by the radiating electrons has an amplitude $\mathbf{E}_{rad}$ and the electric field contributed by the original wave has an amplitude $\mathbf{E}_0$. These must be opposite,

$$\mathbf{E}_{rad} = -\mathbf{E}_0 \tag{19}$$

The field contributed by the electrons can be split into two parts: that due to the electrons in the perforated plate (without the plug) and that in the plug. Thus, Eq. (19) becomes[5]

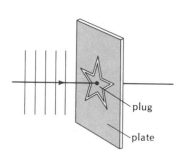

Fig. 39.15 An opaque plate with an aperture. The aperture has been closed with a plug.

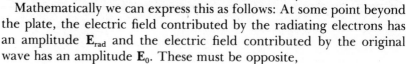
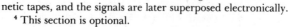

---

[3] For radiotelescopes separated by an intercontinental distance, it is not practical to connect both antennas to the same, single-radio receiver. Instead, each antenna is connected to its own radio receiver, recordings of the received signals are made on magnetic tapes, and the signals are later superposed electronically.

[4] This section is optional.

[5] This equation is not exact. An unperforated plate, made of a single piece of material, is not completely equivalent to a perforated plate plus a separate plug. The motion

$$\boxed{\mathbf{E}_{\text{rad, plate}} + \mathbf{E}_{\text{rad, plug}} = -\mathbf{E}_0} \qquad (20)$$

This relationship between the fields radiated from the perforated plate and the plug is called **Babinet's Principle.**[6] The plug can be regarded as a plate of a shape *complementary* to that of the perforated plate. With this terminology, Babinet's Principle simply states that the sum of the fields radiated from a plate and from its complement equals the negative of the field of the original wave.

With these preliminaries, we are now ready to examine the amplitude beyond the plate when the plug is *absent*. This amplitude has a contribution from the radiating electrons in the perforated plate and a contribution from the original wave:

$$\mathbf{E}_{\text{beyond plate}} = \mathbf{E}_{\text{rad, plate}} + \mathbf{E}_0 \qquad (21)$$

But, in view of the identity (20) the right side can be replaced by $-\mathbf{E}_{\text{rad,plug}}$. We then obtain

$$\mathbf{E}_{\text{beyond plate}} = -\mathbf{E}_{\text{rad, plug}} \qquad (22)$$

Thus, apart from a minus sign, the amplitude beyond the plate happens to be just what we would calculate if we were to pretend that the points in the aperture radiate as point sources (or electrons). Of course, this is exactly what the Huygens–Fresnel Principle tells us to do — we have therefore obtained its justification. The minus sign on the right side of Eq. (22) is left out of account by the Huygens–Fresnel Principle, but this is of no importance because ultimately we are only interested in the *intensity* and this depends only on the square of the electric field.

Finally, let us rephrase the Babinet Principle in terms of the amplitude of the diffracted light. Suppose we remove the perforated plate from the light beam and instead place the plug there, i.e., we replace the plate by the complementary plate. Then an argument similar to that which led to Eq. (22) yields

$$\mathbf{E}_{\text{beyond plug}} = -\mathbf{E}_{\text{rad, plate}} \qquad (23)$$

By substituting Eqs. (22) and (23) into Eq. (20) we obtain

$$\boxed{\mathbf{E}_{\text{beyond plate}} + \mathbf{E}_{\text{beyond plug}} = \mathbf{E}_0} \qquad (24) \qquad \textit{Babinet's Principle}$$

This identity is another form of Babinet's Principle. It states that the sum of the amplitudes of light diffracted by a plate and by its complement equals the amplitude of the original wave.

---

of the electrons on the perforated plate and on the plug is subject to the constraint that the electrons cannot cross the boundary separating the perforated plate and the plug; the motion on the unperforated plate is not subject to this constraint. This difference leads to small deviations from Eq. (20); but these deviations are only important in regions fairly near the plate–plug boundary. As in the preceding sections, we will only deal with points at a large distance from the plate and there Eq. (20) is a good approximation.

[6] For another version of Babinet's Principle, see below.

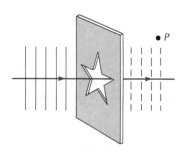

Fig. 39.16 The point *P* is not in the path of the incident light wave.

Equation (24) has an interesting application. Suppose that we want to calculate the intensity of light in some region outside of the path of the original light wave (Figure 39.16). In this region $\mathbf{E}_0 = 0$. Hence Eq. (24) gives

$$\mathbf{E}_{\text{beyond plate}} = -\mathbf{E}_{\text{beyond plug}} \qquad (25)$$

Since the intensity of light depends on the square of the electric field, Eq. (25) shows that a plate and its complement produce the same diffraction pattern! For instance, an arrangement of star-shaped holes in a plate and the complementary arrangement of opaque stars suspended in midair (Figure 39.17) produce the same diffraction pattern. Figure 39.18 shows photographs of these diffraction patterns. Of course they differ in the central region which is in the path of the original wave; but elsewhere they are identical.

Fig. 39.17 (a) An opaque plate with star-shaped holes. (b) The complementary plate consists of opaque stars. (Courtesy C. C. Jones, Union College.)

Fig. 39.18 (a) Photograph of the diffraction pattern of the plate shown in Figure 39.17a. (b) Photograph of the diffraction pattern of the complementary plate shown in Figure 39.17b. The two diffraction patterns are identical, except in the central region, where the undiffracted light beam passes through the complementary plate with high intensity. (Courtesy C. C. Jones, Union College.)

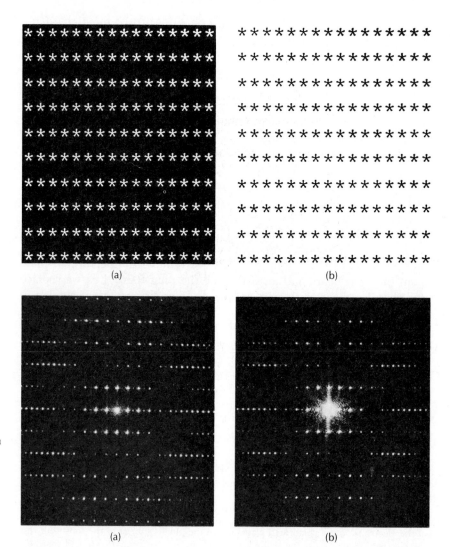

(a)   (b)

(a)   (b)

SUMMARY

Minima for single slit: $a \sin \theta = \lambda,\ 2\lambda,\ 3\lambda, \dots$

First minimum for circular aperture: $a \sin \theta = 1.22\lambda$

Rayleigh's criterion: $\Delta\theta = 1.22\lambda/a$

Babinet's Principle: $\mathbf{E}_{\text{beyond plate}} + \mathbf{E}_{\text{beyond plug}} = \mathbf{E}_0$

QUESTIONS

1. Gratings used for analyzing light are often called diffraction gratings, but it would be more accurate to call them interference gratings. Why?

2. Consider the diffraction pattern shown in Figure 39.7. How would this pattern change if the entire apparatus were immersed in water?

3. Figure 16.28 shows the diffraction of water waves at the entrance of a harbor. Could this diffraction be eliminated by making the entrance smaller?

4. What would you expect the diffraction pattern of a rectangular aperture to look like?

5. In the center of the shadow of a disk or a sphere there is a small bright spot, called the **Poisson spot** (Figure 39.19). This spot is very faint near the disk, but becomes more noticeable at large distances. Qualitatively, explain why the diffraction of the light waves around the edges of the disk gives rise to this spot.

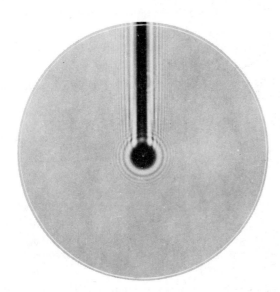

Fig. 39.19 The Poisson spot.

6. Besides good angular resolution, what other advantage does a telescope of large aperture have over a telescope of small aperture?

7. Spy satellites use cameras with lenses of very large diameter, 30 cm or more. Why are such large diameters necessary?

8. What might be the advantages of building a radiotelescope in space, in orbit around the Earth?

9. In order to beam a sound wave sharply in one direction with a loud-hailer, should the horn of the loud-hailer have a small aperture or a large aperture?

10. The manufacturers of the Questar telescope claim that this telescope can distinguish two stars even if their separation is somewhat smaller than that specified by Rayleigh's criterion. Why is this possible?

11. The maximum useful magnification of an optical microscope is determined by diffraction effects at the objective lens. Explain.

Fig. 39.20 Atoms in a barium titanate crystal.

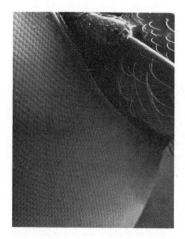

Fig. 39.21 A portion of the compound eye of a dragonfly. This eye consists of about 28,000 ommatidia. (Courtesy R. Olberg, Union College.)

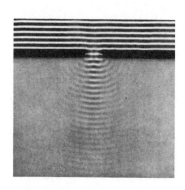

Fig. 39.22

12. Other things being equal, how much resolution can you gain by operating an optical microscope with blue light instead of red light? Why can you not operate the microscope with ultraviolet light?

13. The picture of atoms in Figure 39.20 was taken with an electron microscope using electron waves of extremely short wavelength. Roughly, how short must the wavelength be to make individual atoms visible?

14. The compound eye of insects consists of a large number of small eyes, or ommatidia (Figure 39.21). Each ommatidium is typically 0.03 mm across; it does not form an image, but merely acts as a sensor of the intensity of light arriving from a narrow cone of directions. What are some of the advantages and disadvantages of such a compound eye as compared to the camera eye of vertebrates?

15. The corona formed by diffraction of sunlight by small droplets of water in clouds consists of a bright ring surrounding the Sun. What is the color of the outer edge of the corona? The inner edge?

16. Antennas used for the transmisson of microwaves in communication links consist of metal dishes. What factors determine the size of these dishes?

17. In the joint operation of two radiotelescopes separated by an intercontinental distance, it is not practical to connect both of these antennas to the same, single radio receiver. Instead, each antenna is connected to its own radio receiver, and recordings of the received signals are made on magnetic tapes. How can we extract the interference patterns from the information stored on these tapes?

18. What would the diffraction pattern of a narrow opaque strip look like? Of a small opaque disk?

19. How would the diffraction patterns of Figures 39.18a and b change if all the stars in Figure 39.17 were made smaller?

20. What feature of the plate shown in Figure 39.17a determines the spacing between the dots in Figure 39.18a? What feature determines the gradual modulation of the intensity of these dots?

PROBLEMS

Section 39.1

1. Light of wavelength 6328 Å from a He–Ne laser illuminates a single slit of width 0.10 mm. What is the width of the central maximum formed on a screen placed 2.0 m beyond the slit?

2. Consider the water waves shown in Figure 39.22. With a ruler, measure the wavelength of the waves and the length of the gap; with a protractor, measure the angular positions of the two nodal lines (minima). Check whether these quantities satisfy Eq. (5).

3. A sound wave of frequency 820 Hz passes through a doorway of width 1.0 m. What are the angular directions of the minima of the diffraction pattern?

4. Water waves of wavelength 20 m approach a harbor entrance 50 m across at right angles to their path. What is the angular width of the central beam of diffracted waves beyond the entrance?

5. Light of wavelength λ is obliquely incident on a slit of width a. The angle of incidence of the light on the slit is ϕ (Figure 39.23).
   (a) Show that the diffracted light emerging at an angle θ interferes destructively if

$$a \sin \theta - a \sin \phi = \lambda, \, 2\lambda, \, 3\lambda, \, \ldots$$

(b) Show that, for small values of θ, the angular separation between the directions of destructive interference is independent of ϕ.

6. A slit of width 0.11 mm cut in a sheet of metal is illuminated with light of wavelength 5770 Å from a mercury lamp. A screen is placed 4.0 m beyond the slit.
   (a) Find the width of the central maximum in the diffraction pattern on the screen, i.e., find the distance between the first minimum on the left and on the right.
   (b) Find the width of the secondary maximum, i.e., find the distance between the first minimum and the second minimum on the same side.

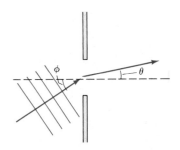

Fig. 39.23

7. Plot intensity vs. angle for the light diffracted by a pair of slits of width $a = 4\lambda$ separated by a center-to-center distance $d = 8\lambda$.

8. Two narrow slits of width a are separated by a center-to-center distance d. Suppose that the ratio of d to a is an integer, $d/a = n$.
   (a) Show that in the diffraction pattern produced by this arrangement of slits, the nth interference maximum (corresponding to $d \sin \theta = n\lambda$) is suppressed because of coincidence with a diffraction minimum. Show that this is also true for the $2n$th, $3n$th, etc., interference maxima.
   (b) How many interference maxima are there between one diffraction minimum on one side and the next on the same side?

9. Figure 38.19 shows the intensity curve for the interference pattern produced by light emerging from three very thin slits (three pointlike sources) separated by distances of 12λ. Suppose the three very thin slits are replaced by three new slits of width 6λ. This will change the intensity curve.
   (a) On top of Figure 38.19 (or on a copy of Figure 38.19), plot the intensity curve for the diffraction pattern produced by *one* of these new slits.
   (b) By suitably combining these two curves, obtain the complete intensity curve for the system of the three new slits.

10. Figure 39.24 is a plot of the intensity of the light on a screen placed beyond an opaque plate with two parallel narrow slits. The distance between the slits is 0.12 mm. Deduce the width of each of the slits.

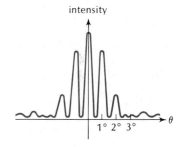

Fig. 39.24 Intensity as a function of θ in the diffraction pattern of two slits.

Section 39.2

11. According to a recent proposal (see Section F.4), solar energy is to be collected by a large power station on an artificial satellite orbiting the Earth at an altitude of 35,000 km. The power is to be beamed down to the surface of the Earth in the form of microwaves. If the microwaves have a wavelength of 10 cm and if the antenna emitting the microwaves is 1.5 km in diameter, what is the angular width of the central beam emerging from this antenna? What will be the transverse dimension of the beam when it reaches the surface of the Earth?

12. When the eye looks at a star (a point of light), diffraction at the pupil spreads the image of the star on the retina into a small disk.
   (a) When opened to maximum size, the diameter of the pupil of a human eye is 7.0 mm. Assuming the starlight has a wavelength of 5500 Å, what is the angular size of the image on the retina?
   (b) The distance from pupil to retina is 23 mm. What is the linear size of the image of the star?
   (c) At the midpoint on the retina (fovea), there are 150,000 light-sensitive cells (rods) per mm². How many of these cells are illuminated when the eye looks at a star?

13. A sailor uses a speaking trumpet to concentrate his voice into a beam. The opening at the front end of the speaking trumpet has a diameter of 25 cm. If the sailor emits a sound of wavelength 15 cm (this is very roughly the wavelength a man emits when yelling "eeeee . . ."), what is the angular width of the central maximum of the beam of sound?

14. The antenna of a small radar transmitter operating at 1.5×10^{10} Hz consists of a circular dish of diameter 1.0 m. What is the angular width of the central maximum of the radar beam? What is the linear width at a distance of 5.0 km from the transmitter?

15. A microwave antenna used to relay communication signals has the shape of a circular dish of diameter 1.5 m. The antenna emits waves with $\lambda = 4.0$ cm.
   (a) What is the width of the central maximum of the beam of this antenna at a distance of 30 km?
   (b) The power emitted by the antenna is 1.5×10^3 W. What is the energy flux directly in front of the antenna? What is the energy flux at a distance of 30 km? Assume that the power is evenly distributed over the width of the central beam.

16. For an optically perfect lens, the size of the focal spot is limited only by diffraction effects. Suppose that a lens of diameter 10 cm and focal length 18 cm is illuminated with parallel light of wavelength 5500 Å. What is the angular width of the central maximum in the diffraction pattern? What is the corresponding linear width at the focal distance?

17. The Space Telescope that will be placed into an orbit above the atmosphere of the Earth has an aperature of 2.4 m. According to Rayleigh's criterion, what angular resolution can this telescope achieve with visible light of wavelength 5500 Å? With ultraviolet light of wavelength 1200 Å? How much better is this than the angular resolution of 0.5 seconds of arc achieved by telescopes on the surface of the Earth?

18. The radiotelescope at Jodrell Bank (England) is a dish with a circular aperture of diameter 76 m. What angular resolution can this radiotelescope achieve when operating at a wavelength of 21 cm?

19. According to newspaper reports, a photographic camera on a "Blackbird" reconnaissance jet flying at an altitude of 17 mi can distinguish detail on the ground as small as the size of a man.
   (a) Roughly, what angular resolution does this require?
   (b) According to Rayleigh's criterion, what minimum diameter must the lens of the camera have?

20. Rumor has it that a photographic camera on a spy satellite can read the license plate of an automobile on the ground.
   (a) If the altitude of the satellite is 100 mi, roughly what angular resolution does the camera need to read a license plate? Assume that the reading requires a linear resolution of about 2 in.
   (b) To attain this angular resolution, what must be the diameter of the aperture of the camera?

21. (a) According to Rayleigh's criterion, what is the angular resolution that the human eye can achieve for light of wavelength 5500 Å? The fully distended pupil of the human eye has a diameter of 7.0 mm.
   (b) Even during steady fixation, the eye has a spontaneous tremor that swings it through angles of 20 or 30 seconds of arc. Compare this angular tremor with the angular resolution that you found in part (a). Would the elimination of the tremor greatly improve the acuity of the eye?

22. At night, on a long stretch of straight road in Nevada, a truckdriver sees the distant headlights of another truck. How close must he be to the other truck in order for his eyes to resolve two headlights? Assume that the pupils of the truckdriver have a diameter of 5.0 mm, that the headlights are separated by 1.8 m, and that the light has a wavelength of 5500 Å.

23. Some spy satellites carry cameras with lenses 30 cm in diameter and with a focal length of 2.4 m.

(a) What is the angular resolution of the camera according to Rayleigh's criterion? Assume that the wavelength of light is 5500 Å.

(b) If such a satellite looks down on the Earth from a height of 150 km, what is the distance between two points on the ground that the camera can barely resolve?

(c) The lens projects images of the two points on a film at the focal plane of the lens. What is the distance between the two images projected on the film?

24. 7×50 binoculars magnify angles by a factor of 7 and their objective lenses have an aperture of 50-mm diameter.

(a) According to Rayleigh's criterion, what is the intrinsic angular resolution of these binoculars? Assume that the light has a wavelength of 5500 Å.

(b) At best, the pupil of your eye has an aperture of 7.0-mm diameter. Compare the angular resolution of your eye divided by a factor of 7 with the intrinsic angular resolution of the binoculars. Which of the two numbers determines the actual angular resolution you can achieve while looking through the binoculars?

25. When exposed to strong light, the pupil of the eye of a cat narrows to a fine slit, about 0.3 mm across. Suppose that the cat is looking at two white mice 20 m away and separated by a distance of 5 cm. Can the cat distinguish one mouse from the other?

26. Show that the optimum diameter for the pinhole of a pinhole camera (see Section 37.6) is approximately $\sqrt{2.44\lambda L}$, where λ is the wavelength of the light, and L is the distance from the pinhole to the film. (Hint: Consider light arriving from a distant point source. If the pinhole is excessively small, diffraction at the pinhole spreads the light over a spot diameter $2.44\lambda L/a$ on the film. If the pinhole is excessively large, then the light from a distant point source illuminates a spot diameter a. What choice of a makes the smallest spot on the film?

Section 39.3

27. A narrow, opaque strip, 0.060 mm across, is placed in the light beam of wavelength 4579 Å from a laser. The diffraction pattern formed by this strip on a screen 4.8 m beyond shows bright and dark fringes. What is the distance between the second and the third dark fringe on one side of the central beam?

28. A circular opaque disk of diameter 0.12 mm is placed in the beam of a He–Ne laser emitting a wavelength of 6328 Å. The disk makes a diffraction pattern of bright and dark rings on a screen placed 5.0 m beyond. What is the diameter of the first dark circular ring in this diffraction pattern?

29. A thin strip of opaque material 1.5 mm wide is illuminated by light of wavelength 6940 Å. At what distance beyond the strip is the width of the central maximum of the diffracted light as large as the width of the geometric shadow (1.5 mm), so that the shadow disappears?

CHAPTER 40

Quanta of Light

In the two preceding chapters we examined the wave properties of light. We saw that light exhibits interference and diffraction, in agreement with Maxwell's theory, according to which light is a wave consisting of electric and magnetic fields with a smooth distribution of energy. In this chapter we will discover that light has particle properties. We will describe experimental evidence that shows that a light beam consists of a stream of particle-like energy packets. These energy packets are called **quanta of light** or **photons.**

The fact that light has the dual attributes of wave and of particle indicates that neither the wave concept of classical physics nor the particle concept of classical physics gives an adequate description of light. We have to think of light as a wave–particle object, sometimes called a **wavicle.** This object sometimes behaves pretty much as a classical wave, sometimes pretty much as a classical particle, and sometimes as a bit of both. Furthermore, we will see in the next chapter that electrons, protons, neutrons, and all the other known ''particles'' also exhibit such dual attributes of wave and of particle — we must regard all of them as wavicles.

40.1 Blackbody Radiation

The first hint of a failure of classical physics emerged around 1900 from the study of thermal radiation. When we heat a body to high temperature, it glows; for instance, when we heat a bar of iron to 1300 K, it glows in a deep red. This glow is thermal radiation (''radiant heat''). The spectrum of thermal radiation emitted by a hot body is continuous — if we analyze the light with a prism, we find that the energy is smoothly distributed over all wavelengths.

A quantitative description of the distribution of energy over the different wavelengths is provided by the **spectral emittance** S_λ. This is the energy flux (or power per unit area) emitted by the surface of the glowing body per unit wavelength interval; thus, $S_\lambda \, d\lambda$ is the flux emitted in a small wavelength interval $d\lambda$. The spectral emittance is a function of wavelength. Measurements of the thermal radiation emitted by a glowing body show that the energy flux at very long and at very short wavelengths is quite small, and that the energy flux has a maximum at some intermediate wavelength. The position of this maximum depends on the temperature. For example, an iron bar at 1300 K has a maximum emittance at 22,000 Å (infrared), whereas the Sun, with a surface temperature of 5800 K, has a maximum emittance at 5000 Å (blue-green). Figure 40.1 is a plot of the emittance of the Sun as a function of the wavelength; this figure may be regarded as a plot of the intensity distribution in the smear of light that an (ideal) prism produces when exposed to sunlight.

Spectral emittance

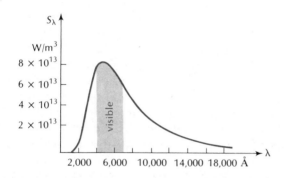

Fig. 40.1 Spectral emittance of the Sun. In this plot the discrete dark spectral lines (Fraunhofer lines), which result from the blocking out of some of the thermal radiation by gas in the solar atmosphere, have been neglected.

The thermal radiation emerging from the surface of a glowing body is generated within the volume of the body by the random thermal motions of atoms and electrons. Before the radiation reaches the surface and escapes, it is absorbed and re-emitted many times and it attains thermal equilibrium with the atoms and electrons. This equilibration process shapes the continuous spectrum of the radiation, completely washing out all of the original spectral features of the radiation.

The flux of thermal radiation emerging from the surface of a glowing body depends to some extent on the characteristics of the surface. The surface usually permits the escape of only a fraction of the flux reaching it from the inside of the body. Correspondingly, if the body is irradiated with an equal flux of thermal radiation from the outside, the surface permits the ingress of only an equal fraction of this flux, reflecting the rest. This equality of the emissive and absorptive characteristics of the surface can be deduced by an argument based on thermodynamics. Thus we are led to a general rule: *a good absorber is a good emitter; and a poor absorber is a poor emitter.* With this rule we can understand how thermos bottles or dewars provide such excellent thermal insulation. These bottles are constructed with a double glass wall, and the space between these walls is evacuated (Figure 40.2). Heat cannot flow across the evacuated space by conduction or convection; it can only flow by radiation. To inhibit radiation, the glass walls

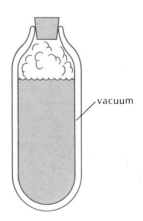

Fig. 40.2 A thermos bottle.

are silvered and thereby made into mirrorlike reflecting surfaces; these highly reflective surfaces are very poor absorbers and emitters of radiation. This keeps the heat transfer between the walls very small.

A body with a perfectly absorbing (and emitting) surface is called a **blackbody;** such a body would look black under illumination from the outside. When a blackbody is hot, its surface emits more thermal radiation than that of any other hot body at the same temperature. In practice the characteristics of an ideal blackbody are most easily achieved by a trick: take a body with a cavity, such as a hollow cube, and drill a small hole in one side of the cube (Figure 40.3). The hole then acts as a blackbody — any radiation incident on the hole from outside will be completely absorbed. Because of this equivalence between a blackbody and a hole in a cavity, the terms *blackbody radiation* and *cavity radiation* are used interchangeably.

The blackbody plays a special role in the study of thermal radiation because its spectral emittance does not depend on the material of which it is made or on any other characteristics of the body — the spectral emittance depends only on the temperature of the body. We can prove this by appealing to the Second Law of Thermodynamics, according to which heat cannot flow spontaneously from a colder body to a warmer body. Consider two cavities at equal temperature with holes of equal size (Figure 40.4). The cavity on the left radiates into the cavity on the right and vice versa. If the flux emitted by the cavity on the left were larger than that emitted by the cavity on the right, the radiative heat transfer would lead to an increase of temperature on the right and a decrease on the left. Heat would then be flowing from a colder to a warmer body, in contradiction to the Second Law. This argument establishes that the fluxes emitted by both cavities are the same. Furthermore, a slight refinement of this argument establishes that the fluxes in any given small wavelength interval $d\lambda$ are also the same. We need only make a slight alteration in the arrangement shown in Figure 40.4 by inserting a filter for light between the two cavities, a filter that only permits the passage of radiation of wavelengths in the interval λ to $\lambda + d\lambda$. Our thermodynamic argument then leads to the conclusion that the fluxes in this chosen wavelength interval must be the same.

Thus for a blackbody, the spectral emittance S_λ is a universal function of the wavelength λ and of the temperature T, and of nothing else. In the last years of the nineteenth century, physicists engaged in an intensive experimental and theoretical effort to find the function S_λ.

Blackbody

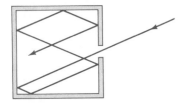

Fig. 40.3 A cavity with a small hole. Any radiation entering the hole is trapped; it will suffer multiple reflections and it will ultimately be absorbed.

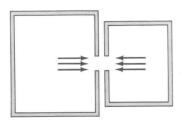

Fig. 40.4 Two cavities with holes of equal size.

40.2 Energy Quanta

One of the first attempts at a theoretical derivation of the blackbody spectrum was made by Lord Rayleigh. Since the energy flux of the radiation emerging from the hole in a cavity is directly proportional to the energy density of the radiation inside the cavity,[1] Rayleigh decided to calculate the latter quantity. He began by noting that the radiation

---

[1] According to Eqs. (36.30) and (36.32), for a plane wave the constant of proportionality between the energy flux (Poynting vector) and the energy density is c, the speed of light. For thermal radiation, with many plane waves traveling in random directions, the constant of proportionality is $c/4$.

in a cavity is made up of a large number of standing waves; Figure 40.5 shows some of these standing waves. Each of these standing waves can be regarded as a mode of vibration of the cavity. Rayleigh then appealed to the equipartition theorem, according to which, at thermal equilibrium, each mode of vibration has an average thermal energy of kT (in Section 19.4 we stated a special case of the equipartition theorem for free translational or rotational motion). Thus, each of the standing waves of Figure 40.5 ought to have an energy kT, and from this one can calculate the spectral emissivity. Although this calculation gave reasonable results at the long-wavelength end of the blackbody spectrum, it gave disastrous results at the short-wavelength end: the number of possible standing-wave modes of very short wavelength is infinitely large and, if each of these modes had an energy kT, the total energy in the cavity would be infinite! This disastrous failure of classical theory has been called the **ultraviolet catastrophe.**

Ultraviolet catastrophe

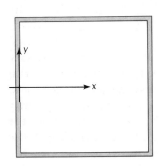

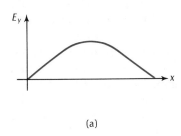 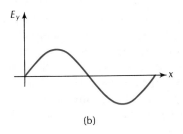 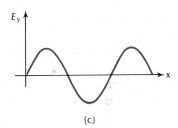

(a) (b) (c)

Fig. 40.5 Some of the possible standing electromagnetic waves in a closed cavity. For the sake of simplicity, only waves with a horizontal direction of propagation are shown. The plots give the electric field as a function of x at one instant of time.

The correct formula for the spectral emittance of a blackbody was finally obtained by Max Planck in 1900. By some inspired guesswork, based on thermodynamics, Planck hit upon the following formula:

$$S_\lambda = \frac{2\pi c^2 h}{\lambda^5} \frac{1}{e^{hc/kT\lambda} - 1}$$

(1) *Spectral emittance of blackbody*

where h is **Planck's constant,**[2]

$$h = 6.63 \times 10^{-34}\, \text{J} \cdot \text{s}$$

(2) *Planck's constant, h*

and k is Boltzmann's constant. Experimental investigation quickly established that the values of the spectral emittance calculated from Eq.

---

[2] See Appendix 8 for a more precise value of Planck's constant.

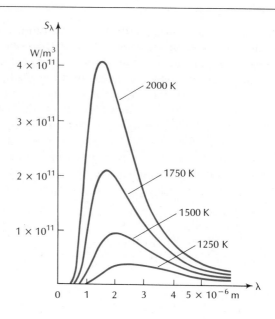

Fig. 40.6 Spectral emittance of a blackbody according to Planck's Law. Each curve is labeled with the appropriate temperature.

Max Planck, *1858–1947, German physicist, professor at Berlin and president of the Kaiser Wilhelm Society for the Advancement of Science (now the Max Planck Institute). Planck made significant contributions to thermodynamics before he became involved with the problem of blackbody radiation. He was a very conservative scientist, and he regarded his quantization postulate as "an act of desperation" into which he was forced because "a theoretical explanation had to be found at any cost, whatever the price." Planck received the Nobel Prize in 1918.*

(1) agreed very precisely with the measured values. Figure 40.6 shows the spectral emittance for several values of the temperature.

At first, Planck had proposed his formula merely as an empirical law that happens to give a good fit to the experimentally measured points. But then he searched for a theoretical justification of this law. The search led Planck to a revolutionary discovery: the quantization of energy. This discovery was to bring about the overthrow of classical Newtonian physics and the rise of quantum physics. For Planck, the postulate of the quantization of energy was "an act of desperation" which he committed because "a theoretical explanation had to be found at any cost, whatever the price."

Planck's derivation of the blackbody radiation formula involves some sophisticated statistical mechanics and we cannot reproduce the details of his derivation here. We will merely give a sketchy outline of this derivation. Planck began by making a theoretical model of the walls of the cavity; he regarded the atoms in the walls as small harmonic oscillators with electric charges. Although this is a rather crude model of an atom, it was adequate for his purposes since, by the thermodynamic argument of the preceding section, the radiation in a cavity is completely independent of the characteristics of the wall. The random thermal motions of the oscillators result in the emission of electromagnetic radiation. This radiation fills the cavity and acts back on the oscillators. When thermal equilibrium is attained, the average rate of emission of radiation energy by the oscillators matches the rate of absorption of radiation energy. Thus, the oscillators share their energy with the radiation in the cavity, and Planck was able to show that, under equilibrium conditions, the average radiation energy at some frequency ν (or at a wavelength $\lambda = c/\nu$) is directly proportional to the average energy of an oscillator of frequency ν. These steps of Planck's calculation involved nothing but classical mechanics. But in the next step of the calculation, Planck departed radically from classical physics. He postulated that the energy of the oscillators is quantized according to the following rule: *in an oscillator of frequency ν, the only permitted values of the energy are*

$$E = 0, \, h\nu, \, 2h\nu, \, 3h\nu, \, \ldots \tag{3}$$

All other values of the energy are forbidden. The constant h in Eq. (3) is Planck's constant, the same as appears in Eqs. (1) and (2). The energy $h\nu$ is called an **energy quantum;** according to the quantization rule, the energy of an oscillator is always some multiple of the basic energy quantum,

Energy quantum

$$E = nh\nu, \qquad n = 0, 1, 2, 3, \ldots \qquad (4)$$

Energy quantization of oscillator

The integer n is called the **quantum number** of the oscillator.

Quantum number

With this quantization condition, Planck calculated the average energy of the oscillators; and from that he derived his radiation formula. Although we cannot go into the details of this derivation, we can achieve a rough understanding of how Planck's calculation avoids the ultraviolet catastrophe. The thermal energy of the walls of the cavity is shared at random among all the oscillators in these walls. Some of these oscillators have high frequencies, some have low frequencies. For an oscillator of very high frequency, the energy quantum $h\nu$ is very large. If this oscillator is initially quiescent ($n = 0$), it cannot begin to move unless it acquires one energy quantum; but since this energy quantum $h\nu$ is very large, the random thermal disturbances will be insufficient to provide it — the oscillator will remain quiescent. Thus, the quantization of energy tends to inhibit the thermal excitation of the high-frequency oscillators. If the high-frequency oscillators remain quiescent, then they will not supply energy to the corresponding high-frequency standing waves in the cavity, and there will be no ultraviolet catastrophe.

Note that for an oscillator with a frequency of $\nu \simeq 10^{15}/\text{s}$, which is typical for atomic vibrations, the energy quantum is $h\nu = 6.6 \times 10^{-34}\,\text{J} \cdot \text{s} \times 10^{15}/\text{s} = 6.6 \times 10^{-19}\,\text{J}$. Since this is a very small amount of energy, quantization does not make itself felt at a macroscopic level. But quantization plays a pervasive role at the atomic level.

Unfortunately, Planck could not offer any basic justification for his postulate of quantization of energy. His postulate took care of the radiation law, but raised serious questions about classical physics. Obviously, quantization of energy makes no sense in classical physics — there is nothing in the laws of Newton that would prevent an oscillator from acquiring energy in any amount whatsoever. Planck could only justify his postulate by its consequences; but (in physics) the end does not justify the means. A deeper explanation of the quantization of energy only emerged much later, with the development of quantum mechanics (see Chapter 41).

From Planck's formula one can show that the spectral emittance of a blackbody has a maximum at a wavelength given by

$$\lambda_{\text{max}} = (2.90 \times 10^{-3}\,\text{m} \cdot \text{K}) \times \frac{1}{T} \qquad (5)$$

Wien's Law

This inverse proportionality of the wavelength of maximum emittance and the temperature is called **Wien's Law.**

By integrating Planck's formula one obtains the total flux, or power per unit area, emitted by the blackbody at all wavelengths:

$$S = \int S_\lambda \, d\lambda = \int \frac{2\pi c^2 h}{\lambda^5} \frac{1}{e^{hc/kT\lambda} - 1} \, d\lambda \qquad (6)$$

One can show that this total flux is proportional to the fourth power of the temperature,

Stefan–Boltzmann Law

$$S = \sigma T^4 \tag{7}$$

where $\sigma = 5.67 \times 10^{-8}$ W/(m² · K⁴).[3] This is called the **Stefan–Boltzmann Law.** The laws of Wien and of Stefan and Boltzmann had both been discovered empirically many years before Planck wrote down his formula.

EXAMPLE 1.   On a clear night, the surface of the Earth loses heat by radiation. Suppose that the temperature of the ground is 10°C and that the ground radiates like a blackbody. What is the rate of loss of heat per square meter?

SOLUTION: The absolute temperature of the ground is 283 K. Hence, the Stefan–Boltzmann Law tells us that the radiated flux, or power per unit area, is

$$S = \sigma T^4 = 5.67 \times 10^{-8} \text{ W/(m}^2 \cdot \text{K}^4) \times (283 \text{ K})^4$$

$$= 364 \text{ W/m}^2$$

This amounts to 87 cal/m² per second.

EXAMPLE 2.   Astronomers sometimes determine the size of a star by a method that relies on the Stefan–Boltzmann Law. Determine the radius of the star Capella from the following data: the flux of the starlight reaching the Earth is 1.2×10^{-8} W/m², the distance of the star is 4.3×10^{17} m, and its surface temperature is 5200 K. Assume the star radiates like a blackbody.

SOLUTION: According to the Stefan–Boltzmann Law, the flux, or power per unit area, emerging from the surface of the star is σT^4. The surface area of the star is $4\pi R^2$ and hence the total emitted power is

$$4\pi R^2 \times \sigma T^4 \tag{8}$$

This must match the total power crossing a sphere of radius 4.3×10^{17} m centered on the star. Since at each point of this sphere the power per unit area is 1.2×10^{-8} W/m², the total power is

$$4\pi(4.3 \times 10^{17} \text{ m})^2 \times 1.2 \times 10^{-8} \text{ W/m}^2 = 2.8 \times 10^{28} \text{ W} \tag{9}$$

Comparing Eqs. (8) and (9), we obtain

$$4\pi R^2 \sigma T^4 = 2.8 \times 10^{28} \text{ W} \tag{10}$$

or

$$R = \left(\frac{2.8 \times 10^{28} \text{ W}}{4\pi \times 5.67 \times 10^{-8} \text{ W/(m}^2 \cdot \text{K}^4) \times (5200 \text{ K})^4} \right)^{1/2}$$

$$= 7.3 \times 10^9 \text{ m}$$

This is about 10 times the radius of the Sun. Capella is a giant star.

---

[3] See Appendix 8 for a more precise value of the Stefan–Boltzmann constant.

40.3 Photons and the Photoelectric Effect

In 1905, Einstein showed that Planck's formula could be understood much more simply in terms of a direct quantization of the energy of the radiation. Planck had postulated that the oscillators in the wall of the cavity have discrete quantized energies, but he had treated the electromagnetic radiation as a smooth, continuous distribution of energy, exactly as it is supposed to be according to classical electromagnetic theory. In contrast, Einstein proposed that electromagnetic radiation consists of particle-like packets of energy. He regarded a wave of some given frequency ν as a stream of more or less localized energy packets, each with one quantum of energy $h\nu$. The wave then has an energy $h\nu$ if it contains only one such quantum, $2h\nu$ if it contains two, etc. The particle-like energy packets in light are called **photons.** The thermal radiation in a cavity, with waves traveling randomly in all directions, can then be regarded as a gas of photons. Einstein applied statistical mechanics to calculate the energy spectrum of this gas and he thereby obtained the energy spectrum of the cavity radiation.

Photon

EXAMPLE 3.   The energy flux of sunlight reaching the surface of the Earth is 1.0×10^3 W/m². How many photons reach the surface of the Earth per square meter per second? For the purposes of this calculation assume that all the photons in sunlight have an average wavelength 5000 Å.

SOLUTION: The energy of a photon of wavelength 5000 Å is

$$E = h\nu = \frac{hc}{\lambda}$$

$$= \frac{6.63 \times 10^{-34} \text{ J} \cdot \text{s} \times 3.00 \times 10^8 \text{ m/s}}{5.0 \times 10^{-7} \text{ m}} = 4.0 \times 10^{-19} \text{ J}$$

The energy incident per square meter per second is 1.0×10^3 J. To obtain the number of photons, we must divide this by the energy per photon, 4.0×10^{-19} J. This gives 1.0×10^3 J$/4.0 \times 10^{-19}$ J $= 2.5 \times 10^{21}$ photons per square meter per second.

Because the number of photons in sunlight and other common light sources is so large, we do not perceive the grainy character of the energy distribution in light at the macroscopic level.

With this concept of light as a stream of photons, Einstein was also able to offer an explanation of the **photoelectric effect.** In the early experiments on the production of radio waves by sparks, Hertz had noticed that light shining on an electrode tended to promote the formation of sparks. Subsequent careful experimental investigations demonstrated that the impact of light on an electrode can eject electrons. The electrons emerge with a kinetic energy that increases directly with the frequency of the light.

Photoelectric effect

Figure 40.7 is a schematic diagram of the apparatus used in the investigation of the photoelectric effect. Light from a lamp illuminates an electrode of metal (C) enclosed in an evacuated tube. Electrons ejected from this electrode travel to the collecting electrode (A), and then flow around the external circuit. A galvanometer (G) detects this

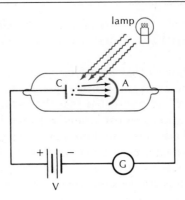

Fig. 40.7 Schematic diagram of the apparatus for the investigation of the photoelectric effect. The light from the lamp ejects electrons from the electrode C (cathode) and they travel to the collecting electrode A (anode).

flow of electrons. The kinetic energy of the ejected photoelectrons can be determined by applying a potential difference between the emitting and the collecting electrodes by means of an adjustable source of emf *(V)*. With the polarity shown in the figure, the collector has a negative potential relative to the emitter, i.e., the collector exerts a repulsive force on the photoelectrons. If the potential energy matches or exceeds the initial kinetic energy of the photoelectrons, then the flow of these electrons will stop. The corresponding potential is called the *Stopping potential* **stopping potential;** the measured value of this stopping potential gives us the kinetic energy of the electrons:

$$K = eV_{\text{stop}} \tag{11}$$

Experimentally, one finds that the kinetic energy determined in this way increases linearly with the frequency of the incident light. For example, Figure 40.8 is a plot of the kinetic energy vs. the frequency of the light for photoelectrons ejected from sodium. Note that if the frequency is below 4.4×10^{14} Hz, then the light is incapable of ejecting *Threshold frequency* electrons. This frequency is called the **threshold frequency.**

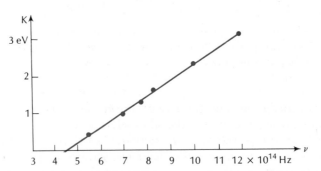

Fig. 40.8 Kinetic energies (in electron-volts) of photoelectrons ejected from sodium by light of different frequencies.

Einstein's quantum theory of light accounts for these experimental observations as follows. The electrons in the illuminated electrode absorb quanta of light, one at a time. When an electron absorbs a quantum, it acquires an energy $h\nu$. But before this electron can emerge from the electrode, it must overcome the restraining forces that bind it to the metal of the electrode. The energy required for this is called *Work function* the **work function** of the metal, designated by ϕ. The energy left with the electron is then $hv - \phi$ and this must be the kinetic energy of the emerging electron,[4]

$$K = hv - \phi \qquad (12)$$

Einstein's photoelectric equation

This is Einstein's photoelectric equation. It shows that the kinetic energy is indeed a linearly increasing function of the frequency, in agreement with the data of Figure 40.8. According to Eq. (12), the slope of the straight line in Figure 40.8 should equal Planck's constant.

Einstein's photoelectric equation was verified in detail by a long series of meticulous experiments by R. A. Millikan (the data in Figure 40.8 are due to him). In order to obtain reliable results, Millikan found it necessary to take extreme precautions to avoid contamination of the surface of the photosensitive electrode. Since the surfaces of metals exposed to air quickly accumulate a layer of oxide, he developed a technique for shaving the surfaces of his metals in a vacuum by means of a magnetically operated knife.

The results of these experiments gave strong support to the quantum theory of light. This success of Einstein's theory was all the more striking in view of the failure of the classical wave theory of light to account for the features of the photoelectric effect. According to the wave theory, the crucial parameter that determines the ejection of a photoelectron should be the intensity of light. If an intense electromagnetic wave strikes an electron, it should be able to jolt it loose from the metal, regardless of the frequency of the wave. Furthermore, the kinetic energy of the ejected electron should be a function of the intensity of the wave. The observational evidence contradicts these predictions of the wave theory: A wave with a frequency below the threshold frequency never ejects an electron, regardless of its intensity. And, furthermore, the kinetic energy depends on the frequency, and not on the intensity. High-intensity light ejects more photoelectrons, but does not give the individual electrons more kinetic energy.

Today, the photoelectric effect finds many practical applications in sensitive electronic devices for the detection of light. For instance, in a photomultiplier tube, an incident photon ejects an electron from an electrode; this electron is accelerated toward a second electrode (called a dynode; see Figure 40.9) where its impact ejects several secondary electrons; these, in turn, are accelerated toward a third electrode where their impact ejects tertiary electrons, etc. Thus, one electron from the first electrode generates an avalanche of electrons. In a high-gain photomultiplier tube, a pulse of 10^9 electrons emerges from the last electrode, delivering a measurable pulse of current to an external circuit. In this way, the photomultiplier tube can detect the arrival of individual photons. Some sensitive television cameras, such as the image orthicon, rely on the same multiplier principle to convert the arrival of a photon at a photosensitive faceplate into a measurable pulse of current.

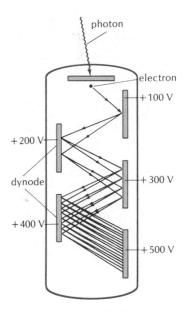

Fig. 40.9 Schematic diagram of a photomultiplier tube. The secondary electrodes are called dynodes. For the purpose of this diagram it has been assumed that each electron impact on a dynode releases two electrons. The arrows show an avalanche of electrons.

EXAMPLE 4. The work function for platinum is 9.9×10^{-19} J. What is the threshold frequency for the ejection of photoelectrons from platinum?

---

[4] Some electrons suffer extra energy losses in collisions within the metal before they emerge. Thus, the expression actually gives the *maximum* kinetic energy with which electrons can emerge.

SOLUTION: If an electron absorbs a photon at the threshold frequency, it will just barely have enough energy to overcome the binding forces holding it in the metal, and it will emerge with zero kinetic energy. By Eq. (12) this corresponds to

$$0 = h\nu - \phi$$

or

$$\nu = \frac{\phi}{h} = \frac{9.9 \times 10^{-19}\,\mathrm{J}}{6.63 \times 10^{-34}\,\mathrm{J \cdot s}} = 1.5 \times 10^{15}\,\mathrm{Hz}$$

40.4 The Compton Effect

Arthur Holly Compton, *1892–1962, American experimental physicist, professor at the University of Chicago. For his discovery of the Compton effect, he received the Nobel Prize in 1927.*

Very clear experimental evidence for the particle-like behavior of photons was uncovered by A. H. Compton in 1922. Compton had been investigating the scattering of X rays by a target of graphite. When he bombarded the graphite with monochromatic X rays, he found that the scattered (deflected) X rays had a wavelength somewhat larger than that of the original X rays. Compton soon recognized that this effect could be understood in terms of collisions of photons with electrons, collisions in which the photons behave like particles.

In this collision the electron of a carbon atom can be regarded as free because the force binding the electron to the atom is insignificant compared to the force exerted by the photon. When the photon bounces off the electron, the electron recoils and thereby picks up some of the photon's energy — the deflected photon is left with reduced energy. Since the energy of the photon is $E = h\nu = hc/\lambda$, a reduction of energy implies an increase of wavelength. Qualitatively, we expect that those photons deflected through the largest angles should lose the most energy and therefore emerge with the longest wavelength. And this is just what Compton found in his experiments.

For a quantitative discussion of the photon–electron collision, we need an expression for the momentum of a photon. From Eq. (36.42) we know that the momentum and the energy of light are related by $p = E/c$. A photon of energy $E = h\nu$ therefore has a momentum

Momentum of photon

$$p = \frac{h\nu}{c} \qquad (13)$$

[Alternatively, we can derive this equation from the relativistic formula for the momentum of a particle. Since the speed of the photon is c, it must be regarded as an ultrarelativistic particle. For such a particle, Eq. (17.53) tells us that $p = E/c$, which leads, again, to $p = h\nu/c$. Incidentally, note that according to Eq. (17.51) the energy of a particle with a speed equal to the speed of light can be finite only if $m = 0$; thus photons must be regarded as particles of zero mass.]

Before the collision, the electron is at rest and the photon has a frequency ν and momentum $h\nu/c$. After the collision, the electron has a recoil momentum $m_e v$ and the photon has a reduced frequency $\nu - \Delta\nu$ and momentum $h(\nu - \Delta\nu)/c$. Figure 40.10a shows the initial momentum vector of the photon and the final momentum vectors of the electron and the photon. Conservation of momentum demands that the sum of the two final momentum vectors equal the initial momentum vector; the three momentum vectors therefore form a triangle (Figure 40.10b). Applying the law of cosines to this triangle, we find

$$(m_e v)^2 = \left(\frac{h\nu}{c}\right)^2 + \left(\frac{h(\nu - \Delta\nu)}{c}\right)^2 - \frac{2h\nu}{c}\frac{h(\nu - \Delta\nu)}{c}\cos\theta \qquad (14)$$

or

$$(m_e v)^2 = \left(\frac{h\nu}{c}\right)^2\left[1 + \left(1 - \frac{\Delta\nu}{\nu}\right)^2 - 2\left(1 - \frac{\Delta\nu}{\nu}\right)\cos\theta\right] \qquad (15)$$

$$= \left(\frac{h\nu}{c}\right)^2\left[2 - 2\frac{\Delta\nu}{\nu} + \left(\frac{\Delta\nu}{\nu}\right)^2 - 2\left(1 - \frac{\Delta\nu}{\nu}\right)\cos\theta\right] \qquad (16)$$

If we neglect the term $(\Delta\nu/\nu)^2$, we can rewrite this as

$$\tfrac{1}{2}m_e v^2 = \frac{1}{m_e}\left(\frac{h\nu}{c}\right)^2\left(1 - \frac{\Delta\nu}{\nu}\right)(1 - \cos\theta) \qquad (17)$$

Conservation of energy demands that the kinetic energy of the electron equals the energy lost by the photon,

$$\tfrac{1}{2}m_e v_e^2 = h\,\Delta\nu \qquad (18)$$

Comparing Eqs. (17) and (18), we see that

$$\Delta\nu = \frac{h\nu^2}{m_e c^2}\left(1 - \frac{\Delta\nu}{\nu}\right)(1 - \cos\theta) \qquad (19)$$

or

$$\frac{c\,\Delta\nu}{\nu(\nu - \Delta\nu)} = \frac{h}{m_e c}(1 - \cos\theta) \qquad (20)$$

The complicated expression on the left side of this equation is simply the change of wavelength:

$$\frac{c\,\Delta\nu}{\nu(\nu - \Delta\nu)} = \frac{c}{\nu - \Delta\nu} - \frac{c}{\nu} = = (\lambda + \Delta\lambda) - \lambda = \Delta\lambda \qquad (21)$$

where we have taken into account that a decreased frequency $\nu - \Delta\nu$ implies an increased wavelength $\lambda + \Delta\lambda$. Our equation for the change of wavelength of the photon as a function of the angle of deflection is then

$$\boxed{\Delta\lambda = \frac{h}{m_e c}(1 - \cos\theta)} \qquad (22)$$

Wavelength shift of photon in Compton effect

This result, first obtained by Compton, remains valid even if the speed

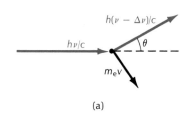

(a)

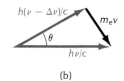

(b)

Fig. 40.10 (a) Momentum vector of photon before the collision and momentum vectors of photon and electron after the collision. Before the collision the photon has a momentum $h\nu/c$, and the electron is at rest; after the collision the photon has a momentum $h(\nu - \Delta\nu)/c$, and the electron has a momentum $m_e v$. (b) The triangle of the momentum vectors.

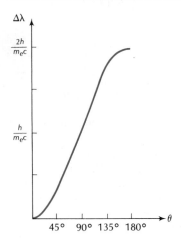

Fig. 40.11 Wavelength shift of the photon as a function of the deflection angle.

Wavicle

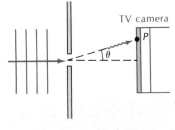

Fig. 40.12 An interference experiment. A light beam is incident from the left on a plate with two narrow slits. A sensitive TV camera detects photons on the right.

of recoil of the electron is relativistic [in fact, the relativistic calculation shows that Eq. (22) is *exact*].

Figure 40.11 is a plot of $\Delta\lambda$ as a function of θ. For $\theta = 180°$ (a head-on collision), the wavelength shift is maximum, $\Delta\lambda = 2h/(m_e c) = 0.0485$ Å; but even this maximum is quite small. Such a small wavelength shift is difficult to detect unless the wavelength of the X rays is itself quite small, so that $\Delta\lambda$ is an appreciable fraction of λ. In his experiments Compton used X rays of a wavelength $\lambda = 0.71$ Å; thus the maximum wavelength shift amounted to about 7% of the wavelength.

40.5 Wave vs. Particle

From the photoelectric effect and the Compton effect we learned that photons have particle properties. On the other hand, we know that light displays interference and diffraction phenomena which prove that the photons also have wave properties. Thus photons are neither classical particles nor classical waves. They are some new kind of object, unknown to classical physics, with a subtle combination of both wave and particle properties. B. Hoffman has coined the name **wavicle** for this new kind of object. It is difficult to achieve a clear understanding of the character of a wavicle because these objects are very remote from our everyday experience. We all have an intuitive grasp of the concept of a classical particle or of a classical wave from our experience with, say, billiard balls and water waves, but we have no such experience with wavicles.

We can gain some insight into the interplay between the particle behavior and the wave behavior of a photon by a new, detailed examination of a simple diffraction experiment. Figure 40.12 shows a light beam striking a plate with a narrow slit. In the usual diffraction experiments described in Chapter 39, we installed a screen or a photographic film at the far right, and with this we recorded the intensity of the light in the diffraction pattern. Instead, in our new arrangement we will install the faceplate of a very sensitive TV camera in place of the customary screen. With this, we can detect the individual photons of the light in the diffraction pattern.

If the incident light beam has a very low intensity, so that there is only one photon passing through the slit at a time, we can watch the photons arriving one by one at the faceplate of our TV camera. Figure 40.13a shows a typical pattern of impacts of 28 photons. The pattern seems quite random. If the photons behaved like classical particles, they would travel along a straight line and they would reach only those points on the faceplate that are within the geometric image of the slit. The widely scattered impacts prove that the photons are certainly not traveling along such straight lines. Figure 40.13b shows the pattern of accumulated impacts for 1000 photons; and Figure 40.13c shows it for 10,000 photons. In these figures we can recognize a tendency of the photons to cluster in bandlike zones. These zones correspond to the maxima of the diffraction pattern predicted by the wave theory of light. Finally, Figure 40.13d shows the pattern of accumulated impacts for a very large number of photons; this is simply the familiar intensity pattern of light diffracted by a slit (see also Figure 39.5).

From this diffraction experiment we learn that the behavior of the photons is governed by a probabilistic law. The point of impact of an

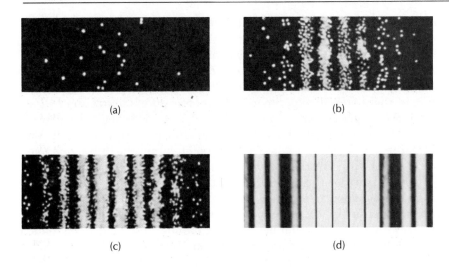

(a)

(b)

(c)

(d)

Fig. 40.13 (a) Pattern of impacts of 28 photons (simulation). (b) Pattern of impacts of 1000 photons (simulation). (c) Pattern of impacts of 10,000 photons (simulation). (d) Intensity pattern of light recorded on a photographic plate by a very large number of photons. The intensity of light at each point is proportional to the number of photons arriving at that point.

individual photon on the faceplate is unpredictable. Only the average distribution of impacts of a large number of photons is predictable: the distribution of photons matches the intensity distribution calculated from the wave theory of light. Thus, the probability that a photon arrives at a given point on the faceplate is proportional to the calculated intensity of the wave at that point, i.e.,

$$[\text{probability for photon at point } P] \propto E^2(P) \tag{23}$$

Probability interpretation of wave

Here we have a connection between the wave and the particle aspects of the photon: *the intensity of the photon wave at some point determines the probability that there is a photon particle at that point.* This probability interpretation of the intensity of the wave was discovered by Max Born.

Our single-slit experiment can also teach us something about the limitations that quantum theory imposes on the ultimate precision of measurement of the position of a wavicle. Suppose we have a light wave consisting of one photon and we want to measure the position of this photon. Of course, the position has x, y, and z components; we will concentrate on the y component, perpendicular to the direction of propagation. Figure 40.14 shows the wave propagating in the horizontal direction; the y direction is vertical. To determine this vertical position of the photon, we use a narrow slit placed in the path of the wave. If the photon succeeds in passing through this slit, then we will have achieved a determination of the vertical position to within an uncertainty

$$\Delta y = a \tag{24}$$

where a is the width of the slit. If the photon fails to pass through this slit, then our measurement is inconclusive and will have to be repeated.[5]

By making the slit very narrow, we can make the uncertainty of our

Max Born, *1882–1970, German, and later British, theoretical physicist, professor at Göttingen and at Edinburgh. He was awarded the Nobel Prize somewhat tardily in 1954 for his discovery of the probabilistic interpretation of quantum waves in 1926.*

---

[5] An alternative method for measuring the position of a photon is to observe its impact on the faceplate of a sensitive TV camera. But this measurement is destructive (the photon is absorbed by the photoelectron and disappears). Besides, the uncertainties in this measurement are not easily analyzed.

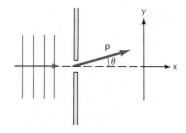

Fig. 40.14 Photon passes through a narrow slit and emerges at an angle θ.

Werner Heisenberg, *1901–1976, German theoretical physicist, professor at Leipzig, and later director of the Max Planck Institute for Physics in Munich. He was one of the founders of the new quantum mechanics, and received the Nobel Prize in 1932.*

Heisenberg's uncertainty relation for y and p_y

determination of the y coordinate very small. But this has a surprising consequence for the y component of the momentum of the photon: if we make the uncertainty in the y coordinate small, we will make the uncertainty in the y component of the momentum large. To see how this comes about, let us recall that according to our preceding discussion of the single-slit experiment, the photon suffers diffraction by the slit and emerges at some angle θ (Figure 40.14). This angle θ is unpredictable; all we can say about the photon after it emerges from the slit is that it will be heading toward some point within the diffraction pattern. Thus, the direction of motion of the photon is uncertain. As a rough measure of the magnitude of this uncertainty in direction, we can take the angular width of the central diffraction maximum (most of the intensity of the photon wave is gathered within the region of this central maximum, and hence the photon is most likely to be found in this region). This estimate of the uncertainty of the angle gives us

$$\Delta\theta \gtrsim \lambda/a \tag{25}$$

The y component of the momentum is $p_y = p \sin \theta$; since we are concerned with a small angle, we can approximate this by $p_y = p\theta$. The uncertainty in p_y is then

$$\Delta p_y = p \, \Delta\theta \gtrsim p\lambda/a \tag{26}$$

But, from Eq. (13),

$$p = \frac{h\nu}{c} = \frac{h}{\lambda} \tag{27}$$

which gives us

$$\Delta p_y \gtrsim h/a \tag{28}$$

Thus, if the slit is very narrow, the uncertainty in the y component of the momentum will be very large! Comparing Eqs. (28) and (24), we find that

$$\boxed{\Delta y \, \Delta p_y \gtrsim h} \tag{29}$$

This equation states that Δy and Δp_y cannot both be small; if one is small then the other must be large, so that their product equals or exceeds Planck's constant.

Equation (29) is one of **Heisenberg's uncertainty relations.** There are corresponding relations for the other components of position and momentum. Although we have obtained the uncertainty relation (29) by examining the special case of a position measurement by means of a slit, it turns out that this relation is actually of general validity for any kind of position measurement. The Heisenberg uncertainty relations tell us that there exist ultimate, insuperable limitations in the precision of our measurements. At the macroscopic level, the quantum uncertanties in our measurements can be neglected. But at the atomic level, these quantum uncertainties are often so large that it is completely meaningless to speak of the position or momentum of a wavicle.

SUMMARY

Spectral emittance of blackbody:

$$S_\lambda = \frac{2\pi c^2 h}{\lambda^5} \frac{1}{e^{hc/kT\lambda} - 1}$$

$$h = 6.63 \times 10^{-34} \, \text{J} \cdot \text{s}$$

Energy quantization of an oscillator: $E = nh\nu$

Wien's Law: $\lambda_{\text{max}} \propto \dfrac{1}{T}$

Stefan–Boltzmann Law: $S \propto T^4$

Energy and momentum of a photon: $E = h\nu$, $\qquad p = h\nu/c$

Kinetic energy of photoelectron: $K = h\nu - \phi$

Wavelength shift of photon (Compton effect):

$$\Delta\lambda = \frac{h}{m_e c} (1 - \cos\theta)$$

Probability interpretation of wave:

[probability for presence of photon] $\propto$ [intensity of wave]

Heisenberg's uncertainty relation for y and p_y:

$$\Delta y \, \Delta p_y \gtrsim h$$

QUESTIONS

1. Is the light emitted by a neon tube thermal radiation? The light emitted by an ordinary incandescent light bulb?

2. Does your body emit thermal radiation?

3. Other things being equal, on a clear night, the ground is likely to become much colder than on a cloudy night. Explain.

4. The insulation used in the walls of homes consists of a thick blanket of fiber glass covered on one side by a thin aluminum foil. What is the purpose of these two layers?

5. Black velvet looks much blacker than black paint. Why?

6. For protection against the heat of sunlight, parts of the Lunar Lander (and some other spacecraft) were wrapped in shiny aluminum foil. Why is shiny foil useful for this purpose?

7. If you look into a kiln containing pottery heated to a temperature equal to that of the walls of the kiln, you can scarcely see the pottery. Explain.

8. According to Figure 40.6, at what wavelength is the spectral emittance maximum for a body at 2000 K? At 1750 K? At 1500 K? At 1250 K? Do these wavelengths satisfy Wien's Law?

9. The quantization of electric charge is consistent with classical physics, but the quantization of energy is not. Does this make sense?

10. Suppose that Planck's constant were much larger than it is, say, 10^{34} times larger. What strange behavior would you notice in a simple harmonic oscillator consisting of a mass hanging on a spring?

11. Consider a seconds pendulum, i.e., a pendulum that has a period of 2 seconds. What is the magnitude of one energy quantum for such a pendulum? Would you expect that quantum effects are noticeable in such a pendulum?

12. Suppose that two stars have the same size, but the temperature of one is twice that of the other. By what factor will the thermal power radiated by the hotter star be larger than that radiated by the cooler star?

13. Why do we not notice the discrete quanta of light when we look at a light bulb?

14. Day-glo paints achieve their exceptionally bright orange or red color by converting short-wavelength photons into long-wavelength (red) photons. Why can we not make such a paint in a blue or violet color?

15. Figure 40.15 shows a plot of current vs. applied potential for the photoelectric current emitted by the surface of a metal illuminated with light of a given wavelength. Qualitatively, explain why the current is zero if $V < -V_{stop}$, explain why the current levels off for a large positive V, and explain why the curves differ for different intensities of the light.

16. When light of a given wavelength ejects photoelectrons from the surface of a metal, why is it that not all of these photoelectrons emerge with the same kinetic energy?

17. Can a particle of mass zero ever be at rest?

18. According to Eq. (22), a photon suffers a maximum change of wavelength in a collision with an electron if it emerges at an angle $\theta = 180°$, and a minimum change of wavelength (no change) if it emerges at an angle $\theta = 0°$. Is this reasonable?

19. Suppose that a photon and an electron have the same momentum. Which has the larger energy, taking into account both the rest-mass energy and the kinetic energy?

20. Can the Compton effect occur with visible light? Would it be observable?

21. Photons of short wavelength are more particle-like than photons of long wavelength. Why?

22. Give an example of an experiment in which photons behave like waves. Give an example of an experiment in which they behave like particles.

23. What happens to the y momentum that a photon gains or loses in the diffraction experiment described in Figure 40.14?

24. If photons were classical particles, what pattern of impact points would we find in the diffraction experiment that led to the results described in Figures 40.13a–d?

25. According to Eq. (23) we can only predict the *probability* that a photon will be found at some given point. Does this mean that quantum physics is not deterministic?

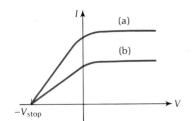

Fig. 40.15 I vs. V for two different values of the intensity of light: (a) high intensity and (b) low intensity.

PROBLEMS

Sections 40.1 and 40.2

1. Incandescent light bulbs have a tungsten filament whose temperature is typically 3200 K. At what wavelength does such a filament radiate a maximum flux? Assume that the filament acts as a blackbody.

2. Interplanetary and interstellar space is filled with thermal radiation of a temperature of 3 K left over from the Big Bang.
(a) At what wavelength is the flux of this radiation maximum?

(b) What is the power incident on the surface of the Earth due to this radiation?

3. The tungsten filament of a light bulb is a wire of diameter 0.080 mm and length 5.0 cm. The filament is at a temperature of 3200 K. Calculate the power radiated by the filament. Assume the filament acts as a blackbody.

4. The spectral emittance of the Sun is maximum at 5000 Å. By what factor is the emittance smaller at 7000 Å? At 4000 Å?

5. Prove that the spectral emissivity given by Eq. (1) does have a maximum at $\lambda_{max} = (2.9 \times 10^{-3} \text{ m} \cdot \text{K})/T$. (Hint: Substitute this wavelength into the mathematical condition for a maximum.)

6. Derive the Stefan–Boltzmann Law

$$S \propto T^4$$

from Planck's Law. [Hint: Consider the integral given in Eq. (6); change the variable of integration to $x = hc/\lambda kT$ and show that the result has the form $S = [\text{constant}] \times T^4$; you do not have to evaluate the constant.]

7. Show that the flux radiated by a blackbody in a frequency interval dv is

$$S_v \, dv = \frac{2\pi h}{c^2} \frac{v^3}{e^{hv/kT} - 1} \, dv$$

Does the maximum of S_v coincide with the maximum of S_λ?

8. The star Procyon B is at a distance of 11 light-years from Earth. The flux of its starlight reaching us is 1.7×10^{-12} W/m² and the surface temperature of the star is 6600 K. Calculate the size of the star.

9. At the Earth, the flux of sunlight per unit area facing the Sun is 1.34×10^3 W/m². The Earth absorbs heat from the sunlight and reradiates heat as thermal infrared radiation. For equilibrium, the power arriving from the Sun must equal the average power radiated by the surface of the Earth. What average surface temperature for the Earth can you deduce from this? Assume that the Earth radiates like a blackbody.

10. Deduce the average surface temperature of Pluto by the method described in Problem 9. You will find data on Pluto in the table printed on the endpapers.

11. If you stand naked in a room, your skin and the walls of the room will exchange heat by radiation. Suppose the temperature of your skin is 33°C; the total area of your skin is 1.5 m². The temperature of the walls is 15°C. Assume your skin and the walls behave like blackbodies.
   (a) What is the rate at which your skin radiates heat?
   (b) What is the rate at which your skin absorbs heat? What is your net rate of loss of heat?

12. An oscillator of frequency 2×10^{15} Hz consists of a mass of 9.1×10^{-31} kg attached to a spring. What is the amplitude of oscillation of this oscillator if its energy of oscillation is one energy quantum? Two energy quanta?

13. We know from Chapter 14 that at small amplitudes a pendulum behaves like an oscillator. Suppose that a pendulum consists of a mass of 0.10 kg attached to a (massless) string of length 1.0 m.
   (a) Taking into account the quantization of energy, what is the least (nonzero) amount of energy that this pendulum can have?
   (b) What is the amplitude of oscillation of the pendulum with this least amount of energy?

*14. In Problem 24.19 you will find a description of the Thomson model of the hydrogen atom.
   (a) What is the frequency of oscillation of the electron in this atom?

(b) The electron can be regarded as an oscillator. Show that if the energy of the electron is one quantum, the amplitude of oscillation exceeds the radius (0.5 Å) of the atom.

Section 40.3

15. Photons of green light have a wavelength of 5500 Å. What is the energy and what is the momentum of one of these photons?

16. Show that if we express the energy of a photon in keV and the wavelength in angstroms, then

$$E = 12.4/\lambda$$

17. A radio transmitter radiates 10 kW at a frequency of 8.0×10^5 Hz. How many photons does the transmitter radiate per second?

18. The energy flux in the starlight reaching us from the bright star Capella is 1.2×10^{-8} W/m². If you are looking at this star, how many photons per second enter your eyes? The diameter of your pupil is 0.70 cm. Assume that the average wavelength of the light is 5000 Å.

19. If you want to make a very faint light beam that delivers only 1 photon per square meter per second, what must be the amplitude of the electric field in this light beam? The wavelength of the light is 5000 Å.

20. The energy density of starlight in intergalactic space is 10^{-15} J/m³. What is the corresponding density of photons? Assume the average wavelength of the photons is 5000 Å.

21. An incandescent light bulb radiates 40 W of thermal radiation from a filament of temperature 3200 K. Estimate the number of photons radiated per second; assume that the photons have an average wavelength equal to the λ_{max} given by Wien's Law.

22. Show that the flux of photons, or the number of photons per unit area and unit time, emitted by a blackbody in the frequency interval dv is

$$\frac{2\pi}{c^2} \frac{v^2}{e^{hv/kT} - 1} dv$$

23. A photon has an energy of 5 eV in the reference frame of the laboratory. What is the energy of this photon in the reference frame of a proton moving through the laboratory at a speed of $\frac{1}{2}c$ in the same direction as the photon?

24. According to Figure 40.8, what is the work function of sodium? Express your answer in electron-volts.

25. The work function of potassium is 2.26 eV. What is the threshold frequency for the photoelectric effect in potassium?

26. The work functions of K, Cr, Zn, and W are 2.26, 4.37, 4.24, and 4.49 eV, respectively. Which of these metals will emit photoelectrons when illuminated with red light ($\lambda = 7000$ Å)? Blue light ($\lambda = 4000$ Å)? Ultraviolet light ($\lambda = 2800$ Å)?

27. By inspection of Figure 40.8, find the slope of the line in eV/Hz. Convert these units into J · s and verify that the slope is the same as Planck's constant.

28. The binding energy of an electron in a hydrogen atom is 13.6 eV. Suppose that a photon of wavelength 400 Å strikes the atom and gives up all of its energy to the electron. With what kinetic energy will the electron be ejected from the atom?

Section 40.4

29. X Rays emitted by molybdenum have a wavelength of 0.72 Å. What are the energy and the momentum of one of the photons in these X rays?

30. In a collision with an initially stationary electron, a photon suffers a wavelength increase of 0.022 Å. What must have been the deflection angle of this photon?

31. What is the maximum energy that a free electron (initially stationary) can acquire in a collision with a photon of energy 4.0×10^3 eV?

32. X rays of wavelength 0.30 Å collide with free electrons at rest. Calculate the wavelength of the X rays that emerge from this collision with a deflection of 60°. Calculate the wavelength of the X rays that emerge from this collision with a deflection of 120°.

33. In a collision with a free electron, a photon of energy 2.0×10^3 eV is deflected by 90°. What energy does the electron acquire in this collision?

34. A photon of energy 1.6×10^8 eV collides with a *proton* initially at rest. The photon is deflected by 45°. What is its new energy?

35. A photon of initial wavelength 0.40 Å suffers two successive collisions with two electrons. The deflection in the first collision is 90° and in the second collision it is 60°. What is the final wavelength of the photon?

*36. A photon of energy 5.0×10^3 eV collides head on with a free electron of energy 2.0×10^3 eV. After the collision the photon moves in a direction opposite to its initial direction. Find the energy of the photon. Find the energy of the electron.

Section 40.5

37. A photon passes through a horizontal slit of width 5×10^{-6} m. What uncertainty in the vertical position will this photon have as it emerges from the slit? What uncertainty in vertical momentum?

38. Consider a radio wave in the form of a pulse lasting 0.001 s. This pulse then has a length of $0.001 \text{ s} \times c = 3 \times 10^5$ m. Since an individual photon of this radio wave can be anywhere within this pulse, the uncertainty in the position of the photon is $\Delta x = 3 \times 10^5$ m along the direction of propagation.
    (a) According to Heisenberg's relation, what is the corresponding uncertainty in the momentum of the photon?
    (b) What is the uncertainty in the frequency of the photon?

CHAPTER 41

Atomic Structure and Spectral Lines

The photographs of the Prelude reproduced in Figure 41.1 give convincing visual evidence that solids, liquids, and gases are made of atoms, small grains of matter with a diameter of about 10^{-10} m. These photographs were prepared with powerful microscopes of a special design. Unfortunately, none of these microscopes is sufficiently powerful to reveal the inside of the atom. Hence, for the exploration of the internal structure of the atom, we still have to rely on the technique developed by Ernest Rutherford and his associates around 1910: bombard the atom with a beam of particles and use this beam as a probe to "feel" the interior of the atom.

By 1910, most physicists had come to believe that atoms are made of some combination of positive and negative electric charges, and that the attractions and repulsions between these electric charges are the basis for all the chemical and physical phenomena observed in solids,

Fig. 41.1 (a) Platinum atoms in the tip of a fine needle as seen with an ion microscope (magnification $\sim 5 \times 10^5 \times$). (b) Uranium atom seen with an electron microscope (magnification $\sim 5 \times 10^6 \times$).

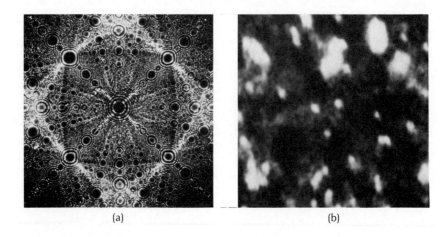

(a)   (b)

liquids, and gases. Since electrons were known to be present in all of these forms of matter, it seemed reasonable to suppose that each atom consists of a combination of electrons and positive charge. The vibrational motions of the electrons within such an atom would then result in the radiation of electromagnetic waves; this was supposed to account for the emission of light by the atom. However, both the arrangement of the electric charges within the atom and the mechanism that accounts for the characteristic colors of the emitted light remained mysteries until Rutherford's discovery of the nucleus and Niels Bohr's discovery of the quantization of atomic states. In this chapter we will look at these two crucial discoveries.

The study of the internal structure of the atom led to the inescapable conclusion that in the atomic realm Newton's laws of motion are not valid. Electrons and other subatomic particles obey new equations of motion that are drastically different from the old equations of motion obeyed by planets, billiard balls, or shotgun pellets. The new theory of motion is called **quantum mechanics;** it rules the realm of the atom. In contrast, the old theory of Newton is called **classical mechanics.** The discovery of an entirely new set of laws of motion was without doubt the greatest scientific revolution of this century.

Quantum mechanics

41.1 Spectral Lines

The earliest attempts at a theory of atomic structure ended in failure — they were not able to explain the characteristic colors of the light emitted by atoms. These colors show up very distinctly when a small sample of gas is made to emit light by the application of heat or of an electric current. For instance, if we put a few grains of ordinary salt into a flame, the sodium vapor released by the salt will glow with a characteristic yellow color. If we put neon gas into an evacuated glass tube and connect the ends of the tube to a high-voltage generator (Figure 41.2), the gas will glow with the familiar orange-red color of neon signs.

The light emitted by an atom can be precisely analyzed with a prism (Figure 41.3); this breaks the light up into its component colors. In the arrangement shown in Figure 41.3, each separate color generates a bright line. These are the **spectral lines.** Each kind of atom has its own discrete spectral lines. The color plate between pages 850–851 shows the spectral lines of hydrogen; the numbers next to the spectral lines give the wavelengths in angstroms. Hydrogen has four spectral lines in the visible region (already mentioned in Chapter 37) and many ultraviolet and infrared lines not visible to the human eye. The color plate

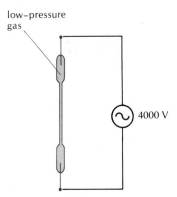

Fig. 41.2 An electric discharge tube. The tube contains gas at a very low pressure. When the terminals are connected to a high-voltage generator, an electric current flows through the gas and makes it glow.

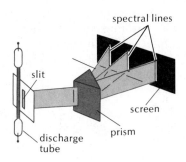

Fig. 41.3 Analysis of light by means of a prism. In this arrangement, each separate color of the light emerging from the slit gives rise to a separate spectral line on the screen.

Spectrum

Joseph von Fraunhofer, *1787–1826, German optician and physicist. Starting as an apprentice to a glazier, he became a famous maker of optical instruments and a member of the Bavarian Academy of Sciences. He rediscovered the dark lines in the solar spectrum, observed previously by W. H. Wollaston, and introduced many improvements in the design of the spectroscope. He was one of the first makers of diffraction gratings.*

Fraunhofer lines

also shows the spectral lines of sodium, helium, and mercury. Obviously, the set of spectral lines, or **spectrum,** belonging to hydrogen is unmistakably different from those of sodium, helium, and mercury — the spectrum of an atom can serve as a fingerprint in its identification.

In spectroscopy laboratories, scientists often perform the quantitative analysis of a sample of atoms with absorption lines rather than emission lines. It so happens that an atom capable of emitting light of a given wavelength is also capable of absorbing light of that wavelength. When we illuminate a sample of atoms with white light (a mixture containing all colors or wavelengths), the atoms will absorb light of their characteristic wavelengths and, upon analyzing the remaining light with a prism, we find dark absorption lines in the uniform background generated by the white light. The last picture in the color plate shows such an absorption spectrum for sodium vapor. The dark lines of this absorption spectrum coincide with the bright lines in the emission spectrum (see the second picture in the color plate).[1]

One advantage of spectroscopy over chemistry is that the analysis can be performed even on minuscule amounts of material. What is more, atoms can be identified at a distance. For example, we can identify the atoms on the surface of the Sun by careful analysis of the distribution of colors in sunlight — we do not need to pluck a sample of atoms from the Sun. The power of this technique is best illustrated by the story of the discovery of helium (the "Sun element"). In 1868, the gas was yet unknown to chemists when an astronomer discovered it on the Sun by means of its light; 30 years later chemists finally found traces of helium in minerals on the Earth. By spectroscopic techniques astronomers can identify atoms in remote stars, clouds of interstellar gas, galaxies, and quasars. Figure 41.4 shows the spectrum of the star Caph in the constellation Cassiopeia; the spectral lines indicate the presence of hydrogen, calcium, iron, manganese, chromium, etc.

Note that only a small part of the light from a star is in the form of discrete spectral lines from individual atoms on the stellar surface. Most of the starlight is white light, a more or less uniform mixture of all colors. This is thermal radiation produced in the stellar interior. As we know from the preceding chapter, this kind of light does not retain the fingerprint of the atoms that produced it. Light originating in the stellar interior cannot escape directly, but is first tossed back and forth (scattered) many times by the restless atoms of the hot stellar gas. The random motion of these atoms communicates random changes of wavelength to the light and what finally emerges from the stellar interior is a continuous mixture of a wide range of wavelengths.

Figure 41.5 displays a portion of the spectrum of the white light from our Sun. White light is a nearly uniform mixture of all colors. However, there are many dark lines in this spectrum, caused by absorption in the gas surrounding the Sun. These are called the **Fraunhofer lines.**

Careful examination of the sets of spectral lines in the color plate between pages 850–851 reveals certain systematic regularities in the spacing of the lines. For instance, in the hydrogen spectrum we find that the spacing of the lines and their intensity decrease systematically

---

[1] The latter spectrum sometimes has extra lines that are absent in the former spectrum. This is so because the absorption process tends to suppress some lines; they become so faint as to be unnoticeable.

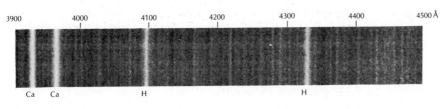

Fig. 41.4 A portion of the spectrum of the star Caph (β Cassiopeiae), from 3900 Å to 4500 Å. The two strong absorption lines on the left are due to ionized calcium. The other two strong lines (middle and right) are due to hydrogen.

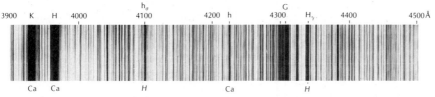

Fig. 41.5 A portion of the spectrum of the Sun, from 3900 Å to 4500 Å. Besides the strong absorption lines of calcium and of hydrogen, similar to those in Figure 41.4, there are also strong lines of iron (close pair, right of center).

as we look at shorter and shorter wavelengths; we will see in the next section that the spacing can be described by a simple mathematical formula. These hydrogen lines are said to form a **spectral series.** In the spectra of other elements, we find similar series; however, the spectra usually contain several overlapping series, and this makes it a bit harder to perceive the regularities in the spacing.

Spectral series

41.2 The Balmer Series and Other Spectral Series

The systematic pattern in the spacing of the spectral lines of hydrogen suggests that the wavelengths of these lines should be described by some simple mathematical formula. Table 41.1 lists the wavelengths of the first few of these spectral lines; there actually is an infinite number of spectral lines, the spacing between them becoming smaller and smaller at shorter wavelengths. In 1885, Johann Balmer toyed with the numbers in such a table and discovered that the wavelengths accurately fit the formula

$$\lambda = 911.76 \text{ Å} \times \frac{4n^2}{n^2 - 4} \tag{1}$$

with $n = 3, 4, 5, 6$, etc. This infinite series of spectral lines is called the **Balmer series.** Note that for $n \to \infty$, the wavelength approaches the asymptotic value $\lambda = 3647.0$ Å; this is called the **series limit.**

Balmer's formula was purely descriptive, or phenomenological; it did not explain the atomic mechanism responsible for the production of the spectral lines. Nevertheless, it proved very fruitful because it led to more general formulas describing other series of spectral lines. It is best to rewrite the formula in terms of the frequency,

$$\nu = \frac{c}{\lambda} = \frac{c}{911.76 \text{ Å}} \left(\frac{1}{4} - \frac{1}{n^2} \right) \tag{2}$$

or

Table 41.1 THE BALMER SERIES IN THE HYDROGEN SPECTRUM

Wavelength λ[a]
6564.7 Å
4862.7
4341.7
4102.9
3971.2
3890.2
3836.5
3799.0
etc.

[a] Wavelengths are measured in vacuum.

Balmer series of hydrogen

$$v = cR_{\mathrm{H}}\left(\frac{1}{2^2} - \frac{1}{n^2}\right) \qquad (3)$$

where R_{H} is the **Rydberg constant,**

Rydberg constant

$$R_{\mathrm{H}} = \frac{1}{911.76 \text{ Å}} = 109{,}678 \text{ cm}^{-1}$$

Balmer proposed that there might be other series in the hydrogen spectrum, with the 2 in Eq. (3) replaced by 1, or 3, or 4, etc. This yields the frequencies

$$v = cR_{\mathrm{H}}\left(\frac{1}{1^2} - \frac{1}{n^2}\right) \qquad n = 2, 3, 4, \dots \qquad (4)$$

$$v = cR_{\mathrm{H}}\left(\frac{1}{3^2} - \frac{1}{n^2}\right) \qquad n = 4, 5, 6, \dots \qquad (5)$$

$$v = cR_{\mathrm{H}}\left(\frac{1}{4^2} - \frac{1}{n^2}\right) \qquad n = 5, 6, 7, \dots \qquad (6)$$

$$v = cR_{\mathrm{H}}\left(\frac{1}{5^2} - \frac{1}{n^2}\right) \qquad n = 6, 7, 8, \dots \qquad (7)$$

These four series of spectral lines were actually discovered many years after Balmer proposed them; they are called, respectively, the **Lyman,** the **Paschen,** the **Brackett,** and the **Pfund series** (Figure 41.6). We can combine all these formulas into the single general formula

Spectral series of hydrogen

$$v = cR_{\mathrm{H}}\left(\frac{1}{n_2^2} - \frac{1}{n_1^2}\right) \qquad (8)$$

where n_1 and n_2 are positive integers and $n_1 > n_2$.

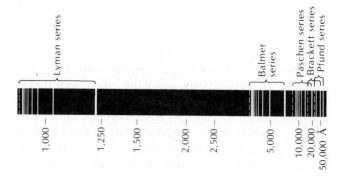

Fig. 41.6 The series of lines in the spectrum of hydrogen.

EXAMPLE 1.   What is the shortest wavelength that a hydrogen atom will emit or absorb?

SOLUTION: To find the shortest wavelength, we must choose n_1 and n_2 in Eq. (8) so as to obtain the highest frequency. Obviously, this demands $n_1 = \infty$ and $n_2 = 1$, which gives

$$\nu = cR_H\left(\frac{1}{1} - \frac{1}{\infty}\right) = cR_H \tag{9}$$

or

$$\lambda = c/\nu = 1/R_H = 911.76 \text{ Å}$$

Note that according to Eq. (8) the frequencies of hydrogen are written as differences between two terms, cR_H/n_2^2 and cR_H/n_1^2. Therein lies a crucial clue to the atomic mechanism responsible for the production of the spectral lines — as we will see in Section 41.4, these term differences correspond to energy differences. Careful examination of the frequencies of the spectral lines of other atoms shows that in all cases we can write the frequencies as differences between two terms, although the terms for these other atoms do not have as simple a mathematical form as those for the hydrogen atom. Furthermore, whenever we take any difference between two terms, there actually exists a spectral line that corresponds to this difference; this rule is called the **Rydberg–Ritz combination principle.**

41.3 The Nuclear Atom

The regularity in the series of spectral lines of the atom must be due to an underlying regularity in the structure of the atom. We may think of an atom as analogous to a musical instrument, such as a flute. The atom can only emit a discrete set of spectral lines, just as the flute can only emit a discrete set of tones which make up a musical scale. The regularity in the spacing of tones in this musical scale is due to an underlying regularity in the structure of the flute — the tube of the instrument has regularly spaced blowholes that determine what kind of standing waves can build up within the tube and what kind of waves will be radiated.

J. J. Thomson, the discoverer of the electron, made one of the first attempts at explaining the emission of light in terms of the structure of the atom. Having established that electrons are a ubiquitous component of matter, Thomson proposed the following picture: An atom consists of a number of electrons, say, Z electrons, embedded in a cloud of positive charge. The cloud is heavy, carrying almost all of the mass of the atom. The positive charge in the cloud is $+Ze$, so that it exactly neutralizes the negative charge $-Ze$ of the electrons. In an undisturbed atom the electrons will sit at their equilibrium positions, where the attraction of the cloud on the electrons balances their mutual repulsion (Figure 41.7). But if the electrons are disturbed by, say, a collision, then they will vibrate around their equilibrium positions and emit light. This model of the atom, called the "plum-pudding model," does yield frequencies of vibration of the same order of magnitude as the frequency of light, but it does not yield the observed spectral series; for instance, on the basis of this model, hydrogen should have only one single spectral line, in the far ultraviolet. And in 1910, experiments by

Sir Joseph John Thomson, *1856–1940, English experimental physicist, director of the Cavendish Laboratory at Cambridge, and president of the Royal Society. His discovery of the electron marks the beginning of modern experimental physics. Thomson received the Nobel Prize in 1906 for his investigations on electric discharges in gases. His experiments with beams of positive ions led to the first separation of the isotopes of a chemical element.*

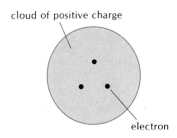

Fig. 41.7 The lithium atom according to the "plum-pudding" model. The three electrons sit at their equilibrium positions.

Sir Ernest Rutherford, *1871–1937, British experimental physicist, professor at McGill and at Manchester, and director of the Cavendish Laboratory at Cambridge, where he succeeded J. J. Thomson. Rutherford identified alpha and beta rays. He founded nuclear physics with his discoveries of the nucleus and of transmutation of elements by radioactive decay; he also produced the first artificial nuclear reaction. He was awarded the Nobel Prize for Chemistry in 1908.*

Rutherford and his collaborators established conclusively that most of the mass of the atom is not spread out over a cloud — instead, the mass is concentrated in a small kernel, or **nucleus,** at the center of the atom.

Rutherford had been studying the emission of alpha particles from radioactive substances. These alpha particles carry a positive charge $2e$ and they have a mass of 6.64×10^{-27} kg, about four times the mass of a proton (alpha particles have the same structure as nuclei of helium atoms; see Section B.1). Some radioactive substances, such as radioactive polonium and radioactive bismuth, spontaneously emit alpha particles with energies of several million electron-volts. These energetic alpha particles readily pass through thin foils of metal, or thin sheets of glass, or other materials. Rutherford was much impressed by the penetrating power of these alpha particles and it occurred to him that a beam of these particles can serve as a probe to "feel" the interior of the atom. When a beam of alpha particles strikes a foil of metal, the alpha particles penetrate the atoms and they are deflected by collisions with the subatomic structures; the magnitude of these deflections gives a clue about the subatomic structures. For example, if the interior of the atom had the "plum-pudding" structure proposed by J. J. Thomson, then the alpha particles would suffer only very small deflections since neither the electrons, with their small masses, nor the diffuse cloud of positive charge would be able to disturb the motion of a massive and energetic alpha particle.

The crucial experiments were performed by H. Geiger and E. Marsden working under Rutherford's direction. They used thin foils of gold and of silver as targets and bombarded these with a beam of alpha particles from a radioactive source. After the alpha particles passed through the foil, they were detected on a zinc sulfide screen which registers the impact of each particle by a faint scintillation (Figure 41.8). To Rutherford's amazement, some of the alpha particles were deflected by such a large angle that they came out backward. In Rutherford's own words: "It was quite the most incredible event that has ever happened to me in my life. It was almost as incredible as if you fired a 15-inch shell at a piece of tissue paper and it came back and hit you." Rutherford immediately recognized that the large deflection must be produced by a close encounter between the alpha particle and a very small but very massive kernel inside the atom. He therefore proposed the following picture: An atom consists of a small nucleus of charge $+Ze$ containing almost all of the mass of the atom; this nucleus is surrounded by a swarm of Z electrons. Thus, the atom is like a solar sys-

Fig. 41.8 Rutherford's apparatus.

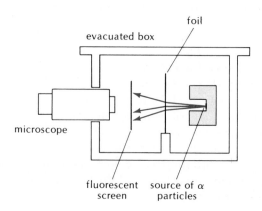

tem — the nucleus plays the role of Sun and the electrons play the role of planets.

On the basis of this nuclear model of the atom, Rutherford calculated what fraction of the beam of alpha particles should be deflected through what angle. If an alpha particle passes close to the nucleus it will experience a large electric repulsion and it will be deflected by a large angle; if it passes far from the nucleus it will only be deflected by a small angle. Figure 41.9 shows the trajectories of several alpha particles approaching a nucleus; these trajectories are hyperbolas. The perpendicular distance between the nucleus and the original (undeflected) line of motion is called the **impact parameter.** In order to suffer a large deflection, the alpha particle must hit an atom with a very small impact parameter, 10^{-13} m or less; since the alpha particles in the beam strike the foil of metal at random, only very few of them will score such a close hit.

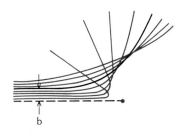

Fig. 41.9 Hyperbolic trajectories of several alpha particles with different impact parameters passing by a nucleus.

EXAMPLE 2. An alpha particle of energy E approaches a nucleus of charge Ze. The impact parameter of the alpha particle is b. What is the distance of the closest approach between the alpha particle and the nucleus?

SOLUTION: Figure 41.10 shows the hyperbolic orbit of the alpha particle and the point of closest approach. When the alpha particle is far away, its velocity is $\mathbf{v}$ and the perpendicular distance between the origin and the line of motion is b; hence the initial angular momentum is mvb. When the alpha particle is at the point of closest approach, its velocity is $\mathbf{v}'$ and its radius vector is $\mathbf{r}'$; since these vectors are perpendicular, the angular momentum is $mv'r'$. Conservation of angular momentum tells us

$$mv'r' = mvb \tag{10}$$

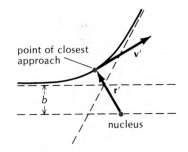

Fig. 41.10 At the point of closest approach, the alpha particle has a velocity $\mathbf{v}'$ perpendicular to its position vector $\mathbf{r}'$.

When the alpha particle is far away its energy is $\frac{1}{2}mv^2$. When the alpha particle is at the point of closest approach its energy is a sum of kinetic and potential energies,

$$\frac{1}{2}mv'^2 + \frac{(2e)(Ze)}{4\pi\varepsilon_0}\frac{1}{r'}$$

Conservation of energy then tells us

$$\frac{1}{2}mv'^2 + \frac{2Ze^2}{4\pi\varepsilon_0}\frac{1}{r'} = \frac{1}{2}mv^2 \tag{11}$$

Using Eq. (10), we can eliminate v' from Eq. (11),

$$\frac{1}{2}mv^2\frac{b^2}{r'^2} + \frac{2Ze^2}{4\pi\varepsilon_0}\frac{1}{r'} = \frac{1}{2}mv^2$$

We can regard this as a quadratic equation for the unknown $1/r'$. The standard formula for the solution of such a quadratic equation then gives

$$\frac{1}{r'} = \frac{-2Ze^2/4\pi\varepsilon_0 \pm \sqrt{(2Ze^2/4\pi\varepsilon_0^2) + m^2v^4b^2}}{mv^2b^2}$$

Only the positive square-root sign makes sense (the negative sign gives a negative value of $1/r'$). In terms of the energy $E = \frac{1}{2}mv^2$ of the alpha particle, we can express the distance of the closest approach as

$$r' = \frac{Eb^2}{-(Ze^2/4\pi\varepsilon_0) + \sqrt{(Ze^2/4\pi\varepsilon_0)^2 + E^2b^2}} \tag{12}$$

For example, if an alpha particle of energy $E = 1.2 \times 10^{-12}$ J approaches a gold nucleus ($Z = 79$) with an impact parameter of 5.0×10^{-14} m, then

$$\frac{Ze^2}{4\pi\varepsilon_0} = \frac{79 \times (1.6 \times 10^{-19} \text{ C})^2}{4\pi\varepsilon_0} = 1.8 \times 10^{-26} \text{ J} \cdot \text{m}^-$$

and the distance of closest approach will be

$$r' = \frac{1.2 \times 10^{-12} \text{ J} \times (5.0 \times 10^{-14} \text{ m})^2}{-1.8 \times 10^{-26} \text{ J} \cdot \text{m} + \sqrt{(1.8 \times 10^{-26} \text{ J} \cdot \text{m})^2 + (1.2 \times 10^{-12} \text{ J} \times 5.0 \times 10^{-14} \text{ m})^2}}$$

$$= 6.7 \times 10^{-14} \text{ m}$$

41.4 Bohr's Theory

Niels Bohr, *1885–1962, Danish theoretical physicist. He worked under J. J. Thomson and Rutherford in England and then became professor at Copenhagen and director of the Institute of Theoretical Physics, for the foundation of which he was largely responsible. After formulating the quantum theory of the atom, he played a leading role in the further development of the new quantum mechanics. He received the Nobel Prize in 1922.*

Rutherford's experiments did reveal the gross arrangement of the electrons in the atom, but not the details of their motion. Since the electrons make up the outer layers of an atom, their arrangement and motion should determine the chemical properties of the atom and the emission of light. But when physicists tried to calculate the electron motion according to the laws of classical mechanics and electromagnetism, they immediately ran into trouble.

To gain some insight into the source of this trouble, let us examine the case of the hydrogen atom. Suppose that the single electron of this atom is moving, according to the laws of classical mechanics, in a circular orbit of radius $\sim 10^{-10}$ m. The electron would then have a centripetal acceleration which is very large, about 10^{23} m/s². Because of this acceleration, the electron would emit high-frequency electromagnetic radiation, i.e., it would emit light. The energy carried away by the light must be supplied by the electron. Hence the emission process has the same effect on the electron as a friction force — it removes energy from the electron. This kind of friction would cause the electron to leave its circular orbit and gradually spiral in toward the nucleus, just like the residual atmospheric friction on an artificial satellite in a low-altitude orbit around the Earth causes it to spiral down toward the ground. A calculation using the laws of classical mechanics and electricity shows that the rate of emission of light by the orbiting electron in a hydrogen atom would be quite large. Correspondingly, the rate of energy loss of the electron would be large — the electron would spiral inward and collide with the nucleus within a time as short as 10^{-10} s!

Thus, our classical calculation leads us to the troublesome conclusion that hydrogen atoms, and other atoms, ought to be unstable — all the electrons ought to collapse into the nucleus almost instantaneously. Furthermore, the light that the electron emits during the spiraling motion ought to be a wave of continually increasing amplitude and increasing frequency (in musical terminology, crescendo and glissando); this is so because the closer the electron comes to the nucleus, the larger its acceleration and the higher its frequency of orbital motion. Hydrogen atoms do not behave as this calculation predicts. Hydrogen atoms are stable and when they do emit light, they emit discrete frequencies (spectral lines) instead of a continuum of frequencies.

These irreconcilable differences between the observed properties of atoms and the calculated properties gave evidence of a serious breakdown of the classical mechanics of Newton and the classical theory of electromagnetism. Although these theories had proved very successful on a macroscopic scale, they were in need of some drastic modification on an atomic scale.

In 1913, Niels Bohr took a bold step toward resolving these difficulties. He made the radical proposal that, at the atomic level, the laws of classical mechanics and of classical electromagnetism must be replaced or supplemented by other laws. Bohr expressed these new laws of atomic mechanics in the form of several postulates:

1. The orbits and the energies of the electrons in an atom are quantized, i.e., only certain discrete orbits and energies are permitted. When an electron is in one of the quantized orbits, it does not emit any electromagnetic radiation; thus, the electron is said to be in a **stationary state.** The electron can make a discontinuous transition, or **quantum jump,** from one stationary state to another. During this transition it does emit radiation.

2. The laws of classical mechanics apply to the orbital motion of the electrons in a stationary state, but these laws do not apply during the transition from one state to another.

3. When an electron makes a transition from one stationary state to another, the excess energy ΔE is released as a single photon of frequency $v = \Delta E / h$.

4. The permitted orbits are characterized by quantized values of the orbital angular momentum. This angular momentum is always an integral multiple of $h/2\pi$:

Bohr's postulates

$$\boxed{L = nh/2\pi \qquad n = 1, 2, 3, \ldots} \tag{13}$$

Quantization of angular momentum

Let us now see how to calculate the stationary states and the spectrum of the hydrogen atom on the basis of these postulates. For the sake of simplicity, we will assume that the electron moves in a circular orbit around the proton which remains at rest (Figure 41.11). Since the electric force of attraction between the electron and the proton is $e^2/4\pi\varepsilon_0 r^2$, the equation of motion for the electron in a circular orbit is

$$\frac{m_e v^2}{r} = \frac{1}{4\pi\varepsilon_0} \frac{e^2}{r^2} \tag{14}$$

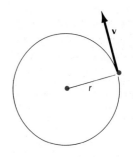

Fig. 41.11 Electron in circular orbit around proton.

According to Bohr's postulate, the orbital angular momentum must be $h/2\pi$ multiplied by an integer,

$$L = m_e v r = nh/2\pi \qquad n = 1, 2, 3, \ldots \tag{15}$$

or

$$m_e v r = n\hbar \tag{16}$$

where $\hbar$ (pronounced "h bar") is Planck's constant divided by 2π:

$$\hbar = h/2\pi = 1.06 \times 10^{-34} \, \text{J} \cdot \text{s}$$

Angular-momentum quantum number

The number n is called the **angular-momentum quantum number.** From Eq. (16)

$$v^2 = \frac{n^2 \hbar^2}{m_e^2 r^2} \tag{17}$$

which, when substituted into Eq. (14), yields an expression for the radius of the orbit,

$$r = \frac{4\pi\varepsilon_0 n^2 \hbar^2}{m_e e^2} \tag{18}$$

Thus, the radius of the smallest permitted orbit ($n = 1$) is

$$r_1 = \frac{4\pi\varepsilon_0 \hbar^2}{m_e e^2} = 0.529 \times 10^{-10} \, \text{m} = 0.529 \, \text{Å}$$

Bohr radius

This is called the **Bohr radius,**[2] usually designated by a_0,

$$\boxed{a_0 = \frac{4\pi\varepsilon_0 \hbar^2}{m_e e^2}} \tag{19}$$

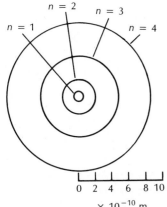

$n = 2$ $n = 3$
$n = 1$ $n = 4$

0 2 4 6 8 10
$\times 10^{-10}$ m

Fig. 41.12 The possible Bohr orbits of an electron in the hydrogen atom.

Figure 41.12 shows the permitted circular orbits, drawn to scale.

The energy of the electron in one of these orbits is a sum of kinetic and potential energies:

$$E = \tfrac{1}{2} m_e v^2 - \frac{e^2}{4\pi\varepsilon_0 r}$$

$$= \tfrac{1}{2} m_e \left(\frac{n^2 \hbar^2}{m_e^2} \right) \left(\frac{m_e e^2}{4\pi\varepsilon_0 n^2 \hbar^2} \right)^2 - \frac{e^2}{4\pi\varepsilon_0} \left(\frac{m_e e^2}{4\pi\varepsilon_0 n^2 \hbar^2} \right) \tag{20}$$

or

Energy of stationary states of hydrogen

$$\boxed{E = -\frac{m_e e^4}{2(4\pi\varepsilon_0)^2 \hbar^2} \frac{1}{n^2}} \tag{21}$$

Thus, the energy of the stationary state of least energy ($n = 1$) is

$$E_1 = -\frac{m_e e^4}{2(4\pi\varepsilon_0)^2 \hbar^2}$$

$$= -2.18 \times 10^{-18} \, \text{J} = -13.6 \, \text{eV} \tag{22}$$

The energies of the other stationary states are

[2] See Appendix 8 for a more precise value of the Bohr radius.

$$E_n = -\frac{13.6 \text{ eV}}{n^2} \qquad (23)$$

Figure 41.13 displays these quantized energies in an **energy-level diagram.** Each horizontal line represents one of the energies given by Eq. (23). According to Bohr's assumptions, the electron radiates when it makes a quantum jump from one stationary state to a lower stationary state. Such quantum jumps have been indicated by arrows in Figure 41.13. The stationary state of lowest energy is called the **ground state;** the next one is called the **first excited state,** etc. Ordinarily, the electron of the hydrogen atom is in the ground state, i.e., the circular orbit of radius a_0. This is the configuration of least energy into which the atom tends to settle when it is left undisturbed. As long as the atom remains in the ground state it does not emit light. To bring about the emission of light, we must first kick the electron into one of the excited states, i.e., a circular orbit of larger radius. We can do this by heating a sample of atoms or by passing an electric current through the sample. Collisions between the atoms will then disturb the electronic motions and occasionally kick an electron into a larger orbit. From there, the electron will spontaneously jump into a smaller orbit, giving off a quantum of light. Note that the quantum jumps shown in Figure 41.13 form several series: one series consists of all those jumps that end in the ground state, another series consists of all those jumps that end in the first excited state, etc. These series of jumps give rise to the series of spectral lines: the Lyman series, the Balmer series, etc.

Energy-level diagram

Ground state and excited states

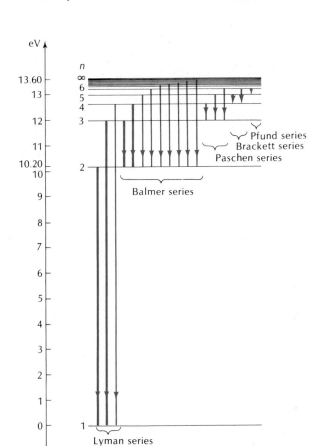

Fig. 41.13 Energy-level diagram for the hydrogen atom. The arrows show the possible quantum jumps for the electron. Note that in this diagram the energies are given relative to the ground state, which is assigned an energy of zero.

Let us now calculate the frequency of the light emitted in a quantum jump from some initial state i to a final state f. In this jump the electron releases an energy

$$\Delta E = E_i - E_f$$

$$= \frac{m_e e^4}{2(4\pi\varepsilon_0)^2\hbar^2} \left(\frac{1}{n_f^2} - \frac{1}{n_i^2} \right) \tag{24}$$

According to Bohr's postulate, this energy is radiated as a single photon of frequency $\nu = \Delta E/h$, i.e.,

Frequency of photon emitted in transition

$$\boxed{\nu = \frac{E_i - E_f}{h} = \frac{m_e e^4}{4\pi(4\pi\varepsilon_0)^2\hbar^3} \left(\frac{1}{n_f^2} - \frac{1}{n_i^2} \right)} \tag{25}$$

This equation looks just like the general formula (8) for the frequencies of the spectral series. Comparison of Eqs. (25) and (8) yields the following theoretical formula for the Rydberg constant:

$$R_H = \frac{m_e e^4}{4\pi(4\pi\varepsilon_0)^2\hbar^3 c} \tag{26}$$

Upon insertion of the accurate values of the fundamental constants given in Appendix 8, we obtain

$$R_H = \frac{9.10953 \times 10^{-31} \text{ kg} \times (1.602189 \times 10^{-19} \text{ C})^4}{4\pi(4\pi \times 8.854178 \times 10^{-12} \text{ F/m})^2 \times (1.054589 \times 10^{-34} \text{ J} \cdot \text{s})^3 \times 2.997925 \times 10^8 \text{ m/s}}$$

$$= 109{,}737 \text{ cm}^{-1}$$

This theoretical value of R_H agrees quite well with the experimental value quoted in Section 41.2.[3]

EXAMPLE 3. Suppose that the atoms in a sample of hydrogen gas are initially in the ground state. If we illuminate these atoms with light (from some kind of lamp), what frequencies will the atoms absorb?

SOLUTION: Absorption of light is the reverse of emission. When an electron in an atom absorbs a photon (supplied by the lamp), it jumps from the initial state to a state of higher energy. The energy of the photon must match the energy difference between the states. Thus, the frequencies of the photons that the electrons can absorb when jumping upward from the ground state are exactly those frequencies that they emit when jumping downward into the ground state, i.e., the frequencies of the Lyman series.

With his theory Bohr attained the goal of explaining the regularities in the spectrum of hydrogen in terms of the regularities of the struc-

---

[3] The small disagreement between the theoretical value of R_H given in Eq. (26) and the experimental value given in Section 41.2 is due to the motion of the nucleus of the atom, which we have neglected in our calculation. A careful calculation that takes into account the motions of electron and proton about their common center of mass eliminates the disagreement.

ture of the atom. By showing that this structure is based on a simple numerical sequence, he fulfilled the ancient dream of Pythagoras of a universe based on simple numerical ratios, a dream that arose from an analogy with musical instruments. Bohr's theory tells us how the atom plays its tune.

41.5 The Correspondence Principle

We saw in Chapter 40 that in an ordinary light wave we usually do not notice the individual quanta of energy — the number of quanta is so large that the energy distribution seems continuous. Likewise, when the electron in a hydrogen atom has a large angular momentum, an increase or decrease of this angular momentum by one quantum $\hbar$ represents such a small fraction of the whole that the change in angular momentum seems continuous. For instance, if an electron in a Bohr orbit with quantum number $n = 5000$ jumps down, step by step, to the next orbit ($n = 4999$), and to the next orbit ($n = 4998$), and etc., the changes of angular momentum, energy, and radius will seem quite continuous because each step is small compared to the remaining angular momentum, energy, and radius. Under these conditions the electron will behave pretty much like a classical particle — the quantized character of the angular momentum and energy, and the discrete character of the orbits will not be very noticeable.

This is an instance of Bohr's **Correspondence Principle.** This principle states that *in the limiting case of large quantum numbers, the results obtained from quantum theory must agree with those obtained from classical theory.*

Correspondence Principle

We can check that the behavior of an electron in a hydrogen atom does obey this principle both qualitatively and quantitatively, by calculating the frequency of the emitted light; we will see that in the limiting case of large n, the results of quantum theory and of classical theory agree. According to Eq. (25), the frequency emitted in a transition from the state n to the state $n - 1$ is

$$\nu = \frac{m_e e^4}{4\pi(4\pi\varepsilon_0)^2\hbar^3}\left[\frac{1}{(n-1)^2} - \frac{1}{n^2}\right] \tag{27}$$

$$= \frac{m_e e^4}{4\pi(4\pi\varepsilon_0)^2\hbar^3}\frac{2n-1}{n^2(n-1)^2} \tag{28}$$

If n is very large, $(2n - 1)/[n^2(n-1)^2] \cong 2n/n^4 = 2/n^3$ so that

$$\nu \cong \frac{m_e e^4}{4\pi(4\pi\varepsilon_0)^2\hbar^3}\frac{2}{n^3} \tag{29}$$

This is the frequency according to quantum theory. To find the frequency according to classical theory, we note that for a charge in accelerated motion, classical electromagnetism predicts that the frequency of the emitted light coincides with the frequency of the motion. For our electron in a circular orbit, the frequency of the motion is $v/2\pi r$. From Eqs. (16) and (18) we then find that the frequency of the emitted light is

$$v_{\text{class}} = \frac{v}{2\pi r} = \frac{n\hbar/m_e r}{2\pi r} = \frac{n\hbar}{2\pi m_e}\frac{1}{r^2}$$

$$= \frac{n\hbar}{2\pi m_e}\left(\frac{m_e e^2}{4\pi\varepsilon_0 n^2 \hbar^2}\right)^2 \tag{30}$$

If we simplify the right side of this equation, we recognize that it coincides with the right side of Eq. (29), i.e., the result of the classical calculation agrees with the result of the quantum-mechanical calculation. Note, however, that this agreement holds only for large values of n; if n is *not* large, then the classical frequency [Eq. (30)] is smaller than the quantum-mechanical frequency [Eq. (28)].

41.6 Quantum Mechanics

Bohr's theory is a hybrid. It relies on some basic classical features (orbits) and grafts onto these some quantum features (quantum jumps, quanta of light). In the 1920s the cooperative efforts of several brilliant physicists — L. de Broglie, E. Schrödinger, W. Heisenberg, M. Born, P. Jordan, P. A. M. Dirac — established that the remaining classical features had to be eradicated from the theory of the atom. Bohr's semiclassical theory had to be replaced by a new quantum mechanics with an entirely different equation of motion.

The basis of the new quantum mechanics was laid by the discovery that electrons — as well as protons, neutrons, and all the other "particles" found in nature — have not only particle properties but also wave properties. When a beam of electrons is made to pass through an extremely narrow slit, the electrons exhibit diffraction. This means that electrons are neither classical particles nor classical waves. Electrons, just like photons, are a new kind of object with a subtle combination of particle and wave properties. Electrons are *wavicles*. The wavelength associated with an electron or some other wavicle is inversely proportional to its momentum:

Louis Victor, prince de Broglie (de broy), *1892– , French theoretical physicist, professor at the University of Paris. He discovered Eq. (41.31) by reasoning that if waves have particle properties, then maybe particles have wave properties. For his discovery of the wave properties of matter he was awarded the Nobel Prize in 1929, after the existence of these wave properties was confirmed experimentally.*

De Broglie wavelength

$$\boxed{\lambda = h/p} \tag{31}$$

This is called the **de Broglie wavelength.**

Schrödinger equation

In quantum mechanics, the motion of an electron is described by a wave equation, the **Schrödinger equation.** This Schrödinger equation plays the same role for electrons as the Maxwell equations play for photons. The intensity of the electron wave at some point determines the probability that there is an electron particle at that point [compare Eq. (40.23)]. Furthermore, as a consequence of their wave properties, electrons obey the Heisenberg uncertainty relations for position and momentum [see Eq. (40.29)]. These quantum uncertainties are of crucial importance for the behavior of an electron inside an atom. For such an electron, the uncertainty in the position is very large — about as large as the size of the atom. This implies that the electron follows no definite orbit. It is therefore not surprising that the Bohr theory should have failed in its attempt at calculating the electron motion in the helium atom and in other atoms with several electrons; what is surprising is that this theory should have succeeded as well as it did in the case of the hydrogen atom.

EXAMPLE 4. Consider an electron in the ground state of hydrogen. Show that a well-defined orbit is inconsistent with the Heisenberg uncertainty relations.

SOLUTION: If the electron is to follow a well-defined orbit, the uncertainty in its momentum (in any direction) must be much smaller than the magnitude of the momentum. According to Eq. (16), the speed of the electron in the smallest circular orbit ($n = 1$) is $v = \hbar/m_e r$ and the magnitude of the momentum is $p = m_e v = \hbar/r$. For a well-defined orbit we therefore require

$$\Delta p_y \ll \hbar/r \tag{32}$$

Furthermore, we require that the uncertainty in the position must be much smaller thanthe size of the orbit,

$$\Delta y \ll r \tag{33}$$

Taking the product of Eqs. (32) and (33), we find

$$\Delta y \, \Delta p_y \ll \hbar \tag{34}$$

This is inconsistent with the Heisenberg uncertainty relation [Eq. (40.29)].

Although electrons are wavicles, they will sometimes behave pretty much like classical particles. Roughly, we can say that classical mechanics will be a good approximation whenever the quantum uncertainties are small compared to the relevant magnitudes of positions and momenta. For instance, for the electrons in the beam of a TV tube, the quantum uncertainty in the momentum is negligible compared with the magnitude of the momentum. Under these conditions, classical mechanics gives an adequate description of the motion of the electrons.

What we have said about the quantum mechanics of electrons also applies to other "particles" found in nature — they all have wave properties and they all have quantum uncertainties in their position and momentum. Strictly speaking, even large macroscopic bodies have wave properties. For example, an automobile is a wavicle and it has some quantum uncertainty in its position. However, it turns out that the quantum uncertainties are very small whenever the mass of the body is large compared to atomic masses — the quantum uncertainty in the position of the body of an automobile is typically no more than about 10^{-18} m, a number that can be ignored for all purposes. Hence, for automobiles and other macroscopic bodies, quantum effects are completely insignificant and classical mechanics gives an excellent description of the motion of these bodies.

Erwin Schrödinger, *1887–1961, Austrian theoretical physicist, professor at Berlin and at Vienna. Another of the founders of the new quantum mechanics, he received the Nobel Prize in 1933.*

SUMMARY

Spectral series of hydrogen:

$$v = cR_H \left(\frac{1}{n_2^2} - \frac{1}{n_1^2} \right); \qquad R_H = 109{,}678 \text{ cm}^{-1}$$

Quantization of angular momentum: $L = n\hbar$

Bohr radius: $a_0 = \dfrac{4\pi\varepsilon_0 \hbar^2}{m_e e^2}$

Energy of stationary states of hydrogen:

$$E = -\frac{m_e e^4}{2(4\pi\varepsilon_0)^2 \hbar^2} \frac{1}{n^2}$$

$$= -\frac{13.6 \text{ eV}}{n^2}$$

Frequency of photon emitted in transition: $\nu = \dfrac{E_i - E_f}{h}$

Correspondence Principle: In the limiting case of large quantum numbers, the results obtained from quantum theory must agree with those obtained from classical theory.

De Broglie wavelength: $\lambda = h/p$

QUESTIONS

1. Do the spectral lines seen in a stellar spectrum (e.g., Figure 41.4) tell us anything about the chemical composition of the stellar interior?

2. The spectrum of hydrogen shown in the color plate between pages 850–851 displays all of the spectral lines simultaneously. Since a hydrogen atom emits only one spectral line at a time, how can all the lines be visible simultaneously?

3. What is the longest wavelength that a hydrogen atom will emit or absorb?

4. The target used in Rutherford's scattering experiment was a very thin foil of metal. What is the advantage of a thin foil over a thick foil in this experiment?

5. How can Rutherford's experiment tell us something about the size of the nucleus?

6. Why is Bohr's postulate of stationary states in direct contradiction with classical mechanics and electromagnetism?

7. If an electron in a hydrogen atom makes a transition from some state to a lower state, does its kinetic energy increase or decrease? Its potential energy? Its orbital angular momentum?

8. At low temperatures, the absorption spectrum of hydrogen displays only the spectral lines of the Lyman series. At higher temperatures, it also displays other series. Explain.

9. The planets move around the Sun in circular orbits. Is their orbital angular momentum quantized?

10. In a muonic atom, a muon orbits around the nucleus. The mass of the muon is 207 times the mass of the electron. What is the Bohr radius for a muonic atom with a hydrogen nucleus?

11. Given that the orbital angular momentum of an atom is quantized, can we conclude that the orbital magnetic moment is also quantized?

Fine-structure constant

12. The quantity $\hbar/(m_e c)$ is called the **Compton wavelength.** The quantity $e^2/(4\pi\varepsilon_0 m_e c^2)$ is called the **"classical electron radius."** Show that the Bohr radius, the Compton wavelength, and the classical electron radius are in the ratio $1:\alpha:\alpha^2$, where $\alpha = e^2/(4\pi\varepsilon_0 \hbar c)$. The quantity α is called the **fine-structure constant.** What is the numerical value of this constant?

13. Would you expect that Bohr's theory of the hydrogen atom can be adapted to the singly ionized helium atom, i.e., the helium atom with one missing electron? To what other ionized atoms can Bohr's theory be adapted?

14. According to the **Complementarity Principle,** formulated by Bohr, a wavicle has both wave properties and particle properties, but these properties are never exhibited simultaneously: if the wavicle exhibits wave properties in an experiment, then it will not exhibit particle properties, and conversely. Give some examples of experiments in which wave or particle properties (but not both simultaneously) are exhibited.

Complementarity Principle

15. Show that photons obey the de Broglie relation.

16. If the de Broglie wavelengths of two electrons differ by a factor of 2, by what factor must their energies differ?

17. According to the de Broglie relation, the wavelength of an electron of very small momentum is very large. Could we take advantage of this to design an experiment that makes the wave properties of the electron obvious?

18. An electron and a proton have the same energy. Which has the longer de Broglie wavelength?

19. Describe the interference pattern expected for an electron wave incident on a plate with two very narrow parallel slits separated by a small distance.

20. Electron microscopes achieve high resolution because they use electron waves of very short wavelength, usually less than 0.1 Å. Why can we not build a microscope that uses *photons* of equally short wavelength?

PROBLEMS

Section 41.2

1. Use Eq. (4) to calculate the wavelengths of the first four lines of the Lyman series.

2. Which of the spectral lines of the Brackett series is closest in wavelength to the first spectral line ($n = 6$) of the Pfund series? By how much do the wavelengths differ?

3. Show that the spectral lines of the Balmer series all have a higher frequency than the spectral lines of the Paschen series. Do the spectral lines of the Paschen series all have a higher frequency than those of the Brackett series?

4. When astronomers examine the light of a distant galaxy, they find that all the wavelengths of the spectral lines of the atoms are longer than those of the atoms here on Earth by a common multiplicative factor. This is the *red shift* of light; it is a Doppler shift caused by the motion of recession of the galaxy, away from the Earth. In the light of a galaxy beyond the constellation Virgo, astronomers find spectral lines of wavelengths 4117 Å and 4357 Å.
    (a) Assume that these are two spectral lines of hydrogen, with the wavelengths multiplied by some factor. Identify these lines. What is the factor by which these wavelengths are longer than the normal wavelengths of the two spectral lines?
    (b) What is the speed of recession of the galaxy?

5. One of the spectral series of the lithium atom is the **principal series,** with the following wavelengths: 6707.9 Å, 3232.6 Å, 2741.3 Å, 2562.5 Å, 2475.3 Å. Show that these wavelengths approximately fit the formula

$$\frac{1}{\lambda} = R\left[\frac{1}{(1 + s)^2} - \frac{1}{(n + p)^2}\right] \qquad n = 2, 3, 4, \ldots$$

where $R = 109,728$ cm^{-1} is the Rydberg constant for lithium, and s and p are constants characteristic of the series. Given that $p = -0.041$, what value of s must you use to make the wavelengths fit the formula?

6. Another of the spectral series of the lithium atom is the **diffuse series,** with the following wavelengths: 6103.5 Å, 4603.0 Å, 4132.3 Å, 3915.0 Å, 3794.7 Å. These wavelengths approximately fit the formula

$$\frac{1}{\lambda} = R\left[\frac{1}{(2+p)^2} - \frac{1}{(n+d)^2}\right] \qquad n = 3, 4, 5, \ldots$$

where, as in the preceding problem, $R = 109,728$ cm^{-1} and p and d are constants.
   (a) Given that $d = -0.0015$, what value of p must you use to make the wavelengths fit the formula?
   (b) The principal series (see Problem 5) and the diffuse series of lithium are analogous to two spectral series of hydrogen. Which two series?

Section 41.3

7. What is the distance of closest approach for a 5.5-MeV alpha particle in a head-on collision with a gold nucleus? With an aluminum nucleus?

8. The nucleus of platinum has a radius 6.96×10^{-15} m and an electric charge of $78e$. What must be the minimum energy of an alpha particle in a head-on collision if it is to just barely reach the nuclear surface? Assume the alpha particle is pointlike.

9. An alpha particle of energy 5.5 MeV is incident on a silver nucleus with an impact parameter 8.0×10^{-15} m. Find the distance of closest approach of the particle. Find the speed at the point of closest approach.

10. Prove that the distance of closest approach given by Eq. (13) can be expressed as

$$r' = \frac{b^2}{-r^*/2 + \sqrt{(r^*/2)^2 + b^2}}$$

where r^* is the distance of closest approach for a head-on collision (with $b = 0$).

*11. A foil of gold, 2.1×10^{-5} cm thick, is being bombarded by alpha particles of energy 7.7 MeV. The particles impact at random over an area of 1 cm^2 of the foil of gold.
   (a) How many atoms are there within the volume 1 cm$^2 \times 2.1 \times 10^{-5}$ cm under bombardment? The density of gold is 19.3 g/cm^2 and the mass of one atom is 3.27×10^{-25} kg.
   (b) It can be shown that to suffer a deflection of more than 30°, an alpha particle must strike within 5.5×10^{-14} m of the center of a gold nucleus. What is the probability for this to happen?
   (c) If 10^{10} alpha particles impact on the foil, how many will suffer deflections of more than 30°?

Section 41.4

12. What is the speed of an electron in the smallest ($n = 1$) Bohr orbit? Express your answer as a fraction of the speed of light.

13. What is the frequency of the orbital motion for an electron in the smallest ($n = 1$) Bohr orbit? In the next ($n = 2$) Bohr orbit? Do either of these frequencies coincide with the frequency of the light emitted during the transition $n = 2$ to $n = 1$?

14. If a hydrogen atom is in the ground state, what is the *longest* wavelength it will absorb?

15. What is the ionization energy of hydrogen (i.e., what energy must you supply to remove the electron from the atom when it is in the ground state)? Express the answer in electron-volts.

16. A hydrogen atom emits a photon of wavelength 1026 Å. From what stationary state to what lower stationary state did the electron jump?

17. Suppose that the electron in a hydrogen atom is initially in the second excited state ($n = 3$). What wavelength will the atom emit if the electron jumps directly to the ground state? What two wavelengths will the atom emit if the electron jumps to the first excited state and then to the ground state?

18. Find the orbital radius, the speed, the angular momentum, and the centripetal acceleration for an electron in the $n = 2$ orbit of hydrogen.

19. If you bombard hydrogen atoms in their ground state with a beam of particles, the collisions will (sometimes) kick atoms into one of their excited states. What must be the minimum kinetic energy of the bombarding particles if they are to achieve such an excitation?

20. A hydrogen atom is initially in the ground state. In a collision with an argon atom, the electron of the hydrogen atom absorbs an energy of 15.0 eV. With what speed will the electron be ejected from the hydrogen atom?

21. The singly ionized helium atom (usually designated HeII) has one electron in orbit around a nucleus of charge $2e$.
   (a) Apply Bohr's theory to this atom and find the energies of the stationary states. What is the value of the ionization energy, i.e, the energy that you must supply to remove the electron from the atom when it is in the ground state? Express the answer in electron-volts.
   (b) Show that for every spectral line of the hydrogen atom, the ionized helium atom has a spectral line of identical wavelength.

22. Doubly ionized lithium (usually designated LiIII) has one electron in orbit around a nucleus of charge $Z = 3e$. What is the radius of the smallest Bohr orbit in doubly ionized lithium? What is the energy of this orbit?

23. The muon (or mu meson) is a particle somewhat similar to an electron; it has a charge $-e$ and a mass 206.8 times as large as the mass of the electron. When such a muon orbits around a proton, they form a **muonic hydrogen atom,** similar to an ordinary hydrogen atom, but with the muon playing the role of the electron. Calculate the Bohr radius of this muonic atom and calculate the energies of the stationary states. What is the energy of the photon emitted when the muon makes a transition from $n = 2$ to $n = 1$?

24. Assume that, as proposed by J. J. Thomson, the hydrogen atom consists of a cloud of positive charge e, uniformly distributed over a sphere of radius R. However, instead of placing the electron in static equilibrium at the center of the sphere, assume that the electron orbits around the center with uniform circular motion under the influence of the electric centripetal force $(e^2/4\pi\varepsilon_0)\,(r/R^3)$. If the angular momentum of this orbiting electron is quantized according to Bohr's theory (so that $L = n\hbar$), what are the radii and the energies of the quantized orbits? What are the frequencies of the photons emitted in transitions from one quantized orbit to another? What must be the value of R if at least two orbits are to fit inside this atom?

*25. In our calculation of the energies of the stationary states of hydrogen we pretended that the proton remains at rest. Actually, both the electron and the proton orbit about their common center of mass. Show that the energies of the stationary states, taking into account this motion of the proton, are given by

$$E_n = -\frac{\mu e^4}{2(4\pi\varepsilon_0)^2\hbar^2}\frac{1}{n^2}$$

where

$$\mu = \frac{m_e m_p}{m_e + m_p}$$

(Hint: the electron and the proton move in circles of radii

$$r_e = r\frac{m_p}{m_p + m_e} \quad \text{and} \quad r_p = r\frac{m_e}{m_p + m_e}$$

where r is the distance between the electron and the proton. According to Bohr's theory, the net angular momentum of this system of two particles is quantized, $L = n\hbar$.)

*26. The atom of **positronium** consists of an electron and a positron (or anti-electron) orbiting about their common center of mass. According to Bohr's theory, the net angular momentum of this system is quantized, $L = n\hbar$. What is the radius of the smallest possible circular orbit of this system? What is the wavelength of the photon released in the transition $n = 2$ to $n = 1$?

Section 41.5

27. In principle, Bohr's theory also applies to the motion of the Earth around the Sun. The Earth plays the role of the electron, the Sun that of the nucleus, and the gravitational force that of the electric force.
  (a) Find a formula analogous to Eq. (18) for the radii of the permitted circular orbits of the Earth around the Sun.
  (b) The actual radius of the Earth's orbit is 1.50×10^{11} m. What value of the quantum number n does this correspond to?
  (c) What is the radial distance between the Earth's actual orbit and the next larger orbit?

28. According to classical electrodynamics, an electron in an elliptical orbit emits not only radiation at the orbital frequency, but also radiation at multiples of the orbital frequency; thus, if the orbital frequency is v, the radiation contains the fundamental frequency v and also the harmonic frequencies $2v$, $3v$, $4v$, etc. Does this contradict the Correspondence Principle? (Hint: Calculate what frequencies the electron will emit in the transition n to $n - 2$, n to $n - 3$, n to $n - 4$, etc.)

29. What must be the energy of an electron if its wavelength is to equal the wavelength of visible light, about 5500 Å?

30. A photon and an electron each have an energy of 6.0×10^3 eV. What are their wavelengths?

31. An electron microscope operates with electrons of energy 40,000 eV. What is the wavelength of such electrons? By what factor is this wavelength smaller than that of visible light?

32. What is the de Broglie wavelength of a tennis ball of mass 0.060 kg moving at a speed of 1.0 m/s?

33. What is the de Broglie wavelength, in angstroms, of an electron in the ground state of hydrogen? In the first excited state?

34. Use Eq. (16) to derive an expression for the de Broglie wavelength of an electron in the nth state of hydrogen. Show that the circumference of the orbit equals n times the de Broglie wavelength.

Section 41.6

35. Interferometric methods permit us to measure the position of a macroscopic body to within $\pm 10^{-12}$ m (see Section 1.3). Suppose we perform a position measurement of such a precision on a body of mass 0.050 kg. What

uncertainty in momentum is implied by the Heisenberg relation? What uncertainty in velocity?

36. Suppose that the velocity of an electron has been measured to within an uncertainty of ± 1 cm/s. What minimum uncertainty in the position of the electron does this imply?

37. If the position of a parked automobile of mass 2×10^3 kg is uncertain by $\pm 10^{-18}$ m, what is the corresponding uncertainty in its velocity?

38. The nucleus of aluminum has a diameter of 7.2×10^{-15} m. Consider one of the protons in this nucleus. The uncertainty in the position of this proton is necessarily less than 7.2×10^{-15} m. What is the minimum uncertainty in its momentum and velocity?

39. Consider an electron in a circular orbit of quantum number n in a hydrogen atom. The orbit is well defined provided that $\Delta p_y \ll p$ *and* $\Delta y \ll r$. Show that if $n \gg 1$, these requirements are *not* in conflict with the Heisenberg uncertainty relations. Thus, *large* orbits in the hydrogen atom are well defined (this is in accord with the Correspondence Principle).

LASER LIGHT[1]

Man is a visual animal. Man's intake of visual information is about nine times larger than his combined intake of all other kinds of information. It is therefore no wonder that, up to a few years ago, we regarded light as illumination — our main direct use of light was to make our environment visible. The most powerful light sources were searchlights and lighthouses with an output of up to 1 kW.

The invention of the laser in 1958 changed all that. We now have available light beams of fearsome intensity — up to 10^{10} kW in short pulses. Light beams from lasers are now used in industry to cut and weld metal, cloth, and plastic; in medicine to cut and cauterize human tissues; and in plasma research to vaporize and ignite the fuel for fusion reactors. Light has become a tool and, in military applications, it has become a weapon.

L.1 STIMULATED EMISSION

Light is emitted by the electrons in the atoms of the light source. As we saw in Chapter 41, the electrons in an atom can only move in certain selected, quantized orbits. As long as the electron stays in one of these orbits, it does not emit light. But when the electron jumps from one orbit to a smaller orbit, it emits a packet of light, or photon, with an energy equal to the energy difference between the two orbits (Figure L.1).

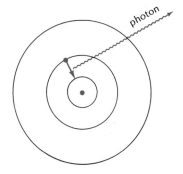

Fig. L.1 The electron in an atom emits a photon during a quantum jump.

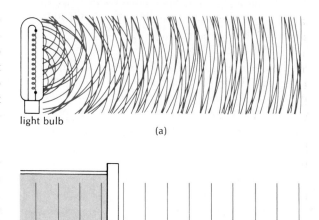

Fig. L.2 (a) Light waves emitted by an ordinary light bulb. (b) Light wave emitted by a laser.

In an ordinary light source, such as a light bulb, the atoms emit their light independently. The atom suffers concussions from thermal collisions and this excites one or more electrons from lower orbits to higher orbits. At some later, unpredictable time the electron spontaneously jumps back down, emitting a pulse of light. Because the atoms emit at random, the light emerging from an ordinary light bulb is a confused combination of many different waves with no special directions of propagation and no special phase relationships (Figure L.2a). On the average, the total intensity of the light is simply the sum of the intensities contributed by the individual atoms; therefore, the total intensity is simply proportional to the number of atoms.

In a laser, the atoms emit their light in unison. The electrons in different atoms either jump down at the same time or else they jump with a time difference of one or several periods of oscillation of the light wave. Furthermore, the electrons emit their light waves in the same direction. The result is that the light emerging from a laser is a **coherent** combination of waves (Figure L.2b). All the light waves from different atoms are in phase — the light wave contributed by each atom combines crest to crest with the light waves contributed by the other atoms. The total amplitude is then proportional to the number of atoms, and therefore the total intensity is proportional to the *square* of the number of atoms. Since the number of atoms in even a fairly small light source is more than 10^{16}, the coherent emission from these atoms can be enormously stronger than the incoherent emission.

What keeps the electrons in different atoms in a laser in step is the phenomenon of **stimulated emission.** Imagine that an excited electron in one of the atoms jumps to its lower orbit, releasing a wave of light. As this wave passes by some other atom with an excited electron, it will jiggle the electron; this causes the electron to resonate and to jump down in unison with the wave instead of waiting to jump spontaneously. The passage of the light wave triggers, or stimulates, the emission of an additional coherent light wave.

The mechanism behind this process involves quantum theory and we cannot present the details here, but it turns out that the additional wave will have the same direction of propagation and the same phase as the original light wave. The two waves therefore combine constructively and they proceed to stimulate the emission of more and more waves from other excited atoms. The process has some features of a chain reaction: whenever an excited atom joins in the emission, it strengthens the wave and thereby increases the probability that the remaining excited atoms will also join in the emission.

The word *laser* is an acronym for *l*ight *a*mplification by *s*timulated *e*mission of *r*adiation. From the above discussion it is obvious that the key to the operation of a laser is the initial presence of a large number of atoms in an excited configuration. This is called a **population inversion** because under normal conditions atoms tend to settle into an unexcited configuration. Indeed, there must be more atoms in the excited configuration than in the unexcited configuration — otherwise the light wave will lose energy by stimulated absorption rather than gain energy by stimulated emission. One method for achieving the required population inversion uses an intense flash of (ordinary) light to lift the electrons into excited orbits; this is called **optical pumping.** Unfortunately, the flash of light that pumps the electrons upward into an excited orbit can also pump them downward; hence direct pumping into an excited orbit does not produce a population inversion. This problem can be circumvented by making the electron jump upward by an indirect route: it first jumps to some high excited orbit and then spontaneously jumps down into a somewhat lower excited orbit, where it remains for some time awaiting stimulated emission (Figure L.3). The lasing action begins

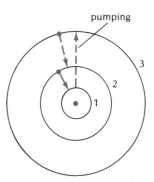

Fig. L.3 Upward and downward jumps of an electron in an atom subjected to optical pumping. A flash of light lifts the electron from the lowest orbit (1) to a high excited orbit (3). From there the electron spontaneously jumps to a somewhat lower excited orbit (2). The lasing action occurs between this orbit and the lowest orbit.

when a sufficiently large number of electrons have accumulated in these lower excited orbits.

The fundamental theoretical principles involved in the operation of a laser were first published by C. H. Townes and A. L. Schawlow.[2] Their speculations on the generation of coherent light were motivated by an earlier discovery of the generation of coherent microwaves in a maser, a device similar to a laser but operating with radiowaves rather than light waves.

L.2 LASERS

The first laser designed according to the above principles was a **ruby laser** built by T. H. Maiman in 1960 (Figure L.4). This kind of laser, still in common use today, generates light in a long cylindrical crystal of synthetic ruby. The crystal contains corundum, an oxide of aluminum, with a few chromium impurities; the red color of ruby is due to these impurities. In the ruby laser, only the chromium atoms lase. To pump the electrons in these atoms into their excited orbits, the crystal is surrounded by a flash lamp (Figure L.4); the high-intensity light from the flash lamp supplies the energy for the initial upward jump of the electrons. The first electron that engages in the spontaneous emission of light will trigger the stimulated emission of light by other electrons.

In order to ensure that all, or almost all, the excited atoms participate in stimulated emission, it is best to make the light wave traverse the ruby rod several times so that any excited atoms not triggered on the first pass are triggered on a later pass. For this purpose,

[2] Charles H. Townes, 1915–, and Arthur L. Schawlow, 1921–, American physicists. Townes shared the Nobel Prize in 1964 with the Soviet physicists Nicolai G. Basov and Alexander M. Prochorov for the discovery of the maser. Schawlow was awarded the 1981 Nobel Prize for his work on lasers.

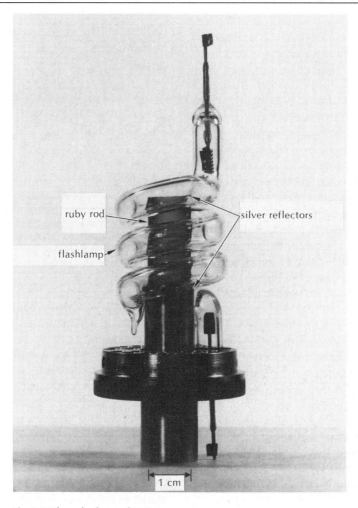

ruby rod

silver reflectors

flashlamp

1 cm

Fig. L.4 The ruby laser of Maiman.

the ends of the rod are polished flat and silvered, making them into mirrors. One end is only partially silvered so that the light can ultimately escape. The back-and-forth reflections drastically enhance the lasing action for those light waves that are emitted exactly parallel to the axis of the rod and that have a wavelength fitting into the length of the rod exactly an integral number of times. For such light waves, the rod acts as a resonant cavity oscillating in one of its normal modes. In this cavity, the light wave is a standing wave and the stimulated emission by the excited atoms feeds more and more energy into this standing wave. Some of the wave leaks out of the partially silvered end of the rod, forming the useful external beam of the laser. The chromium atoms in ruby will lase at one or another of several wavelengths between 6930 Å and 7000 Å, in the red part of the spectrum. The emerging beam is essentially unidirectional since it originates from light waves traveling exactly parallel to the axis of the rod. Typically, the beam of a laser has an angular spread of a minute of arc or less.

The beam from a ruby laser emerges as a pulse last-ing as long as excited atoms remain in the rod, about 10^{-3} s. When the atoms become exhausted, they must be pumped again with a new flash from the flash lamp. For some investigations of the effect of intense light on matter, the concentration of power into a succession of short pulses of laser light is very useful. Neodymium glass lasers designed for thermonuclear fusion experiments generate pulses as short as 10^{-12} s with a peak power of 10^9 kW.

However, for many other applications it is desirable to build lasers with a continuous output of light. Obviously, this requires that, on the average, the electrons in the atoms be pumped up into the excited orbits at the same rate as they jump down by stimulated emission. The most commonly employed laser with a continuous output of light is the **helium–neon laser.** It consists of a glass tube with a mixture of helium and neon at low pressure. The tube has a silvered and a partially silvered mirror at the two ends, and also two electrodes (Figure L.5). The lasing action is due to the neon; the helium merely serves to pump the neon. When the electrodes are connected to a high-voltage

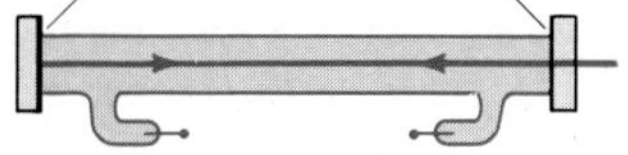

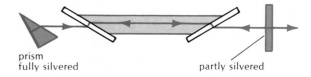

Fig. L.5 Schematic diagram of a helium–neon laser. The terminals are connected to a high-voltage supply.

Fig. L.6 Laser with a prism for the selection of one color.

power supply, a current of electrons flows through the tube. These electrons collide with the helium atoms and kick their atomic electrons into excited orbits. The excited helium atoms in turn collide with the neon atoms. In such a collision it is very likely that the helium atom will transfer its excess internal energy to the neon atom. This pumps the electron in the neon into an excited orbit suitable for stimulated emission. The neon atoms will lase at a wavelength of 6328 Å (red light) and also at several wavelengths in the infrared region of the spectrum (there exist several techniques to restrict the lasing action to just a single wavelength; for instance, see below).

A laser capable of very high power output is the **carbon dioxide–nitrogen laser.** This laser operates in much the same way as the helium–neon laser — the carbon dioxide molecules lase and the nitrogen molecules transfer energy to them. Lasers of this kind can easily deliver 10 kW of continuous power. But they can also be operated in a pulsed fashion with a much higher power concentrated in very short pulses. The carbon dioxide laser emits infrared light at several wavelengths, but no visible light. The high power and relatively high efficiency (∼30%) attained with this laser is largely due to the long wavelength of the emitted infrared light. The downward energy jump during emission by a molecule is small and, correspondingly, the upward energy jump during pumping is also small. The pumping mechanism can supply these small energy quanta much more easily than the larger energy quanta required for lasers operating with shorter-wavelength visible light.

The resonant cavity of a ruby, helium–neon, or carbon dioxide laser has very many different normal modes and, unless special precautions are taken, the atoms will simultaneously lase in all the normal modes whose frequencies coincide with the frequencies of light generated by jumps from the available excited states. This means that the light emerging from the laser is a mixture of several colors. To obtain truly monochromatic light, a special selective device must be attached to the laser. Figure L.6 shows one of the devices that can be used for this purpose. The rear end of the laser tube is left transparent and a prism with one silvered face is placed beyond it. The prism

refracts rays of different colors at different angles. When the multicolored light from the laser penetrates the prism, only one of the rays, of a selected color, strikes the silvered face at a right angle and is reflected back into the laser. Hence only standing waves of one selected color are possible, and lasing will occur only at this one color. The emerging laser beam will then be perfectly monochromatic. Note that by turning the prism through some angle, we can select a different color for lasing. This is the principle of tunable lasers that produce light of adjustable color.

If laser light illuminates a slightly irregular surface — a wall or a sheet of paper — the patch of light on the surface will look grainy to the eye, appearing to have bright specks and dark specks. This is called the **speckle effect** (Figure L.7). The apparent irregularity of the illumination of the wall is an illusion. The laser beam illuminates the entire patch of surface uniformly, but small irregular points of the surface reflect light with somewhat different phases; thus, the incident light is a uniform plane wave, but the reflected light is a superposition of many spherical waves with somewhat different phases. When these waves reach the retina of the eye or a photographic plate, they interfere constructively or destructively, making bright spots and dark spots.

Incidentally, looking directly into a laser beam can damage the eye. This danger is of course obvious in the case of powerful lasers that are capable of burning holes in metals, but the danger subsists even in the

(a)   (b)

Fig. L.7 Bright spot made by a beam of light on a cement surface: (a) ordinary light from a mercury arc and (b) laser light from a helium–neon laser.

case of the small milliwatt laser commonly used for optical experiments. The lens of the eye can focus the laser beam on a single spot of the retina and it will then not take very much power to cause a retinal burn.

L.3 SOME APPLICATIONS

Laser light is useful because of its directionality, intensity, pure color, and coherence. The applications of laser light exploit one or several of these characteristics.

Surveys and Distance Measurements

Because of its sharp directionality, the beam from a laser is a very convenient tool for laying out straight lines over large distances. The traditional and tedious method for doing this involves setting up a row of markers along the line of sight of a survey telescope. If we replace the telescope by a laser, we can dispense with the cumbersome markers since the laser beam itself can serve as a marker. For example, to dig a straight horizontal trench, we can aim a laser beam parallel to the ground and proceed to dig in its direction; we can check the depth of the trench by intercepting the beam with a vertical meter stick resting against the bottom of the trench. Laser beams can similarly be used in the construction of large aircraft and ships to check the alignment of ribs and frames.

Laser pulses have found application in rangefinders, especially for military purposes. The rangefinder consists of a laser and a light detector, both linked to a timing device. The laser sends a short pulse of light out to the target and the target reflects part of this pulse back. The travel time for this round trip indicates the distance.

As was mentioned in Section 37.2, this method has also been used with spectacular success in an accurate determination of the Earth–Moon distance. Both the Apollo 11 and the Apollo 14 astronauts installed corner reflectors on the surface of the Moon during their missions (Figure L.8). By means of a telescope at MacDonald Observatory, Texas, a pulse from a powerful laser was aimed at one of these reflectors. By the time the pulse reached the Moon, it had spread out to a diameter of about 3 km, but enough light was reflected back to the Earth to be picked up by a sensitive phototube. The precise measurement of the elapsed time for the round trip gave the distance to the Moon to within about 15 cm, i.e., about nine significant figures!

Light-Wave Communications

As we saw in Section 37.3, light pulses can be piped through long, thin glass fibers. If the sound of the human voice is encoded as a series of light pulses, then these optical fibers can be used to transmit telephone conversations. In commercial telephone installations of this kind, the light pulses are generated either by small lasers or by light-emitting diodes. The laser light is much more monochromatic than the diode light, and this is an advantage because it helps to preserve the shape of the light pulses. If a light pulse containing a mixture of colors is sent along an optical fiber, the dispersion of the medium will cause the short-wavelength colors to fall behind and the long-wavelength colors to get ahead; this tends to

Fig. L.8 Corner reflector placed on the surface of the Moon by the Apollo 14 astronauts.

Fig. L.9 Solid-state laser.

Fig. L.10 Robot laser welding machine at an automobile assembly line.

Fig. L.11 Laser beam cutting a thick steel plate.

spread the pulse out, making it indistinct. For laser light the spreading effect is about 10 times smaller than for diode light.

The lasers that generate the pulses for such telephone installations are solid-state devices made of a sandwich of semiconductor, roughly the size of a grain of salt (Figure L.9); their power is about 0.5 mW. In an optical telephone line, each laser feeds 5×10^7 pulses per second into its attached optical fiber; this is enough to encode 672 one-way telephone speeches.

Laser as Torch

Many applications take advantage of the high concentration of energy in laser light to melt, weld, cut, or vaporize materials. For industrial applications, a carbon dioxide laser with a power of several kilowatts makes an excellent welding torch; it has a high welding speed, requires little or no filler metal, and produces joints of excellent quality. Figure L.10 shows a laser welding system in service at an automobile factory. High-power lasers are also used to cut holes in very hard materials, e.g., steel (Figure L.11). In the clothing industry, laser beams are now used to cut out patterns in stacks of several hundreds of layers of cloth all at once.

In medicine, lasers have been used for many years for surgery of the eye. A laser beam aimed through the (transparent) lens of the eye and focused on the retina can "weld" a detached patch of retina into its proper position. In some other surgical operations, a laser beam can serve as a scalpel; this is particularly suitable for operations on blood-rich tissues, such as the liver, where the immediate cauterization of the tissues by the laser beam prevents excessive bleeding.

Laser Inteferometry

Lasers make excellent light sources for interference ex-

periments. As we saw in Section 38.4, when an ordinary light source is used to illuminate the slits or holes, a shield with a pinhole must be placed around the light source so as to select waves of a single direction from the confused combination of waves produced by the light source; this, of course, entails a drastic reduction of the light intensity. The light from a laser already has a sharply defined direction and no pinhole is required.

In an interferometer, such as that of Michelson, the great advantage of laser light is its large-scale coherence. An ordinary light beam, even when selected in one direction, is only coherent over a length of a few meters. This is so because the individual wave trains of light emitted by individual atoms are only about that long. Hence portions of the light beam separated by more than a few meters consist of waves produced by different atoms; these waves are not coherent — they have no regular phase relationships and they will not produce interference patterns (Figure L.12a). When the light beam is split into two by a half-silvered mirror, it will display interference patterns only if each portion travels roughly the same distance to the place of recombination, because only then will the wave pulse of a given atom be able to interfere with itself. On the other hand, in the light beam of a laser all the waves are coherent and the entire beam is a single coherent wave (Figure L.12b). Therefore any segment of the wave train will display interference with any other segment when they are brought together.

The light from ordinary lasers is nearly monochromatic, but not perfectly monochromatic because the thermal motion of the emitting atoms gives the light a fluctuating Doppler shift. To eliminate or reduce these Doppler shifts, scientists have developed **stabilized lasers** in which a feedback device monitors the wave-

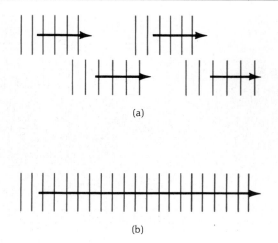

Fig. L.12 (a) The waves of light produced by different atoms are not coherent. (b) The waves produced by a laser are coherent.

Fig. L.13 Several of the lasers used at the National Bureau of Standards for a precise comparison of frequencies. (Courtesy K. M. Evenson, National Bureau of Standards.)

length. A laser of this kind can serve as an excellent standard of length and also of frequency.

As mentioned in Chapter 1, stabilized lasers are used to implement the new definition of the meter. For this purpose, the frequency of the laser light must be measured in terms of the frequency of the cesium atomic clock that serves as our standard of time. The wavelength of the laser light can then be calculated by multiplying the inverse of the frequency by the standard value of the speed of light specified in the definition of the meter, $c = 2.99792458 \times 10^8$ m/s. However, it is difficult to measure the frequency of a laser with precision: even for a long-wavelength laser (infrared laser) the frequency is several orders of magnitude higher than that of the cesium clock, and direct comparison is therefore impossible. Scientists at the National Bureau of Standards overcame this obstacle by using several pairs of lasers of different frequencies and cleverly exploiting beat frequencies and harmonic frequencies generated when the light waves struck suitable detectors of light (diodes). With a chain of four pairs of lasers (Figure L.13), they were able to span the range of frequencies from the cesium atomic clock to a methane-stabilized laser, and thereby determine the frequency of the laser light to nine significant figures.

Laser Spectroscopy
The development of tunable lasers has led to spectacular advances in spectroscopy. **Tunable lasers** operate with molecules of organic dyes, which emit a large number of closely spaced spectral lines; each of the spectral lines contains a spread of frequencies[3] and

---

[3] All spectral lines contain some small spread of frequencies; in organic dyes, this spread is exceptionally large.

therefore overlaps to some extent with the adjacent spectral lines. Thus the frequencies available for lasing in an organic dye span a continuous range, and the selective device attached to the laser — such as the prism described in Section L.2 — can be used to tune the laser to any chosen frequency in this range. When the beam from the laser strikes a sample of atoms, it will trigger resonant quantum jumps of the electrons, provided that the frequency of the laser light coincides with the frequency of a spectral line of these atoms. Thus, the laser light will be strongly absorbed if, and only if, its frequency coincides with that of an atomic spectral line; this provides a sensitive and convenient method for the measurement of the frequency of the spectral line. One of the great advantages of this method is that, with a clever arrangement of two laser beams passing through the sample in opposite directions, it is possible to selectively trigger absorption in only those atoms of the sample that have zero velocity. This eliminates the Doppler shift usually present in spectral lines, and therefore permits an extremely precise measurement of their frequencies. Recent measurements of the frequencies of the spectral lines of hydrogen atoms by this technique have led to the best available determination of the Rydberg constant.

Laser Fusion
The most powerful lasers have been developed in an attempt to extract energy from the thermonuclear fusion of heavy hydrogen (deuterium and tritium). At a temperature of 10^8 K, the violent thermal collisions between the nuclei of heavy hydrogen merge them together into helium. This reaction releases a large amount of heat (see Interlude J for details of this reaction).

In the laser-fusion reactors now under develop-

Fig. L.14 Neodymium lasers at Lawrence Livermore Laboratory.

Fig. L.15 Target chamber at Lawrence Livermore Laboratory. Laser beams, within tubes, converge toward the center of this chamber.

ment, a small pellet of deuterium–tritium mixture is dropped into a combustion chamber where the intense beam of a very powerful laser is suddenly focused on it. The sudden influx of energy not only heats the pellet to 10^8 K, but also compresses its inner core because the sudden vaporization of the outer layers generates high pressure, which causes these layers to explode outward and implode inward. This increases the density of the core by a factor of about 1000. The high temperature and high density ignite thermonuclear reactions within the pellet. These reactions go on for a short time; then the pellet flies apart. Thus, the fusion energy is released in a miniature explosion — the pellet acts as a miniature H-bomb. The essential difference between the H-bomb and the pellet is the method of ignition; fusion in the H-bomb is ignited by an A-bomb, whereas fusion in the pellet is ignited by the laser beam.

An experimental laser fusion device, called **Shiva,** at the Lawrence Livermore Laboratory employs 20 neodymium lasers focused into a target chamber (Figures L.14 and L.15). The lasers deliver a total power of more than 2×10^{10} kW in a short pulse. The radiant energy strikes a deuterium–tritium pellet about 0.1 mm in diameter. Figure L.16 shows some of these pellets; they are small glass balloons filled with deuterium and tritium under high pressure; each holds an energy equivalent to that of a barrel of oil. Figure L.17 shows

Fig. L.16 Small glass balloons, or microballoons, filled with a mixture of deuterium and tritium.

the compression of one of these pellets. The Shiva device is able to trigger miniature thermonuclear explosions, but the energy output is below the energy input required by the lasers. Livermore will build a new device, called **Nova,** with lasers 10 times more powerful. This device is expected to reach the break-even point — the energy output of the explosions is expected to equal the energy input of the lasers. Of course, it will be necessary to go beyond this point and it will also be necessary to build accessories that convert the heat from the fusion into electric energy.

L.4 HOLOGRAPHY

The large-scale coherence and high intensity of laser light has made it possible to take three-dimensional photographic pictures of objects. The photographic plate, or **hologram,** is prepared by illuminating the object with laser light. Figure L.18 shows the arrangement of light source, object, and photographic plate. The laser beam is split into a reference beam and an object beam, and these are allowed to interfere at the photographic plate. The plate records the pattern of interference fringes (Figure L.19). After developing the

plate, we shine laser light on it. If we then view the plate from the far side, we will see an exact replica of the original object. The image is three dimensional; by moving our eye up, down, or sideways we can bring the top, bottom, or sides of the object into view (Figure L.20). But, of course, the image is virtual — if we try to touch the ghostly thing behind the hologram with a finger, we find that nothing is there.

The hologram is produced by the interference between the complicated wave fronts emerging from the

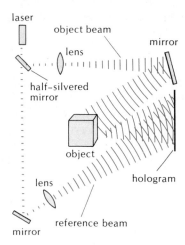

Fig. L.18 Arrangement for making a hologram. All points of the illuminated object act as sources of spherical waves. For the sake of simplicity, only one spherical wave originating at one point of the object has been shown.

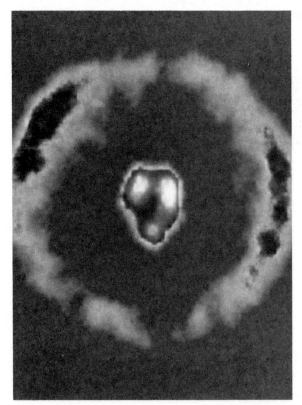

Fig. L.17 X-ray picture of a deuterium–tritium fuel pellet that has been compressed by an intense pulse of light from the Shiva laser at Livermore. The pellet has reached a temperature of 10^7 K and it shines in its own X-ray light. Fusion reactions are occurring within the pellet.

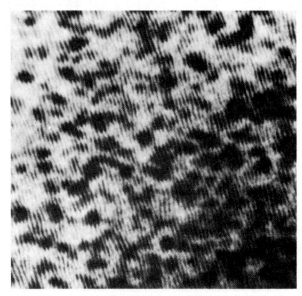

Fig. L.19 A hologram. The dark and the light fringes on this photographic plate record the constructive and the destructive interference of reference and object beams.

illuminated object and the plane wave fronts of the reference beam. To understand how this hologram generates a three-dimensional image, suppose that instead of a complicated object we have a very simple object: a plane mirror (Figure L.21). The wave fronts emerging from the illuminated mirror are then simply plane waves and, at the photographic plate, these plane waves interfere with the plane waves of the reference beam. Obviously, this interference will give a fringe pattern consisting of parallel bright and dark bands. If the angle of incidence of the object beam is α, then the distance d between one dark fringe and the next is related to the wavelength λ by (Figure L.22)

$$d \sin \alpha = \lambda \qquad (1)$$

The developed photographic plate will then look like a grating with a distance d between the slits. If we illuminate this grating with laser light, constructive interference from these slits will yield several diffracted beams of maximum intensity (Figure L.23). The angle of the first-order beam will satisfy the condition [see Eq. (38.21)]

$$d \sin \theta = \lambda \qquad (2)$$

Comparing Eqs. (1) and (2), we recognize that the angle of emergence of the diffracted beam (produced by the illuminated hologram) coincides with the angle of incidence of the original beam (produced by the illuminated mirror). Thus, the hologram *reconstructs* the wave fronts — it generates a light wave that has exactly the same characteristics as the light wave emitted by the object that was photographed. If we place our eyes beyond the hologram (Figure L.20), we will see exactly what we would see if instead of the illuminated hologram we had the illuminated object in front of our eyes.

It can be demonstrated that the same reconstruction of wave fronts obtains if the illuminated object has some complicated shape, so that the emitted wave fronts also have some complicated shape. The illuminated hologram will then generate a wave with exactly

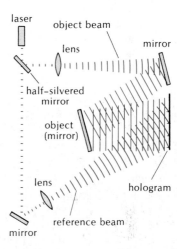

Fig. L.21 Arrangement for making a hologram. The object is a mirror. The marks on the hologram show the points of constructive interference.

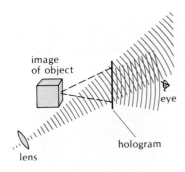

Fig. L.20 Arrangement for viewing a hologram.

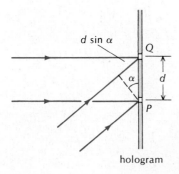

Fig. L.22 The object beam (horizontal) and the reference beam (at an angle α) reaching the photographic plate. If there is constructive interference at P, then there is also constructive interference at Q, provided the extra distance $d \sin \alpha$ equals one wavelength.

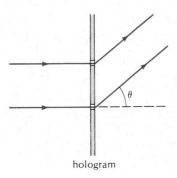

Fig. L.23 Diffracted beam (at an angle θ) emerging from the illuminated hologram.

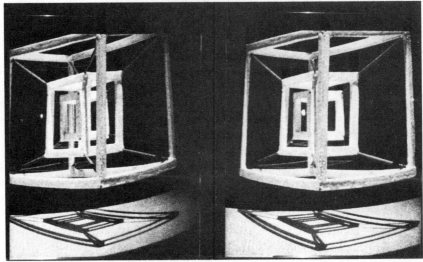

Fig. L.24 The holographic image of a piece of modern sculpture has been photographed from two different directions, giving different points of view. (If you can de-couple your eyes, and view the left image with the left eye and the right image with the right eye, you will be able to perceive these photos as three dimensional.)

the same complicated shape, and our eyes cannot tell the difference between this reconstructed wave and the wave emitted by the object itself — when we view the hologram, we believe we see the object with all its three-dimensional features (Figure L.24).

A remarkable aspect of this imaging process is that each piece of a hologram contains enough information for the reconstruction. We can cut a hologram in two or more pieces and each piece will retain the capability of giving an entire view of the three-dimensional image. However, small pieces provide a view from only a narrow range of directions, like a view through a small window.

Note that, in principle, we could make a hologram with light from an ordinary monochromatic light source (e.g., a mercury lamp) instead of light from a laser. As in the investigation of interference and diffraction by thin slits, such a light source must be placed either at a very large distance from the object or else covered with a shield with a single pinhole (see Section 38.4). But this would reduce the intensity of the light and would require very long and very impractical exposure times. Even with light from a laser, the exposure times for holography are much longer than for photography. Because of this, it is usually necessary to take special precautions to hold the object stationary during exposure.

Holographic images make spectacular displays for demonstrations, but they also have many practical applications. They can be used for sensitive measurements of small deformations and displacements. For

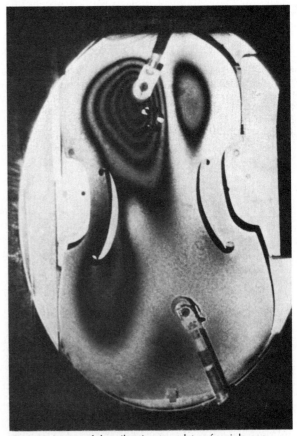

Fig. L.25 Image of the vibrating top plate of a viola generated by a time-exposure hologram.

Fig. L.26 Image of shock waves around a bullet in flight, generated by a doubly exposed hologram.

example, Figure L.25 is a holographic image of the vibrating body of a viola. The interference fringes indicate contours of equal height and therefore reveal the pattern of vibration of the viola. The exposure time in Figure L.25 was much longer than the period of vibration of the viola; since a vibrating body spends most of its time at the turning points of the motion (where the velocity is zero), a time exposure is then essentially equivalent to a double exposure — one exposure at each turning point. When we illuminate such a doubly exposed hologram, it will reconstruct simultaneously the wave fronts emitted by the body at the two separate instants of exposure. These two wave fronts will interfere constructively or destructively, depending on how far the surface of the body has moved between one exposure and the next; thus, in the holographic image, the displacement of the surface becomes visible by interference fringes. Scientists hope that the investigation of the pattern of vibration of the viola will lead to improvements in its loudness, dynamic range, and playing ease.

A double-exposure hologram can also make visible small alterations of the density of air. Figure L.26 shows the shock waves surrounding a bullet speeding through air. The hologram was first exposed before the bullet entered the field of view, and then again when the bullet arrived. The compression and rarefaction of the air near the shock fronts alters the index of refraction and hence changes the phase of light between one exposure and the next. When the reconstructed wave fronts from this hologram interfere, the variations of index of refraction show up as bright and dark zones. Incidentally, the exposures were made

with high-intensity laser pulses of very short duration so as to "stop" the action at one instant.

Another very promising application of holography is information storage. Obviously, we can store the information of, say, a printed page by taking a holographic photograph of the page. The hologram can store the information at high density because it can be made as small or smaller than an ordinary microfilm photograph, and it has the further advantage that specks of dust or small defects in the film will not block out the words on the page (remember that each part of the hologram can reconstruct the entire image). In principle, a three-dimensional hologram would permit the most efficient storage of information. In a three-dimensional medium, such as a thick photographic emulsion or a light-sensitive crystal, many different holograms can be stored successively by means of multiple exposures with light waves incident at different angles. To recover any one of the holographic images, it is necessary only to illuminate the medium with a light wave incident at the same angle as the original wave. About 1000 different holograms could be stored in a layer about 1 mm thick. With the right technology, it should be possible to store an entire 3D movie in a cube of transparent material as small as a cube of sugar.

Holography was originally invented by D. Gabor[4] as a means for increasing the power of microscopes. Gabor intended to photograph holographically the image produced by a microscope with light of short

[4] Dennis Gabor, 1900–1979, British physicist. He received the Nobel Prize in 1971 for the discovery of the principles of holography.

wavelength and afterward illuminate the hologram with light of long wavelength. The reconstructed image would then be larger than the original image in the ratio of the wavelengths. This technique of holographic microscopy was employed to make the photograph of the neon atom shown in Figure 22.2. The atom was first illuminated with electron waves of very short wavelength,[5] which generated a hologram on a photographic plate. The hologram was then illuminated with visible light, of a wavelength much larger than that of the electron waves. This yielded the enormous magnification in Figure 22.2.

Further Reading

Lasers and Holography by W. E. Kock (Doubleday, Garden City, 1969) is an elementary introduction to coherent light, its generation by lasers, and its application in holography. *The Amazing Lazer* by B. Bova (Westminster, Philadelphia, 1971) is another elementary, and very readable, introduction.

Engineering Applications of Lasers and Holography by W. E. Kock (Plenum Press, New York, 1975) and *Laser* by J. Hecht and D. Teresi (Ticknor & Fields, New Haven, 1982) give broad surveys of many practical applications of lasers.

The following magazine articles describe lasers and their applications:

"Laser Light," A. L. Schawlow, *Scientific American,* September 1968

"Applications of Laser Light," D. R. Herriott, *Scientific American,* September 1968

"The Lunar Laser Reflector," J. E. Faller and E. J. Wampler, *Scientific American,* March 1970

"Progress in Holography," E. N. Leith and J. Upatnieks, *Physics Today,* March 1972

"Metal-Vapor Lasers," W. T. Silfvast, *Scientific American,* February 1973

"Ultrafast Phenomena in Liquids and Solids," R. R. Alfano and S. L. Shapiro, *Scientific American,* June 1973

"Laser-induced Thermonuclear Fusion," J. Nuckolls, J. Emmet, and L. Wood, *Physics Today,* August 1973

"Fusion Power by Laser Implosion," J. L. Emmet, J. Nuckolls, and L. Wood, *Scientific American,* June 1974

"X-Ray Lasers," G. Chapline and L. Wood, *Physics Today,* June 1975

"White-Light Holograms," E. N. Leigh, *Scientific American,* October 1980

"Processing Materials with Lasers," E. M. Breinan, B. H. Kear, and C. M. Banas, *Physics Today,* November 1976

"Laser Separation of Isotopes," R. N. Zare, *Scientific American,* February 1977

"Light-Wave Communications," W. S. Boyle, *Scientific American,* August 1977

"Laser Fusion," C. M. Stickley, *Physics Today,* May 1978

"Cosmic Masers," D. F. Dickinson, *Scientific American,* June 1978

"Guided-Wave Optics," A. Yariv, *Scientific American,* January 1979

"Laser Chemistry," A. M. Ronn, *Scientific American,* May 1979

"Counting the Atoms," G. S. Hurst, M. G. Pane, S. D. Kramer, and C. H. Chen, *Physics Today,* September 1980

"Laser Weapons," K. Tsipis, *Scientific American,* December 1981

"Laser Applications in Manufacturing," A. V. La Rocca, *Scientific American,* March 1982

"Laser Annealing of Silicon," J. M. Poate and W. L. Brown, *Physics Today,* June 1978

"The Feasibility of Inertial-Confinement Fusion," J. N. Nuckolls, *Physics Today,* September 1982.

"Lasers and Physics: A Pretty Good Hint," A. L. Schawlow, *Physics Today,* December 1982

The Resource Letter L-1 in the *American Journal of Physics,* October 1981, gives a comprehensive list of sources of information on lasers.

Questions

1. Consider two overlapping waves with equal amplitudes and with a random phase difference. Explain why the superposition of these waves has an average intensity equal to twice the average intensity of each wave.

2. Consider two overlapping waves with equal amplitudes and with equal phases. Explain why the superposition of these waves has an intensity four times as large as the intensity of each individual wave. Does this violate energy conservation?

3. Suppose that a light source consists of 10^{16} atoms. If these atoms radiate coherently, by what factor is the average intensity larger than if they radiate incoherently?

4. Can you guess what the acronym *maser* stands for?

5. Why can the flash of light that pumps electrons upward into excited orbits also pump them downward?

6. It is easiest to attain a population inversion by optical pumping if the lower excited orbit (see orbit 2, Figure L.3) is a **metastable** orbit, i.e., an orbit in which the electron tends to remain for a long time without performing a spontaneous downward jump. Why does this help?

7. Explain how a grating can be used as a color-selective device in a laser, instead of the prism shown in Figure L.6.

8. The beam of a laser can be made visible by blowing chalk dust or smoke in its path. Can this also be done with a beam of ordinary, incoherent light?

9. For which of the applications of laser light described in Section L.3 is the coherence essential? For which is the pure color essential?

10. Because of the pull of gravity, a tightly stretched string

---

[5] As was pointed out in Chapter 41, on the atomic scale electrons behave as waves; their wavelength is usually much shorter than that of light. The operation of electron microscopes takes advantage of these electron waves.

sags downward slightly, whereas a laser beam "sags" upward (however, on the Earth the amount of this "sag" is too small to be measurable). How does this difference arise?

11. The U.S. Department of Defense is trying to develop a laser gun. What might be the advantages and disadvantages of such a gun over a conventional gun? Explain how a cloud would stop the beam of a laser gun, even though it cannot stop the bullet from a conventional gun.

12. One obstacle to shooting intense laser beams through the atmosphere at a distant target is that the beam heats the air, changing its index of refraction. This causes the beam to spread out laterally, an effect called **thermal blooming.** Explain.

13. Could an intense laser beam be used to propel a rocket? For this purpose would a laser beam be any better than a light beam?

14. If the individual wave trains of light emitted by individual atoms have a length of a few meters in space, how long are they in time?

15. The wavelength of the light emitted by a laser can be determined to nine significant figures. If you use the light from this laser in conjunction with an interferometer to measure a displacement of roughly 1 m, what precision can you expect?

Assume that the interferometer introduces no additional uncertainties.

16. The frequency of a methane-stabilized laser is 8.84×10^{13} Hz. By what factor is this larger than the frequency of the cesium atomic clock?

17. Suppose that a hologram has been prepared with green light. If you illuminate this hologram simultaneously with green light and with red light, what do you see?

18. Why is the image produced by a very small piece of a hologram somewhat fuzzy? (Hint: A hologram can be regarded as a grating. According to Section 38.5, the angular resolution of a grating increases in direct proportion with the number of slits, that is, with the size of the grating.)

19. The automatic checkout system found in large supermarkets sweeps many beams of laser light over the bottom and sides of a package, so that at least one of the beams is reflected off the product-code stripes with sufficient intensity to register in the light detector. To generate these many beams of light, the system requires a complicated array of mirrors; but instead of actual mirrors, the system uses a hologram. Explain how you could make a hologram that simulates a complicated array of mirrors.

APPENDIX 1: INDEX TO TABLES

Note: In the two-volume edition, all pages after 530 are in Vol. 2 except those in parentheses, which are in Vol. 1. Interlude F is the last in Vol. 1.

APPENDIX 2: MATHEMATICAL SYMBOLS AND FORMULAS

A2.1 Symbols

$a = b$ means a equals b

$a \neq b$ means a is not equal to b

$a > b$ means a is greater than b

$a < b$ means a is less than b

$a \geq b$ means a is not less than b

$a \leq b$ means a is not greater than b

$a \propto b$ means a is proportional to b

$a \cong b$ means a is approximately equal to b

$a \sim b$ means a is of the order of magnitude of b, i.e., a is within a factor of 10 or so of b

$a \gg b$ means a is much larger than b

$a \ll b$ means a is much less than b

$\Sigma_i a_i$ stands for the sum $a_1 + a_2 + a_3 + a_4 + \cdots$

$n!$ (or "n factorial") stands for the product $1 \cdot 2 \cdot 3 \cdots n$

$\pi = 3.14159 \ldots$

$e = 2.71828 \ldots$

A2.2 The Quadratic Equation

The quadratic equation $ax^2 + bx + c = 0$ has two solutions:

$$x = \frac{-b \pm \sqrt{b^2 - 4ac}}{2a} \tag{1}$$

A2.3 Some Approximations

The following approximations are valid for small values of x, that is, $x \ll 1$:

$$(1 + x)^n \cong 1 + nx \tag{2}$$

$$(1 + x)^{\frac{1}{2}} \cong 1 + \frac{x}{2} \tag{3}$$

$$\frac{1}{(1 + x)} \cong 1 - x \tag{4}$$

$$\frac{1}{(1 + x)^{\frac{1}{2}}} \cong 1 - \frac{x}{2} \tag{5}$$

These approximations can be derived from the binomial expansion

$$(1 + x)^n = 1 + nx + \frac{n(n - 1)}{2!}x^2 + \frac{n(n - 1)(n - 2)}{3!}x^3 + \cdots \tag{6}$$

by neglecting all powers of x, except the first power.

A2.4   The Exponential and the Logarithmic Functions

The **exponential function** $\exp(x)$ is defined by the following infinite series:

$$\exp(x) = 1 + x + \frac{x^2}{2!} + \frac{x^3}{3!} + \frac{x^4}{4!} + \cdots \tag{7}$$

This function is equivalent to raising the constant $e = 2.71828 \ldots$ to the power x,

$$\exp(x) = e^x \tag{8}$$

The **natural logarithmic function** is the inverse of the exponential function, so that

$$x = e^{\ln x} \tag{9}$$

or

$$x = \ln(e^x) \tag{10}$$

Natural logarithms obey the usual rules for logarithms,

$$\ln(x \cdot y) = \ln x + \ln y \tag{11}$$

$$\ln\left(\frac{x}{y}\right) = \ln x - \ln y \tag{12}$$

$$\ln(x^a) = a \ln x \tag{13}$$

Note that

$$\ln e = 1 \tag{14}$$

and

$$\ln 10 = 2.3026 \ldots \tag{15}$$

If we designate the **common,** or base-10, **logarithm** by log x, then the relationship between the two kinds of logarithm is as follows:

$$\ln x = \ln(10^{\log x}) = (\log x)(\ln 10) = 2.3026 \log x \tag{16}$$

APPENDIX 3: PERIMETERS, AREAS, AND VOLUMES

[perimeter of a circle of radius r] $= 2\pi r$
[area of a circle of radius r] $= \pi r^2$
[area of a triangle of base b, altitude h] $= hb/2$
[surface of a sphere of radius r] $= 4\pi r^2$
[volume of a sphere of radius r] $= 4\pi r^3/3$
[curved surface of a cylinder of radius r, height h] $= 2\pi rh$
[volume of a cylinder of radius r, height h] $= \pi r^2 h$

A4.1 Angles

The angle between two intersecting straight lines is defined as the fraction of a complete circle included between these lines (Figure A4.1). To express the angle in **degrees,** we assign an angular magnitude of 360° to the complete circle; any arbitrary angle is then an appropriate fraction of 360°. To express the angle in **radians,** we assign an angular magnitude of 2π radian to the complete circle; any arbitrary angle is then an appropriate fraction of 2π. For example, the angle shown in Figure A4.1 is $\frac{1}{8}$ of a complete circle, that is, 45°, or $\pi/4$ radian. In view of the definition of angle, the length of arc included between the two intersecting straight lines is proportional to the angle θ between these lines; if the angle is expressed in radians, then the constant of proportionality is simply the radius:

$$s = r\theta \tag{1}$$

Since 2π radian $= 360°$, it follows that

$$1 \text{ radian} = \frac{360°}{2\pi} = \frac{360°}{2 \times 3.14159} = 57.2958° \tag{2}$$

Each degree is divided into 60 minutes of arc (arcminutes), and each of these into 60 seconds of arc (arcseconds). In degrees, minutes of arc, and seconds of arc, the radian is

$$1 \text{ radian} = 57° \ 17' \ 44.8'' \tag{3}$$

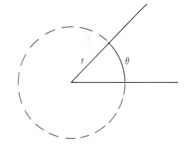

Fig. A4.1

A4.2 The Trigonometric Functions

The trigonometric functions of an angle are defined as ratios of the lengths of the sides of a right triangle erected on this angle. Figure A4.2 shows an acute angle θ and a right triangle, one of whose angles coincides with θ. The adjacent side OQ has a length x, the opposite side

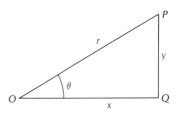

Fig. A4.2

QP a length y, and the hypothenuse OP a length r. The **sine, cosine, tangent, cotangent, secant,** and **cosecant** of the angle θ are then defined as follows:

$$\sin \theta = y/r \tag{4}$$

$$\cos \theta = x/r \tag{5}$$

$$\tan \theta = y/x \tag{6}$$

$$\cot \theta = x/y \tag{7}$$

$$\sec \theta = r/x \tag{8}$$

$$\csc \theta = r/y \tag{9}$$

EXAMPLE 1.   Find the sine, cosine, and tangent for angles of $0°$, $90°$, and $45°$.

SOLUTION: For an angle of $0°$, the opposite side is zero ($y = 0$), and the adjacent side coincides with the hypothenuse ($x = r$). Hence

$$\sin 0° = 0 \quad \cos 0° = 1 \quad \tan 0° = 0 \tag{10}$$

For an angle of $90°$, the adjacent side is zero ($x = 0$), and the opposite side coincides with the hypothenuse ($y = r$). Hence

$$\sin 90° = 1 \quad \cos 90° = 0 \quad \tan 90° = \infty \tag{11}$$

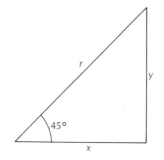

Fig. A4.3

Finally, for an angle of $45°$ (Figure A4.3), the adjacent and the opposite sides have the same length ($x = y$) and the hypothenuse has a length of $\sqrt{2}$ times the length of either side ($r = \sqrt{2}x = \sqrt{2}y$). Hence

$$\sin 45° = \frac{1}{\sqrt{2}} \quad \cos 45° = \frac{1}{\sqrt{2}} \quad \tan 45° = 1 \tag{12}$$

Table A4.1 lists the values of the trigonometric functions sine, cosine, and tangent for angles ranging from $0°$ to $90°$.

The definitions (4)–(9) are also valid for angles greater than $90°$, such as the angle shown in Figure A4.4. In the general case, the quantities x and y must be interpreted as the rectangular coordinates of the point P. For any angle larger than $90°$, one or both of the coordinates x and y are negative. Hence some of the trigonometric functions will also be negative. For instance,

$$\sin 135° = \frac{1}{\sqrt{2}} \quad \cos 135° = -\frac{1}{\sqrt{2}} \quad \tan 135° = -1 \tag{13}$$

Figure A4.5 shows plots of the sine, cosine, and tangent vs. θ for the range $0°$–$360°$.

If the angle θ is small, say, $\theta < 0.2$ radian, or $\theta < 10°$, then the length of the opposite side is approximately equal to the length of the circular arc (Figure A4.6). If we express the angle in radians, this length is [see Eq. (1)]

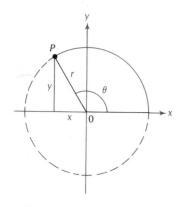

Fig. A4.4

$$y \cong s = r\theta \tag{14}$$

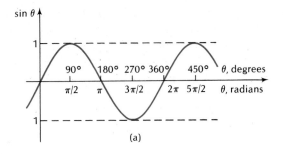

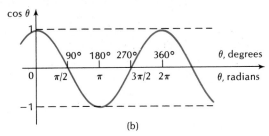

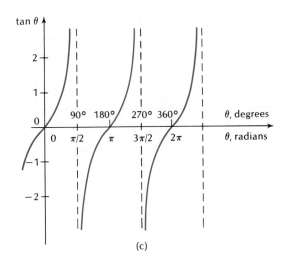

Fig. A4.5

and therefore

$$\sin \theta \cong \theta \tag{15}$$

To obtain a corresponding approximation for $\cos \theta$, we use the Pythagorean theorem to express the adjacent side in terms of the hypotenuse r and the opposite side $y \cong r\theta$,

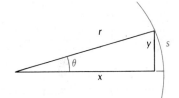

$$x = \sqrt{r^2 - y^2} \cong \sqrt{r^2 - r^2\theta^2} \tag{16}$$

$$\cong r\sqrt{1 - \theta^2} \tag{17}$$

Fig. A4.6

For small θ, Eq. (A2.3) gives us the approximation

$$\sqrt{1 - \theta^2} \cong 1 - \theta^2/2 \tag{18}$$

so that

$$\cos \theta = x/r \cong 1 - \theta^2/2 \tag{19}$$

When using the approximate formulas (15) and (19), we must always remember that the angle θ is expressed in radians!

The inverse trigonometric functions **sin⁻¹**, **cos⁻¹**, and **tan⁻¹** give the angle that corresponds to a specified value of the sine, cosine, or tangent, respectively. Thus, if

$$\sin \theta = u$$

then

$$\theta = \sin^{-1} u$$

Table A4.1 VALUES OF TRIGONOMETRIC FUNCTIONS

Degrees	Radians	Sine	Tangent	Cotangent	Cosine		
0	0	0	0	∞	1.0000	1.5708	90
1	.0175	.0175	.0175	57.290	.9998	1.5533	89
2	.0349	.0349	.0349	28.636	.9994	1.5359	88
3	.0524	.0523	.0524	19.081	.9986	1.5184	87
4	.0698	.0698	.0699	14.301	.9976	1.5010	86
5	.0873	.0872	.0875	11.430	.9962	1.4835	85
6	.1047	.1045	.1051	9.5144	.9945	1.4661	84
7	.1222	.1219	.1228	8.1443	.9925	1.4486	83
8	.1396	.1392	.1405	7.1154	.9903	1.4312	82
9	.1571	.1564	.1584	6.3138	.9877	1.4137	81
10	.1745	.1736	.1763	5.6713	.9848	1.3963	80
11	.1920	.1908	.1944	5.1446	.9816	1.3788	79
12	.2094	.2079	.2126	4.7046	.9781	1.3614	78
13	.2269	.2250	.2309	4.3315	.9744	1.3439	77
14	.2443	.2419	.2493	4.0108	.9703	1.3265	76
15	.2618	.2588	.2679	3.7321	.9659	1.3090	75
16	.2793	.2756	.2867	3.4874	.9613	1.2915	74
17	.2967	.2924	.3057	3.2709	.9563	1.2741	73
18	.3142	.3090	.3249	3.0777	.9511	1.2566	72
19	.3316	.3256	.3443	2.9042	.9455	1.2392	71
20	.3491	.3420	.3640	2.7475	.9397	1.2217	70
21	.3665	.3584	.3839	2.6051	.9336	1.2043	69
22	.3840	.3746	.4040	2.4751	.9272	1.1868	68
23	.4014	.3907	.4245	2.3559	.9205	1.1694	67
24	.4189	.4067	.4452	2.2460	.9135	1.1519	66
25	.4363	.4226	.4663	2.1445	.9063	1.1345	65
26	.4538	.4384	.4877	2.0503	.8988	1.1170	64
27	.4712	.4540	.5095	1.9626	.8910	1.0996	63
28	.4887	.4695	.5317	1.8807	.8829	1.0821	62
29	.5061	.4848	.5543	1.8040	.8746	1.0647	61
30	.5263	.5000	.5774	1.7321	.8660	1.0472	60
31	.5411	.5150	.6009	1.6643	.8572	1.0297	59
32	.5585	.5299	.6249	1.6003	.8480	1.0123	58
33	.5760	.5446	.6494	1.5399	.8387	.9948	57
34	.5934	.5592	.6745	1.4826	.8290	.9774	56
35	.6109	.5736	.7002	1.4281	.8192	.9599	55
36	.6283	.5878	.7265	1.3764	.8090	.9425	54
37	.6458	.6018	.7536	1.3270	.7986	.9250	53
38	.6632	.6157	.7813	1.2799	.7880	.9076	52
39	.6807	.6293	.8098	1.2349	.7771	.8901	51
40	.6981	.6428	.8391	1.1918	.7660	.8727	50
41	.7156	.6561	.8693	1.1504	.7547	.8552	49
42	.7330	.6691	.9004	1.1106	.7431	.8378	48
43	.7505	.6820	.9325	1.0724	.7314	.8203	47
44	.7679	.6947	.9657	1.0355	.7193	.8029	46
45	.7854	.7071	1.0000	1.0000	.7071	.7854	45
		Cosine	Cotangent	Tangent	Sine	Radians	Degrees

A4.3   Trigonometric Identities

From the definitions (4)–(9) we immediately find the following identities:

$$\tan \theta = \sin \theta / \cos \theta \qquad (20)$$

$$\cot \theta = 1/\tan \theta \qquad (21)$$

$$\sec \theta = 1/\cos \theta \qquad (22)$$

$$\csc \theta = 1/\sin \theta \qquad (23)$$

Figure A4.7 shows a right triangle with angles θ and $90° - \theta$. Since the adjacent side for the angle θ is the opposite side for the angle $90° - \theta$ and vice versa, we see that the trigonometric functions also obey the following identities:

$$\sin(90° - \theta) = \cos \theta \qquad (24)$$

$$\cos(90° - \theta) = \sin \theta \qquad (25)$$

$$\tan(90° - \theta) = \cot \theta = 1/\tan \theta \qquad (26)$$

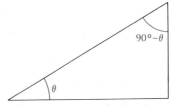

Fig. A4.7

According to the Pythagorean theorem, $x^2 + y^2 = r^2$. With $x = r \cos \theta$ and $y = r \sin \theta$ this becomes $r^2 \cos^2 \theta + r^2 \sin^2 \theta = r^2$, or

$$\cos^2 \theta + \sin^2 \theta = 1 \qquad (27)$$

The following are a few other trigonometric identities, which we state without proof:

$$\sec^2 \theta = 1 + \tan^2 \theta$$

$$\csc^2 \theta = 1 + \cot^2 \theta$$

$$\sin 2\theta = 2 \sin \theta \cos \theta$$

$$\cos 2\theta = 2 \cos^2 \theta - 1$$

$$\sin \tfrac{1}{2}\theta = \sqrt{(1 - \cos \theta)/2}$$

$$\cos \tfrac{1}{2}\theta = \sqrt{(1 + \cos \theta)/2}$$

$$\sin(\alpha + \beta) = \sin \alpha \cos \beta + \cos \alpha \sin \beta$$

$$\cos(\alpha + \beta) = \cos \alpha \cos \beta - \sin \alpha \sin \beta$$

$$\tan(\alpha + \beta) = \frac{\tan \alpha + \tan \beta}{1 - \tan \alpha \tan \beta}$$

$$\sin \alpha + \sin \beta = 2 \sin \tfrac{1}{2}(\alpha + \beta) \cos \tfrac{1}{2}(\alpha - \beta)$$

$$\cos \alpha + \cos \beta = 2 \cos \tfrac{1}{2}(\alpha + \beta) \cos \tfrac{1}{2}(\alpha - \beta)$$

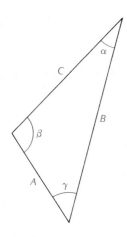

Fig. A4.8

A4.4 The Laws of Cosines and of Sines

In an arbitrary triangle the lengths of the sides and the angles obey the laws of cosines and of sines. The law of cosines states that if the lengths of two sides are A and B and the angle between them is γ (Figure A4.8), then the length of the third side is given by

$$C^2 = A^2 + B^2 - 2AB \cos \gamma \qquad (28)$$

The law of sines states that the sines of the angles of the triangle are in the same ratio as the lengths of the opposite sides (Figure A4.8):

$$\frac{\sin \alpha}{A} = \frac{\sin \beta}{B} = \frac{\sin \gamma}{C} \qquad (29)$$

Both of these laws are very useful in the calculation of unknown lengths or angles of a triangle.

A5.1 Derivatives

We saw in Section 2.3 that if the position of a particle is some function of time, say $x = x(t)$, then the instantaneous velocity of the particle is the derivative of x with respect to t,

$$v = \frac{dx}{dt} \tag{1}$$

This derivative is defined by first looking at a small increment Δx that results from a small increment Δt, and then evaluating the ratio $\Delta x / \Delta t$, in the limit when both Δx and Δt tend toward zero. Thus

$$\frac{dx}{dt} = \lim_{\Delta t \to 0} \frac{\Delta x}{\Delta t} \tag{2}$$

Graphically, in a plot of the worldline, the derivative dx/dt is the slope of the straight line tangent to the (curved) worldline at the time t (see Figure 2.6).

In general, if $f = f(u)$ is some given function of a variable u, the **derivative** of f with respect to u is defined by

$$\frac{df}{du} = \lim_{\Delta u \to 0} \frac{\Delta f}{\Delta u} \tag{3}$$

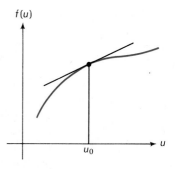

$f(u)$

u_0

u

Fig. A5.1 The derivative of $f(u)$ at u_0 is the slope of the straight line tangent to the curve at u_0.

In a plot of f vs. u, this derivative is the slope of the straight line tangent to the curve representing $f(u)$ (see Figure A5.1).

Starting with the definition (3) we can find the derivative of any function (provided the function is sufficiently smooth so that the derivative exists!). For example, consider the function $f(u) = u^2$. If we increase u to $u + \Delta u$, the function $f(u)$ increases to

$$f + \Delta f = (u + \Delta u)^2 \tag{4}$$

so that

$$\Delta f = (u + \Delta u)^2 - f = (u + \Delta u)^2 - u^2$$

$$= 2u \, \Delta u + (\Delta u)^2 \tag{5}$$

953

The derivative df/du is then

$$\frac{df}{du} = \lim_{\Delta u \to 0} \frac{\Delta f}{\Delta u} = \lim_{\Delta u \to 0} \frac{2u\,\Delta u + (\Delta u)^2}{\Delta u} \tag{6}$$

$$= \lim_{\Delta u \to 0} (2u) + \lim_{\Delta u \to 0} (\Delta u) \tag{7}$$

The second term on the right side vanishes in the limit $\Delta u \to 0$; the first term is simply $2u$. Hence

$$\frac{df}{du} = 2u \tag{8}$$

or

$$\frac{d}{du}(u^2) = 2u \tag{9}$$

This is one instance of the general rule for the differentiation of u^n,

$$\frac{d}{du}(u^n) = nu^{n-1} \tag{10}$$

This general rule is valid for any positive or negative number n, including zero. The proof of this rule can be constructed by an argument similar to that above. Table A5.1 lists the derivatives of the most common functions.

Table A5.1 SOME DERIVATIVES

$$\frac{d}{du}u^n = nu^{n-1}$$

$$\frac{d}{du}\ln u = \frac{1}{u}$$

$$\frac{d}{du}e^u = e^u$$

$$\frac{d}{du}\sin u = \cos u \qquad \text{(where } u \text{ is in } \textit{radians)}$$

$$\frac{d}{du}\cos u = -\sin u \qquad \text{(where } u \text{ is in } \textit{radians)}$$

$$\frac{d}{du}\tan u = \sec^2 u \qquad \text{(where } u \text{ is in } \textit{radians)}$$

$$\frac{d}{du}\cot u = -\csc^2 u \qquad \text{(where } u \text{ is in } \textit{radians)}$$

$$\frac{d}{du}\sec u = \tan u \sec u \qquad \text{(where } u \text{ is in } \textit{radians)}$$

$$\frac{d}{du}\csc u = -\cot u \csc u \qquad \text{(where } u \text{ is in } \textit{radians)}$$

$$\frac{d}{du}\sin^{-1} u = 1/\sqrt{1-u^2} \qquad \text{(where } u \text{ is in } \textit{radians)}$$

$$\frac{d}{du}\cos^{-1} u = -1/\sqrt{1-u^2} \qquad \text{(where } u \text{ is in } \textit{radians)}$$

$$\frac{d}{du}\tan^{-1} u = \frac{1}{1+u^2} \qquad \text{(where } u \text{ is in } \textit{radians)}$$

A5.2 Other Important Rules for Differentiation

i. Derivative of a constant times a function:

$$\frac{d}{du}(cf) = c\,\frac{df}{du} \tag{11}$$

For instance,

$$\frac{d}{du}(6u^2) = 6\,\frac{d}{du}(u^2) = 6 \times 2u = 12u$$

ii. Derivative of the sum of two functions:

$$\frac{d}{du}(f+g) = \frac{df}{du} + \frac{dg}{du} \tag{12}$$

For instance,

$$\frac{d}{du}(6u^2 + u) = \frac{d}{du}(6u^2) + \frac{d}{du}(u) = 12u + 1$$

iii. Derivative of the product of two functions:

$$\frac{d}{du}(f \times g) = g\,\frac{df}{du} + f\,\frac{dg}{du} \tag{13}$$

For instance,

$$\frac{d}{du}(u^2 \sin u) = \sin u\,\frac{d}{du}u^2 + u^2\,\frac{d}{du}\sin u$$

$$= \sin u \times 2u + u^2 \times \cos u$$

iv. Chain rule for derivatives: If f is a function of g and g is a function of u, then

$$\frac{d}{du}f(g) = \frac{df}{dg}\frac{dg}{du} \tag{14}$$

For instance, if $g = 2u$ and $f(g) = \sin g$, then

$$\frac{d}{du}\sin(2u) = \frac{d\sin(2u)}{d(2u)}\frac{d(2u)}{du}$$

$$= \cos(2u) \times 2$$

A5.3 Integrals

We have learned that if the position is known as a function of time, then we can find the instantaneous velocity by differentiation. What about the converse problem: if the instantaneous velocity is known as

$$\frac{d^3}{du^3} \sin u = \frac{d}{du} (-\sin u) = -\cos u = -1$$

$$\frac{d^4}{du^4} \sin u = \frac{d}{du} (-\cos u) = \sin u = 0, \qquad \text{etc.}$$

Hence Eq. (32) gives

$$\sin u = 0 + 1 \times (u - 0) + \frac{1}{2!} \times 0 \times (u - 0)^2 + \frac{1}{3!} \times (-1) \times (u - 0)^3$$

$$+ \frac{1}{4!} \times 0 \times (u - 0)^4 + \cdots$$

$$= u - \tfrac{1}{6} u^3 + \cdots$$

Note that for very small values of u, we can neglect all higher powers of u, so that $\sin u \cong u$, which agrees with the approximation given in Appendix 4.

APPENDIX 6: THE INTERNATIONAL SYSTEM OF UNITS (SI)

A6.1 Base Units

The SI system of units is the modern version of the metric system. The SI system recognizes seven fundamental, or base, units for length, mass, time, electric current, thermodynamic temperature, amount of substance, and luminous intensity.[a] The following definitions of the base units were adopted by the Conférence Générale des Poids et Mesures in the years indicated:

Meter (m) "The metre is the length of the path travelled by light in vacuum during a time interval of 1/299 792 458 of a second." (Adopted in 1983.)

Kilogram (kg) "The kilogram is . . . the mass of the international prototype of the kilogram." (Adopted in 1889 and in 1901.)

Second (s) "The second is the duration of 9 192 631 770 periods of the radiation corresponding to the transition between the two hyperfine levels of the ground state of the cesium-133 atom." (Adopted in 1967.)

Ampere (A) "The ampere is that constant current which, if maintained in two straight parallel conductors of infinite length, of negligible circular cross section, and placed one meter apart in vacuum, would produce between these conductors a force equal to 2×10^{-7} newton per meter of length." (Adopted in 1948.)

Kelvin (K) "The kelvin . . . is the fraction 1/273.16 of the thermodynamic temperature of the triple point of water." (Adopted in 1967.)

Mole "The mole is the amount of substance of a system which contains as many elementary entities as there are atoms in 0.012 kilogram of carbon-12." (Adopted in 1967.)

Candela (cd) "The candela is the luminous intensity, in a given direction, of a source that emits monochromatic radiation of frequency 540×10^{12} Hz and that has a radiant intensity in that direction of $\frac{1}{683}$ watt per steradian." (Adopted in 1979.)

Besides these seven base units, the SI system also recognizes two supplementary units of angle and solid angle:

Radian (rad) "The radian is the plane angle between two radii of a circle which cut off on the circumference an arc equal in length to the radius."

Steradian (sr) "The steradian is the solid angle which, having its vertex in the center of a sphere, cuts off an area equal to that of a [flat] square with sides of length equal to the radius of the sphere."

---

[a]At least two of the seven base units of the SI system are redundant. The mole is merely a certain number of atoms or molecules, in the same sense that a dozen is a number; there is no need to designate this number as a unit. The candela is equivalent to $\frac{1}{683}$ watt per steradian; it serves no purpose that is not served equally well by watt per steradian. Two other base units could be made redundant by adopting new definitions of the unit of temperature and of the unit of electric charge. Temperature could be measured in energy units because, according to the equipartition theorem, temperature is proportional to the energy per degree of freedom. Hence the kelvin could be defined as a derived unit, with 1 K $= \frac{1}{2} \times 1.38 \times 10^{-23}$ joule per degree of freedom. Electric charge could also be defined as a derived unit, to be measured with a suitable combination of the units of force and distance, as is done in the cgs system.

Furthermore, the definitions of the supplementary units — radian and steradian — are gratuitous. These definitions properly belong in the province of mathematics and there is no need to include them in a system of physical units.

A6.2 Derived Units

The derived units are formed out of products and ratios of the base units. Table A6.1 lists those derived units that have been glorified with special names.

(Other derived units are listed in the tables of conversion factors in Appendix A.7.)

Table A6.1 NAMES OF DERIVED UNITS

Quantity	Derived Unit	Name	Symbol
frequency	1/s	hertz	Hz
force	$kg \cdot m/s^2$	newton	N
pressure	N/m^2	pascal	Pa
energy	$N \cdot m$	joule	J
power	J/s	watt	W
electric charge	$A \cdot s$	coulomb	C
electric potential	J/C	volt	V
electric capacitance	C/V	farad	F
electric resistance	V/A	ohm	Ω
conductance	A/V	siemens	S
magnetic flux	$V \cdot s$	weber	Wb
magnetic field	$V \cdot s/m^2$	tesla	T
inductance	$V \cdot s/A$	henry	H
temperature	K	degree Celsius	°C
luminous flux	$cd \cdot sr$	lumen	lm
illuminance	$cd \cdot sr/m^2$	lux	lx
radioactivity	1/s	becquerel	Bq
absorbed dose	J/kg	gray	Gy
dose equivalent	J/kg	sievert	Sv

A6.3 Prefixes

Multiples and submultiples of SI units are indicated by prefixes, such as the familiar *kilo, centi,* and *milli* used in *kilometer, centimeter,* and *millimeter,* etc. Table A6.2 lists all the accepted prefixes. Some enjoy more popularity than others; it is best to avoid the use of uncommon prefixes, such as *atto* and *exa,* since hardly anybody will recognize those.

Table A6.2 PREFIXES FOR UNITS

Multiplication factor	Prefix	Symbol
10^{18}	exa	E
10^{15}	peta	P
10^{12}	tera	T
10^{9}	giga	G
10^{6}	mega	M
10^{3}	kilo	k
10^{2}	hecto	h
10	deka	da
10^{-1}	deci	d
10^{-2}	centi	c
10^{-3}	milli	m
10^{-6}	micro	μ
10^{-9}	nano	n
10^{-12}	pico	p
10^{-15}	femto	f
10^{-18}	atto	a

The units for each quantity are listed alphabetically, except that the SI unit is always listed first. The numbers are based on "American National Standard; Metric Practice" published by the Institute of Electrical and Electronics Engineers, 1982.

ANGLE

1 radian $= 57.30° = 3.438 \times 10^3\, ' = \frac{1}{2\pi}$ rev $= 2.063 \times 10^5\, ''$

1 degree (°) $= 1.745 \times 10^{-2}$ **radian** $= 60' = 3600'' = \frac{1}{360}$ rev

1 minute of arc (') $= 2.909 \times 10^{-4}$ **radian** $= \frac{1}{60}° = 4.630 \times 10^{-5}$ rev $= 60''$

1 revolution (rev) $= 2\pi$ **radian** $= 360° = 2.160 \times 10^4\, ' = 1.296 \times 10^6\, ''$

1 second of arc ('') $= 4.848 \times 10^{-6}$ **radian** $= \frac{1}{3600}° = \frac{1}{60}' =$
7.716×10^{-7} rev

LENGTH

1 meter (m) $= 1 \times 10^{10}$ Å $= 6.685 \times 10^{-12}$ AU $= 100$ cm $= 1 \times 10^{15}$ F $=$
3.281 ft $= 39.37$ in. $= 1 \times 10^{-3}$ km $= 1.057 \times 10^{-16}$ light-year $=$
$1 \times 10^6\, \mu m = 5.400 \times 10^{-4}$ nmi $= 6.214 \times 10^{-4}$ mi $=$
3.241×10^{-17} pc $= 1.094$ yd

1 angstrom (Å) $= 1 \times 10^{-10}$ m $= 1 \times 10^{-8}$ cm $= 1 \times 10^5$ F $=$
3.281×10^{-10} ft $= 1 \times 10^{-4}\, \mu m$

1 astronomical unit (AU) $= 1.496 \times 10^{11}$ m $= 1.496 \times 10^{13}$ cm $=$
1.496×10^8 km $= 1.581 \times 10^{-5}$ light-year $= 4.848 \times 10^{-6}$ pc

1 centimeter (cm) $= 0.01$ m $= 1 \times 10^8$ Å $= 1 \times 10^{13}$ F $= 3.281 \times 10^{-2}$ ft
$= 0.3937$ in. $= 1 \times 10^{-5}$ km $= 1.057 \times 10^{-18}$ light-year $= 1 \times 10^4\, \mu m$

1 fermi (F) $= 1 \times 10^{-15}$ m $= 1 \times 10^{-13}$ cm $= 1 \times 10^5$ Å

1 foot (ft) $= 0.3048$ m $= 30.48$ cm $= 12$ in. $= 3.048 \times 10^5\, \mu m =$
1.894×10^{-4} mi $= \frac{1}{3}$ yd

1 inch (in.) $= 2.540 \times 10^{-2}$ m $= 2.54$ cm $= \frac{1}{12}$ ft $= 2.54 \times 10^4\, \mu m = \frac{1}{36}$ yd

1 kilometer (km) $= 1 \times 10^3$ m $= 1 \times 10^5$ cm $= 3.281 \times 10^3$ ft $= 0.5400$ nmi
$= 0.6214$ mi $= 1.094 \times 10^3$ yd

1 light-year $= 9.461 \times 10^{15}$ m $= 6.324 \times 10^4$ AU $= 9.461 \times 10^{17}$ cm $=$
9.461×10^{12} km $= 5.879 \times 10^{12}$ mi $= 0.3066$ pc

1 micron, or micrometer (μm) $= 1 \times 10^{-6}$ m $= 1 \times 10^4$ Å $= 1 \times 10^{-4}$ cm
$= 3.281 \times 10^{-6}$ ft $= 3.937 \times 10^{-5}$ in.

1 nautical mile (nmi) $= 1.852 \times 10^3$ m $= 1.852 \times 10^5$ cm $= 6.076 \times 10^3$ ft
$= 1.852$ km $= 1.151$ mi

1 parsec (pc) $= 3.086 \times 10^{16}$ m $= 2.063 \times 10^5$ AU $= 3.086 \times 10^{18}$ cm $=$
3.086×10^{13} km $= 3.262$ light-years

1 statute mile (mi) $= 1.609 \times 10^3$ m $= 1.609 \times 10^5$ cm $= 5280$ ft $=$
1.609 km $= 0.8690$ nmi $= 1760$ yd

1 yard (yd) $= 0.9144$ m $= 91.44$ cm $= 3$ ft $= 36$ in. $= \frac{1}{1760}$ mi

TIME

1 second (s) = 1.157×10^{-5} day = $\frac{1}{3600}$ h = $\frac{1}{60}$ min = 1.161×10^{-5} sidereal day = 3.169×10^{-8} yr

1 day = 8.640×10^4 s = 24 h = 1440 min = 1.003 sidereal days = 2.738×10^{-3} yr

1 hour (h) = 3600 s = $\frac{1}{24}$ day = 60 min = 1.141×10^{-4} yr

1 minute (min) = 60 s = 6.944×10^{-4} day = $\frac{1}{60}$ h = 1.901×10^{-6} year

1 sidereal day = 8.616×10^4 s = 0.9973 day = 23.93 h = 1.436×10^3 min = 2.730×10^{-3} yr

1 year (yr) = 3.156×10^7 s = 365.24 days = 8.766×10^3 h = 5.259×10^5 min = 366.24 sidereal days

MASS

1 kilogram (kg) = 6.024×10^{26} u = 5000 carats = 1.543×10^4 grains = 1000 g = 1×10^{-3} t = 35.27 oz.-mass = 2.205 lb-mass = 1.102×10^{-3} short ton-mass = 6.852×10^{-2} slug

1 atomic mass unit (u) = 1.6605×10^{-27} kg = 1.6605×10^{-24} g

1 carat = 2×10^{-4} kg = 0.2 g = 7.055×10^{-3} oz.-mass = 4.409×10^{-4} lb-mass.

1 grain = 6.480×10^{-5} kg = 6.480×10^{-2} g = 2.286×10^{-3} oz.-mass = $\frac{1}{7000}$ lb-mass

1 gram (g) = 1×10^{-3} kg = 6.024×10^{23} u = 5 carats = 15.43 grains = 1×10^{-6} t = 3.527×10^{-2} oz.-mass = 2.205×10^{-3} lb-mass = 1.102×10^{-6} short ton-mass = 6.852×10^{-5} slug

1 metric ton (t) = 1×10^3 kg = 1×10^6 g = 2.205×10^3 lb-mass = 1.102 short ton-mass = 68.52 slugs

1 ounce-mass (oz.-mass) = 2.835×10^{-2} kg = 141.7 carats = 437.5 grains = 28.35 g = $\frac{1}{16}$ lb-mass

1 pound-mass (lb-mass)[b] = 0.4536 kg = 453.6 g = 4.536×10^{-4} t = 16 oz.-mass = $\frac{1}{2000}$ short ton-mass = 3.108×10^{-2} slug

1 short ton-mass = 907.2 kg = 9.07×10^5 g = 0.9072 t = 2000 lb-mass

1 slug = 14.59 kg = 1.459×10^4 g = 32.17 lb-mass.

AREA

1 square meter (m²) = 1×10^4 cm² = 10.76 ft² = 1.550×10^3 in.² = 1×10^{-6} km² = 3.861×10^{-7} mi² = 1.196 yd²

1 barn = 1×10^{-28} m² = 1×10^{-24} cm²

1 square centimeter (cm²) = 1×10^{-4} m² = 1.076×10^{-3} ft² = 0.1550 in.² = 1×10^{-10} km² = 3.861×10^{-11} mi²

1 square foot (ft²) = 9.290×10^{-2} m² = 929.0 cm² = 144 in.² = 3.587×10^{-8} mi² = $\frac{1}{9}$ yd²

1 square inch (in.²) = 6.452×10^{-4} m² = 6.452 cm² = $\frac{1}{144}$ ft²

1 square kilometer (km²) = 1×10^6 m² = 1×10^{10} cm² = 1.076×10^7 ft² = 0.3861 mi²

1 square statute mile (mi²) = 2.590×10^6 m² = 2.590×10^{10} cm² = 2.788×10^7 ft² = 2.590 km²

1 square yard (yd²) = 0.8361 m² = 8.361×10^3 cm² = 9 ft² = 1296 in.²

VOLUME

1 cubic meter (m³) = 1×10^6 cm³ = 35.31 ft³ = 264.2 gal. = 6.102×10^4 in.³ = 1×10^3 liters = 1.308 yd³

1 cubic centimeter (cm³) = 1×10^{-6} m³ = 3.531×10^{-5} ft³ = 2.642×10^{-4} gal. = 6.102×10^{-2} in.³ = 1×10^{-3} liter

1 cubic foot (ft³) = 2.832×10^{-2} m³ = 2.832×10^4 cm³ = 7.481 gal. = 1728 in.³ = 28.32 liters = $\frac{1}{27}$ yd³

1 gallon (gal.)[c] = 3.785×10^{-3} m³ = 0.1337 ft³

1 cubic inch (in.³) = 1.639×10^{-5} m³ = 16.39 cm³ = 5.787×10^{-4} ft³

---

[b] This is the "avoirdupois" pound. The "troy" or "apothecary" pound is 0.3732 kg, or 0.8229 lb avoirdupois.

[c] This is the U.S. gallon; the U.K. and the Canadian gallon are 4.546×10^{-3} m³, or 1.201 U.S. gallons.

1 liter (l) = 1×10^{-3} m³ = 1000 cm³ = 3.531×10^{-2} ft³
1 cubic yard (yd³) = 0.7646 m³ = 7.646×10^5 cm³ = 27 ft³ = 202.0 gal.

DENSITY

1 kilogram per cubic meter (kg/m³) = 1×10^{-3} g/cm³ =
6.243×10^{-2} lb-mass/ft³ = 8.345×10^{-3} lb-mass/gal. =
3.613×10^{-5} lb-mass/in.³ = 8.428×10^{-4} short ton-mass/yd³ =
1.940×10^{-3} slug/ft³

1 gram per cubic centimeter (g/cm³) = 1×10^3 kg/m³ = 62.43 lb-mass/ft³
= 8.345 lb-mass/gal. = 3.613×10^{-2} lb-mass/in.³ =
0.8428 short ton-mass/yd³ = 1.940 slug/ft³

1 lb-mass per cubic foot (lb-mass/ft³) = 16.02 kg/m³ =
1.602×10^{-2} g/cm³ = 0.1337 lb-mass/gal. =
1.350×10^{-2} short ton-mass/yd³ = 3.108×10^{-2} slug/ft³

1 pound-mass per gallon (1 lb-mass/gal.) = 119.8 kg/m³ =
7.481 lb-mass/ft³ = 0.2325 slug/ft³

1 short ton-mass per cubic yard (short ton-mass/yd³) =
1.187×10^3 kg/m³ = 74.07 lb-mass/ft³

1 slug per cubic foot (slug/ft³) = 515.4 kg/m³ = 0.5154 g/cm³ =
32.17 lb-mass/ft³ = 4.301 lb-mass/gal.

SPEED

1 meter per second (m/s) = 100 cm/s = 3.281 ft/s = 3.600 km/h =
1.944 knot = 2.237 mi/h

1 centimeter per second (cm/s) = 0.01 m/s = 3.281×10^{-2} ft/s =
3.600×10^{-2} km/h = 1.944×10^{-2} knot = 2.237×10^{-2} mi/h

1 foot per second (ft/s) = 0.3048 m/s = 30.48 cm/s = 1.097 km/h =
0.5925 knot = 0.6818 mi/h

1 kilometer per hour (km/h) = 0.2778 m/s = 27.78 cm/s = 0.9113 ft/s
= 0.5400 knot = 0.6214 mi/h

1 knot, or nautical mile per hour = 0.5144 m/s = 51.44 cm/s =
1.688 ft/s = 1.852 km/h = 1.151 mi/h

1 mile per hour (mi/h) = 0.4470 m/s = 44.70 cm/s = 1.467 ft/s =
1.609 km/h = 0.8690 knot

ACCELERATION

1 meter per second squared (m/s²) = 100 cm/s² = 3.281 ft/s² =
0.1020 gee

1 centimeter per second squared (cm/s²) = 0.01 m/s² =
3.281×10^{-2} ft/s² = 1.020×10^{-3} gee

1 foot per second squared (ft/s²) = 0.3048 m/s² = 30.48 cm/s² =
3.108×10^{-2} gee

1 gee = 9.807 m/s² = 980.7 cm/s² = 32.17 ft/s²

FORCE

1 newton (N) = 1×10^5 dynes = 0.1020 kp = 0.2248 lb =
1.124×10^{-4} short ton

1 dyne = 1×10^{-5} N = 1.020×10^{-6} kp = 2.248×10^{-6} lb =
1.124×10^{-9} short ton

1 kilopond, or kilogram force (kp) = 9.807 N = 9.807×10^5 dynes =
2.205 lb = 1.102×10^{-3} short ton

1 pound (lb) = 4.448 N = 4.448×10^5 dynes = 0.4536 kp = $\frac{1}{2000}$ short ton
1 short ton = 8.896×10^3 N = 8.896×10^8 dynes = 907.2 kp = 2000 lb

ENERGY

1 joule (J) = 9.478×10^{-4} Btu = 0.2388 cal = 1×10^7 ergs =
6.242×10^{18} eV = 0.7376 ft · lb = 2.778×10^{-7} kW · h

1 British thermal unit (Btu)[d] = 1.055×10^3 J = 252.0 cal =
1.055×10^{10} ergs = 778.2 ft · lb = 2.931×10^{-4} kW · h

---

[d] This is the "International Table" Btu; there are several other Btus.

1 calorie (cal)^e = 4.187 J = 3.968 $\times$ 10^{-3} Btu = 4.187 $\times$ 10^7 ergs = 3.088 ft · lb = 1 $\times$ 10^{-3} kcal = 1.163 $\times$ 10^{-6} kW · h

1 erg = 1 $\times$ 10^{-7} J = 9.478 $\times$ 10^{-7} Btu = 2.388 $\times$ 10^{-8} cal = 6.242 $\times$ 10^{11} eV = 7.376 $\times$ 10^{-8} ft · lb = 2.778 $\times$ 10^{-14} kW · h

1 electron-volt (eV) = 1.602 $\times$ 10^{-19} J = 1.602 $\times$ 10^{-12} erg = 1.182 $\times$ 10^{-19} ft · lb

1 foot-pound (ft · lb) = 1.356 J = 1.285 $\times$ 10^{-3} Btu = 0.3239 cal = 1.356 $\times$ 10^7 ergs = 8.464 $\times$ 10^{18} eV = 3.766 $\times$ 10^{-7} kW · h

1 kilocalorie (kcal), or **large calorie** (Cal) = 4.187 $\times$ 10^3 J = 1 $\times$ 10^3 cal

1 kilowatt-hour (kW · h) = 3.600 $\times$ 10^6 J = 3412 Btu = 8.598 $\times$ 10^5 cal = 3.6 $\times$ 10^{13} ergs = 2.655 $\times$ 10^6 ft · lb

POWER

1 watt (W) = 3.412 Btu/h = 0.2388 cal/s = 1 $\times$ 10^7 ergs/s = 0.7376 ft · lb/s = 1.341 $\times$ 10^{-3} hp

1 British thermal unit per hour (Btu/h) = 0.2931 W = 7.000 $\times$ 10^{-2} cal/s = 0.2162 ft · lb/s = 3.930 $\times$ 10^{-4} hp

1 calorie per second (cal/s) = 4.187 W = 14.29 Btu/h = 4.187 $\times$ 10^7 erg/s = 3.088 ft · lb/s = 5.615 $\times$ 10^{-3} hp

1 erg per second (erg/s) = 1 $\times$ 10^{-7} W = 2.388 $\times$ 10^{-8} cal/s = 7.376 $\times$ 10^{-8} ft · lb/s = 1.341 $\times$ 10^{-10} hp

1 foot-pound per second (ft · lb/s) = 1.356 W = 0.3238 cal/s = 4.626 Btu/h = 1.356 $\times$ 10^7 ergs/s = 1.818 $\times$ 10^{-3} hp

1 horsepower (hp)^f = 745.7 W = 2.544 $\times$ 10^3 Btu/h = 178.1 cal/s = 550 ft · lb/s

1 kilowatt (kW) = 1 $\times$ 10^3 W = 3.412 $\times$ 10^3 Btu/h = 238.8 cal/s = 737.6 ft · lb/s = 1.341 hp

PRESSURE

1 newton per square meter (N/m^2), or **pascal** (Pa) = 9.869 $\times$ 10^{-6} atm = 1 $\times$ 10^{-5} bar = 7.501 $\times$ 10^{-4} cmHg = 10 dynes/cm^2 = 2.089 $\times$ 10^{-2} lb/ft^2 = 1.450 $\times$ 10^{-4} lb/in.^{2} = 7.501 $\times$ 10^{-3} torr

1 atmosphere (atm) = 1.013 $\times$ 10^5 N/m^2 = 76.00 cmHg = 1.013 $\times$ 10^6 dynes/cm^2 = 2.116 $\times$ 10^3 lb/ft^2 = 14.70 lb/in.^{2}

1 bar = 1 $\times$ 10^5 N/m^2 = 0.9869 atm = 75.01 cmHg

1 centimeter of mercury (cmHg) = 1.333 $\times$ 10^3 N/m^2 = 1.316 $\times$ 10^{-2} atm = 1.333 $\times$ 10^{-2} bar = 1.333 $\times$ 10^4 dynes/cm^2 = 27.85 lb/ft^2 = 0.1934 lb/in.^{2} = 10 torr

1 dyne per square centimeter (dyne/cm^2) = 0.1 N/m^2 = 9.869 $\times$ 10^{-7} atm = 7.501 $\times$ 10^{-5} cmHg = 2.089 $\times$ 10^{-3} lb/ft^2 = 1.450 $\times$ 10^{-5} lb/in.^{2}

1 kilopond per square centimeter (kp/cm^2) = 9.807 $\times$ 10^4 N/m^2 = 0.9678 atm = 9.807 $\times$ 10^5 dynes/cm^2 = 14.22 lb/in.^{2}

1 pound per square inch (lb/in.^{2}, or psi) = 6.895 $\times$ 10^3 N/m^2 = 6.805 $\times$ 10^{-2} atm = 6.895 $\times$ 10^4 dynes/cm^2 = 7.031 $\times$ 10^{-2} kp/cm^2

1 torr, or **millimeter of mercury** (mmHg) = 1.333 $\times$ 10^2 N/m^2 = 0.1 cmHg

ELECTRIC CHARGE^g

1 coulomb (C) $\leftrightarrow$ 2.998 $\times$ 10^9 statcoulombs, or esu of charge $\leftrightarrow$ 0.1 abcoulomb, or emu of charge

---

^{e} This is the "International Table" calorie, which equals exactly 4.1868 J. There are several other calories; for instance, the thermochemical calorie, which equals 4.184 J.

^{f} There are several other horsepowers; for instance, the metric horsepower, which equals 735.5 J.

^{g} The dimensions of the electric quantities in SI units, electrostatic units (esu), and electromagnetic units (emu) are different; hence the relationships among these units are correspondences ($\leftrightarrow$) rather than equalities (=).

ELECTRIC CURRENT

1 ampere (A) $\leftrightarrow$ 2.998×10^9 statamperes, or esu of current $\leftrightarrow$ 0.1 abampere, or emu of current

ELECTRIC POTENTIAL

1 volt (V) $\leftrightarrow$ 3.336×10^{-3} statvolt, or esu of potential $\leftrightarrow$ 1×10^8 abvolts, or emu of potential

ELECTRIC FIELD

1 volt per meter (V/m) $\leftrightarrow$ 3.336×10^{-5} statvolt/cm $\leftrightarrow$ 1×10^6 abvolts/cm

MAGNETIC FIELD

1 tesla (T), or **weber per square meter** (Wb/m^2) $\leftrightarrow$ 1×10^4 gauss

ELECTRIC RESISTANCE

1 ohm (Ω) $\leftrightarrow$ 1.113×10^{-12} statohm, or esu of resistance $\leftrightarrow$ 1×10^9 abohms, or emu of resistance

ELECTRIC RESISTIVITY

1 ohm-meter ($\Omega \cdot$ m) $\leftrightarrow$ 1.113×10^{-10} statohm-cm $\leftrightarrow$ 1×10^{11} abohm-cm

CAPACITANCE

1 farad (F) $\leftrightarrow$ 8.988×10^{11} statfarads, or esu of capacitance $\leftrightarrow$ 1×10^{-9} abfarad, or emu of capacitance

INDUCTANCE

1 henry (H) $\leftrightarrow$ 1.113×10^{-12} stathenry, or esu of inductance $\leftrightarrow$ 1×10^9 abhenrys, or emu of inductance

FUNDAMENTAL CONSTANTS

Compiled by E. R. Cohen and B. N. Taylor under the auspices of the CODATA Task Group on Fundamental Constants. This set has been officially adopted by CODATA and is taken from J. Phys. Chem. Ref. Data, Vol. 2, No. 4, p. 663 (1973) and CODATA Bulletin No. 11 (December 1973).

Quantity	Symbol	Numerical Value *	Uncert. (ppm)	SI †	cgs ‡
Speed of light in vacuum	c	299792458(1.2)	0.004	m·s^{-1}	10^2 cm·s^{-1}
Permeability of vacuum	μ_0	4π =12.5663706144		10^{-7} H·m^{-1} 10^{-7} H·m^{-1}	
Permittivity of vacuum, $1/\mu_0 c^2$	ϵ_0	8.854187818(71)	0.008	10^{-12} F·m^{-1}	
Fine-structure constant, $[\mu_0 c^2/4\pi](e^2\hbar c)$	α α^{-1}	7.2973506(60) 137.03604(11)	0.82 0.82	10^{-3}	10^{-3}
Elementary charge	e	1.6021892(46) 4.803242(14)	2.9 2.9	10^{-19} C	10^{-20} emu 10^{-10} esu
Planck constant	h $\hbar = h/2\pi$	6.626176(36) 1.0545887(57)	5.4 5.4	10^{-34} J·s 10^{-34} J·s	10^{-27} erg·s 10^{-27} erg·s
Avogadro constant	N_A	6.022045(31)	5.1	10^{23} mol^{-1}	10^{23} mol^{-1}
Atomic mass unit, 10^{-3}kg·mol$^{-1}N_A^{-1}$	u	1.6605655(86)	5.1	10^{-27} kg	10^{-24} g
Electron rest mass	m_e	9.109534(47) 5.4858026(21)	5.1 0.38	10^{-31} kg 10^{-4} u	10^{-28} g 10^{-4} u
Proton rest mass	m_p	1.6726485(86) 1.007276470(11)	5.1	10^{-27} kg u	10^{-24} g u
Ratio of proton mass to electron mass	m_p/m_e	1836.15152(70)	0.38		
Neutron rest mass	m_n	1.6749543(86) 1.008665012(37)	5.1 0.037	10^{-27} kg u	10^{-24} g u
Electron charge to mass ratio	e/m_e	1.7588047(49) 5.272764(15)	2.8 2.8	10^{11} C·kg^{-1}	10^7 emu·g^{-1} 10^{17} esu·g^{-1}
Magnetic flux quantum, $[c]^{-1}(hc/2e)$	Φ_0 h/e	2.0678506(54) 4.135701(11) 1.3795215(36)	2.6 2.6 2.6	10^{-15} Wb 10^{-15} J·s·C^{-1}	10^{-7} G·cm^2 10^{-7} erg·s·emu^{-1} 10^{-17} erg·s·esu^{-1}
Josephson frequency-voltage ratio	$2e/h$	4.835939(13)	2.6	10^{14} Hz·V^{-1}	
Quantum of circulation	$h/2m_e$ h/m_e	3.6369455(60) 7.273891(12)	1.6	10^{-4} J·s·kg^{-1} 10^{-4} J·s·kg^{-1}	erg·s·g^{-1} erg·s·g^{-1}
Faraday constant, $N_A e$	F	9.648456(27) 2.8925342(82)	2.8 2.8	10^4 C·mol^{-1}	10^3 emu·mol^{-1} 10^{14} esu·mol^{-1}
Rydberg constant, $[\mu_0 c^2/4\pi]^2(m_e e^4/4\pi\hbar^3 c)$	R_∞	1.097373177(83)	0.075	10^7 m^{-1}	10^5 cm^{-1}
Bohr radius, $[\mu_0 c^2/4\pi]^{-1}(\hbar^2/m_e e^2)=\alpha/4\pi R_\infty$	a_0	5.2917706(44)	0.82	10^{-11} m	10^{-9} cm
Classical electron radius, $[\mu_0 c^2/4\pi](e^2/m_e c^2)=\alpha^3/4\pi R_\infty$	$r_e=\alpha\lambda_C$	2.8179380(70)	2.5	10^{-15} m	10^{-13} cm
Thomson cross section, $(8/3)\pi r_e^2$	σ_e	0.6652448(33)	4.9	10^{-28} m^2	10^{-24} cm^2
Free electron g-factor, or electron magnetic moment in Bohr magnetons	$g_e/2=\mu_e/\mu_B$	1.0011596567(35)	0.0035		
Free muon g-factor, or muon magnetic moment in units of $[c](e\hbar/2m_\mu c)$	$g_\mu/2$	1.00116616(31)	0.31		
Bohr magneton, $[c](e\hbar/2m_e c)$	μ_B	9.274078(36)	3.9	10^{-24} J·T^{-1}	10^{-21} erg·G^{-1}
Electron magnetic moment	μ_e	9.284832(36)	3.9	10^{-24} J·T^{-1}	10^{-21} erg·G^{-1}
Gyromagnetic ratio of protons in H_2O	γ'_p $\gamma'_p/2\pi$	2.6751301(75) 4.257602(12)	2.8 2.8	10^8 s^{-1}·T^{-1} 10^7 Hz·T^{-1}	10^4 s^{-1}·G^{-1} 10^3 Hz·G^{-1}
γ'_p corrected for diamagnetism of H_2O	γ_p $\gamma_p/2\pi$	2.6751987(75) 4.257711(12)	2.8 2.8	10^8 s^{-1}·T^{-1} 10^7 Hz·T^{-1}	10^4 s^{-1}·G^{-1} 10^3 Hz·G^{-1}
Magnetic moment of protons in H_2O in Bohr magnetons	μ'_p/μ_B	1.52099322(10)	0.066	10^{-3}	10^{-3}
Proton magnetic moment in Bohr magnetons	μ_p/μ_B	1.521032209(16)	0.011	10^{-3}	10^{-3}
Ratio of electron and proton magnetic moments	μ_e/μ_p	658.2106880(66)	0.010		
Proton magnetic moment	μ_p	1.4106171(55)	3.9	10^{-26} J·T^{-1}	10^{-23} erg·G^{-1}
Magnetic moment of protons in H_2O in nuclear magnetons	μ'_p/μ_N	2.7927740(11)	0.38		
μ'_p/μ_N corrected for diamagnetism of H_2O	μ_p/μ_N	2.7928456(11)	0.38		
Nuclear magneton, $[c](e\hbar/2m_p c)$	μ_N	5.050824(20)	3.9	10^{-27} J·T^{-1}	10^{-24} erg·G^{-1}
Ratio of muon and proton magnetic moments	μ_μ/μ_p	3.1833402(72)	2.3		
Muon magnetic moment	μ_μ	4.490474(18)	3.9	10^{-26} J·T^{-1}	10^{-23} erg·G^{-1}
Ratio of muon mass to electron mass	m_μ/m_e	206.76865(47)	2.3		

Quantity	Symbol	Numerical Value *	Uncert. (ppm)	SI †	← Units →	cgs ‡
Muon rest mass	m_μ	1.883566(11)	5.6	10^{-28} kg		10^{-25} g
		0.11342920(26)	2.3	u		u
Compton wavelength of the electron, $h/m_e c = \alpha^2/2R_\infty$	λ_C	2.4263089(40)	1.6	10^{-12} m		10^{-10} cm
	$\lambdabar_C = \lambda_C/2\pi = \alpha a_0$	3.8615905(64)	1.6	10^{-13} m		10^{-11} cm
Compton wavelength of the proton, $h/m_p c$	$\lambda_{C,p}$	1.3214099(22)	1.7	10^{-15} m		10^{-13} cm
	$\lambdabar_{C,p} = \lambda_{C,p}/2\pi$	2.1030892(36)	1.7	10^{-16} m		10^{-14} cm
Compton wavelength of the neutron, $h/m_n c$	$\lambda_{C,n}$	1.3195909(22)	1.7	10^{-15} m		10^{-13} cm
	$\lambdabar_{C,n} = \lambda_{C,n}/2\pi$	2.1001941(35)	1.7	10^{-16} m		10^{-14} cm
Molar volume of ideal gas at s.t.p.	V_m	22.41383(70)	31	10^{-3} m³·mol⁻¹		10^3 cm³·mol⁻¹
Molar gas constant, $V_m p_0/T_0$ ($T_0 \equiv 273.15$ K; $p_0 \equiv 101325$ Pa ≡ 1atm)	R	8.31441(26)	31	J·mol⁻¹·K⁻¹		10^7 erg·mol⁻¹·K⁻¹
		8.20568(26)	31	10^{-5} m³·atm·mol⁻¹·K⁻¹		10 cm³·atm·mol⁻¹·K⁻¹
Boltzmann constant, R/N_A	k	1.380662(44)	32	10^{-23} J·K⁻¹		10^{-16} erg·K⁻¹
Stefan-Boltzmann constant, $\pi^2 k^4/60\hbar^3 c^2$	σ	5.67032(71)	125	10^{-8} W·m⁻²·K⁻⁴		10^{-5} erg·s⁻¹·cm⁻²·K⁻⁴
First radiation constant, $2\pi hc^2$	c_1	3.741832(20)	5.4	10^{-16} W·m²		10^{-5} erg·cm²·s⁻¹
Second radiation constant, hc/k	c_2	1.438786(45)	31	10^{-2} m·K		cm·K
Gravitational constant	G	6.6720(41)[h]	615	10^{-11} m³·s⁻²·kg⁻¹		10^{-8} cm³·s⁻²·g⁻¹
Ratio, kx-unit to ångström, $\Lambda = \lambda(Å)/\lambda(kxu)$; $\lambda(CuK\alpha_1) \equiv 1.537400$ kxu	Λ	1.0020772(54)	5.3			
Ratio, Å* to ångström, $\Lambda^* = \lambda(Å)/\lambda(Å^*)$; $\lambda(WK\alpha_1) \equiv 0.2090100$ Å*	Λ^*	1.0000205(56)	5.6			

ENERGY CONVERSION FACTORS AND EQUIVALENTS

Quantity	Symbol	Numerical Value *	Units	Uncert. (ppm)
1 kilogram (kg·c²)		8.987551786(72)	10^{16} J	0.008
		5.609545(16)	10^{29} MeV	2.9
1 Atomic mass unit (u·c²)		1.4924418(77)	10^{-10} J	5.1
		931.5016(26)	MeV	2.8
1 Electron mass $m_e \cdot c^2$)		8.187241(42)	10^{-11} J	5.1
		0.5110034(14)	MeV	2.8
1 Muon mass ($m_\mu \cdot c^2$)		1.6928648(96)	10^{-11} J	5.6
		105.65948(35)	MeV	3.3
1 Proton mass ($m_p \cdot c^2$)		1.5033015(77)	10^{-10} J	5.1
		938.2796(27)	MeV	2.8
1 Neutron mass ($m_n \cdot c^2$)		1.5053738(78)	10^{-10} J	5.1
		939.5731(27)	MeV	2.8
1 Electron volt		1.6021892(46)	10^{-19} J	2.9
			10^{-12} erg	2.9
	1 eV/h	2.4179696(63)	10^{14} Hz	2.6
	1 eV/hc	8.065479(21)	10^5 m⁻¹	2.6
			10^3 cm⁻¹	2.6
	1 eV/k	1.160450(36)	10^4 K	31
Voltage-wavelength conversion, hc		1.986478(11)	10^{-25} J·m	5.4
		1.2398520(32)	10^{-6} eV·m	2.6
			10^{-4} eV·cm	2.6
Rydberg constant	$R_\infty hc$	2.179907(12)	10^{-18} J	5.4
			10^{-11} erg	5.4
		13.605804(36)	eV	2.6
	$R_\infty c$	3.28984200(25)	10^{15} Hz	0.075
	$R_\infty hc/k$	1.578885(49)	10^5 K	31
Bohr magneton	μ_B	9.274078(36)	10^{-24} J·T⁻¹	3.9
		5.7883785(95)	10^{-5} eV·T⁻¹	1.6
	μ_B/h	1.3996123(39)	10^{10} Hz·T⁻¹	2.8
	μ_B/hc	46.68604(13)	m⁻¹·T⁻¹	2.8
			10^{-2} cm⁻¹·T⁻¹	2.8
	μ_B/k	0.671712(21)	K·T⁻¹	31
Nuclear magneton	μ_N	5.505824(20)	10^{-27} J·T⁻¹	3.9
		3.1524515(53)	10^{-8} eV·T⁻¹	1.7
	μ_N/h	7.622532(22)	10^6 Hz·T⁻¹	2.8
	μ_N/hc	2.5426030(72)	10^{-2} m⁻¹·T⁻¹	2.8
			10^{-4} cm⁻¹·T⁻¹	2.8
	μ_N/k	3.65826(12)	10^{-4} K·T⁻¹	31

* Note that the numbers in parentheses are the one standard-deviation uncertainties in the last digits of the quoted value computed on the basis of internal consistency, that the unified atomic mass scale $^{12}C \equiv 12$ has been used throughout, that u=atomic mass unit, C=coulomb, F=farad, G=gauss, H=henry, Hz=hertz=cycle/s, J=joule, K=kelvin (degree Kelvin), Pa=pascal=N·m⁻², T=tesla (10^4 G), V=volt, Wb=weber= T·m², and W=watt. In cases where formulas for constants are given (e.g., R_∞), the relations are written as the product of two factors. The second factor, in parentheses, is the expression to be used when all quantities are expressed in cgs units, with the electron charge in electrostatic units. The first factor, in brackets, is to be included only if all quantities are expressed in SI units. We remind the reader that with the exception of the auxiliary constants which have been taken to be exact, the uncertainties of these constants are correlated, and therefore the general law of error propagation must be used in calculating additional quantities requiring two or more of these constants.

† Quantities given in u and atm are for the convenience of the reader; these units are not part of the International System of Units (SI).

‡ In order to avoid separate columns for "electromagnetic" and "electrostatic" units, both are given under the single heading "cgs Units." When using these units, the elementary charge e in the second column should be understood to be replaced by e_m or e_e, respectively.

[h] This value has been superseded by a new, more accurate measurement giving $G = 6.6726(5) \times 10^{-11}$ m³·s⁻²·kg⁻¹. The uncertainty is 65 ppm.

Table A9.1 THE PERIODIC TABLE OF THE CHEMICAL ELEMENTS[i]

Table P.1 THE PERIODIC TABLE OF CHEMICAL ELEMENTS[a]

IA																	0
1 H 1.0079	IIA											IIIA	IVA	VA	VIA	VIIA	2 He 4.00260
3 Li 6.941	4 Be 9.01218											5 B 10.81	6 C 12.011	7 N 14.0067	8 O 15.9994	9 F 18.99840	10 Ne 20.179
11 Na 22.98977	12 Mg 24.305	IIIB	IVB	VB	VIB	VIIB		VIII		IB	IIB	13 Al 26.98154	14 Si 28.0855	15 P 30.97376	16 S 32.06	17 Cl 35.453	18 Ar 39.948
19 K 39.098	20 Ca 40.08	21 Sc 44.9559	22 Ti 47.90	23 V 50.9414	24 Cr 51.996	25 Mn 54.9380	26 Fe 55.847	27 Co 58.9332	28 Ni 58.71	29 Cu 63.546	30 Zn 65.38	31 Ga 69.72	32 Ge 72.59	33 As 74.9216	34 Se 78.96	35 Br 79.904	36 Kr 83.80
37 Rb 85.4678	38 Sr 87.62	39 Y 88.9059	40 Zr 91.22	41 Nb 92.9064	42 Mo 95.94	43 Tc 98.9062	44 Ru 101.07	45 Rh 102.9055	46 Pd 106.4	47 Ag 107.868	48 Cd 112.40	49 In 114.82	50 Sn 118.69	51 Sb 121.75	52 Te 127.60	53 I 126.9045	54 Xe 131.30
55 Cs 132.9054	56 Ba 137.34	57–71 Rare Earths	72 Hf 178.49	73 Ta 180.947	74 W 183.85	75 Re 186.2	76 Os 190.2	77 Ir 192.22	78 Pt 195.09	79 Au 196.9665	80 Hg 200.59	81 Tl 204.37	82 Pb 207.2	83 Bi 208.9804	84 Po (210)	85 At (210)	86 Rn (222)
87 Fr (223)	88 Ra 226.0254	89–103 Acti-nides	104 Rf (257)	105 Ha (260)	106 (263)	107 (262)		109 (266)									

															Rare Earths (Lanthanides)
57 La 138.9055	58 Ce 140.12	59 Pr 140.9077	60 Nd 144.24	61 Pm (145)	62 Sm 150.4	63 Eu 151.96	64 Gd 157.25	65 Tb 158.9254	66 Dy 162.50	67 Ho 164.9304	68 Er 167.26	69 Tm 168.9342	70 Yb 173.04	71 Lu 174.97	

														Actinides
89 Ac (227)	90 Th 232.0381	91 Pa 231.0359	92 U 238.029	93 Np 237.0482	94 Pu (244)	95 Am (243)	96 Cm (247)	97 Bk (247)	98 Cf (251)	99 Es (254)	100 Fm (257)	101 Md (258)	102 No (259)	103 Lr (256)

[i] In each box, the upper number is the *atomic number*. The lower number is the *atomic mass,* i.e., the mass (in grams) of one mole or, alternatively, the mass (in atomic mass units) of one atom. Numbers in parentheses denote the atomic masses of the most stable or best-known isotope of the element; all other numbers represent the average masses of a mixture of several isotopes as found in naturally occurring samples of the element.

Table A9.2 THE CHEMICAL ELEMENTS

Element	Chemical symbol	Atomic number	Atomic mass[j]
Hydrogen	H	1	1.0079 u
Helium	He	2	4.00260
Lithium	Li	3	6.941
Beryllium	Be	4	9.01218
Boron	B	5	10.81
Carbon	C	6	12.011
Nitrogen	N	7	14.0067
Oxygen	O	8	15.9994
Fluorine	F	9	18.99840
Neon	Ne	10	20.179
Sodium	Na	11	22.98977
Magnesium	Mg	12	24.305
Aluminum	Al	13	26.98154
Silicon	Si	14	28.0855
Phosphorus	P	15	30.97376
Sulfur	S	16	32.06
Chlorine	Cl	17	35.453
Argon	Ar	18	39.948
Potassium	K	19	39.098
Calcium	Ca	20	40.08
Scandium	Sc	21	44.9559
Titanium	Ti	22	47.90
Vanadium	V	23	50.9414
Chromium	Cr	24	51.996
Manganese	Mn	25	54.9380
Iron	Fe	26	55.847
Cobalt	Co	27	58.9332
Nickel	Ni	28	58.71
Copper	Cu	29	63.546
Zinc	Zn	30	65.38
Gallium	Ga	31	69.72
Germanium	Ge	32	72.59
Arsenic	As	33	74.9216
Selenium	Se	34	78.96
Bromine	Br	35	79.904
Krypton	Kr	36	83.80
Rubidium	Rb	37	85.4678
Strontium	Sr	38	87.62
Yttrium	Y	39	88.9059
Zirconium	Zr	40	91.22
Niobium	Nb	41	92.9064
Molybdenum	Mo	42	95.94
Technetium	Tc	43	98.9062
Ruthenium	Ru	44	101.07
Rhodium	Rh	45	102.9055
Palladium	Pd	46	106.4
Silver	Ag	47	107.868
Cadmium	Cd	48	112.40
Indium	In	49	114.82
Tin	Sn	50	118.69
Antimony	Sb	51	121.75
Tellurium	Te	52	127.60
Iodine	I	53	126.9045

[j] Numbers in parentheses denote the atomic masses of the most stable or best-known isotope of the element; all other numbers represent the average masses of a mixture of several isotopes as found in naturally occurring samples of the element.

Table A9.2 THE CHEMICAL ELEMENTS

Element	Chemical symbol	Atomic number	Atomic mass
Xenon	Xe	54	131.30
Cesium	Cs	55	132.9054
Barium	Ba	56	137.34
Lanthanum	La	57	138.9055
Cerium	Ce	58	140.12
Praseodymium	Pr	59	140.9077
Neodymium	Nd	60	144.24
Promethium	Pm	61	(145)
Samarium	Sm	62	150.4
Europium	Eu	63	151.96
Gadolinium	Gd	64	157.25
Terbium	Tb	65	158.9254
Dysprosium	Dy	66	162.50
Holmium	Ho	67	164.9304
Erbium	Er	68	167.26
Thulium	Tm	69	168.9342
Ytterbium	Yb	70	173.04
Lutetium	Lu	71	174.97
Hafnium	Hf	72	178.49
Tantalum	Ta	73	180.947
Tungsten	W	74	183.85
Rhenium	Re	75	186.2
Osmium	Os	76	190.2
Iridium	Ir	77	192.22
Platinum	Pt	78	195.09
Gold	Au	79	196.9665
Mercury	Hg	80	200.59
Thallium	Tl	81	204.37
Lead	Pb	82	207.2
Bismuth	Bi	83	208.9804
Polonium	Po	84	(210)
Astatine	At	85	(210)
Radon	Rn	86	(222)
Francium	Fr	87	(223)
Radium	Ra	88	226.0254
Actinium	Ac	89	(227)
Thorium	Th	90	232.0381
Protactinium	Pa	91	231.0359
Uranium	U	92	238.029
Neptunium	Np	93	237.0482
Plutonium	Pu	94	(244)
Americium	Am	95	(243)
Curium	Cm	96	(247)
Berkelium	Bk	97	(247)
Californium	Cf	98	(251)
Einsteinium	Es	99	(254)
Fermium	Fm	100	(257)
Mendelevium	Md	101	(258)
Nobelium	No	102	(259)
Lawrencium	Lr	103	(256)
Rutherfordium	Rf	104	(257)
Hahnium	Ha	105	(260)
?		106	(263)
?		107	(262)
?		109	(266)

Chapters 1–21:

$v = dx/dt$

$a = dv/dt = d^2x/dt^2$

$x = x_0 + v_0 t + \frac{1}{2}at^2$

$a(x - x_0) = \frac{1}{2}(v^2 - v_0^2)$

$A_x = A \cos\theta_x$

$A = \sqrt{A_x^2 + A_y^2 + A_z^2}$

$\mathbf{A} \cdot \mathbf{B} = AB \cos\phi$
$\quad = A_x B_x + A_y B_y + A_z B_z$

$|\mathbf{A} \times \mathbf{B}| = AB \sin\phi$

$a = v^2/r$

$\mathbf{v}' = \mathbf{v} - \mathbf{V}$

$m\mathbf{a} = \mathbf{F}$

$\mathbf{p} = m\mathbf{v}$

$\mathbf{L} = \mathbf{r} \times \mathbf{p}$

$\dfrac{d\mathbf{L}}{dt} = \mathbf{r} \times \mathbf{F}$

$w = mg$

$f_k = \mu_k N$

$f_s \leq \mu_s N$

$F = -kx$

$W = \mathbf{F} \cdot \Delta\mathbf{r}$

$W = \int \mathbf{F} \cdot d\mathbf{r}$

$K = \frac{1}{2}mv^2$

$U = mgz$

$E = K + U = [\text{constant}]$

$U(P) = -\displaystyle\int_{P_0}^{P} \mathbf{F} \cdot d\mathbf{r} + U(P_0)$

$U = \frac{1}{2}kx^2$

$E = mc^2$

$P = dW/dt$

$P = \mathbf{F} \cdot \mathbf{v}$

$\mathbf{r}_{CM} = \dfrac{1}{M}\displaystyle\int \mathbf{r}\rho \, dv$

$\mathbf{I} = \displaystyle\int_0^{\Delta t} \mathbf{F}\, dt$

$v_1' = \dfrac{m_1 - m_2}{m_1 + m_2}\, v_1; \quad v_2' = \dfrac{2m_1}{m_1 + m_2}\, v_1$

$\omega = d\phi/dt$

$\alpha = d\omega/dt = d^2\phi/dt^2$

$v = R\omega$

$K = \frac{1}{2}I\omega^2$

$I = \displaystyle\int \rho R^2 \, dV$

$I_{CM} = MR^2 \text{ (hoop)}; \frac{1}{2}MR^2 \text{ (disk)};$
$\qquad \frac{2}{5}MR^2 \text{ (sphere)}; \frac{1}{12}ML^2 \text{ (rod)}$

$I = I_{CM} + Md^2$

$L_z = I\omega$

$I\alpha = \tau_z$

$P = \tau_z \omega$

$F = GMm/r^2$

$v^2 = GM/r$

$g = GM_E/R_E^2$

$U = -GMm/r$

$x = A\cos(\omega t + \delta)$

$T = 2\pi/\omega; \quad v = 1/T = \omega/2\pi$

$m\, d^2x/dt^2 = -kx$

$\omega = \sqrt{k/m}$

$\omega = \sqrt{g/l}; \; T = 2\pi\sqrt{l/g}$

$\omega = \sqrt{mgl/I}$

$\omega = \sqrt{\kappa/I}$

$y = A\cos k(x - vt) = A\cos(kx - \omega t)$

$\lambda = 2\pi/k; \quad v = v/\lambda; \quad \omega = 2\pi v$

$v = \sqrt{F/\mu}$

$P \propto v\omega^2 A^2$

$v_{\text{beat}} = v_1 - v_2$

$v' = v(1 \pm V_R/v)$

$v' = v/(1 \mp V_E/v)$

$\sin\theta = v/V_E$

$x' = \dfrac{x - Vt}{\sqrt{1 - V^2}}; \quad t' = \dfrac{t - Vx}{\sqrt{1 - V^2}}$

$\Delta t = \dfrac{\Delta t'}{\sqrt{1 - V^2}}$

$\Delta x = \sqrt{1 - V^2}\, \Delta x'$

$v_x' = \dfrac{v_x - V}{1 - v_x V}$

$p = \dfrac{m\mathbf{v}}{\sqrt{1 - v^2/c^2}}; \quad E = \dfrac{mc^2}{\sqrt{1 - v^2/c^2}}$

$p - p_0 = -\rho gz$

$\frac{1}{2}\rho v^2 + \rho gz + p = [\text{constant}]$

$pV = NkT$

$T_C = T - 273.15$

$v_{\text{rms}} = \sqrt{3kT/m}$

$pV^\gamma = [\text{constant}]; \quad \gamma = C_p/C_V$

$\Delta E = \Delta Q - \Delta W$

$e = 1 - T_2/T_1$

$S(A) = \displaystyle\int_{A_0 \text{ rev.}}^{A} \dfrac{dQ}{T} + S(A_0)$

$S(B) - S(A) \geq \displaystyle\int_{A \text{ irrev.}}^{B} \dfrac{dQ}{T}$

$g = 9.81 \text{ m/s}^2$
$G = 6.67 \times 10^{-11} \text{ N} \cdot \text{m}^2/\text{kg}^2$
$M_E = 5.98 \times 10^{24} \text{ kg}$
$R_E = 6.37 \times 10^6 \text{ m}$

$m_e = 9.11 \times 10^{-31} \text{ kg}$
$m_p = 1.67 \times 10^{-27} \text{ kg}$
$c = 3.00 \times 10^8 \text{ m/s}$

$N_A = 6.02 \times 10^{23}/\text{mole}$
$k = 1.38 \times 10^{-23} \text{ J/K}$
$1 \text{ cal} = 4.19 \text{ J}$

Chapters 22–41:

$F = \dfrac{1}{4\pi\varepsilon_0}\dfrac{qq'}{r^2}$

$E = \dfrac{1}{4\pi\varepsilon_0}\dfrac{q'}{r^2}$

$E = \sigma/2\varepsilon_0$

$p = lQ$

$\boldsymbol{\tau} = \mathbf{p} \times \mathbf{E}$

$U = -\mathbf{p}\cdot\mathbf{E}$

$\oint E_n\, dS = \dfrac{Q}{\varepsilon_0}$

$V(P_2) - V(P_1) = -\displaystyle\int_{P_1}^{P_2}\mathbf{E}\cdot d\mathbf{l}$

$V = \dfrac{1}{4\pi\varepsilon_0}\dfrac{q'}{r}$

$\dfrac{\partial V}{\partial x} = -E_x, \quad \dfrac{\partial V}{\partial y} = -E_y, \quad \dfrac{\partial V}{\partial z} = -E_z$

$U = \tfrac{1}{2}Q_1 V_1 + \tfrac{1}{2}Q_2 V_2 + \tfrac{1}{2}Q_3 V_3 + \cdots$

$u = \tfrac{1}{2}\varepsilon_0 E^2$

$C = Q/\Delta V$

$C = \varepsilon_0 A/d$

$E = E_{\text{free}}/\kappa$

$\displaystyle\int \kappa E_n\, dS = \dfrac{Q_{\text{free}}}{\varepsilon_0}$

$u = \tfrac{1}{2}\kappa\varepsilon_0 E^2$

$I = \Delta V/R$

$R = \rho l/A$

$P = I\mathscr{E}$

$P = I\,\Delta V$

$\mathbf{F} = \dfrac{\mu_0}{4\pi}\dfrac{qq'}{r^2}\,\mathbf{v}\times(\mathbf{v}'\times\hat{\mathbf{r}})$

$\mathbf{B} = \dfrac{\mu_0}{4\pi}\dfrac{q'}{r^2}(\mathbf{v}'\times\hat{\mathbf{r}})$

$\mathbf{F} = q\mathbf{v}\times\mathbf{B}$

$d\mathbf{B} = \dfrac{\mu_0}{4\pi}I\dfrac{d\mathbf{l}\times\hat{\mathbf{r}}}{r^2}$

$\oint \mathbf{B}\cdot d\mathbf{l} = \mu_0 I$

$B = \mu_0 I_0 n$

$r = \dfrac{p}{qB}$

$d\mathbf{F} = I\,d\mathbf{l}\times\mathbf{B}$

$\mu = I\cdot[\text{area of loop}]$

$\boldsymbol{\tau} = \boldsymbol{\mu}\times\mathbf{B}$

$U = -\boldsymbol{\mu}\cdot\mathbf{B}$

$\mathscr{E} = vBl$

$\mathscr{E} = -\dfrac{d\Phi_B}{dt}$

$\Phi_B = \displaystyle\int \mathbf{B}\cdot d\mathbf{S}$

$\Phi_B = LI$

$\mathscr{E} = -L\dfrac{dI}{dt}$

$U = \tfrac{1}{2}LI^2$

$u = \dfrac{1}{2\mu_0}B^2$

$\mu = \dfrac{e}{2m_e}L$

$B = \kappa_m B_{\text{free}}$

$\omega_0 = 1/\sqrt{LC}$

$Z = \sqrt{R^2 + \left(\dfrac{1}{\omega C} - \omega L\right)^2}$

$Z = 1\Big/ \sqrt{\dfrac{1}{R^2} + \left(\omega C - \dfrac{1}{\omega L}\right)^2}$

$\mathscr{E}_2 = \mathscr{E}_1 \dfrac{N_2}{N_1}$

$\oint \mathbf{B}\cdot d\mathbf{l} = \mu_0 I + \mu_0\varepsilon_0\dfrac{d\Phi}{dt}$

$E_\theta = \dfrac{1}{4\pi\varepsilon_0}\dfrac{qa\sin\theta}{c^2 r}$

$B = E_\theta/c$

$\mathbf{S} = \dfrac{1}{\mu_0}\mathbf{E}\times\mathbf{B}$

$P_x = U/c$

$v = \sqrt{\dfrac{1 - v/c}{1 + v/c}}\,v_0$

$v = c/n$

$\sin\theta = n\sin\theta'$

$f = \pm\tfrac{1}{2}R$

$\dfrac{1}{s} + \dfrac{1}{s'} = \dfrac{1}{f}$

Minima (interference):

$\quad d\sin\theta = \tfrac{1}{2}\lambda, \tfrac{3}{2}\lambda, \tfrac{5}{2}\lambda, \ldots$

Maxima (interference):

$\quad d\sin\theta = 0, \lambda, 2\lambda, \ldots$

Minima (diffraction):

$\quad a\sin\theta = \lambda, 2\lambda, 3\lambda, \ldots$

$\quad a\sin\theta = 1.22\lambda$

$E = h\nu$

$p = h\nu/c$

$\Delta y\,\Delta p_y \gtrsim h$

$L = n\hbar$

$E = -\dfrac{m_e e^4}{2(4\pi\varepsilon_0)^2\hbar^2}\dfrac{1}{n^2} = -\dfrac{13.6\text{ eV}}{n^2}$

$\lambda = h/p$

$e = 1.60\times10^{-19}$ C
$\varepsilon_0 = 8.85\times10^{-12}$ F/m

$\mu_0 = 1.26\times10^{-6}$ H/m
$c = 1/\sqrt{\mu_0\varepsilon_0} = 3.00\times10^8$ m/s

$m_e = 9.11\times10^{-31}$ kg
$m_p = 1.67\times10^{-27}$ kg
$h = 2\pi\hbar = 6.63\times10^{-34}$ J·s

Chapter 1

2. 6×10^{-5} m
4. 8×10^{-7}, 4×10^{-9}, 3×10^{-13}, and 8×10^{-14} in.
6. 6.9×10^8 m
8. 1.00×10^7 m; 9.01×10^6 m
10. 6.3×10^6 m
12. 1.4×10^{17} s
14. 0.25 minute of arc, 0.29 mi
16. 0.134%, 99.866%
18. 0.021%, 99.979%
20. (a) 8.4×10^{24}; (b) 4.3×10^{46}; (c) 1.6×10^3
22. 9.22×10^{56}
24. 6.7×10^{27}
26. 8.3 light-minutes; 1.3 light-seconds
28. 8.9×10^3 kg/m³, 5.6×10^2 lb-mass/ft³, 0.32 lb-mass/in.³, 17 slug/ft³
30. 73×10^{-3} m³
32. 0.038 m³/s; 38 kg/s
34. 2.3×10^8 tons/cm³
36. 6.0×10^7 tons/cm³
38. 5.4×10^3, 5.2×10^3, 5.5×10^3, 3.9×10^3, 1.2×10^3, 0.63×10^3, 1.3×10^3, 0.17×10^3, and 10×10^3 kg/m³, respectively, for Mercury, Venus, Earth, Mars, Jupiter, Saturn, Uranus, Neptune, and Pluto

Chapter 2

2. 23 mi/h
4. 1.9×10^{10} years
6. 13 m/s, 0.77 s
8. (a) 13.8 s, 388 m; (b) 72 m
10. 6.4 m/s; 0 m/s
12. 32.4 m/s
14. (a) 4.25 m, 3.0 m/s, -6.0 m/s²; (b) 2.0 m, -6.0 m/s, -6.0 m/s²; (c) -1.5 m/s, -6.0 m/s²
16. (b) 5650 ft; (c) approximately 5650 ft
18. (b) 1.6 s and also at times earlier or later than this by a multiple of 3.16 s, ±2.0 m/s, 0 m/s²; (c) 0 s and also at times earlier or later than this by a multiple of 3.16 s, 0 m/s, ±2.0 m/s²

20. 2.4 m/s²
22. 4.0 years; 6.2×10^8 m/s
24. 350 m/s²
26. (a) 65.2 ft/s²; (b) at constant acceleration the distance would have been 345 yd; (c) 319 mi/h
28. 7.1 m/s²
32. 1.60 ft/s²; 13.7 s
34. 76 km/h
36. 66 m
38. 44 m
40. 34 m; 25 liters
42. 1.6×10^4 m/s²
44. 802 m/s; 1.89 s
46. (a) $n\sqrt{2h/g}$; (b) $\frac{3}{4}$ h; (c) $\frac{2}{3}$ h

Chapter 3

2. 11.2 km at 2.3° north of east
6. (b) 2.12×10^{11} m
10. 5550 km
12. 9.2 km; 7.7 km
14. (b) $C_x = 2$ cm, $C_y = 5$ cm
16. $A_x = 4.2$, $A_y = 0.5$, $A_z = \pm4.2$
18. 3.74
20. 5.9; 32°, $-60°$, 80°
22. $17\hat{x} + 3\hat{y} - 16\hat{z}$
24. 2.24×10^6 m²
26. $0.44\hat{x} - 0.22\hat{y} - 0.88\hat{z}$
28. 2.4; 1.8
30. 47.3 m down
32. $-12\hat{x} - 14\hat{y} - 9\hat{z}$
34. $0.45\hat{x} - 0.59\hat{y} - 0.67\hat{z}$
38. (a) 2.80 mi, 4.15 mi; (b) 3.59 mi, 3.48 mi
42. rotate coordinate system by an angle $\theta = -26.6°$
44. (a) $x'' = x'$, $y'' = y'\cos\phi + z'\sin\phi$, $z'' = -y'\sin\phi + z'\cos\phi$; (b) $x'' = x\cos\theta + y\sin\theta$, $y'' = -x\sin\theta\cos\phi + y\cos\theta\cos\phi + z\sin\phi$, $z'' = x\sin\theta\sin\phi - y\cos\theta\sin\phi + z\cos\phi$

Chapter 4

2. 29.9 km/s; 19.0 km/s
4. (a) $8\hat{x} + 10\hat{y}$, 12.8 m/s; (b) $4\hat{x} + 6\hat{y}$, 7.2 m/s²
6. 2.65 ft
8. 89 mi/h
10. (a) 2.5×10^4 m; 5.0×10^4 m; (c) no
12. 216 ft/s; 306 ft/s
14. (a) 7.25°; (b) 13 m
16. 25 ft/s; 63°
18. (a) 22 m/s; (b) 3.5 rev/s
20. 230 ft; 9.4 gal.
22. (a) 6.0 m; (b) 1.5 m; (c) 1.1 s
24. 0.11 m
26. (a) 230 ft/s; (b) 14 ft; (c) 26 ft
28. 43°; 35 ft/s
30. (a) 17.5°; (b) 6 m
32. (a) 270 m; (b) air resistance
34. no; 12.2 m
36. 9.95°; 205 m, or 42 minutes of arc
38. 2.4×10^2 ft/s; 1.8×10^5 ft/s²
40. 8.9 m/s²
42. 7.4×10^3 ft/s²
44. 13 m/s at 40° from vertical
46. 27 m/s at 68° from vertical
48. 329 m/s
50. 27 km/h at 43° east of north; 33°
52. 85°
54. 15 km/h at 15° east of north
56. $v = 4v_0^2 t/(4v_0^2 t^2 + h^2)^{1/2}$
58. 528 km/h

Chapter 5

2. 54 ft/s²
4. 6.6×10^2 kp
6. 1.2×10^4 N
8. 3.31×10^5 lb
10. 6.4×10^2 lb; 4.1×10^2 lb
12. $F = -623 + 66.5t$ measured in lb
14. no, since with 1150 children on each side the tension would be 34,500 lb
16. 4.7×10^{20} N at 25° with the Sun–Moon line
18. 175 lb toward dock
20. 5.9×10^2 N; 7.8×10^2 N
22. 6.9×10^5 N; 2.8×10^3 N
24. 40 lb
26. 8.2×10^2 kp
28. 37 lb at 20° downward from horizontal
30. 150 N
32. 3.6×10^3 slug·ft/s; 5.0 mi/h
34. 1.28 slug·ft/s; 1.96 slug·ft/s
36. 7.4×10^2 N
38. 5.3×10^{-13} kg·m²/s
40. 2.1×10^{12} kg·m²/s in north direction
42. (a) 1.1×10^{14} kg·m²/s; (b) 4.3×10^{36} kg·m²/s
44. (a) 0 kg·m²/s; (b) 2.7×10^{40} kg·m²/s, 5.4×10^{40} kg·m²/s, 2.7×10^{40} kg·m²/s, no
46. 5.26×10^{12} m
48. $2\hbar/[(1 - 2\sqrt{2}/3)m_e a_0] = 7.6 \times 10^7$ m/s; $2\hbar/[(1 + 2\sqrt{2}/3)m_e a_0] = 2.2 \times 10^6$ m/s

Chapter 6

2. (a) 0.018 oz; (b) no, since price is based on oz-mass
4. 9.2×10^2 N; 1.4×10^2 N
6. (a) [weight] = 7.4×10^2 N, [normal force] = 6.0×10^2 N, [resultant] = 4.2×10^2 N; (b) 5.6 m/s²
8. 64 m; 5.1 s
10. 9.8 m/s²
12. 7.0°
14. $a_1 = g(4m_2 m_3 - m_1 m_2 - m_1 m_3)/(4m_2 m_3 + m_1 m_2 + m_1 m_3)$, $a_2 = g(3m_1 m_3 - 4m_2 m_3 - m_1 m_2)/(4m_2 m_3 + m_1 m_2 + m_1 m_3)$, $a_3 = g(3m_1 m_2 - 4m_2 m_3 - m_1 m_3)/(4m_2 m_3 + m_1 m_2 + m_1 m_3)$, $T_1 = 2T_2 = 8m_1 m_2 m_3 g/(4m_2 m_3 + m_1 m_2 + m_1 m_3)$
16. 9.6×10^2 lb
18. 3.9 m/s²
20. 1.3×10^2 ft
22. 53 m
24. 0.54 ft/s²
26. 3.8×10^2 ft; 6.4 s
28. 27°
30. 4.4 m/s², 1.3 N
32. $T = \mu_k mg/\sqrt{1 + \mu_k^2}$
34. $a_1 = F/m_1 - \mu_1 g$, $a_2 = \mu_1 m_1 g/m_2 - \mu_2 g(m_1 + m_2)/m_2$
36. 6.8×10^2 N
38. 74 N
40. 0.15 m
42. 180 N/m, 360 N/m
46. 3.4 s
48. 4.4×10^2 N
50. 7.6×10^2 N; 8.1×10^2 N
52. 3.0 m/s
54. 67°
56. 47 mi/h
58. 6.4°
60. $v = \sqrt{gl \tan \theta \sin \theta}$
62. $T = mv^2/(2\pi r)$
64. $\phi = \tan^{-1}[\tan \theta/(1 - v_E^2/gR_E)] - \theta$ or approximately $(\sin \theta \cos \theta)v_E^2/gR_E$, where R_E and v_E are the equatorial radius and speed of the Earth and θ is the latitude angle; 0.099°

Chapter 7

2. 3.6×10^3 ft·lb
4. 8.8×10^3 J
6. 2.6×10^3 J
8. (a) 1.2×10^4 J; (b) 290 N, 1.2×10^4 J
10. 24 J
12. (a) 7.1×10^3 N; (b) 2.2×10^5 J, 8.1×10^3 N
14. curved ramp: $W = mgR[(1 - \sqrt{2}/2) + \mu_k \sqrt{2}/2]$; straight ramp: $W = mgR(1 - \sqrt{2}/2)[1 + \mu_k \cos(45°/2)]$, which is less
16. 2.2×10^{-18} J
18. (a) 4.0×10^5 J; (b) 2.5×10^4 J; (c) 1.2×10^6 J
20. 7.2×10^6 ft·lb, 4.2×10^6 ft·lb; 3.0×10^6 ft·lb
22. 8.6×10^4 ft·lb, 4.2×10^3 ft·lb; 20 times
24. (a) 1.2×10^4 N; (b) 39 m; (c) 4.7×10^5 J; (d) 4.7×10^5 J

26. 1.4×10^5 ft · lb
28. 1.6×10^9 ft · lb, 4.8×10^9 ft · lb
30. 8.2×10^6 m³
32. 5.1 m; yes, since the jumper does some extra work with his arms
34. 99 m/s; 9.8×10^{10} J; 23 tons
36. (a) 2.5×10^4 N; (b) 1.2×10^7 J; (c) 15 m/s
38. 50 J, 17 J
40. 9.9 m/s; 26 m/s
42. 53° from top

Chapter 8

4. $U = K/(3x^3)$
6. 217 J
8. (a) 47 J; (b) 0.89 m
10. 0.26 m
12. $t = v/\mu_k g$; $x = v^2/2\mu_k g$; $mv^2/2$
14. $\mathbf{F} = (2K/x^3)\hat{\mathbf{x}}$
16. 0.19 Å, 0.80 Å
18. (a) none and 0.2 m for E_1, 3.1 m and 0.3 m for E_2, 1.3 m and 0.5 m for E_3; (b) absolute maximum at 0.9 m, absolute minimum at turning points, local maximum at 2.2 m, local minimum at 1.7 m; (c) unbound for E_1, bound for E_2 and E_3
20. 11 eV
22. 2.2×10^6, 1.1×10^7, 2.7×10^6, 5.8×10^5, 3.6×10^6, and 9.8×10^6 J; bus; snowmobile
24. 1.0×10^6 eV
26. 9.4×10^8 eV
28. 540 kcal
30. 18 kW · h
32. 1.1×10^3 W; 0 W
34. 0.61 hp
36. 550 ft · lb
38. 9.1 gal./h
40. 4.2×10^5 W
42. (a) 1.2×10^{10} ft · lb; (b) 17 min; (c) 17 min, 2 or 3 mi
44. 1.2×10^{-4} W
46. 20 mi/h
48. 53 hp
50. 37%
52. (a) 3.2×10^4 W; (b) 7.8×10^2 W; (c) 3.1×10^4 W
54. 2.5×10^3 km²
56. 2.4×10^5 ft · lb/s
58. 4.3×10^9 kg/s; 1.4×10^{17} kg

Chapter 9

2. 4.3 ft/s
4. 14 mi/h; 26 mi/h and 54 mi/h
6. 150 bullets
8. $\sqrt{2}v$, at 135° with respect to the direction of the other fragments
10. (a) 42 km/h at 56° east of north; (b) 3.6×10^5 J
12. 74 lb
14. (a) 0.10 kg/s; (b) 2.3 kg m/s, 2.3 N
16. 0.62 ft from center of seesaw
18. 2.7×10^{-4} m

20. 2.28 Å from H
22. (950, 180, 820)
24. on diagonal of cube, at $\sqrt{3}/3L$ from vertex
26. $L/(64\pi - 4)$ from center of cube, above hole
30. 950 m
32. 6.5×10^7 ft · lb
34. 2120 lb
36. 1.0 m
38. 7.1×10^5 m from first fragment
40. (a) 6.8×10^6 ft · lb, 0 ft · lb; (b) 6.8×10^8 ft · lb, 1.3×10^6 ft · lb, extra energy comes from explosive chemical reactions
42. (a) 3.4×10^5 J, 3.6×10^5 J; (b) 3.4×10^5 J, 0 J
44. 2.81×10^{34} kg · m²/s; 3.46×10^{32} kg · m²/s
46. $1 - 1/e$

Chapter 10

2. -7.8×10^8 N; 1.1 m/s²
4. 3.4×10^5 ft/s²; 1.5×10^4 lb
6. 14 m/s
8. 2.6 km/h, 12.6 km/h
10. (a) 18 m/s; (b) 30 m/s
12. 39 m/s
14. 0.57 J
16. 1.8×10^4 m/s, at 139°
18. 4.0×10^{-13} J
20. (a) $h/9$, $4h/9$; (b) h, 0
22. $v/3$ to left, $2v/9$ to right, $8v/9$ to right
24. 45 ft/s
26. (a) 3.0 m/s; (b) 2.1×10^2 m/s²
28. 21 m/s
30. $v/5$ to left, $2\sqrt{3}v/5$ up 30° to right, and $2\sqrt{3}v/5$ down 30° to right
32. $v_2' = 7.1 \times 10^6$ m/s, $\theta_2' = 52°$
34. 45° each, 4.0×10^{-13} J each
36. 1.9×10^5 eV, 70°
38. 1.25×10^{-13} J
40. 4.3×10^{-11} J

Chapter 11

2. 7.3×10^{-5} radian/s; 460 m/s, 350 m/s
4. 0.13 in./min; 1.2×10^3 in./min; 5.2×10^2 in./min
6. 81 radian/s; 13 rev/s
8. (b) 5.4×10^{-3} m
10. 12 radian/s²
12. 9.6×10^{-22} radian/s²
14. 1.21×10^{-10} m
16. 6.50×10^{-46} kg · m²
18. $\frac{1}{2}M(R_1^2 + R_2^2)$
20. -1.6×10^{22} kg · m²
22. $0.38M_E^2R_E^2$
24. $\frac{1}{12}ML^2 \sin^2 \theta$
28. $\frac{1}{8}MR^2$
30. $\frac{1}{2}MR^2 - (r^2M/R^2)(\frac{1}{2}r^2 + d^2)$
32. $\frac{1}{2}MR^2 - (2r^2M/R^2)(r^2 + \frac{1}{2}R^2)$
34. $0.338M_E^2R_E^2$
36. (a) 2.2×10^2 kg · m²; (b) 4.4×10^3 J
38. 2.61×10^5 ft · lb; 3.83%

42. 2.74×10^3 J; 13.1 J; 4.74×10^{-3}
44. $\frac{1}{6}Ml^2$
48. 3.8×10^{-2} slug $\cdot$ ft^2/s^2
50. 5.6×10^{41} J $\cdot$ s; 3.1×10^{43} J $\cdot$ s; 1.8%
52. 1.8×10^{22} kg $\cdot$ m^2/s^2

Chapter 12

2. 400 N
4. 1.2×10^4 N; 7.2×10^3 N $\cdot$ m; depress
6. 2.7×10^4 N $\cdot$ m
8. 180 lb
10. 9.7 m/s^2
12. (a) 110 J $\cdot$ s; (b) 34 kg $\cdot$ m^2/s^2; (c) 34 N $\cdot$ m
14. (a) 76°; (b) 290 N, 250 N
16. 1.0×10^{18} J $\cdot$ s; 3.0×10^{12} J $\cdot$ s/s; 3.0×10^{12} N $\cdot$ m
18. (a) $\omega_1 R_1/R_2$; (b) $(T - T')R_1$, $(T - T')R_2$;
    (c) $(T - T')R_1\omega_1$, $(T - T')R_2\omega_2$, yes
20. 27 N $\cdot$ m
22. 4.6 radian/s
24. (a) 8.4×10^2 ft $\cdot$ lb; (b) 6.9 ft/s^2
26. (a) 2.4×10^{19} J $\cdot$ s; (b) 3.0×10^{-19} radian/s
28. 7.2°
30. (a) 5.0 m/s; (b) 9.0×10^{-3} radian/s; (c) 19 times
32. [height of end] $= m^2v^2/[2g(M/3 + m)(M/2 + m)]$
34. $v_{CM} = mv/(m + M)$,
    $\omega = v_{CM}/[\frac{1}{3}l + \frac{1}{4}l/(1 + m/M)^2 + \frac{1}{4}lm/M]$
38. 9.6 m/s^2
40. 100 N in *forward* direction; 300 N in forward direction
42. $2a/7$
44. $a = g(\sin\theta - \mu_k \cos\theta)$; $\alpha = (2/R)g\mu_k \cos\theta$
46. $a = (g\sin\theta)(m + 4M)/(m + 6M)$
48. 0.85 radian/s
50. 22 lb at 23° right of vertical at the top, and 23° left of vertical at the bottom
52. $2mg$, $mg/\sqrt{3}$, $mg/\sqrt{3}$
54. 1.7×10^3 N; 1.5×10^3 N
56. $0.11Mg$
58. (a) $U = 5.3 \times 10^2$ J $\times \{\sin(\theta + \tan^{-1} 2) + $ [constant]\}; (b) 27°; (c) 56 J
60. 38% front, 62% rear
62. 0.31 m; 590 N

Chapter 13

2. 4.7×10^{-35} N
4. 8.88 m/s^2, 3.71 m/s^2, 3.71 m/s^2
6. 4.0×10^{-4} radian/s, or 0.23 rev/h
8. 2×10^8 years, 3×10^5 m/s
10. Lincoln, Nebraska, 22.6° west of New York City
12. $m_1/m_2 = 1.6$
14. 3.0×10^{10} m
16. (a) 7.5×10^3 m/s, 8.32×10^3 m/s;
    (b) 3.94×10^8 J, 4.85×10^8 J
18. 3.5 days
20. 2.66×10^{33} J, -5.31×10^{33} J; -2.66×10^{33} J
22. (a) 6.0×10^{-14} m/s^2, 9.1×10^{-14} m/s^2;
    (b) 1.1×10^5 m/s, 1.6×10^5 m/s;
    (c) 3.4×10^{51} J, 5.1×10^{51} J, 7.7×10^{51} J, yes

24. (a) 1.1×10^4 m/s; (b) 1.2×10^{11} J, 29 tons;
    (c) 1.2×10^5 m/s^2
26. (a) -1.1×10^{11} J, -2.2×10^{11} J, -1.1×10^{11} J;
    (b) yes, yes
28. (a) 1.1×10^4 m/s; (b) 2.4×10^3 m/s
30. (a) I and III elliptical, II circular
32. 2.5×10^6 m; 8.2×10^2 m/s
34. (a) 4.21×10^4 m/s; (b) 7.19×10^4 m/s, 1.23×10^4 m/s
36. 7.78×10^7 m
40. (a) 2.6×10^3 m/s; (b) 2.7×10^3 m/s; (c) 0.40 year; (d) at launch, Venus must be 90° behind Earth as seen from Sun
42. 4.92, 7.67, 10.8, 9.8, 9.9, and 9.8 m/s^2
44. $U = GMm(r^2/R^2 - 3)/2R$

Chapter 14

2. (a) 1.7×10^4 m/s^2, 27 m/s; 2.0×10^4 N
4. (a) $A = 3.0$ m, $\nu = 0.32$ Hz, $\omega = 2.0$ radian/s, $T = 3.1$ s; (b) 0.26 s, 1.05 s (and also at times larger and smaller by a multiple of one-half period)
6. (a) $0.20 \cos 6\pi t + (4.0/6\pi) \sin 6\pi t$; (b) 0.043 s, -1.0×10^2 m/s^2
8. 1.9×10^4 N
10. 1.13×10^{14} Hz
12. (a) 4.5 Hz; (b) 8.5 m/s, strong vibrations
14. (a) $x_1 = -x_2 = (-v_1/\omega) \cos \omega t$ with $\omega = \sqrt{3k/m}$;
    (b) $x_1 = x_2 = (-v_1/\omega) \cos \omega t$ with $\omega = \sqrt{k/m}$
16. (a) 6.6 J; (b) $t = 0.30$ s, $t = 0$; (c) $t = 0.15$ s
18. 18 eV; no
22. (a) 0.73 min; (b) 1.0 mm
24. 0.19 Hz
26. (a) 1.3×10^{-3} J; (b) 0.14 m/s
28. (a) 9.4 J; (b) 1.6×10^{-5} W;
    (c) 6.0×10^{-3} N $\cdot$ m/radian
32. 1.988 s
34. 1.5 s
36. (b) $l = R/\sqrt{2}$
38. 3.6 radian/s; 11 ft/s
40. $2\pi\sqrt{4l/5g}$
42. (a) $T = mg[1 - \frac{1}{2}A^2 + \frac{3}{2}A^2 \sin^2 (\sqrt{g/l}\ t)]$;
    (b) $t = \frac{1}{2}\pi/\sqrt{g/l}$ gives $T = mg(1 + A^2)$
44. $\gamma = 2.5 \times 10^{-3}$/s

Chapter 15

2. 0.114 Hz, 0.716 radian/s, 0.0524 /m, 13.7 m/s
4. (a) 6.0×10^{-3} m, 0.31 m, 20 /m, 0.64 Hz, 4.0 radian/s; (b) -0.26 s
6. 7.8 m/s
8. 5.5×10^{-7} m; 4.2×10^{-7} m
10. 0.69 s
12. 1.6 s
16. 1:2.4
18. $A = 5.0$; $\delta = -\tan^{-1} \frac{4}{3}$
20. 0.007 Hz
22. 1.00/day, 0.92/day; 1.00 day, 1.09 day; first is due to Sun, second due to Moon
24. 392 Hz, 588 Hz, 784 Hz, 980 Hz

26. 1.6 Hz, 3.2 Hz
28. 71 N
30. 3.7 m/s; 6.8×10^3 m/s^2
32. (a) 9.6×10^3 N; (b) 12 Hz
34. 1.25×10^{-7} m; 2.5×10^{-7} m
36. 0.018 J/m $\times (\sin^2 \omega t \cos^2 kx + \cos^2 \omega t \sin^2 kx)$,
    where $\omega = 3.7 \times 10^3$ radian/s and $k = 28$/m;
    location of maxima depends on time: maxima are
    at 0.057 m, 0.17 m, 0.28 m when $\sin^2 \omega t <$
    $\cos^2 \omega t$, and maxima are at 0 m, 0.11 m, 0.23 m,
    0.34 m when $\sin^2 \omega t > \cos^2 \omega t$; 0.018 J/m $\times \sin^2 \omega t$
    or 0.018 J/m $\times \cos^2 \omega t$

Chapter 16

2. 55 Hz, 4187 Hz
4. 1.9, 1.8, 1.7, 1.6, 1.5, 1.5, 1.3, 1.3, 1.2, 1.1, 1.1,
   and 1.0 cm
6. 3.1 m
8. 130 times; 21 dB
10. 9.8 s; use visual signal
12. 150 m
14. (a) 1.2×10^{-4} s; (b) on the glass
18. (a) 0.63 m; (b) 3.5, 3.3, 3.2, 3.0, 2.8, 2.7, 2.5,
    2.4, 2.2, 2.1, 2.0, and 1.9 cm
22. 71 ft/s; 0.22 Hz
24. 594 Hz, 596 Hz
26. 476 Hz
28. 481 Hz
30. (a) 33.4°; (b) 30.4 s
32. 16.1 ft/s; 6.76 ft/s^2
34. 1.7×10^2 m
36. (a) 2.3 /h; (b) 320 km, 740 km/h
38. (a) 31 m/s; (b) 0.11 /h, 9.0 h; (c) probably
40. (b) 1.6 m, 3.2 m
42. 4.0×10^3 km
44. 7.2
46. (a) 6.3 m/s; (b) 1.0 m

Chapter 17

2. (b) 2.4, $v' = 0.42$
4. (a) $x = 2.3 \times 10^8$ light-seconds, $y = 4.0 \times 10^8$
   light-seconds, $z = 0$, $t = -1.7 \times 10^8$ s; (b) before
8. $(1 + 3.4 \times 10^{-10})$; 3.0×10^{-5} s
10. 3.7×10^2 m/s; $(1 + 7.5 \times 10^{-13})$
12. $V = (1 - 1.0 \times 10^{-11})$; 2.2×10^6 years
14. (a) 8.2×10^8 s; (b) 7.6×10^8 s
16. 4×10^{-13}%
18. $0.999c$
20. $0.77c$; 67°
22. $0.87c$
26. 4.0×10^{-3}%
28. 130 m/s; 5.3×10^{-16} kg $\cdot$ m/s
30. 1.4×10^{-15} m/s
32. (a) 9.0×10^{13} m/s^2; (b) 1.8×10^{-24} kg;
    (c) 1.6×10^{-10} N
34. $0.83c$
38. (a) $0.99962c$; (b) $0.973c$; (c) 6.5×10^{-10} J

Chapter 18

2. (a) 11 m; (b) 8.1 cm, 11.7 cm
4. 1.4×10^3 lb; layer of air under the paper exerts
   an opposite pressure force
6. 20 in.^{2}
8. 34.1 lb/in.^{2}
10. 5.1×10^4 N
12. 51 atm
14. (a) 360 N; (b) 330 N
16. 0.86 m
18. 10.3 m
20. (b) 5.0×10^{11} N/m^2
22. (a) 4.7×10^7 m^3; (b) 4.8×10^{10} kg
24. yes
26. (a) 9.5 m; (b) 28.4 m
28. 0.115%; 0.011%; 0.002%
30. $\rho = M/[\frac{4}{3}\pi R^3 + R'^2 (l - h)]$
32. 19 cm
34. 29 N/m^2
36. (a) 51 ft/s; (b) 16 lb/in.^{2} (overpressure)
38. (a) $v = \sqrt{2g(h_2 - h_1)}$; (b) $p_{atm} - \rho g h_2$; (c) $p_{atm}/\rho g$
40. 2.3 hp
44. $(A/A')\sqrt{2l/g}$

Chapter 19

2. 58.0°C = 331.2 K; −88.3°C = 184.9 K
4. 2.69×10^{19} molecules
6. 3.4 atm
8. 1.4×10^{-9} N/m^2
10. 11.6 kg/m^3
12. 3.6×10^6 N/m^2
14. 0.094 kg
16. 6.4×10^3 lb/in.^{2}
18. 96.3 g
20. 2.1×10^3 kg
22. (a) 1.2 m; (b) 2.5 kg at 2.5 atm
24. 29.0 g
28. 5.7×10^{-21} J
30. 6×10^6; 30 cm
32. 0.43%
34. 1.9×10^5 J; $\frac{3}{5}$; $\frac{2}{5}$
36. 284 K

Chapter 20

2. 97 kW
4. 25°C
6. 97 W
8. 4.3 cal/s
10. 3.5 h
12. 1.4 liters/min
14. 0.21 m
16. ±0.08°C
18. (a) 0.43 in.; (b) 1.6 ft
22. (a) 709 kg/m^3; (b) 43¢/kg, 45¢/kg
26. 100.32°C
28. 0.085 kg/h
30. 1.8 cm

32. 0.45 cm/h
36. (a) 1.0×10^{11} kcal; (b) 5 A-bombs
38. 22°C
40. 0.65 kg/h
42. 2.5×10^3 J; 50 m/s
44. 1.19×10^2 kcal
46. 1.5×10^4 kcal
48. (a) 2.62 min; (b) 2.64 min, no
50. 30.8 atm; 507°C
52. −225°C
54. 8.1×10^3 ft · lb

Chapter 21

2. $\Delta Q = 1.96$ J, $\Delta W = 1.96$ J, $\Delta E = 0$
4. 518 kcal/kg
6. 0.19 kg/s
10. (a) 2.0×10^4 J; (b) 3.9×10^4 J, 1.9×10^4 J
12. $1 - 3 \times 10^{-9}$, or 99.9999997%
14. 3.9×10^2 W
16. (a) 3.1×10^2 MW; (b) 1.0×10^3 MW
18. 7.1 kW
20. 1.7×10^2 J
22. 8.2×10^3 cal/K · s
24. 3.7×10^2 cal/K · day
26. -2.9×10^2 cal/K
28. 39 cal/K
30. (a) 6.9×10^3 cal/K; (b) 4.6×10^2 cal/K

Chapter 22

2. 2.89×10^{-9} N
4. 17 N; 2.6×10^{27} m/s²
6. $-6.9 \times 10^{-4}\,\hat{\mathbf{x}} + 1.7 \times 10^{-3}\,\hat{\mathbf{y}}$;
   $6.9 \times 10^{-4}\,\hat{\mathbf{x}} - 1.7 \times 10^{-3}\,\hat{\mathbf{y}}$
8. 10^{-9}; 10^{-5}
10. 2.6×10^{-39} C; $1:1.8 \times 10^5$; attractive
14. 8.4×10^{22}
16. 2.4×10^{28} of each
18. 6.8×10^{32} electrons on Earth and 1.9×10^{32} electrons on Moon
20. 9.4×10^{18}/s
22. 4 electrons

Chapter 23

2. 6.2 m/s² upward
4. 1.3×10^{-13} N; 1.4×10^{17} m/s²
6. 9.0×10^3 N/C
8. 6.2×10^{-7} m
10. 5.1×10^{11} N/C
12. 5.1×10^{12} N/C in positive x direction
14. 2.9×10^4 N/C at 71° below horizontal
16. 2.3×10^4, 1.7×10^4, 0.88×10^4, 0.35×10^4, 0.11×10^4, and 0.017×10^4 N/C
18. $\sqrt{15}\lambda/(8\pi\varepsilon_0 d)$
20. $E_x = -\dfrac{1}{4\pi\varepsilon_0}\dfrac{\lambda}{\sqrt{x^2+y^2}}$, $E_y = \dfrac{1}{4\pi\varepsilon_0}\dfrac{\lambda}{y}\left(1 + \dfrac{x}{\sqrt{x^2+y^2}}\right)$
22. 1.1×10^5 N/C up, 1.1×10^5 N/C down, 1.1×10^5 N/C down, 3.4×10^5 N/C down
24. 2.4×10^5 N/C

26. (a) $E = \dfrac{Q}{4\pi\varepsilon_0}\dfrac{1}{x(x+l)}$ in positive x direction
    (b) $E = \dfrac{Q}{2\pi\varepsilon_0}\dfrac{1}{y}\dfrac{1}{\sqrt{l^2+4y^2}}$ in positive y direction
28. $E = 2Q/(l^2\pi\varepsilon_0)$ toward lower right corner
30. $E = \dfrac{Q}{2\pi\varepsilon_0}\dfrac{1}{R^2}\left(1 - \dfrac{z}{\sqrt{x^2+R^2}}\right)$ along axis
32. $E_x = \dfrac{\lambda}{4\pi\varepsilon_0}\left[\dfrac{1}{x}\left(1 + \dfrac{y}{\sqrt{x^2+y^2}}\right) - \dfrac{1}{\sqrt{x^2+y^2}}\right]$

   $E_y = \dfrac{\lambda}{4\pi\varepsilon_0}\left[\dfrac{1}{y}\left(1 + \dfrac{x}{\sqrt{x^2+y^2}}\right) - \dfrac{1}{\sqrt{x^2+y^2}}\right]$
34. $\dfrac{Q^2}{4\pi\varepsilon_0 L^2}\ln\left[\dfrac{(x+L)^2}{x(x+2L)}\right]$
36. (a) 8.5×10^{-30} C · m; (b) 0 C · m
38. 4.8×10^{-24} N · m

Chapter 24

2. 1.1×10^3 N · m²/C
4. 1.1×10^5 N · m²/C; 0 N · m²/C; 1.1×10^5 N · m²/C
6. 4.4×10^{-11} C/m³
8. 8.3×10^{-9} C
10. 4.8×10^{20} N/C; 3.5×10^3 N; 1.8×10^{28} m/s²
12. 7×10^{-9} C/m
14. $\mathbf{E} = (\rho x/\varepsilon_0)\hat{\mathbf{x}}$ for $x < d/2$; $\mathbf{E} = \pm (\rho d/2\varepsilon_0)\hat{\mathbf{x}}$ for $\pm x > d/2$
16. $E = 0$ for $r < a$, $E = \dfrac{Q}{4\pi\varepsilon_0}\dfrac{r^3 - a^3}{b^3 - a^3}\dfrac{1}{r^2}$ for $a < r < b$, $E = \dfrac{Q}{4\pi\varepsilon_0}\dfrac{1}{r^2}$ for $r > b$
18. $E = \dfrac{Q}{4\pi\varepsilon_0}\left[\dfrac{1}{r^2} - \dfrac{1}{8(r - R/2)^2}\right]$ in radial direction
20. 0.5 Å
22. 1.2×10^7 m/s, 2.2×10^{-13} J, 1.1×10^{-34} J · s, 6.5×10^{20} Hz
26. (a) 1.9×10^{-25} kg · m/s; (b) 2.1×10^5 m/s; (c) 2.1×10^5 m/s
28. 10^{-6} C/$(4\pi\varepsilon_0 r^2)$; 4×10^{-6} C/$(4\pi\varepsilon_0 r^2)$
30. 8.8×10^{-10} C/m²

Chapter 25

2. −6.0 eV
4. 4.5×10^3 V
6. 1.3×10^7 V/m
8. 1.9×10^3 V/m
10. 1.7×10^7 V; 2.5×10^7 V/m
12. 5.8×10^{-12} J
14. 7.5×10^7 V; 1.7×10^7 V
16. (c) −13.6 eV
18. $V = \dfrac{1}{4\pi\varepsilon_0}\dfrac{Q}{l}\ln\left(\dfrac{2+\sqrt{3}}{2-\sqrt{3}}\right)$
20. $V = \dfrac{1}{4\pi\varepsilon_0}\dfrac{Q}{l}\ln\left[\dfrac{x+l}{x}\dfrac{l+\sqrt{x^2+l^2}}{x}\dfrac{l+\sqrt{(x+l)^2+l^2}}{x+l}\dfrac{l+x+\sqrt{(x+l)^2+l^2}}{x+\sqrt{x^2+l^2}}\right]$

22. 3.1×10^7 m/s

24. $V = \dfrac{Q}{2\pi\varepsilon_0} \dfrac{1}{R^2}(\sqrt{R^2 + z^2} - z)$

26. $V = 4k/(\varepsilon_0 r^{1/2})$

28. $E = -8\hat{\mathbf{x}} - 4\hat{\mathbf{y}}$ in volt/meter

30. $V = \dfrac{1}{4\pi\varepsilon_0} \dfrac{Q}{l} \ln\left(\dfrac{x+l}{x}\right)$; $E = \dfrac{1}{4\pi\varepsilon_0} \dfrac{Q}{x(x+l)}$

   in positive x direction

32. (a) $V = \dfrac{2Q}{3\pi\varepsilon_0 R^2} (\sqrt{z^2 + R^2} - \sqrt{z^2 + R^2/4})$

   (b) $E = \dfrac{2Qz}{3\pi\varepsilon_0 R^2}\left(-\dfrac{1}{\sqrt{z^2 + R^2}} + \dfrac{1}{\sqrt{z^2 + R^2/4}}\right)$

   along axis

34. (a) 6.4×10^7 V/m parallel to **p**;
    (b) 3.2×10^7 V/m antiparallel to **p**

36. (b) $E_x = \dfrac{Ql^2}{4\pi\varepsilon_0} \dfrac{x}{r^5}\left[2 + \dfrac{5(2z^2 - x^2 - y^2)}{r^2}\right]$

   $E_y = \dfrac{Ql^2}{4\pi\varepsilon_0} \dfrac{y}{r^5}\left[2 + \dfrac{5(2z^2 - x^2 - y^2)}{r^2}\right]$

   $E_z = \dfrac{Ql^2}{4\pi\varepsilon_0} \dfrac{z}{r^5}\left[-4 + \dfrac{5(2z^2 - x^2 - y^2)}{r^2}\right]$

   where $r^2 = x^2 + y^2 + z^2$

Chapter 26

2. 4.0×10^{-17} J

4. 5.8×10^6 eV

6. -50 eV

8. $U = \dfrac{1}{4\pi\varepsilon_0} \dfrac{Q^2}{l^2}\left[x \ln \dfrac{(x+2l)x}{(x+l)^2} + 2l \ln \dfrac{x+2l}{x+l}\right]$

10. 1.7 J

12. 1.41×10^{-15} m

14. 5.1×10^{31} J/m^3

16. 1.8×10^{31}, 4×10^{16}, 1.6×10^{12}, and 1.1×10^2 J/m^3; 2.0×10^{14}, 0.5, 1.8×10^{-5}, and 1.2×10^{-15} kg/m^3

18. 3.5×10^{-15} m

20. 1.1×10^6 eV

22. (b) 1.2×10^{29} J; $1.5.3 \times 10^{10}$

24. $0.980\ e^2/8\pi\varepsilon_0 R) = 7.1 \times 10^5$ eV

Chapter 27

2. (a) 2.0×10^{-11} F; (b) 4.0×10^{-6} C

4. 9.5 pF

6. 15.5 μF; 1.5 μF

8. 2.4×10^{-5} C, 7.2×10^{-5} C

10. 7.4 m^2

12. $C = \dfrac{\varepsilon_0 A}{d} \dfrac{2\,\kappa_1\kappa_2}{\kappa_1 + \kappa_2}$

16. 0.11%

18. 0.20 C/m^2

20. (a) -2.6×10^{-6} C/m^2, 1.7×10^{-6} C/m^2;
    (b) 1.6×10^5 V/m, 1.1×10^5 V/m;
    (c) 3.0×10^5 V/m

22. 5.0×10^{-2} N

24. $C = \dfrac{2\pi\varepsilon_0 (\kappa_1 + \kappa_2)}{1/R_1 - 1/R_2}$

26. (a) 1.6×10^{-10} F; (b) 3.8×10^2 V;
    (c) 7.5×10^4 V/m; (d) 2.5×10^{-2} J/m^3;
    (e) 1.1×10^{-5} J

28. 0.2 C, 2×10^3 J

30. 8.9×10^{-4} J

32. 1.1×10^8 m^3

Chapter 28

2. 5.7×10^{-14} J$\cdot$s

4. 4.1×10^6 A/m^2; 6.9×10^{-2} V/m

6. (a) 3.6×10^{-2} V/m; (b) 1.7×10^{-4} m/s; 46 years

8. 0.201, 0.256, 0.323, 0.407, and 0.514 Ω

10. 7.9 Ω

12. 2.2 cm

14. 1.0×10^{-7} V

16. 0.88 Ω; 14 A

18. 191 kg vs. 384 kg

20. 0.164 cm

22. 8×10^{-5} A

24. 33° C

26. 3.8 A, 2.2 A

28. (a) 2.1×10^2 A; (b) 2.7×10^{-19} A

30. 2.4%

32. 1.5 km from AB

34. 4.0 A, 2.4 A, 1.5 A; 7.9 A

36. 10.9 A, 7.1 A

38. (a) 1.85 Ω; (b) 6.5 A;
    (c) $\Delta V_1 = \Delta V_2 = \Delta V_3 = 12$ V, $I_1 = 3.0$ A, $I_2 = 2.0$ A, $I_3 = 1.5$ A

40. $(1 + \sqrt{3})\ \Omega$

Chapter 29

2. (a) 6.5×10^3 J, 2.4×10^6 J;
    (b) 2.2×10^3 J/in.^{3}, 3.4×10^3 J/in.^{3};
    (c) 3.4×10^4 J/lb, 4.8×10^4 J/lb

4. 2.9×10^3 J; (b) 48 s

6. 2.0 Ω

8. 8.0 Ω

10. $R_i'\mathscr{E}/(RR_i + RR_i' + R_iR_i')$, $R_i\mathscr{E}/(RR_i + RR_i' + R_iR_i')$

12. 12.5 A, 25.5 A

14. (a) 2.6, 1.7, 1.3, 2.1, and 0.86 A; (b) 2.6 V

16. 6.2×10^{12} protons per second; 7.0×10^2 W

18. 14 A, 8.1 Ω

20. (a) 13.2 V; (b) 7.2 W

22. 6.7 h

24. 0.50 W

26. (a) 1.8×10^2 V; (b) 1.8×10^6 W

28. 1.3×10^2 W; 0.022

32. (a) 1/30; (b) 1/3

34. 19 liters per second

Chapter 30

2. 1.2×10^{-12} N, 0 N

4. (a) $|\mathbf{F}| = 0.89 \times 10^{-13}$ N, $|\mathbf{F'}| = 1.78 \times 10^{-13}$ N;
    (b) 0.89×10^{-13} N in direction of **v**

6. 8×10^{-16} N in direction of $-\mathbf{v} \times \mathbf{B}$

8. 8.2×10^{-8} N and 3.5×10^{-5} N; yes

10. 4.0×10^{-3} T

12. (a) 5.0×10^{-6} T; (b) 16°

14. 1.5×10^{13} m/s²

16. $B = \dfrac{\mu_0}{2\pi} \dfrac{3\sqrt{2}}{2} \dfrac{I}{d}$

18. $B = \dfrac{\mu_0 I}{4\pi} \dfrac{\sqrt{x^2 + y^2} + x + y}{xy}$ in positive z direction

20. $B = \dfrac{\sqrt{5}\,\mu_0 I}{2\pi L}$ into plane

22. $B = (\sqrt{5} + \sqrt{13}/3)\,\dfrac{\mu_0 I}{\pi h}$ out of plane

24. $B = \mu_0 I / 8R$ into plane

26. $B = \dfrac{\mu_0}{\pi} \dfrac{I}{b} \tan^{-1} \dfrac{b}{2z}$ perpendicular to z and to current

28. $B = \dfrac{\mu_0 I}{4\pi R}(3\pi/4 + 2)$ out of plane

30. 1.9 T

32. 2.1×10^9 A; westward

34. $B = \dfrac{4\mu_0}{\pi} \dfrac{IL^2}{(4z^2 + L^2)\sqrt{4z^2 + 2L^2}}$ in positive z direction

36. $Q\omega R^2/4$

38. $B = \dfrac{\mu_0 I}{2L} \left[\dfrac{z + L/2}{\sqrt{(z + L/2)^2 + R^2}} - \dfrac{z - L/2}{\sqrt{(z - L/2)^2 + R^2}} \right]$ in positive z direction

40. $B_x = 0$, $B_y = -\dfrac{\mu_0 I}{2\pi} \dfrac{Rz}{(z^2 + R^2)^{3/2}}$

$B_z = \dfrac{\mu_0 I}{2\pi} \left[\dfrac{\pi}{2} \dfrac{R^2}{(z^2 + R^2)^{3/2}} + \dfrac{R}{R^2 + z^2} \right]$

Chapter 31

2. 6.4×10^{-16} N, 7.0×10^{14} m/s²; into wire

4. 4.0×10^{-8}; 1.9×10^{-18} N

6. 0, $\mu_0 \sigma \hat{\mathbf{x}}$, 0, $-\mu_0 \sigma \hat{\mathbf{x}}$

8. 0.13 T

10. $\mu_0 \sqrt{n^2 I^2 + (I'/2\pi r)^2}$; helical

12. 6.9 T; 2.3 T

14. 1.1×10^{-17} kg·m/s

16. 3.4×10^{-17} kg·m/s

18. 3.3 T

20. 0.39 A

22. 3.6×10^{-2} T

24. 2×10^6 m/s eastward

26. 0.12 N

28. 6.7×10^{-5} N

30. 7.2×10^{-4} N toward the straight wire

32. 3.4×10^{-5} N·m

34. 2.0 N·m; 0.86 hp

36. (a) 4.0×10^{-3} N; (b) 3.2×10^{-4} N·m

Chapter 32

2. 0.75 V

4. 22 rev/s

6. 9×10^{17} V

8. 9.8×10^{-5} V, 7.5×10^{-5} V, 8.7×10^{-5} V

10. 40 rev/s

12. (a) 1.9×10^{-2} T/s; (b) 6.4×10^{-3} V

14. 3.0×10^2 A

16. $\mathscr{E} = \dfrac{\mu_0 I v}{2\pi} \dfrac{l}{(l + d)d}$

18. 45 T

20. (a) 6.0 V/m; (b) 0 V/m; (c) 0.36 V

22. (a) 2.6×10^{-3} V/m; (b) 0 V/m; (c) 9.2×10^{-15} C/m²

24. (a) 0 Wb; (b) 90 V; (c) negative

26. $200\,\mu_0 n\pi R^2$; no

28. (a) 6.3×10^{-3} H; (b) -1.9 V

30. (a) $B = \dfrac{\mu_0}{2\pi} \dfrac{NI}{r}$; (b) $\Phi_B(1) = \dfrac{\mu_0 NI}{2\pi}(R_2 - R_1) \times \ln(R_2/R_1)$, $L = \dfrac{\mu_0 N^2}{2\pi}(R_2 - R_1) \ln(R_2/R_1)$

34. 4×10^{11} J/m³

36. 7×10^{18} J

38. 3.6×10^7 J

40. 9.1×10^3 m³; about 40 m across

42. (a) $u = \dfrac{\mu_0}{8\pi^2} \left(\dfrac{NI}{r}\right)^2$; (b) $U = \dfrac{\mu_0}{4\pi} N^2 I^2 (R_2 - R_1) \times \ln(R_2/R_1)$; (c) $L = \dfrac{\mu_0}{2\pi} N^2 (R_2 - R_1) \ln(R_2/R_1)$

Chapter 33

2. (a) 1.9 T; (b) -1.1×10^{-4} eV, 1.1×10^{-4} eV, parallel

4. (a) $\tan^{-1}[3zx/(2z^2 - x^2)]$; (b) 0°, 0°, 72°

6. 0.24 Hz

8. $(3/14)\,e\hbar/m_n = 2.2 \times 10^{-27}$ A·m²; opposite

10. (a) 13 T; (b) 3.3×10^9 radian/s

12. 3.2×10^{-6} T

14. 1.000912

18. 3.6×10^3 A

20. 2.0×10^{-6} N·m; no

22. 1.6×10^{28} electrons/m³; 0.18 electron/atom

24. 3.4×10^{-9} J·s

26. (a) 4.4×10^{10} radian/s; (b) 2.3 m/s; (c) 4.7×10^{-24} J

Chapter 34

2. 400 W

4. 3.2×10^{-4} m/s; 0.12 m/s²

6. (a) 3.1×10^3 A; (b) less loss in transmission line

8. (a) $Q = 6.5 \times 10^{-8} \sin(120\pi t)$; (b) 1/240 s, 0 s; (c) 5.8×10^{-8} J, 2.9×10^{-8} J

10. $I = \omega C \mathscr{E}_0 \cos \omega t$, $I = -\mathscr{E}_0/(\omega L) \cos \omega t$; $I = (\omega C - 1/\omega L)\,\mathscr{E}_0 \cos \omega t$, $P = (\omega C - 1/\omega L)\,\mathscr{E}_0^2 \cos \omega t \sin \omega t$

12. (a) 240 Ω; (b) 8.3×10^{-4} A; (c) -8.3×10^{-4} A, -5.9×10^{-4} A

14. 6.0×10^{-9} F to 6.6×10^{-10} F

16. 1.4×10^{-3} J; 7.9×10^{-4} s; 15.7×10^{-4} s

18. (a) circuit oscillates at decreased frequency with gradually diminishing amplitude; (b) circuit oscillates at original frequency, but with much diminished amplitude

20. (a) 5.8×10^{-3} J; (b) 40%; (d) 0.035 s, 0.12 s

22. 1.0×10^{-2} H, 2.0×10^{-5} F

24. (a) $0.80 \sin(2\pi \times 2.2 \times 10^4\,t)$; (b) $5.9 \times 10^{-2} \cos(2\pi \times 2.2 \times 10^4\,t)$

26. 13.4×10^{-3} A, 6.6×10^{-3} A
28. (a) 0.20 A; (b) 6.3×10^{-4} s, 6.8×10^{-4} s
30. (a) 5.9×10^{3} Hz; (b) 0.40 A; (c) 0.16 W
32. 0.37 A; 43°; 5.4×10^{-2} W
34. $\frac{1}{2}\mathscr{E}_{max}^2 (\omega C - 1/\omega L) \sin 2\omega t + (\mathscr{E}_{max}^2/R) \sin^2 \omega t$
36. 4550 turns
38. 9.1×10^{4} A; 5.0×10^{3} A
40. 1.96×10^{3} A; 4.35×10^{3} A

Chapter 35

2. (a) 1.0×10^{6} V/m s; (b) 8.3×10^{-13} T, 1.7×10^{-12} T
4. (a) 2.0×10^{-3} A; (b) 1.0 s; (c) 0.89×10^{-9} T, 1.8×10^{-9} T, 2.7×10^{-9} T
6. (a) $(V_0/R) \sin \omega t$; (b) $(\varepsilon_0 A \omega V_0/d) \cos \omega t$; (c) $(V_0/R) \sin \omega t + (\varepsilon_0 A \omega V_0/d) \cos \omega t$; (d) $(\mu_0/2\pi) [(V_0/rR) \sin \omega t + (\varepsilon_0 \pi r \omega V_0/d) \cos \omega t]$
10. $\oint \mathbf{E} \cdot d\mathbf{S} = Q/\varepsilon_0$, $\oint \mathbf{B} \cdot d\mathbf{S} = q_m$, $\oint \mathbf{E} \cdot d\mathbf{l} = -I_m - d\Phi_B/dt$, $\oint \mathbf{B} \cdot d\mathbf{l} = \mu_0 I + \mu_0 \varepsilon_0 d\Phi/dt$
12. (a) $B_{(2)} = -\frac{1}{8}\mu_0^2 \varepsilon_0^2 r^3 \omega^3 E_0 \cos \omega t$; (b) $B = \frac{1}{2}\mu_0 \varepsilon_0 r \omega (1 - \frac{1}{4}\mu_0 \varepsilon_0 \omega^2 r^2) E_0 \cos \omega t$; (c) $B = \sqrt{\mu_0 \varepsilon_0}\, E_0 \cos \omega t$
14. (a) $(2\pi/\mu_0 \omega^2) E_0 \sin \omega t$; (b) $(2\pi/\mu_0 \omega) E_0 \cos \omega t$
16. 0.11 V/m; 6.7×10^{-10} s
18. (a) 7.4×10^{17} m/s²; (b) 7.9×10^{-9} V/m
20. (a) 4.8×10^{-15} m, 6.0×10^{27} m/s²; (b) 3.2×10^{11} V/m; (c) 0 V/m
22. 4.7×10^{4} m/s²
24. (a) 2.5×10^{-12} V/m, 8.3×10^{-21} T
26. (a) 1.1×10^{29} m/s²; (b) 7.1×10^{13} V/m, 2.4×10^{5} T
28. (a) 6.8×10^{9} m/s²; (b) 2.7×10^{-17} V/m, 9.1×10^{-26} T; (c) 6.1×10^{5} V/m, 2.1×10^{-3} T; (d) 3.8×10^{-9} T
30. (a) $q^2 v a/(12\pi\varepsilon_0 c^3)$; (b) $q^2 v a/(12\pi\varepsilon_0 c^3)$

Chapter 36

2. 5.3×10^{-6} A/m
4. (a) vertical; (b) vertical and perpendicular to **B**
6. 2.0×10^{-9} T, northward
8. (a) negative z; (b) 63°, 27°, 90°; (c) $2\hat{\mathbf{x}}(E_0/c) \sin (\omega t + \omega z/c) - \hat{\mathbf{y}}(E_0/c) \sin (\omega t + \omega z/c)$
10. $S = \frac{1}{2}S_0 \cos^2 \theta$
14. 1.85 m
16. 1.0×10^{3} V/m, 3.4×10^{-6} T
18. 6.9%
20. 51
22. 1.9×10^{9} W/m²; 1.2×10^{6} V/m, 4.0×10^{-3} T
24. 5.5×10^{-2} V/m, 1.8×10^{-10} T
26. 2.0×10^{4} W
30. (a) 50 W/m²; (b) 1.9×10^{2} V/m, 6.5×10^{-7} T; (c) 1.7×10^{-7} N/m²
32. (a) 6.2×10^{-18} N; (b) 3.7×10^{-18} N, gravity is larger; (c) 7.7×10^{-19} N, 9.2×10^{-19} N, radiation pressure is larger
34. 5.8×10^{-4} kg·m/s
38. $0.81c$
40. 2203.13 MHz

42. (a) 1.8×10^{-7}; (b) 1.4×10^{3} Hz
44. 2.2×10^{3} m/s, 1.50×10^{5} m/s

Chapter 37

2. 0.50 m × 0.42 m
4. 50°
6. 39° 0′ 50″
8. l, $l + 2d/n$, $l + 4d/n$
10. (a) 0.67 mm; (b) 2.67 mm
12. 47 m
14. (b) 64°
16. 5.7 m
20. R
22. 120 cm; concave
24. (a) 60 cm in front of first mirror; (b) 30 cm in front of second mirror
26. 8.3 mm
28. 21 cm
30. 10 cm
34. (a) 36 cm to right of lens; (b) 130 cm to left of mirror
36. 475 cm to left of lens
38. 49 cm to infinity
40. 0.022 s
42. (a) −22 cm, divergent; (b) 0.90 cm
44. 10
46. 64
48. 1344

Chapter 38

2. 3×10^{8} m/s
4. (b) 9.48×10^{5}, 9.48×10^{5}; (c) 1582 Å, 3164 Å
6. (a) 996 Å; (b) no
8. $2d = \frac{1}{2}\lambda, \frac{3}{2}\lambda, \frac{5}{2}\lambda, \ldots$
10. (c) 1.2 mm
12. 5.8 mm/s
14. 7.4 mm
16. $\lambda/d \cong 0.20$ implies nodal lines at 12°, 24°, 37°, and 53°
20. $\pm 1.3 \times 10^{-4}$ radian, $\pm 2.6 \times 10^{-4}$ radian, $\pm 3.9 \times 10^{-4}$ radian, etc.
22. 61°, 22°, −7°, −39°
24. 5.7°
26. 0.45
28. 1.0 mm; 2.6 mm
30. 8.7×10^{4}
32. 0.3 mm
34. $N - 2 = 8.7 \times 10^{4}$

Chapter 39

2. $\lambda/a \cong 0.44$ implies nodal lines at 26° and 63°
4. 47°
6. (a) 4.2 cm; (b) 2.1 cm
8. (b) $n - 1$
10. 0.040 mm
12. (a) 1.9×10^{-4} radian; (b) 4.4×10^{-3} mm; (c) 2.3
14. 2.8°; 2.4×10^{2} m
16. 1.3×10^{-5} radian; 2.4×10^{-4} cm

18. 3.4×10^{-3} radian, or $12'$
20. (a) 3×10^{-7} radian; (b) 2 m
22. 13 km
24. (a) 1.3×10^{-5} radian; (b) 1.4×10^{-5} radian, eye
28. 6.4 cm

Chapter 40

2. (a) 1 mm; (b) 2×10^9 W
4. 0.78; 0.87
8. 1.3×10^7 m
10. 44 K
12. 1.4×10^{-10} m; 1.9×10^{-10} m
14. (a) 7.2×10^{15} Hz
18. 1.2×10^6/s
20. 2.5×10^3/m³
24. 1.8 eV
26. none; K; K, Cr, Zn
28. 17.5 eV
30. 85°
32. 0.31 Å; 0.34 Å
34. 1.5×10^8 eV
36. 5.8×10^3 eV; 1.2×10^3 eV
38. (a) 2×10^{-39} kg · m/s; (b) 1×10^3 Hz

Chapter 41

2. first ($n = 5$); 34,076 Å
4. (a) 4102.9 Å and 4341.7 Å in Balmer series;
   (b) 1.1×10^6 m/s
6. (a) $p = -0.0409$; (b) Lyman and Balmer
8. 3.22×10^7 eV
12. 7.30×10^{-3} c
14. 1215.7 Å
16. $n = 3$ to $n = 1$
18. 2.12×10^{-10} m, 1.09×10^6 m/s, 2.10×10^{-34} J · s,
    5.66×10^{21} m/s²
20. 7.0×10^5 m/s
22. 0.176 Å; -122 eV
24. $r = (a_0 R^3 n^2)^{1/4}$, $\quad E = \dfrac{n \hbar e}{\sqrt{4\pi\varepsilon_0 m_e R^3}} - \dfrac{e^2}{4\pi\varepsilon_0} \dfrac{3}{2R}$,

$\quad v = \dfrac{n_2 - n_1}{2\pi} \dfrac{e}{\sqrt{4\pi\varepsilon_0 m_e R^3}}$; $4a_0$

26. 1.06 Å, 2431 Å
30. 2.1 Å, 0.16 Å
32. 1.1×10^{-32} m
34. $\lambda = 2\pi r/n$
36. ± 7 cm
38. 9.2×10^{-20} kg · m/s, 5.5×10^7 m/s

PHOTOGRAPH CREDITS

CHAPTER 7

page 154   (James P. Joule) AIP Niels Bohr Library
page 165   (Christiaan Huygens) AIP Niels Bohr Library

CHAPTER 8

Fig. 8.9   U.S. Air Force Photo
page 176   (Joseph Louis LaGrange) The Granger Collection
page 185   (Hermann von Helmholtz) AIP Niels Bohr Library
page 187   (James Watt) NBS Archives, courtesy AIP Niels Bohr Library

CHAPTER 9

Fig. 9.5   *PSSC Physics*, 2nd ed., 1965, D. C. Heath & Co. with the Education Development Center, Newton, Mass.
Fig. 9.11   The Bettmann Archive, Inc.
Fig. 9.13   NASA
Fig. 9.17   NASA
Fig. 9.18   Duomo Photography, Inc.

CHAPTER 10

Fig. 10.1   Mercedes-Benz of North America, Inc.
Fig. 10.9   Reprinted from *Philosophical Magazine*, R. H. Brown et al., vol. 40, 1949
Fig. 10.10   Hans C. Ohanian
Fig. 10.11   W. B. Hamilton, U.S. Geological Survey
Fig. 10.18   C. F. Powell and G. P. S. Occhialini, *Nuclear Physics in Photographs*, Oxford Univ. Press, New York

INTERLUDE B

Fig. B.3   Reprinted from the *Proceedings of the Royal Society*, P. M. S. Blackett and D. S. Lees, 1932
Fig. B.8   Reprinted from the *Proceedings of the Royal Society*, P. M. S. Blackett and D. S. Lees, 1932
Fig. B.13   Nuclear Medicine Section, Dept. of Radiology, Hospital of the Univ. of Pennsylvania
Fig. B.14   Michael G. Velchik, M.D., Nuclear Medicine Section, Dept. of Radiology, Hospital of the Univ. of Pennsylvania
Fig. B.15   Michael G. Velchik, M.D., Nuclear Medicine Section, Dept. of Radiology, Hospital of the Univ. of Pennsylvania
Fig. B.16   Michael G. Velchik, M.D., Nuclear

Medicine Section, Dept. of Radiology, Hospital of the Univ. of Pennsylvania
Fig. B.17   Michael G. Velchik, M.D., Nuclear Medicine Section, Dept. of Radiology, Hospital of the Univ. of Pennsylvania
Fig. B.18   M. D. Anderson Hospital and Tumor Institute, Univ. of Texas, Houston

CHAPTER 11

Fig. 11.1   Dr. Harold E. Edgerton, MIT

CHAPTER 12

Fig. 12.12   Sperry Flight Systems
Fig. 12.15   NASA
Fig. 12.19   *PSSC Physics*, 2nd ed., 1965, D. C. Heath & Co. with the Education Development Center, Newton, Mass.
Fig. 12.32   Wide World Photos
Fig. 12.46   NASA

INTERLUDE C

Fig. C.2   Leybold-Heraeus GMBH & Co.
Fig. C.3   "Lent to Science Museum, London," by the late J. J. Thomson, Trinity College, Cambridge, in the case of Cathode Ray Tube of J. J. Thomson
Fig. C.7   (a) The MIT Museum; (b) Cavendish Laboratory, Cambridge, England
Fig. C.8   Lawrence Berkeley Laboratory, Univ. of California
Fig. C.9   Fermilab Photo
Fig. C.10   Fermilab Photo
Fig. C.11   Fermilab Photo
Fig. C.12   Stanford Linear Accelerator Center
Fig. C.14   CERN
Fig. C.15   Stanford Linear Accelerator Center
Fig. C.16   CERN
Fig. C.17   CERN
Fig. C.18   MIT Bubble Chamber Group
Fig. C.19   CERN
Fig. C.20   CERN
Fig. C.21   Fermilab Photo
Fig. C.24   Brookhaven National Laboratory

CHAPTER 13

Fig. 13.5   From Cavendish, *Philosophical Transactions of the Royal Society 18* (1798): 388.
Fig. 13.14   (a) Sovfoto Agency
Fig. 13.15   NASA
Fig. 13.35   NASA
Fig. 13.38   (a) National Film Board of Canada; (b) National Film Board of Canada
Fig. 13.41   NASA

Fig. 13.45   The Bettmann Archive, Inc.
page 308   (Henry Cavendish) NBS Archives, courtesy AIP Niels Bohr Library
page 309   (Nicolas Copernicus) National Portrait Gallery, Smithsonian Institution
page 312   (Johannes Kepler) AIP Niels Bohr Library

INTERLUDE D

Fig. D.1   (a) Palomar Observatory Photograph; (b) Palomar Observatory Photograph
Fig. D.2   Palomar Observatory Photograph
Fig. D.3   Palomar Observatory Photograph
Fig. D.4   Palomar Observatory Photograph
Fig. D.5   Palomar Observatory Photograph
Fig. D.6   Palomar Observatory Photograph
Fig. D.7   Palomar Observatory Photograph
Fig. D.8   Palomar Observatory Photograph
Fig. D.9   Palomar Observatory Photograph
Fig. D.11   Palomar Observatory Photograph
Fig. D.12   AT&T Bell Laboratories

CHAPTER 14

Fig. 14.6   NASA
Fig. 14.8   From *Energy, A Sequel to IPS* by Uri Haber-Schaim, Prentice-Hall, Englewood Cliffs, N.J., 1983
Fig. 14.14   Hans C. Ohanian
Fig. 14.17   U.S National Bureau of Standards
Fig. 14.20   Dr. J. E. Faller, Univ. of Colorado
Fig. 14.21   U.S. National Bureau of Standards
Fig. 14.25   Photo Researchers, Inc.

CHAPTER 15

Fig. 15.19   *PSSC Physics*, 2nd ed., 1965, D. C. Heath & Co. with the Education Development Center, Newton, Mass.
Fig. 15.20   *PSSC Physics*, 2nd ed., 1965, D. C. Heath & Co. with the Education Development Center, Newton, Mass.
Fig. 15.21   *PSSC Physics*, 2nd ed., 1965, D. C. Heath & Co. with the Education Development Center, Newton, Mass.
Fig. 15.22   UPI © Bashford Thompson, Tacoma, Wash.

page 884 (Augustin Fresnel)
AIP Neils Bohr
Library
page 890 (Rayleigh) The Royal
Society, London,
courtesy AIP Niels
Bohr Library

CHAPTER 40
Fig. 40.13 (a, b, c) E. R.
Huggins, New York,
1968; (d) C. C. Jones,
Union College
page 913 (Max Born) AIP
Niels Bohr Library
page 910 (Arthur Compton)
AIP Niels Bohr
Library
page 914 (Werner Heisenberg)
Bainbridge
Collection, AIP Niels
Bohr Library
page 904 (Max Planck) AIP
Niels Bohr Library

CHAPTER 41
Fig. 41.9 Hale Observatories
Photograph

page 928 (Niels Bohr)
Princeton University,
courtesy AIP Niels
Bohr Library
page 934 (Louis Victor, Prince
de Broglie) AIP
Meggers Gallery of
Nobel Laureates
page 922 (Joseph von
Fraunhofer) NMAH
Archives Center,
Smithsonian
Institution
page 926 (Ernest Rutherford)
AIP Niels Bohr
Library
page 935 (Erwin Schrödinger)
Ullstein, courtesy
AIP Niels Bohr
Library
page 925 (Joseph John
Thomson) Bell
Telephone
Laboratories,
courtesy AIP Niels
Bohr Library

INTERLUDE L
Fig. L.4 Hughes Aircraft

Fig. L.7 The Warder
Collection
Fig. L.8 NASA
Fig. L.9 Bell Laboratories
Fig. L.10 Arnold Zann © 1980
Black Star
Fig. L.11 United Technologies
Fig. L.13 U.S. National Bureau
of Standards
Fig. L.14 Lawrence Livermore
National Laboratory
Fig. L.15 Lawrence Livermore
National Laboratory
Fig. L.16 Lawrence Livermore
National Laboratory
Fig. L.17 Lawrence Livermore
National Laboratory
Fig. L.19 From H. M. Smith,
*Principles of
Holography,* 2nd ed.,
© John Wiley &
Sons, Inc., 1975
Fig. L.24 Museum of
Holography, New
York City, © Ronald
R. Erickson, 1983
Fig. L.25 Karl Stetson
Fig. L.26 TRW

In the two-volume edition, all pages after 530 are in Volume Two, except those in parentheses, which are in Volume One. Interlude F is the last one in Volume One.

In the two-volume edition, all pages after 530 are in Volume Two, except those in parentheses, which are in Volume One. Interlude F is the last one in Volume One.

In the two-volume edition, all pages after 530 are in Volume Two, except those in parentheses, which are in Volume One. Interlude F is the last one in Volume One.

In the two-volume edition, all pages after 530 are in Volume Two, except those in
parentheses, which are in Volume One. Interlude F is the last one in Volume One.

In the two-volume edition, all pages after 530 are in Volume Two, except those in parentheses, which are in Volume One. Interlude F is the last one in Volume One.

In the two-volume edition, all pages after 530 are in Volume Two, except those in parentheses, which are in Volume One. Interlude F is the last one in Volume One.

In the two-volume edition, all pages after 530 are in Volume Two, except those in parentheses, which are in Volume One. Interlude F is the last one in Volume One.

In the two-volume edition, all pages after 530 are in Volume Two, except those in parentheses, which are in Volume One. Interlude F is the last one in Volume One.

In the two-volume edition, all pages after 530 are in Volume Two, except those in
parentheses, which are in Volume One. Interlude F is the last one in Volume One.

In the two-volume edition, all pages after 530 are in Volume Two, except those in parentheses, which are in Volume One. Interlude F is the last one in Volume One.

In the two-volume edition, all pages after 530 are in Volume Two, except those in
parentheses, which are in Volume One. Interlude F is the last one in Volume One.

Speed of light	$c = 3.00 \times 10^8$ m/s
Planck's constant	$h = 6.63 \times 10^{-34}$ J · s
	$\hbar = h/2\pi = 1.05 \times 10^{-34}$ J · s
Gravitational constant	$G = 6.67 \times 10^{-11}$ N · m^2/kg^2
Permeability constant	$\mu_0 = 1.26 \times 10^{-6}$ H/m
Permittivity constant	$\varepsilon_0 = 8.85 \times 10^{-12}$ F/m
Boltzmann constant	$k = 1.38 \times 10^{-23}$ J/K
Electron charge	$-e = -1.60 \times 10^{-19}$ C
Electron mass	$m_e = 9.11 \times 10^{-31}$ kg
Proton mass	$m_p = 1.673 \times 10^{-27}$ kg
Neutron mass	$m_n = 1.675 \times 10^{-27}$ kg
Rydberg constant	$R_H = 1.10 \times 10^7$/m
Bohr radius	$4\pi\varepsilon_0 \hbar^2/m_e e^2 = 5.29 \times 10^{-11}$ m
Compton wavelength	$\hbar/m_e c = 3.86 \times 10^{-13}$ m

Acceleration of gravity	1 gee $= 9.81$ m/s$^2 = 32.2$ ft/s^2
Atomic mass unit	1 u $= 1.66 \times 10^{-27}$ kg
Avogadro's number	$N_A = 6.02 \times 10^{23}$/mole
Density of dry air	1.29 kg/m^3 (0°C, 1 atm)
Molecular mass of air	28.98 g/mole
Speed of sound in air	331 m/s (0°C, 1 atm)
Density of water	1000 kg/m^3
Heat of vaporization of water	539 kcal/kg
Heat of fusion of ice	79.7 kcal/kg
Mechanical equivalent of heat	1 cal $= 4.19$ J
Solar constant	1.4 kW/m^2
Index of refraction of water	1.33

THE PLANETS[a]

Planet	Mean distance from Sun	Period of revolution	Mass	Equatorial radius	Surface gravity	Period of rotation
Mercury	57.9×10^6 km	0.241 year	3.30×10^{23} kg	2,439 km	0.38 gee	58.6 days
Venus	108	0.615	4.87×10^{24}	6,052	0.91	243
Earth	150	1.00	5.98×10^{24}	6,378	1.00	0.997
Mars	228	1.88	6.42×10^{23}	3,397	0.38	1.026
Jupiter	778	11.9	1.90×10^{27}	71,398	2.53	0.41
Saturn	1,430	29.5	5.67×10^{26}	60,000	1.07	0.43
Uranus	2,870	84.0	8.70×10^{25}	25,400	0.92	0.65
Neptune	4,500	165	1.03×10^{26}	24,300	1.19	0.77
Pluto	5,910	248	6.6×10^{23}	2,500	0.72	6.39

[a] Based on The Astronomical Almanac, 1984.